THE CHEMISTRY OF
PLATINUM AND PALLADIUM

Consulting Editor

P. L. ROBINSON

*Emeritus Professor of Chemistry in the University of Durham
and the University of Newcastle upon Tyne*

THE CHEMISTRY OF PLATINUM AND PALLADIUM

With Particular Reference to Complexes of the Elements

F. R. HARTLEY

M.A., D.Phil., A.R.I.C.

University of Southampton, Great Britain

A HALSTED PRESS BOOK

JOHN WILEY & SONS
New York—Toronto

PUBLISHED IN THE U.S.A. AND CANADA BY
HALSTED PRESS
A DIVISION OF JOHN WILEY & SONS, INC., NEW YORK

Library of Congress Cataloging in Publication Data

Hartley, F. R.

The chemistry of platinum and palladium.

'A Halsted Press book.'

1. Platinum. 2. Palladium. 3. Complex compounds. I. Title.
QD181.P8H37   546'.645'2   72-11319
ISBN 0-470-35658-8

WITH 59 ILLUSTRATIONS AND 129 TABLES

© APPLIED SCIENCE PUBLISHERS LTD 1973

Printed in Great Britain by J. W. Arrowsmith Ltd., Bristol, England.

Preface

This monograph has been written for three reasons: they are first, the author's general interest in these two metals and his special interest in certain aspects of their chemistry; secondly, the large output of original research from many centres since the Second World War; and thirdly, the absence of a recent survey covering the literature dealing with platinum and palladium. Within the limits of the space available, the present text constitutes a reasonably comprehensive account of the subject matter. It more than adequately meets the probable needs of students and teachers and, because of the extensive documentation, should provide a convenient place of first reference for those already involved in research. To those embarking upon such work, the author hopes it offers both suggestion and guidance. All concerned with laboratory investigation should find the two appendices of special service: one gives essential details of the preparation and properties of the compounds of platinum and palladium usually employed as starting materials for further experimentation; the other sets out a complete, up-to-date list of the published structures of the compounds of these elements arranged so as to serve as an immediate source of data for correlation purposes.

Platinum and palladium have attracted continuing attention largely because they have been the source of an increasing stream of new compounds of high intrinsic interest, particularly with respect to bonding and structure. The kinetics of the reactions involved in their formation and utilisation have been the object of much enquiry. Their organometallic chemistry is extensive; and they have excited interest as homogeneous and heterogeneous catalysts; and the breakthrough to the fixation of some of the noble gases initiated by the preparation of xenon hexafluoroplatinate(V) has led to notable results. In the monograph emphasis has been laid on structure and bonding, and an attempt has been made to evaluate critically the role of the physical techniques currently used to elucidate these problems. Attention has been called to the need to appreciate the scope and limitations of present theories of bonding as they are applicable to the complexes of these elements.

· A brief run-down of the contents of the monograph may help the reader by showing how the subject-matter has been subdivided and grouped. Beyond the introductory chapter the oxidation states are summarised and their class 'b' or 'soft' character is discussed. Then follow through several chapters an extended consideration of zerovalent, divalent and tetravalent states. The treatment is comparative and designed to display similarities and differences between the respective chemistries of platinum and palladium; as a help to this end much use has been made of tables. Throughout, the diagnostic techniques employed have been stressed and structure and bonding given a prominent place. From this the monograph passes on to deal with substitution reactions, associative mechanism in reactions of square-planar complexes, and isomerism in compounds of the elements in a divalent state. Then follow accounts of metal–carbon σ-bonded compounds, olefin and acetylene complexes, and finally the π-allyl complexes which are of interest both for their bonding and their significance in reactions catalysed by palladium(II) salts.

I am indebted to Professor L. M. Venanzi for first introducing me to inorganic chemistry in general and the chemistry of these two metals in particular, and to the late Professor Sir Ronald Nyholm FRS for later fostering this interest. I should like to thank Professor P. L. Robinson for critically reading the drafts of each chapter and for the many helpful suggestions which he made for their improvement. Finally I should like to thank my wife both for her patience during the writing of this book and for her help in handling the large amount of information which had to be sifted before this book could be written.

F. R. HARTLEY
Southampton

Contents

CHAPTER 1

Introduction

This chapter is in two parts. The first provides an introductory survey of the scope of the monograph which, as the sub-title indicates, and as the character of the chemistry of platinum and palladium demands, is mainly about the complexes of these elements. The second part gives brief notes on the occurrence, extraction and properties of the elements and their uses, especially that of the metals as such.

INTRODUCTORY SURVEY

A systematic study of the chemistry of platinum began soon after its arrival in Europe in 1741 and by 1830 not only had many of the inorganic compounds been made but the first organometallic derivative $K[Pt(C_2H_4)Cl_3].H_2O$ had actually been prepared.[1] Palladium was discovered in 1803, and the chemistry of both metals developed during the rest of the nineteenth century and, with increasing acceleration, in the twentieth century, in the course of which the empirical facts and the theoretical significance of covalence were recognised. There were three factors which attracted interest to and sustained research on these two metals.

First, in the divalent oxidation state both readily form complexes with ligands containing donor atoms from most groups of the periodic table. Following from this the nineteenth and early twentieth centuries saw the preparation of many crystalline platinum derivatives as a routine method of characterising organic compounds; from this much information on the co-ordination chemistry of platinum flowed.

Secondly, there came the recognition of the square–planar geometry of the divalent oxidation states, which opened up the possibility of cis- trans-isomerisation in such complexes. From this followed the discovery of the trans-effect in platinum(II) complexes in the 1920's (pp. 299–301).[2] Its elucidation enabled the systematic synthesis of any desired platinum(II) complex to be accomplished. These extensive researches reached their peak with the synthesis of all the three possible geometric isomers of $[Pt(pyr)(NH_3)BrCl]$.[3] From 1930 to 1950 was a period of consolidation in which a great deal of the coordination chemistry of the two metals was systematised.

Thirdly, on the basis of this not inconsiderable achievement, but much stimulated by advancing theory, there followed a tremendous growth in interest in platinum and palladium triggered off by the burgeoning of their organometallic chemistry. This is a vast subject of great intrinsic interest. It has applications in the field of homogeneous catalysis in the reactions of organic compounds, and also provides some basis for a discussion of the mechanisms of organic reactions that are heterogeneously catalysed by the metals themselves and by their oxides.

The commonest oxidation state for both metals is $+2$. There are also many compounds with the elements in the zero and $+4$ oxidation states, although these are both much commoner for platinum than palladium. For instance on heating in chlorine, platinum forms a mixture of $PtCl_2$, $PtCl_3$ and $PtCl_4$[4,5] whereas palladium forms only the dichloride $PdCl_2$.[6] In the $+4$ oxidation state both metals exhibit octahedral coordination. The greater reluctance of palladium to form $+4$ compounds, owing to the higher ionisation energy needed to produce Pd^{4+} than Pt^{4+} ions (p. 13) has two important consequences: the first, and most apparent (p. 251) is that there is a much wider range of platinum(IV) than palladium(IV) compounds; the second, and less obvious consequence, is that whereas many complexes of platinum(II) undergo reactions by an oxidative-addition mechanism (pp. 83–91 and pp. 339–340) such a mechanism is generally not available to palladium(II) complexes. Either the palladium(II) complexes do not react or more vigorous conditions are required to promote reaction by an alternative, higher energy, path.

In their zerovalent oxidation states both metals form a range of complexes (pp. 27–47). Although these might possibly be prepared by treating the metal with the ligand, in practice they are usually obtained from the divalent compounds with the aid of a suitable reducing agent.

Although compounds of the metals in the zero, $+2$ and $+4$ oxidation states are common, compounds in the $+1$ oxidation-state (pp. 16–17) are very difficult to make. Compounds in the $+3$ state have not been unambiguously identified for either metal. Moreover, platinum alone forms a fluoride and oxyfluoride in the $+5$ state as well as the hexafluoride PtF_6, in the $+6$ oxidation state. Reasons for each of the metals forming compounds in a particular oxidation state are examined in Chapter 2 and in the same place a general survey of all the oxidation states is given. It is also shown that in oxidation states below $+4$ the two metals are class 'b' or 'soft' acids (pp. 13–15).

In Chapters 4 to 9 compounds and complexes of the elements in the divalent condition are considered; the donor atoms bound directly to the metal are classified according to their positions in the periodic table. Chapter 10, dealing with compounds and complexes of the metals in the tetravalent state, brings out the important point that there is sufficient similarity between the square-planar platinum(II) and octahedral platinum(IV) complexes to enable many of the properties of platinum(IV) complexes to be understood by looking upon them as square-planar platinum(II) complexes with two extra ligands bound one above and one below the plane.

Some topics could not be conveniently covered within the scope of Chapters 4 to 9. Of these the most important is the mechanism of the reactions, particularly the substitution reactions, of the two elements; this subject is enlarged upon in Chapter 11. Most of the information available relates to platinum complexes, because these react much more slowly than their palladium counterparts and are generally amenable to the simple experimental techniques devised to follow slow reactions. On the other hand, palladium complexes, with certain exceptions, call for the use of more expensive equipment designed to measure the rates of fast reactions. However, conclusions involving platinum complexes have generally been assumed to apply to corresponding reactions of palladium complexes. When tested experimentally these assumptions have generally been found to be valid.

A current major growth area in the study of platinum and palladium has been that of their organometallic chemistry. Some dramatic results in the homogeneous catalysis of the reactions of organic compounds, particularly the successful commercial exploitation of the Wacker one-stage process for the homogeneous catalytic oxidation of ethylene to acetaldehyde in the presence of palladium(II) chloride (pp. 388–389), have contributed to this interest. The organometallic chemistry might have been included in the earlier chapters; this was not done because in this instance the oxidation state of the metal is of less importance than the nature of the ligand. Accordingly, the final three chapters deal, respectively, with metal–carbon σ-bonded compounds, olefin and acetylene compounds, and π-allylic compounds. In these chapters structure and bonding has been especially emphasised because this knowledge is essential to an understanding of the chemistry of the organometallic compounds. A feature which emerges in Chapters 12 to 14 is the complementary nature of the properties of platinum and palladium in organometallic compounds. Thus platinum forms more stable compounds with alkyl and olefinic ligands than palladium, but the reverse is true of their behaviour with π-allylic ligands. Although this cannot as yet be fully explained, some tentative suggestions towards this end are advanced in Chapter 14 (p. 441).

RESUMÉ OF THE OCCURRENCE, EXTRACTION AND PROPERTIES OF PLATINUM AND PALLADIUM

Platinum, discovered in the sixteenth century in the Choco district of Columbia, was originally called '*platina del Pinto*' or 'little silver of the Pinto River'.[7] Charles Wood, an Assay Master of Jamaica, brought it to Europe in 1741 and passed the specimen by way of William Brownrigg FRS to Sir William Watson FRS who described it to the Royal Society in 1750.[8] A fairly definitive account of platinum was published in 1755 and 1757 by William Lewis FRS, a medical practitioner of Kingston-upon-Thames.[9]

Palladium was discovered in 1803, some three hundred years after platinum, by W. H. Wollaston FRS when he was investigating the refining of platinum and it was named after the recently discovered asteroid Pallas.[10]

Occurrence

Platinum occurs as metal, and before 1914 most of it came from Russia and Columbia; today the sources of native platinum are Alaska and Columbia, but their contribution represents only a very minor part of world production. Nearly all platinum is derived from the copper–nickel ores of Canada, South Africa and Russia, and they contain less than an ounce of platinum per ton.

Palladium occurs in association with platinum; the Canadian ores have more palladium than platinum, whereas the reverse is true of the South African ores. Incidentally, the native platinum found in Columbia contains between 0·5 and 3 % of palladium.

Extraction

The platinum and palladium in cupronickel ores are recovered along with other noble metals from the anode slimes produced in the electrolytic recovery of the copper and nickel or from the involatile residues from the Mond carbonyl process for refining nickel. Figure 1 outlines the treatment of the concentrates.[11] This begins with aqua-regia which is used to dissolve the platinum, palladium and gold. Because platinum is more easily oxidised

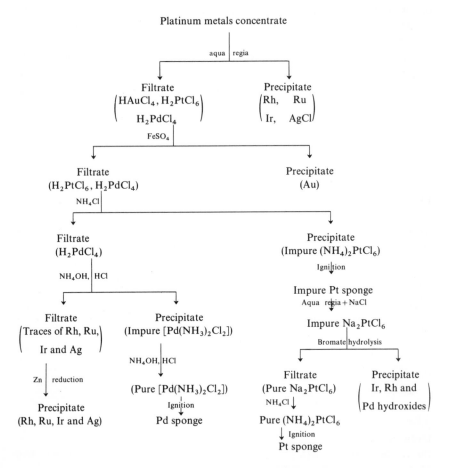

Fig. 1. Scheme for the extraction of platinum and palladium.

than palladium (p. 13) the former reaches the $+4$ and the latter only the $+2$ oxidation state. This solution is treated with either ferrous sulphate or chloride which reduces chloroauric acid to gold, leaving platinum and palladium in solution (eq. 1). The addition of ammonium chloride precipitates orange-

$$HAuCl_4 + 3FeSO_4 \rightarrow \downarrow Au + Fe_2(SO_4)_3 + FeCl_3 + HCl \qquad (1)$$

yellow ammonium hexachloroplatinate(IV), which is filtered off and ignited to give an impure platinum sponge. This crude sponge is dissolved in aquaregia, filtered and evaporated with sodium chloride and hydrochloric acid to remove nitric acid and nitrosyl compounds and convert the platinum to water-soluble sodium hexachloroplatinate(IV). The sodium hexachloroplatinate(IV) solution is then treated with sodium bromate to oxidise the remaining impurities (Ir, Rh and Pd) to valence states from which, by a careful addition of sodium bicarbonate the metals are quantitatively precipitated as dark slimy hydroxides. The pure filtrate is then boiled with hydrochloric acid to destroy the excess of bromate and treated with ammonium chloride to precipitate platinum as ammonium hexachloroplatinate(IV); this is filtered off, washed, dried and finally heated very slowly up to 1000°C to give $> 99.9\%$ pure platinum sponge.

The impure palladium salt (H_2PdCl_4) is treated with dilute ammonia (eq. 2) and precipitated as $[Pd(NH_3)_2Cl_2]$ by hydrochloric acid (eq. 3). The process is repeated to yield pure $[Pd(NH_3)_2Cl_2]$, which is filtered off and

$$H_2PdCl_4 + 6NH_4OH \rightarrow [Pd(NH_3)_4]Cl_2 + 6H_2O \qquad (2)$$

$$[Pd(NH_3)_4]Cl_2 + 2HCl \rightarrow [Pd(NH_3)_2Cl_2] + 2NH_4Cl \qquad (3)$$

ignited to palladium sponge. To avoid the formation of an oxide film the sponge is cooled in hydrogen. The product is $> 99.9\%$ palladium.

Physical and chemical properties

Platinum and palladium are grey-white, lustrous metals which are sufficiently ductile and malleable to be drawn into wire, rolled into sheet or formed by spinning and stamping. They have high melting-points and considerable resistance to corrosion. Their main physical properties are summarised in Table 1 which gives the currently acceptable values, some of which are critically dependent on purity.

Both platinum and palladium are noble metals owing to a combination of high sublimation energy and high ionisation potential (p. 9). Of the two metals, platinum is slightly the less reactive. It is not attacked by any single mineral acid but readily dissolves in aqua-regia, whereas palladium, even in the compact state, is attacked by hot concentrated nitric and sulphuric acids, particularly in the presence of oxygen and the oxides of nitrogen. In powder form palladium is slowly dissolved by hydrochloric acid in the presence of oxygen. Platinum is not oxidised when heated in air whereas palladium is oxidised to palladium(II) oxide in air at 700°C. Above 875°C the oxide dissociates to the free metal and oxygen.[15] Palladium, but not platinum, is attacked by moist chlorine and bromine at room temperature.

TABLE 1

The physical properties of platinum and palladium (compiled from references 12–14)

Property	Platinum	Palladium
Atomic number	78	46
Atomic weight (relative to ^{12}carbon $= 12$)	195·09	106·4
Density (g/cc at 20°C)	21·45	12·02
Crystal lattice	Face-centered cubic	Face-centered cubic
Lattice cell (Å)	3·9158	3·8825
Atomic radius (Å)	1·387	1·375
Allotropic forms	None known	None known
Melting point (°C)	1773·5	1554
Boiling point (°C, estimated)	4530	3980
Thermal conductivity (cal/cm/cm^2/sec/°C)	0·17	0·17
Linear thermal coefficient of expansion at 0°C (per °C)	$8 \cdot 9 \times 10^{-6}$	$11 \cdot 67 \times 10^{-6}$
Specific heat at 0°C (cal/g/°C)	0·0314	0·0584
Electrical resistivity at 20°C (micro ohm-cm)	10·6	10·8
Hardness (annealed—Brinell Hardness number)	42	46
Tensile strength (annealed—ton/in^2)	9	13·8
Young's modulus (annealed–ton/in^2)	$1 \cdot 1 \times 10^4$	$(8 – 8 \cdot 8) \times 10^3$

Uses

The properties of platinum make it of use for laboratory ware subject to high temperatures and certain corrosive materials, and in industry for the lining of crucibles used in the melting of special glasses and of orifices through which glass fibre is extruded. The metal is employed as a surface catalyst in the oxidation of ammonia. Other uses are found in contact-breakers, electrical furnaces and as electrodes in laboratory apparatus and industrial plant. The pure metal is relatively soft and for many purposes is alloyed with either gold or rhodium—with the latter for spinnerets in nylon fibre manufacture, since the alloy resists both acid and alkali attack. Platinum is employed in the jewellery trade.

Both metals, when dispersed on a suitable porous solid, are useful heterogeneous catalysts: examples are platinum in the production of high-octane fuels and palladium to promote liquid-phase hydrogenation reactions in the general chemical, dyestuffs and pharmaceutical industries. The promise of their salts as homogeneous catalysts is dealt with in some detail in this monograph (pp. 386–395, 443–448).

REFERENCES

1   W. C. Zeise, *Mag. Pharm.*, **35** (1830) 105.
2   I. I. Chernyaev, *Ann. Inst. Platine USSR*, **4** (1926) 243; *Chem. Abs.*, **21** (1927) 2620.
3   A. D. Gel'man, E. F. Karandashova and L. N. Essen, *Dokl. Akad. Nauk. SSSR*, **63** (1948) 37; *Chem. Abs.*, **43** (1949) 1678i.
4   L. Wöhler and F. Martin, *Ber.*, **42** (1909), 3958.
5   L. Wöhler and S. Streicher, *Ber.*, **46** (1913) 1591.
6   E. H. Keiser and M. B. Breed, *J. Amer. Chem. Soc.*, **16** (1894) 20.

7   D. McDonald, *A History of Platinum*, Johnson Matthey and Co. Ltd., London, 1960.
8   W. Watson, *Phil. Trans.*, **46** (1751) 584.
9   W. Lewis, *Phil. Trans.*, **48** (1755) 638 ; **50** (1757) 148.
10 W. H. Wollaston, *Phil. Trans.*, **95** (1805) 316.
11 C. Johnson and R. H. Atkinson, *Trans. Inst. Chem. Engrs. (London)*, **15** (1937) 131.
12 R. F. Vines and E. M. Wise, *The Platinum Metals and Their Alloys*, International Nickel Co., New York, 1941.
13 E. M. Wise in *Metals Handbook*, ed. by T. Lyman, American Society of Metals, Cleveland, 1968.
14 F. E. Beamish, W. A. E. McBryde and R. R. Barefoot, *Rare Metals Handbook*, ed. by C. A. Hampel, Reinhold, New York, 1954, p. 291.
15 L. Wöhler, *Z. Elektrochem.*, **11** (1905) 836.

CHAPTER 2

Oxidation States and Stereochemistry

Here is given a general survey of the oxidation states and stereochemistries of the compounds and complexes of platinum and palladium. We start by considering some of the fundamental properties, such as atomic radii and ionisation potentials, and then discuss their influence on the oxidation states exhibited by the two metals. Then we consider the oxidation-reduction potentials of the metals in solution: this introduces solution properties and leads to a discussion of the class 'b' or 'soft' characters of platinum and palladium. We pass to a survey of the oxidation states exhibited by the two metals and describe the chemistry of those particular oxidation states that do not receive attention later in the book.

ATOMIC RADII

The radii of platinum and palladium in a number of environments have been either measured or estimated. The values (Table 2) indicate that except for the free metals their radii are essentially equal. This equality, common among pairs of second and third transition series metals, arises from the long recognised 'lanthanide' contraction of the atomic radius due to imperfect shielding of the outer electrons from the nuclear charge by the intermediately placed $4f$ subshell of electrons. It is noteworthy that in complexes the metal atoms in the zerovalent, divalent and tetravalent states all possess the same covalent radius.

TABLE 2
Atomic radii of platinum and palladium

Radius	Palladium (Å)	Platinum (Å)	Reference
M(0) (in metal)	1·375	1·387	1
M(0) (covalent)		1·31[a]	2, 3
M(II) (covalent, square-planar)	1·31	1·31	4, 5
M(IV) (covalent, octahedral)	1·31	1·31	4, 5
M^{4+} (ionic, octahedral)	0·62	0·65	6, 7

[a] Mean value determined from the Pt–P bond lengths in $[(Ph_3P)_2Pt(PhC \equiv CPh)]$ (ref. 2) and $[(Ph_3P)_2Pt(OC(CH_3)_2OO)]$ (ref. 3).

IONISATION POTENTIALS

It is interesting that the second, third and fourth ionisation potentials of platinum are all lower than the corresponding ones for palladium (Table 3) indicating that a $5d$-electron on platinum experiences a lower effective nuclear charge than a $4d$-electron on palladium. This is because the inner shell of f-electrons provides a better shield for the outer d-electrons of platinum from the nuclear charge than is experienced by the outer d-electrons of palladium. The first ionisation potential of platinum is greater than that of palladium because for platinum ionisation represents the loss of an s-electron (Pt^0 has the outer electron configuration d^9s^1); by contrast for palladium the electron lost during the first ionisation is a d-electron (Pd^0 has the outer electron configuration $d^{10}s^0$).

TABLE 3
Ionisation potentials of platinum and palladium (eV)

Ionisation potential	Palladium	Platinum	Reference
$M^0 \rightarrow M^+$	8·33	9·0	8
$M^+ \rightarrow M^{2+}$	19·42	18·56	8
$M^{2+} \rightarrow M^{3+}$	32·92	28·5	8, 9
$M^{3+} \rightarrow M^{4+}$	48·8	41·1	9
$M^{4+} \rightarrow M^{5+}$	66	55	9
$M^0 \rightarrow M^{2+}$	27·75	27·56	
$M^0 \rightarrow M^{3+}$	60·67	56·06	
$M^0 \rightarrow M^{4+}$	109·47	97·16	
$M^0 \rightarrow M^{5+}$	175·47	152·16	

SUBLIMATION ENERGIES

The sublimation energy of platinum (121·6 kcal/mol at 25°C[10]) is rather greater than that of palladium (92·96 kcal/mol at 25°C[10]; this is consistent with the lower melting- and boiling-points of palladium (Table 1, p. 6).

ELECTRONEGATIVITIES

The Pauling electronegativities of platinum and palladium are identical (Table 4) and even the optical electronegativities of the divalent and tetravalent ions, deduced from electron-transfer spectra, are extremely similar.

TABLE 4
Electronegativities of palladium and platinum

	Palladium	Platinum	Reference
Pauling electronegativity	2·2	2·2	4
Optical electronegativity of M(II)	2·2[a]	2·3	11
Optical electronegativity of M(IV)	2·7	2·7	11

[a] The value of 2·2 was obtained by combining the results from electron-transfer spectra (ref. 11) with some n.q.r. data (ref. 12).

Thus it is not surprising to find considerable similarities in the nature of the bonds formed between platinum and palladium and other elements.

FORMATION AND STABILITY OF PLATINUM AND PALLADIUM COMPOUNDS

The influence of the above properties on the formation and stability of a given compound are most easily understood with the aid of the Born–Haber

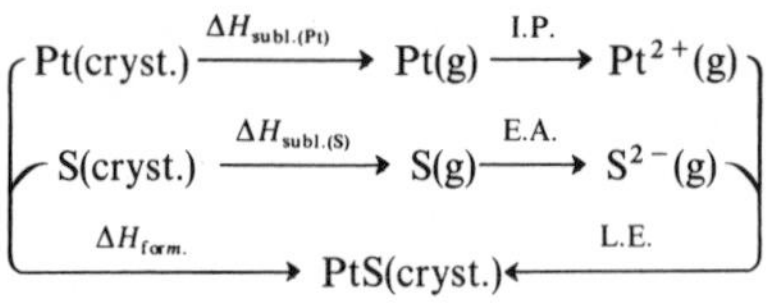

Fig. 2. Born–Haber cycle for the formation of crystalline platinum(II) sulphide.

cycle. This is applied in Fig. 2 to the formation of crystalline platinum(II) sulphide.

$\Delta H_{\text{subl.(Pt)}}$ (sublimation energy of platinum) = 121·6 kcal/mol[10]

I.P. (sum of first and second ionisation potentials of platinum) = 635·5 kcal/mol[8]

$\Delta H_{\text{subl.(S)}}$ (heat of formation gaseous sulphur) = 53·25 kcal/mol[13]

E.A. (electron affinity of sulphur) = 71·95 kcal/mol[13]

ΔH_{form} (heat of formation of platinum(II) sulphide) = -26 kcal/mol[14]

The Born–Haber cycle can be used to calculate an experimental lattice energy (L.E.) since,

$$\text{L.E.} = \Delta H_{\text{subl.(Pt)}} + \text{I.P.} + \Delta H_{\text{subl.(S)}} + \text{E.A.} - \Delta H_{\text{form}}$$

Appropriate substitution gives an experimental lattice energy of platinum(II) sulphide of 908 kcal/mol.

Although in the case of ionic compounds the Born–Haber cycle can often be used to estimate enthalpies of formation by calculating the lattice energy of the compound, this is not possible for platinum and palladium compounds for two reasons. (i) The expressions used for calculating lattice energies,[15] for instance eqn. 4, consider the attractive force as arising solely from a coulombic term. The important, though difficult to calculate, contribution

$$\text{Lattice energy} = \frac{NA(z+)(z-)e^2(1 - 1/n)}{r} \qquad (4)$$

(N = Avogadro's number, A = Madelung constant, $z+$ = numerical charge on the cation, e = electronic charge, r = interatomic distance, n = Born exponent.)

of the covalent bonding to the total lattice energy is thus completely neglected. (ii) A further complication that arises with many platinum and palladium compounds is the evaluation of the Madelung constant. The structures of most of these compounds are far from simple, whereas Madelung constants are only available for those compounds of the two metals that crystallise in a simple structure; for instance PtS_2 which crystallises in a cadmium iodide structure.

The contribution of covalent bonding to the total lattice energy is well illustrated by comparing the calculated and experimental values of the lattice energy of platinum(II) sulphide. As shown above the experimental lattice energy of platinum(II) sulphide is 908 kcal/mol. The lattice energy calculated from eqn. 4, with a Madelung constant of 1·58[15] a Born exponent of 11[4] and a platinum–sulphur bond length of 3·31 Å,[16] is 607 kcal/mol, indicating the importance of the covalent bonding term. An even more dramatic difference is found between the enthalpy of formation of platinum(II) sulphide calculated from the Born–Haber cycle (+275 kcal/mol) and the experimental value (−26 kcal/mol). The vast difference emphasises an important point, namely that the enthalpy of formation is a small term in relation to both ionisation potential and lattice energy.

Thus in platinum and palladium chemistry the Born–Haber cycle can provide only qualitative evaluation and, moreover, its use is restricted to the comparison of two compounds of the same structure (for example PtS and PdS or PtO and PdO). Within these strict limitations, it can be used to find which fundamental parameters are dominant in determining the relative enthalpies of formation of a pair of platinum and palladium compounds. This is illustrated below for some compounds of both the divalent and tetravalent metals.

(i) Compounds of the divalent metals

Table 5 shows that the enthalpies of formation of a number of palladium(II) salts are higher than those of the corresponding platinum(II) salts. The heat of sublimation of palladium is lower than that of platinum (p. 9) and this favours a larger enthalpy of formation for palladium compounds. The sum of the first two ionisation potentials (Table 3) is only slightly lower for platinum but this will favour a slightly larger enthalpy of formation for platinum(II) compounds. The almost identical radii of the divalent metals gives rise to

TABLE 5
Enthalpies of formation of some typical compounds of divalent palladium and platinum

Compound	$\Delta H°$ (*form.*) *of Pt compound* (*kcal/mol*)	*Reference*	$\Delta H°$ (*form.*) *of Pd compound* (*kcal/mol*)	*Reference*
$M(OH)_2$	−84	17	−88	17, 18
MCl_2	−33	19	−39	20–22
MBr_2	−15	23	−25	18
K_2MCl_4	−255	17, 24	−259	17

similar lattice energies, as calculated from an ionic model. However, the greater spacial extension of the platinum $5d$-electrons as compared to the palladium $4d$-electrons should give a slightly greater overlap of the ligand orbitals and hence slightly higher lattice energies for the platinum compounds. Thus the greater enthalpies of formation of the palladium(II) compounds as compared to the platinum(II) compounds in Table 5 arises largely from the lower heat of sublimation of palladium.

(ii) Compounds of the tetravalent metals

Enthalpies of formation have been determined for only two pairs of compounds containing the tetravalent metals (Table 6) and in both cases the enthalpy of formation of the platinum compound is greater than that of the palladium compound. The effects of the relative heats of sublimation and

TABLE 6
Enthalpies of formation of compounds containing tetravalent platinum and palladium

Compound	$\Delta H°$ (form.) of Pt compound (kcal/mol)	Reference	$\Delta H°$ (form.) of Pd compound (kcal/mol)	Reference
MS_2	-26	25, 26	-19	27
K_2MCl_6	-296	28, 17	-277	22

covalent radii will be identical with those for the divalent compounds. However, the sum of the first four ionisation potentials is rather less for platinum than palladium (see Table 3) so that 284 kcal/mol less energy is expended in the formation of platinum(IV) than palladium(IV) ions. This is sufficient to give platinum(IV) compounds larger enthalpies of formation than palladium(IV) compounds in spite of a difference of about 29 kcal/mol in the sublimation energies of the two metals.

OXIDATION–REDUCTION POTENTIALS

Although heats of formation give important information about the overall thermodynamic stability of compounds with respect to their decomposition into their elements, the oxidation potentials give a measure of their stability to oxidation or reduction. Some important oxidation–reduction potentials are given in Table 7. These values illustrate a number of important properties of platinum and palladium in aqueous solution.

(i) The standard oxidation potentials for equilibria involving the zero-valent and divalent oxidation states are less (i.e. more negative) for platinum than palladium. In other words it is more difficult to oxidise zerovalent platinum to the divalent oxidation state and easier to reduce divalent platinum than divalent palladium. These properties, as stated in detail in connection with the heats of formation of compounds of the divalent metals (pp. 11–12), are due principally to the lower sublimation energy of palladium. This accounts also for platinum being a more noble metal than palladium as already mentioned in the introduction (p. 5).

TABLE 7
Oxidation potentials of platinum and palladium

Oxidation state change	Equilibrium	Oxidation potential (volts)		Reference	
		Pt	Pd	Pt	Pd
$M^0 \rightleftharpoons M^{II}$	$M \rightleftharpoons M^{2+} + 2\varepsilon$	$-1\cdot2$	$-0\cdot92$	29	30
	$M + 2OH^- \rightleftharpoons M(OH)_2 + 2\varepsilon$	$-0\cdot14$	$-0\cdot07$	31	29
	$M + 4Cl^- \rightleftharpoons MCl_4^= + 2\varepsilon$	$-0\cdot75$	$-0\cdot59$	32, 33	29
	$M + 4Br^- \rightleftharpoons MBr_4^= + 2\varepsilon$	$-0\cdot67$	$-0\cdot49$	32	29
	$M + 4I^- \rightleftharpoons MI_4^= + 2\varepsilon$	$-0\cdot40$	$-0\cdot18$	32, 34	29
$M^{II} \rightleftharpoons M^{IV}$	$MCl_4^= + 2Cl^- \rightleftharpoons MCl_6^= + 2\varepsilon$	$-0\cdot77$	$-1\cdot26$	29	35
	$MBr_4^= + 2Br^- \rightleftharpoons MBr_6^= + 2\varepsilon$	$-0\cdot64$	$-0\cdot99$	34	36
	$MI_4^= + 2I^- \rightleftharpoons MI_6^= + 2\varepsilon$	$-0\cdot39$	$-0\cdot48$	32, 34	36

(ii) By contrast, the standard oxidation potentials for equilibria involving the divalent and tetravalent oxidation states are less (i.e. more negative) for palladium than for platinum; and platinum(II) compounds are easier to oxidise than palladium(II) compounds. This occurs, as mentioned under heats of formation (p. 12) because of the lower ionisation potential of platinum(IV) and the greater covalent bond energy, arising from greater orbital overlap in platinum(IV) compounds, outweighing the greater sublimation energy of platinum compared with palladium. It is of interest, in this connection, to note that the platinum(II) complex ion $[PtCl_4]^{2-}$ is unstable, disproportionating to the free metal and the tetravalent complex (equilibrium 5). Thus at 60°C the equilibrium constant $[PtCl_4^{2-}]^2/[PtCl_6^{2-}][Cl^-]^2$-

$$2[PtCl_4]^{2-} \rightleftharpoons Pt(0) + [PtCl_6]^{2-} + 2Cl^- \tag{5}$$

[Pt(0)] in a 3 M chloride or perchlorate medium is $(20 \pm 3) \times 10^{-3}$ when the cations are mainly protons and $(45 \pm 7) \times 10^{-3}$ when the cations are mainly sodium ions.[37]

(iii) Examination of both the M(0)/M(II) and M(II)/M(IV) oxidation potentials (Table 7) indicates that, as the halogen increases in atomic weight oxidation becomes more favourable. This observation, which is in direct contrast to that expected from the relative oxidation potentials of the halogens[13] ($E°$ $Cl^-/Cl_2 = -1\cdot36$ V; $E°$ $Br^-/Br_2 = -1\cdot09$ V; $E°$ $I^-/I_2 = -0\cdot54$ V), indicates a greater stability in aqueous solution of the complexes of the heavier halides. Indeed, this observation is the basis of Chatt's description of these metals as of class 'b' character[38] and Pearson's description of them as 'soft'[39] (see below).

CLASS 'b' OR 'SOFT' ACID CHARACTER†

Chatt[38] showed that metal ions could be conveniently divided into two groups, which he labelled as class 'a' and class 'b'. Class 'a' metals are those

† For reviews on this topic see References 38–43.

that prefer to form complexes with elements of lower atomic number within a given group of the Periodic Table, whereas class 'b' metals prefer to form complexes with elements of higher atomic number in that group. Thus with class 'b' metals the stabilities of the halide complexes decrease in the order $I^- > Br^- > Cl^- \gg F^-$. The Pearson theory[39] is very similar in that it states that 'hard' acids form their stronger complexes with the harder bases and soft acids form their stronger complexes with the softer bases. The properties 'hard' and 'soft' are related to polarisability so that the halide ions increase in softness with increasing atomic number (i.e. with increasing polarisability). Thus a Pearson 'soft acid' is equivalent to a Chatt class 'b' metal.

When Chatt said that platinum(II) and palladium(II) were class 'b' metal ions he had to rely on stability constants for the equilibria 6[44] and 7[45]. However, since then more experimental data has become available[29]

$$trans\text{-}[Pt(C_2H_4)(H_2O)Cl_2] + X^- \xrightleftharpoons{\text{(X = halide)}}$$
$$trans\text{-}[Pt(C_2H_4)XCl_2] + H_2O \quad (6)$$

$$[Pd(n\text{-}C_8H_{17}NH_2)_2Cl_2] + P(n\text{-}C_4H_9)_3 \rightleftharpoons$$
$$[P(n\text{-}C_4H_9)_3Pd(n\text{-}C_8H_{17}NH_2)Cl_2] + n\text{-}C_8H_{17}NH_2 \quad (7)$$

so that instead of comparing stability constants (i.e. free energies of formation) it is now possible to compare the free energies and also the enthalpies of formation of the halogen complexes of both divalent and tetravalent platinum and palladium.

(i) Relative free energies and enthalpies of formation of the divalent halogen complex ions $[MX_4]^{2-}$

The class 'b' character of the divalent metals could best be demonstrated by comparing both the standard free energies and the enthalpies for reaction 8. Although these have not been reported they can be evaluated from the free energies and enthalpies of reactions 9–11 since $\Delta H_1^\circ = \Delta H_4^\circ + \Delta H_2^\circ - 4(\Delta H_3^\circ)$

$$M^{2+}(aq.) + 4X^-(aq.) \rightleftharpoons MX_4^{2-}(aq.) \quad \Delta G_1^\circ, \Delta H_1^\circ \quad (8)$$

and the fact that $\Delta G_1^\circ = \Delta G_4^\circ + \Delta G_2^\circ - 4(\Delta G_3^\circ)$. Unfortunately accurate values are not available for the standard free energies and enthalpies of formation

$$M(cryst.) \rightarrow M^{2+}(aq.) + 2\varepsilon \quad \Delta G_2^\circ, \Delta H_2^\circ \quad (9)$$

$$\tfrac{1}{2}X_2(g) + \varepsilon \rightarrow X^-(aq.) \quad \Delta G_3^\circ, \Delta H_3^\circ \quad (10)$$

$$M(cryst.) + 2X_2(g) + 2\varepsilon \rightarrow MX_4^{2-}(aq.) \quad \Delta G_4^\circ, \Delta H_4^\circ \quad (11)$$

of the divalent metal ions in aqueous solution (ΔG_2° and ΔH_2°).[29] However, this is not serious since values of $\Delta H_1^\circ + \Delta H_2^\circ$ (Table 8) for a given metal with a series of halide ions show trends in the ΔH_1° values. Table 8 indicates that the free energy and enthalpy of formation decreases (the most favourable reaction is that for which the value of ΔG or ΔH is least positive) as the atomic number of the halogen increases: thus both free energies and enthalpies confirmed that platinum(II) and palladium(II) are both class 'b' or 'soft' acceptors.

TABLE 8

Relative free energies and enthalpies of formation of the halide complex anions $[MX_4]^{2-}$ in aqueous solution $(\Delta H_{aq.})^a$

Halogen	$\Delta G_{aq.}$ (kcal/mol)		$\Delta H_{aq.}$ (kcal/mol)	
	Pt	Pd	Pt	Pd
Cl	+34·40	+27·40	+38·08	+41·08
Br	+31·80	+23·80	+26·60	+23·60
I	+18·40	+8·40	+9·48	—

^{a} The value of $\Delta H_{aq.}$ is numerically equal to that of $\Delta H_1^0 + \Delta H_2^0$ (see text) which is evaluated from the equation $\Delta H_1^0 + \Delta H_2^0 = \Delta H_4^0 - 4\Delta H_3^0$. ΔH_3^0 values were obtained from reference 13 and ΔH_4^0 values from reference 29. ΔG values are defined and evaluated similarly.

Although the classification into class 'a' and 'b' character is useful, its evaluation on the basis of a reversal of the stability of the halide complexes from $F \gg Cl > Br > I$ to $F \ll Cl < Br < I$ on going from a class 'a' to a class 'b' acceptor is valid only for experiments in aqueous solution.[46] This is because the free energy and enthalpy of hydration of the halide ion decreases in the order $F^- > Cl^- > Br^- > I^-$ and also because for class 'b' metals, the free energies and enthalpies of hydration of the lighter halides are sufficient to render the metal complexes of these ions in aqueous solution less stable than the complexes of the heavier halides.

(ii) Relative free energies and enthalpies of formation of the tetravalent halogen complex anions $[PtX_6]^{2-}$

Although there is insufficient data available for the hexahalogenopalladate(IV) complex ions, entirely analogous calculations to those for the divalent complex anions may be carried out for the hexahalogenoplatinate(IV) anions. The results (Table 9) indicate that platinum(IV) fits Chatt's concept of a class 'b' metal ion.

Since class 'b' or 'soft' acid character increases with increasing numbers of d-electrons and also decreasing charge on the metal,[39] we may confidently predict, although no quantitative evidence is available, that since platinum(IV), platinum(II) and palladium(II) are all class 'b' in character both platinum(0) and palladium(0) will also be class 'b'. This is in agreement with qualitative evidence from the properties of the complexes of the zerovalent metals.

SURVEY OF THE OXIDATION STATES EXHIBITED BY PLATINUM AND PALLADIUM

Before surveying the oxidation states exhibited by platinum and palladium, we must define what we mean by oxidation state. It is to be taken as 'the formal charge left on the metal atom after all the ligands have been removed in their closed-shell configurations (for instance chlorine as Cl^-) and all metal–metal bonds have been cleaved homolytically (i.e. metal–metal bonds

The Chemistry of Platinum and Palladium

TABLE 9

**Relative free energies and enthalpies of formation of
$[PtX_6]^{2-}$ anions in aqueous solution**[a]

Halogen	$\Delta G_{aq.}$ (kcal/mol)	$\Delta H_{aq.}$ (kcal/mol)
Cl	+70·1	+78
Br	+58·5	+59
I	+37·1	+30

[a] $\Delta G_{aq.}$ is numerically equal to $\Delta G_5^0 + \Delta G_6^0$ in aqueous
solution, which is evaluated from the equation $\Delta G_5^0 +
\Delta G_6^0 = \Delta G_8^0 - 6\Delta G_7^0$. The ΔH values correspond to
the ΔG values.

$$Pt^{4+}(aq.) + 6X^- \rightarrow PtX_6^{2-}(aq.) \quad \Delta G_5^0, \Delta H_5^0$$

$$Pt(cryst.) \rightarrow Pt^{4+}(aq.)$$
$$+ 4\varepsilon \qquad \Delta G_6^0, \Delta H_6^0$$

$$\tfrac{1}{2}X_2(g) + \varepsilon \rightarrow X^-(aq.) \qquad \Delta G_7^0, \Delta H_7^0$$

$$Pt(cryst.) + 3X_2(g) + 2\varepsilon \rightarrow PtX_6^{2-}(aq.) \quad \Delta G_8^0, \Delta H_8^0$$

Values of ΔG_7^0 and ΔH_7^0 were obtained from reference
13 and values of ΔG_8^0 and ΔH_8^0 were obtained from
reference 29.

are ignored in the computation of oxidation states).'[47] There are, however,
some compounds for which it is difficult to ascribe an oxidation state to the
metal atom but these exceptions will be considered as they arise.

We shall now rapidly review platinum and palladium oxidation states,
stopping only to consider in detail those that are not covered in later chapters.

Negative oxidation states
Neither platinum nor palladium complexes in which the metals exhibit
negative oxidation states have been reported.

Zero oxidation state
Zerovalent platinum and palladium complexes have coordination numbers
between four and two (see Table 10) owing to the ease with which they give
rise to coordinately unsaturated complexes. More about these complexes
and also a number of cluster compounds of platinum(0) containing carbonyl
ligands appears in Chapter 3 (pp. 27–47).

+1 Oxidation state
Platinum and palladium in the +1 oxidation state is extremely rare. Plati-
num(I) monochloride, which may be trimeric Pt_3Cl_3,[48] has been reported
to be formed in the gas phase on heating platinum(II) dichloride above 330°C;
it decomposes at still higher temperatures to platinum metal and chlorine.[19]
The only other platinum(I) complex is the cluster compound $[Pt(SnCl_3)_2$-
$(C_8H_{12})_3]$ whose structure is shown on p. 94 (structure **22**). If the three
platinum–platinum bonds are considered to be single bonds the formal

TABLE 10

Co-ordination numbers and stereochemistries of zerovalent platinum and palladium complexes

Co-ordination number	Structure	Examples
4[a]	Tetrahedral[a]	$[Pt(PF_3)_4]$, $[(PPh_3)_3Pt(CO)]$, $[Pt\{P(OR)_3\}_4]$
3	Planar	$[Pt(PPh_3)_3]$
	Distorted planar	$[(PPh_3)_2PtX]$ where X = unsaturated ligand such as O_2, C_2H_4, CF_3CN, CS_2 etc.
2	Probably linear	$[Pt(PPh_3)_2]$
2(?)	Possibly linear	$K_2[Pd(C\equiv CR)_2]$
2(?)	Unknown	$[Pd(CNR)_2]$

[a] Most common.

oxidation state of platinum is $+\frac{2}{3}$, however these bonds are significantly shorter $(2·58(1)\ Å^{(49)})$ than those in other platinum cluster compounds such as $[Pt(PPhMe_2)(CO)_5]$, where the shortest platinum–platinum bond is $2·750(6)\ Å$.[50] This has led to the suggestion that the platinum–platinum bond order in $[Pt(SnCl_3)_2(C_8H_{12})_3]$ is between one and two and therefore that the platinum is in an approximately $+1$ oxidation state.[51]

Two palladium(I) complexes have been reported. One of these, $[Pd(C_6H_6)\text{-}(H_2O)(ClO_4)]_n$, has been prepared by reducing palladium(II) acetate in a benzene–acetic acid solvent containing perchloric acid with either benzene, hex-1-ene or phenylboronic acid.[51] The product is an explosive purple solid of unknown structure, which should only be prepared in small amounts. On treatment with halide ions it disproportionates precipitating palladium metal and leaving $[PdCl_4]^{2-}$ in solution. Oxygen, bromine, and permanganate ion all oxidise palladium(I) to palladium(II). The only other palladium(I) compound, prepared by treating palladium(II) dichloride in benzene with aluminium and aluminium trichloride, is $[PdAl_2Cl_7(C_6H_6)]_2$, which has been well-characterised by x-ray diffraction[52] (Appendix II, structure 50). It is an unusual compound, in which a dimeric Pd_2^{2+} unit lies sandwiched between two benzene rings.

Since only four complexes of platinum(I) and palladium(I) have been reported, this is clearly a rare oxidation state for these metals. In fact, neither element readily assumes an odd-oxidation state in its compounds. This is probably because such compounds involve an unpaired electron in an outer orbital and when such orbitals are exposed the situation is less stable than one with a pair of electrons in that outer orbital. Thus in the second and third transition series even-electron configurations are generally more stable than odd-electron configurations.[53] In connection with this instability arising from a single unpaired electron in an outer orbital it is noteworthy that at least two of the univalent complexes described above involve metal–metal bonds which enable these unpaired electrons to become paired.

+2 Oxidation state

The $+2$ oxidation state is easily the commonest state for both platinum and palladium. The co-ordination numbers and stereochemistries that have been

TABLE 11

Co-ordination numbers and stereochemistries of divalent platinum and palladium compounds and complexes

Co-ordination number	Structure	Examples
4[a]	Square-planar[a]	All well authenticated four-co-ordinate platinum(II) and palladium(II) compounds and complexes e.g. MO, MCl_2, $[MCl_4]^{2-}$ etc.
4	Tetrahedral	No well authenticated examples but bis(N,N'-ethylenedimorpholine)-palladium(II) diiodide may involve tetrahedral palladium(II).
5	Trigonal bipyramidal	$[Pt(SnCl_3)_5]^{3-}$, $[M(QAs)X]^{+}$ [b]
5	Square pyramidal	$[Pd(TPAs)Cl]^{+}$ [c]
5	Distorted square pyramid	$[Pd(PPI)_3Br_2]$ [d]
6	Octahedral	$K_2[Pt(NO)Cl_5]$, PdF_2

[a] Most common.
[b] QAs = **57** on p. 147.
[c] TPAs = **59** on p. 147.
[d] PPI = **63** on p. 149.

observed for divalent platinum and palladium are summarised in Table 11. No compounds or complexes of these divalent metals with a coordination number less than four are known, indeed most of the divalent compounds are four-co-ordinate. These four-co-ordinate complexes are all square-planar, or slight distortions of this (e.g. $[(PMe_3)_2PtCl_2]$[54]—Appendix II, structure 143). Only one tetrahedral complex of either palladium(II) or platinum(II) has been reported, although this is still tentative since it rests on the observation that a solution thought to contain bis (N,N'-ethylenedimorpholine)-palladium(II) diiodide is paramagnetic[55] and the complex itself has not yet been isolated (further details are on p. 117). A number of five-co-ordinate complexes of the divalent metals have been made. All involve ligands that are considered to be π-acceptors, such as $SnCl_3^-$, and ligands containing phosphorus, arsenic or antimony donor atoms. This can be rationalised by considering the effect of adding a fifth ligand to a four-co-ordinate metal ion. When the fifth ligand is added the energy of the σ-antibonding orbitals is raised and the energy of the π-bonding orbitals is lowered as a consequence of a reduction of the positive charge on the metal atom by the fifth ligand. In cases where all the ligands are purely σ-donors the bonding of a fifth ligand will be resisted as the overall strength of the complex is weakened. However, when the ligand is capable of π-bonding, five-co-ordination will be possible since it lowers the energy of the metal π-bonding orbitals. Thus the relative abilities of the group VB elements to stabilise five-co-ordination is N < P < As < Sb, which parallels the ability of these elements to take part in π-bonding (i.e. N ≪ P < As < Sb).[56]

Octahedral complexes of the divalent metals should be paramagnetic and indeed palladium difluoride, which has a distorted rutile structure in which palladium is surrounded by six fluorine atoms, four at 2·155 Å and two at 2·171 Å,[57] is indeed paramagnetic.[58] However, the only octahedral complexes of platinum(II) are the diamagnetic nitric oxide complexes in which nitric oxide is formally an NO^+ ligand. It has been suggested that the bonding in octahedral platinum(II) and palladium(II) complexes involves the use of four $nd(n + 1)s(n + 1)p^2$ hybrid orbitals in a plane and two $(n + 1)p(n + 1)d$ hybrid orbitals normal to that plane. This type of bonding should lead to a distorted octahedral structure due to the repulsion of the axial ligands by the filled nd_{z^2} orbital.[59] In agreement with this, both PdF_2 and $[Pd(diars)_2I_2]$ have long axial bonds (see Appendix II, structures 7 and 73 respectively).

Like five-co-ordinate complexes, the known six-co-ordinate complexes (except PdF_2) involve strong π-acceptor ligands such as NO^+ or tertiary arsines. This π-back-donation is essential in order to delocalise the excessive negative charge that would otherwise build up on the metal, so raising the energy of the σ-bonding orbitals and reducing the overall bonding energy in the complex. The only six-co-ordinate compound that does not involve a strong π-acceptor group is palladium difluoride and here the high electronegativity of fluorine prevents an excessive build-up of negative charge on the palladium. In this connection it is relevant to observe that palladium only very reluctantly forms five- and six-co-ordinate complexes $[PdCl_5]^{3-}$ and $[PdCl_6]^{4-}$ with the less electronegative chlorine.

The approximate relative energies of the d-orbitals in crystal fields of different symmetries are shown in Fig. 3, following the calculations of Hush.[60] Bearing in mind that the divalent metals ions have eight d-electrons, it is apparent why square-planar geometry is that most favoured by platinum(II) and palladium(II) which both exhibit large crystal-field splittings, since it is possible to leave one d-orbital (the $d_{x^2-y^2}$ orbital) empty. It is also apparent that both the five-co-ordinate structures should be observed since, again, these gave rise to a considerable ligand-field stabilisation if the highest energy d-orbital is left empty. Furthermore, octahedral stereochemistry would be expected to be rare, as in fact it is, and might, moreover, be expected to be distorted towards a square-planar structure since this would raise the energy of the $d_{x^2-y^2}$ orbital, which could then be left empty to give a diamagnetic product, and in addition lower the energy of the d_{z^2} orbital, thus giving an overall stabilisation relative to a regular octahedral structure.

+3 Oxidation state

There is no unequivocal evidence for either platinum or palladium occurring in the +3 oxidation state; this supports the observation that odd-electron configurations are much rarer than even-electron configurations for the ions of these elements (p. 17). A number of compounds and complexes, such as PdF_3 (see p. 276), $[M(am)_2X_3]$, where M = Pd or Pt, am = amine and X = halide (pp. 253–254) and $[Pt(am)_4X]^{2+}$ (pp. 253–254), which were originally thought to involve the trivalent metals are now known to contain two metal atoms, one in the divalent and the other in the tetravalent oxidation state. The 'trivalent' oxides M_2O_3 of both platinum[61] and

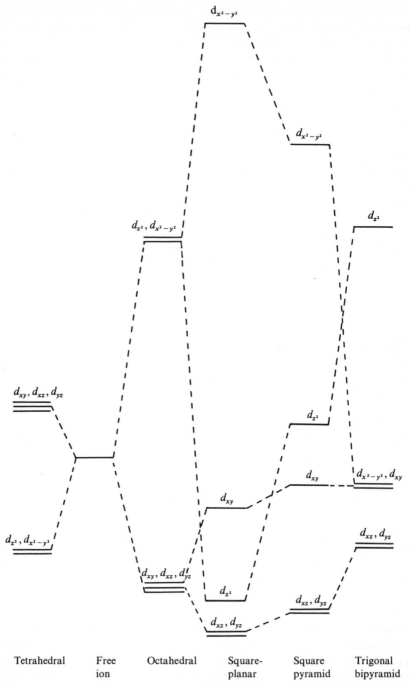

Fig. 3. Relative energies of the *d*-orbitals in different environments.

palladium[62,63] have been described, and although neither compound has been investigated by x-ray diffraction, Pt_2O_3 has been found to be dia-magnetic,[64] suggesting that it contains a mixture of platinum(II) and platinum(IV) ions rather than platinum(III) ions alone, when it would be paramagnetic. The halides $PtCl_3$[63] and $PtBr_3$[65] have both been described. Although their structures are unknown their very dark colours ($PtCl_3$ is almost black whereas $PtCl_2$ is yellow and $PtCl_4$ is red-brown) and methods for their preparation are consistent with the presence of both platinum(II) and platinum(IV) ions. Platinum triiodide, however, which is best prepared[66,67] by adding potassium iodide to a solution of hexachloroplatinic acid and heating the precipitate to 200°C is, from x-ray evidence, a true compound, not a mixture of PtI_2 and PtI_4. It may, however, be similar to PdF_3 (i.e. Pt^{II} [$Pt^{IV}I_6$]), which would be consistent with its observed diamagnetism.[66] On heating to 230°C platinum triiodide dissociates into the diiodide and iodine.

The green compound [$Pt(NH_3)_2I(SCN)_2$] formed on treating either *cis*-or *trans*-[$Pt(NH_3)_2(SCN)_2$] with iodine is thought, on the basis of esr spec-troscopy, to be a platinum(III) complex.[68] If it is, it is the first of this class and, accordingly, merits a detailed investigation. There is some evidence (sum-marised in Chapter 11, pp. 319–320) that platinum(III) complexes are formed as intermediates during both the chemical and electrochemical oxidation of platinum(II), and also reduction of platinum(IV) complexes.

+4 Oxidation state

Platinum and palladium form a number of compounds and complexes in which the metals are formally tetravalent. As discussed above (p. 13 and later in Chapter 10, pp. 251–252), there are many fewer palladium(IV) than platinum(IV) compounds, largely because of the much higher ionisation energy needed to produce 'Pd^{4+}' than 'Pt^{4+}' ions. All the compounds and complexes of the tetravalent metals have an octahedral or almost octahedral arrangement of groups around the metal atoms. The tetravalent compounds are all diamagnetic because of the considerable ligand field splitting which renders these d^6 metal ions 'low spin' with empty $d_{x^2-y^2}$ and d_{z^2} orbitals (see Fig. 3). A detailed description of the chemistry of the tetravalent state appears in Chapter 10 (pp. 251–285).

+5 Oxidation state

Palladium does not form any compounds in the +5 oxidation state, whereas platinum forms PtF_5, $PtOF_3$ and a series of hexafluoroplatinates(V) of which the most famous is the xenon compound $Xe^+[PtF_6]^-$. The most notable feature of platinum(V) chemistry is undoubtedly the ability of the hexafluoroplatinate(V) ion to stabilise compounds and complex ions of the noble gas xenon. Indeed the discovery of this compound led to the opening up of the field of noble gas chemistry. All the platinum(V) compounds of which the structures are known have octahedral co-ordination about the platinum atom (the stereochemistry of platinum in $PtOF_3$ is unknown), and those investigated are paramagnetic, with a single unpaired electron. Since the platinum(V) compounds are not discussed elsewhere in this book, a brief summary of the properties and structures of the more important ones is given below.

PtF_5

This is prepared by fluorinating $PtCl_2$ at about 350°C and is a deep-red solid which melts at 80°C to a viscous red liquid.[69] Moreover, it is paramagnetic with a moment of 2·05 *B.M.* and is isostructural with ruthenium pentafluoride, indicating that it contains slightly distorted tetrameric Pt_4F_{20} units (**1**).[70] On heating, PtF_5 disproportionates into PtF_4 and PtF_6. It reacts with xenon

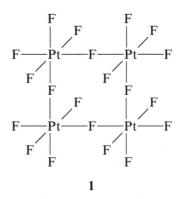

1

in the presence of excess fluorine at 80 lb/in² and 200°C to give $[XeF_5]^+[PtF_6]^-$ (see below).[70]

$PtOF_3$

$PtOF_3$ is a light-brown solid prepared by the fluorination of platinum dioxide at 200°C.[69]

$M[PtF_6]$ (M = alkali metal)

Pure alkali metal hexafluoroplatinates(V) are best made by treating dioxygenylhexafluoroplatinate(V) in iodine pentafluoride with an alkali metal fluoride.[69] Impure hexafluoroplatinates(V) results from the action of fluorine on a mixture of the platinum halide and alkali metal halide at 300°C.[71] The reaction product is contaminated with hexafluoroplatinate(IV) indicating how difficult it is to oxidise that complex. Potassium hexafluoroplatinate(V) is paramagnetic with a room temperature magnetic moment of 0·87 *B.M.*[69] X-ray diffraction has shown that the alkali metal hexafluoroplatinates(V) contain discrete octahedral $[PtF_6]^-$ anions.[69] The hexafluoroplatinates(V) all react violently with water, yielding chiefly platinum dioxide, hexafluoroplatinate(IV) and ozone.[69]

$(O_2^+)PtF_6^-$

Dioxygenyl hexafluoroplatinate(V) was first prepared by passing fluorine over platinum dichloride at 350°C in a silica apparatus.[72] A detailed examination showed that the compound was paramagnetic with a room temperature magnetic moment of 2·46 *B.M.*; this is consistent with two unpaired electrons one associated with the dioxygenyl cation and the other with the hexafluoroplatinate(V) anion. The x-ray diffraction pattern conformed with the formulation $(O_2^+)[PtF_6]^-$ (for further details see Appendix II structure 105).[73,74] The compound, which has also been synthesised by treating platinum(VI) hexafluoride with oxygen at room temperature, is very reactive; it forms oxygen and a bromine trifluoride–platinum tetrafluoride adduct with bromine trifluoride at room temperature and oxidises selenium tetrafluoride to selenium hexafluoride with the liberation of oxygen.[69,74] Both iodine pentafluoride and chlorine trifluoride react with dioxygenyl hexafluoroplatinate(V) to form their respective 1:1 adducts with platinum(V) pentafluoride.[69,74]

$(Xe^+)[PtF_6]^-$

Since xenon and oxygen have very similar oxidation potentials, Bartlett believed that if platinum(VI) hexafluoride could oxidise the oxygen molecule it should also be able to oxidise xenon. In agreement with this he found that xenon was oxidised by platinum(VI) hexafluoride, although the reaction was complex and yielded both $XePtF_6$ and $Xe(PtF_6)_2$.[70,75] These two compounds which can also be prepared by heating platinum wire in a mixture of xenon and fluorine[76,77] are deep-red solids which decompose at about 165°C with the evolution of xenon tetrafluoride, and are hydrolysed by water with the liberation of xenon and oxygen[75] (reaction 12). The infrared spectra and

$$2XePtF_6 + 6H_2O \rightarrow \uparrow 2Xe + \uparrow O_2 + PtO_2 + 12HF \qquad (12)$$

chemical reactions of the two compounds are consistent with their being $(Xe^+)[PtF_6]^-$ and $(Xe^{++})[PtF_6]_2^-$ respectively.[78]

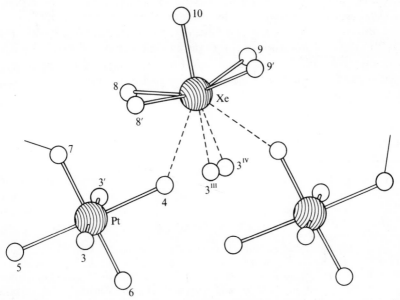

Fig. 4. Structure of $[XeF_5]^+[PtF_6]^-$ (the fluorine atoms have been displaced slightly for clarity) (Reproduced with permission from ref. 79).

$[XeF_5]^+[PtF_6]^-$

This stable, yellow crystalline compound, is formed when equimolar amounts of xenon and platinum(V) pentafluoride react with fluorine at 80 lb/in^2 in a bomb at about 200°C. A single crystal x-ray diffraction study indicated that the structure (Fig. 4) consists of approximately square pyramidal penta-fluoroxenon(VI) and octahedral hexafluoroplatinate(V) groups held together by weak fluorine bridges[79] (for further details see Appendix II, structure 106).

+6 Oxidation state

Only platinum forms a compound in which it exhibits the +6 oxidation state, this is the fluoride PtF$_6$. The compound PtOF$_4$, reported to be formed along with platinum(V) pentafluoride by fluorination of platinum sponge or anhydrous platinum(II) dichloride in a silica vessel at 350°C,[72] was subsequently shown to be dioxygenyl hexafluoroplatinate(V).[73]

PtF$_6$, an extremely reactive, thermally unstable compound, is formed by electrically heating a platinum wire in fluorine close to a surface cooled by liquid nitrogen.[80,81] It is a dark, red-black solid that melts at 61·3°C to give a dark-red liquid which boils at 69·14°C to give a monomeric brown vapour. It is thermally unstable, producing initially fluorine and platinum(V) pentafluoride; this disproportionates to the hexafluoride and tetrafluoride so that the overall decomposition products are fluorine and platinum(IV) tetrafluoride. Platinum(VI) hexafluoride is isostructural with the hexa-fluorides of tungsten, osmium and iridium[80] and its infra-red spectrum implies perfect octahedral symmetry[82] and indicates a platinum–fluorine stretching force constant of 4·52 mdyn/Å.[83] Platinum(VI) hexafluoride is an extremely powerful oxidising agent with an electron affinity in excess of 170 kcal/mol.[75] It reacts with oxygen and xenon at room temperature to give dioxygenyl hexafluoroplatinate(V) and a mixture of XePtF$_6$ and Xe[PtF$_6$]$_2$.[74,75] With nitrosylfluoride and nitric oxide it gives the para-magnetic platinum(V) compound NOPtF$_6$ although in the presence of excess nitric oxide some of the diamagnetic platinum(V) compound (NO)$_2$PtF$_6$ is formed.[84,85]

Oxidation states higher than +6

Neither platinum nor palladium form any compounds in which they exhibit oxidation states in excess of +6.

REFERENCES

1   *Interatomic Distances*, ed. by L. E. Sutton, Chemical Society Special Publications, Nos. 11 and 18, 1958 and 1965.

2   J. O. Glanville, J. M. Stewart and S. O. Grim, *J. Organometal. Chem.*, 7 (1967) p. 9.

3   R. Ugo, F. Conti, S. Cenini, R. Mason and G. B. Robertson, *Chem. Comm.*, (1968) 1498.

4   L. Pauling, *The Nature of the Chemical Bond*, Cornell University Press, 3rd edn., 1960.

5   J. D. Bell, D. Hall and T. N. Waters, *Acta Cryst.*, 21 (1966), 440.

6   L. H. Ahrens, *Geochim. et Cosmochim. Acta.*, 2 (1952) 155.

7   R. D. Shannon and C. T. Prewitt, *Acta Cryst.*, B25 (1969), 925.

8   C. E. Moore, *Atomic Energy Levels*, National Bureau of Standards Circular C467, vol. 3 (1958).

9   F. A. Cotton and G. Wilkinson, *Advanced Inorganic Chemistry*, Interscience, 2nd edn., 1966, p. 910.

10 Landolt-Börnstein, *Eigenschaften Der Materie in Ihren Aggregatzustanden*, Springer Verlag, 1961, vol. 4, pp. 238–9.

11 C. K. Jørgenson, *Orbitals in Atoms and Molecules*, Academic Press, London, 1962, p. 95.

12 W. van Bronswijk and R. S. Nyholm, *J. Chem. Soc.* (*A*), (1968) 2084.

13 W. M. Latimer, *The Oxidation States of the Elements and their Potentials in Aqueous Solution*, Prentice-Hall, New York, 2nd edn., 1952.

14 F. Grønvold, T. Thurmann-Moe, E. F. Westrum and E. Chang, *J. Chem. Phys.*, **35** (1961), 1665.

15 T. C. Waddington, *Adv. Inorg. Radiochem.*, **1** (1959), 157.

16 F. Grønvold, H. Haraldsen and A. Kjekshus, *Acta. Chem. Scand.*, **14** (1964), 1879.

17 J. Thomsen, *Thermochemistry*, Longmans Greens and Co., London, 1908.

18 Joannis, *Compt. Rend.*, **95** (1882) 295.

19 S. A. Shchukarev, M. A. Oranskaya and T. S. Shemyakina, *Zhur. Neorg. Khim.*, **1** (1956) 17; *Chem. Abs.*, **50** (1956) 9833i.

20 W. E. Bell and M. Tagami, *J. Phys. Chem.*, **70** (1966) 3736.

21 M. A. Oranskaya and N. A. Mikhailova, *Russ. J. Inorg. Chem.*, **5** (1961) 5.

22 F. Puche, *Ann. Chim.*, **9** (1938), 233; *Chem. Abs.*, **32** (1938), 5322.

23 S. A. Shchukarev, T. A. Tolmacheva, M. A. Oranskaya and L. V. Komandrovskaya, *Zhur. Neorg. Khim.*, **1** (1956) 8; *Chem. Abs.*, **50** (1956) 9833g.

24 V. A. Palkin, N. N. Kuz'mina and I. I. Chernyaev, *Russ J. Inorg. Chem.*, **10** (1965) 978.

25 W. Biltz and R. Juza, *Z. anorg. allgem. Chem.*, **190** (1930) 161.

26 E. F. Westrum, H. G. Carlson, F. Grønvold and A. Kjekshus, *J. Chem. Phys.*, **35** (1961) 1670.

27 W. Biltz and J. Laar, *Z. anorg. allgem. Chem.*, **238** (1936) 257.

28 L. V. Coulter, K. S. Pitzer and W. M. Latimer, *J. Amer. Chem. Soc.*, **62** (1940) 2845.

29 R. N. Goldberg and L. G. Hepler, *Chem. Rev.*, **68** (1968) 229.

30 R. M. Izatt, D. J. Eatough, C. E. Morgan and J. J. Christensen, *J. Chem. Soc.* (*A*), (1970) 2514.

31 D. T. Sawyer and L. V. Interrante, *J. Electroanal Chem.*, **2** (1961) 310.

32 A. A. Grinberg and M. I. Gel'fman, *Dokl. Akad. Nauk. SSSR*, **133** (1960) 1081.

33 H. Yamamoto, S. Tanaka, T. Nagai and T. Takei, *Denki Kagaku*, **32** (1964) 43; *Chem. Abs.*, **61** (1964) 10317d.

34 A. A. Grinberg, B. V. Ptitsyn and V. N. Lavrent'ev, *Zh. Fiz. Khim.*, **10** (1937) 661; *Chem. Abs.*, **32** (1938) 5719.

35 H. B. Wellman, *J. Amer. Chem. Soc.*, **52** (1930) 985.

36 A. A. Grinberg and A. S. Shamsiev, *Zh. Obsch. Khim.*, **12** (1942) 55; *Chem. Abs.*, **37** (1943) 1915.

37 O. Ginstrup and I. Leden, *Acta. Chem. Scand.*, **21** (1967) 2689.

38 St. Ahrland, J. Chatt and N. R. Davies, *Quart. Rev.*, **12** (1958) 265.

39 R. G. Pearson, *J. Amer. Chem. Soc.*, **85** (1963) 3533.

40 R. G. Pearson, *Science*, **151** (1966) 172; *Chemistry in Britain*, **3** (1967) 103.

41 R. J. P. Williams and J. D. Hale, *Structure and Bonding*, **1** (1966) 249.

42 St. Ahrland, *Structure and Bonding*, **1** (1966) 207; **5** (1968) 118.

43 C. K. Jørgenson, *Structure and Bonding*, **3** (1967) 106.

44 I. Leden and J. Chatt, *J. Chem. Soc.*, (1955) 2936.

45 B. Meddings and A. R. Burkin, *J. Chem. Soc.*, (1956) 1115.

46 F. Basolo and R. G. Pearson, *Mechanisms of Inorganic Reactions*, John Wiley, 1st edn., 1958, p. 180.

47 R. S. Nyholm and M. L. Tobe, *Adv. Inorg. Radiochem.*, **5** (1963) 1.

48 A. Landsberg and J. L. Schaller, *J. Less-Common Metals*, **23** (1971) 195.

49 L. J. Guggenberger, *Chem. Comm.*, (1968) 512.

50 R. G. Vranka, L. F. Dahl, P. Chini and J. Chatt, *J. Amer. Chem. Soc.*, **91** (1969) 1574.

51 J. M. Davidson and C. Triggs, *J. Chem. Soc. (A)*, (1968) 1324.
52 G. Allegra, A. Immirzi and L. Porri, *J. Amer. Chem. Soc.*, **87** (1965) 1394.
53 C. S. G. Phillips and R. J. P. Williams, *Inorganic Chemistry*, Oxford, 1966, vol. 2, p. 294.
54 G. G. Messmer, E. L. Amma and J. A. Ibers, *Inorg. Chem.*, **6** (1967) 725.
55 A. L. Lott and P. G. Rasmussen, *J. Amer. Chem. Soc.*, **91** (1969) 6502.
56 A. D. Westland, *J. Chem. Soc.*, (1965) 3060.
57 N. Bartlett and R. Maitland, *Acta Cryst.*, **11** (1958) 747.
58 N. Bartlett and M. A. Hepworth, *Chem. Ind. (London)*, (1956) 1425.
59 C. M. Harris and R. S. Nyholm, *J. Chem. Soc.*, (1956) 4375.
60 N. S. Hush, *Austral. J. Chem.*, **15** (1962) 378.
61 L. Wöhler and F. Martin, *Z. anorg. allgem. Chem.*, **57** (1908) 398; *Chem. Abs.*, **2** (1908) 2054.
62 W. L. Dudley, *J. Amer. Chem. Soc.*, **28** (1902) 59.
63 L. Wöhler and F. Martin, *Ber.*, **42** (1909) 3958.
64 Ya. K. Syrkin and V. I. Belova, *Zhur. Fiz. Khim.*, **23** (1949) 664; *Chem. Abs.*, **43** (1949) 7277g.
65 L. Wöhler and F. Müller, *Z. anorg. allgem. Chem.*, **149** (1925) 377; *Chem. Abs.*, **20** (1926) 718.
66 G. R. Argue and J. J. Banewicz, *J. Inorg. Nucl. Chem.*, **25** (1963) 923.
67 S. A. Shchukarev, T. A. Tolmacheva and G. M. Slavutskaya, *Russ. J. Inorg. Chem.*, **9** (1964) 1351.
68 G. S. Muraveiskaya, G. M. Larin and V. F. Sorokina, *Russ. J. Inorg. Chem.*, **13** (1968) 771.
69 N. Bartlett and D. H. Lohmann, *J. Chem. Soc.*, (1964) 619.
70 N. Bartlett, F. Einstein, D. F. Stewart and J. Trotter, *Chem. Comm.*, (1966) 550.
71 N. Bartlett, *Preparative Inorganic Reactions*, **2** (1965) 301.
72 N. Bartlett and D. H. Lohmann, *Proc. Chem. Soc.*, (1960) 14.
73 N. Bartlett and D. H. Lohmann, *Proc. Chem. Soc.*, (1962) 115.
74 N. Bartlett and D. H. Lohmann, *J. Chem. Soc.*, (1962) 5253.
75 N. Bartlett, *Proc. Chem. Soc.*, (1962) 218.
76 F. Mahieux, *Compt. Rend.*, **257** (1963) 1083.
77 F. Mahieux, *Compt. Rend.*, **258** (1964) 3497.
78 N. Bartlett and N. K. Jha, *Noble Gas Compounds*, ed. by H. H. Hyman, Univ. of Chicago Press, 1963, p. 23.
79 N. Bartlett, F. Einstein, D. F. Stewart and J. Trotter, *J. Chem. Soc. (A)*, (1967) 1190.
80 B. Weinstock, H. H. Claasen and J. G. Malm, *J. Amer. Chem. Soc.*, **79** (1957) 5832.
81 B. Weinstock, J. G. Malm and E. E. Weaver, *J. Amer. Chem. Soc.*, **83** (1961) 4310.
82 B. Weinstock, H. H. Claasen and J. G. Malm, *J. Chem. Phys.*, **32** (1960) 181.
83 B. Weinstock and G. L. Goodman, *Adv. Chem. Phys.*, **9** (1965) 169.
84 N. Bartlett, S. P. Beaton and N. K. Jha, *Chem. Comm.*, (1966) 168.
85 F. P. Gortsema and R. H. Toeniskoetter, *Inorg. Chem.*, **5** (1966) 1217.

CHAPTER 3

Complexes of the Zerovalent Metals

The compounds and complexes of platinum and palladium in their zero oxidation states (see Chapter 2, pp. 16–17) all involve d^{10} metal ions. Since the fifties these compounds have been of particular interest partly because they are considered to be models for the 'compounds' formed when small molecules are chemisorbed on to the bulk metal during heterogeneous catalysis and partly because of the intrinsic interest of compounds in which the metal is in a very low oxidation state. Zerovalent complexes of platinum and palladium have been prepared with phosphine,[1,2] arsine,[1,2] phosphite,[1] isocyanide,[3,4] cyanide,[5] acetylide,[6] ammonia[7] and nitric oxide[8] ligands. In the present chapter the preparation of these compounds is considered first and then, after a discussion of their structures, the electronic properties of the ligands that stabilise these complexes are considered. It is concluded that whereas the stabilisation of low oxidation states of the transition metals is generally associated with ligands that are strong π-acceptors, such as carbon monoxide, the stabilisation of zerovalent platinum and palladium complexes requires strong σ-donor ligands that may also be capable of π-acceptance. The chapter concludes with a discussion of the chemical properties of zerovalent complexes, and in the case of the tertiary phosphine complexes these are considered in terms of the two main classes of reactions that these complexes undergo, namely reactions involving co-ordinative dissociation in which the zerovalent metal is oxidised to the +2 oxidation state during the course of the reaction, and reactions involving co-ordinative association in which the metal retains its zero oxidation state throughout the reaction.

PREPARATION OF THE ZEROVALENT COMPLEXES

(a) Phosphine and arsine complexes

Phosphine and arsine complexes of platinum(0) and palladium(0) were first synthesised in 1957–8 by reduction of the corresponding platinum(II) and palladium(II) complexes with hydrazine or alcoholic potassium hydroxide.[1,9] The original preparations, together with more recent procedures involving the use of sodium borohydride,[10] trifluorophosphine[11] or triphenylphosphine itself[1] as reducing agents, are given in Appendix I (pp. 452–454). Although most of the preparative routes have used complexes

27

of the divalent metals, it has recently been found that $[Pd(PPh_3)_2]$ can be formed by heating metallic palladium with triphenylphosphine in the presence of excess triethylsilane.[12] Soon after zerovalent complexes were first reported divalent hydrido-complexes were prepared by a very similar method.[13] This, together with the remarkable stability of the zerovalent complexes, led to the suggestion that the original zerovalent complexes were in fact hydrido-complexes.[14] However, a number of authors independently demonstrated that the complexes claimed to be zerovalent were really so.[15–17] The reason for the early confusion has recently become apparent. In the first stage of the reduction of platinum(II) complexes hydrido-complexes are formed, which can be dehydrogenated in the strongly basic mother media to give zerovalent complexes.[18] Thus whether the product be a hydrido-platinum(II) complex or a platinum(0) complex depends critically on the temperature, concentration of the reducing agent, reaction time and the nature of the tertiary phosphine.

Both the hydrazine and the ethanolic potassium hydroxide reductions are extremely complex. The hydrazine reduction is thought to follow the reaction scheme in Fig. 5, giving a mixture of the two nitrogen-bridged complexes

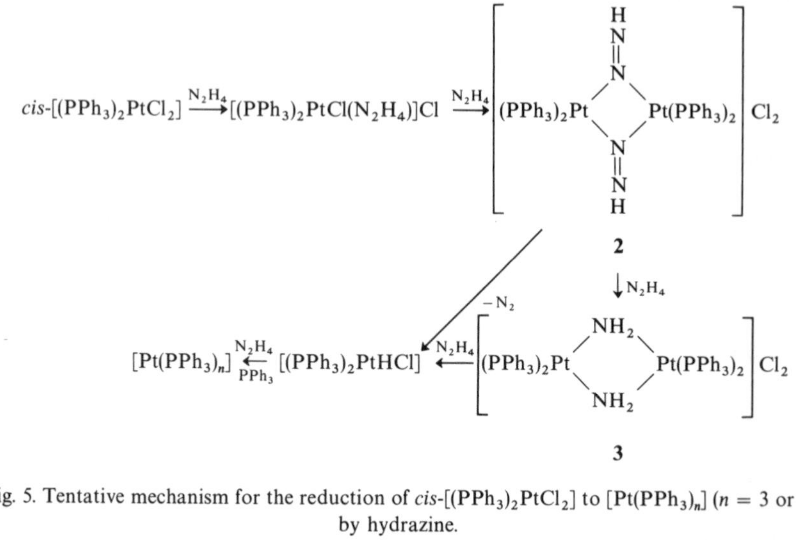

Fig. 5. Tentative mechanism for the reduction of *cis*-$[(PPh_3)_2PtCl_2]$ to $[Pt(PPh_3)_n]$ ($n = 3$ or 4) by hydrazine.

(**2** and **3**) which have been isolated and examined by x-ray diffraction[19] (for further details see Appendix II, structure 212). The alcohol reduction is thought to occur by one of two routes (Fig. 6), depending on the nature of the alcohol, the time and temperature of the reaction and the stoichiometry of the products. Route (a) involves the phosphine as the reducing agent and is characterised by the formation of triphenylphosphine oxide whereas route (b) involves the alcohol acting as the reducing agent and is characterised by the formation of carbonyl complexes.[20] This mechanism was supported by the observation that on treating $[(PPh_3)_2MCl_2]$, where M = Pt or Pd, with

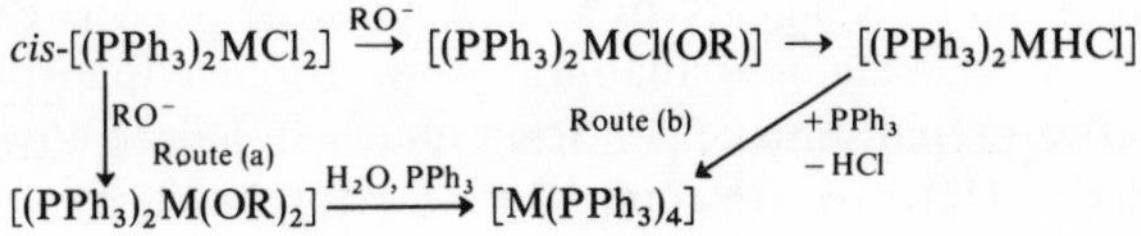

$$cis\text{-}[(PPh_3)_2MCl_2] \xrightarrow{RO^-} [(PPh_3)_2MCl(OR)] \rightarrow [(PPh_3)_2MHCl]$$

$$\Big\downarrow \begin{matrix} RO^- \\ Route\ (a) \end{matrix} \qquad Route\ (b) \qquad \begin{matrix} +PPh_3 \\ -HCl \end{matrix}$$

$$[(PPh_3)_2M(OR)_2] \xrightarrow{H_2O,\ PPh_3} [M(PPh_3)_4]$$

Fig. 6. Mechanism for the reduction of *cis*-[(PPh$_3$)$_2$MCl$_2$] by alcoholic potassium hydroxide.

potassium phenoxide in benzene solution [(PPh$_3$)$_2$M(OPh)$_2$] is formed. The bis(phenoxide) platinum(II) complex reacts with excess triphenylphosphine at 25°C to give [Pt(PPh$_3$)$_4$] in quantitative yields whereas the palladium(II) complex [(PPh$_3$)$_2$Pd(OPh)$_2$] only reacts with triphenylphosphine to give [Pd(PPh$_3$)$_4$] in the presence of water.

The first zerovalent complexes that were prepared had tertiary aryl phosphine ligands. The aryl groups used were either phenyl or substituted phenyl groups.[1,9] Since then the range of phosphines has been extended to include trifluorophosphine,[21] mixed alkylfluorophosphines,[22] mono-alkyl-di-arylphosphines,[23] such as PMePh$_2$, as well as chelating phosphines and arsines, such as Ph$_2$PCH$_2$CH$_2$PPh$_2$, *o*-C$_6$H$_4$(AsMe$_2$)$_2$ and Me$_2$PCH$_2$CH$_2$PMe$_2$.[24] The fully alkylated phosphine, PEt$_3$, has only been reported[25] in the zerovalent complex [(PEt$_3$)$_2$Pt(TCNE)] (TCNE = tetracyanoethylene), where the stronger σ-donor and weaker π-acceptor properties of the trialkyl- relative to the triaryl-phosphine are acceptable because of the very strong π-acceptor ability of the tetracyanoethylene.

(b) Phosphite complexes

Phosphite complexes of platinum(0) and palladium(0) have been prepared by hydrazine reduction of the divalent phosphite complexes[9] (reaction 13) or by displacing the phosphine[9] (reaction 14) or isocyanide[1] (reaction 15) ligands

$$[\{P(OPh)_3\}_2PtCl_2] + N_2H_4 + P(OPh)_3 \xrightarrow{C_2H_5OH} [Pt\{P(OPh)_3\}_{3\ or\ 4}] \qquad (13)$$

from their zerovalent complexes.

$$[Pt(PPh_3)_4] + 4P(OPh)_3 \rightarrow [Pt\{P(OPh)_3\}_4] + 4PPh_3 \qquad (14)$$

$$[Pd(p\text{-}CH_3C_6H_4NC)_2] + 4P(OPh)_3 \rightarrow$$
$$[Pd\{P(OPh)_3\}_4] + 2p\text{-}CH_3C_6H_4NC \qquad (15)$$

(c) Isocyanide complexes

No platinum(0) isocyanide complexes have been prepared.[4] Palladium(0) isocyanide complexes are prepared by reducing the palladium(II) isocyanide complexes with isocyanide in the presence of strong alkali (reaction 16).[26,27] It is of interest how specific the conditions of this reaction are, since palladium(II) isocyanide complexes cannot be reduced to the zerovalent state

$$[PdI_2(CNR)_2] + 2KOH \rightarrow 2KI + [Pd(OH)_2(CNR)_2] \xrightarrow{3RNC}$$
$$[Pd(CNR)_4] + RNCO + H_2O$$
$$\downarrow$$
$$[Pd(CNR)_2] + 2RNC \qquad (16)$$

even by strong reducing agents in acid, neutral or even slightly alkaline solution.

An alternative preparation of $[Pd(CNR)_2]$ which involves treating either $[(\pi-C_5H_5)Pd(\pi-C_6H_9)]$ or $[(\pi-C_5H_5)Pd(\pi-C_3H_5)]$ with the appropriate isocyanide has also been described.[28]

(d) Carbonyl complexes

Platinum(0)–carbonyl complexes containing no other ligands can be prepared either by the action of water on a benzene solution of $[PtCl_2(CO)_2]$ under carbon monoxide or by the action of carbon monoxide on an ethanolic solution of Na_2PtCl_4.[29,30] The product, which is a brown crystalline material that is sensitive to air, is polymeric with an empirical formula $[Pt(CO)_2]_n$. It is insoluble in all single solvents.

Monomeric platinum(0)–carbonyl complexes have been prepared by treating platinum(0)–phosphine complexes in hydrocarbon suspension under pressure with carbon monoxide[9] (reaction 17), but when tetrahydrofuran or acetone are used $[Pt(PPh_3)_3CO]$ is formed very rapidly at atmospheric

$$[Pt(PPh_3)_{3 \text{ or } 4}] + 2CO \rightarrow [Pt(PPh_3)_2(CO)_2] \xrightarrow[PPh_3]{1 \text{ mol of}} [Pt(PPh_3)_3CO] \quad (17)$$

pressure.[31] The trimeric $[Pt_3(PPh_3)_4(CO)_3]$ results from boiling an ethanolic solution of $[Pt(PPh_3)_2(CO)_2]$[32] or by treating a suspension of $[(Pt(CO)_2]_n$ in acetone with triphenylphosphine.[29] Treatment of $[Pt_3(PPh_3)_4(CO)_3]$ with further carbon monoxide at room temperature and atmospheric pressure yields the tetrameric complex $[Pt_4(PPh_3)_4(CO)_5]$.[32]

(e) Other complexes of the zerovalent metals

A zerovalent cyanide complex of palladium(0), $K_4[Pd(CN)_4]$ has been prepared by treating $K_2(Pd(CN)_4]$ with potassium in liquid ammonia.[5] The ammine $[Pt(NH_3)_4]$ was prepared similarly from $[Pt(NH_3)_4]Br_2$.[7] Acetylide complexes of both platinum(0)[6] and palladium(0)[33] have been prepared by treating the divalent cyanides with potassium acetylide in liquid ammonia and then treating the resulting divalent cyano–acetylide complexes with potassium in liquid ammonia (reaction 18; R = H, CH_3 or C_6H_5). A preliminary account has appeared of the preparation of a neutral diolefin

$$Pd(CN)_2 + 2KC \equiv CR \xrightarrow{\text{liq. NH}_3} K_2[Pd(CN)_2(C \equiv CR)_2] \xrightarrow[\text{liq. NH}_3]{+2K \text{ in}}$$
$$K_2[Pd(C \equiv CR)_2] + 2KCN \quad (18)$$

complex of palladium(0), $[Pd(PhCH=CHCOCH=CHPh)_2]$, by treating a hot methanolic solution of Na_2PdCl_4 containing excess dibenzylidene-acetone with sodium acetate.[34] Nitric oxide complexes of zerovalent palladium are discussed in Chapter 7 (pp. 127–128).

STRUCTURE OF THE ZEROVALENT COMPLEXES

(a) Phosphine complexes

Both the four-co-ordinate phosphine complexes which have been examined, $[Pt(PF_3)_4]$[35] and $[(PPh_3)_3Pt(CO)]$,[36,37] have been found to be tetrahedral (for further details see Appendix II, structure 210). The tetrakistrifluoro-phosphine complexes are stable in solution and do not lose trifluorophos-phine.[22] However, the triphenylphosphine complexes all dissociate in solution to give co-ordinatively unsaturated species (reaction 19). In the case of the palladium(0) complexes this dissociation has been inferred from the very low molecular weights found in benzene solution[1] and the isolation of

$$[M(PPh_3)_4] \overset{(M\,=\,Pd,Pt)}{\rightleftharpoons} [M(PPh_3)_3] \rightleftharpoons [M(PPh_3)_2] \rightleftharpoons [M(PPh_3)]_n \qquad (19)$$

trico-ordinated complexes such as $[Pd\{(p\text{-}ClC_6H_4)_3P\}_3]$ and $[Pd\{(p\text{-}CH_3C_6H_4)_3P\}_3]^{[1]}$ and dico-ordinated complexes such as $[Pd(PPh_3)_2]$ and $[Pd\{P(cyclohexyl)_3\}_2]$.[38] For the platinum(0) complexes the evidence in favour of reaction 19 is more detailed. Stable trico-ordinated species have been isolated with a number of phosphine ligands[9] and for $[Pt(PPh_3)_3]$ an x-ray diffraction study showed that the packing of the phenyl rings in the crystal leads to a distorted planar structure with the three phosphorus atoms in a plane and the platinum atom lying 0·1 Å above that plane[39] (for further details see Appendix II, structure 209). $[Pt(PPh_3)_3]$ and $[Pt(PPh_3)_4]$ can be distinguished by their photoluminescence since the former appears red and the latter bright orange under a 'black light' source.[40] The dico-ordinated species $[Pt(PPh_3)_2]$ has been isolated as a yellow compound that is stable for a few hours in the solid state,[41,42] but unstable in solution where it dissociates further into the stable diamagnetic tetrameric complex $[Pt(PPh_3)]_4$.[43] This is in contrast to $[Pd(PPh_3)_2]$ which disproportionates in solution to $[Pd(PPh_3)_4]$ and palladium metal.[38]

The relatively high stabilities of the co-ordinatively unsaturated complexes suggests that the phosphine ligands are strong σ-donors but relatively poor π-acceptor ligands so that there is a build-up of negative charge on the metal. It has been suggested that phosphine ligands can dissipate this excess negative charge by π-back acceptance more effectively in the trigonal and linear complexes than in a tetrahedral complex.[44] The decrease of the excess negative charge with decreasing co-ordination number is demonstrated by the tendency of the low co-ordination number complexes to increase their co-ordination number. This arises from the electrostatic repulsion of the σ-electron pair of the entering ligands by the negative charge on the metal.

(b) Phosphite complexes

The tetrakis(triethylphosphite) complexes of platinum(0) and palladium(0) are not appreciably dissociated in solution.[45] This reflects the weaker σ-donor properties and better π-acceptor properties of the phosphite ligands relative to the phosphine ligands, which prevents a build-up of negative charge on the metal in the phosphite complexes. The infra-red spectra of the phosphite complexes are consistent with tetrahedral co-ordination around the metal.[46]

(c) Isocyanide complexes

The structures of the palladium(0) isocyanide complexes, $[Pd(CNR)_2]$, are unknown. The two most plausible suggestions that have been put forward are:

  (i) They are monomeric with linear C–Pd–C bonds. This is supported
      by the strong σ-donor and weak π-acceptor properties of isocya-
      nides.[3]

 (ii) They are polymeric with palladium–palladium bonding.[27] This is
      supported by the insolubility of the complexes.

ELECTRONIC PROPERTIES OF ZEROVALENT COMPLEXES

Whereas the stabilisation of low oxidation states is commonly associated
with ligands such as carbon monoxide, which is a relatively poor σ-donor but
strong π-acceptor, the stabilisation of zerovalent platinum and palladium
complexes is associated with strong σ-donor ligands that are also capable of
some π-acceptance. The reduced importance of π-back-donation in the
platinum(0) and palladium(0) complexes parallels the high ionisation
potentials of these metals[47] (8·20 and 8·33 eV respectively), which are in
contrast to the lower ionisation potential of nickel(0) (5·83 eV). It is significant
that of the three metals only nickel(0) forms a simple monomeric complex
with the strong π-acceptor ligand carbon monoxide. The force constants for
the complexes $[M\{P(OEt)_3\}_4]$ increase in the order Ni(0) < Pd(0) <
Pt(0).[46] Since this is the order of decreasing π-bonding ability, it supports
the contention made above that the σ-bond is the factor primarily responsible
for the stability of the zerovalent complexes. Similarly platinum alone forms
zerovalent complexes with the purely σ-donor ligands ammonia[7] and
ethylenediamine.[48]

The photoelectron spectrum of $[Pt(PF_3)_4]$ suggests that there is little
net migration of charge to or from platinum.[49,50] This, combined with the
n.m.r. data on a series of phosphine and phosphite complexes,[51] demonstrates that the σ-donor ability increases from left to right and the π-acceptor
capacity decreases from left to right across the series PF_3, $P(OR)_3$, PR_3. It
is noteworthy that the tendency of the metal(0) complexes towards dissociation in solution parallels the increase in σ-donor ability. The platinum
$4f_{7/2}$ electron binding energies in a series of platinum complexes were studied
by photoelectron spectroscopy and the ESCA shifts assumed to be proportional to the charge on the platinum atom.[52] In this way it was shown that
the net positive charge on the platinum atom in $[(PPh_3)_2PtX]$ complexes
increased across the series X = $2PPh_3$ < $PhC\equiv CPh$ < C_2H_4 < CS_2 <
O_2 < Cl_2. This can be understood in terms of the molecular-orbital description of the bonding in π-complexes of zerovalent platinum (pp. 400–403)
if it is assumed that the π-back-donation from the metal to olefin or acetylene
is more important than the σ-donation from olefin or acetylene to metal.
This assumption is supported by a number of observations (see especially
pp. 406–407) and the present results further suggest that the carbon disulphide
and oxygen complexes of platinum(0) involve substantial π-back-donation
of charge from the metal to the π^*-(antibonding) orbitals of the ligand and
rather less σ-donation of charge from the ligand to the metal.

CHEMICAL PROPERTIES OF THE PHOSPHINE COMPLEXES

Of the various zerovalent complexes those containing tertiary phosphine ligands have been the most widely investigated. This is partly due to their versatility and stability and partly because, in contrast to the isocyanides, they are soluble in a number of common organic solvents. It has been demonstrated conclusively that the reactions of triphenylphosphine platinum(0) complexes occur by an initial rate-determining dissociation reaction yielding $[Pt(PPh_3)_2]$ which then reacts rapidly with the incoming ligand.[53-55]

The reactions of $[Pt(PPh_3)_2]$ with an incoming ligand fall into two classes:

(i) In the first group the incoming molecule is dissociated into two fragments both of which co-ordinate to the metal and the metal is oxidised to the divalent state (reaction 20).

$$[M(PPh_3)_2] + XY \rightleftharpoons [(PPh_3)_2M(X)(Y)] \qquad (20)$$

(ii) In the second group the incoming molecule does not dissociate so that there is no change in the formal oxidation state of the metal (reaction 21)

$$[M(PPh_3)_2] + XY \rightleftharpoons [(PPh_3)_2M(XY)] \qquad (21)$$

The first step of both reactions 20 and 21 involves donation of electron density from the incoming XY molecule to an empty s or sp hybrid orbital on the metal. At the same time some of the excess negative charge on the metal is transferred to the XY molecule. If the XY molecule has no low-energy antibonding orbitals (e.g. HCl or CH_3Cl) then it dissociates and the excess electron density of the metal is transferred to the more electronegative fragment. The transfer of further electron density from the metal to the other fragment results in the co-ordination of both fragments and the oxidation of the metal to the divalent state. If, however, the XY molecule has low-energy antibonding orbitals (such as olefins, acetylenes or carbon monoxide) then this molecule can delocalise the excess negative charge on the metal by means of a π-mechanism and so form a stable zerovalent addition product.

The most important electronic factor that favours initial co-ordination of XY is the ability of the empty np_z orbital on the metal to hybridise with the filled $(n - 1)d_{z^2}$ orbital to provide two dp hybrid orbitals one of which can interact with the incoming XY molecule. Thus the activation energy of the association reaction will be to a large extent related to the $(n - 1)d$ to np promotion energy. This is less for platinum(0) (3.28 eV[47]) than for palladium(0) (4.23 eV[47]) and so the co-ordinative reactivity of palladium(0) might be expected to be less than that of platinum(0). Whilst it is certainly true that the co-ordinative reactivity of platinum(0) has been investigated more extensively than that of palladium(0), this is probably less a consequence of electronic factors than of the fact that palladium(0) complexes are less stable and more air sensitive than their platinum(0) analogues.

(a) Reactions involving co-ordinative dissociation
(i) Acids
Acids react with platinum(0) complexes to give oxidative addition reactions yielding hydrido-complexes. The overall reaction, which is reversible, can be represented by the scheme in Fig. 7.[56] The product obtained depends on the

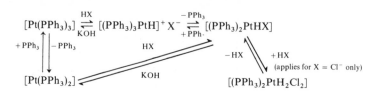

Fig. 7. The interaction of acids with platinum(0)–phosphine complexes.

nature of the anion X. Thus when $X = ClO_4$, HSO_4 or BF_4, the product obtained is $[(PPh_3)_3PtH]^+X^-$, whereas when $X = CN$,[56] CNS,[56] a dehydroimide[57] (see below), $SCOCH_3$,[58] $SCOC_6H_5$,[58] $C_7H_5S_2N$,[58] PhS[59] or $PhSe$[59] the product is *trans*-$[(PPh_3)_2PtHX]$. When $X = Cl$, however, the ionic hydride was isolated in polar solvents such as water or aqueous ethanol, whereas the covalent hydride was isolated in non-polar solvents such as benzene. The platinum(IV) dihydride was only obtained in a non-polar solvent with $X = Cl$. An investigation of a series of acids has shown that the essential requirements for an acid to undergo oxidative addition with a platinum(0) complex to yield a hydrido-platinum(II) complex are that the acid should either be a strong acid, or if it is not that its conjugate base should be a good ligand for platinum(II). Thus, for example, the very weak acid HCN (pK_a 8-12) gives $[(PPh_3)_2PtH(CN)]$[56] whereas the much stronger hydrofluoric acid (pK_a 3·5) does not react with $[Pt(PPh_3)_4]$ to form $[(PPh_3)_2PtHF]$.[60] Ligands such as fluoride ions which are poor ligands for platinum(II) lead to unstable complexes largely because of the very strong *trans*-influence of the hydrido-ligand which further weakens the platinum(II)–fluoride bond.

When palladium(0) complexes are treated with acids a similar series of reactions probably occur. However, because palladium(II)–hydride complexes are unstable both thermally and to excess acid, the products isolated are the di-halo-complexes $[(PPh_3)_2PdX_2]$. This, together with the observation that hydrogen is evolved, suggests that the reaction occurs as in 22.

$$[Pd(PPh_3)_3] + HCl \rightleftharpoons [(PPh_3)_2PdHCl] \xrightarrow{HCl} [(PPh_3)_2PdH_2Cl_2] \rightleftharpoons$$
$$[(PPh_3)_2PdCl_2] + \uparrow H_2 \quad (22)$$

Imides such as succimides, phthalimide, saccharin and parabanic acid react with platinum(0) complexes in a similar way to acids to give hydrido–platinum(II) complexes (reaction 23; L = As or P), whereas with palladium(0) the diimide complex is obtained (reaction 24).[57] Substituted tetrazoles react in a similar way with both platinum(0) and palladium(0) complexes to give the bis(tetrazolato) complexes[61] (reaction 25).

$$Pt(LPh_3)_4 \; + \begin{array}{c} CH_2-CO \\ | \qquad\qquad NH \\ CH_2-CO \end{array} \longrightarrow \left[(LPh_3)_2PtH\left(-N \begin{array}{c} CO-CH_2 \\ | \\ CO-CH_2 \end{array}\right) \right] \qquad (23)$$

$$Pd(PPh_3)_4 \; + \begin{array}{c} CH_2-CO \\ | \qquad\qquad NH \\ CH_2-CO \end{array} \longrightarrow \left[(PPh_3)_2Pd\left(-N \begin{array}{c} CO-CH_2 \\ | \\ CO-CH_2 \end{array}\right)_2 \right] \; +\uparrow H_2 \qquad (24)$$

$$M(PPh_3)_4 \; + \begin{array}{c} X \\ \| \\ C \\ N \diagdown \quad \diagup N-H \\ | \qquad | \\ N=\!=\!N \end{array} \longrightarrow cis\text{-} \left[(PPh_3)_2M \left(\begin{array}{c} X \\ \| \\ C \\ N \diagdown \quad \diagup N \\ | \qquad | \\ N=\!=\!N \end{array} \right)_2 \right] + \uparrow H_2 \qquad (25)$$

(ii) Alkyl halides

Alkyl,[41,62,63] aryl,[64] benzyl,[65] acyl,[63] aroyl,[63,65] sulphonyl[63] and perfluoroalkyl halides[67] as well as carbon tetrachloride[68] react with $[Pt(PPh_3)_n]$ ($n = 2, 3$ or 4) to cleave the carbon–halogen bond, or in the case of acyl, aroyl and sulphonyl halides the oxygen–halogen bond, to yield *cis*-$[(PPh_3)_2Pt(R)X]$. This reaction can be used to prepare *cis*-platinum(II) monalkyl complexes (pp. 327–328). $[Pd(PPh_3)_4]$ reacts similarly with alkyl halides, acetyl chloride, ethylchloroformate and iodobenzene,[69] but because of the ease of *cis-trans*-isomerisation of palladium(II) complexes the product is the *trans-complex*.

Other organic molecules containing halogens react in a similar way. For example, ethyldithiochloroformate (ClCSSEt) undergoes oxidative addition with $[Pt(PPh_3)_3]$ to give $[(PPh_3)_2Pt(-CSSEt)Cl]$.[70] However, allyl chloride and allyl isothiocyanate both react with $[Pt(PPh_3)_3]$ to give five-co-ordinate π-allyl–platinum(II) complexes $[(PPh_3)_2Pt(\pi\text{-}C_3H_5)X]$ (X = Cl, Br or SCN), which are stable both in the solid state and in chloroform solution, but which dissociate into square-planar ionic complexes $[(PPh_3)_2\text{-}Pt(\pi\text{-}C_3H_5)]^+X^-$ in polar solvents such as nitrobenzene.[66] By contrast methallyl chloride $(CH_2=C(CH_3)CH_2Cl)$ reacts with $[Pd(PPh_3)_4]$ to give a σ-allyl complex $[(PPh_3)_2PdCl(-CH_2C(CH_3)=CH_2)]$ as opposed to a π-allyl complex.[69] Chloroolefins also react differently with platinum(0) and palladium(0) complexes. In non-polar solvents such as benzene chloro-olefins give platinum(0)–π-olefin complexes by non-dissociative addition to $[Pt(PPh_3)_4]$,[71] whereas in polar solvents such as ethanol they form platinum(II)–σ-alkenyl complexes by dissociative addition.[72] $[Pd(PPh_3)_4]$ always reacts with chloroolefins by a dissociative mechanism to give palladium(II)–σ-alkenyl complexes.[73] These reactions are discussed further in Chapter 13 (pp. 398 and 410–411).

(iii) Metal halides

Some metal halide derivatives react with platinum(0) and palladium(0) complexes by an oxidative addition reaction to give complexes containing metal–metal bonds. Complexes containing Pt–Pt,[74] Pt–Ni,[74] Pt–Cu,[74] Pt–Au,[74] Pt–Hg,[74] Pt–Sn,[55,74] Pt–B[75] and Pd–Hg[76] bonds have been prepared in this way (for instance reactions 26, 27 and 28). Although these

$$[Pt(PPh_3)_4] + [(PPh_3)_2PtCl_2] \rightarrow [(PPh_3)_2ClPt-PtCl(PPh_3)_2] \quad (26)$$

$$[Pt(PPh_3)_4] + [ClCu(PPh_3)_3] \rightarrow [(PPh_3)_2ClPt-Cu(PPh_3)_3] \quad (27)$$

$$[Pd(PPh_3)_4] + HgCl_2 \rightarrow [(PPh_3)_2ClPd-HgCl] \quad (28)$$

reactions occur readily, they are by no means universal, since R_3SiCl and R_3PbCl both react with $[Pt(PPh_3)_4]$ to give $[(PPh_3)_2PtCl_2]$,[2] and the Grignard reagent C_6F_5MgBr gave no isolable product.[77]

(iv) Halogens

Bromine[14] and iodine[15,17] both react with $[Pt(PPh_3)_4]$ by oxidative addition to give the bis–halogen complexes (reaction 29). The Ph_3PI_2 formed in this reaction reacts with water to give hydrogen iodide (reaction 30),

$$[Pt(PPh_3)_4] + 3I_2 \xrightarrow{\text{benzene}} [(PPh_3)_2PtI_2] + 2Ph_3PI_2 \quad (29)$$

$$Ph_3PI_2 + 2H_2O \longrightarrow Ph_3P(OH)_2 + 2HI \quad (30)$$

a reaction that led to some confusion in the early sixties when there was doubt as to whether or not the complexes were truly complexes of platinum(0) or were hydridoplatinum(II) complexes.[14,15] Cyanogen reacts with both platinum(0) and palladium(0) triphenylphosphine complexes to give the bis–cyano complexes of the divalent metals[78,79] (reaction 31).

$$[Pt(PPh_3)_4] + C_2N_2 \xrightarrow{\text{benzene}} \text{cis-}[(PPh_3)_2Pt(CN)_2] \quad (31)$$

There is, however, some doubt about the ease of this reaction, one research group finding that it occurs rapidly at room temperature,[79] and the other that it requires 3 hours at 100°C.[78]

(v) Hydrogen

Although it has been claimed that hydrogen reacts with $[Pt(PPh_3)_n]$ ($n = 3$ or 4) to give $[(PPh_3)_2PtH_2]$,[14,15] recent work has in fact shown that no reaction occurs when pure hydrogen is used at 400 atm. and 50°C, and further that the product obtained previously was actually $[(PPh_3)_2Pt(CO_3)]$ formed by the traces of carbon dioxide and oxygen present in commercial hydrogen.[80]

(vi) Miscellaneous

Trifluoromethyl nitrosyl reacts[81] with platinum(0) complexes to give the oxidative addition product $[(PPh_3)_2Pt(CF_3)(NO)]$. Nitrous oxide reacts with both platinum(0) and palladium(0) complexes to give an oxidative-addition reaction in which a metal–metal bond is formed[82] (reaction 32) whereas with cyanogen[78,79] and 1,1,1-tricyanoethane[83] a carbon–carbon bond cleavage occurs (reactions 31 and 33). Azides such as $RCON_3$, where $R = Ph$, $p\text{-}NO_2C_6H_4$ or EtO, react with $[Pt(PPh_3)_2]$ to give

$$2[M(PPh_3)_2] + 2NO \xrightarrow[(M = Pd,Pt)]{} [(PPh_3)_2(NO)M—M(NO)(PPh_3)_2] \quad (32)$$

$$[Pt(PPh_3)_4] + CH_3C(CN)_3 \rightarrow \left[(PPh_3)_2Pt(CN)(C{\overset{\displaystyle CH_3}{\underset{\displaystyle CN}{—}}}CN) \right] \quad (33)$$

$[(PPh_3)_2Pt(N_3)_2]$.[84] With sulphonyl azides (e.g. $PhSO_2N_3$) the products are more complex due to reactions subsequent to the initial oxidative-addition occurring. The main products are $[(PPh_3)_2Pt(OH)(NHSO_2Ph)]$ when the reaction is carried out in the presence of traces of water and $[(PPh_3)_2Pt(N_3)-(NHSO_2Ph)]$ in the presence of a large amount of water.

Ortho-quinones such as *o*-quinone,[85] tetrachloro-*o*-quinone[86] and 9,10-phenanthrenequinone[87] react with zerovalent platinum and palladium complexes to give complexes of the divalent metals (reaction 34). By contrast

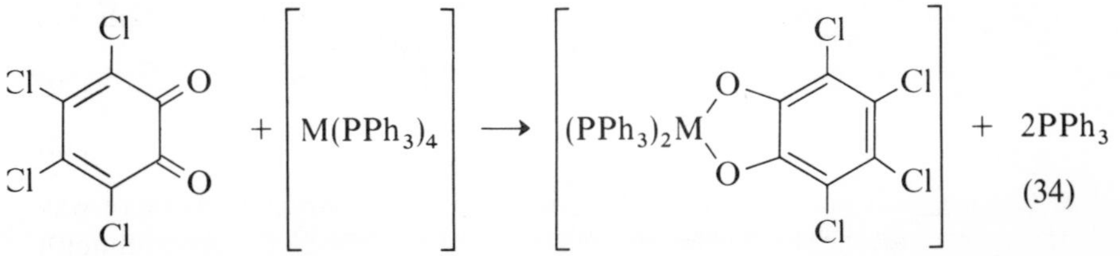

$$(34)$$

p-quinones add non-oxidatively to $[Pt(PPh_3)_4]$ to form diolefin complexes of platinum(0)[85] (reaction 35).

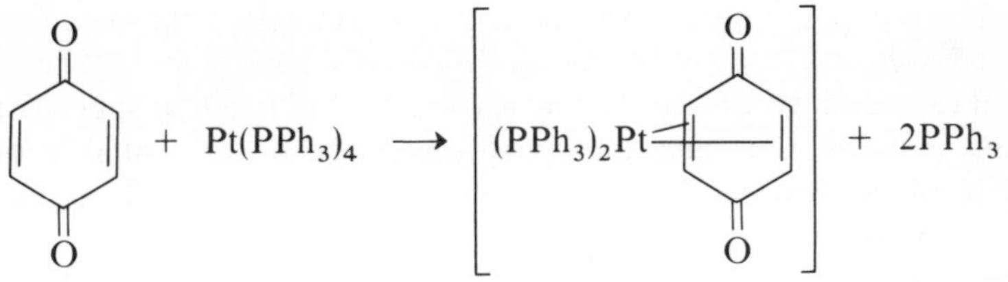

$$(35)$$

Aldehydes and formic acid esters were originally reported to react with $[Pt(PPh_3)_4]$, by carbon–hydrogen bond-cleavage reactions, yielding acyl complexes[88] $[(PPh_3)_2Pt(COR)_2]$, where $R = CH_3$, C_2H_5, C_6H_5 and OC_2H_5. However, a re-investigation of this reaction has shown that some oxygen has been taken up and that the products are not acyls but carboxylates,[89] $[(PPh_3)_2Pt(OCOR)_2]$, probably formed via the intermediate peroxo-compounds (see p. 43). These reactions are somewhat exceptional; they are not typical of carbonyl compounds since ketones give non-oxidative addition (see p. 39), nor are they an inevitable consequence of the presence of an activated hydrogen, since both acetylacetone and diethylmalonate react without carbon–hydrogen bond cleavage to give non-oxidative addition.[88]

(b) Reactions involving co-ordinative addition

(i) Olefins, acetylenes and benzene derivatives

The reactions of olefins and acetylenes with platinum(0) and palladium(0) complexes and the properties of the resulting products are discussed in detail in Chapter 13 (pp. 398–411). Whereas the zerovalent triphenylphosphine complexes dissolve in benzene without any change, other than dissociation, [Pt(PPh$_3$)$_3$] was found to react with 1,3,5-trinitrobenzene to give presumably [(PPh$_3$)$_2$Pt(C$_6$H$_3$(NO$_2$)$_3$)], although the structure of the product is not yet certain.[90]

(ii) Reactions with $>$C$=$S *and* $>$C$=$O *Bonds*

Carbon disulphide adds non-oxidatively to platinum(0) triphenylphosphine complexes to give [(PPh$_3$)$_2$Pt(CS$_2$)][66,91] An x-ray investigation of this complex showed that the bonding involved a π-bond between the metal and one C$=$S double-bond (see Fig. 8 and Appendix II structure 197),[92,93] which is similar to the bonding in metal(0)–olefin complexes (see pp. 400–406).

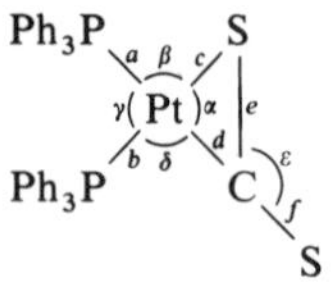

a	2·346 Å	α	45·5°
b	2·240 Å	β	99·7°
c	2·328 Å	γ	107°
d	2·063 Å	δ	108°
e	1·72 Å	ε	136°
f	1·54 Å		

Fig. 8. Structure of [(PPh$_3$)$_2$Pt(CS$_2$)] (from ref. 93).

The S–C–S angle is 136° in contrast to the angle of 180° in free carbon disulphide, but very close to the angle of 135° found in the lowest excited state ($^3A_2$) of carbon disulphide.[94] The mean C–S bond length in the co-ordinated CS$_2$ (1·63 Å) is very close to that in the $^3A_2$ excited state (1·64 Å). This similarity of co-ordinated CS$_2$ to the $^3A_2$ excited state of free CS$_2$ suggests that bonding between CS$_2$ and platinum(0) involves a substantial π-back-donation of electron density from platinum to the π-antibonding orbitals of the C$=$S double-bond.

Other compounds containing carbon–sulphur double-bonds that add non-oxidatively to [Pt(PPh$_3$)$_n$] are carbonylsulphide,[66] alkyl and aryl isothiocyanates,[66] perfluorothioacetone,[66] carbon subsulphide[95] and bis(trifluoromethyl)thioketen ({CF$_3$}$_2$C$=$C$=$S).[96] In all cases the metal–ligand bonding is similar to that in the carbon disulphide complex. On heating the carbonylsulphide complex [(PPh$_3$)$_2$Pt(COS)] in chloroform, carbon monoxide is lost and a dimeric complex [(PPh$_3$)$_2$PtSPt(CO)(PPh$_3$)], which involves a platinum–platinum bond of length 2·647 Å and a bridging sulphide group (mean Pt–S bond length = 2·22 Å), is formed[97,98] (for further details see Appendix II, structure 211).

Carbon dioxide will not add non-oxidatively to platinum(0) or palladium(0) triphenylphosphine complexes.[99] An earlier report suggesting that carbon dioxide did add to these zerovalent complexes[91] has been found to be incorrect,[99] the discrepancy arising because of inadequate care to rigorously exclude oxygen in the earlier work. Although the carbon–oxygen double-bonds in both carbon dioxide and acetone do not co-ordinate to platinum(0) or palladium(0) by a π-interaction the carbonyl group in perfluoroacetone will, giving $[(PPh_3)_2M(\{CF_3\}_2CO)]$.[23] These perfluoro-acetone complexes are believed to have structures analogous to the carbon disulphide complex shown in Fig. 8. The electronegative trifluoromethyl groups activate the $>C=O$ bond by withdrawing electron density from it and so lowering the energy of the π-antibonding orbital of the $>C=O$ bond so that its energy is suitable for π-interaction with the zerovalent metals. The interaction of keten dimer $(CH_3-C=CH$ equivalent to 2 × $O-C=O$

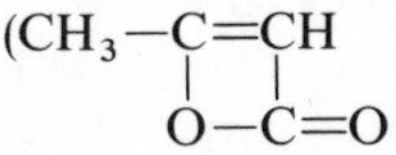

$CH_2=C=O)$ with $[Pt(PPh_3)_4]$ gives a non-oxidative addition product. However, the bonding in this complex (4) involves σ-bonds from platinum to

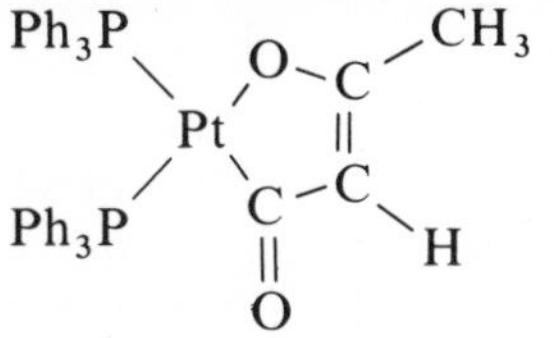

4

both carbon and oxygen rather than π-interaction between a carbon–oxygen double-bond and the metal.[100]

(iii) *Reaction with* $-C\equiv N$ *and* $>C=N$-*Bonds*

Although trifluoromethyl cyanide will displace *trans*-stilbene from $[(PPh_3)_2Pt(PhCH=CHPh)]$ to give complex **5** in which the carbon–nitrogen triple-bond is bound to platinum(0) by a π-interaction, trifluoromethyl cyanide reacts with $[Pt(PPh_3)_4]$ to give a ring compound, **6**, in which π-interaction is avoided.[101] The existence of the extra $>N-H$ group, the

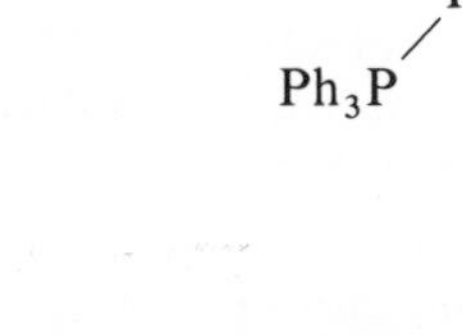

5 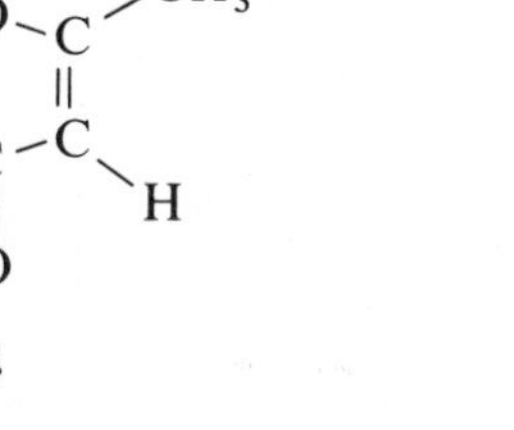   **6**

nitrogen of which was confirmed by x-ray diffraction (for further details see Appendix II, structure 204), was assumed to result from hydrolytic degradation of trifluoromethylcyanide by traces of water present in the reaction mixture.

Displacement of *trans*-stilbene from $[(PPh_3)_2Pt(PhCH=CHPh)]$ by hexafluoroisopropylideneamine and its *N*-methyl analogue yields complexes in which the $>C=N=$ group is π-bonded to platinum(0) (7).[102]

$$
\begin{array}{c}
CF_3 \\
| \diagup CF_3 \\
Ph_3P \diagdown \quad C \\
\quad \diagdown Pt \diagup \quad | \\
Ph_3P \diagup \quad \diagdown N \\
\quad \diagdown R
\end{array}
\qquad R = H \text{ or } CH_3
$$

7

(iv) Reaction with group VIB donors

The reaction of triphenylphosphine platinum(0) complexes with sulphur dioxide to give a dark-purple product occurs almost instantaneously. However, the formulation of the product has been somewhat variable. Thus when $[Pt(PPh_3)_n]$ (n = 3 or 4) is treated in benzene with sulphur dioxide the initial red-brown solution yields dark-purple needles of $[(PPh_3)_3Pt(SO_2)]$. $1\frac{1}{2}C_6H_6$[103] in which the sulphur dioxide is bound via a platinum(0)– sulphur bond. This compound readily loses a molecule of triphenylphosphine and all the benzene to yield $[(PPh_3)_2Pt(SO_2)]$.[74] Moreover, when $[(PPh_3)_2Pt(C_2H_4)]$ is treated with sulphur dioxide the initial product is a brown crystalline material, probably $[(PPh_3)_2Pt(SO_2)_2]$; this on warming to 50°C loses one molecule of sulphur dioxide to yield the green air-stable complex $[(PPh_3)_2Pt(SO_2)]$.[104] The sulphur dioxide complexes, all of which appear to involve direct platinum(0)–sulphur bonding, thus illustrate the ability of platinum(0) to exhibit co-ordinative unsaturation. On heating in a vacuum at 100–120°C $[(PPh_3)_2Pt(SO_2)]$ slowly loses sulphur dioxide. The palladium(0) compound, $[Pd(PPh_3)_4]$, reacts similarly with sulphur dioxide, but the only compound so far reported is $[(PPh_3)_3Pd(SO_2)]$.[103] Whether this reflects a lower tendency in palladium(0) to exhibit co-ordinative unsaturation or whether it is because the reaction has been insufficiently investigated is uncertain. Sulphur dioxide complexes of both platinum(0) and palladium(0) react with oxygen to form the metal(II)-sulphate complexes $[(PPh_3)_2M(SO_4)]$ (M = Pt, Pd).[103,104] This reaction is of interest because it supports the suggestion that the catalytic oxidation of sulphur dioxide by platinum metal involves chemisorption of the gas on to the metal.[105]

Since hydrogen sulphide and hydrogen selenide both poison metal surfaces their interaction with platinum(0) complexes has been investigated. The reaction occurs by a two-stage mechanism in which complex **8** is first formed and this then slowly rearranges to **9** (reaction 36).[106] The structures of **8** and **9** were determined by proton n.m.r. spectroscopy. The compound **9** is a

$$
Pt(PPh_3)_n + H_2M \underset{\text{heat}}{\rightleftharpoons}
\begin{array}{c}
H \diagdown \quad \diagup H \\
M \\
| \\
Pt \\
Ph_3P \diagup \quad \diagdown PPh_3
\end{array}
\longrightarrow
\begin{array}{c}
\quad \diagup H \\
M \cdots \diagup PPh_3 \\
\diagdown Pt \\
Ph_3P \diagup \quad \diagdown H
\end{array}
\qquad (36)
$$

$(n = 2, 3) \quad (M = S, Se)$

8   **9**

typical platinum(II) hydride with a hydride chemical shift of 18·8 p.p.m. (M = Se) or 19·2 p.p.m. (M = S) relative to tetramethylsilane (10 p.p.m.).

(v) Reaction with oxygen

The zerovalent complexes of both platinum and palladium react with oxygen to give addition compounds (reaction 37; M = Pd, Pt; $n = 3, 4$).[99,107,108] The palladium(0)–oxygen complex $[(PPh_3)_2PdO_2]$ is unstable above 20°C,

$$[M(PPh_3)_n] + O_2 \rightarrow [(PPh_3)_2MO_2] + (n - 2)PPh_3 + \text{some } Ph_3PO \quad (37)$$

decomposing to the free metal according to reaction 38. The platinum

$$[(PPh_3)_2PdO_2] \rightarrow Pd + 2Ph_3PO \quad (38)$$

complex is more stable and does not decompose until about 120°C when it eliminates oxygen and reverts to $[Pt(PPh_3)_4]$[2] (reaction 39). The oxygen

$$4[(PPh_3)_2PtO_2] \rightarrow [Pt(PPh_3)_4] + 2O_2 + 4Ph_3PO \quad (39)$$

molecule in these complexes is very reactive, so that the complexes can be used to catalyse the oxidation of triphenylphosphine to triphenylphosphine oxide[107,109] (reaction 37 and 38), isocyanides to isocyanates,[109] hydrocarbons to hydrocarbon hydroperoxides[110] and the oxidation of cyclohexene to cyclohexen-3-one and cylohexene oxide.[111,112] A detailed investigation[54,113] of the kinetics and mechanism of the oxidation of triphenylphosphine catalysed by $[Pt(PPh_3)_3]$ has suggested that it occurs by reactions 40 and 41, in which the triphenylphosphine molecule oxidised is not co-ordinated to the platinum.

$$[Pt(PPh_3)_3] + O_2 \rightarrow [(PPh_3)_2PtO_2] + PPh_3 \quad (40)$$

$$[(PPh_3)_2PtO_2] + PPh_3 \rightarrow [(PPh_3)_3PtO_2] \xrightarrow[\text{fast}]{+2PPh_3}$$
$$[Pt(PPh_3)_3] + 2Ph_3PO \quad (41)$$

The structure of $[(PPh_3)_2PtO_2]$ has been determined by x-ray diffraction by two groups of workers independently[114,115] (see Fig. 9 and Appendix II, structure 194). Both groups found similar results, which makes them significant in spite of the low stability of the complex to x-rays. In particular they found the platinum, phosphorus and oxygen atoms to be virtually coplanar.

The complexes $[(PPh_3)_2MO_2]$ (M = Pd, Pt) react with small molecules such as sulphur dioxide,[104,116,117] nitrogen dioxide,[106,116,117] nitric oxide[111,116,117] and carbon dioxide[80,99] to give the corresponding sulphato, dinitrato, dinitro and carbonato complexes, as exemplified in reactions 42 and 43. The carbonato complex $[(PPh_3)_2Pt(CO_3)]$ was originally thought to be the bis–hydrido-platinum(II) complex $[(PPh_3)_2Pt(H)_2]$.

$$[(PPh_3)_2MO_2] + SO_2 \rightarrow [(PPh_3)_2M(SO_4)] \quad (42)$$

$$[(PPh_3)_2MO_2] + 2NO_2 \rightarrow [(PPh_3)_2M(NO_3)_2] \quad (43)$$

This carbonato-complex, although the first product to be isolated, is not the first product to be formed; this has now been shown to be a

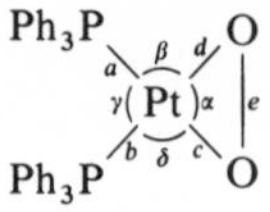

Bond/angle	$[(PPh_3)_2PtO_2].1.5C_6H_6$	$[(PPh_3)_2PtO_2].C_6H_5CH_3$
a	2·28 Å	2·28 Å
b	2·25 Å	2·22 Å
c	2·01 Å	1·99 Å
d	2·01 Å	1·90 Å
e	1·45 Å	1·26 Å
α	43°	38°
β	108·5°	109°
γ	101·2°	100°
δ	107·8°	113°

Fig. 9. Structure of $[(PPh_3)_2PtO_2]$. (a) In $[(PPh_3)_2PtO_2].1.5C_6H_6$ (from ref. 114). (b) In $[(PPh_3)_2 PtO_2].C_6H_5CH_3$ (from ref. 115).

peroxycarbonato complex.[89,118] The overall reaction scheme is that shown in equation 44:

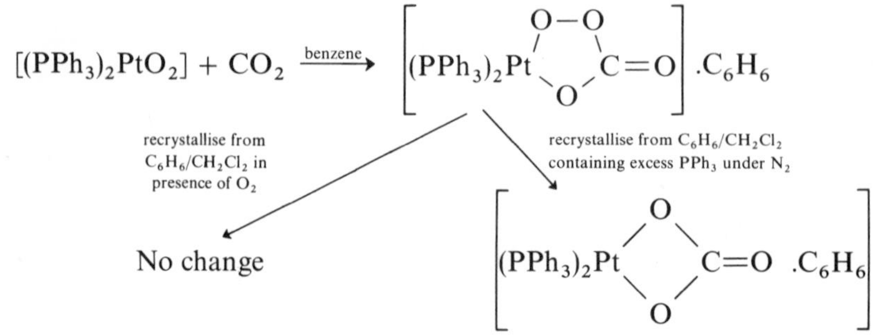

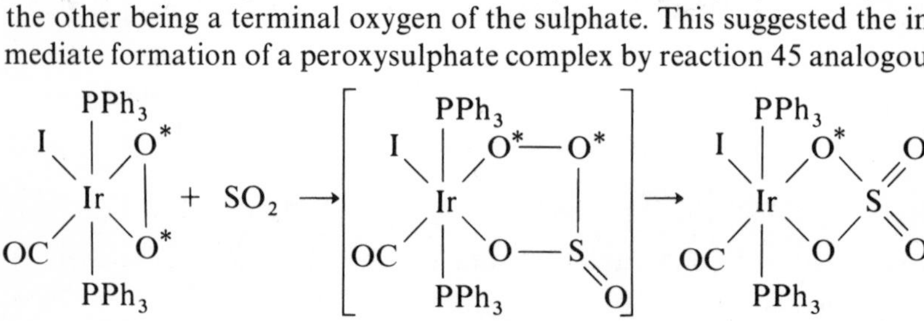

An attempt to determine the mechanism of the reaction of sulphur dioxide with the platinum(0)–oxygen complex by means of isotopically labelled oxygen and infra-red spectroscopy was unsuccessful due to a combination of two causes. First, an uncertainty whether the sulphato-group was bidentate in $[(PPh_3)_2Pt(SO_4)]$ and, secondly, the poor resolution of the infra-red bands which made it difficult to distinguish between ^{16}O and ^{18}O.[119] However, when the corresponding iridium(I)–oxygen complex $[(PPh_3)_2Ir(CO)I(O_2)]$ was treated with sulphur dioxide, only one of the original oxygen atoms from the iridium(I) complex was bound directly to iridium(III) in the product, the other being a terminal oxygen of the sulphate. This suggested the intermediate formation of a peroxysulphate complex by reaction 45 analogous to the other peroxy complexes discussed below. Cleavage of the platinum–oxygen bond must also occur during the reaction of nitric oxide with $[(PPh_3)_2PtO_2]$ since the product is reported to be the bis-nitro complex $[(PPh_3)_2Pt(NO_2)_2]$ containing N-bonded nitro-groups[111] (see also p. 214).

In addition to carbon dioxide and sulphur dioxide, carbon disulphide, aldehydes and ketones all react with $[(PPh_3)_2Pt(O_2)]$ to give peroxy-compounds; these are not particularly stable and lose the peroxy-group on recrystallisation in the absence of oxygen[89,118] (reactions 46 and 47). The reactivity of the carbonyl group in these reactions is related to its polarity.

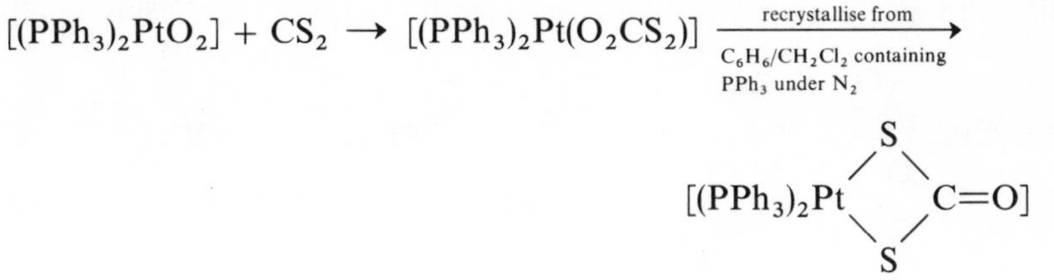

$$[(PPh_3)_2PtO_2] + CS_2 \rightarrow [(PPh_3)_2Pt(O_2CS_2)] \xrightarrow[\substack{C_6H_6/CH_2Cl_2 \text{ containing} \\ PPh_3 \text{ under } N_2}]{\text{recrystallise from}}$$

$$[(PPh_3)_2Pt\underset{S}{\overset{S}{\diagdown}}C=O] \quad (46)$$

$$[(PPh_3)_2PtO_2] + (CH_3)_2CO \rightarrow (PPh_3)_2Pt\underset{\substack{O}}{\overset{O-O}{\diagup}}C(CH_3)_2 \xrightarrow[\substack{N_2 \text{ in } CH_2Cl_2}]{\text{bubble in } H_2 \text{ or}}$$

10

$$[(PPh_3)_2Pt\underset{O}{\overset{O}{\diagdown}}C(CH_3)_2] \quad (47)$$

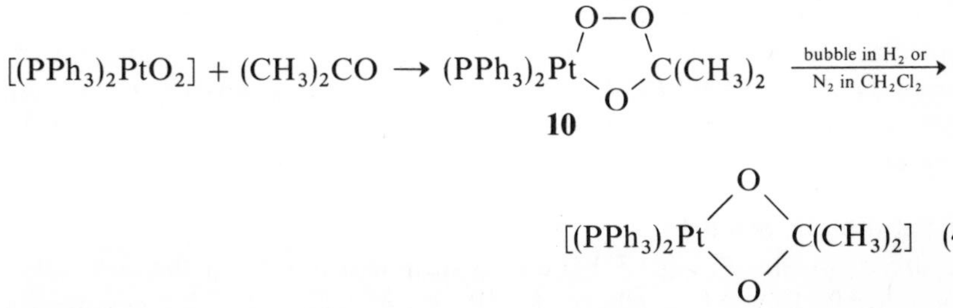

Thus, for instance, amides and esters do not react at all, and aldehydes react faster than ketones.[120] Fluoroketones such as hexafluoroacetone react with $[(PPh_3)_2PtO_2]$ to give the perfluoroanalogue of **10** (**11**). However, with excess hexafluoroacetone this gives a 1:2 adduct (**12**) with a seven-membered ring (reaction 48). Recrystallisation of **12** gives **11** back again and reduction of both **11** and **12** with triphenylphosphine gives the four-membered ring complex

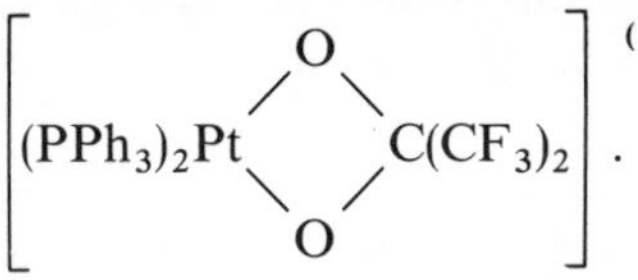

$$\left[(PPh_3)_2Pt\underset{O}{\overset{O}{\diagdown}}C(CF_3)_2 \right] . \quad (121)$$

$$\left[(PPh_3)_2Pt\underset{O}{\overset{O-O}{\diagup}}C(CF_3)_2 \right] + (CF_3)_2CO \rightarrow \left[(PPh_3)_2Pt\underset{O————C(CF_3)_2}{\overset{O-O-C(CF_3)_2}{\diagup}}O \right] \quad (48)$$

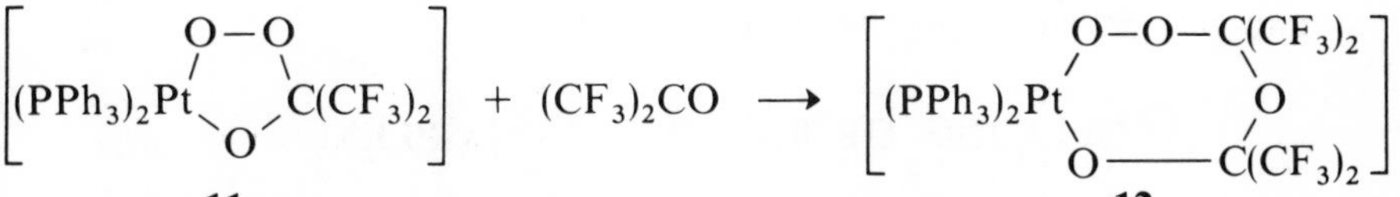

11   **12**

When α-diketones react with $[(PPh_3)_2PtO_2]$ the initial product

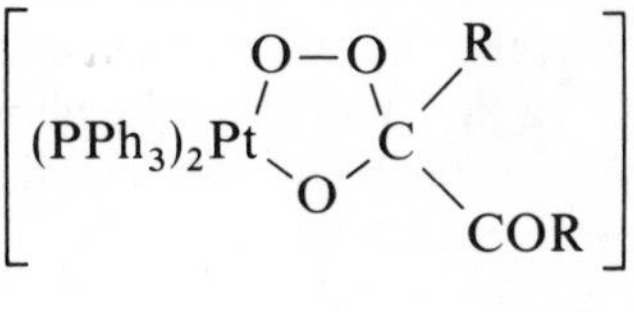

$$\left[(PPh_3)_2Pt\underset{O}{\overset{O-O}{\diagup}}C\underset{COR}{\overset{R}{\diagdown}} \right]$$

can react with further $[(PPh_3)_2PtO_2]$ to give

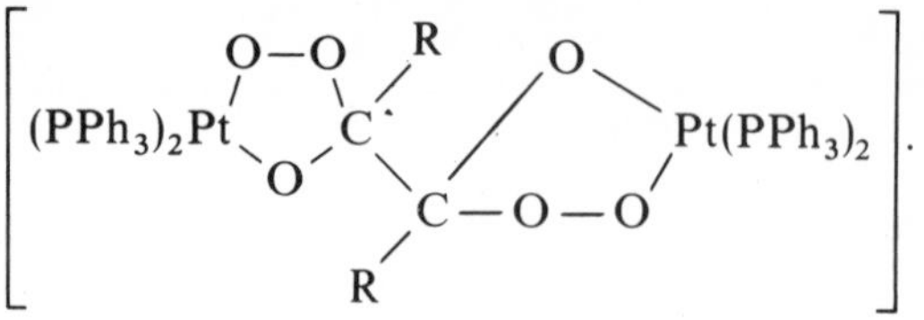

$$\left[(PPh_3)_2Pt \underset{O}{\overset{O-O}{\diagdown}} C \underset{C-O-O}{\overset{R}{\diagup}} \overset{O}{\diagdown} Pt(PPh_3)_2 \right].$$

The initial $1:1$ adduct (13) on standing in dichloromethane solution for some time or even in the solid state rearranges via two, as yet unidentified, intermediates to give a bis-carboxylato platinum(II) complex[122] (reaction 49).

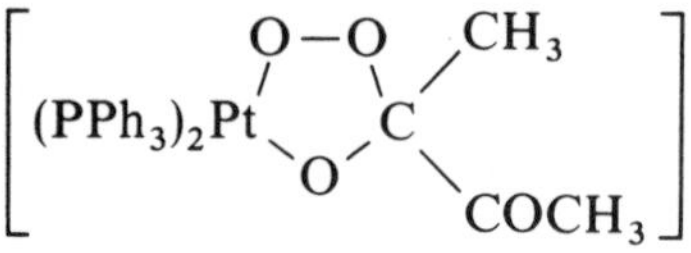

$$\left[(PPh_3)_2Pt \underset{O}{\overset{O-O}{\diagdown}} C \underset{COCH_3}{\overset{CH_3}{\diagup}} \right] \longrightarrow [(PPh_3)_2Pt(OCOCH_3)_2] \quad (49)$$

$$\textbf{13}$$

The palladium complex $[(PPh_3)_2PdO_2]$ appears to react similarly but intermediate peroxy-compounds have not been detected.[89] An x-ray diffraction study of compound 10 has shown that co-ordination about the platinum atom is almost completely planar[120] (for further details see Appendix II, structure 201).

(c) Substitution reactions

As already mentioned (p. 33) the substitution reactions of the zerovalent complexes $[(PPh_3)_2ML_n]$, where $M = Pt$ or Pd and $n = 1$ or 2 depending on the nature of the ligand L, take place by a dissociative mechanism yielding $[M(PPh_3)_2]$ as the reactive intermediate.[53-55] Similarly the substitution reactions of $[Pt(PF_3)_4]$ involve the initial formation of the reactive intermediate $[Pt(PF_3)_3]$.[123] The main exceptions to this are (a) the addition reactions of oxygen complexes which appear to involve an initial rearrangement of the metal–oxygen π-bond to a σ-bond followed by reaction with the incoming ligand (pp. 41–44) and (b) the reactions of acids with π-bonded olefin and acetylene complexes which involve initial co-ordination of the acid to the metal (pp. 409–410).

Methods for the preparation of complexes of ligands that co-ordinate too weakly to be capable of displacing phosphine ligands from zerovalent platinum and palladium are discussed in Chapter 13 (pp. 398–399).

PROPERTIES OF PHOSPHITE COMPLEXES

The kinetics of ligand exchange between free $(EtO)_3P$ and $[M\{(EtO)_3P\}_4]$, where $M = Pd$ or Pt, indicate that the exchange proceeds by a dissociative (S_N1) mechanism.[45] Furthermore, the palladium(0) complex reacts faster than the platinum(0) complex; this is consistent with higher bond strengths in the platinum(0) than in the palladium(0) phosphite complexes, as is indicated by the force constants of the metal–phosphorus symmetric stretching modes in the infrared.[46,124]

PROPERTIES OF FLUOROPHOSPHINE COMPLEXES

Tetrakis(fluorophosphine)platinum(0) complexes PtL_4, where $L = PF_3$, CF_3PF_2 or $(CF_3)_2PF$, in contrast to triphenylphosphine complexes do not react with alkylhalides, hydrogen chloride, ethylene or carbon disulphide.[11] This, together with the absence of any evidence for the formation of tris- and bis-(fluorophosphine)platinum(0) complexes, is probably due to the stronger π-acceptor capacity and the weaker σ-donor ability of the fluorophosphine ligands as compared with triphenylphosphine, due to the inductive effect of fluorine bound directly to phosphorus. This suggestion is consistent with the fact that $[Pt(CF_3PF_2)_4]$ readily reacts with tertiary phosphines to give $[Pt(CF_3PF_2)_2(PR_3)_2]$ complexes, where $PR_3 = PPh_3$, PPh_2Me or $PPhMe_2$, where the σ-donor and π-acceptor capacities of the two different ligands are balanced.

PROPERTIES OF CARBONYL COMPLEXES

Although carbon monoxide forms only a polymeric complex with platinum(0) in the absence of other ligands, it forms a variety of monomeric and polymeric complexes with phosphine complexes of platinum(0). This is because of the strong σ-donor ability of the phosphine ligands, which by donating electron density to the metal increase the metal to carbon monoxide π-back-dona-tion.[47] The relationship between the various mixed carbonyl–phosphine complexes of platinum(0) is summarised in Fig. 10 and the preparative details

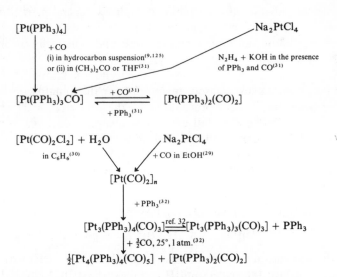

Fig. 10. The interrelation of platinum(O) complexes containing carbon monoxide and tertiary phosphine ligands.

are given on p. 30. The structures of $[Pt(PPh_3)_3CO]$, which is tetrahedral,[36,37] $[Pt_4(PPh(CH_3)_2)_4(CO)_5]$, $[Pt_3(PPh_3)_4(CO)_3]$ and $[Pt\{PPh_2(CH_2C_6H_5)\}_3$-$(CO)_3]$, which are shown in Fig. 11, have been determined by x-ray diffraction

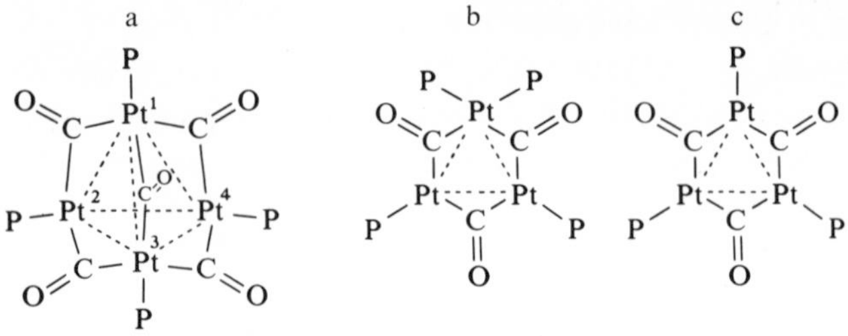

Pt^2–Pt^3 and Pt^3–Pt^4 = 2.75 Å (bonding)
Pt^1–Pt^3 = 2·79 Å (bonding)
Pt^2–Pt^4 = 3·54 Å (non-bonding)

Fig. 11. Structures of (a) $[Pt_4\{PPh(CH_3)_2\}_4(CO)_5]$ (from ref. 126). (b) $[Pt_3(PPh_3)_4(CO)_3]$ (from refs. 29 and 126) and (c) $[Pt_3\{PPh_2(CH_2C_6H_5)\}_3(CO)_3]$ (from refs. 29 and 126).

(see Appendix II, structures 210 and 198). The preference of carbon monoxide to act as a bridging ligand rather than as a terminal ligand in platinum(0) complexes is apparent. Carbon monoxide can be displaced from $[Pt(PPh_3)_2(CO)_2]$ by phenylacetylene[127] (reaction 50).

$$[Pt(PPh_3)_2(CO)_2] + PhC{\equiv}CH \rightarrow [(PPh_3)_2Pt(PhC{\equiv}CH)] + 2CO \quad (50)$$

PROPERTIES OF ISOCYANIDE COMPLEXES

The properties of palladium(0) isocyanide complexes depend markedly on their method of preparation (see p. 29). Thus Malatesta's products are brown-black crystalline substances, stable to air and practically insoluble in all solvents except nitrobenzene and pyridine in which they decompose, and soluble in liquid aryl isocyanides to give yellow solutions that are unstable in the absence of excess isocyanide.[26] The palladium(0) isocyanide complexes prepared by Fischer's procedure have the same empirical formula as Malatesta's (i.e. $[Pd(CNR)_2]$) but are yellow complexes soluble in most organic solvents.[28] Their molecular weights in benzene are time dependent, the minimum value corresponding to a dimeric molecule. At room temperature the yellow solids darken to give compounds identical to those obtained by Malatesta.

$$[(p\text{-}CH_3C_6H_4NC)_2Pd] + 3[(p\text{-}ClC_6H_4O)_3Pd] \rightarrow$$

$$[(p\text{-}CH_3C_6H_4NC)Pd\{(p\text{-}ClC_6H_4O)_3P\}_3] + p\text{-}CH_3C_6H_4NC \quad (51)$$

In alcoholic suspension palladium(0) isocyanide complexes react with one mole of iodine to give the palladium(II) complexes $[PdI_2(CNR)_2]$.[26] On treatment with triarylphosphines or triarylphosphites the isocyanide ligands can be completely or partially displaced to give new zerovalent palladium derivatives[1] as exemplified in reaction 51. The compound $[Pd(CN^tBu)_2]$ absorbs oxygen to give $[Pd(CN^tBu)_2O_2]$ which has an infrared spectrum consistent with the presence of a π-bonded oxygen ligand.[128,129] The analogous platinum(0) oxygen complexes $[Pt(RNC)_2O_2]$, where R =

cyclohexyl or t-butyl, have been found to be good catalysts for the oxidation of isocyanides to isocyanates.[129] In addition to reacting with oxygen [Pd(Bu^{t}NC)$_2$] reacts with tetracyanoethylene in toluene at between 0 and $-20°$C to give the zerovalent palladium–olefin complex [Pd(Bu^{t}NC)$_2$-(TCNE)].[129]

REFERENCES

1   L. Malatesta and M. Angoletta, *J. Chem. Soc.*, (1957) 1186.
2   R. Ugo, *Coord. Chem. Rev.* **3** (1968) 319.
3   L. Malatesta, *Prog. Inorg. Chem.*, **1** (1959) 283.
4   L. Malatesta and F. Bonati, *Isocyanide Complexes of Metals*, John Wiley, New York, 1969.
5   J. J. Burbage and W. C. Fernelius, *J. Amer. Chem. Soc.*, **65** (1943) 1484.
6   R. Nast and W. Hoerl, *Chem. Ber.*, **95** (1965) 1478.
7   G. W. Watt, M. T. Walling and P. I. Mayfield, *J. Amer. Chem. Soc.*, **75** (1953) 6175.
8   W. P. Griffith, J. Lewis and G. Wilkinson, *J. Chem. Soc.*, (1959) 1775.
9   L. Malatesta and C. Cariello, *J. Chem. Soc.*, (1958) 2323.
10   D. T. Rosevear and F. G. A. Stone, *J. Chem. Soc. (A)*, (1968) 164.
11   J. F. Nixon and M. D. Sexton, *J. Chem. Soc. (A)*, (1970) 321.
12   M. Hara, K. Ohno and J. Tsuji, *Chem. Comm.*, (1971) 247.
13   J. Chatt, L. A. Duncanson and B. L. Shaw, *Proc. Chem. Soc.*, (1957) 343.
14   J. A. Chopoorian, J. Lewis and R. S. Nyholm, *Nature*, **190** (1961) 528.
15   L. Malatesta and R. Ugo, *J. Chem. Soc.*, (1963) 2080.
16   J. Chatt and G. A. Rowe, *Nature*, **191** (1961) 1191.
17   A. D. Allen and C. D. Cook, *Proc. Chem. Soc.*, (1962) 218.
18   R. Ugo, G. La Monica, F. Conti and S. Cenini, unpublished results quoted in reference 2.
19   G. C. Dobinson, R. Mason, G. B. Robertson, R. Ugo, F. Conti, D. Morelli, S. Cenini and F. Bonati, *Chem. Comm.*, (1967) 739.
20   P. Roffia, F. Conti and G. Gregorio, *Chim. Ind. (Milan)*, **53** (1971) 361.
21   T. Kruck and K. Baur, *Angew. Chem. Int. Ed.*, **4** (1965) 521.
22   J. F. Nixon and M. D. Sexton, *Inorg. Nucl. Chem. Lett.*, **4** (1968) 275.
23   B. Clarke, M. Green, R. B. L. Osborn and F. G. A. Stone, *J. Chem. Soc. (A)*, (1968) 167.
24   J. Chatt, F. A. Hart and H. R. Watson, *J. Chem. Soc.*, (1962) 2537.
25   W. H. Baddley and L. M. Venanzi, *Inorg. Chem.*, **5** (1966) 33.
26   L. Malatesta, *J. Chem. Soc.*, (1955) 3924.
27   L. Malatesta, *Rec. Trav. Chim.*, **75** (1956) 644.
28   E. O. Fischer and H. Werner, *Chem. Ber.*, **95** (1962) 703.
29   G. Booth, J. Chatt and P. Chini, *Chem. Comm.*, (1965) 639.
30   G. Booth and J. Chatt, *J. Chem. Soc. (A)*, (1969) 2131.
31   P. Chini and G. Longoni, *J. Chem. Soc. (A)*, (1970) 1542.
32   J. Chatt and P. Chini, *J. Chem. Soc. (A)*, (1970) 1538.
33   R. Nast and W. Hoerl, *Chem. Ber.*, **95** (1962) 1470.
34   Y. Takahashi, T. Ito, S. Sakai and Y. Ishii, *Chem. Comm.*, (1970) 1065.
35   J. C. Marriott, J. A. Salthouse, M. J. Ware and J. M. Freeman, *Chem. Comm.*, (1970) 595.
36   V. G. Albano, P. L. Bellon and M. Sansoni, *Chem. Comm.*, (1969) 899.
37   V. G. Albano, G. M. Basso Ricci and P. L. Bellon, *Inorg. Chem.*, **8** (1969) 2109.
38   R. van der Linde and R. O. der Jongh, *Chem. Comm.*, (1971) 563.
39   V. G. Albano, P. L. Bellon and V. Scatturin, *Chem. Comm.*, (1966) 507.
40   R. F. Ziolo, S. Lipton and Z. Dori, *Chem. Comm.*, (1970) 1124.
41   R. Ugo, F. Cariati and G. La Monica, *Chem. Comm.*, (1966) 868.
42   R. Ugo, G. La Monica, F. Cariati, S. Cenini and F. Conti, *Inorg. Chim. Acta*, **4** (1970), 390.
43   R. D. Gillard, R. Ugo, F. Cariati, S. Cenini and F. Bonati, *Chem. Comm.*, (1966) 869.
44   L. Malatesta, R. Ugo and S. Cenini, *Adv. Chem. Ser.*, **62** (1966) 318.
45   M. Meier, F. Basolo and R. G. Pearson, *Inorg. Chem.*, **8** (1969) 795.

46   V. G. Myers, F. Basolo and K. Nakamoto, *Inorg. Chem.*, **8** (1969) 1204.
47   R. S. Nyholm, *Proc. Chem. Soc.*, (1961) 273.
48   G. W. Watt, R. E. McCarley and J. W. Dawes, *J. Amer. Chem. Soc.*, **79** (1957) 5163.
49   J. C. Green, D. I. King and J. H. D. Eland, *Chem. Comm.*, (1970) 1121.
50   I. H. Hillier, V. R. Saunders, M. J. Ware, P. J. Bassett, D. R. Lloyd and N. Lynaugh, *Chem. Comm.*, (1970) 1316.
51   J. F. Nixon and A. Pidcock, *Ann. Rev. NMR Spectroscopy*, **2** (1969) 346.
52   C. D. Cook, K. Y. Wan, U. Gelius, K. Hamrin, G. Johansson, E. Olsson, H. Siegbahn, C. Nordling and K. Siegbahn, *J. Amer. Chem. Soc.*, **93** (1971) 1904.
53   A. D. Allen and C. D. Cook, *Can. J. Chem.*, **42** (1964) 1063.
54   J. P. Birk, J. Halpern and A. L. Pickard, *J. Amer. Chem. Soc.*, **90** (1968) 4491.
55   J. P. Birk, J. Halpern and A. L. Pickard, *Inorg. Chem.*, **7** (1968) 2672.
56   F. Cariati, R. Ugo and F. Bonati, *Inorg. Chem.*, **5** (1966) 1128.
57   D. M. Roundhill, *Inorg. Chem.*, **9** (1970) 254.
58   D. M. Roundhill, P. B. Tripathy and B. W. Rencoe, *Inorg. Chem.*, **10** (1971) 727.
59   R. Ugo, G. La Monica, S. Cenini, A. Segre and F. Conti, *J. Chem. Soc. (A)*, (1971) 522.
60   J. McAvoy, K. C. Moss and D. W. A. Sharp, *J. Chem. Soc.*, (1965) 1376.
61   J. H. Nelson, D. L. Schmitt, R. A. Henry, D. W. Moore and H. B. Jonassen, *Inorg. Chem.*, **9** (1970) 2678.
62   P. Heimbach, *Angew. Chem. Int. Ed.*, **3** (1964) 648.
63   C. D. Cook and G. S. Jauhal, *Can. J. Chem.*, **45** (1967) 301.
64   P. Fitton and E. A. Rick, *J. Organometal. Chem.*, **28** (1971) 287.
65   P. Fitton, J. E. McKeon and B. C. Ream, *Chem. Comm.*, (1969) 370.
66   M. C. Baird and G. Wilkinson, *J. Chem. Soc. (A)*, (1967) 865.
67   M. Green, R. B. L. Osborn, A. J. Rest and F. G. A. Stone, *Chem. Comm.*, (1966) 502.
68   W. J. Bland and R. D. W. Kemmitt, *J. Chem. Soc. (A)*, (1968) 1278.
69   P. Fitton, M. P. Johnson and J. E. McKeon, *Chem. Comm.*, (1968) 6.
70   D. Commereuc, I. Douek and G. Wilkinson, *J. Chem. Soc. (A)*, (1970) 1771.
71   W. J. Bland and R. D. W. Kemmitt, *Nature*, **211** (1966) 963.
72   W. J. Bland, J. Burgess and R. D. W. Kemmitt, *J. Organometal. Chem.*, **14** (1968) 201.
73   P. Fitton and J. E. McKeon, *Chem. Comm.*, (1968) 4.
74   A. J. Layton, R. S. Nyholm, G. A. Pneumaticakis and M. L. Tobe, *Chem. Ind. (London)*, (1967) 465.
75   G. Schmid and H. Nöth, *Z. Naturforsch.*, **B20** (1965) 1008.
76   M. Braithwaite and R. S. Nyholm, Personal Communication.
77   D. M. Roe and A. G. Massey, *J. Organometal. Chem.*, **23** (1970) 547.
78   B. J. Argento, P. Fitton, J. E. McKeon and E. A. Rick, *Chem. Comm.*, (1969) 1427.
79   M. Bressan, G. Favero, B. Corain and A. Turco, *Inorg. Nucl. Chem. Lett.*, **7** (1971) 203.
80   C. J. Nyman, C. E. Wymore and G. Wilkinson, *Chem. Comm.*, (1967) 407.
81   M. Green, R. B. L. Osborn, A. J. Rest and F. G. A. Stone, *J. Chem. Soc. (A)*, (1968) 2525.
82   G. A. Pneumaticakis, *Chem. Comm.*, (1968) 275
83   J. L. Burmeister and L. M. Edwards, *J. Chem. Soc. (A)*, (1971), 1663.
84   W. Beck, M. Bauder, G. La Monica, S. Cenini and R. Ugo, *J. Chem. Soc. (A)*, (1971) 113.
85   S. Cenini, R. Ugo and G. La Monica, *J. Chem. Soc. (A)*, (1971), 416.
86   Y. S. Sohn and A. L. Balch, *J. Amer. Chem. Soc.*, **93** (1971) 1290.
87   J. S. Valentine and D. Valentine, *J. Amer. Chem. Soc.*, **92** (1970) 5795.
88   I. Harvie and R. D. W. Kemmitt, *Chem. Comm.*, (1970) 198.
89   P. J. Hayward, D. M. Blake, G. Wilkinson and C. J. Nyman, *J. Amer. Chem. Soc.*; **92** (1970) 5873.
90   M. C. Baird, *J. Organometal. Chem.*, **16** (1969) P16.
91   M. C. Baird and G. Wilkinson, *Chem. Comm.*, (1966) 514.
92   M. C. Baird, G. Hartwell, R. Mason, A. I. M. Rae and G. Wilkinson, *Chem. Comm.*, (1967) 92.
93   R. Mason and A. I. M. Rae, *J. Chem. Soc. (A)*, (1970) 1767.
94   B. Kleman, *Can. J. Phys.*, **41** (1963) 2034.
95   A. P. Ginsberg and W. E. Silverthorn, *Chem. Comm.*, (1969) 823.

96   M. Green, R. B. L. Osborn and F. G. A. Stone, *J. Chem. Soc. (A)*, (1970) 944.
97   A. C. Skapski and P. G. H. Troughton, *Chem. Comm.*, (1969) 170.
98   A. C. Skapski and P. G. H. Troughton, *J. Chem. Soc. (A)*, (1969) 2772.
99   C. J. Nyman, C. E. Wymore and G. Wilkinson, *J. Chem. Soc. (A)*, (1968) 561.
100 T. Kobayashi, Y. Takahashi, S. Sakai and Y. Ishii, *Chem. Comm.*, (1968) 1373.
101 W. J. Bland, R. D. W. Kemmitt, I. W. Nowell and D. R. Russell, *Chem. Comm.*, (1968) 1065.
102 J. Ashley-Smith, M. Green and F. G. A. Stone, *J. Chem. Soc. (A)*, (1970) 3161.
103 J. J. Levison and S. D. Robinson, *Chem. Comm.*, (1967) 198.
104 C. D. Cook and G. S. Jauhal, *J. Amer. Chem. Soc.*, **89** (1967) 3066.
105 J. K. Dixon and J. E. Longfield, *Catalysis*, ed. by P. H. Emmett, Reinhold, New York, 1960, vol. 7, p. 333.
106 D. Morelli, A. Segre, R. Ugo, G. La Monica, S. Cenini, F. Conti and F. Bonati, *Chem. Comm.*, (1967) 524.
107 G. Wilke, H. Schott and P. Heimbach, *Angew. Chem. Int. Ed.*, **6** (1967) 92.
108 C. D. Cook and G. S. Jauhal, *Inorg. Nucl. Chem. Lett.*, **3** (1967) 31.
109 S. Takahashi, K. Sonagashira and N. Hagihara, *J. Chem. Soc. Japan*, **87** (1966) 610; *Chem. Abs.*, **65** (1966) 14485.
110 E. W. Stern, *Chem. Comm.*, (1970) 736.
111 J. P. Collman, M. Kubota and J. W. Hosking, *J. Amer. Chem. Soc.*, **89** (1967) 4809.
112 A. Fusi, R. Ugo, F. Fox, A. Pasini and S. Cenini, *J. Organometal. Chem.*, **26** (1971) 417.
113 J. Halpern and A. L. Pickard, *Inorg. Chem.*, **9** (1970) 2798.
114 T. Kashiwagi, N. Yasuoka, N. Kasai, M. Kakudo, S. Takahashi and N. Hagihara, *Chem. Comm.*, (1969) 743.
115 C. D. Cook, P.-T. Cheng and S. C. Nyburg, *J. Amer. Chem. Soc.*, **91** (1969) 2123.
116 J. J. Levison and S. D. Robinson, *Inorg. Nucl. Chem. Lett.*, **4** (1968) 407.
117 J. J. Levison and S. D. Robinson, *J. Chem. Soc. (A)*, (1971) 762.
118 P. J. Hayward, D. M. Blake, C. J. Nyman and G. Wilkinson, *Chem. Comm.*, (1969) 987.
119 R. W. Horn, E. Weissberger and J. P. Collman, *Inorg. Chem.*, **9** (1970) 2367.
120 R. Ugo, F. Conti, S. Cenini, R. Mason and G. B. Robertson, *Chem. Comm.*, (1968) 1498.
121 P. J. Hayward and C. J. Nyman, *J. Amer. Chem. Soc.*, **93** (1971) 617.
122 P. J. Hayward, S. J. Saftich and C. J. Nyman, *Inorg. Chem.*, **10** (1971) 1311.
123 R. D. Johnston, F. Basolo and R. G. Pearson, *Inorg. Chem.*, **10** (1971) 247.
124 R. L. Keiter and J. G. Verkade, *Inorg. Chem.*, **9** (1970) 404.
125 F. Cariati and R. Ugo, *Chim. Ind. (Milan)*, **48**, (1966) 1288; *Chem. Abs.*, **66** (1967) 65618.
126 R. G. Vranka, L. F. Dahl, P. Chini and J. Chatt, *J. Amer. Chem. Soc.*, **91** (1969) 1574.
127 F. Cariati and R. Ugo, quoted in reference 44.
128 K. Hirota, M. Yamamoto, S. Otsuka, A. Nakamura and Y. Tatsuno, *Chem. Comm.*, (1968) 533.
129 S. Otsuka, A. Nakamura and Y. Tatsuno, *J. Amer. Chem. Soc.*, **91** (1969) 6994.

CHAPTER 4

Hydride Complexes of the Divalent Metals

Hydride complexes of platinum and palladium are a recent discovery, the first, trans-$[(PEt_3)_2PtHCl]$, being reported in 1957.[1] Since then many more platinum(II) complexes have been prepared as well as a smaller number of hydrido-palladium(II) complexes. The rapid expansion of the field has been greatly assisted by the increasing availability of commercial n.m.r. spectrometers, since transition metal hydride complexes exhibit large high-field chemical shifts; this enables them to be readily detected in the presence of compounds containing hydrogen atoms bound to main group elements, since these atoms show low-field chemical shifts (relative to tetramethyl-silane). This has allowed not only stable but also unstable hydride complexes which are formed as reaction intermediates to be detected. Hydride complexes have attracted research for two main reasons: first the simplicity of the hydride ligand lends itself to theoretical treatment and secondly the complexes readily undergo insertion reactions whereby olefins are inserted into the metal-hydride bond as well as the corresponding elimination reactions. These two reactions, in succession, lead to olefin isomerisation, which becomes catalytic if the hydride species is formed only transiently, and this is important in the petrochemicals industry.

Here we shall consider the preparation and then the physical properties whereby hydride complexes may be characterised. This will enable us to consider the bonding and so lead on to a discussion of the chemical reactions of these complexes. The principal reviews that may be used to supplement the material in this chapter are given in Table 12.

TABLE 12
Reviews on metal hydride complexes

Title	Reference
Hydrido- and related organo-complexes of transition metals	2
Hydride complexes of the transition metals	3, 4
σ-Complexes of platinum(II) with hydrogen, carbon and other elements of group IV	5
Hydride complexes	6

50

PREPARATION

Because hydride complexes are very reactive and palladium(II)–hydride complexes are unstable in both acidic and alkaline media a wide range of methods have been devised for their preparation. The method of choice will depend on the hydride complex to be prepared.

(a) Hydrazine reduction

Hydride complexes are formed[7] when cis-[(PR$_3$)$_2$PtCl$_2$] complexes (R = Et, Ph, etc.) are reduced with hydrazine in dilute aqueous or alcoholic solution (eq. 52). The reaction involves the intermediate formation of

$$cis\text{-}[(PR_3)_2PtCl_2] \longrightarrow$$
$$trans\text{-}[(PR_3)_2PtHCl] + NH_4Cl + \uparrow N_2 + NH_3 \quad (52)$$

hydrazine complexes (Fig. 5) and, when the phosphine is triphenylphosphine, can easily lead to the isolation of zerovalent platinum complexes (see p. 28). The arsine complex trans-[(AsEt$_3$)$_2$PtHCl] is only formed in good yield when the medium is dilute aqueous ammonia.[7] This stabilises the hydride as [(AsEt$_3$)$_2$PtH(NH$_3$)]$^+$Cl$^-$ which can be converted to the hydridochloride by cautious acidification at 0°C.

trans-Dichloroplatinum(II) complexes are not reduced by hydrazine nor can hydrazine reduction be used to prepare the corresponding hydrido-palladium(II) complexes as these are unstable in both acidic and basic media.[8]

(b) Alcoholic base

Treatment of cis-[(PEt$_3$)$_2$PtCl$_2$] with base in the presence of primary or secondary alcohols yields a hydride complex[7,9,10] (reaction 53). The original suggestion[7] that this reaction proceeds through an intermediate alkoxy-

$$cis\text{-}[(PEt_3)_2PtCl_2] + C_2H_5OH + KOH \rightarrow$$
$$trans\text{-}[(PEt_3)_2PtHCl] + CH_3CHO + KCl \rightarrow H_2O \quad (53)$$

complex (reaction 54) now appears even more likely, first because of the

$$CH_3CH_2OH + OH^- + Cl-\overset{|}{\underset{|}{Pt}}- \rightarrow$$

$$Cl^- + H_2O + CH_3-\overset{H}{\underset{H}{\overset{|}{C}}}\overset{O}{\diagdown}\overset{|}{\underset{|}{Pt}}- \rightarrow H-\overset{|}{\underset{|}{Pt}}- + CH_3CHO \quad (54)$$

isolation of stable alkoxy–platinum(II) complexes [(PEt$_3$)$_2$Pt(OR)(GePh$_3$)], where R = H, CH$_3$, C$_2$H$_5$ and ^{i}C$_3$H$_7$,[11] and secondly because deuteration studies have shown that in the corresponding iridium reaction the hydride ligand results from one of the hydrogen atoms bound to the α-carbon atom of the alcohol.[12] This method cannot be used to prepare hydrido–palladium(II) complexes because they are sensitive to basic media.[8] As already noted (pp. 28–29) triphenylphosphineplatinum(II) complexes tend to be reduced to the platinum(0) complexes, particularly in the presence of an excess of triphenylphosphine.[13]

(c) Metal hydrides

Although lithium aluminium hydride reduces both *cis*- and *trans*-platinum(II) complexes to hydrido-complexes it is not a convenient reagent because it reduces much of the platinum(II) to the free metal.[7] Similarly sodium borohydride is too powerful, reducing $[(Ph_2PCH_2CH_2PPh_2)PtCl_2]$ to $[(Ph_2PCH_2CH_2PPh_2)_2Pt]$[14] and other palladium(II) salts to the free metal.[15] However, the stable nickel borohydride complex *trans*-$[(PCy_3)_2$-$NiH(BH_4)]$, where Cy = cyclohexyl, reduces *trans*-$[(PR_3)_2PdCl_2]$ to a mixture of *trans*-$[(PCy_3)_2PdHCl]$ and *trans*-$[(PCy_3)_2NiHCl]$ and these can be partly separated by recrystallisation.[16,17] Further treatment of the recrystallised material with sodium borohydride yields $[(PCy_3)_2PdH(BH_4)]$ which can reduce more $[(PR_3)_2PdCl_2]$.[15] Repetition of this process, coupled with recrystallisation, eventually yields the pure palladium(II)–hydride complex *trans*-$[(PCy_3)_2PdHCl]$.

The most useful hydrides for the preparation of hydride complexes of platinum(II) and palladium(II) are those of silicon and germanium. For instance, the silanes R_3SiH convert platinum(II) chloro-complexes to hydrido-complexes probably by an addition-elimination mechanism[18] (reaction 55). The germane, Ph_3GeH, reacts similarly with *trans*-$[(PEt_3)_2PtBr_2]$ to give

$$\text{\textit{cis}-}[(PR_3')_2PtCl_2] + R_3SiH \longrightarrow [(PR_3')_2Pt(SiR_3)HCl_2] \longrightarrow$$

$$\text{\textit{trans}-}[(PR_3')_2PtHCl] + R_3SiCl \quad (55)$$

three volatile hydride complexes, but these have not yet been separated.[11] Me_3GeH reacts with *trans*-$[(PEt_3)_2PdCl_2]$ under mild conditions,[19] forming the rather unstable palladium hydride $[(PEt_3)_2PdHCl]$ (reaction 56). The products suggest that this reaction occurs

$$\text{\textit{trans}-}[(PEt_3)_2PdCl_2] + Me_3GeH \text{ (excess)} \xrightarrow[\text{vacuo}]{40°C \text{ in}}$$

$$\text{\textit{trans}-}[(PEt_3)_2PdHCl] + H_2 + Me_3GeCl + Me_6Ge_2 + \text{trace Pd} \quad (56)$$

by a free-radical mechanism.[8]

(d) Hydrogen

Hydride complexes can be prepared with gaseous hydrogen. The conditions vary from severe to very mild (see reactions 57–61). The ability of

$$\text{\textit{cis}-}[(PEt_3)_2PtCl_2] + H_2 \xrightarrow[\text{ref. 7}]{95°C/50 \text{ atm}}$$

$$\text{\textit{trans}-}[(PEt_3)_2PtHCl] + HCl \quad (57)$$

$$\text{\textit{trans}-}[(PEt_3)_2PtPhCl] + H_2 \xrightarrow[\text{ref. 7}]{20°C/1 \text{ atm}}$$

$$\text{\textit{trans}-}[(PEt_3)_2PtHCl] + C_6H_6 \quad (58)$$

$$[(PEt_3)_2Pt(GePh_3)_2] + H_2 \xrightarrow[\text{refs. 11 \& 20}]{25°C/20 \text{ mm}}$$

$$[(PEt_3)_2PdH(GePh_3)] + Ph_3GeH \quad (59)$$

$$trans\text{-}[(PEt_3)_2Pd(GePh_3)_2] + H_2 \xrightarrow[\text{ref. 21}]{25°C/100\,atm}$$

$$trans\text{-}[(PEt_3)_2PdH(GePh_3)] + Ph_3GeH \quad (60)$$

$$(Me_4N^+)[Pt(SnCl_3)_5]^- + H_2 \xrightarrow[\text{ref. 22}]{25°C/3\,atm}$$

$$(Me_4N^+)[PtH(SnCl_3)_4]^- \quad (61)$$

platinum(II) complexes to undergo addition-elimination reactions is illustrated by 59; this occurs rapidly with an activation energy of only 9 kcal/mol,[11] whereas palladium(II) complexes, which do not readily undergo addition, require far more forcing conditions to get the same effect (60).

Hydrogenation of $[(Ph_2PCH_2CH_2PPh_2)PtCl(SiMe_3)]$ at 60°C and 1 atm. has produced the only authentic *cis*-hydride complex of platinum(II), $[(Ph_2PCH_2CH_2PPh_2)PtHCl]$.[23] Earlier claims[24] for *cis*-$[(PPh_3)_2PtHCl]$ have since been discredited, the product being shown to be a crystalline modification of the *trans*-isomer.[23,25]

(e) Grignard and lithium reagents

Certain Grignard reagents give hydride complexes with platinum(II) complexes. For example, cyclohexyl magnesium bromide[7] and styryl magnesium bromide[26] both react with *cis*-$[(PEt_3)_2PtCl_2]$ to give *trans*-$[(PEt_3)_2PtHBr]$, and methyl magnesium chloride reacts with $[(QAs)PtCl]^+Cl^-$ (QAs = tris-(*o*-diphenylarsinophenyl)-arsine) to give $[(QAs)PtH]^+Cl^-$.[27] These reactions probably involve platinum(II)—magnesium intermediates[28] (reaction 62). With the triethylphosphine complex the driving force probably arises from the relative instability of the

$$cis\text{-}[(PEt_3)_2PtBr_2] + C_6H_{11}MgBr \longrightarrow$$

$$[(PEt_3)_2Pt(MgBr)Br] \xrightarrow{H_2O} trans\text{-}[(PEt_3)_2PtHBr]$$

$$\downarrow D_2O$$

$$trans\text{-}[(PEt_3)_2PtDBr] \quad (62)$$

cyclohexyl- and styryl-platinum(II) bonds; because alkylation by the Grignard reagent is reversible (see pp. 325–326) instability of the alkyl complex will not lead to overall decomposition but rather to regeneration of the starting material. The course of the reaction with the QAs complex is probably governed by the steric hindrance of the phenyl groups of the QAs ligand which prevent the more bulky methyl group binding to platinum but allow the linear –MgCl moiety to react.

Metal–halogen exchange reactions between triphenylgermyl lithium and platinum(II) halide complexes yield platinum(II)–lithium complexes which give hydride complexes on hydrolysis[28] (63 and 64).

$$[(PEt_3)_2PtI_2] + Ph_3GeLi \longrightarrow$$

$$Ph_3GeI + [(PEt_3)_2PtLiI] \xrightarrow{H_2O} [(PEt_3)_2PtHI] + LiOH \quad (63)$$

$$\downarrow Ph_3GeLi$$

$$Ph_6Ge_2$$

$[(PEt_3)_2PtI(GePh_3)] + Ph_3GeLi \rightarrow$

$$[(PEt_3)_2Pt(GePh_3)_2] + [(PEt_3)_2PtLi(GePh_3)]$$

$$\downarrow {\scriptstyle H_2O}$$

$$[(PEt_3)_2PtH(GePh_3)] + LiOH \qquad (64)$$

(f) Cleavage of platinum–germanium and platinum–silicon complexes

Treatment of $[(PR_3)_2Pt(GePh_3)_2]$[20] and $trans$-$[(PEt_3)_2PtCl(SiMe_3)]$[29] with hydrogen chloride yields $trans$-$[(PEt_3)_2PtHCl]$ probably by an addition-elimination reaction mechanism (p. 84).

(g) Oxidative-addition to zerovalent complexes

Oxidative-addition of hydrogen chloride to the palladium(0) complexes $[Pd(CO)(PPh_3)_3]$, $[Pd(PPh_3)_4]$ and $[Pd\{P(cyclohexyl)_3\}_2]$ has been used to prepare $trans$-$[(PR_3)_2PdHCl]$[30,31] and imides[32] such as succinimide, phthalimide, saccharin and tetrazoles[33] react with platinum(0) complexes to give hydride complexes (65 and 66). When a solution of $[Pt(PPh_3)_4]$ in

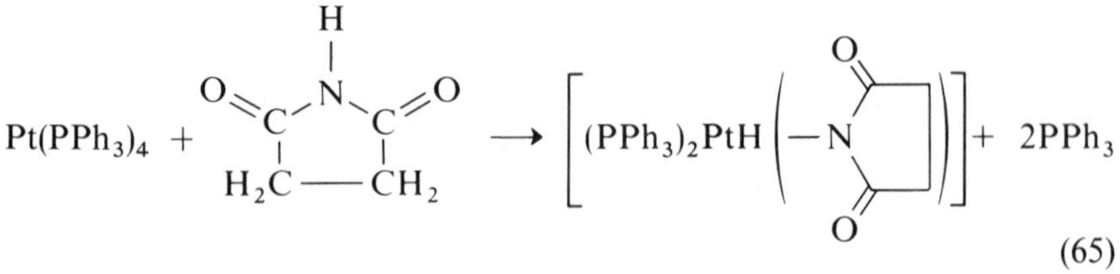

$$(65)$$

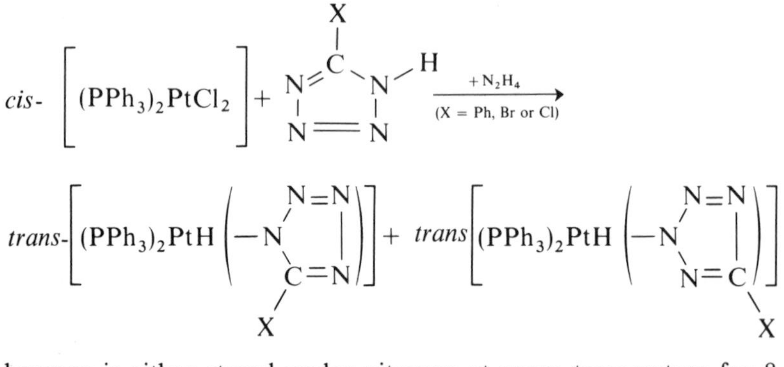

$$(66)$$

benzene is either stored under nitrogen at room temperature for 8 days or treated with hydrogen at 200 atm. for 5 hours $[(PPh_3)_2Pt(CO_3)]$ is formed by reaction of $[Pt(PPh_3)_2]$ with traces of oxygen and carbon dioxide present in the reaction vessel.[34] An earlier suggestion[35] that the product is $[(PPh_3)_2PtH_2]$ is incorrect.[34]

(h) Via carbonyl complexes

The cationic carbonyl complexes $trans$-$[(PR_3)_2PtCl(CO)]^+BF_4^-$, where R = Et or Ph, on boiling in water yield hydride complexes[36-38] (reaction 67). Deuteration has shown that protons in the water form the hydride

$$trans\text{-}[(PR_3)_2PtCl(CO)]^+ + H_2O \xrightarrow{\text{reflux}} trans\text{-}[(PR_3)_2PtHCl] + CO_2 \quad (67)$$

ligand. A remarkable complex $(AsPh_4^+)[Pd(CO)HCl_2]^-$ has been prepared by bubbling carbon monoxide through a solution of palladium(II) chloride in 2-methoxyethanol.[39]

(i) Formic acid
Hot formic acid reacts with cis-$[(PEt_3)_2PtCl_2]$ to form a hydride complex[7] (reaction 68). It seems, at the moment that a Pt–COOH intermediate is

$$cis\text{-}[(PEt_3)_2PtCl_2] + HCOOH \rightarrow trans\text{-}[(PEt_3)_2PtHCl] + CO_2 \quad (68)$$

involved, as tentatively suggested[37,38] for the hydrolysis of carbonyl complexes (see p. 106).

(j) Pyrolysis of platinum(II)–alkyl complexes
Heating trans-$[(PEt_3)_2Pt(C_2H_5)Cl]$ eliminates ethylene and forms the hydride[7] (reaction 69). The reaction is reversible. Deuterium labelled

$$trans\text{-}[(PEt_3)_2Pt(C_2H_5)Cl] \underset{}{\overset{heat}{\rightleftharpoons}} trans\text{-}[(PEt_3)_2PtHCl] + C_2H_4 \quad (69)$$

complexes indicated that the hydrogen atoms attached to either carbon atom of the ethyl group provide the hydride ligand.[40] This type of equilibrium may be present in other systems, for example the reduction of cis-$[(PPr_3)_2$-$PtCl_2]$ by phenylacetylene to give a platinum(II)–styryl complex may involve an intermediate hydrido–platinum(II) complex.[41]

(k) Metathetical replacement reactions
Metathetical replacement reactions by which the halide *trans* to the hydride ligand is replaced by a bromide, iodide, cyanide, cyanate, thiocyanate or nitro-ligand (reaction 70) are greatly facilitated by the high *trans*-effect of the hydride ligand[7] (pp. 299–301, Table 90). The *trans*-chloride ligand can

$$trans\text{-}[(PEt_3)_2PtHCl] + KSCN \xrightarrow{\text{acetone}}$$
$$trans\text{-}[(PEt_3)_2PtH(SCN)] + KCl \quad (70)$$

also be replaced by neutral ligands such as ammonia,[7] diphenylphosphine,[42] tertiary phosphines,[43–45] tertiary arsines,[44] tertiary phosphites,[44] carbon monoxide[44] and iso-cyanides[44] to give cationic hydride complexes that can be isolated with the aid of large anions.

PHYSICAL PROPERTIES OF HYDRIDE COMPLEXES

(a) Structure
All the known hydride complexes of platinum(II) and palladium(II) are square-planar, and all but $[(Ph_2PCH_2CH_2PPh_2)PtHCl]$[23] have a *trans*-configuration. Although x-ray diffraction investigations of trans-$[(PEt_3)_2PtHBr]$[46] and trans-$[(PPh_2Et)_2PtHCl]$[47] did not locate the hydride ligands, single crystal Raman studies show that they lie in the square-plane.[48] The long platinum–halogen bond lengths (Table 13) indicate the high *trans*-influence of the hydride ligand.

TABLE 13
Platinum(II)–Halogen bond lengths in hydride complexes

Bond	Bond length (Å)	Reference	Comparison Pt–X bond	Bond length (Å)	Reference
Pt–Br in trans-[(PEt$_3$)$_2$PtHBr]	2·56(7)	46	Pt–Br in trans-[(PEt$_3$)$_2$PtBr$_2$]	2·428(2)	49
Pt–Cl in trans-[(PPh$_2$Et)$_2$PtHCl]	2·42(1)	47	Pt–Cl in trans-[Pt(NH$_3$)$_2$Cl$_2$]	2·32(1)	50

(b) Infra-red spectra

Hydride complexes generally show a band in the region 1900–2250 cm^{-1} identifiable by comparison with the deuteride complexes ($v_{M-H}/v_{M-D} = 1\cdot4$, in accord with simple theory). The metal–halogen stretching frequency depends strongly on the solvent and a change from chloroform to hexane produces a shift of about 30 cm^{-1} when the hydride is trans to a halogen.[51] Solid state effects are also large, indeed the palladium–hydrogen stretching frequency in [Pd(CO)HCl$_2$]$^-$ lies between 1960 cm^{-1} and 2000 cm^{-1} depending on the cation.[39] The platinum–hydrogen stretching frequency (Table 14) depends greatly on the nature of the trans-ligand, decreasing as the trans-effect of this ligand increases[52,53] (p. 299). There is also a small cis-

TABLE 14
Dependence of the platinum(II)–hydrogen stretching frequencies in the complexes trans-[(PEt$_3$)$_2$PtHX] on the trans-ligand
(data from ref. 53)

X	v_{Pt-H} (cm^{-1})
NO$_3$	2242
Cl	2183
Br	2178
I	2156
NO$_2$	2150
SCN	2112
CN	2041

TABLE 15
Dependence of the platinum(II)–hydrogen stretching frequencies in the complexes trans-[L$_2$PtHX] at 20°C on the cis ligand (data from ref. 53)

L	Solvent	v_{Pt-H} (cm^{-1}) for X = Cl	v_{Pt-H} (cm^{-1}) for X = I
AsEt$_3$	hexane	2174	2139
PMe$_3$	CCl$_4$	2182	
PEt$_3$	hexane	2183	2156
PPr$_3$	hexane	2183	
PPhEt$_2$	hexane	2199	2179
PPh$_2$Et	benzene	2210	
PPh$_3$	benzene	2224	2190

effect in which an increase in the platinum(II)–hydrogen stretching frequency parallels an increase in the electronegativity of the *cis*-ligands (Table 15).

Where comparable complexes are available the palladium(II)–hydrogen stretching frequency is generally lower than the platinum(II)–hydrogen stretching frequency—for example, *trans*-[(PEt$_3$)$_2$MHCl] $v_{Pt-H} =$ 2183 cm^{-1}[53] and $v_{Pd-H} = 2053$ cm^{-1};[2] *trans*-[(PPr$_3^n$)$_2$MHCl], $v_{Pt-H} =$ 2183 cm^{-1}[53] and $v_{Pd-H} = 2010$ cm^{-1}.[15] Since the atomic weight of palladium is less than that of platinum these results indicate a lower stretching force constant for the palladium–hydrogen bond, which is consistent with the lower thermal and chemical stability of hydrido–palladium(II) complexes. The stretching frequencies of platinum–chlorine bonds *trans* to hydride ligands show that the latter has a high *trans*-influence, comparable to that of alkyl and aryl ligands.[54]

(c) Nuclear magnetic resonance
(i) *Proton magnetic resonance*
The proton magnetic resonance spectra of hydride complexes always show a very large high-field chemical shift for the hydride ligand. Typically these occur in the range 17·6 to 33·6 p.p.m. relative to tetramethylsilane at 10 p.p.m. This large chemical shift ensures that the resonance of the hydridic proton is far removed from the proton resonances of the organic part of the complex so that the detection of hydride complexes by p.m.r. is easy. This largely accounts for the rapid growth in metal–hydride chemistry. In platinum(II) complexes the hydride proton is split by ^{195}Pt (natural abundance 33·7%) with a coupling constant in excess of 1000 Hz. It is also split by the ^{31}P nuclei with a coupling constant of about 15 Hz, to give an overall spectrum of the type shown in Fig. 12. Theoretical interpretations of both the chemical shifts of the hydride protons and the large metal–hydrogen coupling constants have been advanced.

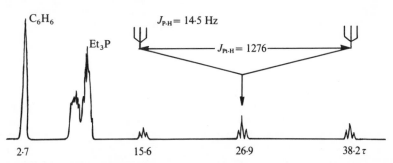

Fig. 12. Proton magnetic resonance spectrum of *trans*-[(PEt$_3$)$_2$PtHCl] in benzene (Reproduced with permission from ref. 2).

Chemical shift of the hydride proton. The large chemical shifts for hydridic protons are ascribed to a shielding of the proton by the non-bonding electrons of the metal. Initially the chemical shifts were thought to result from non-bonding electrons in ground-state orbitals.[55] but later the calculations were refined by considering contributions from excited states as well as the ground-state.[56] These extra contributions are those responsible for temperature independent paramagnetism and lead to the introduction of a paramagnetic shielding term, which according to the position of the hydride proton may either increase or decrease the total shielding. Subsequent work has suggested that for platinum(II)–hydride complexes this paramagnetic shielding of the hydridic proton by the 5d-electrons of platinum is probably the dominant term in determining the chemical shift of the proton.[57] This paramagnetic shielding is very sensitive to the platinum–hydrogen bond length, varying approximately as the inverse cube of its value.[58] This theory is supported[58] by the linear correlation found between the hydride chemical shift and the platinum–hydrogen stretching frequency in the series of complexes *trans*-[(PEt$_3$)$_2$PtHL], where L = *o*-, *m*- or *p*-substituted benzoate, moreover, the platinum–hydrogen stretching frequency is also proportional to the inverse cube of the platinum–hydrogen bond length.[59] In addition to changes in the platinum–hydrogen bond length, the increase in the hydride chemical shift in *trans*-[(PEt$_3$)$_2$MHX], where M = Pd or Pt, in the order X = I < Br < Cl is partly due to the nephelauxetic effect of the halogen ligand which decreases in the order I > Br > Cl[60] allowing greater interaction between the platinum and hydrogen in the chloride complex.[61]

Platinum–Hydrogen Coupling Constants. The platinum–hydrogen coupling constants in complexes of the type *trans*-[(PEt$_3$)$_2$PtHX] are strongly dependent on the nature of the *trans*-ligand X (Table 16). The dominant term in

TABLE 16

Platinum–hydrogen coupling constants for *trans*-[(PEt$_3$)$_2$HX] (data from ref. 58)

X	J_{Pt-H} (Hz)
I	1369
Br	1346
NO$_3$	1322
Cl	1275
—SCN	1233
CNO	1080
NO$_2$	1003
CN	778

determining this variation is the Fermi contact interaction,[62] which in turn depends on the s-electron densities at the two nuclei, the degree of s-character in the platinum(II)–hydrogen bond, and the mean excitation energy. This last term represents a simplification of the perturbation theory used to derive the coupling constant in which the electronic energies of all the triplet states interacting with the ground state are replaced by an average. This mean excitation energy is dependent on the ligand field splitting parameter and a plot of J_{Pt-H}[58] for a series of complex *trans*-$[(PEt_3)_2PtHX]$, where X = CN, NO_2, CNO, Cl, Br and I, against the reciprocal of the first ligand field transition[63] in $[Co(NH_3)_5X]^{2+}$ is approximately a straight line (Fig. 13). However, this mean excitation energy is certainly not the sole contributor to the Fermi contact interaction as shown by both the scatter of the points in Fig. 13 and the variations in J_{Pt-H} for a series of complexes *trans*-$[(PEt_3)_2PtHL]$, where L is an *o*-, *m*- or *p*-substituted benzoate ligand, which are too great for this term alone to explain.[58]

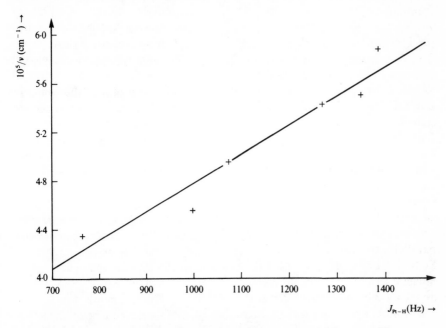

Fig. 13. Plot of the reciprocal of the first ligand field transition $(10^5/v)$ in $[Co(NH_3)_5X]^{2+}$ against J_{Pt-H} in *trans*-$[(PEt_3)_2PtHX]$.

Another important term in the Fermi contact interaction is the *s*-electron density at the nucleus[64] which depends on the effective nuclear charge of the platinum atom and the electronegativity of the ligands around the platinum atom. However, a change in the electronegativity of these ligands has a number of opposing effects. (a) An increase in electronegativity increases the effective nuclear charge on the platinum so that J_{Pt-H} should increase as the ligand is changed in the order I < Br < Cl. (b) The polar nature of the Pt–X bond leads to electron density being withdrawn from platinum to the halogen, so that the *s*-electron density at platinum is reduced by increasing the electronegativity of the halogen. Thus J_{Pt-H} should increase as the ligand changes in the order Cl < Br < I. (c) The more electronegative the ligand *trans* to the platinum–hydrogen bond the greater should be the *s*-character of that bond. Thus J_{Pt-H} should increase in the order I < Br < Cl. Effects *a* and *b* are isotropic whereas *c* is anisotropic. Since the effect on

J_{Pt-H} of changing a *cis*-ligand is the reverse of changing a *trans*-ligand term *c* must be important.[44] A similar conclusion was obtained from the platinum–hydrogen coupling constants for hydride ligands *trans* to a substituted benzoate ligand.[58]

In conclusion, where *trans*-ligands are sufficiently different to give substantial differences in the mean excitation energy term, then this term dominates the observed platinum–hydrogen coupling constants (Fig. 13). However, when these differences are small, differences in the s-characters of the platinum–hydrogen bonds determine the magnitude of the platinum–hydrogen coupling.

(ii) ^{195}Platinum magnetic resonance

The ^{195}platinum chemical shifts in the complexes *trans*-[(PEt$_3$)$_2$PtHX] increase in the order X = substituted benzoate < NO$_3^-$ < NO$_2^-$ < Cl$^-$ < SCN$^-$ < Br$^-$ < CN$^-$ < I$^-$.[65] This is, with the exception of the cyanide ion, the order of increasing π-covalency in the Pt–X bond and arises as follows. The shielding of a nucleus is commonly divided into a diamagnetic and a paramagnetic contribution[66] of which the latter is more important for our present purposes because it is sensitive to the wave function of the valence electrons. This paramagnetic contribution is inversely proportional to the mean excitation energy.[66] The fact that the dependence of the ^{195}Pt chemical shift on the ligand X does not parallel the spectrochemical series suggests that the σ-covalency of the Pt–X bond does not dominate the ^{195}Pt chemical shift. However the ^{195}Pt chemical shift order does closely parallel the nephelauxetic order[60] of benzoate < Cl$^-$ < CN$^-$ < SCN$^-$ < Br$^-$ < I$^-$ except for the position of the cyanide ion. The nephelauxetic series represents the change in interelectron repulsion consequent to an increase in the π-covalency of the Pt–X bond. The anomalous position of the cyanide ligand is probably a consequence of its extreme position in the spectrochemical series. Thus variations in the ^{195}Pt chemical shifts are largely determined by changes in the π-covalency of the platinum–ligand bonds.

(d) Electronic spectra

No detailed studies of the electronic spectra of platinum(II) or palladium(II) hydride complexes have been made, but the hydride ligand appears to have a high ligand-field strength relative to chloride since hydrido-complexes are much paler than their dichloro-analogues.[7] The *d–d* transitions in hydride-complexes are in the ultra-violet region of the spectrum where they are obscured by charge-transfer bands. Because of the impossibility of π-bonding and because of the high polarisability of the hydride ion the ligand-field strength of the hydride ion is more dependent on its environment in a particular complex than the ligand-field strengths of most other ligands.[4] This in complexes with purely σ-donor ligands the hydride ligand should carry a larger negative charge and hence exhibit a weaker ligand-field strength than in complexes which contain ligands capable of π-acceptance as well as σ-donation.

(e) Dipole moment studies

In the early work on hydride complexes dipole moment studies were used extensively.[2] However, their interpretation was difficult because a simple model showed the hydrogen atom to be at the positive end of the dipole. This arises, first because of distortions from the perfect square-planar configuration (p. 55), and secondly because under the influence of the strongly σ-donating hydride ligand the *trans*-ligand becomes more negative and the *trans*-platinum–ligand bond is lengthened.

(f) Mass spectra

The mass spectra of *trans*-[(PEt$_3$)$_2$PtHX] complexes, where X = Cl, Br, CN, or CNO, are dependent on the nature of X.[67] In all cases the parent ion was observed, but with X = Cl, or CNO no peak corresponding to the parent minus the hydridic proton was observed, so that the technique is not always diagnostic for the presence of a hydride ligand.

STABILITY AND BONDING IN HYDRIDE COMPLEXES

There is a marked increase in the stability of transition metal hydride complexes on descending a given group. Thus, for example, *trans*-[(PEt$_3$)$_2$PtHCl] is sufficiently stable to be distilled under high vacuum (at 130°C at 0·01 mm Hg).[7] The corresponding palladium compound can be isolated as a solid but is always contaminated by decomposition products,[52] and the nickel analogue has never been isolated although it can be detected in solution by its proton resonance spectrum.[68] Ligands with high ligand-field strengths such as carbon or phosphorus donor ligands are most effective at stabilising hydride complexes. Ligand effectiveness then falls off in the approximate order tertiary arsines > organic sulphides > amines > oxyacid anions > halide ions.[6] The most effective of the tertiary phosphine ligands is the very bulky tri-cyclohexyl-phosphine which has been used to isolate a series of stable palladium(II)–hydride complexes *trans*-[{P(cyclohexyl)$_3$}$_2$PdHX]. The decomposition of these hydride complexes probably occurs through a tetrahedral intermediate and the role of the bulky phosphine is to stabilise the *trans*-square-planar configuration with respect to a tetrahedral configuration.[17]

Both the properties of the metal and of the ligand point to the need for a high ligand-field stabilisation energy in the complex preventing dissociation to give either hydride ions or hydride radicals both of which would be immediately destroyed by air or moisture. This explanation is similar to that discussed in detail in Chapter 12 for the stabilisation of metal–carbon σ-bonds (pp. 331–334).

CHEMICAL PROPERTIES OF HYDRIDE COMPLEXES

The reactions of platinum(II) and palladium(II) hydrides fall into two classes: first those that involve the hydrogen directly and these are discussed in this section, and secondly reactions such as isomerisation and substitution of the ligand *trans* to the hydride ligand which are discussed in Chapter 11. Whereas the platinum(II)–hydride complexes are stable to air and water, the palladium(II) complexes are sensitive to air,[8,39] although they are stable to water.[39]

(a) Hydrogen chloride

Dry hydrogen chloride reacts with *trans*-[(PEt$_3$)$_2$PtHCl] in ether to give an unstable addition product (reaction 71) which readily loses hydrogen

$$trans\text{-}[(PEt_3)_2PtHCl] + HCl \rightleftharpoons [(PEt_3)_2PtH_2Cl_2] \qquad (71)$$

chloride to revert to the original complex.[7] Palladium complexes as noted elsewhere (pp. 2, 90, 340) do not readily form stable palladium(IV) complexes, so that treatment of *trans*-[(PR$_3$)$_2$PdHCl] with hydrogen chloride results in the formation of *trans*-[(PR$_3$)$_2$PdCl$_2$].[8,30]

(b) Alkyl halides

Alkyl halides such as methyl iodide, ethyl iodide, *n*-butyl bromide, carbon tetrachloride and chloroform react with *trans*-[(PEt$_3$)$_2$PtHCl] to give the dihalo-complex, although the organic products containing the hydride ligand were not isolated.[7] Carbon tetrachloride reacts similarly with *trans*-[(PR$_3$)$_2$PdHCl] and in this case chloroform is formed in quantitative yield.[8,30]

(c) Metal halides

Mercury(II) chloride, copper(II) chloride and boron trichloride all react with *trans*-[(PPh$_3$)$_2$PtHCl] in acetone at 0°C to give *trans*-[(PPh$_3$)$_2$PtCl$_2$] and, with mercury(II) chloride, metallic mercury is also formed.[69] Since neither lithium nor cadmium chloride react, complexes containing Pt–Hg, Pt–Cu or Pt–B bonds are probably involved as intermediates.

(d) Halogens

Halogens cleave the platinum(II)–hydrogen bond to yield the dihalo-platinum(II) complex and presumably the hydrogen halide.[5]

(e) Unsaturated organic compounds

Some of the most interesting and important reactions of the hydrides involve insertion of an unsaturated organic compound into the metal–hydride bond. The reaction of *trans*-[(PEt$_3$)$_2$PtHCl] with ethylene (reaction 69) has already been discussed (p. 55). Reactions of this type are involved in the homogeneous hydrogenation and hydrosilation of olefins as well as the catalytic isomerisation of olefins by both platinum(II) and palladium(II) complexes. The subject has recently been reviewed by the present author.[70] A recent development in this field is that ethylene can react with a hydride-complex containing a weakly co-ordinating ligand, such as nitrate, to displace that ligand and form a hydride–olefin complex[71] (reaction 72) rather than giving the normal insertion product.

$$\text{trans-}[(PEt_3)_2PtH(NO_3)] + C_2H_4 \longrightarrow$$
$$\text{trans-}[(PEt_3)_2Pt(C_2H_4)H]^+NO_3^- \quad (72)$$

Perfluoro-olefins including tetrafluoroethylene react with hydride complexes to give complexes containing platinum(II)–carbon σ-bonds (pp. 343–344). With tetracyanoethylene hydrogen chloride is eliminated and a platinum(0)-olefin complex is formed[72] (reaction 73). However, tetrachloroethylene behaves like a saturated hydrocarbon and gives $[(PEt_3)_2PtCl_2]$.[73]

$$\text{trans-}[(PEt_3)_2PtHCl] + (NC)_2C{=}C(CN)_2 \longrightarrow$$
$$\left[(PEt_3)_2Pt{-}\begin{array}{c}C(CN)_2\\ \|\\ C(CN)_2\end{array}\right] + HCl \quad (73)$$

Aryldiazonium salts insert into platinum(II)–hydrogen bonds giving arylazo complexes[74] (reaction 74), which can be used to prepare platinum(II)–aryl complexes (see p. 329). Similarly diazomethane and azides such as

$$PhN{\equiv}N^+ + [(PEt_3)_2PtHCl] \longrightarrow \left[\begin{array}{c}\quad H\quad PEt_3\\ \quad |\quad\quad |\\ PhN{=}N{-}Pt{-}Cl\\ \quad\quad\quad |\\ \quad\quad\quad PEt_3\end{array}\right]^+ \xrightarrow{\text{KOH}}$$

$$[(PEt_3)_2Pt(N{=}NPh)Cl] \quad (74)$$

p-toluene-sulphonylazide[75,76] insert into platinum(II)–hydrogen bonds with elimination of nitrogen to give methyl and amido complexes respectively (reactions 75 and 76).

$$CH_2N_2 + [(PEt_3)_2PtHCl] \longrightarrow [(PEt_3)_2Pt(CH_3)Cl] + \uparrow N_2 \quad (75)$$

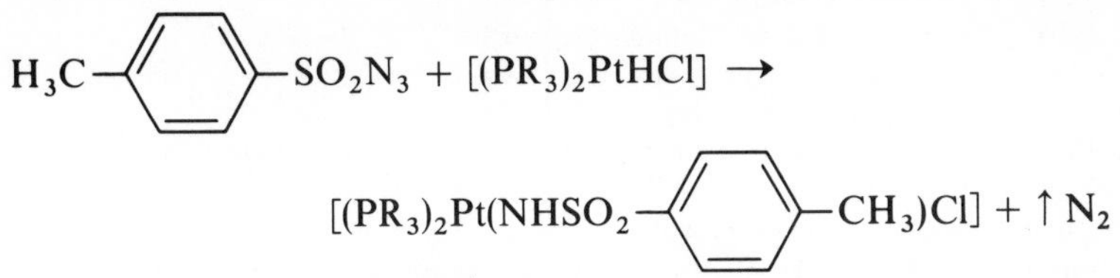

$$H_3C{-}\!\!\left\langle\!\!\bigcirc\!\!\right\rangle\!\!{-}SO_2N_3 + [(PR_3)_2PtHCl] \longrightarrow$$

$$[(PR_3)_2Pt(NHSO_2{-}\!\!\left\langle\!\!\bigcirc\!\!\right\rangle\!\!{-}CH_3)Cl] + \uparrow N_2 \quad (76)$$

(f) Condensation with elimination of hydrogen
Condensation with elimination of hydrogen is a very common reaction of metal hydrides. An example is the reaction of platinum(II) hydrides with group IVB metal hydrides described in Chapter 6 (p. 78) for the preparation of platinum(II) complexes with group IVB metal ligands.

(g) Deuterium exchange
Little reaction takes place when $\text{trans-}[(PEt_3)_2PtHCl]$ is treated with D_2O,[7,36] D_2[77] or C_2D_4.[77] If catalytic amounts of hydrogen chloride are

added, however, a ready exchange occurs in D_2O yielding *trans*-[(PEt$_3$)$_2$-PtDCl].[7] The acid catalyst promotes the formation of platinum(IV) intermediates such as [(PEt$_3$)$_2$PtHDCl$_2$] which undergo rapid exchange with D_2O before losing hydrogen chloride to give the deutero–platinum(II) complex.[78]

REFERENCES

1   J. Chatt, L. A. Duncanson and B. L. Shaw, *Proc. Chem. Soc.*, (1957) 343.
2   J. Chatt, *Proc. Chem. Soc.*, (1962) 318.
3   A. P. Ginsberg, *Trans. Metal Chem.*, **1** (1965) 111.
4   M. L. H. Green and D. J. Jones, *Adv. Inorg. Radiochem.*, **7** (1965) 115.
5   R. J. Cross, *Organometal. Chem. Rev.*, **2** (1967) 97.
6   J. Chatt, *Science*, **160** (1968) 723.
7   J. Chatt and B. L. Shaw, *J. Chem. Soc.*, (1962) 5075.
8   E. H. Brooks and F. Glockling, *J. Chem. Soc. (A)*, (1967) 1030.
9   J. Chatt and B. L. Shaw, *Chem. Ind. (London)*, (1960) 931.
10 J. Chatt and B. L. Shaw, *Chem. Ind. (London)*, (1961) 290.
11 R. J. Cross and F. Glockling, *J. Chem. Soc.*, (1965) 5422.
12 L. Vaska and J. W. DiLuzio, *J. Amer. Chem. Soc.*, **84** (1962) 4989
13 L. Malatesta and C. Cariello, *J. Chem. Soc.*, (1958) 2323.
14 J. Chatt and G. A. Rowe, *Nature*, **191** (1961) 1191.
15 H. Munakata and M. L. H. Green, *Chem. Comm.*, (1970) 881.
16 M. L. H. Green, H. Munakata and T. Saito, *Chem. Comm.*, (1969) 1287.
17 M. L. H. Green, H. Munakata and T. Saito, *J. Chem. Soc. (A)*, (1971) 469.
18 A. J. Chalk and J. F. Harrod, *J. Amer. Chem. Soc.*, **87** (1965) 16.
19 E. H. Brooks and F. Glockling, *Chem. Comm.*, (1965) 510.
20 R. J. Cross and F. Glockling, *Proc. Chem. Soc.*, (1964) 143.
21 F. Glockling and E. H. Brooks, *Abstracts 157th Amer. Chem. Soc. Meeting*, April 1969, Petr. 43.
22 R. D. Cramer, R. V. Lindsey, C. T. Prewitt and U. G. Stolberg, *J. Amer. Chem. Soc.*, **87** (1965) 658.
23 A. F. Clemmit and F. Glocking, *J. Chem. Soc. (A)*, (1969) 2163.
24 J. C. Bailar and H. Itatani, *Inorg. Chem.*, **4** (1965) 1618.
25 I. Collamati, A. Furlani and G. Attioli, *J. Chem. Soc. (A)*, (1970) 1694.
26 J. Chatt and B. L. Shaw, *J. Chem. Soc.*, (1959) 4020.
27 F. R. Hartley and L. M. Venanzi, Unpublished results.
28 R. J. Cross and F. Glockling, *J. Organometal. Chem.*, **3** (1965) 253.
29 F. Glockling and K. A. Hooton, *Chem. Comm.*, (1966) 218.
30 K. Kudo, M. Hidai, T. Murayama and Y. Uchida, *Chem. Comm.*, (1970) 1701.
31 R. Van der Linde and R. O. de Jongh, *Chem. Comm.*, (1971) 563.
32 D. M. Roundhill, *Chem. Comm.*, (1969) 567.
33 J. H. Nelson, D. L. Schmitt, R. A. Henry, D. W. Moore and H. B. Jonassen, *Inorg. Chem.*, **9** (1970) 2678.
34 C. J. Nyman, C. E. Wymore and G. Wilkinson, *Chem. Comm.*, (1967) 407.
35 L. Malatesta and R. Ugo, *J. Chem. Soc.*, (1963) 2080.
36 H. C. Clark, K. R. Dixon and W. J. Jacobs, *Chem. Comm.*, (1968) 548.
37 H. C. Clark, K. R. Dixon and W. J. Jacobs, *J. Amer. Chem. Soc.*, **91** (1969) 1346.
38 H. C. Clark and W. J. Jacobs, *Inorg. Chem.*, **9** (1970) 1229.
39 J. V. Kingston and G. R. Scollary, *Chem. Comm.*, (1969) 455.
40 J. Chatt, R. S. Coffey, A. Gough and D. T. Thompson, *J. Chem. Soc. (A)*, (1968) 190.
41 J. Chatt and G. A. Rowe, *International Conference on Coordination Chemistry*, London, 1959, *Chem. Soc. Special Publ.*, No. 13, p. 117.
42 J. Chatt and J. M. Davidson, *J. Chem. Soc.* (1964) 2433.
43 M. Giustiniani, G. Dolcetti and U. Belluco, *J. Chem. Soc. (A)*, (1969) 2047.
44 M. J. Church and M. J. Mays, *J. Chem. Soc. (A)*, (1970) 1938.

45 M. L. H. Green and H. Munakata, *Chem. Comm.*, (1971) 549.
46 P. G. Owston, J. M. Partridge and J. M. Rowe, *Acta Cryst.*, **13** (1960) 246.
47 R. Eisenberg and J. A. Ibers, *Inorg. Chem.*, **4** (1965) 773.
48 I. R. Beattie and K. M. S. Livingston, *J. Chem. Soc. (A)*, (1969) 2201.
49 G. G. Messmer and E. L. Amma, *Inorg. Chem.*, **5** (1966) 1775.
50 G. H. W. Milburn and M. R. Truter, *J. Chem. Soc. (A)*, (1966) 1609.
51 D. M. Adams, *Proc. Chem. Soc.*, (1961) 431.
52 J. Chatt, L. A. Duncanson and B. L. Shaw, *Chem. Ind. (London)*, (1958) 859.
.53 J. Chatt, L. A. Duncanson, B. L. Shaw and L. M. Venanzi, *Disc. Faraday Soc.*, **26** (1958) 131.
54 D. M. Adams, J. Chatt, J. Gerratt and A. D. Westland, *J. Chem. Soc.*, (1964) 734.
55 R. M. Stevens, C. W. Kern and W. N. Lipscomb, *J. Chem. Phys.*, **37** (1962) 279.
56 A. D. Buckingham and P. J. Stephens, *J. Chem. Soc.*, (1964) 2747.
57 A. D. Buckingham and P. J. Stephens, *J. Chem. Soc.*, (1964) 4583.
58 P. W. Atkins, J. C. Green and M. L. H. Green, *J. Chem. Soc. (A)*, (1968) 2275.
59 P. J. Stephens, D. Phil. Thesis, Oxford, 1964.
60 C. K. Jørgensen, *Absorption Spectra and Chemical Bonding In Complexes*, Pergamon, Oxford, 1962.
61 E. R. Birnbaum, *Inorg. Nucl. Chem. Lett.*, **7** (1971) 233.
62 J. A. Pople and D. P. Santry, *Mol. Phys.*, **8** (1964) 1.
63 R. Tsuchida, *Bull. Chem. Soc. Japan*, **13** (1938) 388 and 436.
64 W. G. Schneider and A. D. Buckingham, *Disc. Faraday Soc.*, **34** (1962) 147.
65 R. R. Dean and J. C. Green, *J. Chem. Soc. (A)*, (1968) 3047.
66 N. F. Ramsey, *Phys. Rev.*, **78** (1950) 699.
67 B. F. G. Johnson, J. Lewis and P. W. Robinson, *J. Chem. Soc. (A)*, (1970) 1684.
68 M. L. H. Green, C. N. Street and G. Wilkinson, *Z. Naturforsch*, **14B** (1959) 738; *Chem. Abs.*, **54** (1960) 16441b.
69 A. D. Allen and M. C. Baird, *Chem. Ind. (London)*, (1965) 139.
70 F. R. Hartley, *Chem. Rev.*, **69** (1969) 799.
71 A. J. Deeming, B. F. G. Johnson and J. Lewis, *Chem. Comm.*, (1970) 598.
72 W. H. Baddley and L. M. Venanzi, *Inorg. Chem.*, **5** (1966) 33.
73 H. C. Clark and W. S. Tsang, *Chem. Comm.*, (1966) 123.
74 G. W. Parshall, *J. Amer. Chem. Soc.*, **87** (1965) 2133.
75 W. Beck, M. Bauder, W. P. Fehlhammer, P. Pöllmann and H. Schächl, *Inorg. Nucl. Chem. Lett.*, **4** (1968) 143.
76 W. Beck and M. Bauder, *Chem. Ber.*, **103** (1970) 583; *Chem. Abs.*, **72** (1970) 74237.
77 R. A. Schunn, *Inorg. Chem.*, **9** (1970) 2567.
78 C. D. Falk and J. Halpern, *J. Amer. Chem. Soc.*, **87** (1965) 3523.

CHAPTER 5

Complexes of the Divalent Metals with Ligands from Groups IA, IIA, IIIB and the Transition Metals

COMPLEXES WITH GROUP IA AND IIA ELEMENTS

No complexes in which either group IA or group IIA elements are bound directly to platinum(II) or palladium(II) have been isolated. However, substantial evidence that $[(PEt_3)_2Pt(Li)I]$ and $([PEt_3)_2Pt(MgBr)Br]$ are formed as intermediates during the preparation of platinum(II)–hydride complexes from Ph_3GeLi and grignard reagents respectively has been obtained (pp. 53–54).[1]

COMPLEXES WITH GROUP IIIB ELEMENTS

Stable platinum(II) complexes containing boron ligands have been prepared by the addition of a boron halide to a platinum(II) hydride complex[2] (reaction 77) and by the oxidative addition of a boron halide to a platinum(0) complex[3] (reaction 78). The platinum(II)-borane complexes are stable in

$$[(PEt_3)_2PtHCl] + Ph_2BCl \rightarrow [(PEt_3)_2Pt(BPh_2)Cl] + HCl \quad (77)$$

air and $[(PPh_3)Pt(BPh_2)Br]$ melts without decomposition at 198–200°C.

$$[Pt(PPh_3)_4] + Y_2BX \xrightarrow[\substack{(X = Cl, Br; Y = Ph, Br)}]{cyclohexane} [(PPh_3)_2Pt(BY_2)X] + 2PPh_3 \quad (78)$$

The platinum(II)–boron bond can be cleaved by bromine[2] (reaction 79). Although no complexes with single palladium(II) boron bonds have been

$$[(PPh_3)_2Pt(BPh_2)Br] + Br_2 \rightarrow [(PPh_3)_2PtBr_2] + Ph_2BBr \quad (79)$$

reported, palladium(II) complexes of bis-(dimethylamino)boron, in which the ligand acts as a π-ligand have been prepared[4] (reaction 80)

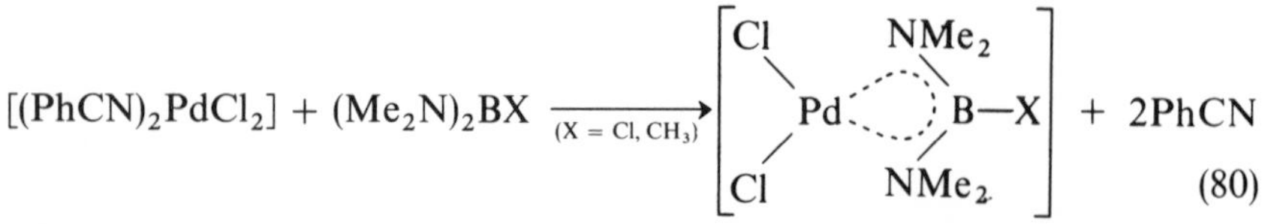

$$[(PhCN)_2PdCl_2] + (Me_2N)_2BX \xrightarrow[(X = Cl, CH_3)]{} \begin{bmatrix} Cl & NMe_2 \\ & \diagdown \\ Pd & \cdots B-X \\ \diagup & \diagup \\ Cl & NMe_2 \end{bmatrix} + 2PhCN \quad (80)$$

A number of compounds in which palladium(II) and platinum(II) are bound directly to more than one atom of a borane ligand are known. These complexes are considered in order of increasing numbers of boron atoms in the borane.

(a) $B_3H_7^{2-}$

Platinum(II) complexes containing $B_3H_7^{2-}$ ligands, prepared by reaction 81, are sensitive to air but fairly stable in aqueous solution.[5] The n.m.r. spectra

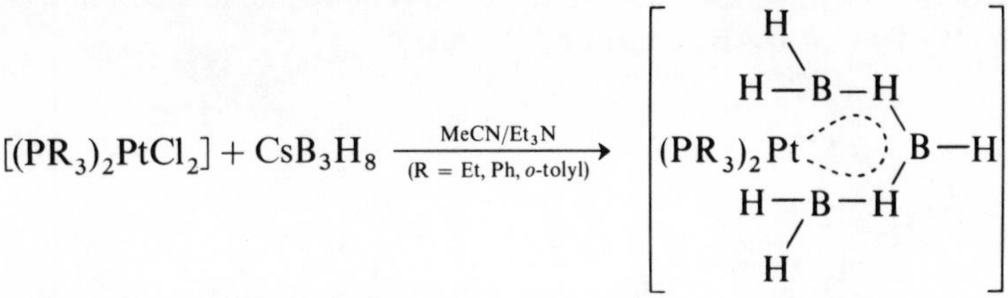

$$[(PR_3)_2PtCl_2] + CsB_3H_8 \xrightarrow[\text{(R = Et, Ph, o-tolyl)}]{\text{MeCN/Et}_3\text{N}} \left[(PR_3)_2Pt \begin{smallmatrix} H \\ \backslash \\ H-B-H \\ \cdots \\ B-H \\ / \\ H-B-H \\ / \\ H \end{smallmatrix} \right] \quad (81)$$

suggest a π-borallyl structure and the x-ray photoelectron spectra are consistent with platinum being in the $+2$ oxidation state: the binding energies of the platinum $4f_{7/2}$ electrons are 72·9 eV.

(b) $B_8H_{12}^{2-}$

A platinum(II) complex of the anionic borane $B_8H_{12}^{2-}$, $[(PEt_3)_2Pt(B_8H_{12})]$, is formed[6] by alcohol degradation of $[(PEt_3)_2Pt\{B_9H_{11}P\,(p\text{-}C_6H_4CH_3)_3\}]$ (p. 68). The n.m.r. spectrum is consistent with structure **14** in which the $B_8H_{12}^{2-}$ acts as a bidentate ligand bonding to the metal through the B_1–B_5 and B_4–B_5 bonds.

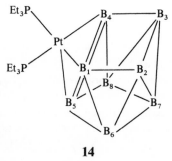

14

(c) $B_9H_{10}S^-$ and $B_9H_{11}L^{2-}$

Treatment of *trans*-$[(PEt_3)_2PtHCl]$ with $CsB_9H_{12}S$ yields the hydrido–platinum(II) borane complex $[(PEt_3)_2PtH(B_9H_{10}S)]$, which x-ray diffraction has shown[6] involves square-pyramidal co-ordination about the platinum(II) ion (**15**) (for further details see Appendix II structure 164). A series of

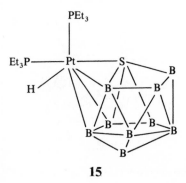

15

complexes $[(PEt_3)_2Pt(B_9H_{11}L)]$ are formed by treating *trans*-$[(PEt_3)_2PtHCl]$ with CsB_9H_{12} in the presence of organic bases (L) such as amines, nitriles, phosphines and sulphides.[6] The n.m.r. spectra of these complexes are consistent with structure **16** in which the borane ligand is bidentate forming bonds between the B_2–B_5 and B_2–B_7 edges and the metal.

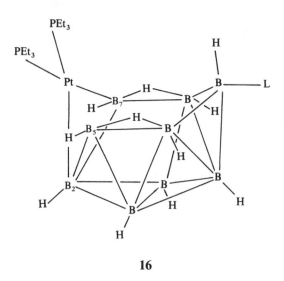

16

(d) $B_9C_2H_{11}^{2-}$ and $B_9C_2H_9(CH_3)_2^{2-}$

Palladium complexes of 1,2-dicarbollide ions ($B_9C_2H_9R_2^{2-}$, where R = H or CH_3) have been prepared. Both the bis(1,2-dicarbollyl)palladium complexes[7,8] and $[(\pi-C_4Ph_4)Pd(\pi-B_9C_2H_9Me_2)]$[9] have sandwich structures as in Fig. 14. A remarkable feature of these complexes, which may be written

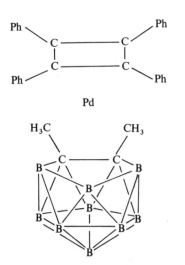

Fig. 14. The structure of $[Pd\{\pi-(C_6H_5)_4C_4\}\{\pi-B_9C_2H_9(CH_3)_2\}]$ (from ref. 9).

$[Pd^{n+}(1,2\text{-}B_9C_2H_{11})_2]^{(n-4)+}$ is their ability readily to undergo one-electron transfer reactions to form complexes in which the palladium atom is formally in the $+2$, $+3$ or $+4$ oxidation state.[7] The structures of these complexes, shown schematically in Fig. 15, have been determined by a combination of x-ray diffraction and n.m.r. studies.[8]

Pd^{II}—'slipped symmetrical sandwich' structure

Pd^{III}—'non-slipped symmetrical sandwich' structure

Pd^{IV}—'cisoid sandwich' structure

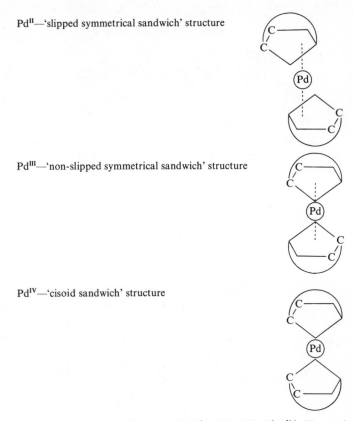

Fig. 15. Schematic structures of the anions $[Pd^{n+}(1,2\text{-}B_9C_2H_{11})]^{(n-4)+}$ (To emphasise the position of the carbon atoms with respect to the palladium atom the dicarbollyl cage shown in full in Fig. 14 is shown schematically here).

The only platinum(II) complex containing the 1,2-dicarbollide anion prepared so far is $[(1,5\text{-cyclo-octadiene})Pt(\pi\text{-}1,2\text{-}B_9C_2H_{11})]$. It has a sandwich structure.[8] This complex reacts with excess dicarbollide ion to precipitate platinum metal, as has occurred in all other attempts to prepare bis–dicarbollyl complexes of platinum.

(e) $B_{10}H_{12}^{2-}$ and $B_{10}H_{10}S^{2-}$

Mono- and bis-complexes of platinum(II) and palladium(II) with the anionic borane $B_{10}H_{12}^{2-}$ have been prepared.[10] The mono–platinum(II) complexes are prepared either by treating the dichloroplatinum(II) complexes with $NaB_{10}H_{13}$ (reaction 82) or by treating a hydride–platinum(II) complex with

$B_{10}H_{14}$ (reaction 83), whereas the bis-complex is prepared from the diolefin complex norbornadieneplatinum(II) dichloride (reaction 84). The mono and bis-complexes of palladium(II) are prepared by reactions 82 and 84 except that treatment of $[(PBu_3)_2PdCl_2]$ with $NaB_{10}H_{13}$ unexpectedly gives the bis-complex $(PBu_3H)_2[Pd(B_{10}H_{12})_2]$. The n.m.r. spectra of these complexes

$$[(PR_3)PtCl_2] + 2NaB_{10}H_{13} \xrightarrow{\text{(R = Et, Bu or Ph)}}$$

$$[(PR_3)_2Pt(B_{10}H_{12})] + B_{10}H_{14} + 2NaCl \quad (82)$$

$$2[(PEt_3)_2PtHCl] + B_{10}H_{14} \rightarrow$$

$$[(PEt_3)_2Pt(B_{10}H_{12})] + \uparrow 2H_2 + [(PEt_3)_2PtCl_2] \quad (83)$$

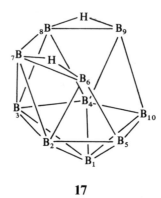

indicate that the $B_{10}H_{12}^{2-}$ anion (**17**) bonds to the metal as a bidentate ligand

17

through its B_5–B_6 and B_9–B_{10} bonds.[9] A platinum(II)–thiaborane complex, $[(PEt_3)_2Pt(B_{10}H_{10}S)]$, has been prepared in which the platinum atom is bound to the sulphur and the four boron atoms which form the face of the thiaborane.[11] Thiaborane itself is isostructural with the dicarbollide anion.

COMPLEXES WITH TRANSITION ELEMENTS

Only a few complexes involving platinum(II) or palladium(II) bound directly to other transition metals have been prepared. It is apparent from Fig. 16 that these metals lie to the right-hand side of the transition series. They form two types of complexes: first those in which the second transition metal is co-ordinated to platinum(II) or palladium(II) through one of the co-ordination sites in the square-plane and secondly those in which the second transition metal is bound axially with respect to the square-plane.

Sc	Ti	V	Cr	Mn	**Fe**	Co	*Ni*	*Cu*	Zn
Y	Zr	Nb	Mo	Tc	Ru	Rh	**Pd**	Ag	Cd
La	Hf	Ta	W	Re	Os	Ir	**Pt**	*Au*	**Hg**

Fig. 16. Transition metal complexes of platinum(II) and palladium(II). (Transition metals known to form complexes involving a direct metal–metal bond with platinum(II) are italicised, and with both palladium(II) and platinum(II) are shown in bold type.)

(i) Complexes in which the co-ordinating transition metal lies in the square-plane

Complexes containing Pt–Pt,[12] Pt–Ni,[12] Pt–Cu,[12] Pt–Hg,[12] and Pd–Hg[13] bonds have been prepared by treating a zerovalent complex of platinum or palladium with the metal halide (for example reactions 85 and 86 and also reactions 26–28 on p. 36), but their properties have not yet been

$$[Pt(PPh_3)_4] + [(PPh_3)_2NiI_2] \rightarrow [(PPh_3)_2IPt–NiI(PPh_3)_2] \qquad (85)$$

$$[Pt(PPh_3)_4] + HgCl_2 \rightarrow [(PPh_3)_2ClPt–HgCl] \qquad (86)$$

reported. Palladium(II)–iron complexes have been obtained by treating π-allylpalladium chloride with iron carbonyl complexes; with iron enneacarbonyl the π-allyl group is transferred to the iron atom via a palladium(II)–iron complex (reaction 87), which can be detected by n.m.r. but not isolated.[14] A palladium(II)–iron complex was actually isolated from a mixture of π-allylpalladium chloride and diphenylphosphine iron tetracarbonyl[15]

$$[(\pi\text{-}C_3H_5)PdCl]_2 + Fe_2(CO)_9 \rightarrow [(\pi\text{-}C_3H_5)Pd(Fe(CO)_4)Cl] \rightarrow$$
$$[(\pi\text{-}C_3H_5)Fe(CO)_3Cl] + Pd + Fe + CO \qquad (87)$$

(reaction 88). This unusual complex undergoes normal bridge-splitting reactions with reagents such as *p*-toluidine. An x-ray diffraction study[16] has confirmed the structure and shown the following bond lengths: Pd–Fe =

$$2[(PPh_2H)Fe(CO)_4] + [(\pi\text{-}C_3H_5)PdCl]_2 \rightarrow$$

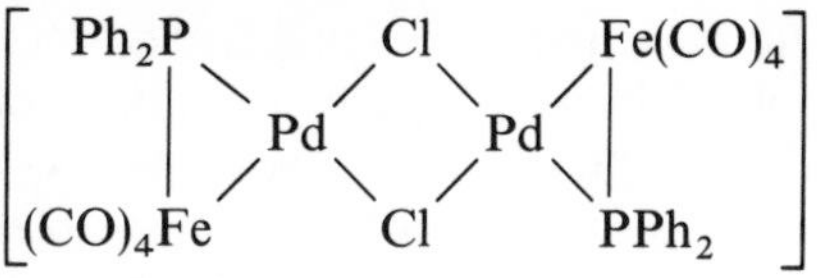

$$+ 2C_3H_6 \qquad (88)$$

2·59(1) Å, Pd–P = 2·15(1) Å, Pd–Cl (*trans* to Fe) = 2·42(1) Å, Pd–Cl (*trans* to P) = 2·45(1) Å. Both Pd–Cl bonds are long compared to bridging chlorine atoms *trans* to another chlorine (e.g. Pd–Cl *trans* to Cl = 2·31 Å in PdCl$_2$[17] and 2·32 Å in [(PhCH=CH$_2$)PdCl$_2$]$_2$[18] but comparable to bridging chlorine atoms *trans* to π-allyl or olefin groups (e.g. Pd–Cl = 2·41 Å in [(π-C$_3$H$_5$)-PdCl]$_2$[19] and 2·40 Å in [(π-C$_4$H$_7$)PdCl]$_2$[20] and 2·41 Å *trans* to PhCH=CH$_2$ in [(PhCH=CH$_2$)PdCl$_2$]$_2$[18]).

(ii) Complexes involving transition metals at the axial positions of the square-plane

Many complexes are known in which square-planar units are stacked vertically with short axial metal–metal bonds (Table 17). These complexes have

TABLE 17
Metal–metal distances in stacked molecules

Molecule	M–M (Å)	Reference
[Pd (dimethylglyoxime)$_2$]	3·26	21–23
[Pt (dimethylglyoxime)$_2$]	3·23	22, 24
α-[Pd(MeS)$_2$]	3·28	25
[Pt(NH$_3$)$_4$][PtCl$_4$]	3·23	26, 28
[Pd(NH$_3$)$_4$][PtCl$_4$]	3·25	27
[Cu(NH$_3$)$_4$][PtCl$_4$]	3·23	27, 29
[Pt(MeNH$_2$)$_4$][PtCl$_4$]	3·25	27, 30
[Pt(EtNH$_2$)$_4$][PtCl$_4$]	3.40	31
[Pd(NH$_3$)$_4$][PdCl$_4$]	3·25	27
[Pt(NH$_3$)$_4$][Pt(SCN)$_4$]	3·35	32
[Pt(en)Cl$_2$]	3·39	33
M^{II}[Pt(CN)$_4$].nH$_2$O M = group IIA metal, n = 2–7	3·09–3·60	34
Ca[Pt(C$_2$O$_4$)$_2$].4H$_2$O red modification)	3·18	35

been extensively studied because of their interesting colours, which are pressure dependent, and also because of their unusual conductivity properties.[31,32,36—38]

Casual observation suggests that in these complexes a band is present in the visible spectrum of the solid that is absent from the solution spectrum. This, coupled with the chain structure of these complexes in which the metal atoms are stacked vertically, has led to speculation that direct metal–metal interaction occurs in the solid.[26,33,37] This view is reinforced by the fact that the 'new' absorption is strongest when the incident radiation is polarised such that its electric vector is parallel to the metal chains.[28] However, more detailed studies[31,38] have shown that this new band is actually a highly perturbed single-molecule transition that moves from the near ultra-violet into the visible and increases sharply in intensity under the influence of the axial perturbation present in the crystal. The pressure dependence of these spectra[39—41] results from an increase in this axial perturbation as the metal–metal bond distance increases. When the ammonia ligands in Magnus' green salt—[Pt(NH$_3$)$_4$][PtCl$_4$]—are replaced by alkylamines, such as ethylamine, that are sufficiently bulky to increase the platinum–platinum bond length, the spectrum gradually changes to one more closely resembling that of K$_2$PtCl$_4$[42] (Pt–Pt = 4·13 Å[43]).

Complexes like Magnus' green salt which contain alternating anions and cations exhibit semiconductivity[42] and photoconductivity.[44] Furthermore the conductivity increases with increasing pressure and is greater when measured along the metal–metal axis of a single crystal than on a polycrystalline sample.[40,45] When the platinum–platinum bond length is increased by substituting bulky amines for the ammonia in Magnus' green salt the conductivity decreases,[42] which is consistent with the suggestion that overlap of the p_z and d_z^2 platinum orbitals occurs in the crystal giving rise to filled and empty bands which are responsible for semiconduction.[37] This band model is supported by molecular orbital calculations which show that there is negligible metal–metal bonding, the platinum–platinum bond having a net order of 0.039 as compared with 0.40 for the platinum–chlorine bond in the $[PtCl_4]^{2-}$ anions.[46]

REFERENCES

1 R. J. Cross and F. Glockling, *J. Organometal. Chem.*, **3** (1965) 253.
2 G. Schmid and H. Nöth, *Z. Naturforsch.*, **B20** (1965) 1008.
3 G. Schmid, W. Petz, W. Arloth and H. Nöth, *Angew Chem. Int. Ed.*, **6** (1967) 696.
4 G. Schmid and L. Weber, *Z. Naturforsch*, **B25** (1970) 1083.
5 A. R. Kane and E. L. Muetterties, *J. Amer. Chem. Soc.*, **93** (1971) 1041.
6 A. R. Kane, L. J. Guggenberger and E. L. Muetterties, *J. Amer. Chem. Soc.*, **92** (1970) 2571.
7 L. F. Warren and M. F. Hawthorne, *J. Amer. Chem. Soc.*, **90** (1968) 4823.
8 L. F. Warren and M. F. Hawthorne, *J. Amer. Chem. Soc.*, **92** (1970) 1157.
9 P. A. Wegner and M. F. Hawthorne, *Chem. Comm.*, (1966) 861.
10 F. Klanberg, P. A. Wegner, G. W. Parshall and E. L. Muetterties, *Inorg. Chem.*, **7** (1968) 2072.
11 W. R. Hertler, F. Klanberg and E. L. Muetterties, *Inorg. Chem.*, **6** (1967) 1696.
12 A. J. Layton, R. S. Nyholm, G. A. Pneumaticakis and M. L. Tobe, *Chem. Ind.* (*London*), (1967) 465.
13 M. Braithwaite and R. S. Nyholm, Personal communication.
14 A. N. Nesmeyanov, S. P. Gubin and A. Z. Rubezhov, *J. Organometal Chem.*, **16** (1969) 163.
15 B. C. Benson, R. Jackson, K. K. Joshi and D. T. Thompson, *Chem. Comm.*, (1968) 1506.
16 B. T. Kilbourn and R. H. B. Mais, *Chem. Comm.*, (1968) 1507.
17 A. F. Wells, *Z. Krist.*, **100** (1939) 189; *Chem. Abs.*, **33** (1939) 2387.
18 J. R. Holden and N. C. Baenziger, *J. Amer. Chem. Soc.*, **77** (1955) 4987.
19 A. E. Smith, *Acta Cryst.*, **18** (1965) 331.
20 R. Mason and A. G. Wheeler, *J. Chem. Soc.* (*A*), (1968) 2549.
21 L. E. Godycki and R. E. Rundle, *Acta Cryst.*, **6** (1953) 487.
22 D. E. Williams, G. Wohlauer and R. E. Rundle, *J. Amer. Chem. Soc.*, **81** (1959) 755.
23 S. Yamada and R. Tsuchida, *J. Amer. Chem. Soc.*, **75** (1953) 6351.
24 C. Panattoni, E. Frasson and R. Zannetti, *Gazz. Chim. Ital.*, **89** (1959) 2132; *Chem. Abs.*, **55** (1961) 5081d.
25 B. Meuthen and M. von Stackelberg, *Z. anorg. u. allgem. Chem.*, **305** (1960) 279.
26 M. Atoji, J. W. Richardson and R. E. Rundle, *J. Amer. Chem. Soc.*, **79** (1957) 3017.
27 J. R. Miller, *Proc. Chem. Soc.*, (1960) 318.
28 S. Yamada, *J. Amer. Chem. Soc.*, **73** (1951) 1579.
29 M. Bukovska and M. A. Porai-Koshits, *Kristallografiya*, **5** (1960) 137.
30 S. Yamada and R. Tsuchida, *Bull. Chem. Soc. Japan*, **31** (1958) 813.
31 P. Day, A. F. Orchard, A. J. Thomson, and R. J. P. Williams, *J. Chem. Phys.*, **43** (1965) 3763.
32 J. R. Miller, *J. Chem. Soc.*, (1961) 4452.
33 D. S. Martin, R. A. Jacobson, L. D. Hunter and J. E. Benson, *Inorg. Chem.*, **9** (1970) 1276.
34 K. Krogmann, *Angew. Chem. Int. Ed.*, **8** (1969) 35.
35 K. Krogmann, *Z. Naturforsch.*, **B23** (1968) 1012.
36 P. Day, A. F. Orchard, A. J. Thomson and R. J. P. Williams, *J. Chem. Phys.*, **42** (1965) 1973.
37 J. R. Miller, *J. Chem. Soc.*, (1965) 713.
38 B. G. Anex, M. E. Ross and M. W. Hedgcock, *J. Chem. Phys.*, **46** (1967) 1090.
39 J. C. Zahner and H. G. Drickamer, *J. Chem. Phys.*, **33** (1960) 1625.
40 L. V. Interrante and F. P. Bundy, *Chem. Comm.*, (1970) 584.
41 L. V. Interrante and F. P. Bundy, *Inorg. Chem.*, **10** (1971) 1169.
42 L. Atkinson, P. Day and R. J. P. Williams, *Nature*, **218** (1968) 668.
43 R. G. Dickinson, *J. Amer. Chem. Soc.*, **44** (1922) 2404.

44   J. P. Collman, L. Slifkin, L. F. Ballard, L. K. Monteith and C. G. Pitt, *International Symposium On Decomposition of Organometallic Compounds to Refractory Ceramics, Metals and Metal Alloys*, (1967) 269; *Chem. Abs.*, **72** (1970) 60322.
45   C. N. R. Rao and S. N. Bhat, *Inorg. Nucl. Chem. Letters*, **5** (1969) 531.
46   L. V. Interrante and R. P. Messmer, *Inorg. Chem.*, **10** (1971) 1174.

CHAPTER 6

Complexes with Group IVB Elements

Here complexes containing silicon, germanium, tin and lead bound directly to either platinum or palladium are considered. Although these do have similarities to complexes in which carbon is bound directly to a transition metal, the differences between them are in general greater than the similarities, and, for this reason, organometallic complexes will be considered only for the purposes of comparison. They are dealt with in Chapters 12 (Organometallic Complexes involving metal–carbon σ-bonds), 13 (Olefin and Acetylene Complexes) and 14 (π-Allyl Complexes). At the end of this chapter complexes containing the carbon donor ligands, cyanide, isocyanide, carbon monoxide and π-cyclopentadienyl are considered.

The differences between complexes containing σ-bonded carbon ligands and ligands containing other group IVB elements arise from:

(i) The potential availability of low energy empty d-orbitals in the heavier group IVB elements; although, as will be seen, it is uncertain whether these d-orbitals are actually involved either in bonding to platinum and palladium or in the reactions of these complexes.

(ii) The unique ability of carbon to bond to itself which leads to a range of complexes of homologous ligands such as CH_3^-, $C_2H_5^-$, $C_3H_7^-$; an ability the heavier group IVB elements do not possess. This is accompanied by a reluctance of carbon to form stable complexes of groups such as the t-butyl group that would be analogous to the complexes of the $E(CH_3)_3$ groups.†

One of the problems in discussing complexes containing metal–metal bonds is to assign oxidation states to the two metals. Lewis and Nyholm[1] have suggested that the oxidation state can be defined as 'the formal charge left on an atom when all the ligands are removed in their closed-shell configurations and any element–element bonds are broken homolytically so as to leave an equal number of electrons on each atom'. By this definition $[(PEt_3)_2PtCl(GeMe_3)]$ contains platinum(I) and germanium(III). Nevertheless, current opinion would assign oxidation states of $+2$ to platinum and $+4$ to germanium by analogy with $[(PEt_3)_2PtCl(CH_3)]$. However, in contrast, both platinum and tin would usually be assigned to the $+2$ oxidation state in $[PtCl_2(SnCl_3)_2]^{2-}$; an assignment that is supported by the chemical

† Throughout this Chapter E is used to represent a group IVB element.

shifts of the Moessbauer spectra of both this complex and its palladium analogue.[2] Although the precise description of the oxidation states of the two metals is probably a question of semantics, it is possible that the co-ordinate link between tin and platinum in $[PtCl_2(SnCl_3)_2]^{2-}$ is predominantly a π-bond with the electron pair from tin that would form the σ-bond remaining close to the tin atom. This is consistent with the observation that with the first row transition metals, where the π-bond from the transition metal to tin will be weaker, the equivalent complexes behave as tin(IV) complexes.[3] Thus in the present chapter at least, it will be found that oxidation states, which are required for a number of purposes such as the interpretation of electronic spectra, are best determined by chemical observation rather than by the application of hard and fast rules.

Complexes in which group IVB elements are bound directly to both platinum(II) and palladium(II) have recently attracted much attention. There are probably three main reasons for this: first their potential use as homogeneous catalysts for the reactions of organic compounds (see pp. 393–394), secondly the current interest in the chemistry of compounds possessing metal–metal bonds, and thirdly these complexes have very interesting chemical properties which show many similarities, and some important differences, to complexes containing metal–carbon σ-bonds. In view of the considerable differences between complexes containing ER_3 ligands (where R is an alkyl or aryl group) and EX_3 ligands (where X = halogen), the preparation and properties of the two series of compounds are treated separately. However, when discussing the bonding they are considered together, because a comparison of the ER_3 and EX_3 complexes is valuable in discussing the importance of π-bonding in metal–group IVB element bonds. The principal reviews that may be used to supplement the material in this chapter are given in Table 18.

TABLE 18
Reviews on complexes containing metal–group IVB element bonds

Title	Reference
σ-Complexes of platinum(II) with hydrogen, carbon and other elements of group IV	4
Metal–metal bonds in transition metal compounds	5
Transition metal complexes with group IVB elements	3
Organometallic compounds with metal–metal bonds between different metals	6
Organogermanium–transition metal complexes	7
Preparation and properties of compounds containing platinum–silicon bonds	8
Group IVB metal derivatives of the transition elements	9

PREPARATION OF COMPLEXES CONTAINING ER$_3$ LIGANDS

A number of methods have been used to prepare platinum and palladium complexes with ligands containing group IVB metals. Some of these have their counterparts in the preparation of the organometallic complexes of these metals which contain metal–carbon σ-bonds, but most are peculiar to the heavier group IVB elements. As yet germanium and lead are the only group IVB elements that appear to give triorgano-group IVB–palladium complexes. This is consistent with the alternation effect observed in the preparation of the platinum complexes (see below p. 82).

(i) Interaction of metal chlorides and triorgano–group IVB–lithium compounds
The interaction of transition metal chlorides with triorgano–group IVB–lithium compounds is by no means simple or readily predictable. Thus $Ph_2MeSiLi$ and $PhMe_2SiLi$ react with cis-$[L_2PtCl_2]$ (where L = tertiary phosphine) to give the cis-$[L_2Pt(SiMe_nPh_{3-n})_2]$ complex[8] (where $n = 1$ or 2), whereas the completely aromatic silyl compound Ph_3SiLi reacts[10] with either cis- or trans-$[(PEt_3)_2PtCl_2]$ to give the unstable monosilyl complex $[(PEt_3)_2PtH(SiPh_3)]$. The hydride ligand in this complex probably arises from the formation of a platinum lithium intermediate ($[(PEt_3)_2PtLi(SiPh_3)]$) as in the corresponding germyl complex (see Scheme 1 below).

Both platinum(II)– and palladium(II)–germyl complexes have been prepared with germyl–lithium reagents. In the case of palladium(II) only the triphenylgermyl complex trans-$[(PEt_3)_2Pd(GePh_3)_2]$ has been prepared[11,12] (reaction 89). The product is difficult to isolate since the palladium–germanium

$$\text{trans-}[(PEt_3)_2PdCl_2] + 2Ph_3GeLi \xrightarrow{\text{monoglyme}}$$

$$\text{trans-}[(PEt_3)_2Pd(GePh_3)_2] + 2LiCl \quad (89)$$

bond is unstable to water and oxygen above $-40°C$. When other palladium(II) complexes such as $[(PPh_3)_2PdCl_2]$, $[(bipyr)PdCl_2]$ or $[(Ph_2PCH_2CH_2PPh_2)PdCl_2]$ were treated with Ph_3GeLi the expected colour change occurred but decomposition occurred during the isolation of the product.[12] Platinum(II)–germyl complexes are readily prepared[13,14] with the lithium germyl reagents (reaction 90). Although the reaction is a two-stage process the second chlorine is replaced considerably faster than the

$$\text{cis- or trans-}[(PR_3)_2PtCl_2] + 2Ph_3GeLi \rightarrow$$

$$\text{cis- or trans-}[(PR_3)_2Pt(GePh_3)_2] \quad (90)$$

first, reflecting the high trans-effect of the triphenylgermyl group, so that if equimolar ratios of reactants are used, the products are unreacted $[(PR_3)_2PtCl_2]$ and $[(PR_3)_2Pt(GePh_3)_2]$. The bis–bromo complex $[(PPh_3)_2PtBr_2]$ on treatment with Ph_3GeLi gives only polymeric material,[14] but the bis-iodo complex $[(PEt_3)_2PtI_2]$ gives a complex series of products[15] that can be explained[7] in terms of extensive halogen–metal exchange according to Scheme 1.

Treatment of cis-$[(PPh_3)_2PtCl_2]$ with Ph_3SnLi produces the mono–tin complex $[(PPh_3)_2PtCl(SnPh_3)]$ although this complex is better prepared by treating trans-$[(PPh_3)_2PtHCl]$ with triphenyltin nitrate.[10] Both platinum(II) and palladium(II) chloro-complexes react with $PbPh_3Li$ in anhydrous ether to form the bis–plumbyl complexes[16,17] trans-$[(PEt_3)_2M(PbPh_3)_2]$ (M = Pt or Pd). Although the initial product is probably the kinetically

SCHEME 1

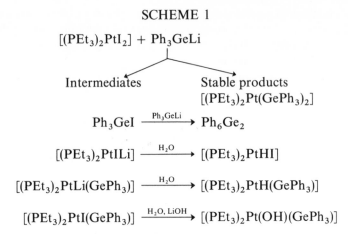

$[(PEt_3)_2PtI_2] + Ph_3GeLi$

Intermediates

Stable products
$[(PEt_3)_2Pt(GePh_3)_2]$

$Ph_3GeI \xrightarrow{Ph_3GeLi} Ph_6Ge_2$

$[(PEt_3)_2PtILi] \xrightarrow{H_2O} [(PEt_3)_2PtHI]$

$[(PEt_3)_2PtLi(GePh_3)] \xrightarrow{H_2O} [(PEt_3)_2PtH(GePh_3)]$

$[(PEt_3)_2PtI(GePh_3)] \xrightarrow{H_2O, LiOH} [(PEt_3)_2Pt(OH)(GePh_3)]$

favoured *cis*-isomer, after standing for some time it is the *trans*-isomer which is recovered.

(ii) Interaction of metal chlorides with $(R_3E)_2Hg$
Treatment of *cis*-$[(PEt_3)_2PtCl_2]$ with $(R_3E)_2Hg$ has been used[18–21] to prepare trimethylsilyl, trimethylgermyl and triphenyllead complexes of platinum (reaction 91) by reactions that are almost quantitative when

$$cis\text{-}[(PEt_3)_2PtCl_2] + (R_3E)_2Hg \rightarrow$$
$$trans\text{-}[(PEt_3)_2PtCl(ER_3)] + R_3ECl + Hg \quad (91)$$

carried out in refluxing benzene. The bis complexes $[(PEt_3)_2Pt(EMe_3)_2]$ do not result when an excess of the mercurial is used; such an excess produces colourless intractable oils. Mercurials have not been used to prepare the corresponding palladium(II) complexes or complexes of either metal with tin or lead. The mercury and lithium salt routes to group IVB metal complexes are complementary in that the former has been used exclusively to prepare complexes containing trialkyl–silyl and –germyl ligands whereas the latter has been used when an aryl group is bound to the group IVB metal.

(iii) Interaction of group IVB metal hydrides with hydridochloroplatinum(II) complexes
The reaction of group IVB metal hydrides such as R_3EH with *trans*-$[(PEt_3)_2PtHCl]$ (reaction 92) has been used to prepare platinum(II)–

$$trans\text{-}[(PEt_3)_2PtHCl] + R_3EH \rightleftharpoons trans\text{-}[(PEt_3)_2PtCl(ER_3)] + \uparrow H_2 \quad (92)$$

silicon[8,22,23] (R = phenyl and substituted phenyl), platinum(II)–germanium[22–5] (R = CH_3 or C_6H_5) and platinum(II)–tin[22] (R = C_2H_5) complexes. As the reaction is reversible it is essential that the hydrogen be allowed to escape from the system. The presence of electron-withdrawing groups bound to the group IVB metal appears to assist the reaction, presumably by facilitating the formation of H^+ which is known to react with platinum(II)–hydrides to liberate hydrogen.[26] A similar reaction (eqn. 93) has been used to prepare platinum complexes in which hydrogen is bound

$$trans\text{-}[(PEt_3)_2PtHCl] + H_3ECl \xrightarrow{\text{benzene}}$$

$$trans\text{-}[(PEt_3)_2PtCl(EH_2Cl)] + \uparrow H_2 \quad (93)$$

directly to both silicon and germanium.[27]

(iv) Interaction of diorgano–group IVB metal amides with platinum hydrides
Amides of both germanium[7] and tin[28,29] have been found to react with
hydridoplatinum(II) complexes (reaction 94) to replace the platinum(II)–

$$R_3ENR'_2 + [(PEt_3)_2PtHCl] \rightarrow [(PEt_3)_2PtCl(ER_3)] + R'_2NH \quad (94)$$

hydride bond by a platinum(II)–germanium or –tin bond. A very closely
related reaction is that of silanes with dichloroplatinum(II) complexes in
refluxing benzene in the presence of triethylamine to give a platinum(II)–
silicon complex[8] (reaction 95). Even in the presence of excess silane it was not

$$R_3SiH + cis\text{-}[(PMe_2Ph)_2PtCl_2] + Et_3N \xrightarrow[\text{reflux}]{C_6H_6,}$$

$$trans\text{-}[(PMe_2Ph)_2PtCl(SiR_3)] + Et_3NHCl \quad (95)$$

possible to replace the second chlorine atom by this method. Reaction 95 is
most successful when R is either an aryl group or an aryl group substituted
with an electron-withdrawing group, thus facilitating release of H^+ from the
silane. For alkyl groups more forcing conditions, such as refluxing xylene
and tri-*n*-butylamine, are necessary to effect reaction.[22] Triphenylgermane
reacts analogously.

In the absence of a base both triphenylgermane and trimethylgermane
react with the palladium(II) complex $trans\text{-}[(PEt_3)_2PdCl_2]$ to give the
hydrido–palladium(II) complex $trans\text{-}[(PEt_3)_2PdHCl]$.[30] The reaction with
trimethylgermane (reaction 96) provides a convenient synthetic route to

$$trans\text{-}[(PEt_3)_2PdCl_2] + \text{excess } Me_3GeH \xrightarrow{40°C}$$

$$trans\text{-}[(PEt_3)_2PdHCl] + H_2 + Me_3GeCl + Me_6Ge_2 + \text{trace Pd} \quad (96)$$

$trans\text{-}[(PEt_3)_2PdHCl]$ since, when the two components are kept in an
evacuated sealed tube at 40°C for several days, the hydrido-complex separ-
ates out as large colourless prisms.

**(v) Interaction of hydridochloroplatinum(II) complexes with triphenyltin
nitrate and triphenyllead nitrate**
Both triphenyltin nitrate and triphenyllead nitrate react[10] with *trans*-
$[(PPh_3)_2PtHCl]$ to give the monosubstituted complexes $[(PPh_3)_2PtCl(EPh_3)]$
where E = Sn or Pb. The lead complex, however, cannot be obtained pure
because it readily disproportionates to the phenyl complex $[(PPh_3)_2\text{-}
PtCl(C_6H_5)]$.

(vi) Oxidative addition to platinum(0) complexes

Some group IVB metal complexes have been prepared by oxidative addition to platinum(0) complexes as in reactions 97–103. Reactions 101 and 102 are

$$[Pt(Ph_2PCH_2CH_2PPh_2)_2] + Ph_2SiH_2 \xrightarrow{\text{ref. 31, 32}}$$

$$[(Ph_2PCH_2CH_2PPh_2)Pt(SiHPh_2)_2] + Ph_2PCH_2CH_2PPh_2 + \uparrow H_2 \quad (97)$$

$$[Pt(PPh_3)_4] + (C_6H_4X)_3SiH \xrightarrow{\text{ref. 31, 32}}$$

$$[(PPh_3)_2Pt(Si(C_6H_4X)_3)_2] + 2PPh_3 + \uparrow H_2 \quad (98)$$

$$(X = m\text{- and } p\text{-F or -CF}_3)$$

$$[Pt(PPh_3)_4] + Ph_3SnCl \xrightarrow{\text{ref. 33, 34}}$$

$$[(PPh_3)_2PtCl(SnPh_3)] + 2PPh_3 \quad (99)$$

$$[Pt(PPh_3)_2(C_2H_4)] + MeCl_2SiH \xrightarrow{\text{ref. 35}}$$

$$[(PPh_3)_2Pt(SiMeCl_2)_2] \quad (100)$$

$$[Pt(PPh_3)_2(C_2H_4)] + Ph_3SnCl \xrightarrow{\text{ref. 34}}$$

$$[(PPh_3)_2PtCl(SnPh_3)] + C_2H_4 \quad (101)$$

$$[Pt(PPh_3)_2(C_2H_4)] + Me_6Sn_2 \xrightarrow{\text{ref. 36}}$$

$$\textit{trans-}[(PPh_3)_2Pt(SnMe_3)_2] + C_2H_4 \quad (102)$$

$$[Pt(Ph_2PCH_2CH_2PPh_2)_2] + 3Me_3SnH \xrightarrow{\text{ref. 21}}$$

$$\uparrow H_2 + Ph_2PCH_2CH_2PPh_2 + [(Ph_2PCH_2CH_2PPh_2)Pt(SnMe_3)_3H]$$

recrystallise
in benzene

$$[(Ph_2PCH_2CH_2PPh_2)Pt(SnMe_3)_2] + SnMe_3H \quad (103)$$

particularly useful as the conditions are mild and the yields good. The only byproduct is ethylene, a gas, so that purification of the product is simple.

(vii) Exchange between group IVB ligands

Trimethylstannyl complexes of platinum(II) have been prepared by treating the corresponding trimethyl–silyl or –germyl complexes with trimethyl-stannane[21,37] (reactions 104 and 105). Although these were the only

$$[(Ph_2PCH_2CH_2PPh_2)PtCl(EMe_3)] + Me_3SnH \xrightarrow[\text{(E = Si or Ge)}]{}$$

$$[(Ph_2PCH_2CH_2PPh_2)PtCl(SnMe_3)] + Me_3EH \quad (104)$$

$$[(Ph_2PCH_2CH_2PPh_2)Pt(EMe_3)_2] + 2Me_3SnH \xrightarrow[\text{(E = Si or Ge)}]{}$$

$$[(Ph_2PCH_2CH_2PPh_2)Pt(SnMe_3)_2] + 2Me_3EH \quad (105)$$

reactions of preparative value it was found that treatment of any complex

$[(Ph_2PCH_2CH_2PPh_2)PtCl(EMe_3)]$ or $[(Ph_2PCH_2CH_2PPh_2)Pt(EMe_3)_2]$
with $E'Me_3$ (where $E \neq E' = Si,Ge,Sn$) led to an equilibrium being set up
within 12 hours at 40°C. This equilibrium favoured tin very much more than
germanium and germanium slightly more than silicon. This order could
either reflect a decreasing order of platinum–group IVB element bond
strengths of Sn > Ge > Si or it could be the result of changing hydrogen–
group IVB element bond strengths.

PROPERTIES OF COMPLEXES CONTAINING ER₃ LIGANDS

The complexes of platinum and palladium with group IVB metal ligands
that have been prepared in a pure state are all crystalline. Although it is
difficult to separate stability and reactivity (for a more detailed discussion
see Chapter 12, p. 331), it appears that (i) the complexes of the heavier group
IVB metals are usually more stable than their methyl analogues, (ii) the
platinum complexes are more stable than the palladium complexes, (iii) the
order of stability for the group IVB ligands is $Cl_3E > Ph_3E > Me_3E$ and
(iv) the stability order of the platinum complexes is approximately Sn ~
Ge > Si > Pb whereas for the palladium complexes it is Ge > Pb ≫ Sn ~
Si.

With the exception of one or two silicon complexes such as *trans*-[(PEt₃)₂-
PtCl(SiMe₃)][18] and *trans*-[(PEt₃)₂PtH(SiPh₃)][10] all the solid complexes
are stable in air; however, their solutions are rather less stable to air. Thus
although solutions of the chloromonosilyl–platinum(II) complexes[8] and
the platinum–germanium complexes[14] are stable in air the bissilyl–platin-
um(II) complexes are stable in solution only in the absence of air.[8] The
triphenyltin and triphenyllead complexes disproportionate in polar solvents[10] (eqn. 106). For the tin complex reaction 106 requires refluxing in

$$\begin{array}{ccc}
\text{Ph} & \text{PPh}_3 & \text{PPh}_3 \\
\diagdown \quad | & & | \\
\text{Ph—E—Pt—Cl} & \xrightarrow[\text{(E = Sn or Pb)}]{} \text{Ph}_2\text{E} + \text{Ph—Pt—Cl} & \quad (106) \\
\diagup \quad | & & | \\
\text{Ph} \quad \text{PPh}_3 & & \text{PPh}_3
\end{array}$$

acetone. The lead complex disproportionates so readily at room temperature
that it cannot be isolated in a pure state. The palladium complexes are much
less stable in solution than their platinum analogues, both palladium–
germanium[12] and palladium–lead[17] complexes decomposing above
−20°C.

Although evidence for the formation of labile platinum(IV) species
containing group IVB metal ligands is often strong, only a few such complexes
have been isolated (see pp. 257–259). All the other complexes so far isolated
contain square-planar platinum(II) and palladium(II). *Cis-* and *trans*-isomers
are known for both bissilyl– and bisgermyl–platinum(II) complexes, the *cis*-
complexes being paler in colour than the *trans*-complexes: for example,
cis-[(PEt₃)₂Pt(GePh₃)₂] is white whereas the *trans*-isomer is pale yellow.[38]
The corresponding monosilyl and monogermyl complexes are generally
trans except the complexes containing the chelating diphosphine

$Ph_2PCH_2CH_2PPh_2$, which are necessarily *cis*. All the platinum–tin, platinum–lead, palladium–germanium and palladium–lead complexes so far isolated are *trans*-isomers.

There is some evidence of the 'alternation effect'[39] among these complexes. Thus palladium forms complexes only with carbon, germanium and lead, although this may be because of insufficient experimental effort with silicon and tin. A rather more tenuous† example of the alternation effect is that when dichloroplatinum(II) complexes are treated with Ph_3ELi only the monosilyl and monostannyl complexes are formed whereas the bisgermyl and bis-plumbyl complexes are formed.

ER_3 ligands have very high *trans*-effects as deduced from their abilities to direct substitution reactions to the position *trans* to themselves. These ligands also have very high *trans*-influences as is indicated by the platinum–chlorine stretching frequencies in the infrared spectra of the $[(PR_3)_2PtCl(ER_3)]$ complexes (Table 19). However, although the *trans*-

TABLE 19

Platinum–chlorine stretching frequencies in the infra-red spectra of *trans*-$[(PR_3)_2PtClY]$

R	Y	$v(Pt–Cl)(cm^{-1})$	Reference
Et	Cl	340	40
Et	$PbPh_3$	285	16
Et	CH_3	274	40
Et	C_6H_5	270	40
Et	H	269	40
Et	$SiMe_3$	238	19
Et	$GeMe_3$	235	19
Ph	Cl	345	10
Ph	$SnCl_3$	315	10
Ph	$SnPh_3$	298	10
Ph	$SnMe_3$	$\begin{cases} 296 \\ 278 \end{cases}$	36 / 28
Ph	H	287	10
Ph	$PbPh_3$	286	10

influences of the trimethylsilyl and trimethylgermyl ligands are the highest yet observed, those of the tin ligands $SnCl_3^-$, $SnPh_3$ and $SnMe_3$ and the lead ligand $PbPh_3$ are slightly less than that of a hydride ligand. Reasons for the lower *trans*-influences of $SnPh_3$, $SnMe_3$ and $PbPh_3$ are at present uncertain, although the low *trans*-influence but high *trans*-effect of the $SnCl_3^-$ ligand is known to result from the π-bonding in the tin-metal bond. This π-bonding causes the metal–ligand bond *trans* to tin not to be weakened as much as it would have been if the tin had been bound by a simple σ-bond to the metal.

† Tenuous because $Ph_2MeSiLi$ and $PhMe_2SiLi$ both yield the bissilyl complexes.[8]

Chemical properties of complexes containing the ER_3 group

It is convenient to consider the chemical properties of the group IVB metal complexes of platinum and palladium in terms of the reagent involved, since this affords a good means of comparing variation in reactivity with variation in the group IVB metal. Some reagents, such as the hydrogen halides, give significantly different reactions with different M–E bonds, whereas others such as hydrogen show no such difference.

All reagents that split Pt–E bonds are now considered to react by initial co-ordination to platinum rather than to the group IVB metal, although for water this is based on a single experimental observation. Reagents such as the hydrogen halides, halogens, alkyl halides and molecular hydrogen which cleave Pt–E bonds are all believed to react by a *cis*-addition to the platinum(II)–group IVB metal complex to give a labile platinum(IV) intermediate which decomposes by a platinum–group IVB metal bond cleavage to yield the final platinum(II) product. Accordingly, it is not surprising to find that these reagents cleave Pd–E bonds differently from Pt–E bonds since palladium does not form a six-co-ordinate tetravalent species as readily as platinum. Although the evidence for the formation of a labile platinum(IV) species is often very strong, only a few platinum(IV) complexes containing group IVB metal ligands have been isolated (pp. 257–259).

A. Reagents which cleave the M–E bond
(i) *Hydrogen halides*
Anhydrous hydrogen chloride cleaves both Pt–E and Pd–E bonds to give products which depend on the group IVB metal and also on the ligand *trans* to the M–E bond.

(a) *Pt–Si and Pt–Ge bonds.* When the chelate complexes *cis*-$[(Ph_2PCH_2CH_2PPh_2)Pt(ER_3)_2]$ where ER_3 is $SiMe_3$,[21] $SiMePh_2$,[8] $GeMe_3$[21] or $GePh_3$[38] are cleaved by hydrogen chloride, the chloride is bound to the platinum and the hydrogen to the group IVB metal (reaction 107). An excess of hydrogen chloride is needed to cleave the germanium

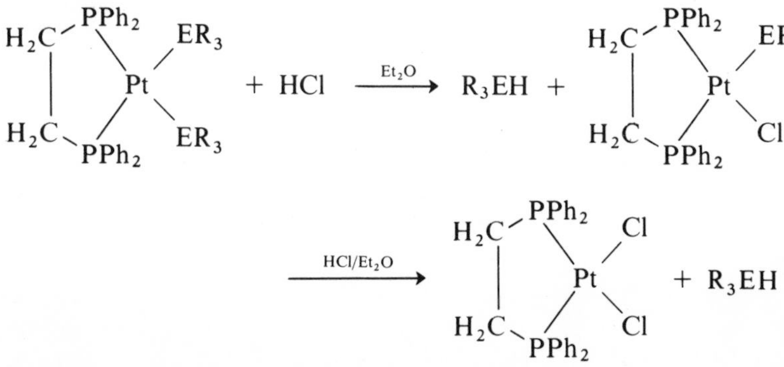

$$\text{(107)}$$

complex and the intermediate mono-chloro-complex cannot be isolated.[38] When the bis-tertiary phosphine complex *cis*-$[(PMe_2Ph)_2Pt(SiMePh_2)_2]$ is treated with hydrogen chloride the first platinum–silicon bond (*trans* to the tertiary phosphine with its high *trans*-effect) is cleaved to give the silane and *trans*-[(PMe$_2$Ph)$_2$PtCl(SiMePh$_2$)]. By contrast, the second platinum–silicon bond (*trans* to the chloride ligand with its low *trans*-effect) gives the chlorosilane[8] (reaction 108). The corresponding germyl complex reacts similarly.[8]

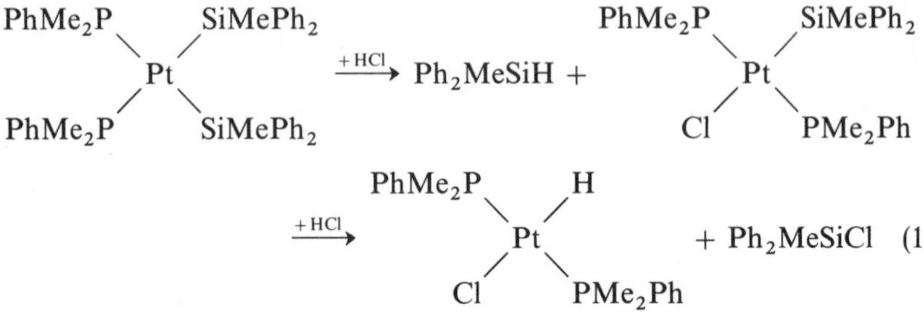

PhMe$_2$P   SiMePh$_2$ PhMe$_2$P   SiMePh$_2$

$\diagdown$ $\diagup$ $\diagdown$ $\diagup$

Pt $\xrightarrow{+\,HCl}$ Ph$_2$MeSiH +   Pt

$\diagup$ $\diagdown$ $\diagup$ $\diagdown$

PhMe$_2$P   SiMePh$_2$ Cl   PMe$_2$Ph

PhMe$_2$P   H

$\diagdown$ $\diagup$

$\xrightarrow{+\,HCl}$ Pt   + Ph$_2$MeSiCl   (108)

$\diagup$ $\diagdown$

Cl   PMe$_2$Ph

By contrast the platinum–germanium complex *trans*-[(PEt$_3$)$_2$Pt(GePh$_3$)$_2$] reacts in the opposite sense giving [(PEt$_3$)$_2$PtH(GePh$_3$)] and Ph$_3$GeCl as products in the first stage and [(PEt$_3$)$_2$PtHCl] and Ph$_3$GeH in the second stage.[14] However, this behaviour is exceptional for germanium since the complex *trans*-[(PEt$_3$)$_2$Pt(GePh$_3$)(GeMe$_3$)] is cleaved according to eqn. 108 giving *trans*-[(PEt$_3$)$_2$Pt(GePh$_3$)Cl] as the initial product.[24]

The hydrogen chloride cleavages of these Pt–E bonds are believed[8,38] to occur by an addition-elimination reaction identical to that which has been shown to account for the hydrogen chloride cleavage of platinum–carbon σ-bonds.[41] *Cis*-addition of hydrogen chloride, the hydrogen going in *trans* to the ligand of highest *trans*-effect, gives the platinum(II) intermediates **18** (Scheme 2) and **19** (Scheme 3) which then revert to platinum(II) complexes by *cis*-elimination of R$_3$EH (Scheme 2) or R$_3$ECl (Scheme 3). Cleavage of this first Pt–E bond is influenced kinetically by the electronegativity of the R$_3$E group

SCHEME 2

Hydrogen chloride cleavage of Pt–E bonds in which the group IVB metal product of the first stage is R$_3$EH

Stage 1

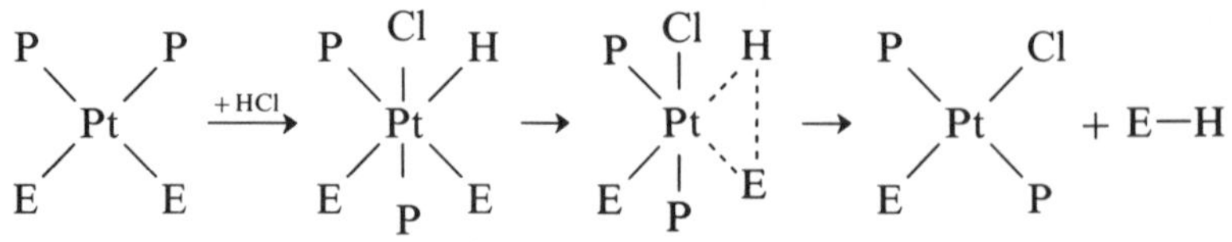

18

Stage 2

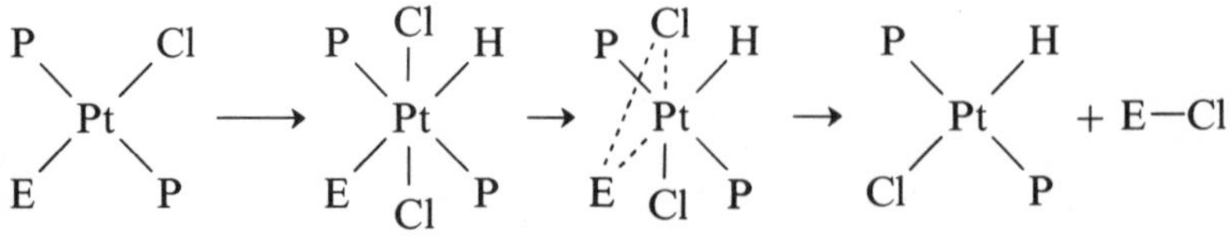

SCHEME 3

Hydrogen chloride cleavage of Pt–E bonds in which the group IVB metal product of the first stage is R_3ECl

Stage 1

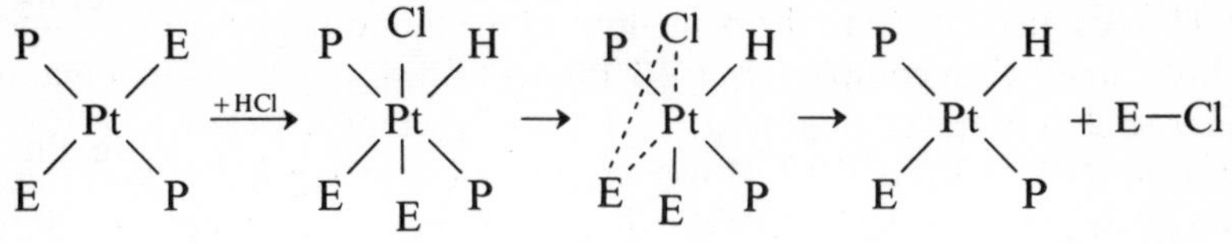

19

Stage 2

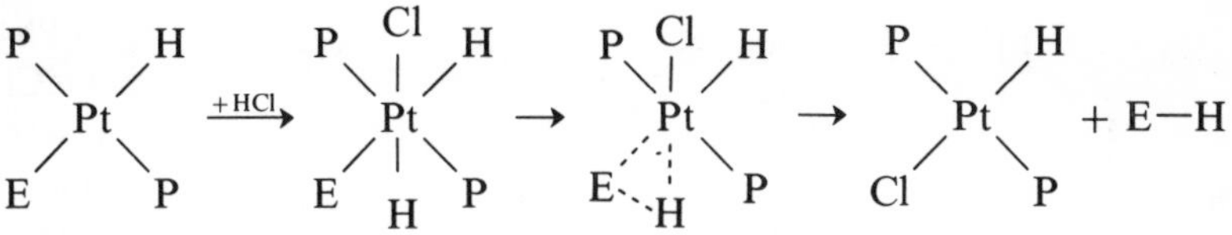

since one of the R_3E groups is *cis* to both H and Cl. Thus at this first stage it is to be expected that in some complexes (e.g. $[(PMe_2Ph)_2Pt(GeMePh_2)_2])\,R_3EH$ is formed, whereas in others (e.g. *trans*-$[(PEt_3)_2Pt(GePh_3)_2])\,R_3ECl$ is formed. In the second stage, *cis*-addition of hydrogen chloride with the hydrogen *trans* to the ligand of highest *trans*-influence occurs. In complexes that follow Scheme 2 this necessarily leaves the group IVB metal *trans* to hydrogen so that *cis*-elimination leads to the formation of R_3ECl, whereas in Scheme 3 the group IVB metal is *cis* to both chlorine and hydrogen. Kinetic factors lead to the elimination of R_3EH so that the final platinum(II) product is identical in each scheme. The postulation of labile platinum(IV) intermediates is strongly supported by the isolation[24] of the six-co-ordinate platinum(IV) complex $[(Ph_2PCH_2CH_2PPh_2)(PEt_3)PtHCl(GeMe_3]^+Cl^-$ formed during the course of reaction 109.

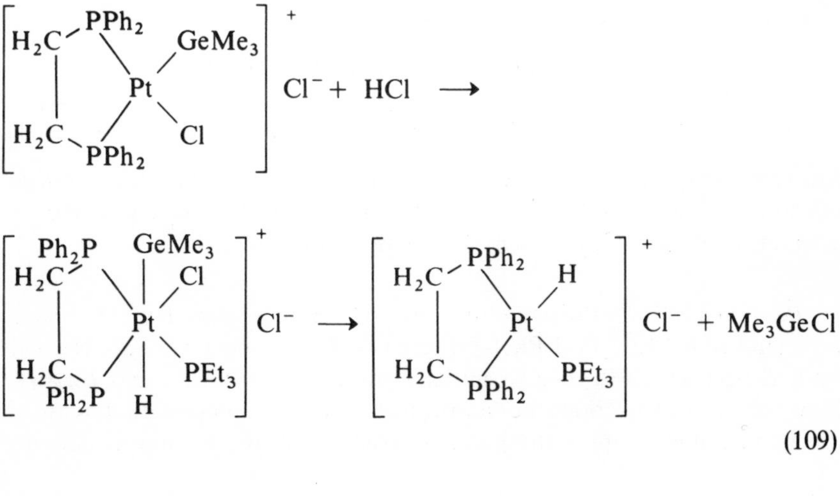

$$(109)$$

(b) *Pt–Sn bonds.* The triphenyltin complex *trans*-[(PPh$_3$)$_2$PtCl(SnPh$_3$)] is cleaved by anhydrous hydrogen chloride[10] to give a mixture of *cis*- and *trans*-[(PPh$_3$)$_2$PtCl$_2$]. This is exactly the reverse of the cleavage in the equivalent Pt–Si and Pt–Ge complexes (eqn. 108 and Scheme 2, stage 2), and illustrates the selectivity of hydrogen chloride attack on platinum–group IVB metal bonds. In contrast to the behaviour of the mono-triphenyltin complex, the bis–trimethyltin complex *trans*-[(PPh$_3$)$_2$Pt(SnMe$_3$)$_2$] reacts in a manner similar to that of the analogous germyl and silyl complexes (see eqn. 108) to give *trans*-[(PEt$_3$)$_2$PtHCl].[36] However, instead of giving R$_3$EH and R$_3$ECl as the other products the bulk of the tin is left as hexamethylditin with only a trace of Me$_3$SnCl. Whether this reaction occurs by a two-stage mechanism such as that in Scheme 2, with the subsequent formation of hexamethylditin, or whether it is a one-stage process (eqn. 110) with the direct formation of

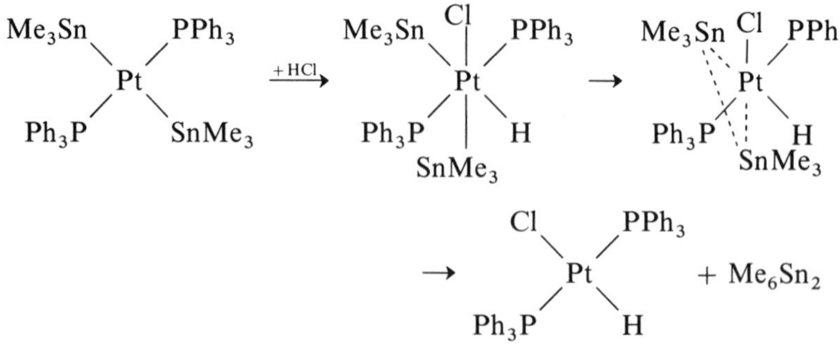

hexamethylditin is at present uncertain. The *cis*-bistrimethyltin complex [(Ph$_2$PCH$_2$CH$_2$PPh$_2$)Pt(SnMe$_3$)$_2$] is cleaved in a similar way to its germyl and silyl analogues to give [(Ph$_2$PCH$_2$CH$_2$PPh$_2$)PtCl$_2$] and trimethyl-stannane.[21]

(c) *Pt–Pb bonds.* Both the platinum–lead bonds in the complex *trans*-[(PEt$_3$)$_2$Pt(PbPh$_3$)$_2$] are cleaved by hydrogen chloride to form platinum–chloride and lead-hydrogen bonds[16,42] (eqn. 111). The second stage of

trans-[(PEt$_3$)$_2$Pt(PbPh$_3$)$_2$] + HCl →

(Ph$_3$PbH) + *trans*-[(PEt$_3$)$_2$PtCl(PbPh$_3$)]

↓ HCl

trans-[(PEt$_3$)$_2$PtCl$_2$] + (Ph$_3$PbH)   (111)

this reaction is identical to that for the corresponding triphenyltin complex (see above). Owing to the instability of plumbane the final reaction mixture contains a mixture of hexaphenyldilead and lead dichloride.

(e) *Pd–E bonds.* Both the palladium–group IVB metal complexes that have been studied ([(PEt$_3$)$_2$Pd(EPh$_3$)$_2$] where E = Ge[12] and Pb[17]) are cleaved by hydrogen chloride to give initially *trans*-[(PEt$_3$)$_2$PdCl$_2$] and Ph$_3$EH. However, in both reactions the end-products are more complex. In the case of the germanium complex the germane reacts with the dichloropalladium complex on heating to give a hydride complex which decomposes to palladium metal[12] (eqn. 112). In the case of the lead complex the triphenylplumbane formed is unstable and decomposes to give hydrogen and Pb_2Ph_6

$$Ph_3GeH + trans\text{-}[(PEt_3)_2PdCl_2] \rightarrow$$

$$Ph_3GeCl + trans\text{-}[(PEt_3)_2PdHCl] \rightarrow \text{Pd metal} \quad (112)$$

which, in the presence of an excess of hydrogen chloride, reacts to give Ph_3PbCl, Ph_2PbCl_2 and $PbCl_2$ which are the identified products.[17] Although detailed reaction mechanisms have not been proposed for these hydrogen chloride cleavages it has been suggested that, in view of the greater reluctance of palladium(II) than platinum(II) to undergo oxidative addition to palladium(IV), the reactions occur by direct attack of hydrogen chloride on the Pd–E bonds.[12]

(ii) Other acids

There are no systematic studies of the cleavage of M–E bonds by acids other than the hydrogen halides. However, thiophenol has been shown to cleave[8] the Pt–Si bonds in $cis\text{-}[(PMe_2Ph)_2Pt(SiMePh_2)_2]$ to give $trans\text{-}[(PMe_2Ph)_2Pt(SPh)_2]$ which is in contrast to the hydrogen chloride cleavage described in eqn. 108 (p. 84). The reaction of phenylacetylene with $trans\text{-}[(PEt_3)_2PtCl(SiMe_3)]$ was investigated[19] because the reaction was unlikely to involve primary attack on silicon. The complex mixture of products appeared to confirm this in that they could be explained in terms of an initial *cis*-addition of phenylacetylene to platinum(II) to give a labile platinum(IV) intermediate (eqn. 113). This provides support for an analogous *cis*-addition-

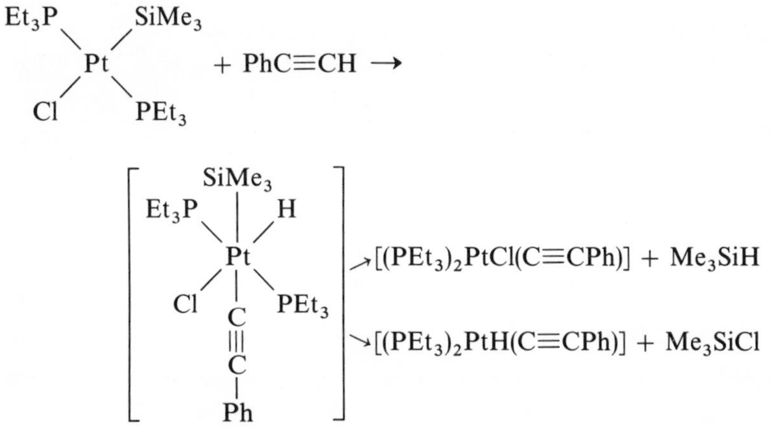

$$(113)$$

elimination mechanism for hydrogen chloride cleavage of the Pt–Si bond (p. 83–85). The reaction of the palladium–germyl complex $[(PEt_3)_2PdH(GePh_3)]$ with phenylacetylene is complex and the products have not been fully characterised, although none of them contain a palladium–hydrogen bond.[30]

(iii) Water

All the complexes containing triphenyl group IVB metal groups are stable to water,[8,10,14,16] the bistriphenylgermyl complexes being stable to alcoholic potassium hydroxide as well.[14] However, the trimethylgermyl and trimethylsilyl complexes $[(PEt_3)_2PtCl(EMe_3)]$ are both hydrolysed (eq. 114),

$$2\,trans\text{-}[(PEt_3)_2PtCl(EMe_3)] + H_2O \rightarrow$$
$$2\,trans\text{-}[(PEt_3)_2PtHCl] + (Me_3E)_2O \quad (114)$$

the former rather slowly and the latter very rapidly.[19] The great difference in rate, together with the fact that the carbon analogues are not hydrolysed (p. 339) might be thought to indicate that primary attack by water occurs at the group IVB metal atom. Since, however, the rate of hydrolysis of the anionic complexes[24] $[(Ph_2PCH_2CH_2PPh_2)(PEt_3)Pt(EMe_3)]^+Cl^-$ is very much lower than that of the neutral complexes $[(PEt_3)_2PtCl(EMe_3)]$ the rate-determining step probably involves proton attack on the platinum atom.

(iv) Halogens

The reactions of the halogens with the platinum complexes of the group IVB metals are all somewhat similar in that the platinum–group IVB metal bond is cleaved and a platinum–halogen bond is formed. The reactions, however, are far from simple as is shown by the yields, which are sometimes above theoretical, and by the nature of the group IVB metal products. Thus treatment of cis-$[(PMe_2Ph)_2Pt(SiMePh_2)_2]$ in benzene solution with 0·50 mol of chlorine or iodine in benzene yields 0·51 mol of $trans$-$[(PMe_2Ph)_2PtCl$-$(SiMePh_2)]$ or 0·58 mol of $trans$-$[(PMe_2Ph)_2PtI(SiMePh_2)]$ respectively together with Ph_2MeSiH.[8] Further halogen cleaves the second Pt–Si bond to give $[(PMe_2Ph)_2PtX_2]$. On the other hand 0·50 mol of bromine reacts with cis-$[(PMe_2Ph)_2Pt(SiMePh_2)_2]$ in benzene solution to give 0·56 mol of $trans$-$[(PMe_2Ph)_2PtBr(SiMePh_2)]$ together with the bromosilane $Ph_2MeSiBr$ instead of the hydrosilane obtained with chlorine and iodine.[8] The nature of the products suggests that some of the platinum–silicon bond cleavage is effected by hydrogen halide, formed by halogen attack on the solvent benzene, rather than by direct halogen attack on the platinum–silicon bond. The above theoretical yields can also be accounted for by secondary reactions involving the halobenzene.

Platinum–germanium,[14] platinum–tin[36] and platinum–lead[16] bonds are all cleaved similarly (eqn. 115), although in the case of lead the products

$$[(PR_3)_2Pt(ER_3')_2] + 2X_2 \rightarrow [(PR_3)_2PtX_2] + 2R_3'EX \quad (115)$$

are complicated by the further reaction of Ph_3PbBr with bromine to give Ph_2PbBr_2 and bromobenzene. The three reactions probably occur by electrophilic attack of the halogen on the platinum(II) complex to give a six-co-ordinate platinum(IV) intermediate which then breaks down to give the observed platinum(II) complex and the halogen adduct of the group IVB metal.

The interaction of halogens with the palladium complexes of group IVB metals has not been investigated.

(v) Alkyl halides

Only the cleavage of platinum–silicon[8] and platinum–germanium[13] complexes by alkyl halides has been investigated. Both reactions follow the route shown in eqn. 116, although the reaction of the silyl complexes is

$$[(PR_3)_2Pt(ER'_3)_2] + CH_3I \rightarrow [(PR_3)_2PtI_2] + R'_3ECH_3 \qquad (116)$$

slightly less straightforward than that of the germyl complexes in that a little R'_3SiI is also formed. Both reactions probably occur by a *cis*-addition-elimination mechanism involving a six-co-ordinate platinum(IV) intermediate. The slightly greater complexity of the products from silicon complexes, already instanced for reactions with the halogens, is indicative of the smaller kinetic differences between loss of R'_3SiCH_3 and R'_3SiI from the labile platinum(IV) intermediate than from the corresponding germanium intermediate. Platinum–germanium bonds are cleaved by carbon tetrachloride.[14] Although the products have not been completely analysed, the germanium is left as Ph_3GeCl and the platinum as $[(R_3P)_2PtCl_2]$.

1,2-Dibromoethane cleaves both palladium–germanium[11,12] and platinum–germanium[13-15] bonds according to the general reaction 117. These

$$[(R_3P)_2M(GeR'_3)_2] + C_2H_4Br_2 \rightarrow [(PEt_3)_2MBr_2] + 2R'_3GeBr + 2C_2H_4$$
$$(117)$$

cleavages probably involve an octahedral platinum(IV) intermediate[21] (eqn. 118).

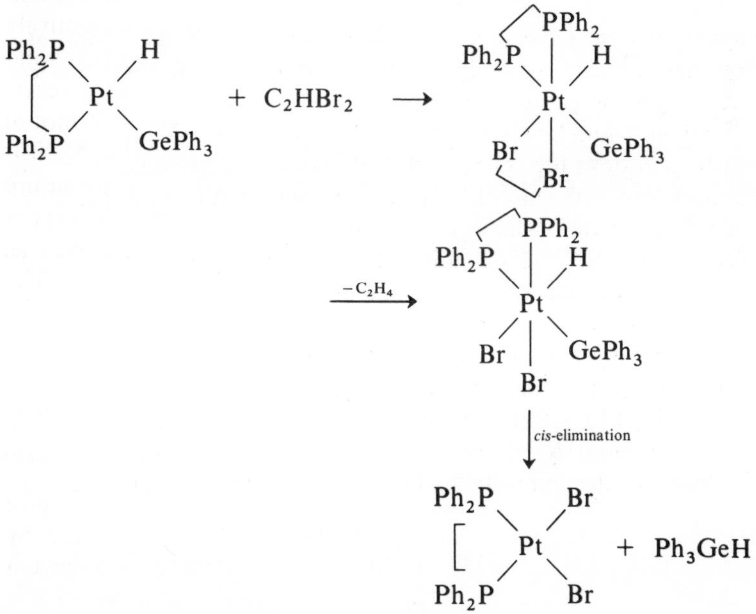

(vi) Molecular hydrogen

Molecular hydrogen cleaves platinum–silicon bonds by a reversible reaction[8] (eqns. 119 and 120) (this reaction in reverse provides one of the standard methods for preparing these complexes, eqn. 92, p. 78). The hydrogenation

$$cis\text{-}[(PMe_2Ph)_2Pt(SiMePh_2)_2] + H_2 \underset{}{\overset{THF}{\rightleftharpoons}}$$

$$trans\text{-}[(PMe_2Ph)_2PtH(SiMePh_2)] + Ph_2MeSiH \quad (119)$$

$$trans\text{-}[(PMe_2Ph)_2PtCl(SiMePh_2)] + H_2 \underset{}{\overset{THF}{\rightleftharpoons}}$$

$$trans\text{-}[(PMe_2Ph)_2PtHCl] + Ph_2MeSiH \quad (120)$$

of platinum–germanium complexes proceeds extremely smoothly at room temperature and pressure. The bis–triphenylgermyl complex $[(PEt_3)_2Pt\text{-}(GePh_3)_2]$ yields the very stable $[(PEt_3)_2PtH(GePh_3)]$[19] (eqn. 121), which is not reduced further by molecular hydrogen. Incidentally the second

$$[(PEt_3)_2Pt(GePh_3)_2] + H_2 \rightarrow [(PEt_3)_2PtH(GePh_3)] + Ph_3GeH \quad (121)$$

platinum–germanium bond can be cleaved by lithium aluminium hydride.[14] Trimethylgermyl groups bound to platinum are cleaved reversibly[19] (eqn. 122) and even more readily than triphenylgermyl groups[19] as is

$$trans\text{-}[(PEt_3)_2PtCl(GeMe_3)] + H_2 \rightleftharpoons trans\text{-}[(PEt_3)_2PtHCl] + Me_3GeH$$

$$(122)$$

illustrated by the complex $trans\text{-}[(PEt_3)_2Pt(GeMe_3)(GePh_3)]$ (eqn. 123).

$$trans\text{-}[(PEt_3)_2Pt(GeMe_3)(GePh_3)] + H_2 \rightarrow$$

$$trans\text{-}[(PEt_3)_2PtH(GePh_3)] + Me_3GeH \quad (123)$$

These hydrogen cleavage reactions are thought to occur by a *cis*-addition–elimination mechanism involving a six-co-ordinate platinum(IV) intermediate. Although the postulated intermediate has not been isolated, the evidence in favour of its involvement includes: (1) the reversibility of some of these hydrogenations (e.g. eqns. 119, 120 and 122); (2) the fact that the hydrogenations proceed smoothly at room temperature and pressure; (3) The very low activation energy of about 9 kcal/mol for reaction 121; (4) the observation that the palladium–germanium complex $[(PEt_3)_2Pd\text{-}(GePh_3)_2]$, which because of the greater reluctance of palladium than platinum to form a six-co-ordinate M(IV) species is unlikely to react via a *cis*-addition–elimination mechanism, requires 100 atm. pressure of hydrogen to effect hydrogenation;[12] and (5) the fact that whilst more than 90% of the hydrogenation of $[(PEt_3)_2Pt(GePh_3)_2]$ occurs as shown in eqn. 121, about 5% of the germanium is recovered as hexaphenyldigermane,[19] which could arise from the decomposition of a platinum(IV) intermediate such as

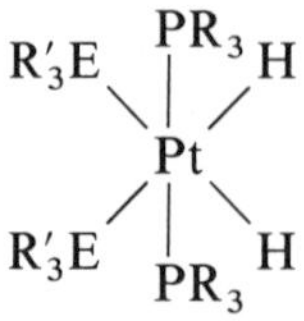

$$\begin{array}{c} R'_3E \diagdown \quad \overset{PR_3}{|} \diagup H \\ \diagup Pt \diagdown \\ R'_3E \diagup \quad \overset{|}{PR_3} \diagdown H \end{array}$$

20. Since the *cis*-complexes $[(Ph_2PCH_2CH_2PPh_2)Pt(ER_3)_2]$, where $ER_3 =$ SiMePh$_2$,[8] or GePh$_3$,[19] are only hydrogenated with difficulty the initial labile platinum(IV) intermediate formed probably has structure **20**, which, by loss of R'_3EH, would give the observed *trans*-$[(PR_3)_2PtH(ER'_3)]$. Structure **20** also accounts for the observation[19] of traces of hexaphenyldigermane during the hydrogenation of $[(PEt_3)_2Pt(GePh_3)_2]$.

The hydrogenation of platinum–tin[36] (eqn. 124) and platinum–lead[14]

$$[(PPh_3)_2Pt(SnMe_3)_2] + H_2 \longrightarrow trans\text{-}[(PPh_3)_2PtH(SnMe_3)] + Me_3SnH$$

$$(124)$$

bonds (eqn. 125) occurs analogously to that of platinum–silicon and

$$[(PEt_3)_2Pt(PbPh_3)_2] + H_2 \longrightarrow trans\text{-}[(PEt_3)_2PtH(PbPh_3)] + Ph_3PbH$$

$$(125)$$

platinum–germanium bonds. There is no mention of the tin reactions being reversible and the lead reaction is not reversible due to the instability of triphenylplumbane.

Neither $[(PEt_3)_2Pd(GePh_3)_2]$ nor $[(Ph_2PCH_2CH_2PPh_2)Pd(GePh_3)_2]$ react with hydrogen at atmospheric pressure although both react readily at 100 atm. pressure[12] to give $[(PEt_3)_2PdH(GePh_3)]$ and $[(Ph_2PCH_2CH_2\text{-}PPh_2)PdH(GePh_3)]$. Whereas *trans*-$[(PEt_3)_2Pt(PbPh_3)_2]$ forms *trans*-$[(PEt_3)_2PtH(PbPh_3)]$ on hydrogenation, the analogous palladium complex reacts instantly with hydrogen in dichloromethane solution even at 0°C to give palladium metal, triethylphosphine and hexaphenyldiplumbane.[17]

(vii) Pyrolysis
Very little systematic work on the pyrolysis of platinum and palladium complexes containing group IVB metals has been reported. However, platinum–silicon complexes are certainly much less stable to heat than platinum–germanium complexes.[10] As with corresponding carbon complexes, palladium–germanium complexes are less stable to heat than platinum–germanium complexes.[38] Where detailed analyses of the products of heating in vacuo have been made the results are consistent with free-radical cleavage reactions (eqns. 126–9). The ethylene formed in some of

$$[(PEt_3)_2Pt(GePh_3)_2] \xrightarrow[\text{ref. 14}]{220°C} PEt_3 + GePh_4 + Ge_2Ph_6 + C_2H_4$$
$$+ C_6H_6 + \text{black residue} \qquad (126)$$

$$[(PEt_3)_2Pd(GePh_3)_2] \xrightarrow[\text{ref. 12}]{97\text{-}107°C} PEt_3 + GePh_4 + Ge_2Ph_6 + C_2H_4$$
$$+ C_6H_6 + H_2 + Pd \qquad (127)$$

$$[(PEt_3)_2Pt(PbPh_3)_2] \xrightarrow[\text{ref. 16}]{\text{pyrolysis}} PEt_3 + PbPh_4 + Pb_2Ph_6 + Pt$$
$$+ \text{other identified products} \qquad (128)$$

$$[(PEt_3)_2Pd(PbPh_3)_2] \xrightarrow[\text{ref. 17}]{110°C} PEt_3 + PbPh_4 + Pb_2Ph_6 + Pd \qquad (129)$$

these reactions results from decomposition of the phosphine, as is shown by the fact that $[(P^nPr_3)_2Pt(GePh_3)_2]$ behaves similarly to its triethylphosphine analogue except in giving propene instead of ethylene.[14]

B. Reagents which do not cleave the M–E bond

(i) 1,2-diphenylphosphinoethane

The chelating diphosphine $Ph_2PCH_2CH_2PPh_2$ reacts with $[(PR_3)_2Pt(ER'_3)_2]$ complexes to displace the two monodentate phosphine ligands and give $[(Ph_2PCH_2CH_2PPh_2)Pt(ER'_3)_2]$ where E = Si,[8] Ge[38] or Pb.[16] Surprisingly, bipyridyl does not replace triethylphosphine from these complexes. Reactions of the monosilyl and monogermyl complexes *trans*-$[(PEt_3)_2$-$PtCl(EMe_3)]$ with $Ph_2PCH_2CH_2PPh_2$ unexpectedly gives the ionic complexes $[(Ph_2PCH_2CH_2PPh_2)(PEt_3)Pt(EMe_3)]^+Cl^-$.[24] Moreover Ph_2-$PCH_2CH_2PPh_2$ displaces both the triethylphosphine ligands from the platinum–lead[12] and palladium–lead[17] complexes *trans*-$[(PEt_3)_2$-$M(PbPh_3)_2]$ to give *cis*-$[(Ph_2PCH_2CH_2PPh_2)M(PbPh_3)_2]$. All these reactions demonstrate the considerable strength of the platinum–group IVB and palladium–group IVB metal bonds relative to the platinum- and palladium–triethylphosphine bonds. However, the strength of the M–E bond depends on the presence of phosphine or arsine ligands since neither platinum– nor palladium–group IVB metal complexes have been prepared without tertiary phosphine or arsine ligands, with the exception of $K_2[M(CN)_2(GePh_3)_2]$ mentioned below. The palladium–germanium complex $[(PEt_3)_2Pd(GePh_3)_2]$ is the only complex in this group that fails to give a stable *cis*-complex with $Ph_2PCH_2CH_2PPh_2$.[12]

(ii) Cyanide

Two moles of cyanide displace both of the triethylphosphine ligands from $[(PEt_3)_2M(GePh_3)_2]$ where M = Pt[38] or Pd[12] to give $K_2[M(CN)_2$-$(GePh_3)_2]$. With an excess of cyanide both complexes give $K_2M(CN)_4$. The complexity of the reaction of potassium cyanide with the corresponding trimethylgermyl complex *trans*-$[(PEt_3)_2PtCl(GeMe_3)]$ indicates that in this complex the Pt–GeMe$_3$ and Pt–PEt$_3$ bonds are of comparable strength, since 1 mol of potassium cyanide gives a mixture of *cis*- and *trans*-$[(PEt_3)_2PtCN(GeMe_3)]$, whereas 3 mol give a mixture of $[(PEt_3)_2Pt(CN)_2]$ and $K_2[Pt(CN)_3(GeMe_3)]$.[19] Thus reaction with potassium cyanide suggests a bond strength series approximating to

$$\text{Pt–Cl} < \text{Pt–PEt}_3 \sim \text{Pt–GeMe}_3 < \text{Pt–GePh}_3.$$

When the platinum–lead complex *trans*-$[(PEt_3)_2Pt(PbPh_3)_2]$ is treated with potassium cyanide under conditions identical with those used for the germanium analogue, the platinum–lead bonds are cleaved[16] (eqn. 130) indicating that the Pt–GePh$_3$ bond is stronger than the Pt–PbPh$_3$ bond. As expected, the palladium–lead complex reacts in an identical manner (eqn. 130).

$$\textit{trans-}[(PEt_3)_2M(PbPh_3)_2] + 4CN^- \rightarrow [M(CN)_4]^{2-} + Pb_2Ph_6 + 2PEt_3$$

$$(\text{M = Pd or Pt}) \hspace{2cm} (130)$$

(iii) *Other reagents*

Only in the case of platinum–germanium complexes have other reagents that react with the complex without cleaving the platinum–germanium bond been studied extensively. The chloride ligand in *trans*-[(PEt$_3$)$_2$PtCl-(GeMe$_3$)] can be replaced by bromide, iodide or thiocyanate by means of the appropriate alkali metal salt[19] or by a phenyl or triphenylgermyl group derived from phenyllithium or triphenylgermyllithium.[19]

PREPARATION AND PROPERTIES OF COMPLEXES CONTAINING EX$_3$ LIGANDS

Although it has long been known that platinum(II) and palladium(II) salts form intensely coloured solutions on treatment with tin(II) chloride, bromide and iodide, only recently has a real understanding of the nature of the products formed when the trichlorostannato ion co-ordinates to platinum been determined; and it is even more recently that this work has been extended to other trichloro–group IVB metal ions.

Platinum(II) complexes containing trichlorosilyl ligands have been prepared from trichlorosilane by two main routes. The first uses platinum(0) complexes[31] (eqn. 131) and the second hydrido–platinum(II) complexes[8]

$$[Pt(PPh_3)_4] + 2HSiCl_3 \rightarrow cis\text{-}[(PPh_3)_2Pt(SiCl_3)_2] + 2PPh_3 + \uparrow H_2 \quad (131)$$

as starting materials (eqn. 132). The reaction of the platinum(0) complex

$$[(PMe_2Ph)_2PtHCl] + Cl_3SiH \rightarrow$$

$$trans\text{-}[(PMe_2Ph)_2PtCl(SiCl_3)] + \uparrow H_2 \quad (132)$$

[Pt(Ph$_2$PCH$_2$CH$_2$PPh$_2$)$_2$] with trichlorosilane depends upon the temperature.[31] At room temperature [(Ph$_2$PCH$_2$CH$_2$PPh$_2$)PtH(SiCl$_3$)] is formed whereas under reflux at 35°C [(Ph$_2$PCH$_2$CH$_2$PPh$_2$)Pt(SiCl$_3$)$_2$] is formed. All the trichlorosilyl complexes are unstable to air and moisture.

In contrast to the very few trichlorosilyl complexes recognised, a wide range of trichlorogermyl and trichlorostannato complexes are known. Although many of the complexes of these two ligands with platinum are similar to one another, where there are differences then GeCl$_3^-$ generally leads to the higher oxidation state (Pt^{II} or Pt^{IV}) and SnCl$_3^-$ to the lower oxidation state (Pt^0 or Pt^{II}). Thus, for example, treatment of H$_2$PtCl$_4$ with the germanium(II) compound HGeCl$_3$ in concentrated hydrochloric acid solution followed by addition of tetramethylammonium chloride yields the platinum(IV) complex (Me$_4$N)$_2$[PtH(GeCl$_5$)],[43] whereas the preparation of a platinum(IV)–SnCl$_3$ complex requires the use of tin tetrachloride[10] (reaction 133). Further examples[43] arise in the treatment of platinum(II)

$$trans\text{-}[(PPh_3)_2PtHCl] + 2SnCl_4 \rightarrow$$

$$[(PPh_3)_2PtCl_2(SnCl_3)_2] + HCl \quad (133)$$

complexes with a germanium(II) compound to give platinum(IV) products.

First, treatment of $[(PPh_3)_2PtCl_2]$ with $HGeCl_3$ yields $[(PPh_3)_2PtCl_3$-$(GeCl_3)]$ after some hours; and, secondly, treatment of one equivalent of K_2PtCl_4 with five equivalents of $HGeCl_3$ in the presence of triphenyl-phosphine yields the phosphonium salt $(Ph_3PH)_2[PtCl_4(GeCl_3)_2]$. These reactions should be contrasted with treatment of the platinum(IV) complex K_2PtCl_6 with stannous chloride in the presence of triphenylphosphine which gives the platinum(II) complex $[(PPh_3)_2Pt(SnCl_3)_2]$,[44] and treatment of chloroplatinic acid with six equivalents of stannous chloride which gives the trigonal bipyramidal platinum(II) complex $[Pt(SnCl_3)_5]^{3-}$.[44,45]

In spite of the differences emphasised above there are many examples where analogous complexes of tin and germanium are formed. Thus treat-ment of K_2PtCl_4 with the divalent compounds $HGeCl_3$[43] or $SnCl_2$[46] in ethanol in the presence of triphenylphosphine yields the platinum(II) complexes $[(PPh_3)_2PtCl(ECl_3)]$, where E = Ge or Sn. Similarly, both *cis*- and *trans*-isomers of the complex ions $[PtCl_2(ECl_3)]^{2-}$, where E = Ge[43] or Sn,[46] are known. However, these complex ions react differently with an excess of the group IVB halide; the germanium complex cannot be reduced with an excess of trichlorogermane,[43] whereas the tin complex is reduced by an excess of stannous chloride to give the platinum(0) cluster compound $[Pt_3Sn_8Cl_{20}]^{4-}$.[47] This compound is assumed to have structure **21** since on treatment with 1,5-cyclo-octadiene it gives $[(C_8H_{12})_3Pt_3Sn_2Cl_6]$ which has been shown[48] by x-ray diffraction to have structure **22**.

21

22

Most of the complexes containing $GeCl_3^-$ and $SnCl_3^-$ ligands described above are stable to air, although the five-co-ordinate $[Pt(SnCl_3)_5]^{3-}$ ion is not.[47] The fact that where differences arise between the $GeCl_3^-$ and $SnCl_3^-$ complexes the tin complexes are found to be in the lower oxidation state suggests that there may be rather more π-bonding in the Pt–Sn than Pt–Ge bond. This is consistent with the well-established preference of tin, as com-pared to germanium, to retain electron density in its vicinity in what is often described as an orbital containing a lone-pair of electrons. This is well illustrated by the standard redox potentials for the two elements[49] (eqns. 134 and 135). The addition of tin(II) chloride to platinum(II) chloride in

$$2H_2O + Ge^{2+} \rightleftharpoons GeO_2 + 4H^+ + 2\varepsilon \qquad E° \sim +0.3 \text{ V} \qquad (134)$$

$$Sn^{2+} \rightleftharpoons Sn^{4+} + 2\varepsilon \qquad\qquad E° = -0.15 \text{ V} \quad (135)$$

hydrochloric acid gives an orange-red solution and sets up a complex series of equilibria[46] (eqns. 136–40). When tin(II) chloride is added dropwise to

$$SnCl_2 + Cl^- \rightleftharpoons SnCl_3^- \qquad\qquad (136)$$

$$[PtCl_4]^{2-} + SnCl_3^- \rightleftharpoons [PtCl_3(SnCl_3)]^{2-} + Cl^- \tag{137}$$

$$[PtCl_3(SnCl_3)]^{2-} + SnCl_3^- \rightleftharpoons trans\text{-}[PtCl_2(SnCl_3)_2]^{2-} + Cl^- \tag{138}$$

$$trans\text{-}[PtCl_2(SnCl_3)_2]^{2-} \rightleftharpoons cis\text{-}[PtCl_2(SnCl_3)_2]^{2-} \tag{139}$$

$$cis\text{-} \text{ and } trans\text{-}[PtCl_2(SnCl_3)_2]^{2-} + 3SnCl_3^-$$

$$\rightleftharpoons [Pt(SnCl_3)_5]^{3-} + 2Cl^- \tag{140}$$

palladium(II) chloride in hydrochloric acid solution a series of colour changes from yellow-orange through brown, purple, blue-green, olive-green and finally to almost colourless in the presence of a large excess of tin(II) chloride, occur.[3,50,51] As yet the only complexes isolated from these solutions are $(Ph_4As)_4[PdCl(SnCl_3)_2]_2$ isolated from the red solution[52] and impure $(Ph_4As)_2[Pd(SnCl_3)_4]$ isolated from the purple solution.[53] Treatment of a solution of $[Pd(1,5\text{-cyclooctadiene})Cl_2]$ in methanol with stannous chloride yields $[Pd_3(C_8H_{12})_3(SnCl_3)_2]$ which probably has structure **22**.[53] Stable palladium(II) complexes, $[Pd(C_6H_5NC)_2(ECl_3)_2]$, where E = Ge and Sn, have recently been reported.[54]

CATALYTIC BEHAVIOUR OF GROUP IVB METAL COMPLEXES

Group IVB metal ligands have been found to be very useful in conjunction with platinum(II) salts in catalysing the homogeneous hydrogenation and isomerisation of olefins. This subject is covered more fully in Chapter 13. Platinum–silicon and platinum–germanium complexes are also involved in the homogeneous hydrosilation and hydrogermanation of olefins in which silanes and germanes are added across the double-bond of an olefin in the presence of chloroplatinic acid (eqn. 141). These reactions are also discussed in more detail in Chapter 13.

$$>C=C< + R_3SiH \xrightarrow{H_2PtCl_6} R_3Si-\overset{|}{\underset{|}{C}}-\overset{|}{\underset{|}{C}}-H \tag{141}$$

BONDING IN PLATINUM AND PALLADIUM COMPLEXES OF GROUP IVB METALS

The bonding in platinum(II) and palladium(II) complexes of the group IVB metals is best understood in terms of the molecular orbital theory, as used to explain the bonding in metal–carbon σ-bonds.[55] This theory, described in detail in Chapter 12 (pp. 330–334), is applicable because of the general similarity between the carbon complexes and the complexes of the heavier group IVB elements exemplified by the following experimental observations:

(i) The presence of tertiary phosphine or arsine ligands appears to be virtually essential for the formation of stable complexes in which the group IVB elements, including carbon, are bound to platinum or palladium. (The complexes in which one or more ECl_3^- ligands are bound to platinum or palladium are, however, an exception to this rule).

(ii) Platinum complexes of all the group IVB elements, including carbon, are more stable than their palladium analogues.

(iii) The decomposition products obtained on pyrolysis are consistent with a free radical cleavage mechanism as required by the theory of Chatt and Shaw.

Rather greater effort has been expended on investigating whether or not π-bonding from the metal to the group IVB element is involved in the M–E bond. The evidence for and against π-bonding is as follows:

(i) Chemical evidence

The 'alternation effect' (p. 82), if indeed it is a real effect, could be explained in terms of π-bonding in the M–E bond, since germanium and lead have less tendency to lose electrons and presumably, therefore, more readily take part in π-bonding, whereas silicon and tin have a greater tendency to lose electrons and may be less ready to take part in π-bonding.

(ii) Infra-red evidence

The platinum–chlorine stretching frequencies in the infrared spectra have been determined for a series of *trans*-$[(PR_3)_2PtClY]$ complexes (Table 19, p. 82). The two series for the *trans*-weakening of the platinum–chlorine bond abstracted from Table 19, namely $Cl < PbPh_3 < CH_3 < C_6H_5 < H < SiMe_3 < GeMe_3$ and $Cl < SnCl_3 < SnPh_3 < SnMe_3 \sim H < PbPh_3$, are in agreement with the *trans*-weakening predicted on the basis of electro-negativity values for a purely inductive mechanism for Cl, CH_3, C_6H_5, $SiMe_3$ and $GeMe_3$ since the Pauling electronegativities of the donor atoms[56] in these ligands decrease in the order $Cl(3\cdot0) > C(2\cdot5) > Si(1\cdot8) \sim Ge(1\cdot8)$. However, the ligands $SnCl_3$, $SnPh_3$, $SnMe_3$ and $PbPh_3$ all give platinum–chlorine stretching frequencies that are higher than predicted on the basis of a purely inductive mechanism. This could indicate some π-bonding in the Pt–Sn and Pt–Pb bonds, as this would strengthen the *trans* Pt–Cl bond and hence increase the platinum–chlorine stretching frequency.[40] However, the normal modes leading to the observed vibrational frequencies in these complexes involve significant contributions from other metal–ligand bonds besides the platinum–chlorine bond, so that these infra-red stretching frequencies do not provide definite evidence for or against π-bonding in the Pt–E bond.[57]

TABLE 20

Tin–chlorine stretching frequencies for a series of trichlorostannato complexes

Complex	$v(Sn–Cl)(cm^{-1})$		Reference
	asym.	sym.	
$SnCl_3^-$	289	252	58
$Cl_3B.SnCl_3^-$	284	255	58
$F_3B.SnCl_3^-$	294	267	58
$(Ph_3P)_2HPt.SnCl_3$	331	309	10
$(Ph_3P)_2ClPt.SnCl_3$	341	325	10
$(Ph_3P)_2PhPt.SnCl_3$	341	318	59
$CH_3.SnCl_3$	384	366	60

The tin–chlorine stretching frequencies in the platinum(II) complexes (Table 20) lie between those in the boron halide adducts, where there is a single σ-bond from tin to boron, and those in trichloromethyltin, where there is a single σ-bond to tin from carbon. This intermediate position of the platinum(II) complexes provides evidence in support of π-bonding from platinum to tin.

(iii) N.m.r. evidence

The ^{19}F n.m.r. chemical shifts of a series of complexes *trans*-[(PEt$_3$)$_2$PtXY]

where $Y = F\text{---}$ or and X = '*trans*-ligand', have been

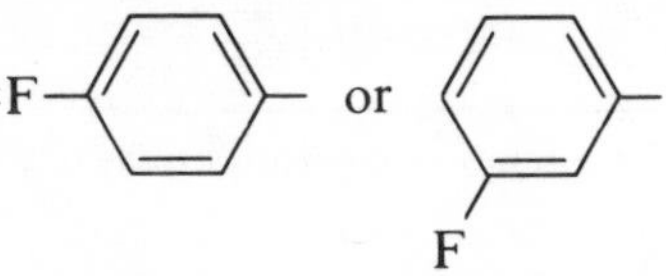

assumed to depend only on the inductive effect of the *trans*-ligand in the *meta*-fluorophenyl complexes and on both the inductive and resonance effects of the *trans*-ligands in the *para*-fluorophenyl complexes.[59,61] Thus the difference in the ^{19}F chemical shifts of the *para*- and *meta*-fluorophenyl complexes, which is proportioned to the π-electron density in the aromatic ring, as the *trans*-ligand X is altered, is taken to reflect the π-acceptor or σ-donor capacity of X. The results (Table 21) indicate that the π-acceptor capacity of SnCl$_3^-$ is comparable to that of the cyanide ion. Although this technique is open to some criticism[62] (pp. 335–336) it is certainly consistent with the infrared technique, discussed above, in predicting the presence of π-bonding in the platinum–trichlorostannato bond.

TABLE 21

^{19}F n.m.r. shielding parameters for complexes *trans*-[(PEt$_3$)$_2$PtXY] where Y =

F--- or (from ref. 59)

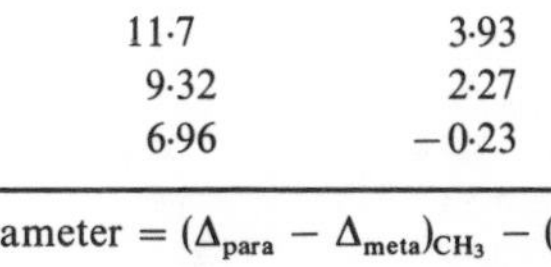

Ligand X	^{19}F Shielding parameters (Δ)		π-Acceptor parameter[a]
	para	meta	
Cl	10·1	2·11	−0·2
CH$_3$	11·7	3·93	0
CN	9·32	2·27	0·7
SnCl$_3$	6·96	−0·23	0·6

[a] π-Acceptor parameter $= (\Delta_{para} - \Delta_{meta})_{CH_3} - (\Delta_{para} - \Delta_{meta})_X$.

The ^{195}platinum–^{31}phosphorus coupling constants in a series of complexes *cis*-[(PR$_3$)$_2$PtXY] depend on the covalency of the platinum–phosphorus bond and on the s-character of the platinum orbital used in the bond.[63] Hence the coupling is small when the phosphorus atom is *trans* to a ligand of high inductive *trans*-influence because the platinum–phosphorus bond is thus made less covalent, and the couplings are large when the phosphorus atom is *trans* to a ligand of low inductive *trans*-influence or high mesomeric *trans*-influence. The very low ^{195}Pt–^{31}P coupling constant obtained for the silicon complex *cis*-[(PhMe$_2$P)$_2$Pt(SiMePh$_2$)$_2$] (Table 22)

TABLE 22
^{195}Pt–^{31}P coupling constants in complexes
cis-[(PR$_3$)$_2$PtX$_2$] (from ref. 57)

PR$_3$	Trans atom (X)	$J_{195_{Pt}-31_P}(Hz)$
PBu$_3$	Cl	3502
PBu$_3$	Br	3477
PBu$_3$	I	3345
	P^a	2824–2265
PEt$_3$	C(phenyl)	1704
PEt$_3$	Si(SiMePh$_2$)	1559

^{a} Values were obtained for a series of complexes trans-[(PR$_3$)$_2$PtX$_2$].

indicates that the silicon atom has a very strong inductive trans-influence. Although this does not rule out π-bonding in the platinum–silicon bond it does indicate that either the amount of π-bonding and hence its associated mesomeric trans-influence is small, or that the inductive trans-influence of silicon is very substantially greater than that of carbon. The latter seems rather unlikely.

(iv) Evidence from X-ray diffraction studies

Pt–E bond lengths determined from x-ray diffraction could give evidence concerning the presence of π-bonding in the Pt–E bond, although the method is not very sensitive due to its inherent inaccuracies. It involves comparing the experimental bond length with that estimated from the sum of the covalent radii of the Pt and E atoms which itself has a certain uncertainty. The available data (Table 23) indicate that the M–E bond lengths are all

TABLE 23
M–E bond lengths

Complex	M–E(Å) (experimental)	M–E(Å) (calculated)^a	Number of structure in Appendix II	Reference
[Pt(SnCl$_3$)$_5$]$^{3-}$	2·54	2·71	103	45
[(π-C$_3$H$_5$)Pd(PPh$_3$)(SnCl$_3$)]	2·563	2·71	74	64
trans-[(PPhMe$_2$)$_2$PtCl(SiMePh$_2$)]	2·29	2·48	192	65

^{a} Calculated from the sum of the covalent radii taken from reference 56.

shorter than predicted from the sum of the covalent radii. This shortening could be due to π-bonding, which for the trichlorostannato ligand would be in agreement with the n.m.r. and infrared evidence mentioned above. However, although π-bonding may be present in the Pt–Si bond of trans-[(PPhMe$_2$)PtCl(SiMePh$_2$)] this bond is essentially a σ-bond since the trans

Pt–Cl bond in this complex is the longest ever recorded (2·45 Å), indicating a very strong inductive *trans*-influence.

In summary, the evidence for π-bonding in the $SnCl_3^-$ complexes of platinum(II) appears to be substantial and this is consistent with assigning the $+2$ oxidation state to tin in these complexes since on cleavage the tin would tend to retain the electron density that it has withdrawn from platinum via π-bonding. There is some rather uncertain infrared evidence for π-bonding in Pt–ER_3 bonds, where ER_3 is $SnPh_3$, $SnMe_3$ and $PbPh_3$, that needs further confirmation before being fully acceptable. In the other Pt–E bonds little, if any, real evidence for π-bonding exists. In particular the *trans*-influences of the germanium and silicon complexes so far investigated can all be understood in terms of purely σ-bonding in the Pt–Ge and Pt–Si bonds.

COMPLEXES WITH CARBON LIGANDS

In the remainder of this chapter complexes of cyanide, isocyanide, carbon monoxide and π-cyclopentadienyl ligands are considered. The organometallic complexes that might have been expected to be found in this section are considered in Chapters 12, 13 and 14. While the bonding of cyanide, isocyanide, carbon monoxide and π-cyclopentadienyl ligands to metals are all similar in that they all involve both σ-donation and π-back-donation of charge, their detailed chemistries are sufficiently different to warrant considering each ligand separately.

Cyanide complexes

The cyanide ion, which is isoelectronic with carbon monoxide, binds to metals by σ-donation of a pair of electrons in an sp hybrid orbital on the carbon atom. This is complemented by π-back-donation from filled orbitals on the metal to the empty π^*-(antibonding) orbitals on the cyanide ion. Because of its negative charge, the cyanide ion is a poorer π-acceptor than carbon monoxide. Nevertheless it forms very strong complexes with both platinum(II) and palladium(II) and together with carbon monoxide exerts one of the strongest ligand fields known.[66]

The preparation and properties of the commoner cyanides of platinum(II) and palladium(II) are given in Table 24. The cyano-complexes are extremely stable, which is responsible for their main application, namely their formation on addition of an excess of potassium cyanide to any platinum(II) or palladium(II) complex with the liberation of the ligands originally bound to the metal which can be separated off and identified. The stability constant $[M(CN)_4^{2-}.aq]/[M^{2+}.aq][CN^-.aq]^4$ is 10^{42} for palladium[84] and about 10^{41} for platinum,[85] although the value for platinum may be in error by several powers of ten as it depends on the value for the $Pt^{2+}.aq/Pt$ standard potential which is not reliably known. Such high stability constants make the preparation of complexes of the type $[M(CN)_2L_2]$ difficult. However, treatment of $[PtX_2L_2]$, where $X = Cl$, NO_2 or NO_3 and $L = NH_3$ or $L_2 = en$, with a stoichiometric amount of potassium cyanide has led to the

TABLE 24

Preparation and properties of the divalent cyanides

Compound	Preparation	Properties	Infra-red ($\nu_{C\equiv N}$)
$Pd(CN)_2$	(1) Treat a Pd^{II} salt with $Hg(CN)_2$[67] (2) Warm $K_2Pt(CN)_4$ with HCl[68]	Diamagnetic yellowish-white solid. Forms adducts with NH_3 and organic bases[69–71]	$2220\,cm^{-1}$,[72] suggests bridging cyanide[73]
$M_2Pd(CN)_4$	Dissolve $Pd(CN)_2$ in aqueous MCN[74] (M = Alkali metal)	Diamagnetic colourless solids which react with potassium acetylide in liquid ammonia to give $K_2[Pd(CN)_2(C\equiv CR)_2]$[75]	$2140\,cm^{-1}$[72,76] Raman in ref. 77 Pd–C force constant = 3·12 mdyn/Å[78]
$H_2Pd(CN)_4$	Treat $M_2Pd(CN)_4$ with HCl in ether[79]	White powder	$2202\,cm^{-1}$[79]
$Pt(CN)_2$	(1) Treat $K_2Pt(CN)_4$ with HCl[80] (2) Heat $(NH_4)_2Pt(CN)_4$ at $300°C$[80]	Yellow compound, insoluble in water	Not recorded, but reported to be consistent with bridging cyanide ligands[81]
$M_2Pt(CN)_4$	(1) Treat a Pt^{II} salt with aqueous MCN[82] (2) Heat Pt sponge in conc. aqueous MCN (M = alkali metal)	Diamagnetic yellow solids	$2150\,cm^{-1}$[76,83] Raman in ref. 77 Pt–C force constant = 3·42 mdyn/Å[78]
$H_2Pt(CN)_4$	Treat $M_2Pt(CN)_4$ with HCl in ether[79]	Yellow powder	$2203\,cm^{-1}$[79]

isolation of $[Pt(NH_3)_2(CN)_2]$[86] and $[Pt(en)(CN)_2]$;[87] the products demonstrate the very high *trans*-effect of the cyanide ligand. Mono-cyano complexes such as $[(PPh_3)_2Pt(CN)Cl]$ can be prepared by refluxing the methyl cyanide complex $[(PPh_3)_2Pt(MeCN)Cl]^+Cl^-$ in benzene when methyl chloride is expelled[88] (see p. 126).

There have been a number of attempts to assign the very intense absorption bands in the ultra-violet spectra of the $[M(CN)_4]^{2-}$ complexes[89–93] and although the relative energies of the d-orbitals in $[Pd(CN)_4]^{2-}$ is still uncertain due to the problem of overlapping bands in the spectrum, recent magnetic circular dichroism results[93] coupled with earlier absorption spectroscopy strongly suggest that the relative energies in $[Pt(CN)_4]^{2-}$ are $d_{xy} < d_{xz}$, $d_{yz} \gtrsim d_{z^2} \ll d_{x^2-y^2}$. The principal difference between this order and the currently accepted order for $[PtCl_4]^{2-}$ ($d_{z^2} < d_{xz}, d_{yz} < d_{xy} < d_{x^2-y^2}$)[94] lies in the very low position of the d_{xy} orbital in the $[Pt(CN)_4]^{2-}$ ion, which is a consequence of the stabilisation of this orbital by π-back-donation to the π^*-(antibonding) orbitals of the cyanide ligands. Very thin films of $(NBu_4)_2$-$[Pt(CN)_4]$ that have been evaporated on to suprasil plates show circular dichroism effects which, since both ions are symmetrical, must be due to the $[Pt(CN)_4]^{2-}$ ions being trapped in some asymmetric environment which has not yet been identified.[95]

Isocyanide complexes

Palladium(II) halides react readily with isocyanides to yield stable orange crystalline products, $[PdX_2(CNR)_2]$ where R = aryl or cyclohexyl,[96,97] that are diamagnetic and monomeric in solution. They dissolve in dichloromethane, chloroform, benzene and nitrobenzene and are slightly soluble in alcohol and ether. No complexes containing the $[Pd(CNR)_4]^{2+}$ cation have been prepared although $[Pt(CNR)_4]^{2+}$ is known. The isocyanide ligands are displaced from palladium(II) by tertiary phosphines and arsines but not by pyridine or bipyridyl indicating the considerable strength of the Pd–CNR bond.[97]

Treatment of a solution of M_2PtX_4, where M is an alkali metal and X is Cl, Br, I, NO_2 or CN, with a slight excess of an alkyl-[98,99] or aryl-isocyanide [100,101] yields very stable compounds of the general formula $[PtX_2 . 2CNR]$. Two forms are obtained with X = Cl, Br, I or CN: first intensely coloured ionic compounds $[Pt(CNR)_4]^{2+}[PtX_4]^{2-}$, that are almost insoluble in organic solvents and secondly pale-yellow or colourless neutral *cis*-compounds[102] $[Pt(CNR)_2X_2]$ that are soluble in most organic solvents but insoluble in water. The nitro-complex is only found in the neutral form. The ionic complexes become neutral on prolonged heating at 110–150°C or on boiling in chloroform. Only a few neutral bis-isocyanide–bis-cyano complexes are known. The methyl isocyanide complex can be prepared by methylation of silver tetracyanoplatinate(II)[103] (reaction 142) and the t-butylisocyanide complex can be prepared either by standing the ionic

$$Ag_2[Pt(CN)_4] + 2CH_3I \rightarrow [Pt(CNCH_3)_2(CN)_2] + 2AgI \quad (142)$$

$[Pt(t\text{-}C_4H_9NC)_4][Pt(CN)_4]$ in water, alcohol or chloroform for a long time or more rapidly by boiling $[Pt(t\text{-}C_4H_9NC)_4][PtCl_4]$ in water in the presence of excess t-butylisocyanide when two of the co-ordinated isocyanide ligands are hydrolysed to t-butylalcohol and cyanide ions[103] (reactions 143 and 144).

$$[Pt(CNBu^i)_4][PtCl_4] + 4Bu^iNC \rightarrow 2[Pt(CNBu^i)_4]Cl_2 \quad (143)$$

$$[Pt(CNBu^i)_4]Cl_2 + 2H_2O \xrightarrow{\text{boil}}$$

$$[Pt(CNBu^i)_2(CN)_2] + 2Bu^iOH + 2HCl \quad (144)$$

Isocyanide complexes of both platinum(II)[104–7] and palladium(II)[97,105] are attacked by nucleophiles such as alcohols, amines, hydroxide, hydrosulphide and amide ions to form carbene complexes (reactions 145 and 146; see also pp. 348–350).

$$[(PEt_3)Pt(CNPh)Cl_2] + C_2H_5OH \rightarrow$$

$$\left[(PEt_3)Pt \left(-C \begin{array}{c} OC_2H_5 \\ \\ NHPh \end{array} \right) Cl_2 \right] \quad (145)$$

$$\text{trans-}[(PPh_3)_2Pt(CNMe)_2]^{2+}(BF_4^-)_2 + OH^- \xrightarrow[25°C]{MeCN/H_2O}$$

$$\text{trans-}\left[(PPh_3)_2Pt(CNMe)\left(C\begin{smallmatrix} \nearrow O \\ \searrow NHMe \end{smallmatrix}\right)\right]^+ BF_4^- + BF_4^- \quad (146)$$

Fluoroboric acid reacts with the carboxamido complex formed in reaction 146 to liberate carbon monoxide and form a mixed nitrile–isonitrile complex (reaction 147).

$$\left[(PPh_3)_2Pt(CNMe)\left(C\begin{smallmatrix} \nearrow O \\ \searrow NHMe \end{smallmatrix}\right)\right]^+ BF_4^- \xrightarrow[CH_3CN/H_2O]{HBF_4 \text{ in}}$$

$$[(PPh_3)_2Pt(CNMe)(NCMe)]^{2+}(BF_4^-)_2 + \uparrow CO \quad (147)$$

Isocyanide ligands are isoelectronic with carbon monoxide and like carbon monoxide bind to platinum(II) and palladium(II) through the carbon atom. The isocyanide molecule has a carbon–nitrogen triple bond made up of one σ and two π-components. The σ and one of the π-components involve an overlap of carbon and nitrogen atomic orbitals containing one electron each and the other π-component involves overlap of an empty carbon $2p$ atomic orbital with a doubly filled nitrogen $2p$ orbital, so that the overall bonding may be represented schematically as $C\overline{\overline{\equiv}}N-R$. Isocyanide ligands bond to metals by σ-donation of the pair of electrons in the sp-hybrid orbital of the carbon atom. This σ-donation, which is stronger than in the case of carbon monoxide, causes an unfavourable build-up of negative charge on the metal, which is relieved by π-back-donation of electron density from filled orbitals on the metal to empty π^* (antibonding) orbitals on the isocyanide. A consequence of this bonding scheme is that a linear M–C–N unit would be anticipated. Although no platinum(II) or palladium(II) isocyanide complexes have been studied in detail by x-ray crystallography (see, however, structure 168 in Appendix II), linear bonding has been found in the cobalt(I) complex $[Co(CNCH_3)_5]^+ClO_4^-$.[108] The effect of this bonding scheme on the C–N stretching frequency of isocyanides on co-ordination will be twofold. Firstly, the stretching frequency would be expected to decrease on co-ordination due to π-back donation of electron density from the metal to the π^*-(antibonding) isocyanide orbitals. However, this effect will be opposed by the fact that σ-donation of electron density from the carbon to the metal will enhance the donation of electron density from nitrogen to carbon within the isocyanide molecule so strengthening and increasing the stretching frequency of the C–N bond. The result of these two opposing effects is that in both platinum(II) and palladium(II) complexes there is an increase in the C–N stretching frequency of between 50 and 100 cm^{-1} on co-ordination to the metal.[109] The platinum(II)–chlorine stretching frequencies in the infra-red in a series of complexes $\text{trans-}[(PEt_3)_2PtLCl]^+ClO_4^-$ suggests that the *trans*-influence of isocyanides is slightly greater than that of carbon monoxide but considerably less than that of tertiary phosphines.[62]

Carbonyl complexes

The carbonyl complexes of divalent platinum and palladium have been studied extensively for three main reasons. First, carbon monoxide has shown a remarkable ability for stabilising low oxidation states in transition metals which has provoked a general interest in metal–carbonyl chemistry. Secondly, palladium(II)–carbonyl complexes are implicated in the homogeneous catalysis by palladium salts of the carbonylation of olefins to give acid chlorides and esters (pp. 390–391) and this has led to a general interest in the possibility of the carbonyl complexes of platinum and palladium being of use as homogeneous catalysts.

Preparation of carbonyl complexes

Four main methods have been described for the preparation of carbonyl complexes of platinum(II) and palladium(II).

(*i*) *Direct reaction.* Treatment of $PtCl_3$ (p. 21) with carbon monoxide under 40–120 atm. at 110°C yields $[Pt(CO)_2Cl_2]$ and phosgene.[110] The corresponding bromo- and iodo-complexes are prepared by treating H_2PtBr_6 and PtI_2 with carbon monoxide at 180–210 atm.[111] These monomeric complexes are all thermally unstable and lose carbon monoxide to form the halogen-bridged complexes (reaction 148); the iodo-complex at room temperature,[111] and the chloro-complex either on standing in vacuo or under

$$2[Pt(CO)_2X_2] \rightarrow [Pt(CO)X_2]_2 + 2CO \qquad (148)$$

nitrogen at high temperature.[110] $[Pd(CO)Br_2]_2$ can be formed by passing carbon monoxide over solid palladium(II) bromide at 180°C.[112] $[Pt(CO)X_3]^-$ anions, together with by-products, are prepared by treating solid H_2PtCl_6, K_2PtCl_4, H_2PtBr_6 and H_2PtI_6[111] and aqueous acid solutions of Na_2PtCl_6[113] with carbon monoxide. $(NH_4)[Pd(CO)Cl_3]$, originally thought to be $(NH_4)_2[Pd_2(CO)_2Cl_4]$,[114] is prepared by passing carbon monoxide into a saturated solution of $(NH_4)_2PdCl_4$.[115]

Treatment of palladium(II) chloride with methanol saturated with carbon monoxide at room temperature yields $[Pd(CO)Cl_2]_n$, which is probably a dimer with $n = 2$.[116] However, when the solution is acidic the anionic hydridocarbonyl species $[PdH(CO)Cl_2]^-$ is formed.[117]

(*ii*) *Displacement of halide ion by carbon monoxide.* When carbon monoxide is passed into an acetone solution of $[(PEt_3)_2PtCl_2]$, containing excess sodium perchlorate, under ambient conditions, the carbon monoxide displaces a chloride ligand to give a cationic carbonyl complex[118] (reaction 149). Similarly $[(PPh_3)_2PtH(CO)]^+BPh_4^-$ is formed when a solution of

$$cis\text{- or } trans\text{-}[(PEt_3)_2PtCl_2] + CO + NaClO_4 \rightarrow$$

$$trans\text{-}[(PEt_3)_2Pt(CO)Cl]^+ ClO_4^- + NaCl \qquad (149)$$

trans-$[(PPh_3)_2PtHCl]$ containing sodium tetraphenylborate is treated with carbon monoxide in the presence of silver nitrate which acts as a chloride abstracting reagent[119] (reaction 150). A similar reaction occurs when

$$[(PPh_3)_2PtHCl] + CO + NaBPh_4 + AgNO_3 \rightarrow$$

$$[(PPh_3)_2PtH(CO)]BPh_4 + AgCl + NaNO_3 \quad (150)$$

$[(PEt_3)_2PdCl_2]$ is treated with carbon monoxide in the presence of boron trifluoride, although in this case the reaction occurs in two stages (reactions 151 and 152) since $[(PEt_3)_4Pd_2Cl_2](BF_4)_2$ formed in the first stage can be isolated.[116]

$$2[(PEt_3)_2PdCl_2] + 4BF_3 \rightarrow$$

$$[(PEt_3)_4Pd_2Cl_2](BF_4)_2 + 2BF_2Cl \quad (151)$$

$$[(PEt_3)_4Pd_2Cl_2](BF_4)_2 + 2CO \rightarrow$$

$$2 \text{ } trans\text{-}[(PEt_3)_2Pd(CO)Cl](BF_4) \quad (152)$$

When a refluxing methanolic solution of Na_2PtCl_6 and $SnCl_2$, which contains $[Pt(SnCl_3)_2Cl_2]^{2-}$ anions,[46] is treated with carbon monoxide at atmospheric pressure the red colour deepens and addition of tetraethylammonium chloride precipitates $(NEt_4)[Pt(CO)(SnCl_3)_2Cl]$.[120] The corresponding palladium(II) complexes can be prepared by treating a solution of palladium(II) chloride (one equivalent) and stannous chloride (ten equivalents) with carbon monoxide followed by addition of tetraethylammonium chloride.[120]

(iii) Cleavage of halogen bridged complexes. Treatment of $[Pt(CO)Cl_2]_2$ with hydrogen chloride cleaves the chloride bridge to form anionic $[Pt(CO)Cl_3]^-$.[121] Carbon monoxide itself can cleave chloride bridges as in reaction 153.[122]

$$[(PR_3)_2Pt_2Cl_4] + CO \rightarrow cis\text{-}[(PR_3)Pt(CO)Cl_2] \quad (153)$$

(iv) Displacement of olefins by carbon monoxide. Carbon monoxide can readily displace olefins and acetylenes from their platinum(II) complexes (reactions 154 and 155).[123,124]

$$[Pt(C_2H_4)Cl_3]^- + CO \rightarrow [Pt(CO)Cl_3]^- + C_2H_4 \quad (154)$$

$$trans\text{-}[Pt(C_2H_4)(RNH_2)Cl_2] + CO \rightarrow trans\text{-}[Pt(CO)(RNH_2)Cl_2] + C_2H_4$$

$$\downarrow 120°C$$

$$cis\text{-}[Pt(CO)(RNH_2)Cl_2] \quad (155)$$

Physical properties of carbonyl complexes

The bonding in metal–carbonyl complexes consists of a σ-bond formed by overlap of a filled sp-hybrid orbital on the carbon atom with a vacant hybrid orbital on the metal, complemented by π-back donation of electron density from filled metal d or dp hybrid orbitals to empty π^*-(antibonding) orbitals on the carbon monoxide. This two-way electron transfer process effectively raises carbon monoxide to an electronically excited state.[125]

The presence of electrons in the π^*-(antibonding) orbitals of carbon monoxide weakens the C–O bond and hence lowers the C–O stretching frequency, which in free carbon monoxide occurs at 2155 cm^{-1}. A steady decrease in the C–O stretching frequency in platinum(II) halogen carbonyl complexes occurs as the atomic number of the halogen is increased (Table 25), reflecting

TABLE 25
Carbon–oxygen stretching frequencies in platinum(II)–carbonyl complexes

Complex	X = Cl	X = Br	X = I	Reference
Cs$^+$[PtX$_3$(CO)]	2132 cm^{-1}	2112 cm^{-1}	2088 cm^{-1}	126
[PtX$_2$(CO)]$_2$	2152 cm^{-1}	2130 cm^{-1}	2112 cm^{-1}	127
[(dipyr)PtX(CO)]	2145 cm^{-1}	2132 cm^{-1}	2120 cm^{-1}	115
cis-[(PEt$_3$)PtX$_2$(CO)]	2100 cm^{-1}	2094 cm^{-1}	2085 cm^{-1}	128

the increasing importance of π-back donation from platinum(II) to carbon monoxide as the electronegativity of the halogen decreases. The carbonyl stretching frequency in a series of complexes *trans*-[Pt(CO)LCl$_2$] where L is a 4-substituted pyridine or a 4-substituted pyridine-N-oxide increases as the electron-withdrawing power of the 4-substituent increases.[129,130]

There is a decrease in the metal–carbon stretching frequency in Cs$^+$-[PtX$_3$(CO)]$^-$ with increasing halogen atomic number[126] ($v_{Pt-C} = 498$ cm^{-1} (Cl), 495 cm^{-1}(Br) and 486 cm^{-1}(I)). However, the assignment of this band is not universally agreed upon (see ref. 131), nor should it be expected to vary very much with charge in halogen because of the presence of two opposing effects: first a decrease in the electronegativity of the halogen should increase the strength of the platinum–carbon bond and hence increase v_{Pt-C} in the order Cl < Br < I, and secondly an increase in the mass of the halogen should decrease v_{Pt-C} in the order Cl < Br < I.

Carbon monoxide has no *trans*-influence (see pp. 301–303 and Table 91 as well as ref. 62). Its high position in the *trans*-effect series (p. 299) is therefore due to substantial π-bonding in the metal–carbonyl bond. This strong π-bonding causes [Pt(CO)$_2$Cl$_2$][132] and [(PEt$_3$)Pt(CO)X$_2$][104,122,128] to be *cis* rather than *trans* since this ensures that three rather than two d-orbitals are involved in the π-bonding.

Chemical properties of carbonyl complexes

(i) *Hydrolysis.* The resistance of the platinum(II)–carbonyl complexes to hydrolysis decreases in the order [PtX$_3$(CO)]$^-$ > [PtX$_2$(CO)]$_2$ > [PtX$_2$-(CO)$_2$].[133] The corresponding order for palladium(II) complexes is unknown. However, [PdCl$_2$(CO)]$_2$ is slowly decomposed by moisture at room temperature.[116] When *trans*-[(PEt$_3$)$_2$Pt(CO)Cl]BF$_4$ is dissolved in aqueous acetone the resulting solution is acidic. This may be due to reaction156[134] since, although no direct evidence for the carboxylate complex has been

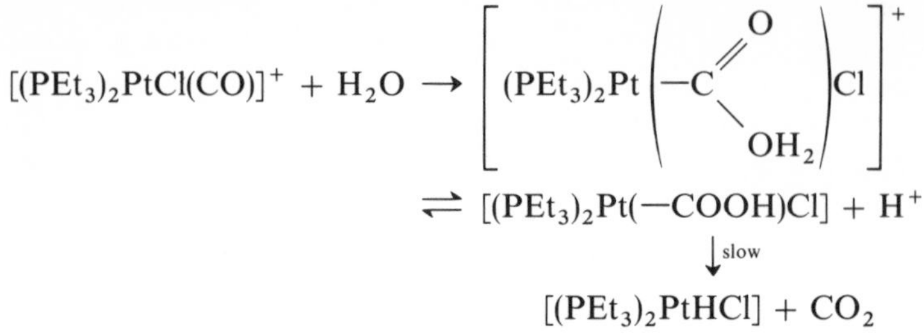

$$[(PEt_3)_2PtCl(CO)]^+ + H_2O \rightarrow \left[(PEt_3)_2Pt \left(-C \underset{OH_2}{\overset{O}{\diagup}} \right) Cl \right]^+$$

$$\rightleftharpoons [(PEt_3)_2Pt(-COOH)Cl] + H^+$$

$$\downarrow \text{slow}$$

$$[(PEt_3)_2PtHCl] + CO_2 \qquad (156)$$

obtained, carboxylic acid esters can be isolated when *trans*-[(PEt$_3$)$_2$PtCl-(CO)]BF$_4$ is treated with methanol or ethanol at 25°C[135] (reaction 157). These ester complexes exchange ester groups with other alcohols (reaction 158) and yield hydrido-complexes with water in the presence of an ammonium

$$\textit{trans-}[(PEt_3)_2PtCl(CO)]^+ + ROH \underset{k_{-1}}{\overset{k_1}{\rightleftharpoons}}$$

$$\textit{trans-}[(PEt_3)_2Pt(-COOR)Cl] + H^+ \quad (157)$$

$$[(PPh_3)_2Pt(-COOR)Cl] + R'OH \rightleftharpoons$$

$$[(PPh_3)_2Pt(-COOR')Cl] + ROH \quad (158)$$

or potassium halide salt as a catalyst[136] (reaction 159). The kinetics of the reversible reaction 157 suggest that the mechanism involves attack of the

$$[(PPh_3)_2Pt(-COOR)Cl] + H_2O \xrightarrow[\text{catalyst}]{+ NH_4Cl \text{ or } KCl}$$

$$\textit{trans-}[(PPh_3)_2PtHCl] \quad (159)$$

alcohol on the co-ordinated carbon monoxide ligand (reaction 160).[137] The k_{-1} rate constants for reaction 157 are almost independent of the

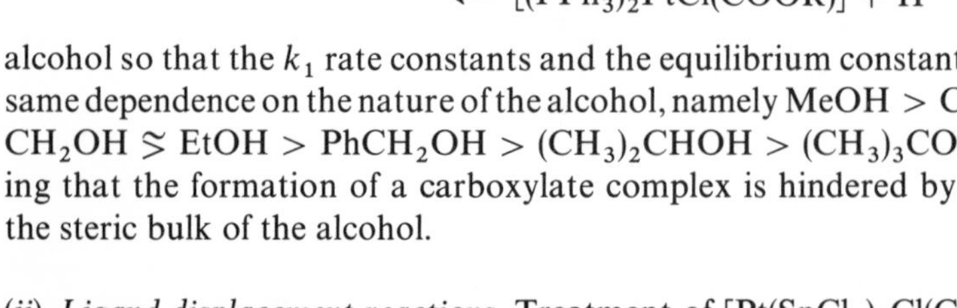

$$[(PPh_3)_2PtCl(CO)]^+ + ROH \rightleftharpoons \left[(PPh_3)_2ClPt -C \underset{\underset{H}{\overset{|}{O-R}}}{\overset{O}{\diagup}} \right]^+$$

$$\rightleftharpoons [(PPh_3)_2PtCl(COOR)] + H^+ \qquad (160)$$

alcohol so that the k_1 rate constants and the equilibrium constants show the same dependence on the nature of the alcohol, namely MeOH > CH$_3$OCH$_2$-CH$_2$OH $\gtrsim$ EtOH > PhCH$_2$OH > (CH$_3$)$_2$CHOH > (CH$_3$)$_3$COH, indicating that the formation of a carboxylate complex is hindered by increasing the steric bulk of the alcohol.

(ii) *Ligand displacement reactions.* Treatment of [Pt(SnCl$_3$)$_2$Cl(CO)]$^-$ complexes with triphenylphosphine, 1,10-phenanthroline and pyridine yields [(PPh$_3$)$_2$Pt(SnCl$_3$)$_2$(CO)], [Pt(SnCl$_3$)$_2$(phen)(CO)] and [Pt(SnCl$_3$)$_2$(pyr)(CO)] respectively by displacement of the chloride ligand.[120] Although it was not reported whether an excess of triphenylphosphine had been tried, triphenylphosphine did react with $(NEt_4)^+[Pd(SnCl_3)_2Cl(CO)]^-$ to give $[(PPh_3)_2$-$PdCl_2]$, which is similar to the reaction observed when platinum(II)–carbonyl complexes are treated with excess tertiary arsine[127] (reaction 161).

$$[(AsMePh_2)PtCl_2(CO)] + AsMePh_2 \rightarrow [(AsMePh_2)_2PtCl_2] + CO \quad (161)$$

(iii) Halogen exchange reactions. The complexes $[PtX_2(CO)]_2$ and $[PtX_3$-$(CO)]^-$, where X = bromide or iodide, can be obtained by treating the corresponding chloro-complexes with the appropriate hydrogen halide.[121]

(iv) Bridge cleavage reactions. Although bridge cleavage reactions with monodentate ligands such as *p*-toluidine, ammonia, pyridine, phosphorus trichloride and tertiary arsines occur symmetrically as shown in reaction 162, unsymmetrical cleavage of the type shown in reaction 163 occurs with

$$[PtCl_2(CO)]_2 + 2L \rightarrow 2[PtCl_2L(CO)] \quad (162)$$

$$[PtCl_2(CO)]_2 + dipyr \rightarrow$$

$$[PtCl(dipyr)(CO)]^+[PtCl_3(CO)]^- \quad (163)$$

bidentate ligands such as 2,2'-dipyridyl.[127,138]

π-Cyclopentadienyl complexes

π-Cyclopentadienyl complexes of platinum and palladium are relatively rare. They can be prepared by treating halogen bridged platinum(II) or palladium(II) complexes with sodium, thallium or mercury cyclopentadienides[139–41] (reactions 164–6). The monomeric complexes $[(PR_3)_2MX_2]$, where M = Pd or Pt, R = Et or Ph and X = Cl, Br or I all give decomposi-

$$[(PPh_3)_2Pd_2Br_4] + 2C_5H_5Tl \xrightarrow[\text{30 min.}]{\text{THF 20°C}}$$

$$2[(\pi\text{-}C_5H_5)(PPh_3)PdBr] + 2TlBr \quad (164)$$

$$[(PEt_3)_2Pt_2Br_2Ph_2] + 2C_5H_5Tl \xrightarrow[\text{30 min.}]{\text{THF 20°C}}$$

$$2[(\pi\text{-}C_5H_5)(PEt_3)PtPh] + 2TlBr \quad (165)$$

$$[(\pi\text{-}C_3H_5)PdCl]_2 + 2C_5H_5Na \xrightarrow{\text{THF/benzene}}$$

$$2[(\pi\text{-}C_3H_5)Pd(\pi\text{-}C_5H_5)] + 2NaCl \quad (166)$$

tion although $[Pt(SMe_2)_2Cl_2]$ does react with sodium cyclopentadienide to yield *cis*- and *trans*-$[(\pi\text{-}C_5H_5)_2Pt(SMe_2)_2]$.[142] Treatment of the mixture of platinum carbonyl chlorides, obtained by heating platinum(II) chloride in a stream of carbon monoxide, with sodium cyclopentadienide yields $[(\pi\text{-}C_5H_5)Pt(CO)]_2$[143] and treatment of this dimeric complex with iodine in the absence of air gave $[(\pi\text{-}C_5H_5)PtI(CO)]$.[140] $[(\pi\text{-}C_5H_5)Pt(NO)]$ is prepared by treating the product obtained by stirring $[Pt_2(CO)_2Cl_4]$ under nitric oxide at 50°C for a week with sodium cyclopentadienide.[144]

The nitrosyl and carbonyl complexes are unstable in air and the thioether complexes are unstable in solution at room temperature even in the absence of air. The halides, however, are indefinitely stable in air at room temperature;

the palladium halides being green and the platinum analogues orange. The mass spectrum of the complex $[(\pi\text{-}C_5H_5)Pd(\pi\text{-}C_3H_5)]$, which is very stable, has been reported,[145] but the remaining properties of this complex await general publication.

All the π-cyclopentadienyl complexes exhibit a single proton resonance in the n.m.r. spectrum between 3.71 and 5.40 p.p.m. downfield from TMS,[139,142] which is consistent with symmetrical π-bonding between the cyclopentadienyl ring and the metal. Coupling to the ^{195}platinum nucleus splits the proton resonance with a coupling constant of between 12 and 20 Hz.[139] In those complexes that contain phosphorus, $^{31}P\text{-}^1H$ coupling further splits the proton resonance with a coupling constant of about 2 Hz.[139] It is very likely that many more π-cyclopentadienyl complexes of platinum(II) and palladium(II) will be reported in the near future because a large number of nickel(II) complexes are known and preliminary work has suggested that both the thermal and chemical stabilities of π-cyclopentadienyl complexes increase in the order Ni < Pd < Pt.[139]

REFERENCES

1 J. Lewis and R. S. Nyholm, *Sci. Progr. (London)*, **52** (1964) 557.
2 V. I. Baranovskii, V. P. Sergeev and B. E. Dzevitskii, *Dokl. Akad. Nauk. SSSR*, **184** (1969) 632; *Chem. Abs.*, **70** (1969) 82777.
3 J. F. Young, *Adv. Inorg. Radiochem.*, **11** (1968) 91.
4 R. J. Cross, *Organometal Chem. Rev.*, **2** (1967) 97.
5 M. C. Baird, *Prog. Inorg. Chem.*, **9** (1968) 1.
6 N. S. Vyazankin, G. A. Razuvaev and O. A. Kruglaya, *Organometal. Chem. Rev.*, **A3** (1968) 323.
7 F. Glockling, *The Chemistry of Germanium*, Academic Press, London, 1969, chapter 8.
8 J. Chatt, C. Eaborn, S. D. Ibekwe and P. N. Kapoor, *J. Chem. Soc. (A)*, (1970) 1343.
9 E. H. Brooks and R. J. Cross, *Organometal Chem. Rev.*, **A6** (1970) 227.
10 M. C. Baird, *J. Inorg. Nucl. Chem.*, **29** (1967) 367.
11 E. H. Brooks and F. Glockling, *Chem. Comm.*, (1965) 510.
12 E. H. Brooks and F. Glockling, *J. Chem. Soc. (A)*, (1966) 1241.
13 R. J. Cross and F. Glockling, *Proc. Chem. Soc.*, (1964) 143.
14 R. J. Cross and F. Glockling, *J. Chem. Soc.*, (1965) 5422.
15 R. J. Cross and F. Glockling, *J. Organometal Chem.*, **3** (1965) 253.
16 G. Deganello, G. Carturan and U. Belluco, *J. Chem. Soc. (A)*, (1968) 2873.
17 G. Carturan, G. Deganello, T. Boschi and U. Belluco, *J. Chem. Soc. (A)*, (1969) 1142.
18 F. Glockling and K. A. Hooton, *Chem. Comm.*, (1966) 218.
19 F. Glockling and K. A. Hooton, *J. Chem. Soc. (A)*, (1967) 1066.
20 G. Deganello, G. Carturan and P. Uguagliati, *J. Organometal. Chem.*, **17** (1969) 179.
21 A. F. Clemmit and F. Glockling, *J. Chem. Soc. (A)*, (1971) 1164.
22 J. Chatt, C. Eaborn, S. Ibekwe and P. N. Kapoor, *Chem. Comm.*, (1967) 869.
23 J. E. Bentham, S. Cradock and E. A. V. Ebsworth, *J. Chem. Soc. (A)*, (1971) 587.
24 F. Glockling and K. A. Hooton, *J. Chem. Soc. (A)*, (1968) 826.
25 J. E. Bentham and E. A. V. Ebsworth, *Inorg. Nucl. Chem. Lett.*, **6** (1970) 671.
26 J. Chatt and B. L. Shaw, *J. Chem. Soc.*, (1962) 5075.
27 J. E. Bentham, S. Cradock and E. A. V. Ebsworth, *Chem. Comm.*, (1969) 528.
28 D. J. Cardin and M. F. Lappert, *Chem. Comm.*, (1966) 506.
29 D. J. Cardin, S. A. Keppie and M. F. Lappert, *J. Chem. Soc. (A)*, (1970) 2594.
30 E. H. Brooks and F. Glockling, *J. Chem. Soc. (A)*, (1967) 1030.
31 J. Chatt, C. Eaborn and P. N. Kapoor, *J. Chem. Soc. (A)*, (1970) 881.

32 J. Chatt, C. Eaborn and P. N. Kapoor, *J. Organometal. Chem.*, (1968) 13, P21.
33 A. J. Layton, R. S. Nyholm, G. A. Pneumaticakis and M. L. Tobe, *Chem. Ind. (London)*, (1967) 465.
34 J. P. Birk, J. Halpern and A. L. Pickard, *Inorg. Chem.*, 7 (1968) 2672.
35 K. Yamamoto, T. Hayashi and M. Kumada, *J. Organometal. Chem.*, 28 (1971) C37.
36 M. Akhtar and H. C. Clark, *J. Organometal. Chem.*, 22 (1970) 233.
37 A. F. Clemmit and F. Glockling, *Chem. Comm.*, (1970) 705.
38 E. H. Brooks, R. J. Cross and F. Glockling, *Inorg. Chim. Acta.*, 2 (1968) 17.
39 C. S. G. Phillips and R. J. P. Williams, *Inorganic Chemistry*, Oxford, 1966, vol. 1, chapter 8; vol. 2, chapter 30.
40 D. M. Adams, J. Chatt, J. Gerratt and A. D. Westland, *J. Chem. Soc.*, (1964) 734.
41 U. Belluco, U. Croatto, P. Uguagliati and R. Pietropaolo, *Inorg. Chem.*, 6 (1967) 718.
42 G. Deganello, G. Carturan and P. Uguagliati, *J. Organometal. Chem.*, 17 (1969) 179.
43 J. K. Wittle and G. Urry, *Inorg. Chem.*, 7 (1968) 560.
44 R. D. Cramer, E. L. Jenner, R. V. Lindsey and U. G. Stolberg, *J. Amer. Chem. Soc.*, 85 (1963) 1691.
45 R. D. Cramer, R. V. Lindsey, C. T. Prewitt and U. G. Stolberg, *J. Amer. Chem. Soc.*, 87 (1965) 658.
46 J. F. Young, R. D. Gillard and G. Wilkinson, *J. Chem. Soc.*, (1964) 5176.
47 R. V. Lindsey, G. W. Parshall and U. G. Stolberg, *Inorg. Chem.*, 5 (1966) 109.
48 L. J. Guggenberger, *Chem. Comm.*, (1968) 512.
49 W. M. Latimer, *The Oxidation States of the Elements and their Potentials in Aqueous Solutions*, Prentice-Hall Inc., New York, 2nd edn., 1952, chapter 8.
50 G. H. Ayres and J. H. Alsop, *Anal. Chem.*, 31 (1959) 1135.
51 V. I. Shlenskaya, A. A. Biryukov and L. N. Moryakova, *Russ. J. Inorg. Chem.*, 14 (1969) 255.
52 M. A. Khattak and R. J. Magee, *Chem. Comm.*, (1965) 400.
53 G. E. Batley and J. C. Bailar, *Inorg. Nucl. Chem. Lett.*, 4 (1968) 577.
54 B. Crociani, T. Boschi and U. Belluco, *Proc. Third Int. Inorg. Chim. Acta Symposium*, Venice, (1970) B.1.
55 J. Chatt and B. L. Shaw, *J. Chem. Soc.*, (1959) 705.
56 L. Pauling, *The Nature of the Chemical Bond*, Cornell University Press, 3rd. ed., (1960).
57 B. T. Heaton and A. Pidcock, *J. Organometal. Chem.*, 14 (1968) 235.
58 M. P. Johnson, D. F. Shriver and S. A. Shriver, *J. Amer. Chem. Soc.*, 88 (1966) 1588.
59 G. W. Parshall, *J. Amer. Chem. Soc.*, 88 (1966) 704.
60 P. Taimsalu and J. L. Wood, *Spectrochim Acta*, 20 (1964) 1043.
61 G. W. Parshall, *J. Amer. Chem. Soc.*, 86 (1964) 5367.
62 M. J. Church and M. J. Mays, *J. Chem. Soc. (A)*, (1968) 3074.
63 A. Pidcock, R. E. Richards and L. M. Venanzi, *J. Chem. Soc. (A)*, (1966) 1707.
64 R. Mason and P. O. Whimp, *J. Chem. Soc. (A)*, (1969) 2709.
65 R. Mason, Personal communication.
66 C. K. Jørgensen, *Absorption Spectra and Chemical Bonding in Complexes*, Pergamon, Oxford, 1962.
67 G. H. Bailey and T. Lamb, *J. Chem. Soc.*, 61 (1892) 745.
68 M. Frenkel, *Z. anorg. allgem. Chem.*, 1 (1892) 217.
69 H. Fehling, *Liebig's Ann. Chem.*, 39 (1841) 110.
70 F. Feigl and G. B. Heisig, *J. Amer. Chem. Soc.*, 73 (1951) 5631.
71 R. B. Jones, *J. Amer. Chem. Soc.*, 57 (1935) 471.
72 M. F. A. El-Sayed and R. K. Sheline, *J. Inorg. Nucl. Chem.*, 6 (1958) 187.
73 D. A. Dows, A. Haim and W. K. Wilmarth, *J. Inorg. Nucl. Chem.*, 21 (1961) 33.
74 J. H. Brigelow, *Inorg. Synth.*, 2 (1946) 245.
75 R. Nast and W. Hörl, *Chem. Ber.*, 95 (1962) 1470.
76 J.-P. Mathieu and H. Poulet, *Compt. Rend.*, 248 (1959) 2315.
77 J.-P. Mathieu and S. Cornevin, *J. Chem. Phys.*, 36 (1939) 271.
78 C. W. F. T. Pistorius, *Z. Physik. Chem. (Leipzig)*, 23 (1960) 206.
79 D. F. Evans, D. Jones and G. Wilkinson, *J. Chem. Soc.*, (1964) 3164.

80   A. A. Grinberg, *Izv. Inst. Izuch Platiny*, **6** (1928) 155.
81   Yu. Ya. Kharitonov, O. N. Evstaf'eva, I. B. Baranovskii, G Ya Mazo and A. V. Babaeva, *Russ. J. Inorg. Chem.*, **11** (1966) 927.
82   B. M. Chadwick and A. G. Sharpe, *Adv. Inorg. Radiochem.*, **8** (1966) 83.
83   D. M. Sweeney, I. Nakagawa, S. Mizushima and J. V. Quagliano, *J. Amer. Chem. Soc.*, **78** (1956) 889.
84   R. M. Izatt, G. D. Watt, D. Eatough and J. J. Christenson, *J. Chem. Soc. (A)*, (1967) 1304.
85   A. A. Grinberg and M. I. Gel'man, *Dokl. Akad. Nauk. SSSR*, **133** (1960) 1081; *Chem. Abs.*, **54** (1960) 23632.
86   I. I. Chernyaev, N. N. Zheligovskaya and A. V. Babkov, *Russ. J. Inorg. Chem.*, **6** (1961) 1328.
87   I. I. Chernyaev, A. V. Babkov and N. N. Zheligovskaya, *Russ. J. Inorg. Chem.*, **8** (1963) 703.
88   P. M. Treichel and R. W. Hess, *Chem. Comm.*, (1970) 1626.
89   H. B. Gray and C. J. Ballhausen, *J. Amer. Chem. Soc.*, **85** (1963) 260.
90   C. Moncuit, *J. Chim. Phys.*, **64** (1967) 494; *Chem. Abs.*, **67** (1967) 37812.
91   P. J. Stephens, A. J. McCaffery and P. N. Schatz, *Inorg. Chem.*, **7** (1968) 1923.
92   W. R. Mason and H. B. Gray, *J. Amer. Chem. Soc.*, **90** (1968) 5721.
93   S. B. Piepho, P. N. Schatz and A. J. McCaffery, *J. Amer. Chem. Soc.*, **91** (1969) 5994.
94   P. Day, M. J. Smith and R. J. P. Williams, *J. Chem. Soc. (A)*, (1968) 668.
95   J. L. Glatch and W. R. Mason, *Inorg. Nucl. Chem. Lett.*, **6** (1970) 75.
96   M. Angoletta, *Ann. Chim. (Italy)*, **45** (1955) 970; *Chem. Abs.* **50** (1956) 8362h.
97   B. Crociano, T. Boschi and U. Belluco, *Inorg. Chem.*, **9** (1970) 2021.
98   L. Chugaev, *Compt. Rend.*, **159** (1914) 188.
99   L. Chugaev and P. Teearu, *Chem. Ber.*, **47** (1914) 568.
100 K. A. Hofmann and G. Bugge, *Chem. Ber.*, **40** (1907) 1772.
101 L. Ramberg, *Chem. Ber.*, **40** (1907) 2578.
102 K. A. Jenson, *Z. anorg. allgem. Chem.*, **231** (1937) 365.
103 L. Malatesta and F. Bonati, *Isocyanide Complexes of Metals*, John Wiley, London, (1969) p. 169.
104 E. M. Badley, J. Chatt, R. L. Richards and G. A. Sim, *Chem. Comm.*, (1969) 1322.
105 F. Bonati, G. Minghetti, T. Boschi and B. Crociani, *J. Organometal. Chem.*, **25** (1970) 255.
106 E. M. Badley, J. Chatt and R. L. Richards, *J. Chem. Soc. (A)*, (1971) 21.
107 W. J. Knebel and P. M. Treichel, *Chem. Comm.*, (1971) 516.
108 F. A. Cotton, T. G. Dunne and J. S. Wood, *Inorg. Chem.*, **4** (1965) 318.
109 F. A. Cotton and F. Zingales, *J. Amer. Chem. Soc.*, **83** (1961) 351.
110 J. M. Lutton and R. W. Parry, *J. Amer. Chem. Soc.*, **76** (1954) 4271.
111 L. Malatesta and L. Naldini, *Gazz. Chim. Ital.*, **90** (1960) 1505.
112 W. Pullinger, *Chem. Ber.*, **24** (1891) 229.
113 A. D. Gel'man, *Dokl. Akad, Nauk SSSR.*, **31** (1941) 761; *Chem. Abs.*, **37** (1943) 843.
114 A. D. Gel'man and E. Meilakh, *Dokl. Akad. Nauk. SSSR.*, **36** (1942) 171; *Chem. Abs.*, **37** (1943) 2678.
115 R. J. Irving and E. A. Magnusson, *J. Chem. Soc.*, (1958) 2283.
116 W. Manchot and J. König, *Chem. Ber.*, **59** (1926) 883.
117 J. V. Kingston and G. R. Scollary, *Chem. Comm.*, (1969) 455.
118 M. J. Church and M. J. Mays, *Chem. Comm.*, (1968) 435.
119 H. C. Clark and K. R. Dixon, *J. Amer. Chem. Soc.*, **91** (1969) 596.
120 J. V. Kingston and G. R. Scollary, *Chem. Comm.*, (1970) 362.
121 F. Mylius and F. Foerster, *Chem. Ber.*, **24** (1891) 2424.
122 A. C. Smithies, M. Rycheck and M. Orchin, *J. Organometal Chem.*, **12** (1968) 199.
123 R. Ellis, T. A. Weil and M. Orchin, *J. Amer. Chem. Soc.*, **92** (1970) 1078.
124 T. Theophanides and P. C. Kong, *Can. J. Chem.*, **48** (1970) 1084.
125 E. W. Abel and F. G. A. Stone, *Quart. Rev.*, **23** (1969) 325.
126 M. J. Cleare and W. P. Griffith, *J. Chem. Soc. (A)*, (1970) 2788.
127 R. J. Irving and E. A. Magnusson, *J. Chem. Soc.*, (1956) 1860.

128   J. Chatt, N. P. Johnson and B. L. Shaw, *J. Chem. Soc.*, (1964) 1662.

129   A. R. Brause, M. Rycheck and M. Orchin, *J. Amer. Chem. Soc.*, **89** (1967) 6500.

130   T. A. Weil, P. Schmidt, M. Rycheck and M. Orchin, *Inorg. Chem.*, **8** (1969) 1002.

131   R. G. Denning and M. J. Ware, *Spectrochim. Acta*, **24A** (1968) 1785.

132   J. Chatt and A. A. Williams, *J. Chem. Soc.*, (1951) 3061.

133   F. Calderazzo in *Halogen Chemistry*, ed. by V. Gutmann, Academic Press, London, vol. 3, (1967) p. 383.

134   H. C. Clark and W. J. Jacobs, *Inorg. Chem.*, **9** (1970) 1229.

135   H. C. Clark, K. R. Dixon and W. J. Jacobs, *Chem. Comm.*, (1968) 548.

136   H. C. Clark, K. R. Dixon and W. J. Jacobs, *J. Amer. Chem. Soc.*, **91** (1969) 1346.

137   J. E. Byrd and J. Halpern, *J. Amer. Chem. Soc.*, **93** (1971) 1634.

138   R. J. Irving and E. A. Magnusson, *J. Chem. Soc.*, (1957) 2018.

139   R. J. Cross and R. Wardle, *J. Organometal. Chem.*, **23** (1970) C4.

140   B. L. Shaw, *Proc. Chem. Soc.*, (1960) 247.

141   E. O. Fischer and H. Werner, *Chem. Ber.*, **95** (1962), 695.

142   H. P. Fritz and K. E. Schwarzhans, *J. Organometal. Chem.*, **5** (1966) 181.

143   E. O. Fischer, H. Schuster-Woldan and K. Bittler, *Z. Naturforsch.*, **18B** (1963) 429.

144   E. O. Fischer and H. Schuster-Woldan, *Z. Naturforsch.*, **19B** (1964) 766.

145   N. V. Zakurin, S. P. Gubin and V. P. Bochin, *J. Organometal. Chem.*, **23** (1970) 535.

CHAPTER 7

Complexes with Group VB Elements

Here complexes containing nitrogen, phosphorus, arsenic and antimony bound directly to platinum(II) and palladium(II) are considered. The most important features of the group VB ligands are: first, the considerable ability of nitrogen in almost any environment to bind strongly to both metals which has led to the preparation of complexes with a wide range of nitrogen ligands; secondly, the key position that the tertiary phosphine ligands have played in the development of the organometallic chemistry of platinum and palladium through their ability to form very stable complexes with both metals, and thirdly, whilst all four group VB elements form strong σ-bonds, phosphorus, arsenic and antimony are also potentially capable of accepting back-donation of electron density from the metal through π-bonds possibly involving the empty d-orbitals of these ligands.

This possibility of π-bonding accounts for the considerable differences in the properties of nitrogen ligands as compared with ligands containing the heavier group VB elements, which has dictated that this chapter be split to enable the nitrogen complexes to be discussed separately. This is done in the first part of the chapter where each class of nitrogen ligand is considered with the emphasis being placed on the structure and bonding in their complexes. In the second part of the chapter the complexes of phosphorus, arsenic and antimony are considered, and since it is currently the subject of much controversy, the evidence concerning π-bonding in these metal–ligand bonds is considered in some detail. The final part of the chapter considers complexes in which the group VB ligand acts as a bidentate ligand with one of the carbon atoms of the ligand being the second donor atom.

NITROGEN COMPLEXES

Amines
The bis–amine complexes, $[PtL_2X_2]$, where L is ammonia or an amine and X is a halogen, which exist as *cis*- and *trans*-isomers, are prepared in accordance with the *trans*-effect series (p. 299).[1,2] Thus heating $[Pt(NH_3)_4]Cl_2$ yields *trans*-$[Pt(NH_3)_2Cl_2]$[3] (reaction 167), whereas treating a tetrachloroplatinate(II) salt with a buffered ammonia solution yields the *cis*-isomer[4] (reaction 168).

112

$$\left[\begin{array}{c} H_3N \quad\quad NH_3 \\ \diagdown\;Pt\;\diagup \\ \diagup\quad\quad\diagdown \\ H_3N \quad\quad NH_3 \end{array}\right]^{2+} 2Cl^- \xrightarrow[-NH_3]{\text{heat}} \left[\begin{array}{c} H_3N \quad\quad Cl \\ \diagdown\;Pt\;\diagup \\ \diagup\quad\quad\diagdown \\ H_3N \quad\quad NH_3 \end{array}\right]^{+} Cl^-$$

$$\xrightarrow[-NH_3]{\text{heat}} \left[\begin{array}{c} H_3N \quad\quad Cl \\ \diagdown\;Pt\;\diagup \\ \diagup\quad\quad\diagdown \\ Cl \quad\quad NH_3 \end{array}\right] \tag{167}$$

$$[PtCl_4]^{2-} + NH_3 \rightarrow \left[\begin{array}{c} Cl \quad\quad Cl \\ \diagdown\;Pt\;\diagup \\ \diagup\quad\quad\diagdown \\ Cl \quad\quad NH_3 \end{array}\right]^{-} \xrightarrow{NH_3} \left[\begin{array}{c} Cl \quad\quad NH_3 \\ \diagdown\;Pt\;\diagup \\ \diagup\quad\quad\diagdown \\ Cl \quad\quad NH_3 \end{array}\right] + Cl^-$$

$$\tag{168}$$

Palladium(II) complexes undergo *cis-trans*-isomerisation more rapidly than platinum(II) complexes (p. 313). As a result, until recently, only *trans*-$[Pd(NH_3)_2X_2]$, where $X = Cl$, Br or I, could be prepared pure.[5,6] However, the pure *cis*-isomer can be prepared[7] by leaving an aqueous solution of $[Pd(NH_3)_4](ClO_4)_2$ to stand in excess aqueous perchloric acid, when two molecules of ammonia are replaced by water giving *cis*-$[Pd(NH_3)_2(H_2O)_2]^{2+}$. This is then treated with sodium halide to precipitate *cis*-$[Pd(NH_3)_2X_2]$. The *cis*-dichloride and *cis*-dibromide complexes so prepared are stable for 14 months, whereas the *cis*-diiodide complex had begun to isomerise after a month and after 14 months had isomerised completely. *Cis*-$[Pd(NH_3)_2(SCN)_2]$ and *cis*-$[Pd(NH_3)_2(NO_2)_2]$ complexes could not be prepared indicating that *cis*-$[Pd(NH_3)_2L_2]$ complexes are stable with ligands that lie low in the *trans*-effect series (e.g. $L = H_2O$, Cl and Br), whereas with ligands that lie higher in the *trans*-effect series they are either formed reluctantly ($L = I$) or not at all ($L = SCN$ or NO_2).

In addition to the *cis*- and *trans*-isomers of $[ML_2X_2]$ where $M = Pd$ or Pt, $L = $ amine and $X = $ halogen, ionic complexes of the type $[ML_4]^{2+}[MX_4]^{2-}$ are also known.[8] They are prepared by treating $[M(NH_3)_4]X_2$ with M'_2MX_4 where $M = Pd$ or Pt and M' is an alkali metal,[9,10] and possess anomalous spectral and conductivity properties (pp. 71–73).

Complexes containing only one amino-ligand are prepared by treating the halogen-bridged complexes with amines[11-13] (reaction 169). Equilibrium (169) lies to the right for $X = Cl$ and Br, but to the left for $X = I$

$$[L_2M_2X_4] + 2 \text{ amine} \rightleftharpoons 2[ML(\text{amine})X_2] \tag{169}$$
$$\text{(M = Pd or Pt; L = neutral}$$
$$\text{ligand; X = halide)}$$

Equilibrium (169) lies to the right for $X = Cl$ and Br, but to the left for $X = I$ so that isolation of the $[ML(\text{amine})I_2]$ complexes is difficult even in the presence of excess amine. The product of reaction 169 is always *trans* for platinum(II) complexes except for the special case in which L is an amine when a *cis-trans* mixture of products is obtained. With palladium(II) complexes a product [PdL(amine)X$_2$], which is always *trans*, can only be isolated when the ligand L has a phosphorus or arsenic donor atom. When L contains sulphur, selenium or tellurium donor atoms [PdL(amine)Cl$_2$] disproportionates to [PdL$_2$Cl$_2$] and [Pd(amine)$_2$Cl$_2$].

Amines bond to platinum(II) and palladium(II) by donation of a pair of electrons in an sp^3 hybrid orbital on the nitrogen to an empty orbital on the metal. There are no low-energy empty orbitals on the nitrogen suitable for π-back-donation so that the metal–nitrogen bond is a pure σ-bond. Hence ligands capable of π-bonding enhance the strength of the metal–nitrogen bond, and in the case of primary amines facilitate deprotonation of the co-ordinated amine.[14,15] A second consequence of the bonding in metal–amine complexes is that amines lie very low in both the *trans*-effect and *trans*-influence series, in direct contrast to the ligands of the other group VB elements. A further consequence of the metal–nitrogen bond being purely a σ-bond is that amines lie high in the spectrochemical series but low in the nephelauxetic series where π-bonding is important.[16] The increase in the electronic transition energies on replacing two chloride ligands in [MCl$_4$]$^{2-}$ by either two ammonia ligands or a diamine ligand is apparent from Table 26.

TABLE 26
Electronic spectra of platinum(II) and palladium(II) diamine complexes

cis-[Pt(NH$_3$)$_2$Cl$_2$]^a	trans-[Pt(NH$_3$)$_2$Cl$_2$]^a	[Pt(l-pn)Cl$_2$]^b	K$_2$PtCl$_4$^{b}	Assignment^c
29,800 cm^{-1}	26,600 cm^{-1}	27,300 cm^{-1}	21,000 cm^{-1}	Singlet $\rightarrow$ triplet
32,450 cm^{-1}	31,600 cm^{-1}	33,100 cm^{-1}	25,700 cm^{-1}	$d_{xy} \rightarrow d_{x^2-y^2}$
36,610 cm^{-1}	36,610 cm^{-1}	$\sim$37,000 cm^{-1}	30,500 cm^{-1}	$d_{yz}, d_{xz} \rightarrow d_{x^2-y^2}$

[Pd(en)Cl$_2$]^d	[Pt(l-pn)Cl$_2$]^b	K$_2$PdCl$_4$^{b}	Assignment^c
21,277 cm^{-1}	22,500 cm^{-1}	16,500 cm^{-1}	Singlet $\rightarrow$ triplet
27,030 cm^{-1}	26,200 cm^{-1}	21,000 cm^{-1}	$d_{xy} \rightarrow d_{x^2-y^2}$
29,998 cm^{-1}	28,700 cm^{-1}	23,700 cm^{-1}	$d_{yz}, d_{xz} \rightarrow d_{x^2-y^2}$
35,994 cm^{-1}	$\sim$39,000 cm^{-1}	30,500 cm^{-1}	$d_{z^2} \rightarrow d_{x^2-y^2}$
45,959 cm^{-1}	>46,000 cm^{-1}	35,700 cm^{-1}	Cl$_{(p_\pi)} \rightarrow d_{x^2-y^2}$
		45,190 cm^{-1}	Cl$_{(p_\sigma)} \rightarrow d_{x^2-y^2}$

^{a} Reference 17. ^{b} Reference 18 *l*-pn = *l*-propylenediamine. ^{c} Reference 18. ^{d} Reference 19.

Assignment[18,20] of these spectra, achieved with the aid of circular dichroism studies on the *l*-propylenediamine complexes, suggests that the relative energies of the *d*-orbitals in diamino–platinum(II) and –palladium(II) complexes are $d_{x^2-y^2} > d_{xy} > d_{xz}, d_{yz} > d_{z^2}$ which is identical to that found in the tetrachloro-complexes.[21]

An asymmetric centre in a co-ordinated amine perturbs the metal and gives rise to circular dichroism in the *d–d* transitions.[22] The platinum(II) complexes of bidentate asymmetric secondary amines such as H$_2$NCH$_2$CH$_2$-NHCH$_3$ have been resolved into their optically active forms.[23] Although optically active diamines form bis-complexes in which the ligands can, in principle, be bound in a number of conformations, an n.m.r. study of the bis(l-propylenediamine) complexes of platinum(II) and palladium(II) indicated that the chelate rings had a fixed λ-gauche conformation in aqueous solution at room temperature.[24]

The metal–nitrogen stretching frequencies in the infra-red spectra of *cis*- and *trans*-[M(am)$_2$X$_2$][6,7,25,26] and [ML(am)X$_2$][27–9] (Table 27) are con-

TABLE 27
Metal–nitrogen stretching frequencies in amine complexes (from refs. 25(Pd) and 26(Pt))

Trans-[M(NH$_3$)$_2$X$_2$]

X	ν_{Pt-N} (cm^{-1})	ν_{Pd-N} (cm^{-1})
Cl	509	496
Br	502	490
I	498	480

Cis-[M(NH$_3$)$_2$X$_2$]

X	ν_{Pt-N} (cm^{-1})		ν_{Pd-N} (cm^{-1})	
	sym.	*asym.*	*sym.*	*asym.*
Cl	508	517	476	495
Br	499	513	460	480
I	477	491	—	—

sistently higher for the platinum than the palladium complexes, indicating that the stretching force constants of the former are higher. The –NH$_3$ anti-symmetric deformation vibrations at about 1610 cm^{-1} are split by between 65 and 100 cm^{-1} in the platinum(II) complexes[26] and slightly broadened in the palladium(II) complexes.[6] Two suggestions, which cannot be distinguished at present, have been advanced to explain this: first this broadening or splitting could arise from steric hindrance to free rotation of the –NH$_3$ group about the metal–nitrogen bond,[30] and secondly this splitting could be due to ligand–ligand interactions through a filled d-orbital on the metal.[26] The latter suggestion can readily account for the greater splitting in the platinum as compared with the palladium complexes, since the $5d$ orbitals on platinum are spatially more extensive than the $4d$-orbitals on palladium giving rise to greater ligand–ligand interactions.

A considerable number of x-ray structures of palladium(II) and platinum(II) amine complexes have been determined (Appendix II, Structures 10, 11, 20, 29, 37, 46, 93, 94, 96, 117, 118, 122, 123, 124, 132, 137 and 174).

Although little stability constant data for amines is available, the equilibrium constant for reaction 170, K ($= [(C_2H_4)PtLCl_2][Cl^-]/[(C_2H_4)PtCl_3^-][L])$

$$[(C_2H_4)PtCl_3]^- + L \rightleftharpoons [(C_2H_4)PtLCl_2] + Cl^- \qquad (170)$$

in 0.2 M perchloric acid varied with L as follows:[31] L = NH$_3$, $K = 2 \times 10^5$;

L = $MeNH_2$, K = 13 × 10^5; L = Me_2NH, K = 3.4 × 10^5; L = Me_3N, $K \approx 10^3$; L = piperidine, K = 5 × 10^5; L = 1-methylpiperidine, K = 0.2 × 10^5. The stability order NH_3 < $MeNH_2$ > Me_2NH ≫ Me_3N arises from three effects. First the basic strengths of the amines, which reflect their electron donor abilities, vary in the order NH_3 ≪ $MeNH_2$ < Me_2NH ≫ NMe_3.[31] Secondly the stability of those complexes in which the amino-groups carry hydrogen atoms is probably enhanced by interaction of these hydrogen atoms either with the metal *d*-orbitals or with the chloride ligands giving a stability order of NH_3 > $MeNH_2$ > Me_2NH > Me_3N, and thirdly the stability of the trimethylamine complex is probably depressed by steric hindrance.

A wide range of multidentate amine ligands co-ordinate to platinum and palladium. Bidentate amines can give rise to several conformations of the ring formed on co-ordination. Variable temperature n.m.r. studies have indicated that for five- and six-membered rings conformational equilibrium is achieved rapidly.[32,33] Two potentially tetradentate ligands (**23** where

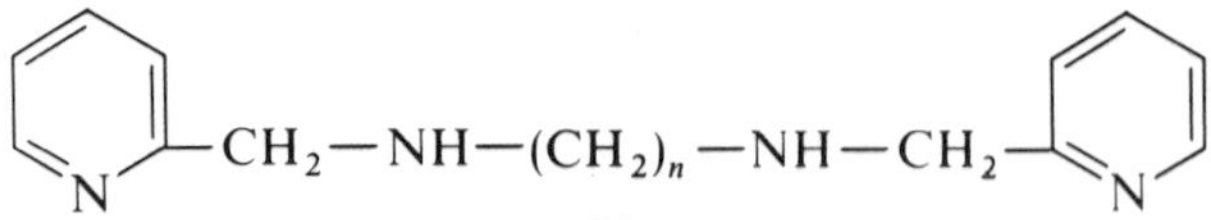

$$CH_2-NH-(CH_2)_n-NH-CH_2$$

23

n = 2 or 3) have been prepared and shown to act as tetradentate ligands towards palladium(II) when only perchlorate or hexafluorophosphate anions are present.[34] With halide ions, however, one of the pyridine rings is displaced so that the ligand only acts as a tridentate ligand. As a result of the high stability of amine complexes ethylenediaminetetraacetic acid normally binds to palladium(II) as a bidentate ligand via the nitrogen atoms.[35-8] Only when the concentrations of all other co-ordinating ligands, including hydroxide ions, is reduced below equimolar amounts is it possible for EDTA to act as either a tridentate (two nitrogen and one oxygen donor atoms) or quadridentate (two nitrogen and two oxygen donor atoms) ligand[35,36] Some special bidentate ligands such as the diazapropellanes **24** and **25** that co-ordinate to palladium but not to platinum have been described.[38a,38b] The stability constants (K = $[PdLCl_2]/[PdCl_2][L]$) were 8·25 × 10^9 and 2·16 × 10^{10} for **24** and **25** respectively. *N,N'*-ethylenedimorpholine (EDM-**26**) also forms complexes with palladium(II) but not with platinum(II).[39] Besides forming normal diamagnetic square-planar com-

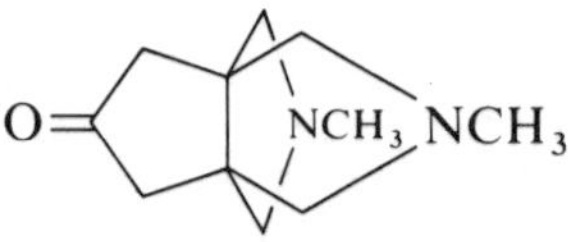

24 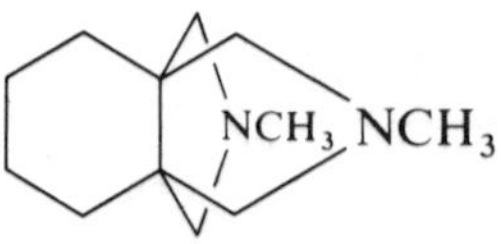   **25**

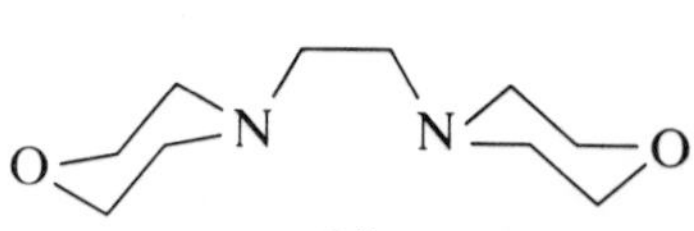

26

plexes $[Pd(EDM)X_2]$, where X = Cl, Br or I, a paramagnetic iodide complex, possibly $[Pd(EDM)_2]I_2$ with tetrahedral palladium(II), that is a 2:1 electrolyte, is formed.[39]

Aromatic heterocyclic nitrogen ligands

Aromatic heterocyclic ligands such as pyridine, bipyridyl and 1,10-phenanthroline form complexes with both platinum(II) and palladium(II) that are very similar to those formed by amines. However, these ligands have π-electron systems associated with their aromatic rings and therefore in addition to the σ-component there is the possibility of a π-component in the metal–nitrogen bond, so that these ligands show some properties that are reminiscent of tertiary phosphine ligands. For example, both pyridine[40] and bipyridyl[41] can stabilise metal(II)–carbon σ-bonds (p. 325).

Most of the complexes are planar, although recently complexes of the very hindered ligand 2,9-dimethyl-1,10-phenanthroline[42,43] **(27)** and the rigid ligand dipyrromethene[44] **(28)** have been prepared in which the plane

H₃C **27** CH₃   **28**

of the ligand lies out of the square-plane around the metal atom (Appendix II, structures 57 and 85).

Normally when a bidentate ligand such as ethylenediamine reacts with $[PtCl_4]^{2-}$ or $[PdCl_4]^{2-}$, one chloride ligand is substituted by one end of the bidentate ligand in a slow rate-determining step and then the other end of the bidentate ligand displaces a second chloride ion rapidly,[45,46] both steps occurring by a mechanism similar to that for a unidentate ligand, the greater speed of the second step being the result of a statistical advantage. Bipyridyl reacts with $[PtCl_4]^{2-}$ in this manner, although the rate is 33 times slower than for ethylene diamine due to the steric crowding in the initial product in which bipyridyl acts as a unidentate ligand.[47] However, 1,10-phenanthroline is too rigid a ligand to be able to form a complex in which it is unidentate and so it reacts with $[MCl_4]^{2-}$ ions by a different mechanism. The kinetics of this reaction indicate that it is first order in both 1,10-phenanthroline and $[MCl_4]^{2-}$ ions. No term that was first-order in $[MCl_4]^{2-}$ but zero order in 1,10-phenanthroline was found.[47-9] Although the detailed mechanism is uncertain the scheme in reaction 171 fits the data for both $[PtCl_4]^{2-}$ and $[PdCl_4]^{2-}$. The calculated ratio $k_1:k_2:k_3$ is 1:20:50 for palladium(II) and 1:4:10 for platinum(II),[49] which is in accordance with a ligand leaving ability of $OH^- > H_2O > Cl^-$.

$$[MCl_4]^{2-} \rightleftharpoons [M(H_2O)Cl_3]^- \rightleftharpoons \text{hydroxo complex(es), e.g.} \\ [M(OH)Cl_3]^{2-}$$

$$k_1 \downarrow + \text{phen} \qquad k_2 \downarrow + \text{phen} \qquad k_3 \downarrow + \text{phen}$$

$$[M(phen)Cl_2] \qquad [M(phen)Cl_2] \qquad [M(phen)Cl_2]$$

(171)

Hydrazine

Although it is well known that hydrazine reacts with platinum(II) and palladium(II) complexes, the products of these reactions are far from well understood. *Cis*-[(PPh$_3$)$_2$PtCl$_2$] reacts with hydrazine to give a product which x-ray diffraction has shown to be **29** contaminated with **30** and possibly **31**.[50]

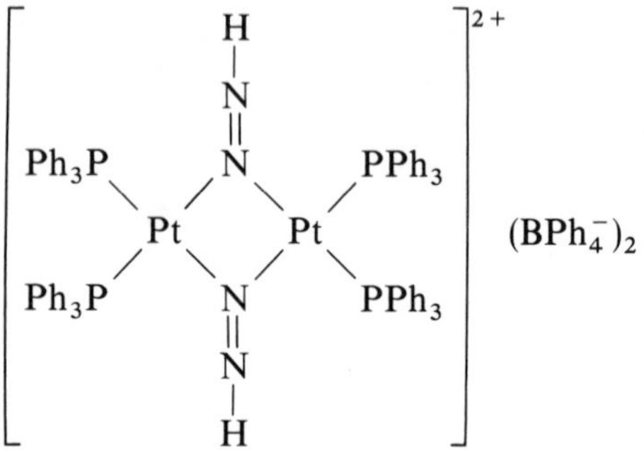

29

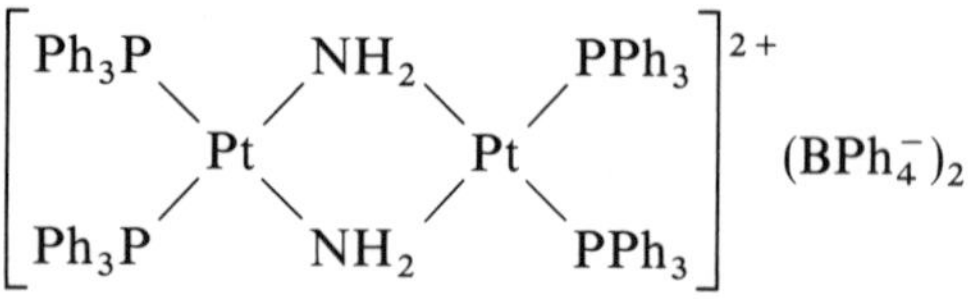

30

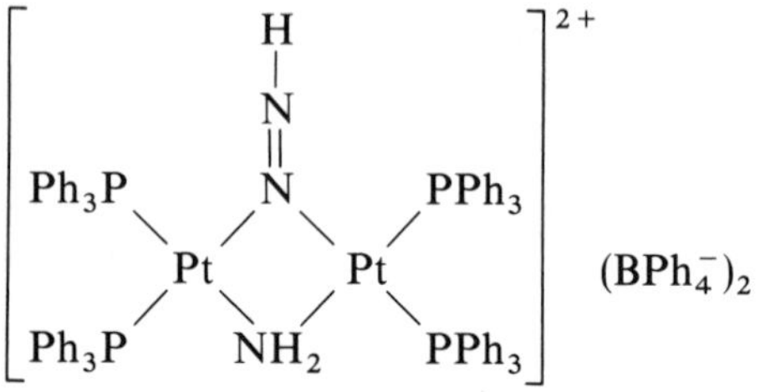

31

A series of complexes, originally formulated as

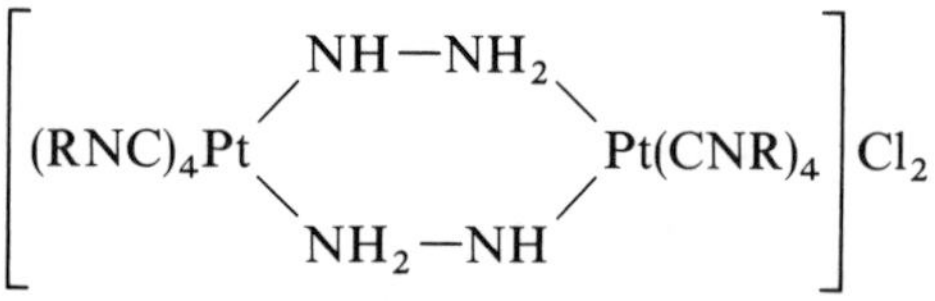

with octahedral platinum(II), are obtained by treating a concentrated aqueous solution of Na$_2$PtCl$_4$ or K$_2$PtCl$_4$ first with an alkyl isocyanide and then with hydrazine.[51] Recent n.m.r., conductivity and x-ray diffraction measurements (see Appendix II, structure 24) indicate that their structure is either **32**[52,53] or **33**[54] in which the platinum is four-co-ordinate.

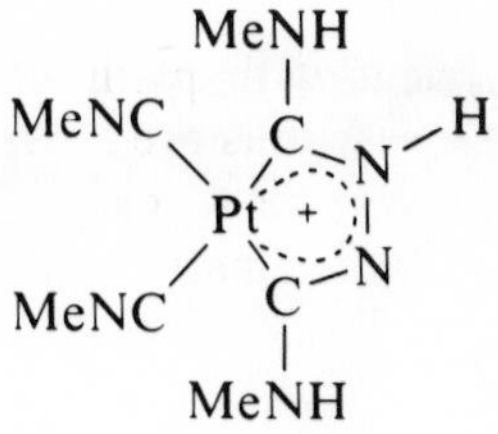

32

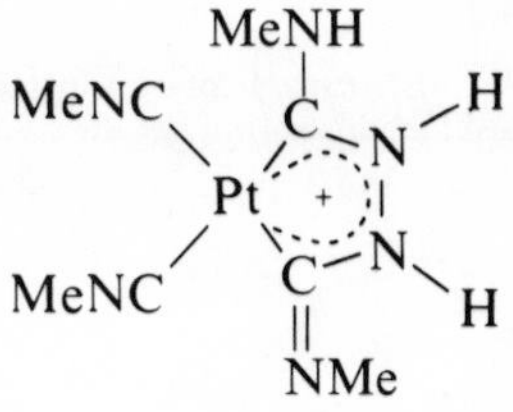

33

Salicylaldimines, salicylaldoximes and 8-hydroxyquinoline

Salicylaldimines, salicylaldoximes and 8-hydroxyquinoline are all bidentate ligands containing one oxygen and one nitrogen donor atom. Bis(8-hydroxy-quinolato)palladium(II) has been extensively studied both for its own conducting properties[55] and for the remarkable series of charge transfer compounds it forms with molecules such as tetracyanobenzene and chloranil in which a 1 : 1 molecular complex is formed in which alternate palladium(II) complex molecules and organic molecules are stacked one above the other.[56–8] These molecular complexes exhibit extra bands in their reflection spectra due to intermolecular charge transfer transitions. The structure of bis(8-hydroxyquinolato)palladium(II) is planar with Pd–N and Pd–O bond lengths of 2·02 Å.[59]

The ultra-violet spectra of palladium(II)–salicylaldimine complexes (34) are altered significantly when a *n*-alkyl substituent (R in 34) is replaced by a

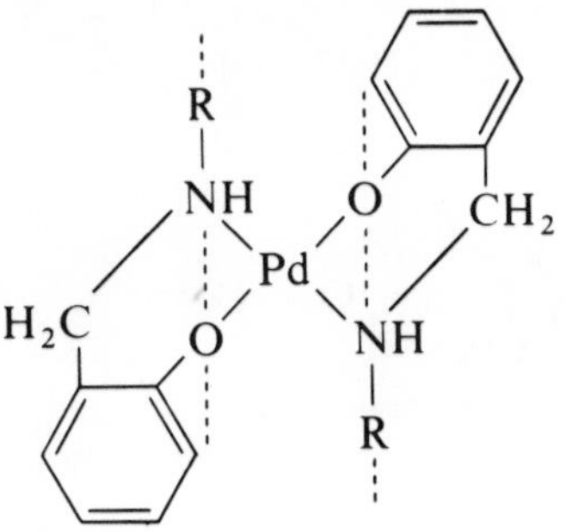

34

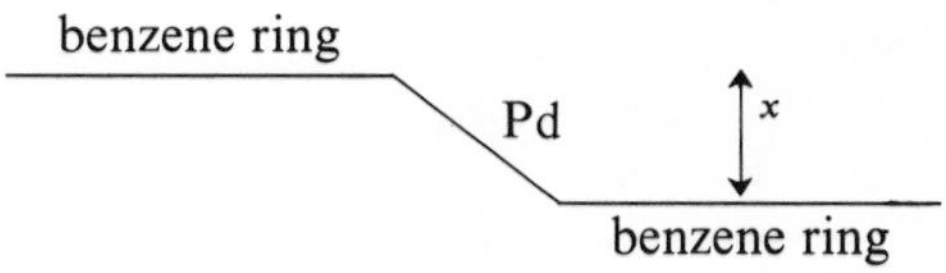

35

branched alkyl substituent. X-ray diffraction studies have shown that this is due to the flexibility of these molecules which enables the complex to be bent into a chair form (35) by folding along the dashed lines in 34. This folding releases the internal strain in the complex, although it does decrease the stabilisation afforded by the delocalised π-bonding present in the planar molecule. The amount of folding (Table 28) increases as the alkyl substituent is altered across the series *n*-alkyl < *i*-propyl < *t*-butyl. As the molecule folds there is very little change in the Pd–O bond length, but a substantial increase in the Pd–N bond length. The *N*-substituted salicylideneimines (36) give similar complexes which also show a change in electronic spectrum[65] when the substituent on the nitrogen is changed from a normal to a branched

TABLE 28
Structural data for bis(salicylaldoximato)palladium(II) complexes

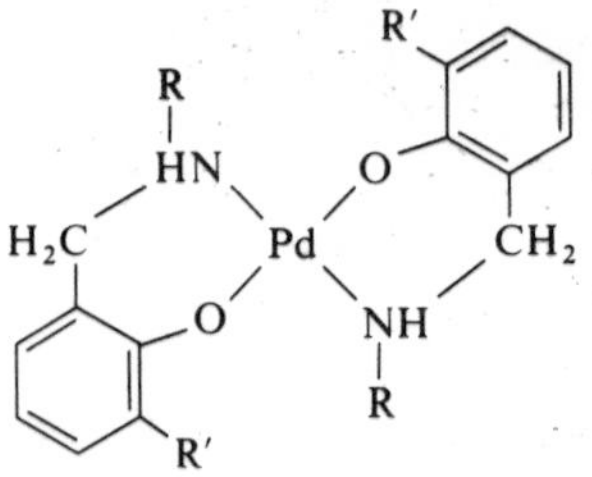

R	R'	*distance x in* **35**	*Pd–N*	*Pd–0*	*∠0–Pd–N*	*Reference*
n-Et	H	0·29 Å	1·86 Å	1·94 Å	91°	60
n-Bu	H	~0 Å	2·01 Å	2·00 Å	104°	61
i-Pr	CH_3 tetragonal form	0·512 Å	2·032 Å	1·987 Å	91·1°	62
i-Pr	CH_3 monoclinic form	0·424 Å	2·019 Å	1·988 Å	91·3°	62
i-Pr	C_2H_5	0·42 Å	2·031 Å	1·991 Å	91·3°	63
t-Bu	H	1·72 Å	2·059 Å	1·982 Å	88·3°	64

alkyl group. Only a preliminary report has appeared of a salicylideneimine complex in which the two nitrogen atoms are linked (**37**) forcing a *cis*-configuration.[66]

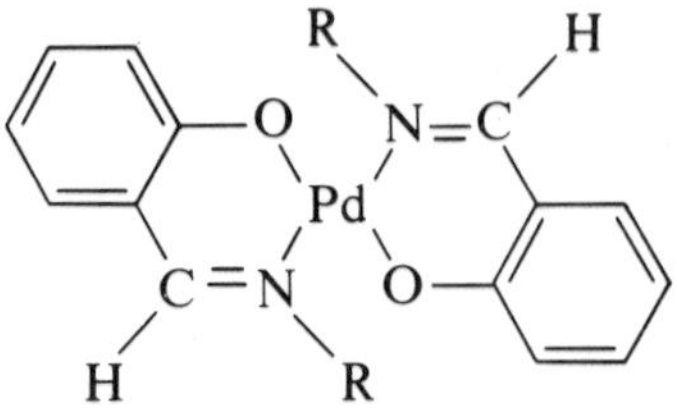

36

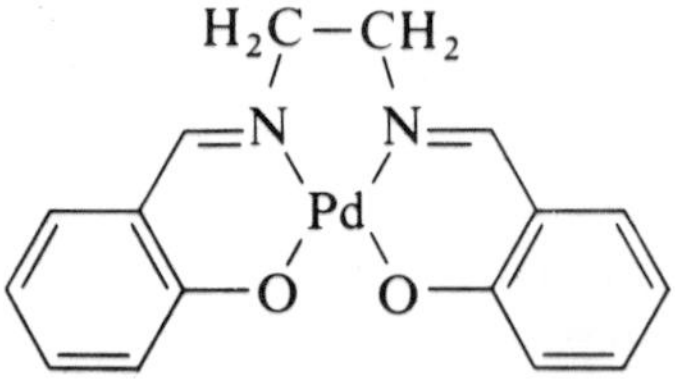

37

This bis(salicylaldoximato)palladium(II) complex (**38**) has a similar structure[67] to the salicylaldimine complexes discussed above in that it forms a chair structure (as in **35**) with folds across the dashed lines drawn in **38**. The step, however, is quite shallow with x (in **35**) equal to 0·16 Å. The Pd–N and Pd–O distances are 1·96 Å and 1·98 Å respectively.

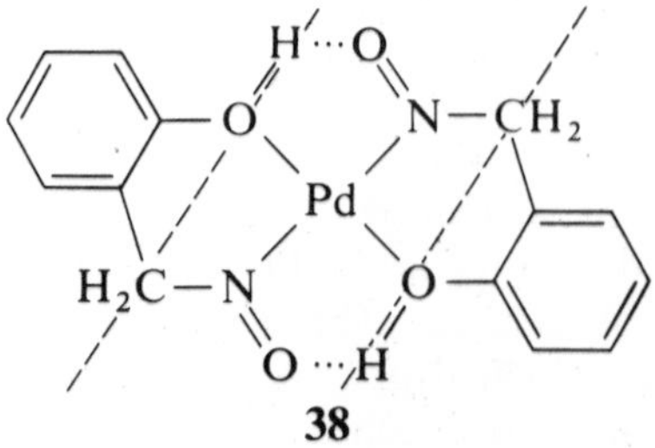

38

Oximes

Both glyoxime and dimethylglyoxime form bis-complexes with platinum(II) and palladium(II) in which the oxime is bound to the metal by metal–nitrogen bonds.[68–72] These complexes (39) are planar and involve two strong hydrogen bridges. In the related complex bis–(2-amino-2-methyl-2-butanone oximato)platinum(II) chloride (40) only one hydrogen bridge is

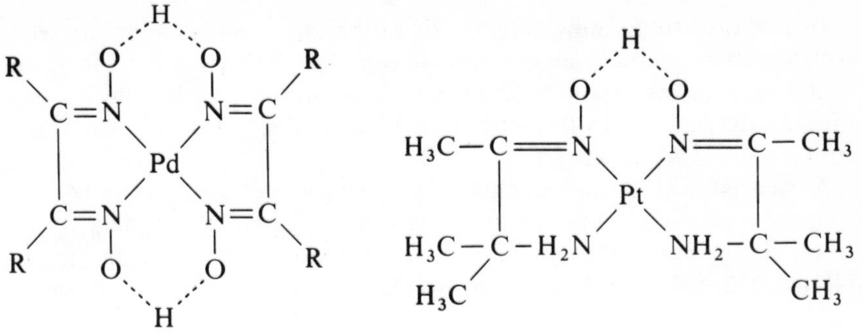

39

40

present. The metal–N–O bond angles in both the one and the two hydrogen bridged complexes are all about 120° which is consistent with an sp^2 hybrid-ised nitrogen atom in an unstrained ring. Further details on these structures are given in Appendix II, structures 22, 43, 128, 146 and 152.

Stable bis(monoxime) complexes of platinum(II) and palladium(II) are formed when potassium tetrachloroplatinate(II)[73] or palladium(II) chloride[74] react with the monoxime directly. Bis(acetoxime)platinum(II) dichloride is mainly formed as the *cis*-isomer although under certain conditions some of the *trans*-isomer can also be prepared,[73] whereas bis(cyclo-hexanone oxime)palladium(II) dichloride is formed as the *trans*-isomer. In addition to direct methods, monoxime complexes of palladium(II) have been prepared from both π-allyl and acetylacetonato complexes of palladium(II)[75] (reactions 172 and 173).

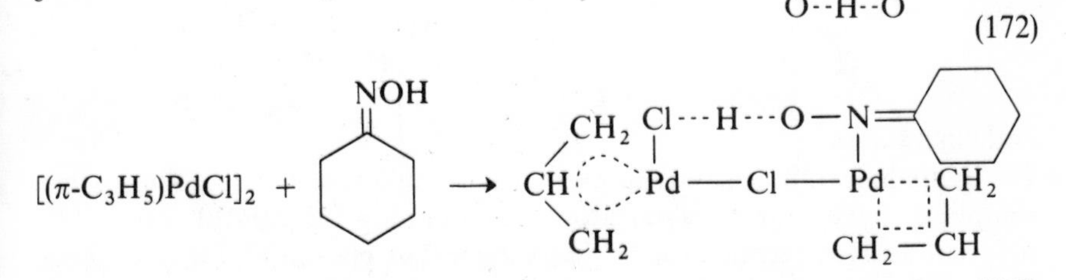

(172)

(173)

The oxime in these monoxime complexes has been shown by x-ray diffraction to be bound to the metal through the nitrogen atom[74] (for further details see Appendix II, structure 52). The oxime hydrogen atom of one acetoxime ligand in *cis*-$[Pt\{(CH_3)_2NOH\}_2Cl_2]$ can be removed by treating the complex with potassium hydroxide to give

$$K[Pt\{(CH_3)_2CNOH\}Cl\{(CH_3)_2CNO\}].[73]$$

In addition to forming monomeric complexes, acetoxime reacts with palladium(II) acetate to give a trimeric complex $[Pd(OAc)(ONCMe_2)]_3$ in which each oxime ligand bridges two palladium atoms, binding to one through oxygen and to the other through nitrogen[76] (for full details see structure 61 in Appendix II).

As well as forming simple complexes involving unidentate nitrogen bonded ligands, aromatic oximes can act as bidentate ligands by bonding to the metal via the nitrogen atom and the ortho-carbon atom of the aryl group (**41**) (pp. 152, 154–155). Aryl azo-oximes (**42** where R = H, alkyl or aryl) react

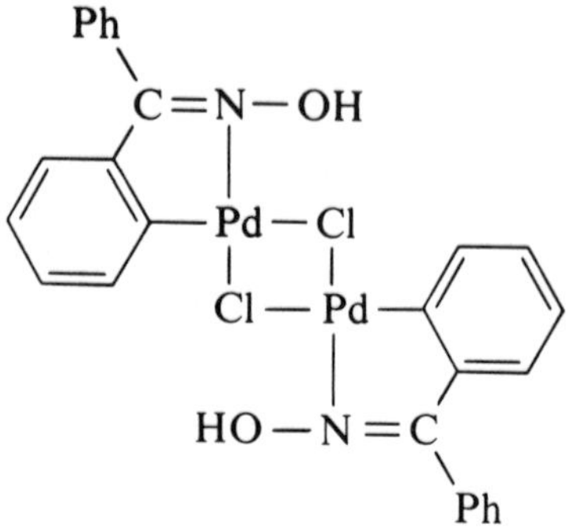

41

with sodium tetrachloropalladate(II) to form chloride bridged complexes (**43**) in which the ligand is a bidentate nitrogen donor.[77]

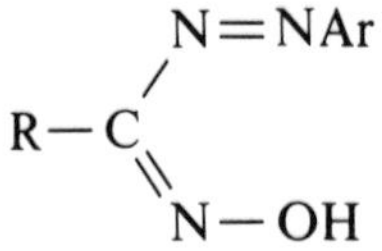

42 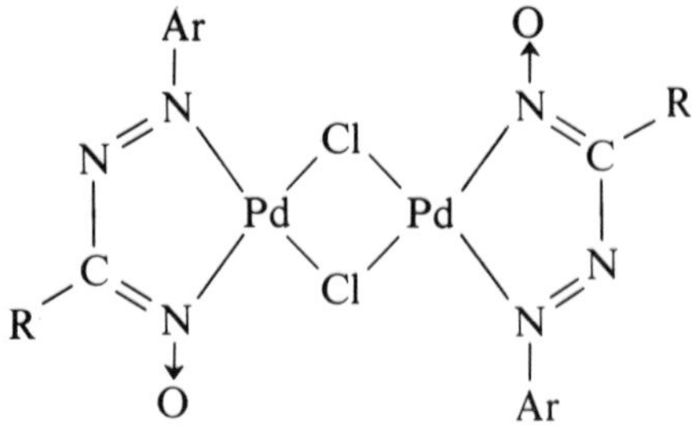   **43**

Phthalocyanines

The phthalocyanine ligand forms planar complexes that like the oxime complexes have four nitrogen atoms surrounding the central metal (**44**). A recent x-ray determination[78] has shown that platinum phthalocyanine forms two distinct types of crystal each of which involves two different

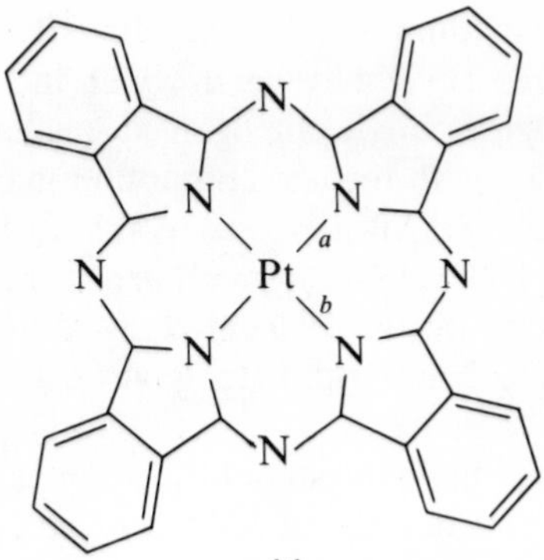

44

platinum–nitrogen bond lengths (*a* and *b* in **44**). In the α-polymorph, which is the less distorted, $a = 1.97$ Å and $b = 1.99$ Å whereas in the γ-polymorph $a = 2.02$ Å and $b = 1.95$ Å.

Azo complexes

The low basicity of azobenzene prevents it readily forming complexes with platinum(II) and palladium(II)[79,80] so that when K_2PtCl_4 or K_2PdCl_4 are treated with azobenzene in aqueous dioxane solution chloro-bridged complexes $[(PhN{=}NPh)MCl]_2$ in which the azobenzene acts as a bidentate ligand (**45**) are formed[81] (pp. 150, 153–155). When these complexes are treated

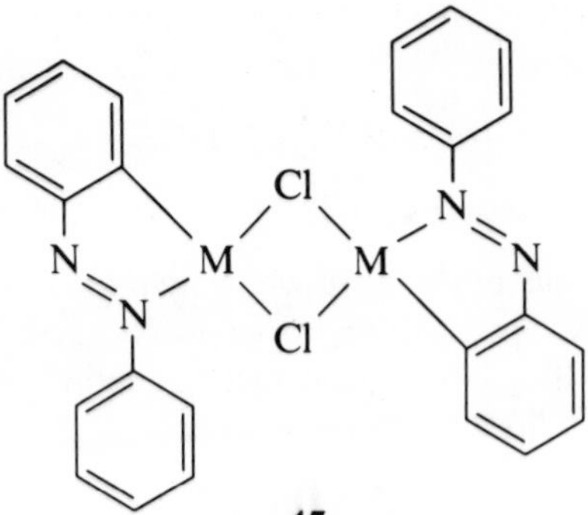

45

with triethylphosphine the chloride bridge and the metal–nitrogen bond are cleaved (reaction 174) emphasising the weakness of that bond.[82]

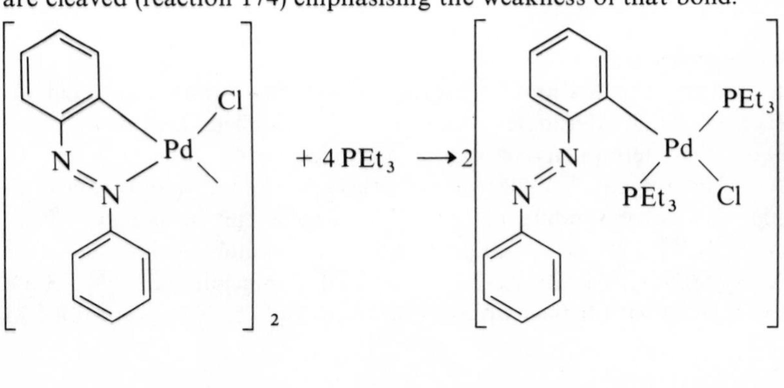

$$(174)$$

Azo complexes of palladium, trans-[Pd(RN=NR)$_2$Cl$_2$] where R = CH$_3$, Ph, substituted phenyl and cyclohexene, in which the azo group is bound to the metal atom through a single nitrogen atom, have been prepared by treating [Pd(PhCN)$_2$Cl$_2$] with the azo compound in a non-aqueous solvent such as dry benzene.[83–5] Attempts to displace benzonitrile from the platinum(II) complex [Pt(PhCN)$_2$Cl$_2$] with azo compounds have all failed. The azo ligand is readily displaced from trans-[Pd(RN=NR)$_2$Cl$_2$] by triphenylphosphine and 2,2'-bipyridyl to give trans-[(PPh$_3$)$_2$PdCl$_2$] and [Pd(bipyridyl)Cl$_2$] respectively.

The structural data for bis–azo-palladium complexes is incomplete. They have been assigned a trans-configuration because they exhibit only a single palladium–chlorine stretching frequency in their infra-red spectra.[84] However, it is uncertain whether the azo ligands themselves are in the cis- or the trans-configuration. Until the full structure is known the details of the bonding in the metal–nitrogen bond and the reason for the slight increase in the nitrogen–nitrogen stretching frequency on coordination[84,85] are unlikely to be known.

Palladium(II) complexes of two of the oxidation products of azobenzene, namely azoxybenzene and nitrosobenzene, have been prepared,[84] although, apart from their empirical formulae of [Pd(C$_6$H$_4$—N(O)=N—C$_6$H$_5$)Cl] and [Pd(C$_6$H$_5$NO)$_2$Cl$_2$] respectively, little is known of their properties.

Diimide complexes

Treatment of trans-[(PEt$_3$)$_2$PtHCl] with a diazonium salt yields a yellow, crystalline diimide complex[85a,86] (reaction 175), which has been described

$$(p\text{-FC}_6\text{H}_4\text{N}{=}\text{N})^+\text{BF}_4^- + trans\text{-}[(\text{PEt}_3)_2\text{PtHCl}] \rightarrow$$

$$trans\text{-}[(\text{PEt}_3)_2\text{Pt}(\text{HN}{=}\text{N}{-}\text{C}_6\text{H}_4\text{F-}p)\text{Cl}]^+\text{BF}_4^- \quad (175)$$

as an inorganic analogue of the nitrogen reductase enzyme[249] because it is rapidly reduced by gaseous hydrogen to give trans-[(PEt$_3$)$_2$Pt(p-FC$_6$H$_4$NHNH$_2$)Cl]$^+$BF$_4^-$. Prolonged hydrogenation regenerates the original hydride (reaction 176).

$$\left.\begin{array}{l} trans\text{-}[(\text{PEt}_3)_2\text{Pt}(\text{NH}{=}\text{N}{-}\text{C}_6\text{H}_4\text{F})\text{Cl}] \\[6pt] trans\text{-}[(\text{PEt}_3)_2\text{Pt}(\text{NH}_2\text{NHC}_6\text{H}_4\text{F})\text{Cl}] \end{array}\right\} \xrightarrow{\text{H}_2} trans\text{-}[(\text{PEt}_3)_2\text{PtHCl}]$$

$$+ \; p\text{-FC}_6\text{H}_4\text{NHNH}_2 \quad (176)$$

Azide complexes

When solutions of palladium(II) nitrate and sodium azide are mixed palladium(II) azide, which detonates when touched, is formed. Treatment of the solution with tetraphenylarsonium chloride yields a precipitate of (Ph$_4$As$^+$)$_2$[Pd$_2$(N$_3$)$_6$]$^{2-}$[87] (46) which, although it burns rapidly, does not explode. The corresponding platinum(II) complex can be prepared from (NH$_4$)$_2$PtCl$_4$.[88] If a 25-fold excess of sodium azide is used (Ph$_4$As$^+$)$_2$[M(N$_3$)$_4$]$^{2-}$, where M = Pt and Pd, is precipitated.[88,89] These complexes react with triphenylphosphine, bipyridyl or o-phenanthroline to

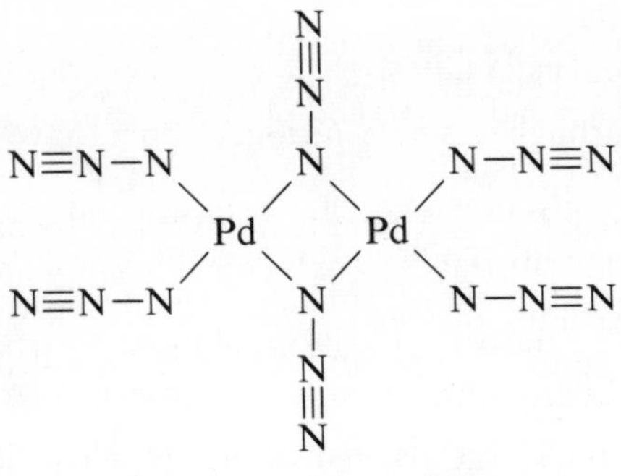

$$
\begin{array}{c}
\text{N} \\
\text{|||} \\
\text{N} \\
\text{|} \\
\text{N} \equiv \text{N} - \text{N} \qquad \text{N} \qquad \text{N} - \text{N} \equiv \text{N} \\
\diagdown \quad \diagup \quad \diagdown \\
\text{Pd} \qquad \text{Pd} \\
\diagup \quad \diagdown \quad \diagup \\
\text{N} \equiv \text{N} - \text{N} \qquad \text{N} \qquad \text{N} - \text{N} \equiv \text{N} \\
\text{|} \\
\text{N} \\
\text{|||} \\
\text{N}
\end{array}
$$

46

form neutral $[L_2M(N_3)_2]$ complexes.[88,89] The infra-red spectrum and preliminary x-ray diffraction data indicate that the two azide ligands in $[(PPh_3)_2Pd(N_3)_2]$ are equivalent in the solid state,[90] although infra-red spectroscopy indicates that they are not equivalent in solution. Until the full solid state structure is available it is impossible to comment on the structural changes that may occur on dissolving the solid. The electronic spectra of azide complexes[88,91] indicate that azide is comparable to iodide in the nephelauxetic series[16] and just above S-bonded thiocyanate in the spectrochemical series.[16]

Co-ordinated azide ions undergo a number of reactions that do not involve metal–nitrogen bond fission. Thus they react with carbon monoxide in non-alcoholic solvents to liberate nitrogen and form isocyanate ligands[92-4] (reaction 177 where M = Pd, Pt). In alcoholic solvents a carboxylate complex is formed[94] (reaction 178). With carbon disulphide

$$[(PPh_3)_2M(N_3)_2] + 2CO \xrightarrow[\text{in CHCl}_3]{\text{ambient conditions}} [(PPh_3)_2M(NCO)_2] + \uparrow 2N_2 \tag{177}$$

and trifluoroacetonitrile cyclic complexes in which the palladium–nitrogen

$$[(PPh_3)_2Pt(N_3)_2] + 2CO \xrightarrow[\text{in EtOH}]{\text{ambient conditions}} [(PPh_3)_2Pt(NCO)(COOEt)] \tag{178}$$

bond is retained are formed[92] (reactions 179 and 180).

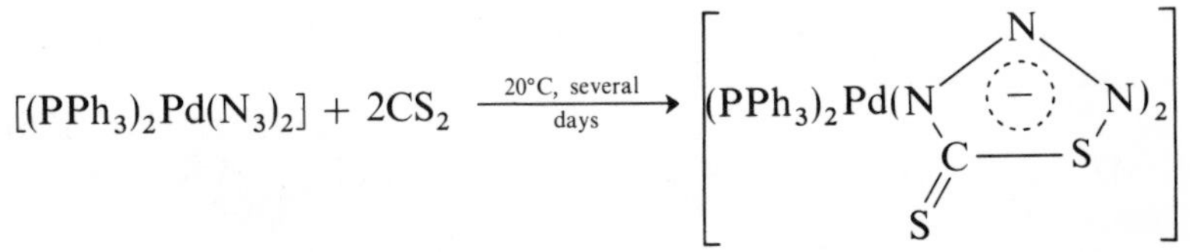

$$[(PPh_3)_2Pd(N_3)_2] + 2CS_2 \xrightarrow[\text{days}]{20°C, \text{ several}} \left[(PPh_3)_2Pd(N \underset{\substack{C \\ \diagup \\ S}}{\overset{\substack{N \\ \diagup \diagdown}}{\underbrace{(-)}}} N)_2 \right] \tag{179}$$

$$[(PPh_3)_2Pd(N_3)_2] + 2CF_3CN \xrightarrow[\text{1 atm., 0°C for 1 hour}]{\text{CH}_2\text{Cl}_2 \text{ solvent};} \left[(PPh_3)_2Pd(N \underset{\substack{C \\ \diagup \\ CF_3}}{\overset{\substack{N \\ \diagup \diagdown}}{\underbrace{(-)}}} N)_2 \right] \tag{180}$$

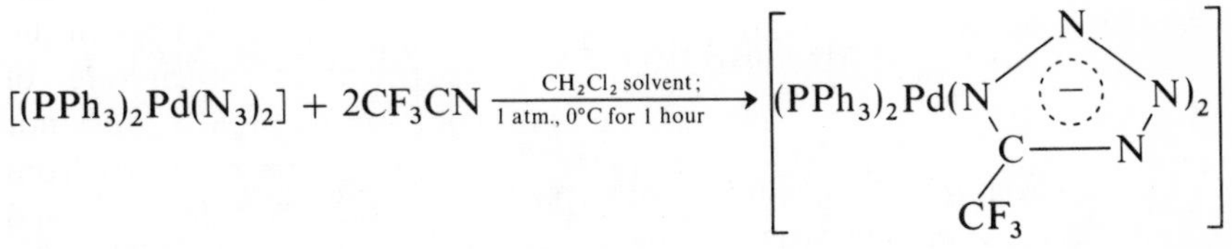

Alkyl and aryl cyanides

Both platinum(II) and palladium(II) halides react with alkyl and arylcyanides. The bis(benzonitrile)palladium(II) complexes[95,96] are particularly useful as starting materials for the preparation of other palladium(II) complexes because they are easily prepared, stable in air and soluble in non-polar solvents such as benzene, chloroform, acetone and benzonitrile.[97] With olefins in benzene solution they form olefin complexes as crystalline precipitates (eqn. 181) leaving the benzonitrile in solution. A partial crystal structure of $[Pd(PhCN)_2Cl_2]$ has shown it to have a *trans*-planar configuration.[98]

$$2[Pd(PhCN)_2] + 2C_2H_4 \rightarrow [Pd(C_2H_4)Cl_2]_2 + 4PhCN \qquad (181)$$

Pale-yellow platinum(II) complexes, $[Pt(RCN)_2Cl_2](R = Me$ or $Ph)$, first prepared by treating potassium tetrachloroplatinate(II) with the appropriate cyanide,[99,100] are better prepared from platinum(II) chloride or bromide.[101] These complexes have been assigned a *cis*-configuration on the basis of their chemical properties,[102,103] their dipole moments[104] and their infrared spectra in which the C–N stretching vibration is split.[105]

Treatment of the chloro- or bromo-complexes $[(PPh_3)_2PtX_2](X = Cl$ or $Br)$ with methyl cyanide yields $[(PPh_3)_2Pt(CNMe)X]^+X^-$, whereas the corresponding iodo-complex gives the five-co-ordinate adduct $[(PPh_3)_2-Pt(CNMe)_2I^+]I^-$. All these complexes eliminated methyl halide on refluxing in benzene (reactions 182 and 183) to yield cyano-complexes.[106] A similar

$$\begin{bmatrix} MeNC & PPh_3 \\ & Pt & \\ Ph_3P & X \end{bmatrix}^+ X^- \xrightarrow[(X = Cl, Br)]{reflux\ in\ C_6H_6} \begin{bmatrix} NC & PPh_3 \\ & Pt & \\ Ph_3P & X \end{bmatrix} + MeX$$

$$(182)$$

$$[(PPh_3)_2Pt(CNMe)_2I]^+I^- \xrightarrow{reflux\ in\ C_6H_6}$$

$$[(PPh_3)_2Pt(CNMe)(CN)]^+I^- + (trace)[(PPh_3)_2Pt(CN)_2] \quad (183)$$

reaction is probably responsible for the unexpected formation of $[(PPh_3)_2Pt(C_6F_5)(CN)]$ when $[(PPh_3)_2Pt(C_6F_5)Br]$ is heated with methyl cyanide.[107] Co-ordinated perfluorobenzonitrile ligands react with alcohols to form imino–ether complexes, which can be isolated in the presence of tetrafluoroborate anions[108] (reaction 184).

$$trans\text{-}[(PMe_2Ph)_2PtCH_3Cl] + C_6F_5CN + AgBF_4 + ROH \xrightarrow[(R\ =\ Me,\ Et,\ Pr)]{25°C}$$

$$trans\text{-}\begin{bmatrix} (PMe_2Ph)_2Pt \begin{pmatrix} NH{=}C \overset{OR}{\underset{C_6F_5}{\diagup\diagdown}} \end{pmatrix} Cl \end{bmatrix}^+ BF_4^- + AgCl \quad (184)$$

The bonding in alkyl cyanide complexes involves σ-donation of the lone pair of electrons in the nitrogen sp hybrid orbital to an empty orbital on the metal. This is complemented by back-donation of electron density from a filled orbital on the metal to the π^*-(antibonding) orbitals of the alkyl cyanide. As a result of this bonding description the C–C–N–M unit in the complex should be linear, as has been found for all the complexes whose structures have been determined, with the exception of certain copper(II) complexes.[109] The heats of formation of some platinum and palladium alkyl and aryl cyanide complexes have been determined using a differential scanning calorimeter[110] (reactions 185 and 186). The low value for the *cis*-

$$PdCl_2(cryst) + 2RCN(g) \rightarrow trans\text{-}[Pd(RCN)_2Cl_2]$$

$$\Delta H_{CH_3CN} = -25 \cdot 2 \text{ kcal/mol}$$

$$\Delta H_{PhCN} = -23 \cdot 2 \text{ kcal/mol}$$

(185)

$$PtCl_2(cryst) + 2RCN(g) \rightarrow cis\text{-}[Pt(RCN)_2Cl_2]$$

$$\Delta H_{CH_3CN} = -29 \cdot 2 \text{ kcal/mol}$$

$$\Delta H_{PhCN} = -17 \cdot 5 \text{ kcal/mol}$$

(186)

$[Pt(PhCN)_2Cl_2]$ complex (186) as compared to the other complexes probably arises from steric interactions between the bulky *cis*-ligands.

Amine-alkylcyanide complexes of platinum(II), prepared by treating the dichloro-complexes with ammonia or amines,[111] are anomalous in that the two chloride ligands are replaced by four amine molecules. For a time they were considered to be six-co-ordinate complexes of platinum(II),[112] but more recent work, following a suggestion of Grinberg,[113] has shown them to be four-co-ordinate complexes containing two amidine and two amine ligands[114–7]—$[Pt(RC(NH)=NHR')_2(R'NH_2)_2]X_2$ (**47**).

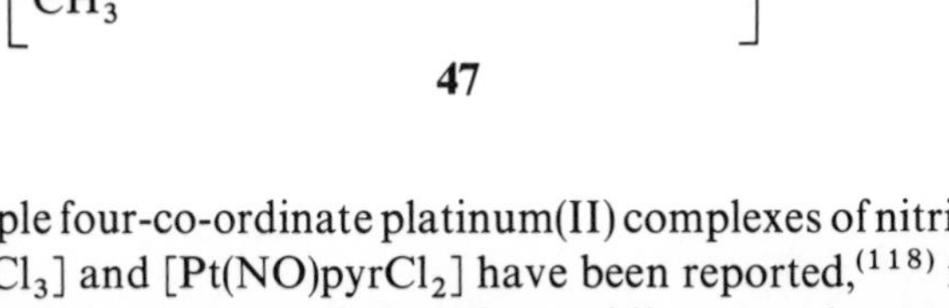

47

Nitric oxide
Although simple four-co-ordinate platinum(II) complexes of nitric oxide such as $K[Pt(NO)Cl_3]$ and $[Pt(NO)pyrCl_2]$ have been reported,[118] repetition of this earlier work has suggested that the pyridine complex may contain a nitrito ligand and that $K[Pt(NO)Cl_3]$, obtained by treating a neutral solution of potassium tetrachloroplatinate(II) with nitric oxide, may well be polymeric.[119] Several six-co-ordinate platinum(II) complexes containing nitric oxide have been prepared by treating platinum(II) complexes with nitrosyl chloride[119] (reaction 187). In addition to reaction 187 $[Pt(NO)(NH_3)_4Cl]$ has also been prepared by treating a solution of $[Pt(NH_3)_4]Cl_2$ in fairly strong

$$K_2PtX_4 + NOCl \xrightarrow[\text{or } \frac{1}{2}\text{en})]{(X = Cl^-, NO_2^-, NH_3} K_2[Pt(NO)X_4Cl]$$

(187)

hydrochloric acid with nitric oxide[120,121] and $K_2[M(NO)(NO_2)_4(NO_3)]$, where M = Pd or Pt, has been prepared by treating $K_2[M(NO_2)_4]$ with nitric acid.[119,122] These complexes appear to be genuine six-co-ordinate complexes of platinum(II) or palladium(II). The oxidation state of the metal is based on the co-ordinated nitric oxide ligand being NO^+, which in turn is based on the diamagnetism of the complexes and the observation of an N–O stretching frequency of about 1800 cm^{-1} in the infra-red.[119]

In addition to these nitric oxide complexes of the divalent metals, nitric oxide complexes of zerovalent palladium such as $[Pd(NO)_2Cl_2]$, $[Pd(NO)Cl]_n$ and $[Pd(NO)_2SO_4]$ have been prepared by passing nitric oxide into palladium(II) chloride or sulphate solutions in either methanol (monomeric products) or water (polymeric products).[123–5] The metal was assigned to the zero oxidation state because the complexes were diamagnetic and exhibited N–O stretching frequencies around 1800 cm^{-1} in their infra-red spectra,[123] which is typical of co-ordinated NO^+.[126]

Dialkylnitrogen oxide radicals

The organic free radical di-t-butylnitrogen oxide reacts with palladium(II) chloride or bromide to give curious diamagnetic halogen bridged complexes $[Pd(Bu_2^tNO)X]_2$, where X = Cl or Bu, which may have structure **48**.[127]

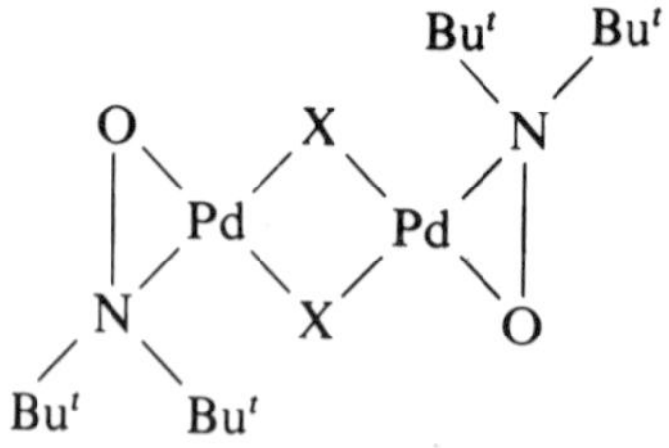

48

Thiazoles

Platinum(II) and palladium(II) chlorides react with thiazoles (**49**) to form $[M(thiazole)_2Cl_2]$ complexes, where M = Pd or Pt, in which the thiazole

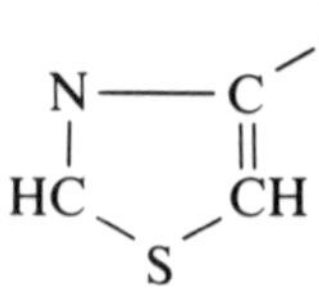

(R = H or CH_3)

49

is bonded to the metal via the nitrogen atom only.[128–30]

COMPLEXES WITH GROUP VB ELEMENTS OTHER THAN NITROGEN

Complexes of platinum(II) and palladium(II) containing ligands which have phosphorus, arsenic or antimony donor atoms are extremely numerous, although they have only been reviewed twice.[131,132] In the present section a description of the preparation of the complexes is followed by their physical properties which enables us to discuss the nature of the bonding in metal–Group VB element bonds. The majority of these complexes involve planar four-co-ordinate metal ions, but at the end of the section the few known five-co-ordinate complexes are considered.

(a) Preparation

A wide range of different Group VB ligands form complexes with platinum(II) and palladium(II) and their preparative methods are sufficiently different to make it worth considering each class of ligand separately, rather than attempting to give a general account of all the ligands.

(i) ER_3 complexes ($E = P$, As or Sb; $R =$ alkyl or aryl)

$[(ER_3)_2MX_2]$ complexes are formed when solutions of $[MX_4]^{2-}$, where $M = Pd$ or Pt and $X =$ halogen, are treated with two equivalents of a tertiary phosphine, arsine or stibine (reaction 188).[133,134] Platinum(II) first forms ionic complexes, $[Pt(PR_3)_4]^{2+}[PtCl_4]^{2-}$, with phosphines

$$[MX_4]^{2-} + 2ER_3 \rightarrow [(ER_3)_2MX_2] + 2X^- \qquad (188)$$

containing alkyl groups lower than *n*-butyl; these form $[(PR_3)_2PtCl_2]$ slowly.[134,135] When R is an alkyl group the platinum(II) complex formed is a mixture of *cis*- and *trans*-isomers which can be separated by fractional recrystallisation.[134] However, triphenylphosphine only forms a *cis*-platinum(II) complex unless the reaction is carried out in the presence of ultra-violet light (*ca.* 366 mμ) when a *cis-trans*-mixture is formed.[136] Detailed accounts of the preparations of monomeric tertiary phosphine complexes are given in Appendix I (pp. 456–459).

Dimeric halogen-bridged complexes of the type $[(ER_3)PtX_2]_2$ (where $M = Pd$ or Pt and $E = P$, As or Sb) can be prepared by fusing together the mononuclear complex $[(ER_3)_2PtX_2]$ and the platinum(II) halide.[137,138] However, this method depends on maintaining the mixture in a molten state, which with high melting compounds demands a high degree of thermal stability from both reactants and products. Accordingly the method cannot be used where the products decompose on melting or with $[(PPh_3)_2PtCl_2]$ or $[(PMe_3)_2PtCl_2]$ both of which decompose before melting. This difficulty, however, can be overcome by heating a finely divided slurry of the reactants either in a high boiling hydrocarbon solvent[139] or in tetrachloroethylene.[140] The solvent assists the reaction in three ways: first it keeps the reactants mobile, secondly it lowers their melting-points and thirdly it may dissolve them slightly. The dimeric halogen bridged palladium(II) complexes are prepared rather more easily than their platinum(II) analogues by boiling a mixture of the mononuclear complex $[(ER_3)_2PdCl_2]$ in a solvent such as ethanol or chloroform with sodium tetrachloropalladate(II) in ethanol.[141] The dimeric tri-n-propylstibine platinum(II) complex $[(SbPr_3^n)PtCl_2]_2$ is best prepared from the dimeric ethylene complex $[(C_2H_4)PtCl_2]_2$[137]

(reaction 189). Although this reaction is potentially suitable for preparing

$$[(C_2H_4)PtCl_2]_2 + 2SbPr_3^n \rightarrow [(SbPr_3^n)PtCl_2]_2 + \uparrow 2C_2H_4 \quad (189)$$

the phosphine and arsine complexes it is not used because better methods are available since the yields are low because the monomeric $[(ER_3)_2PtCl_2]$ complexes are also formed in substantial amounts. Detailed accounts of the preparations of the dimeric tertiary phosphine complexes are given in Appendix I (pp. 459–461).

(ii) PR_2H complexes

Palladium(II) halides react with dialkyl or dicycloalkyl phosphines to form bis-phosphine complexes[142,143] (reaction 190). These complexes react with

$$PdX_2 + 2PEt_2H \rightarrow [(PEt_2H)_2PdX_2] \quad (190)$$

base to eliminate HX and form a dialkylphosphido-bridged complex[143–5] (reaction 191); the ease of elimination of HX decreases in the order X = Cl <

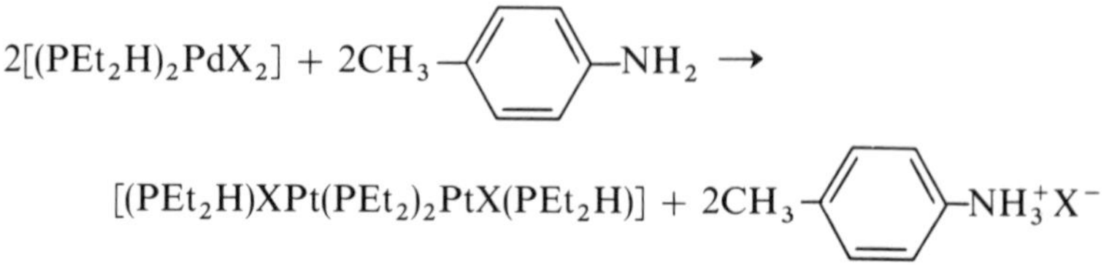

$$2[(PEt_2H)_2PdX_2] + 2CH_3{-}\langle\!\bigcirc\!\rangle{-}NH_2 \rightarrow$$

$$[(PEt_2H)XPt(PEt_2)_2PtX(PEt_2H)] + 2CH_3{-}\langle\!\bigcirc\!\rangle{-}NH_3^+X^- \quad (191)$$

Br < I. Diarylphosphines yield four end products[146] (reaction 192), the actual product obtained depends on: (a) The elimination of HX to form

$$Na_2PdX_4 + 2Ph_2PH \rightarrow [(PPh_2H)_2PdX_2]$$

$$\nearrow \quad \mathbf{51} \quad \searrow$$

$$[(PPh_2H)_3PdX]^+X^- \qquad \tfrac{1}{2}[(PPh_2H)PdX(PPh_2)]_2 + HX$$

$$\mathbf{52} \qquad\qquad\qquad\qquad \Big\Vert_{+2PPh_2H} \quad \mathbf{50} \qquad (192)$$

$$[(PPh_2H)_2Pd(PPh_2)_2] + HX$$

$$\mathbf{53}$$

products 50 and 53 decreases with increasing halogen atomic weight, indeed $[(PPh_2H)_2PdCl_2]$ is too unstable with respect to loss of hydrogen chloride to be isolated. (b) The elimination of HX also depends on the phosphine, decreasing in the order $PPh_2H > PEtPhH > PEt_2H$. This is probably because progressive substitution of ethyl groups by the more electronegative phenyl groups causes a drift of electrons away from the phosphorus atom into the aromatic system, which results in a decrease of electron density in the P–H bond and hence a more acidic proton. (c) The presence of HX favours product 52. (d) The use of an alcohol as a solvent favours the elimination of HX and hence the formation of products 50 and 53, whereas with benzene ionic complexes, 52, are formed. This may be due to the heats of solvation of hydrogen halides being greater in alcohols than in benzene.

(iii) *PH$_3$ complexes*

Polymeric phosphine complexes of platinum(II) and palladium(II) are prepared by treating bis–tertiary phosphine complexes with PH$_3$[147] (reactions 193 and 194). They have not yet been characterised beyond

$$[(PPh_3)_2PtI_2] \xrightarrow{PH_3} [(PPh_3)_3Pt_3(PH_3)_3I_2] \qquad (193)$$

$$[(PPh_3)_2PdCl_2] \xrightarrow{PH_3} [(PPh_3)Pd(PH_3)Cl]_4 \qquad (194)$$

determining their molecular weights.

(iv) *ER$_2$Cl complexes*

When dimeric tertiary phosphine or arsine complexes of platinum(II) are treated with ER$_2$Cl (ER$_2$Cl = PPh$_2$Cl, PEt$_2$Cl, AsPh$_2$Cl or AsMe$_2$Cl), the halogen bridge is cleaved and a monomeric *cis*-product is formed[148] (reaction 195).

$$[(E'R_3')PtX_2]_2 + 2ER_2Cl \rightarrow 2 \, cis\text{-}[(E'R_3')(ER_2Cl)PtX_2] \qquad (195)$$

(v) *PX$_3$ where X is a halogen*

Trifluoro- and trichlorophosphine complexes are formed when platinum(II) chloride is treated with the appropriate trihalophosphine[149,150] (reaction 196). The trichlorophosphine complexes can also be prepared from platinum

$$4PX_3 + 3PtCl_2 \xrightarrow[(X = F \text{ or } Cl)]{} [(PX_3)_2PtCl_2] + [(PX_3)PtCl_2]_2 \qquad (196)$$

metal using phosphorus pentachloride[151] (reaction 197). By contrast treatment of palladium(II) chloride with either phosphorus pentachloride or

$$Pt \xrightarrow{+PCl_5} [(PCl_3)_2PtCl_2] + [(PCl_3)PtCl_2]_2 \qquad (197)$$

phosphorus pentabromide yields only the dimeric [(PX$_3$)PdX$_2$]$_2$, where X = Cl or Br.[152]

(vi) *P(OR)$_3$ complexes*

Complexes of the type [{P(OR)$_3$}MX$_2$]$_2$, where X = Cl or Br, M = Pd or Pt and R = H, CH$_3$ or C$_2$H$_5$, can be prepared by solvolysis of the analogous halogen complex [(PX$_3$)MX$_2$]$_2$ in water or the appropriate alcohol.[152] However, a better method for the trimethylphosphite complexes of platinum(II), starting from the bis–benzonitrile complexes,[153] (reaction 198) should be applicable to other trialkylphosphites as well as to palladium(II)

$$trans\text{-}[PtX_2(PhCN)_2] + 2P(OMe)_3 \xrightarrow[\substack{(X = Cl, Br, I, \\ NCO, N_3 \text{ and } Ph)}]{}$$

$$[\{P(OMe)_3\}_2PtX_2] + 2PhCN \qquad (198)$$

complexes, which are currently prepared by treating an aqueous solution of sodium tetrachloropalladate(II) with the trialkyl-phosphite[154] (reaction 199).

$$2Na_2PdCl_4 + 4P(OEt)_3 \rightarrow 4NaCl + [Pd\{P(OEt)_3\}_4][PdCl_4] \xrightarrow{\text{HCl}}$$

$$2[\{P(OEt)_3\}_2PdCl_2] \quad (199)$$

(vii) Phosphite ($\{RO\}_2PO^-$) complexes
Diethyl hydrogen phosphite, which exists in solution as a tautomeric equilibrium (200) displaced largely to the right, reacts with potassium

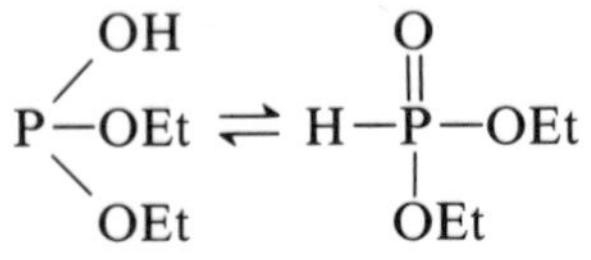

$$P\overset{OH}{\underset{OEt}{\diagup}}-OEt \rightleftharpoons H-\overset{O}{\underset{OEt}{\overset{||}{P}}}-OEt \qquad (200)$$

Diethyl   Diethyl
hydrogen   phosphonate
phosphite

tetrachloropalladate(II)[154] to form a dimeric chloro-bridged complex in which both tautomers are co-ordinated[154,155] (reaction 201). Similarly sodium tetrachloroplatinate(II) reacts with dimethyl- or diphenyl-

$$2Na_2PdCl_4 + 4HOP(OEt)_2 \rightleftharpoons [\{P(OEt)_2OH\}\{P(OEt)_2O\}PdCl]_2$$

$$+ 2HCl + 4NaCl \quad (201)$$

phosphonate in ethanol to yield a complex containing both tautomers[156] (reaction 202). Monomeric complexes containing the phosphonate tautomer

$$Na_2PtCl_4 + 4HOP(OR)_2 \xrightarrow{\text{EtOH}}$$

$$[Pt\{(RO)_2PO\}_2\{(RO)_2POH\}_2] + 2HCl + 2NaCl \quad (202)$$

only can be prepared using sodium diethylphosphite[156] (reaction 203). If sodium diphenylphosphite is used, phosphite bridged complexes are formed

$$cis\text{-}[(PEt_3)_2PtCl_2] + 2NaOP(OEt)_2 \xrightarrow[\text{THF}]{\text{reflux in}}$$

$$trans\text{-}[(PEt_3)_2Pt\{OP(OEt)_2\}_2] + 2NaCl \quad (203)$$

instead of the monomeric complexes (reaction 204).

$$2\ cis\text{-}[(PEt_3)_2PtCl_2] + 2NaOP(OPh)_2 \xrightarrow[\text{THF}]{\text{reflux in}}$$

$$[(PEt_3)_2Pt\{OP(OPh)_2\}_2Pt(PEt_3)_2]Cl_2 + 2NaCl \quad (204)$$

(b) Properties

The tertiary phosphine, arsine and stibine complexes of platinum(II) and palladium(II) are stable in air, soluble in organic solvents, readily recrystallised and of well-defined melting point. Their general stability decreases in the order $PR_3 > AsR_3 > SbR_3$.[131] The monomeric platinum(II) complexes generally exist as *cis*- and *trans*-isomers although the *trans*-isomer of $[(PF_3)_2PtCl_2]$ has not yet been reported and preparation of the *trans*-isomer of $[(PPh_3)_2PtCl_2]$ requires ultra-violet irradiation of the *cis*-isomer.[136]

By contrast only the *trans*-isomer of $[\{P(C_6F_5)_3\}_2PtCl_2]$ has so far been prepared.[157,158] The ratio of *cis*- to *trans*-$[(ER_3)_2PtCl_2]$ present in benzene solution increases as the group VB element, E, is replaced in the order As < P ≪ Sb,[159,160] due to the smaller entropy difference between the *cis*- and *trans*-stibine complexes than between the *cis*- and *trans*-isomers of the corresponding phosphine and arsine complexes (pp. 313–314). The monomeric palladium(II) complexes with tertiary phosphine and arsine ligands are always *trans*,[131] although the tertiary stibine complexes contain up to 40% of the *cis*-isomer in equilibrium and only the crystalline *cis*-isomers can be separated from solutions of $[(SbR_3)_2PdCl_2]$ because their solubilities are so much smaller than those of the *trans*-isomers.[161] The general solubility and colour properties of the tertiary phosphine complexes are described in Appendix I (pp. 458–459).

The dimeric halogen[137,138,141,162-4] and diphenylphosphide[144] bridged complexes have *trans*-structures (**54**), although the thiol-bridged palladium(II) complexes[165] have *cis*-structures (**55**) and the thiol-bridged

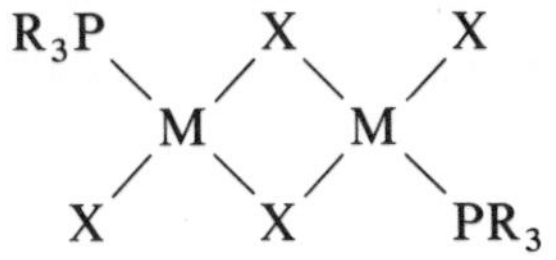

54

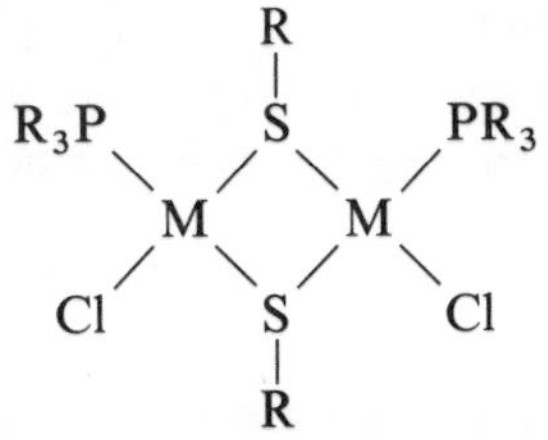

55

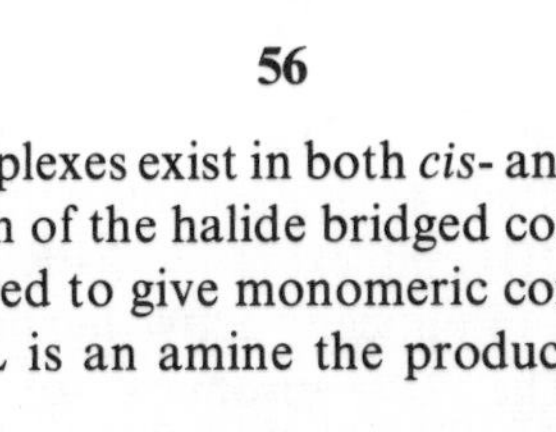

56

platinum(II)[166,167] complexes exist in both *cis*- and *trans*-structures (**55** and **56**). The main application of the halide bridged complexes is in reactions in which the bridge is cleaved to give monomeric complexes[168,169] (reaction 205). When the ligand L is an amine the product is invariably the *trans*-

$$[(PR_3)MCl_2]_2 + 2L \xrightarrow{\text{(M = Pd, Pt)}} 2[(PR_3)MLCl_2] \qquad (205)$$

isomer[168,169] whereas with carbon monoxide or olefins it is the *cis*-isomer.[170]

The product obtained when the monomeric tertiary phosphine complexes are treated with alkali is difficult to predict: *cis*-$[(PEt_3)_2PtCl_2]$ reacts with alcoholic potassium hydroxide to form the hydrido–platinum(II) complex *trans*-[(PEt$_3$)$_2$PtHCl][171] (p. 51), whereas *cis*-[(PPh$_3$)$_2$PtCl$_2$] reacts with alcoholic potassium hydroxide in the presence of excess triphenylphosphine to form the zerovalent complexes [Pt(PPh$_3$)$_n$], where $n = 3$ or 4[172] (pp. 27–29). The tertiary phosphine complexes of palladium(II) generally decompose in strong alkali giving palladium metal, but potentiometric titration indicates that the secondary phosphine complex [(PHEt$_2$)$_2$PdX$_2$] reacts with aqueous alcoholic sodium hydroxide in two stages, possibly reactions 206 and 207.[143]

$$2[(PHEt_2)_2PdX_2] + 2NaOH \rightarrow$$

$$[(PHEt_2)Pd(PEt_2)X]_2 + 2NaX + 2H_2O \quad (206)$$

$$[(PHEt_2)Pd(PEt_2)X]_2 + 2NaOH \rightarrow$$

$$2[Pd(PEt_2)_2] + 2NaX + 2H_2O \quad (207)$$

The hydrolysis of platinum(II) complexes *cis*-[(E'R$_3'$)(ER$_2$Cl)PtX$_2$], where X is a halogen, containing monochlorophosphine and monochloroarsine ligands (reaction 208) involves initial hydrolysis of the chloro-phosphine (or -arsine) to a hydroxyphosphine giving complexes which exist as strongly

$$\textit{cis-}[(E'R_3')(ER_2Cl)PtX_2] \xrightarrow[C_6H_6]{H_2O\ in} \textit{cis-}[(E'R_3')(ER_2OH)PtX_2] \quad (208)$$

hydrogen-bonded dimers and which readily lose a molecule of hydrogen halide to give binuclear phosphinato- or arsinato-bridged complexes (reaction 209).[148] The arsinato-complexes are stable and can be hydrolysed no further, but the phosphinato complexes are further hydrolysed by alkali

$$2\ \textit{cis-}[(E'R_3')(ER_2OH)PtX_2] \rightarrow [(E'R_3')XPt(R_2EO)_2PtX(E'R_3')] + 2HX$$
$$(209)$$

to give very stable hydroxy complexes (reaction 210).

$$[(E'R_3')XPt(R_2EO)_2PtX(E'R_3')] \xrightarrow{OH^-}$$

$$[(E'R_3')(HO)Pt(R_2EO)Pt(OH)(E'R_3')] + 2X^- \quad (210)$$

Both tertiary phosphines[173] and tertiary arsines[174] of the type ER^{1}R^{2}R^3 have been resolved into their separate stereoisomers by preparing the diastereoisomeric platinum(II) complexes *trans*-[(ER^{1}R^{2}R^3)PtLCl$_2$], where L is the pure stereoisomer of an optically active ligand such as stilbenediamine or deoxyephedrine. The stereoisomers of the chelating ligand MePhAsCH$_2$CH$_2$AsPhMe have similarly been separated by preparing the diastereosiomers of [(MePhAsCH$_2$CH$_2$AsPhMe)PdCl$_2$] and separating them chromatographically.[174] An x-ray diffraction study of [Pd(PhMePCH$_2$CH$_2$PMePh)$_2$]Cl$_2$ has shown that both the phenyl groups of one ligand are above and both the phenyl groups of the other ligand are below the square-plane containing the palladium atom[175] (for further details see Appendix II, structure 84).

TABLE 29
Palladium(II)–phosphorus and –arsenic bond lengths

Complex	Atom or group trans to group VB element	Pd–P bond length (Å)	No. of structure in Appendix II	Reference
trans[(PPh$_3$)$_2$PdCl- (—C$_6$H$_4$N=NC$_6$H$_5$)]	PPh$_3$	2·306	88	176
[(π-C$_4$H$_7$)PdCl(PPh$_3$)]	π-allyl	2·31	75	177
[(π-C$_3$H$_5$)Pd(SnCl$_3$)(PPh$_3$)]	π-allyl	2·317	74	178, 179
[Pd(PhMePCH$_2$CH$_2$PPhMe)$_2$]Cl$_2$	PPhMeCH$_2$-	2·333	84	175
[{EtN(PPh$_2$)}$_2$PdCl$_2$]	Cl	2·367	81	180
trans-[(PMe$_2$Ph)$_2$PdI$_2$]	PMe$_2$Ph	2·61 (mean of 2 isomers)	63	181, 182
		Pd–As bond length (Å)		
[(AsMe$_3$)PdBr$_2$]$_2$	Br(bridging)	2·50	39	162

TABLE 30
Platinum(II)–phosphorus and –arsenic bond lengths

Complex	Atom or group trans to group VB element	Pt–P bond length (Å)	No. of structure in Appendix II	Reference
trans-[(PPr$_3$)$_2$PtCl$_2$]	PPr$_3$	2.230	178	183
[(PPh$_3$)$_2$Pt(CO$_3$)]	O	2·24	196	184
cis-[(PEt$_3$)Pt(—C(OEt)(NPh))Cl$_2$]	Cl	2·24	172	185
[(PPr$_3^n$)ClPt(SCN)$_2$PtCl(PPr$_3^n$)]	N	2·244	182	186
cis-[(PMe$_3$)$_2$PtCl$_2$]	Cl	2·247 (mean)	143	187
trans-[(PEt$_3$)$_2$PtHBr]	PEt$_3$	2·26	159	188
[(PPr$_3^n$)ClPt(SCN)$_2$PtCl(PPr$_3^n$)]	S	2·262	182	186
[(PPh$_3$)$_2$Pt(O$_3$CMe$_2$)]	O	2·265	201	189
trans-[(PPh$_2$Et)$_2$PtHCl]	PPh$_2$Et	2·268	191	190
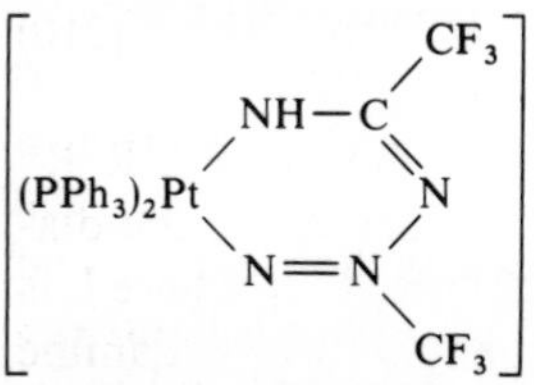 (PPh$_3$)$_2$Pt[NH—C(CF$_3$)=N—N=N—C(CF$_3$)]	N	2·295	204	191
trans-[(PMe$_2$Ph)$_2$PtCl(CH$_2$SiMe$_3$)]	PMe$_2$Ph	2·30	180	192
trans-[(PEt$_3$)$_2$PtCl$_2$]	PEt$_3$	2·30	157	193
trans-[(PEt$_3$)$_2$PtBr$_2$]	PEt$_3$	2·315	156	193
trans-[(PEt$_3$)$_2$Pt(CO)Cl]$^+$BF$_4^-$	PEt$_3$	2·345	169	194
		Pt–As bond length (Å)		
[(Me$_3$As)PtCl$_2$]$_2$	Cl (bridging)	2·308	142	195, 196

(c) X-ray diffraction studies

Examination of the M–E bond lengths in square-planar complexes of palladium(II) and platinum(II) (Tables 29 and 30) shows that there is no correlation between either the Pd–P or the Pt–P bond lengths and the *trans*-influence of the ligand in the *trans*-position. Such a correlation is found in the case of metal–chlorine bond lengths (pp. 302–303) and may indicate that metal(II)–phosphorus bonds are not simple σ-bonds, but are more complex and therefore affected by the *trans*-ligand in a more complex way than metal(II)–chlorine bonds.

If the covalent tetrahedral radii of phosphorus and arsenic are taken as 1·10 and 1·18 Å respectively[197] then the M–P and M–As bond lengths might be expected to be 2·41 Å and 1·49 Å respectively for both palladium(II) and platinum(II). With one notable exception—*trans*-[$(PMe_2Ph)_2PdI_2$]—all the metal(II)–group VB element bond lengths in Tables 29 and 30 fall below these values, which may indicate the presence of π-bonding in these bonds. However, it is possible that the one exception is significant and that to conclude from these bond lengths that π-bonding is present may be dangerous.

(d) Nuclear magnetic resonance studies

The results of n.m.r. spectroscopy have been crucial to an understanding of the bonding in complexes of platinum(II) and palladium(II) with ligands containing phosphorus and arsenic donor atoms.

(i) ^{31}P chemical shift data

In spite of the large amount of ^{31}P chemical shift data available for co-ordination compounds,[198] the interpretation of this data is still very uncertain, because although the main factors that contribute to the chemical shifts have been described[199] there are uncertainties in the magnitudes and occasionally even in the signs of some of these factors. The chemical shifts of tertiary phosphines move downfield on co-ordination indicating that the phosphorus atom is less shielded from the magnetic field in the complex than in the free ligand.[200] However, exactly the opposite is true for tri-methyl- and triphenyl-phosphite complexes where an upfield shift is observed on co-ordination.[201,202] Where halide ligands are also present the ^{31}P chemical shift moves upheld as the atomic number of the halogen increases.[200,203] This is consistent with either a decrease in the σ-component or an increase in the π-component of the platinum(II)–phosphorus bond as a result of an increase in the covalency of the platinum(II)–halogen bond as the halogen increases in atomic number.[203]

The ^{31}P chemical shifts of the *cis*-isomers of tertiary phosphine and tri-phenylphosphite complexes are upfield of the *trans*-isomers.[202,204] This is consistent with the simple idea that strong σ-donation by the phosphorus atom gives a large downfield chemical shift whereas platinum(II)–phosphorus π-back donation gives an upfield chemical shift,[205] since the opportunities for π-bonding are greater in the *cis*- than in the *trans*-isomers.[206]

(ii) $^{31}P-^{31}P$ coupling constants

Phosphorus–phosphorus coupling constants have been widely used to distinguish between *cis*- and *trans*-isomers because the $^{31}P-^{31}P$ coupling is weak in *cis*-bis(phosphine) complexes and fairly strong in the *trans*-complexes.[207-9] Although this has been known for some time,[210,211] only recently have quantitative values of these coupling constants been obtained.[212-4] Double-resonance experiments have shown that the phosphorus–phosphorus coupling constants in the one *cis*- and the one *trans*-bis(tertiary phosphine) complex so far examined are positive in sign.[201,215]

The magnitudes of the phosphorus–phosphorus coupling constants depend on four factors. (i) The stereochemistry of the complex has the largest effect since *cis*-complexes have coupling constants in the range 0–80 Hz, whereas *trans*-complexes exhibit values in the range 551–829 Hz.[215] (ii) The nature of the groups bound to the phosphorus atom have a strong influence on the coupling constants. In the *trans*-complexes the coupling constants increase with the increasing electronegativity of the substituents on the phosphine,[202,212] although in the *cis*-complexes there is a weak tendency for the reverse of this to occur.[153] This may be due to the presence of a strong contact interaction term $((\alpha_p^2)|\psi_{p(3s)}(0)|^2$, where α_p^2 is the s-character of the hybrid orbital used by phosphorus in the platinum–phosphorus bond and $\psi_{p(3s)}(0)$ is the electron density of the 3s orbital of phosphorus at the phosphorus nucleus) in the *trans*-complexes,[212] but that for some, as yet unexplained, reason this term is weak in the *cis*-complexes. (iii) A change of metal atom from platinum(II) to palladium(II) results in an increase in the phosphorus–phosphorus coupling constant.[202] (iv) The ligands other than phosphorus also affect the coupling constant, which increases as the halogen is altered from iodide to bromide to chloride.[215]

A detailed understanding of the factors responsible for the observed phosphorus–phosphorus coupling constants is not yet possible. However, following Pople and Santry's analysis discussed in detail below (pp. 138–140), it has been suggested that the sign of the phosphorus–phosphorus coupling constants is determined by the symmetries of the electronic transitions between orbitals having some phosphorus s-character, an electronic excitation involving orbitals of opposite symmetry giving a positive contribution to the coupling constant and one involving orbitals of the same symmetry a negative contribution.[198,201,215] The magnitude of each of these contributions depends on at least three factors:[216] (i) The coupling constant depends directly on the value of the valence s-orbitals at the phosphorus nucleus. A recent structural review[217] has suggested that the s-character of the metal–phosphorus bonds increases as the electronegativity of substituents on the phosphorus increases, which is again consistent with the observation that coupling constants increase with the increasing electronegativity of the substituents. (ii) The coupling constant depends inversely on the energy difference of the transition, which depends on the energies of the molecular orbitals involved, which in turn depend on the effective nuclear charges of the metal and phosphorus nuclei. (iii) The coupling constant depends on the degree of phosphorus 3s-character in the two orbitals.

(iii) ^{195}Pt *chemical shifts*
Very few studies of the ^{195}platinum chemical shifts of platinum(II) complexes
have been reported because of the difficulties of detecting the signal, due to
the low natural abundance (34 %) and low magnetogyric ratio of ^{195}Pt as well
as the unfavourable relaxation times that are frequently observed. However,
three studies[203,218,219] have shown that the ^{195}Pt chemical shifts extend
over a wide range (> 1600 p.p.m.), suggesting that they are dominated by the
paramagnetic shielding term. A simplification of Ramsey's equation for the
^{195}Pt shielding constant in square-planar platinum(II) complexes has led
to an equation in which the paramagnetic chemical shift is proportional to a
linear combination of the wavelengths of two transitions in the absorption
spectra of the complexes. The proportionality factor depends on the covalency
of the metal–ligand bonds and the radial functions of the platinum 5d orbitals.
The ^{195}Pt chemical shifts in a series of *cis*- and *trans*-[L$_2$PtX$_2$] complexes
depend approximately on the wavelength of the electronic transitions, that
is on the position of the ligand L in the spectrochemical series, which is
related to the covalency of the Pt–L bond. However, for a given ligand L the
chemical shift depends on the position of the halide ligand in the nephelauxetic series. This series represents the fall in electron–electron repulsion
due to lowering of the charge on the metal by increasing the π-covalency of
the metal–ligand bond. Thus the ^{195}Pt chemical shifts appear to be due to
variations in the paramagnetic shielding which arise from changes in both
the σ- and π-covalency of the metal–ligand bond.[203,219]

(iv) ^{195}Pt–^{31}P *coupling constants*
Platinum–phosphorus coupling constants in square-planar platinum(II)–
tertiary phosphine complexes have been widely investigated because of their
potential ability to provide information about the nature of the platinum–
phosphorus bond. Although coupling constants can only give information
about σ-bonding directly, if π-bonding has the usually assumed synergic
effect on the σ-bonds then the coupling constants will also be sensitive to π-
bonding.[202]

Platinum–phosphorus coupling arises from the covalent part of the
platinum–phosphorus bond, so that any change in this covalency (i.e. change
in λ in the valence bond expression 211 for the wavefunction of the platinum–
phosphorus bond) will modify the platinum–phosphorus coupling. The

$$\psi_{Pt-P} = \psi_{covalent} + \lambda \psi_{ionic} \qquad (211)$$

covalent contribution to the platinum–phosphorus coupling can be examined
by considering each term in Pople and Santry's expression (212) for the
Fermi contact term.[220]

$$J_{Pt-P} \propto \gamma_{Pt}\gamma_P(\Delta E^{-1})\alpha_{Pt}^2\alpha_P^2|\psi_{Pt(6s)}(0)|^2|\psi_{P(3s)}(0)|^2 \qquad (212)$$

γ_{Pt} and γ_P = magnetogyric ratios for the nuclei with spin quantum number $\frac{1}{2}$

ΔE = average excitation energy

α_X^2 = s-character of the hybrid orbital used by atom X in the platinum–phosphorus bond

$|\psi(0)|^2$ = electron density of the indicated orbital at the parent nucleus.

(a) ΔE^{-1}. The terms involved in ΔE are singlet–triplet excitation energies. Complexes with the lowest ligand fields should give the lowest ΔE values and hence the largest coupling constants. In fact the coupling constants in *trans*-$[(PBu_3)_2PtX_2]$ decreases as X is changed in the order Cl > NCS > CN, which is opposite to that expected from the order of these ligands in the spectrochemical series, so that any variations in ΔE^{-1} must be dominated by changes in the other parameters in eqn. 212.

(b) α_P^2 and $|\psi_{P(3s)}(0)|^2$. Alteration of the groups bound directly to the phosphorus atom changes α_P^2 and $|\psi_{P(3s)}(0)|^2$ and causes a large change in the observed coupling constant $-J_{Pt-P} = 3508$ Hz in *cis*-$[(PBu_3)_2PtCl_2]$ and 5698 Hz in *cis*-$[\{P(OEt)_3\}_2PtCl_2]$. However, gross changes in the acceptor atom to which the phosphorus is bound seem to have little influence on α_P^2 and $|\psi_{P(3s)}(0)|^2$ and on the metal–phosphorus coupling constants. Thus the ratio $(J_{M-phosphine})/(J_{M-phosphite})$ has a value of 1·62 in *cis*-$[L_2PtCl_2]$ and 1·53 for L-BH_3 where L is a tertiary phosphine or a tertiary phosphite.[221]

(c) α_{Pt}^2 and $|\psi_{Pt(6s)}(0)|^2$. Variations in $|\psi_{Pt(6s)}(0)|^2$, even when quite profound changes are made in the complex, seem to be small and so can be neglected in considering the platinum–phosphorus coupling constant. Thus if it is assumed that, in comparing *cis*- and *trans*-$[(PBu_3)_2PtCl_2]$ with *cis*- and *trans*-$[(PBu_3)_2PtCl_4]$, changes in $|\psi_{Pt(6s)}(0)|^2$ are negligible, a ratio of the coupling constants $(J_{Pt^{IV}-P})/(J_{Pt^{II}-P})$ of 0·667 would be predicted from the ratio of the s-orbital contributions, since in platinum(IV) the s-orbital is shared by six ligands, whereas in platinum(II) it is only shared by four ligands. The experimental coupling constant ratios are 0·59 (*cis*-isomers) and 0·61 (*trans*-isomers).[200]

Heteronuclear magnetic double resonance experiments indicate that the coupling constants in *cis*- and *trans*-$[(PEt_3)_2PtCl_2]$ are positive and similar in magnitude to the constants in other bis–tertiary phosphine platinum(II) complexes.[222] If it is assumed that all platinum–phosphorus coupling constants are positive then the principle factors that influence the platinum(II)–phosphorus coupling constants are the covalency of the platinum(II)–phosphorus bond and the s-character of the platinum orbital.[200] This conclusion is strongly supported by the large differences in the platinum–phosphorus coupling constants for the two non-equivalent phosphorus atoms in complexes such as $[(PBu_3)_3PtCl]^+$ and *cis*-$[(PEt_3)_2Pt(CH_3)Cl]$ where the ΔE^{-1} and $|\psi_{Pt(6s)}(0)|^2$ terms must make identical contributions to the two different coupling constants.[223]

One explanation for the larger platinum–phosphorus coupling constants in *cis*- than in *trans*-complexes considered the effect of π-bonding in the platinum–phosphorus bond.[205] This will be greater in the *cis*-complex because there the two phosphorus atoms (considered to be in the xy-plane of the complex) can use three d-orbitals (d_{xy}, d_{xz}, and d_{yz}) for π-bonding, whereas in the *trans*-complex only the d_{xz} and d_{xy} orbitals are available for π-bond formation.[206] If π-bonding is greater in the *cis*-complex, then since σ- and π-bonding are synergically linked, σ-bonding should also be stronger accounting for the greater platinum–phosphorus coupling constants in *cis*-complexes. Doubt was cast on the validity of this explanation by the observation that the ratio ($J_{Pt–P}(cis)/(J_{Pt–P}(trans)$)) for the complexes [(PBu$_3$)$_2$PtCl$_2$] and [(PBu$_3$)$_2$PtCl$_4$] was 1·41 for the platinum(IV) and 1·47 for the platinum(II) complex suggesting that the σ-bonding was similar in the two oxidation states.[200] Since π-bonding from platinum to phosphorus should be much less important in the higher than in the lower oxidation state it has been suggested that neither oxidation state involves any π-bonding contibution.[132,200] This certainly provoked a great deal of interest and emphasised that the σ-donor ability of tertiary phosphines had previously been underestimated. However, it was probably unnecessarily sweeping since not only the π-acceptor ability but also the σ-donor ability of phosphine ligands would be expected to be greater in *cis*-complexes than in *trans*-complexes, because from the σ-*trans*-effect theory (see pp. 300–301) a strong σ-donor ligand should weaken the bond *trans* to itself. Therefore the similarity of the ($J_{Pt–P}(cis)$)/ ($J_{Pt–P}(trans)$) ratios for platinum(II) and platinum(IV) could arise from a decrease in both the σ- and π-bonding in the platinum(II) case and from a decrease in only the σ-bonding in the platinum(IV) case on going from the *cis*- to the *trans*-isomer. Thus the observation that the ratio ($J_{Pt–P}(cis)$)/ ($J_{Pt–P}(trans)$) is virtually independent of the oxidation state of the platinum atom does not necessarily indicate whether or not π-bonding is present in the lower oxidation state.

Platinum–phosphorus coupling constants are strongly dependent on the ligand *trans* to the phosphorus atom, due to the inductive effect of that ligand on the platinum–phosphorus bond.[202,223,224] Ligands with a very strong σ-inductive character reduce the positive charge on the platinum atom and thus weaken the overlap of the phosphorus and metal orbitals relative to *trans*-ligands which have weaker σ-inductive effects. This has been used to deduce that the σ-inductive ability or *trans*-influence of ligands is in the order MePh$_2$Si > Ph > Me ≫ R$_3$P > (PhO)$_3$P ∼ CN > Et$_3$As > NO$_2$ > amine ∼ N$_3$ ∼ NCO ∼ NCS > Cl ∼ Br ∼ I > ONO$_2$[202,223,224] (the atom italicised in an ambidentate ligand is the donor atom). This series is comparable to that deduced from the metal–chlorine bond length *trans* to a given ligand (pp. 302–303). If further determinations of the *trans*-influence of ligands by other means are found to support the n.m.r.-based *trans*-influence series in cases where this series differs from that based on x-ray diffraction results, then this technique would be preferred to x-ray diffraction because of the greater sensitivity relative to the potential precision with which n.m.r. can determine the *trans*-influence of a ligand.

(e) Photoelectron spectroscopy

Photoelectron spectroscopy can be used to study the effects of σ-donation and π-back-donation in metal–ligand bonds since changes in the electron density about an atom are reflected in changes in the binding energies of core electrons, larger binding energies being associated with smaller electron densities.[225] Unfortunately phosphorus photoelectron spectroscopy has been disappointing in that the binding energies of the phosphorus $2p$ electrons are hardly altered on co-ordination (Tables 31 and 32). However, the observation that the chlorine binding energies are greater and the phosphorus binding energies slightly less in *trans*- than in *cis*-$[(PR_3)_2MCl_2]$ (Table 32) whilst the metal binding energies are identical in both isomers is consistent with the σ-*trans*-effect (pp. 300–301) of tertiary phosphine ligands, which leads to an increase in the negative charge on the ligand *trans* to phosphorus and hence a decrease in the binding energies of the core electrons of that ligand.

TABLE 31

Phosphorus $2p$ electron binding energies in triphenylphosphine and some of its complexes (data from ref. 226)

Compound	Binding energy of $P_{(2p)}$ in eV
PPh_3	131.9 ± 0.4
$[(PPh_3)_2NiCl_2]$	131.6 ± 0.3
$[(PPh_3)_2PdCl_2]$	131.7 ± 0.1
$[(PPh_3)_2CdCl_2]$	131.6 ± 0.2

TABLE 32

Electron binding energies in some tertiary phosphine complexes (data from ref. 227)

Compound	Binding energies (eV)				
	$Pt_{5f_{7/2}}$	$Pd_{3d_{5/2}}$	$Cl_{2p_{3/2}}$	$I_{3d_{5/2}}$	$P_{2p_{3/2}}$
Free metal	71.1	366.3			
cis-$[(PBu_3)_2PtCl_2]$	72.0		198.1		131.1
trans-$[(PBu_3)_2PtCl_2]$	72.0		199.3		130.7
cis-$[(PPh_3)_2PtCl_2]$	72.2		198.0		131.0
cis-$[(PMe_3)_2PtI_2]$	72.9			621.4	132.6
trans-$[(PMe_3)_2PtI_2]$	72.9			622.1	132.1
cis-$[(PEt_2Ph)_2PdCl_2]$		338.0	197.4		130.9
trans-$[(PEt_2Ph)_2PdCl_2]$		338.0	198.2		130.6
trans-$[(PPr_3)_2Pt_2Cl_4]$	71.9		$\begin{cases} 199.1 \text{ terminal} \\ 198.1 \text{ bridging} \end{cases}$		130.8
trans-$[(PBu_3)_2PdCl_4]$		337.7	$\begin{cases} 198.7 \text{ terminal} \\ 197.7 \text{ bridging} \end{cases}$		130.7

(f) Vibrational spectroscopy

Many authors have reported and assigned the infra-red spectra of tertiary phosphine complexes of platinum(II) and palladium(II).[228–40] Prior to the work of Nakamoto *et al.*[240] all the metal–phosphorus stretching vibrations were assigned intuitively without solid experimental evidence, so that there was considerable disagreement over the assignments. However, in 1970

Nakamoto used the palladium-104 (89·75% pure) and palladium-110 (96·98% pure) isotopes to prepare trans-[(PEt$_3$)$_2$PdCl$_2$] and trans-[(PPh$_3$)$_2$PdCl$_2$]. Each complex exhibited only two bands that showed large isotope shifts, one at about 360 cm^{-1} and one at about 190 cm^{-1}. The former was assigned to the metal–halogen stretching frequency and the latter to the metal–phosphorus stretching frequency.[240] Metal–phosphorus or metal–arsenic stretching modes that have been identified in this way are given in Table 33.

TABLE 33

Metal–phosphorus and metal–arsenic stretching frequencies assigned as a result of isotopic substitution studies (data from ref. 240)

| | v_{Pd-P} or v_{Pd-As} (cm^{-1}) | |
Complex	Infra red	Raman
trans-[(PEt$_3$)$_2$PdCl$_2$]	182	
trans-[(PPh$_3$)$_2$PdCl$_2$]	191	152
trans-[(PPh$_3$)$_2$PdBr$_2$]	185	150
trans-[(AsPh$_3$)$_2$PdCl$_2$]	180	132
trans-[(AsPh$_3$)$_2$PdBr$_2$]	180	130

The differences between these values and some earlier assignments, such as the band at 410 cm^{-1} in trans-[(PEt$_3$)$_2$PdCl$_2$] to the infra-red active Pd–P stretching mode,[239] makes it apparent that all stretching modes that have been assigned without the aid of isotopic substitution should be regarded with suspicion. A great deal of experimental work is necessary to straighten out this field.

(g) Electronic spectroscopy

The paler colour of the tertiary phosphine and arsine complexes [(ER$_3$)$_2$MCl$_2$] than the tetrachloro anions [MCl$_4$]$^{2-}$ indicates that tertiary phosphine and arsine ligands have higher ligand-field strengths than chloride ligands. This is supported by the occurrence of the first allowed d–d band in [PtCl$_4$]$^{2-}$ at 25,500 cm^{-1} (Table 65, p. 245) and at 38,000 cm^{-1} in trans-[(PBu$_3$)$_2$PtCl$_2$].[219] This high ligand-field strength means that the d–d transitions are largely obscured by the more intense charge transfer bands. However, tentative assignments of the spectra of platinum(II) complexes indicate that the ligand-field strengths of group VB ligands increase in the order amines < R$_3$As < R$_3$P,[219] which is consistent with earlier work on chromium(III) complexes.[241] The first allowed transition in trans-[Pt(piperidine)LCl$_2$] complexes decreases in energy across the series

L = P(OMe)$_3$ > PPr$_3$ > piperidine > AsPr$_3$ > Et$_2$S > Et$_2$Se > Et$_2$Te.[242]

(h) Nuclear quadrupole resonance spectroscopy

The ^{35}Cl n.q.r. frequencies in cis- and trans-[L$_2$PtCl$_2$] complexes, where L = Me$_2$NH, pyridine and tri-n-butylphosphine, are lower in the cis than the *trans* complexes, and the difference $v_{trans} - v_{cis}$ decreases in the order $PBu_3 > pyr > Me_2NH$ (Table 34).[243] This order, which parallels the *trans*-effect of these ligands, can be explained by assuming that the charge donating properties of L have a larger effect on the ^{35}Cl resonance frequency

TABLE 34
^{35}Cl n.q.r. frequencies in *cis*- and *trans*-$[L_2PtCl_2]$ complexes (data from ref. 243)

| L | ^{35}Cl resonance frequencies (MHz) | | |
	trans-$[L_2PtCl_2]$	cis-$[L_2PtCl_2]$	v_{trans}-v_{cis}
Me_2NH	18·16	17·21	0·95
pyr	19·64	17·65	1·98
PBu_3	20·9 (mean)	17·8 (mean)	3·1

than does the *trans* bond weakening influence of L. Thus the resonance frequencies in the *trans*-$[L_2PtCl_2]$ complexes reflect the charge donating ability of L which decreases in the order $PBu_3 > pyr > MeNH_2$, and $v_{trans} - v_{cis}$ reflects the *trans*-bond weakening effect of L. The results for the *cis*-$[L_2PtCl_2]$ complexes represent the balance of these two effects and thus show very little dependence on the nature of L.

(i) Bonding in phosphine, arsine and stibine complexes

Metal(II)–group VB element bonds are formed by σ-donation of a pair of electrons from the group VB element to an empty orbital on the metal. Phosphorus, arsenic and antimony all have empty *d*-orbitals which are potentially capable of accepting electron density from suitable π-symmetry orbitals on the metal. This π-back donation has been the subject of a great deal of controversy, some authors claiming it is of negligible importance[132,200] and others claiming that it is of significance in the overall bonding pattern of the group VB elements.[244]

Whilst π-back donation from the metal to the group VB element provides an attractive explanation for many of the properties of these complexes, unequivocal evidence to support it is rather limited. Metals with nearly full *d*-orbitals have been described as of class 'b' character, which was taken to be indicative of the ability of the metal to back-donate electron density from its nearly full *d*-orbitals.[245] At first this back-donation was considered to be to a definite orbital, such as a 3*d*-orbital on phosphorus, but recently Jørgenson has suggested that it may be to the continua of orbitals which lie at energies above the first ionisation energy.[246] Subsequently the term 'soft' or readily polarised was introduced[247-9] and consideration of this led to a drift away from the concept of π-back donation and the suggestion that soft–soft interactions are essentially covalent, resulting from favourable enthalpy terms, whereas hard–hard interactions, which are basically ionic, result from favourable entropy terms.[250,251] Let us then examine the evidence for back-donation in metal(II)–group VB element bonds.

(i) Preparative results

The high electronegativity of fluorine should make trifluorophosphine a weaker σ-donor and a stronger π-acceptor than a trialkylphosphine. In agreement with this trifluorophosphine only forms the *cis*-isomer of $[(PF_3)_2PtCl_2]$ where the opportunities for π-back donation are greater than in the *trans*-form.[206] The fact that trifluorophosphine does not form complexes with the boron halides is also indicative of the importance of π-bonding in the formation of stable platinum(II) and palladium(II) complexes.[149] By contrast trichlorophosphine, on account of the lower electronegativity of chlorine than fluorine, should be a better σ-donor and a weaker π-acceptor than trifluorophosphine and in agreement with this PCl_3 forms complexes with boron halides.[252]

Triphenylphosphine is also reluctant to form *trans*-dihalo–platinum(II) complexes, which might be indicative of π-bonding in the platinum(II)–phosphorus bond. By contrast perfluorotriphenylphosphine, which ^{19}F n.m.r. studies suggest does not take part in π-bonding with platinum(II), only forms a *trans*-complex with platinum(II).[158]

The high stability of complexes of tertiary phosphines, arsines and stibines relative to tertiary amines could be the result of π-bonding in the former since the σ-bond strength should decrease in the order $N > P > As > Sb$.[160] Similarly the relative abilities of the group VB elements to promote five-co-ordination in palladium(II) complexes, $(N \ll P < As < Sb)$ can be understood if π-back-acceptance increases in the order $N \ll P < As < Sb$, since the greater the delocalisation of charge away from palladium(II) the easier it is for the metal to accept a fifth ligand.[253] The decreasing reactivity and increasing decomposition temperatures of the platinum(II) complexes $[(EPh_3)_2PtR_2]$, where R = alkyl or aryl, as the group VB element increases in atomic number is consistent with π-back acceptance in the order $PPh_3 < AsPh_3 < SbPh_3$, if the electronic arguments (pp. 332–334) concerning the stability of alkyl and aryl complexes are correct.[254]

(ii) Dipole moments

All bonding descriptions based on dipole moment studies must be regarded with scepticism because these measurements are sensitive to the geometry of the complex. Small distortions from the assumed perfect geometry can cause changes in the dipole moments that are as large as those anticipated from the addition of a small amount of π-back donation in the metal–group VB element bond.

Since *cis*-$[(PF_3)_2PtCl_2]$ and *cis*-$[(CO)_2PtCl_2]$ have very similar dipole moments $(4.4 \pm 0.5$ and 4.65 ± 0.5 Debye respectively)[149] it has been suggested that the $Pt–PF_3$ bond is similar to the $Pt–CO$ bond and involves both σ-donation and π-back donation. The dipole moments for the complexes *cis*-$[(EEt_3)_2PtCl_2]$ $(E = P, 10.9 D; As, 10.9 D; Sb, 10.45 D)$[160] and *cis*-$[(EEt_3)_2PtPh_2]$ $(E = P, 7.2 D; As, 7.2 D; Sb, 6.3 D)$[255] could result from a decrease in σ-donation from ligand to metal in the order $P > As > Sb$ and an approximately constant π-back-donation across this series. However, until exact structural data become available the possibility of the trend in the measured dipole moments being due to small distortions in the geometries of these complexes cannot be ruled out.

(iii) X-ray diffraction data

Pt–P, Pt–As, Pd–P and Pd–As bond lengths found by x-ray diffraction are, with only one exception, less than the sum of the covalent radii of the two atoms (pp. 135–136). This strongly suggests the presence of π-bonding, although these are solid state bond lengths and therefore will be susceptible to modification by crystal forces. It is just conceivable that the one exception is an indication of the sort of effect that crystal forces can have, so that it may not be valid to conclude that these bonds involve π-bonding.

(iv) N.m.r. spectroscopy

Although a study of platinum–phosphorus coupling constants should be able to give information concerning the nature of the bonding in platinum–phosphorus bonds, the results obtained so far are not entirely unequivocal, although they do emphasise the importance of the σ-component and tend to suggest that the π-component is negligible (pp. 136–140).

An alternative application of n.m.r. is the extension of Parshall's ^{19}n.m.r. method[256,257] (pp. 335–336) to the cis-$[(EEt_3)_2Pt(m$- or p-F-$C_6H_4)_2]$ complexes.[255] In this method the difference in the ^{19}F n.m.r. chemical shifts of the para- and meta-fluorosubstituted aryl complexes, which is proportional to the π-electron density in the aryl ring, is taken to reflect the π-acceptor capacity of the trans-ligand. The chemical shifts suggested that the π-bonding in the Pt–E bond decreased in the order $SbEt_3 > AsEt_3 > PEt_3$, although, since the exact geometry of these complexes is not known, it is possible that the orientation of the meta C–F bond may vary from one complex to another, so that the observed chemical shifts may not really reflect π-bonding differences at all.[255]

(v) Infra-red spectroscopy

The $C{\equiv}C$ stretching frequency of $Ph_2PC{\equiv}CPPh_2$ increases on co-ordination to platinum(II) and palladium(II), which is consistent with π-back donation from the metal to the phosphorus d-orbitals decreasing the interaction between the p_π-orbital of the acetylene and the phosphorus d-orbitals so raising the $C{\equiv}C$ bond order. Thus the greater the increase in the $C{\equiv}C$ stretching frequency the greater the π-back donation from the metal to the phosphorus atoms. The results ($v_{C{\equiv}C} = 2097$ cm^{-1} in $Ph_2PC{\equiv}CPPh_2$, 2136 cm^{-1} in $[Pt(Ph_2PC{\equiv}CPPh_2)_2Cl_2]$ and 2135 cm^{-1} in $[Pd(Ph_2PC{\equiv}CPPh_2)Cl_2]$) were considered to indicate that the π-back-donation to platinum(II) was marginally greater than the palladium(II),[257a] although the present author would be very wary of distinguishing between the platinum(II) and palladium(II) cases on the basis of one wavenumber difference.

It is generally assumed in metal–carbonyl complexes that any multiple bond character between the metal and the ligand trans to carbon monoxide will reduce the ability of the metal to undergo π-back-donation to the carbon monoxide so increasing the C–O stretching force constant and the C–O stretching frequency. The C–O and N–O stretching frequencies in a series of cobalt complexes $[LCo(CO)_2NO]$ suggested that the π-bonding ability of a series of ligands L decreased in the order $L = NO \sim CO \sim PF_3 > PCl_3 > P(OR)_3 > PPh_3 > SR_2 > PR_3 > pyr > NR_3$.[257b] A similar but more extensive series of phosphine ligands was obtained from the C–O stretching frequencies of $[Ni(CO)_3L]$ complexes, where L is a phosphine ligand.[258]

(vi) Electronic spectroscopy
The higher ligand field strength of $P(OMe)_3$ than PPr_3[242] suggests the presence of π-bonding in trimethylphosphite complexes because replacement of an alkyl group by an alkoxy group should decrease the σ-donor ability of the phosphorus atom, so that the higher ligand field strength of the alkoxy derivative must result from its higher π-acceptor ability.

(vii) Thermodynamic properties
The heat of formation of cis-$[(EEt_3)_2PtCl_2]$, where $E = P$, As or Sb, is greater than that of the *trans*-isomer.[159,160,259,260] This could be due to π-bonding which should be greater in the *cis*-complex where the two phosphorus atoms (considered to be on the xy-plane of the molecule) can use three d-orbitals (d_{xy}, d_{xz} and d_{yz}) for π-bonding, whereas in the *trans*-complex only the d_{xz} and d_{xy} orbitals are available for π-bond formation.[206]

(viii) Quantum mechanical calculations
When experimental chemists are unable to resolve problems such as the importance of π-bonding in metal–group VB element bonds they often turn to the theoretical chemists for help. However, at present quantum mechanics cannot treat the many-electron problem involved here with sufficient accuracy for a crucial experiment to be designed.[261]

(ix) Conclusion
In conclusion it would appear that group VB ligands are powerful σ-donors and that this σ-donor ability decreases on descending the group ($N > P > As > Sb$). Furthermore, the replacement of alkyl groups (R) by other groups decreases the σ-donor ability in the order $PR_3 > PPh_3 > P(OR)_3 > PCl_3 > PF_3$. Although a detailed analysis of each of the techniques discussed above shows that there is no unequivocal evidence in support of π-bonding in metal–group VB element bonds it would seem to the present author that in the absence of a completely definitive experiment there is on balance more evidence in favour of than against π-bonding and that the order of decreasing π-bonding is $CO \sim PF_3 > PCl_3 > P(OR)_3 > PPh_3 > PR_3 > pyr > NR_3$. This rather cautious conclusion is in line with a recent conclusion about the occurrence of multiple bonding throughout the periodic table.[262]

(j) Five-co-ordinate complexes
Platinum(II) and palladium(II) form complexes in which the metal is five-co-ordinate. There are two idealised geometries for these complexes, namely the trigonal bipyramid of D_{3h} symmetry and the square (or tetragonal)

pyramid of C_{4v} symmetry. These, together with a commonly observed distorted form of the square pyramid, are shown in Fig. 17. The energy difference between the two extreme configurations (the distorted form can be considered as being intermediate) is small[263] and is probably often less than the packing forces in the solid state and the energy of solvation in solution.

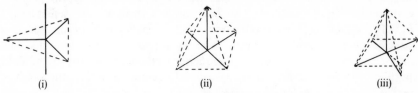

(i)   (ii)   (iii)

Fig. 17. Structures found in five co-ordinate complexes. (i) Trigonal bipyramid, (ii) square pyramid, (iii) a commonly observed distorted form of the square pyramid with one of the ligands displaced below the basal plane.

However, by designing a semi-rigid multidentate ligand[264] such as QAs[265] or QP[266] (**57**) which form trigonal bipyramidal complexes[265,267] (**58**)

$$Ph_2E$$

$$E$$

$$Ph_2E \qquad EPh_2$$

57   (E = P or As)

$$\begin{array}{c} As \quad As \\ As - M \\ \quad As \\ X \end{array}$$

58

it is possible to ensure that a fixed geometry exists both in the solid state[268] and in solution.[269] Electronic spectroscopy has been used to confirm the identity of the solid and solution structures. The *d–d* transitions give unusually intense bands with molar extinction coefficients in the range 500–5000 in these spectra.[270] The solution spectra have an unusual temperature dependence in that the lowest energy band becomes more symmetrical on lowering the temperature.[271] This could be due to either a temperature dependent static distortion of the ground state or a Jahn–Teller distortion of the doubly degenerate excited state.

A slightly less rigid tetra–arsine (TPAs) (**59**) forms a square-pyramidal complex[272] [Pd(TPAs)Cl]$^+$ClO$_4^-$ (**60**). The reasons for the formation of a

$$Me \qquad Me$$

$$As \qquad As$$

$$AsMe_2 \qquad Me_2As$$

59

$$\begin{array}{c} As \\ As - As \\ Pd \\ As \quad Cl \end{array}$$

60

square-pyramidal structure in this case were considered in some detail as the TPAs ligand is sufficiently flexible to form either a square-pyramidal or trigonal bipyramidal structure. The important factors that determine which structure is formed are (a) repulsions between bonding pairs of electrons,[273] (b) the π-bonding requirements of the ligand,[274] (c) ligand-field stabilisation effects,[274] (d) the steric requirements of the ligands[264] and (e) the packing requirements in the crystal. The lack of distortion in the structure of [Pd(TPAs)Cl]$^+$ClO$_4^-$ suggests that both the repulsions between bonding pairs of electrons and the crystal packing forces are small and that the observed geometry is largely a consequence of the fact that the ligand field stabilisation energy of the square pyramidal structure is greater than that of the trigonal bipyramidal structure.[274]

The bidentate tertiary arsine ligand, diars (**61**), forms *bis*-complexes

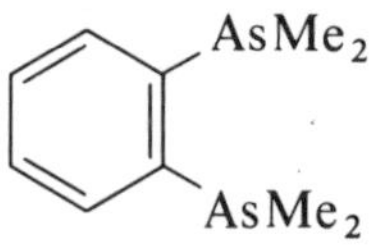

61

[M(diars)$_2$]$^{2+}$, where M = Pd or Pt,[275,276] which on conductimetric titration with halide ions give sharp end-points corresponding to the addition of one equivalent of halide ion. However, crystal-structure analysis of [Pd(diars)$_2$I$_2$] has shown that the palladium atom can be considered as either four- or six-co-ordinate, since the axial palladium–iodine bonds are very long, but not five-co-ordinate[277] (**62**). This emphasises the importance of both solvent interactions and crystal forces in determining the structure

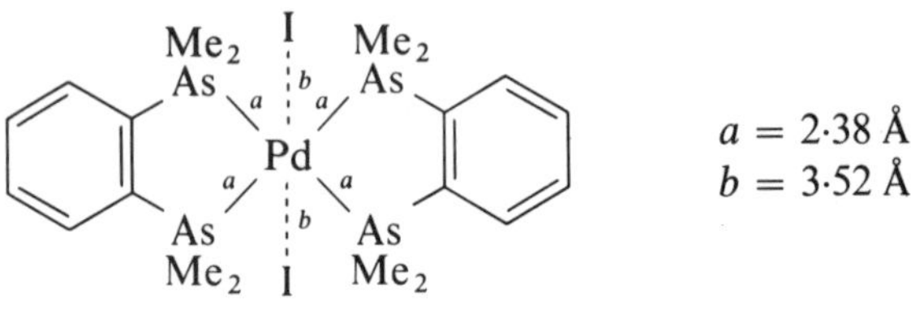

a = 2·38 Å
b = 3·52 Å

62

of these complexes. The stability constant (K = [Pt(diars)$_2$X]/[Pt(diars)$_2$][X]) for the four-co-ordinate/five-co-ordinate equilibrium (213) in methanol

$$[Pt(diars)_2]^{2+} + X^m \xrightleftharpoons{\text{MeOH}} [Pt(diars)_2X]^{2+m} \qquad (213)$$

increased in the order X = thiourea < N$_3^-$ < Cl$^-$ < SCN$^-$ < Br$^-$ < I$^-$ whilst the enthalpy of formation of the five-co-ordinate complexes increased in the order N$_3^-$ < Cl$^-$ < Br$^-$ < SCN$^-$ < I$^-$ < thiourea. The anomalous position of thiourea in the stability constant series may be due to its uncharged character which leads to an unfavourable solvation entropy change. The order of the ligands in terms of increasing enthalpy of formation of the five-co-ordinate complexes follows the order of these ligands in the nephelauxetic series,[16] reflecting the importance of the ability of a ligand to delocalise charge away from the metal if that ligand is to stabilise five-co-ordinate structures. This was also indicated by the increase in ability of group VB ligands to stabilise five-co-ordinate platinum(II) and palladium(II) complexes in the order N < P < As < Sb.[253] This order could arise from a combination of the σ-donor abilities of N > P > As > Sb and π-acceptor capacities of N ≪ P < As < Sb. When a fifth ligand is added to a four-co-ordinate metal the energy of the σ-bonding orbitals is raised and the energy of the π-bonding orbitals is lowered as a consequence of the reduction of the positive charge on the metal atom by the fifth ligand. In cases where σ-bonding is very important (e.g. with nitrogen donors) the bonding of the fifth ligand will be resisted as the overall strength of the complex is weakened. However, if π-bonding is important relative to σ-bonding the fifth ligand will be welcomed as in $[(SbPh_3)_4Pt(NO_3)]^+NO_3^-$.[253]

The third type of structure shown in Fig. 17 is exhibited by the tris-(phosphine)palladium(II) dibromide complex of the phosphine 2-phenyliso-phosphindoline (63)[278] (structure 87 in Appendix II) in which one of the

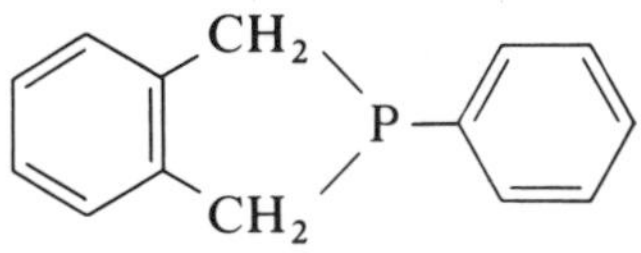

63

bromine atoms is distorted 15° out of the basal plane. It is uncertain why this distortion occurs, although it may well be due to the steric size of the bromine atom, since there is spectroscopic evidence that whereas in $[Pd(TPAs)Cl]^+$ the chloride ligand is in the basal plane (60), in $[Pd(TPAs)I]^+$ the bulky iodide ligand is in axial position.[272]

COMPLEXES WITH MULTIDENTATE LIGANDS CONTAINING BOTH A GROUP VB AND A CARBON DONOR ATOM

Some phosphorus and nitrogen ligands, instead of being unidentate as expected are occasionally bidentate with one of the carbon atoms of the ligand acting as the second donor (Table 35). Many of these complexes involve metallation of a phenyl ring, but a significant number involve the metallation of alkyl groups. Further, with only one exception, all the complexes formed involve five-membered rings.

A classic example of this metallation occurs in the reaction of azobenzene with $[MCl_4]^{2-}$ salts, where M = Pt or Pd[81] (reaction 214). The structure of the product was established first by cleaving the metal–carbon σ-bond with lithium aluminium deuteride which yielded *o*-deuterioazobenzene and secondly by treatment with triethylphosphine, which cleaved the chloride bridge and the palladium–nitrogen bond to yield *trans*-$[(PEt_3)_2Pd(C_6H_5N=NC_6H_4)Cl]$, in which palladium is bound to the *ortho*-carbon atom of one of the phenyl rings[82,293] (see structure 88 in Appendix II). When substituted azobenzenes are used the preference for the metal to become attached to the substituted ring decreases as the substituent is altered in the order OCH_3 >

TABLE 35
Some examples of complexes of nitrogen and phosphorus ligands which
contain intramolecular metal–carbon bonds

Ligand	Complex	Reference
N,N-dialkyl-benzylamine	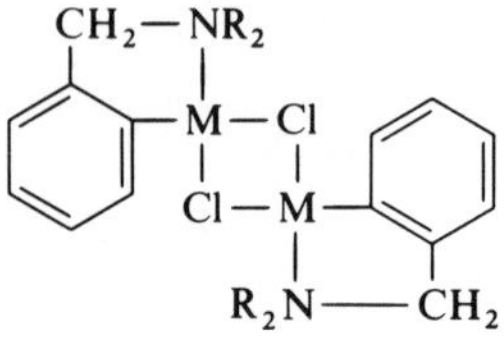 (M = Pd, Pt)	279
N,N-diethylamino-*p*-xylylene	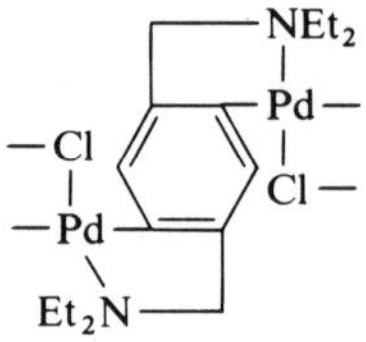 and 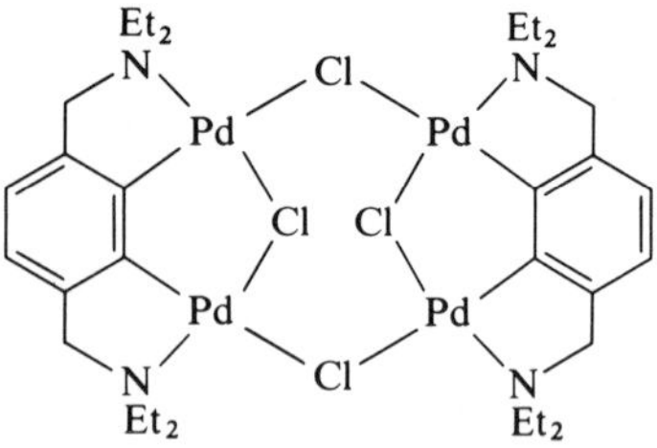	280
1-phenylpyrazole	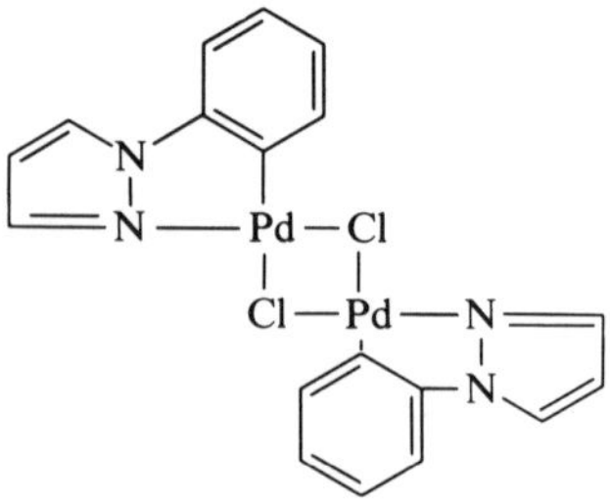	281
N,N-dimethyl-1-naphthylamine	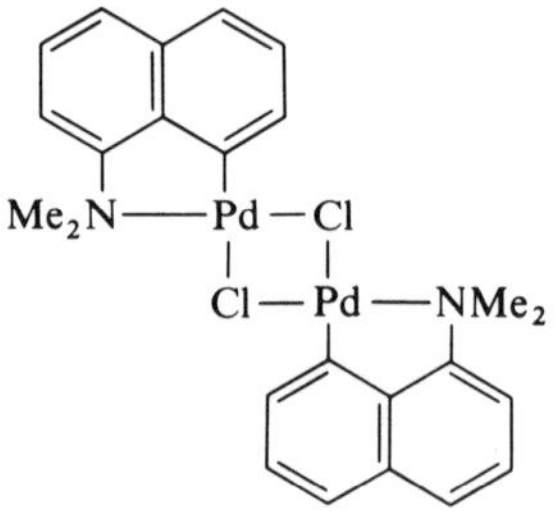	282
Azobenzene	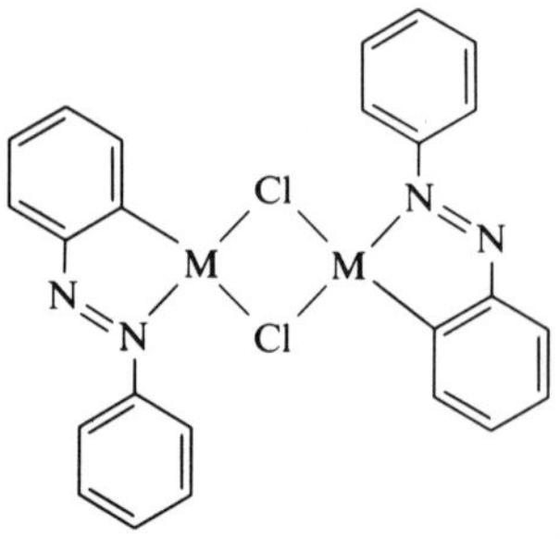 (M = Pd, Pt)	81, 82

TABLE 35 (contd.)

Ligand	Complex	Reference
Substituted azobenzenes	See text for a discussion of the position of the substituent in relation to the phenyl ring that is *ortho*-metallated.	81, 283, 284
Schiff's bases	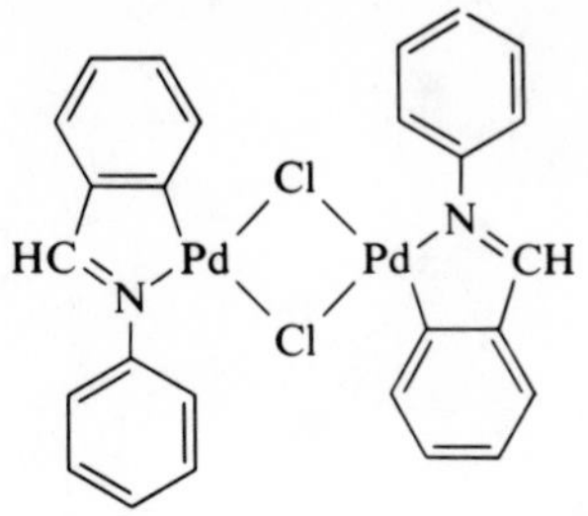	285
2-phenyl-pyridine	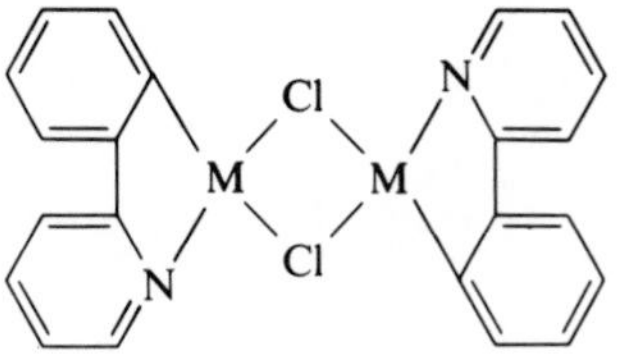 (M = Pd, Pt)	285, 287
2-phenyl-quinoline	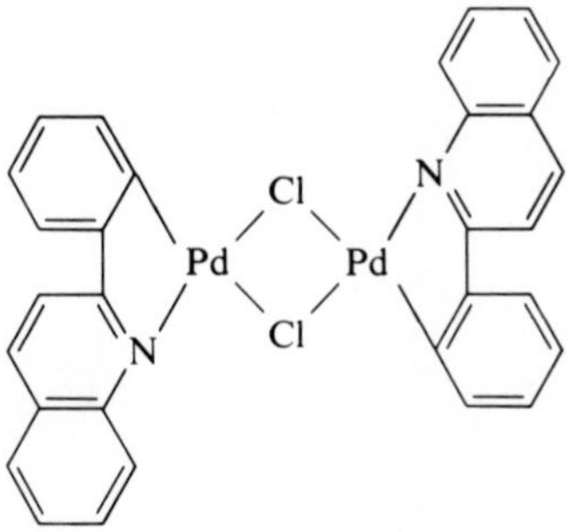	286
Benzo-(h)-quinoline	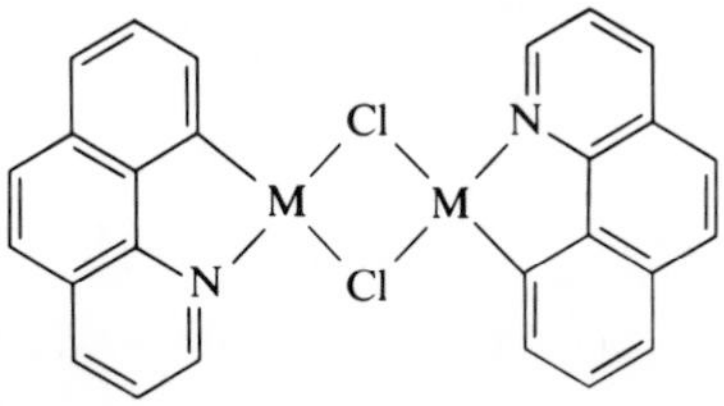 (M = Pd, Pt)	287
8-methyl-quinoline	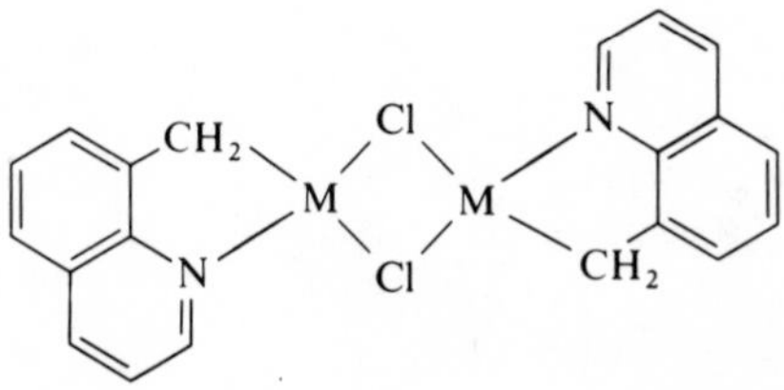 (M = Pd, Pt)	287

TABLE 35 (contd.)

Ligand	Complex	Reference
Benzophenone oxime	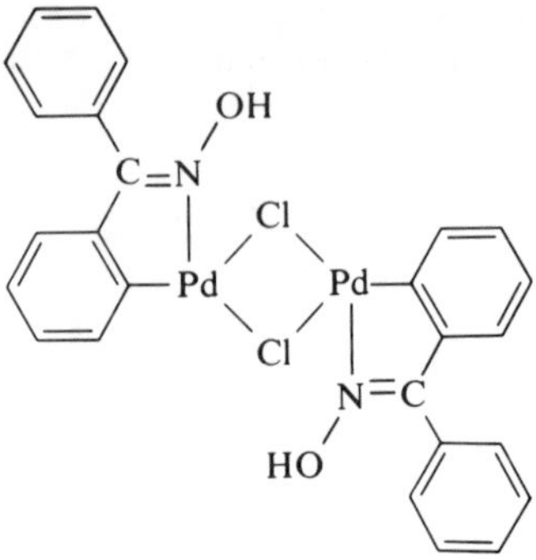	288
Acetophenone oxime	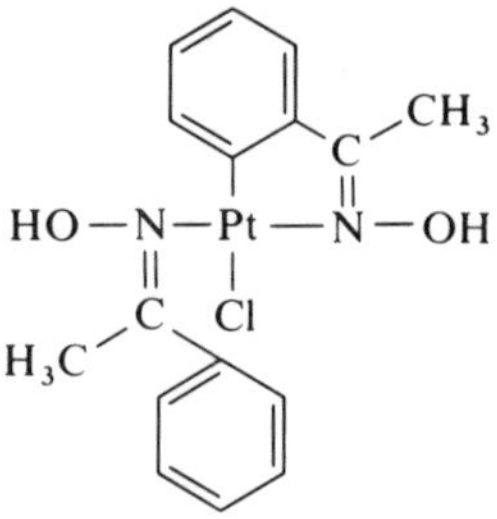	288
PBu$^t_2$Ph	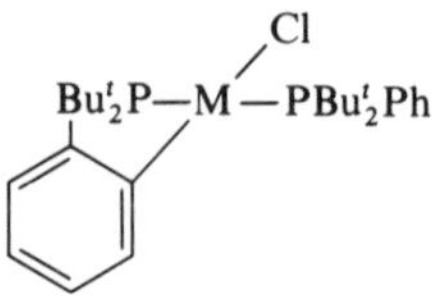 (M = Pd, Pt)	289
PBu$^t_2$Pr^n	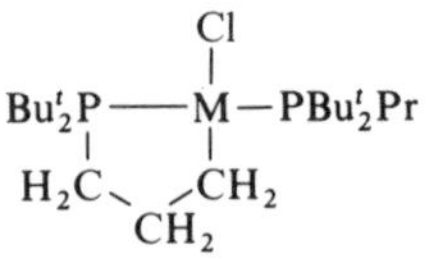 (M = Pd, Pt)	289
PBu$^t_2$(o-tolyl)	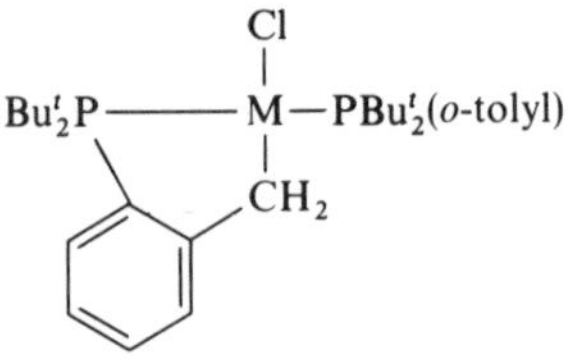 (M = Pd, Pt)	290
PPh$_2$(o-tolyl)	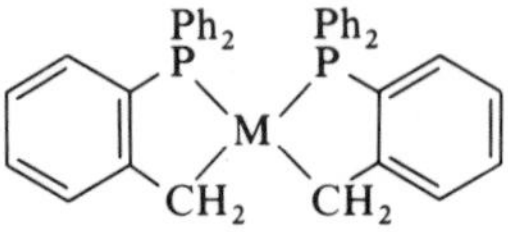 (M = Pd, Pt)	291

TABLE 35 (contd.)

Ligand	Complex		Reference
$P(OC_6H_4X)_3$	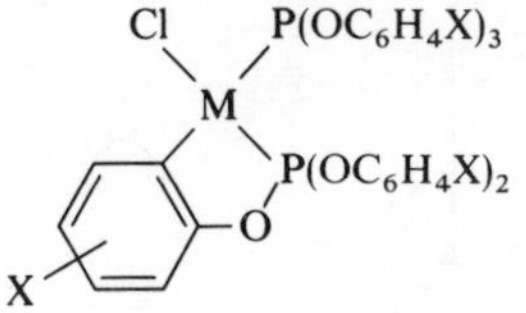	(M = Pd, Pt)	292

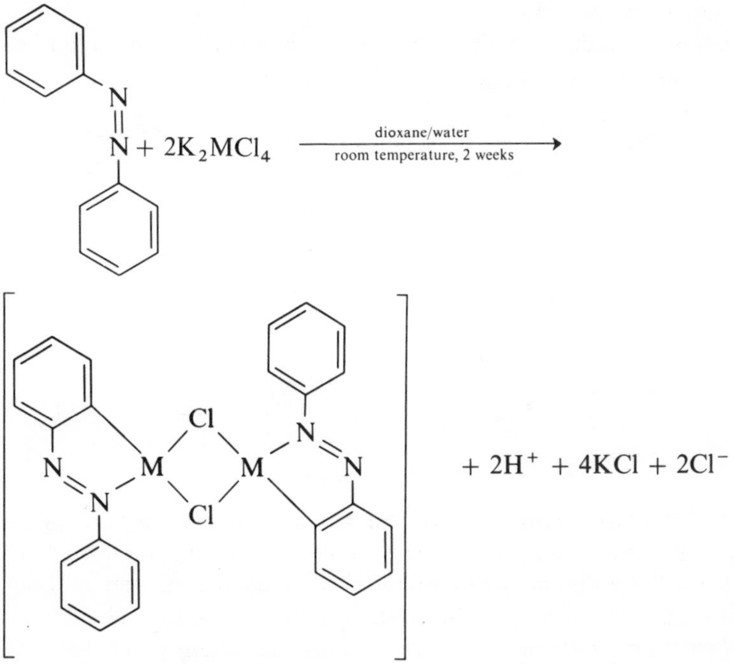

$$(214)$$

$$+ 2H^+ + 4KCl + 2Cl^-$$

$CH_3 > H > Cl$. Thus with 4-methoxyazobenzene only the methoxy substituted ring was attacked whereas with 4-chloroazobenzene the product involving attack in the unsubstituted ring predominated by a 3:1 margin.[283] The electronic effect of these substituents suggests that the aromatic ring has undergone an electrophilic substitution reaction by palladium. The detailed steps for this are probably:

(i) Formation of a complex in which azobenzene is unidentate and bound to palladium through nitrogen (reaction 215) (see p. 124).

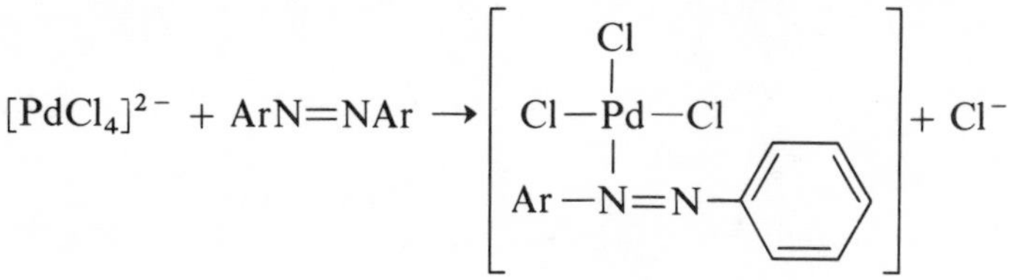

$$[PdCl_4]^{2-} + ArN=NAr \rightarrow \qquad + Cl^- \qquad (215)$$

(ii) The formation of a π-arene complex by the elimination of a further chloride ion (reaction 216). Although there is no direct evidence for the

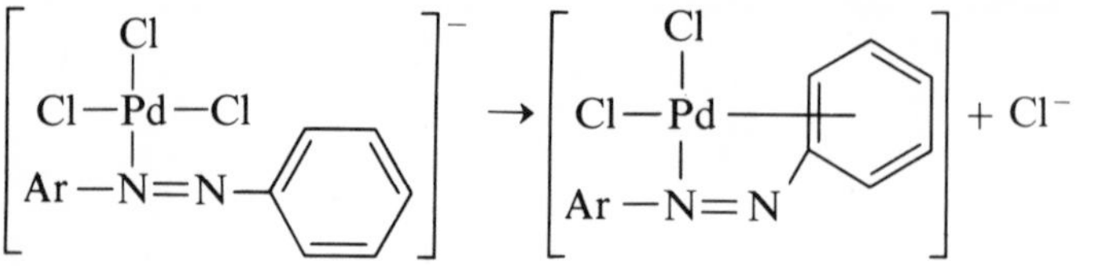

$$
\left[\begin{array}{c} Cl \\ | \\ Cl-Pd-Cl \\ | \\ Ar-N{=}N- \bigcirc \end{array}\right]^{-} \rightarrow \left[\begin{array}{c} Cl \\ | \\ Cl-Pd- \|\ \bigcirc \\ | \\ Ar-N{=}N \end{array}\right] + Cl^{-} \quad (216)
$$

formation of palladium π-arene complexes other than $[Pd_2Cl_{14}Al_4(C_6H_6)_2]$ (p. 17), there is considerable circumstantial evidence that palladium can form transient π-arene complexes.

(iii) Intramolecular electrophilic attack on the benzene ring by palladium leads to the final metallated product (reaction 217).

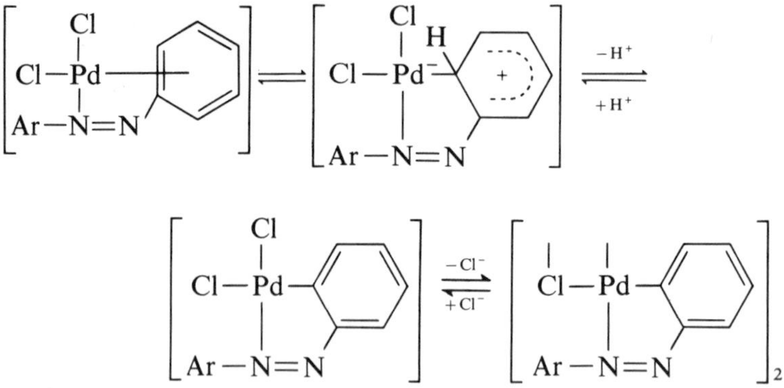

$$
\left[\begin{array}{c} Cl \\ | \\ Cl-Pd- \|\ \bigcirc \\ | \\ Ar-N{=}N \end{array}\right] \rightleftharpoons \left[\begin{array}{c} Cl \ \ H \\ | \\ Cl-Pd^{-} \bigcirc \\ + \\ | \\ Ar-N{=}N \end{array}\right] \underset{+H^{+}}{\overset{-H^{+}}{\rightleftharpoons}}
$$

$$
\left[\begin{array}{c} Cl \\ | \\ Cl-Pd- \bigcirc \\ | \\ Ar-N{=}N \end{array}\right] \underset{+Cl^{-}}{\overset{-Cl^{-}}{\rightleftharpoons}} \left[\begin{array}{c} | \ \ | \\ Cl-Pd- \bigcirc \\ | \\ Ar-N{=}N \end{array}\right]_2 \quad (217)
$$

The overall reaction involves loss of a proton and in many cases metallation is accelerated by the addition of base, although some ligands such as 1-phenylpyrazole[281] and 8-methylquinoline[287] are metallated even in dilute hydrochloric acid. Many ligands, however, are protonated in acid solution and then form salts with the $[MCl_4]^{2-}$ anions. In certain cases, such as benzophenone oxime, a different product is formed in the presence (reaction 218) and absence (reaction 219) of base.[288]

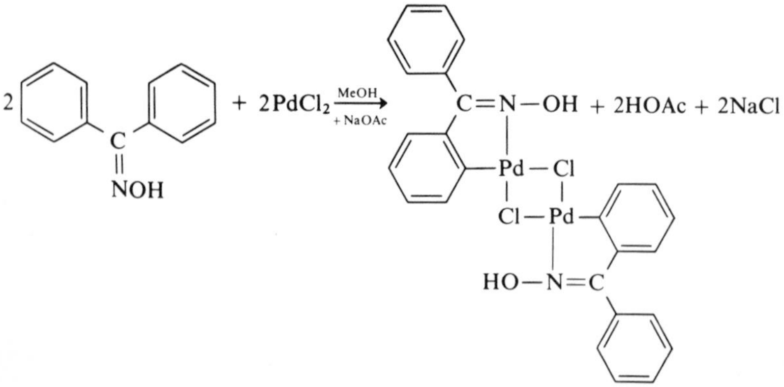

$$
2\ \bigcirc-\underset{\underset{NOH}{\|}}{C}-\bigcirc + 2PdCl_2 \xrightarrow[+\ NaOAc]{MeOH} \quad C{=}N-OH + 2HOAc + 2NaCl
$$

(218)

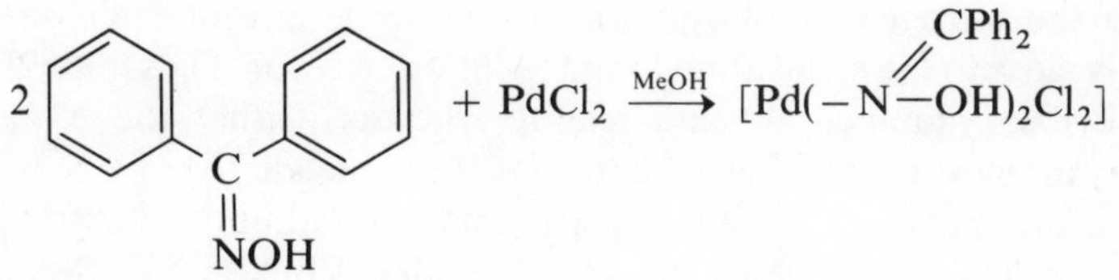

$$2 \text{(structure)} + PdCl_2 \xrightarrow{MeOH} [Pd(-N-OH)_2Cl_2] \qquad (219)$$

Azobenzene–palladium(II) complexes undergo two reactions that are useful in synthetic organic chemistry. Chlorination of azobenzene in the presence of palladium(II) chloride yields 2-chloro- (12%), 2,6-dichloro- (22%), 2,2'-dichloro- (30%), 2,6,2'-trichloro- (33%) and 2,6,2',6'-tetrachloro-azobenzene (3%).[294,295] Carbonylation of azobenzene–palladium(II) complexes in a protic solvent such as ethanol or water yields indazolinones together with metallic palladium[283] (reaction 220) by the insertion of carbon monoxide into the palladium–carbon bond followed by displacement of the palladium.

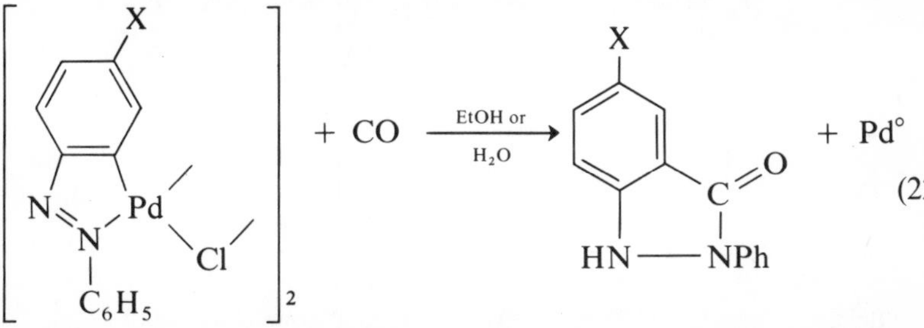

$$\left[\text{(structure)} \right]_2 + CO \xrightarrow[H_2O]{EtOH \text{ or}} \text{(structure)} + Pd^\circ \qquad (220)$$

Many tertiary amine, phosphine or phosphite ligands can be metallated (Table 35) and many of the products are stable to ring cleavage. For example treatment of the chloro-bridged palladium(II) complexes of benzophenone oxime, N,N-dimethylbenzylamine and 2-methoxy-3-N,N-dimethylamino-propane with triphenylphosphine cleaves the chlorobridge but leaves the five-membered ring intact.[288,296] Similarly treatment of the acetophenone complex of platinum(II) with triphenylphosphine displaces the monodentate nitrogen bonded acetophenone oxime ligand but leaves the metallated acetophenone oxime ligand unaffected[288] (reaction 221).

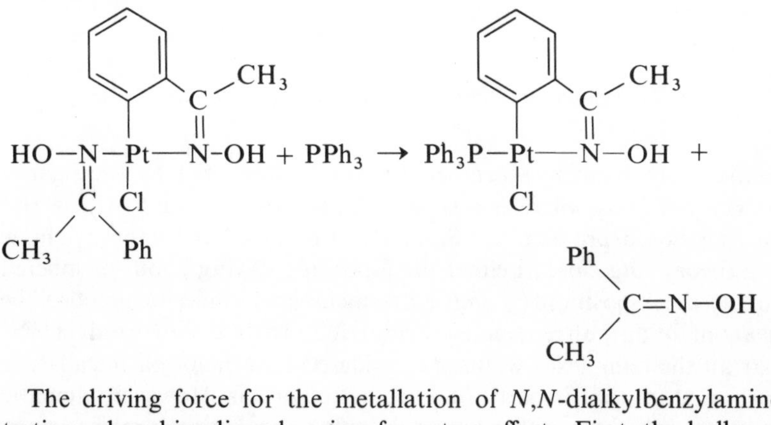

$$\text{(structure)} + PPh_3 \rightarrow \text{(structure)} + \text{(structure)} \qquad (221)$$

The driving force for the metallation of N,N-dialkylbenzylamine and tertiary phosphine ligands arises from two effects. First, the bulky groups

bound to the nitrogen or phosphorus atoms force one of their number
sufficiently close to the metal atom for metallation to occur. Thus for example
N,N-dialkylbenzylamines alkylate readily whereas neither the *N*-mono-
alkylbenzylamines nor free benzylamine itself are metallated.[279] Similarly
PMe_2Ph is not metallated by either platinum or palladium and PBu^tMePh
is metallated less readily than the very crowded PBu_2^tPh.[289] Secondly,
when the groups bound to the nitrogen or phosphorus atom are small the
molecule has a large rotational entropy, since there is virtually free rotation
about the metal–nitrogen or metal–phosphorus bond. However with bulky
amines and phosphines the rotational entropy is low so that the entropy of
activation for the metallation is also low.

Most ring-closed complexes involve five-membered rings (Table 35)
because of the considerable strain in four-membered rings[297] (see Fig. 18).

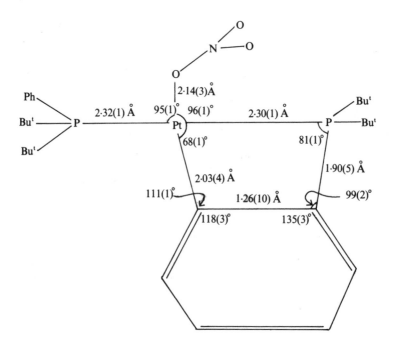

Fig. 18. The structure of $[(PPhBu_2^t)Pt(PBu_2^tC_6H_4)(NO_3)]$ (from ref. 297)

The tendency to form five-membered rings is illustrated by the ligands
$C_6H_5(CH_2)_nN(CH_3)_2$ where $n = 0 - 3$, since only the ligand with $n = 1$
gives a ring-closed product.[279] Similarly *N,N*-dimethyl-1-naphthylamine
could in theory ring close at either the 2-position (giving a four-membered
ring) or at the 8-position (giving a five-membered ring); in practice the
complex containing the five-membered ring is found to be the sole product.[282]

So far all the complexes we have considered have involved metallation
of a carbon atom that is part of an aromatic nucleus. However, aliphatic
carbon atoms, which are either bound directly to an aromatic nucleus as in

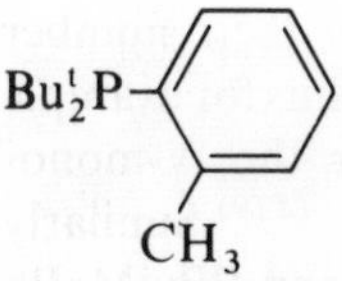

$Bu_2^t P$—⟨phenyl with CH_3⟩ or are part of an aliphatic chain as in $Bu_2^t P$—$CH_2CH_2CH_3$

can also be metallated[289] (Table 35). When non-bulky tertiary phosphines are used, metallation can be forced by first preparing the lithium derivative of the tertiary phosphine[291] (reaction 222).

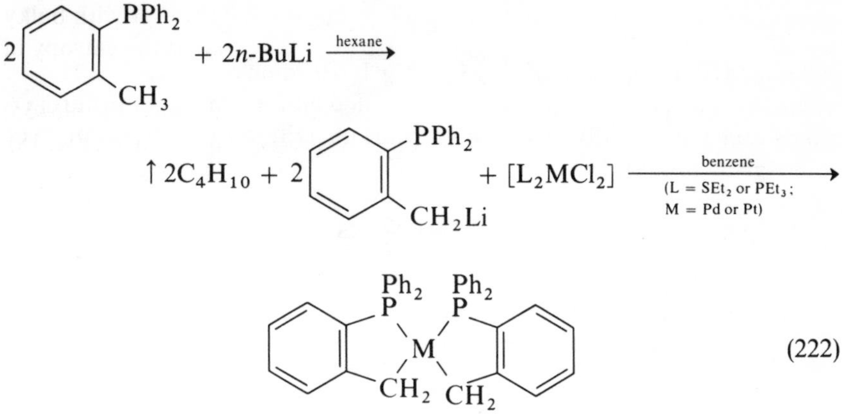

$$2 \text{⟨phenyl with } PPh_2, CH_3⟩} + 2n\text{-BuLi} \xrightarrow{\text{hexane}}$$

$$\uparrow 2C_4H_{10} + 2 \text{⟨phenyl with } PPh_2, CH_2Li⟩} + [L_2MCl_2] \xrightarrow[\substack{(L = SEt_2 \text{ or } PEt_3; \\ M = Pd \text{ or } Pt)}]{\text{benzene}}$$

$$\text{⟨complex structure with } \begin{array}{cc} Ph_2 & Ph_2 \\ P & P \end{array} \text{ bonded to M, } CH_2 \; CH_2⟩ \tag{222}$$

Bis–triarylphosphite complexes eliminate hydrogen halide on boiling in decalin to yield five-membered ring complexes[292] (reaction 223). This reaction is much easier for platinum(II) than for palladium(II) complexes, which may indicate that it occurs by an oxidative-addition of the C–H group to the metal[282] since palladium(II) is known to be reluctant to undergo such reactions (see p. 83 and p. 340).

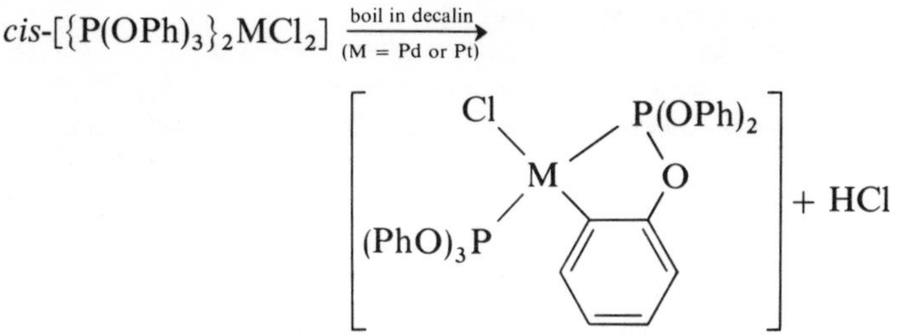

$$cis\text{-}[\{P(OPh)_3\}_2MCl_2] \xrightarrow[\substack{(M = Pd \text{ or } Pt)}]{\text{boil in decalin}}$$

$$\left[\begin{array}{c} \text{⟨Cl, } P(OPh)_2, M, O, (PhO)_3P \text{ structure⟩} \end{array} \right] + HCl \tag{223}$$

The effect of halogens on the ease of ring formation is difficult to rationalise, since for phosphine complexes the rate of ring closure increases as the halogen increases in atomic number[289] (Cl < Br < I) whereas for phosphite complexes the reverse order is observed[292] (Cl > Br > I).

The tris–olefin–phosphine, –arsine and –stibine ligands **64** react to form platinum(II) complexes [PtLBr$_2$], in which infra-red and Raman studies indicate that two of the olefinic double-bonds are free and one is co-ordin-ated.[298] However, the proton n.m.r. spectrum indicates that all three double-bonds are equivalent in solution suggesting that a rapid bonded/non-bonded equilibrium is set up reminiscent of the behaviour of the

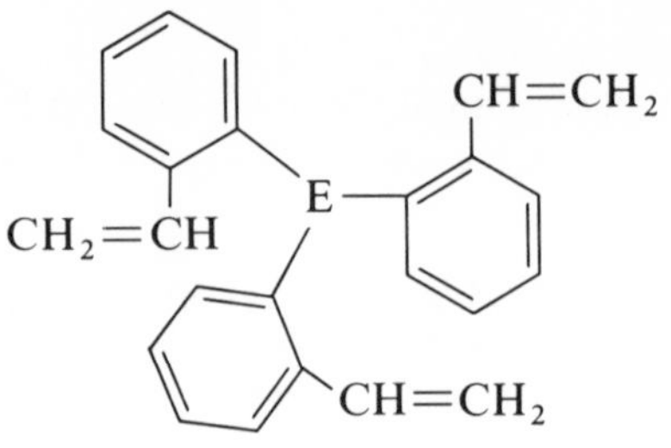

64

palladium(II) complexes of $(o\text{-}Me_2NC_6H_4)_2PPh$ and $(o\text{-}Me_2NC_6H_4)_3P$, where the phosphorus atoms are co-ordinated all the time and the nitrogen atoms exchange rapidly at room temperature, with only one co-ordinated at any given instant.[299]

REFERENCES

1   F. Basolo and R. G. Pearson, *Prog. Inorg. Chem.*, **4** (1962) 381.
2   G. B. Kauffman, *Inorg. Synth.*, **7** (1963) 249.
3   J. Reiset, *Compt. Rend.*, **18** (1844) 1103.
4   M. Peyrone, *Ann. Chem. Liebigs*, **51** (1845) 15.
5   A. A. Grinberg and V. M. Shulman, *Compt. Rend. Acad. Sci. URSS.*, (1933), 215; *Chem. Abs.*, **28** (1934) 1624.
6   R. Layton, D. W. Sink and J. R. Durig, *J. Inorg. Nucl. Chem.*, **28** (1966) 1965.
7   J. S. Coe and J. R. Lyons, *Inorg. Chem.*, **9** (1970) 1775.
8   Magnus, *Pogg. Ann.*, **14** (1828) 204.
9   H. D. K. Drew, F. W. Pinkard, G. H. Preston and W. Wardlaw, *J. Chem. Soc.*, (1932) 1895.
10   E. G. Cox, F. W. Pinkard, W. Wardlaw and G. H. Preston, *J. Chem. Soc.*, (1932) 2527.
11   J. Chatt and L. M. Venanzi, *J. Chem. Soc.*, (1955) 3858.
12   J. Chatt and L. M. Venanzi, *J. Chem. Soc.*, (1957) 2445.
13   J. Chatt and F. G. Mann, *J. Chem. Soc.*, (1939) 1622.
14   G. W. Watt and J. E. Cuddeback, *J. Inorg. Nucl. Chem.*, **33** (1971) 259.
15   G. W. Watt and J. E. Cuddeback, *Inorg. Chem.* **10** (1971) 947.
16   C. K. Jørgensen *Absorption Spectra and Chemical Bonding in Complexes*, Pergamon, Oxford (1962).
17   A. V. Babaeva and R. I. Rudyi, *Zhur. Neorg. Khim.*, **1** (1956) 921; *Chem. Abs.*, **51** (1957) 4140.
18   H. Ito, J. Fujita and K. Saito, *Bull. Chem. Soc. Japan*, 40 (1967) 2584.
19   F. R. Hartley, *J. Organometal. Chem.*, **21** (1970) 227.
20   H. Ito, J. Fujita and K. Saito, *Bull. Chem. Soc. Japan*, **42** (1969) 2863.
21   P. Day, M. J. Smith and R. J. P. Williams, *J. Chem. Soc. (A)*, (1968) 668.
22   B. Bosnich, *J. Chem. Soc. (A)*, (1966) 1394.
23   D. A. Buckingham, L. G. Marzilli and A. M. Sargeson, *J. Amer. Chem. Soc.*, **91** (1969) 5227.
24   S. Yano, H. Ito, Y. Koike, J. Fujita and K. Saito, *Bull. Chem. Soc. Japan*, **42** (1969) 3184.
25   C. H. Perry, D. P. Athans, E. F. Young, J. R. Durig and B. R. Mitchell, *Spectrochim. Acta.*, **23A** (1967) 1137.
26   K. Nakamoto, P. J. McCarthy, J. Fujita, R. A. Condrate and G. T. Behnke, *Inorg. Chem.*, **4** (1965) 36.
27   J. Chatt, L. A. Duncanson and L. M. Venanzi, *J. Chem. Soc.*, (1958) 3203.
28   J. Chatt, L. A. Duncanson and L. M. Venanzi, *J. Chem. Soc.*, (1955) 4461.
29   J. Chatt, L. A. Duncanson and L. M. Venanzi, *J. Chem. Soc.* (1956) 2712.

30 R. C. Leech, D. B. Powell and N. Sheppard, *Spectrochim. Acta*, **21** (1965) 559.
31 J. Chatt and G. A. Gamlen, *J. Chem. Soc.*, (1956) 2371.
32 S. Yano, H. Ito, Y. Koike, J. Fujita and K. Saito, *Chem. Comm.*, (1969) 460.
33 T. G. Appleton and J. R. Hall, *Inorg. Chem.*, **9** (1970) 1807.
34 J. C. Gibson and E. D. McKenzie, *J. Chem. Soc. (A)*, (1971) 1666.
35 Y. O. Aochi and D. T. Sawyer, *Inorg. Chem.*, **5** (1966) 2085.
36 B. B. Smith and D. T. Sawyer, *Inorg. Chem.*, **8**, (1969) 1154.
37 D. J. Robinson and C. H. L. Kennard, *Chem. Comm.*, (1967) 1236.
38 D. J. Robinson and C. H. L. Kennard, *J. Chem. Soc. (A)*, (1970) 1008.
38a J. Altman, E. Babad, J. Itzchaki and D. Ginsburg, *Tetrahedron*, (1966) Suppl. no. 8, 279.
38b J. Altman, E. Babad, J. Pucknat, N. Reshef and D. Ginsburg, *Tetrahedron*, **24** (1968) 975.
39 A. L. Lott and P. G. Rasmussen, *J. Amer. Chem. Soc.*, **91** (1969) 6502.
40 C. R. Kistner, J. H. Hutchinson, J. R. Doyle and J. C. Storlie, *Inorg. Chem.*, **2** (1963) 1255.
41 G. Calvin and G. E. Coates, *J. Chem. Soc.* (1960) 2008.
42 R. A. Plowman and L. F. Power, *Aust. J. Chem.*, **24** (1971) 303.
43 L. F. Power, *Inorg. Nucl. Chem. Lett.*, **6** (1970) 791.
44 F. C. March, D. A. Couch, K. Emerson, J. E. Fergusson and W. T. Robinson, *J. Chem. Soc. (A)*, (1971) 440.
45 P. Haake and P. A. Cronin, *Inorg. Chem.*, **2** (1963) 879.
46 F. Basolo and R. G. Pearson, *Mechanism of Inorganic Reactions*, 2nd edn., Wiley, New York, 1967, pp. 216–18.
47 F. A. Palocsay and J. V. Rund, *Inorg. Chem.*, **8**, (1969) 524.
48 J. V. Rund and F. A. Palocsay, *Inorg. Chem.*, **8** (1969) 2242.
49 J. V. Rund, *Inorg. Chem.*, **9** (1970) 1211.
50 G. C. Dobinson, R. Mason, G. B. Robertson, R. Ugo, F. Conti, D. Morelli, S. Cenini and F. Bonati, *Chem. Comm.*, (1967) 739.
51 L. Chugaev, M. Skanavy Grigorieva and A. Posniak, *Z. anorg. allgem. Chem.*, **148** (1925) 37.
52 G. Rouschias and B. L. Shaw, *Chem. Comm.*, (1970) 183.
53 G. Rouschias and B. L. Shaw, *J. Chem. Soc. (A)*, (1971) 2097.
54 A. Burke, A. L. Balch and J. H. Enemark, *J. Amer. Chem. Soc.*, **92** (1970) 2555.
55 C. K. Prout, R. J. P. Williams and J. D. Wright, *J. Chem. Soc. (A)*, (1966) 747.
56 A. S. Bailey, R. J. P. Williams and J. D. Wright, *J. Chem. Soc.*, (1965) 2579.
57 B. Kamenar, C. K. Prout and J. D. Wright, *J. Chem. Soc.*, (1965) 4851.
58 B. Kamenar, C. K. Prout and J. D. Wright, *J. Chem. Soc. (A)*, (1966) 661.
59 C. K. Prout and A. G. Wheeler, *J. Chem. Soc. (A)*, (1966) 1286.
60 E. Frasson, C. Panattoni and L. Sacconi, *Acta Cryst.*, **17** (1964) 85.
61 E. Frasson, C. Panattoni and L. Sacconi, *Acta Cryst.*, **17** (1964) 477.
62 P. C. Jain and E. C. Lingafelter, *Acta Cryst.*, **23** (1967) 127.
63 R. L. Braun and E. C. Lingafelter, *Acta Cryst.*, **22** (1967) 787.
64 V. W. Day, M. D. Glick and J. L. Hoard, *J. Amer. Chem. Soc.*, **90** (1968) 4803.
65 S. Yamada and K. Yamanouchi, *Bull. Chem. Soc. Japan*, **42** (1969) 2543.
66 K. S. Patel and J. C. Bailar, *J. Inorg. Nucl. Chem.*, **33** (1971) 1399.
67 C. E. Pfluger, R. L. Harlow and S. H. Simonsen, *Acta Cryst.*, **B26** (1970) 1631.
68 M. Calleri, G. Ferraris and D. Viterbo, *Inorg. Chim. Acta*, **1** (1967) 297.
69 G. Ferraris and D. Viterbo, *Acta Cryst.*, **B25** (1969) 2066.
70 E. Frasson, C. Panattoni and R. Zannetti, *Acta Cryst.*, **12** (1959) 1027.
71 C. Panattoni, E. Frasson and R. Zannetti, *Gazz. Chim. Ital.*, **89** (1959) 2132.
72 D. E. Williams, G. Wohlauer and R. E. Rundle, *J. Amer. Chem. Soc.*, **81** (1959) 755.
73 A. V. Babaeva and M. A. Mosyagina, *Dokl. Akad. Nauk. SSSR.*, **89** (1953), 293; *Chem. Abs.*, **47** (1953) 10392.
74 M. Tanimura, T. Mizushima and Y. Kinoshita, *Bull. Chem. Soc. Japan*, **40** (1967) 2777.
75 S. Imamura, T. Kajimoto, Y. Kitano and J. Tsuji, *Bull. Chem. Soc. Japan*, **42** (1969) 805.
76 A. Mawby and G. E. Pingle, *Chem. Comm.*, (1970) 560.

77   K. C. Kalia and A. Chakravorty, *Inorg. Chem.*, **8** (1969) 2586.
78   C. J. Brown, *J. Chem. Soc. (A)*, (1968) 2494.
79   R. H. Nuttall, E. R. Roberts and D. W. A. Sharp, *J. Chem. Soc.* (1962) 2854.
80   J. J. Porter and J. L. Murray, *J. Amer. Chem. Soc.*, **87** (1965) 1628.
81   A. C. Cope and R. W. Siekman, *J. Amer. Chem. Soc.*, **87** (1965) 3272.
82   R. W. Siekman and D. L. Weaver, *Chem. Comm.*, (1968) 1021.
83   R. Murray, *Inorg. Nucl. Chem. Letters*, **5** (1969) 811.
84   A. L. Balch and D. Petridis, *Inorg. Chem.*, **8** (1969) 2247.
85   L. Caglioti, L. Cattalini, F. Gasparrini, G. Marangoni and P. A. Vigato, *J. Chem. Soc. (A)*, (1971) 324.
85a G. W. Parshall, *J. Amer. Chem. Soc.*, **87** (1965) 2133.
86   G. W. Parshall, *J. Amer. Chem. Soc.*, **89** (1967) 1822.
87   W. Beck, K. Feldl and E. Schuierer, *Angew. Chem. Int. Ed.*, **4** (1965) 439.
88   W. Beck, W. P. Fehlhammer, P. Pöllmann, E. Schuierer and K. Feldl, *Chem. Ber.*, **100** (1967) 2335.
89   W. Beck, E. Schuierer and K. Feldl, *Angew. Chem. Int. Ed.*, **5** (1966) 249.
90   K. Bowman and Z. Dori, *Inorg. Chem.* **9** (1970) 395.
91   H.-H. Schmidtke and D. Garthoff, *J. Amer. Chem. Soc.*, **89** (1967) 1317.
92   W. Beck and W. P. Fehlhammer, *Angew. Chem. Int. Ed.*, **6** (1967) 169.
93   W. Beck, W. P. Fehlhammer, P. Pöllmann and H. Schächl, *Chem. Ber.*, **102** (1969) 1976.
94   W. Beck, M. Bauder, G. La Monica, S. Cenini and R. Ugo, *J. Chem. Soc. (A)*, (1971) 113.
95   M. S. Kharasch, R. C. Seyler and F. R. Mayo, *J. Amer. Chem. Soc.*, **60** (1938) 882.
96   J. R. Doyle, P. E. Slade and H. B. Jonassen, *Inorg. Synth.*, **6** (1960) 218.
97   F. R. Hartley, *Organometal. Chem. Rev.*, **A6** (1970) 119.
98   J. R. Holden and N. C. Baenziger, *Acta Cryst.*, **9** (1956) 194.
99   K. A. Hofmann and G. Bugge, *Chem. Ber.*, **40** (1970) 1772.
100 L. Ramberg, *Chem. Ber.* **40** (1907) 2578.
101 M. S. Kharasch and T. A. Ashford, unpublished results quoted in reference 95.
102 V. V. Lebedinskii and V. A. Golovnya, *Izvest. Sekt. Platiny*, **18** (1945) 38; *Chem. Abs.*, **41** (1947) 6187.
103 V. V. Lebedinskii and V. A. Golovnya, *Izvest. Sekt. Platiny*, **21** (1948) 32; *Chem. Abs.*, **44** (1950) 10566.
104 K. A. Jensen, *Z. anorg. allgem. Chem.*, **231** (1937) 365.
105 R. D. Gillard and G. Wilkinson, *J. Chem. Soc.*, (1964) 2835.
106 P. M. Treichel and R. W. Hess, *Chem. Comm.*, (1970) 1626.
107 P. M. Treichel and R. W. Hess, *J. Amer. Chem. Soc.*, **92** (1970) 4731.
108 H. C. Clark and L. E. Manzer, *Chem. Comm.*, (1971) 387.
109 R. A. Walton, *Quart. Rev.*, **19** (1965) 126.
110 G. Beech, G. Marr and S. J. Ashcroft, *J. Chem. Soc. (A)*, (1970) 2903.
111 V. V. Lebedinskii and V. A. Golovnya, *Izvest. Sekt. Platiny*, **16** (1939) 57; *Chem. Abs.*, **34** (1940) 4685.
112 C. M. Harris and N. C. Stephenson, *Chem. Ind. (London)*, (1957) 426.
113 A. A. Grinberg and Kh. I. Gil'dengershel', *Izvest. Sekt. Platiny*, **26** (1951) 115; *Chem. Abs.*, **48** (1954), 13517b.
114 V. A. Golovnya and Ni. Chia-Chien, *Russ. J. Inorg. Chem.*, **6** (1961) 64.
115 Yu. Ya. Kharitonov, Ni. Chia-Chien and A. V. Babaeva, *Russ. J. Inorg. Chem.*, **7** (1962) 513.
116 Yu. Ya. Kharitonov, Ni. Chia-Chien and A. V. Babaeva, *Russ. J. Inorg. Chem.*, **8** (1963) 17.
117 N. C. Stephenson, *J. Inorg. Nucl. Chem.*, **24** (1962) 801.
118 A. D. Gelman and Z. P. Maximova, *Compt. Rend. Acad. Sci. URSS.*, **24** (1939) 748; *Chem. Abs.*, **34** (1940) 1930.
119 W. P. Griffith, J. Lewis and G. Wilkinson, *J. Chem. Soc.*, (1961) 775.
120 E. A. Hadow, *J. Chem. Soc.*, **19** (1866) 345.
121 K. A. Jensen, *5th Nord. Kemikermøde Forh.*, (1939) 200; *Chem. Abs.*, **38** (1944) 2283.

122 Tscherjajev and Hening, *Izvest. Inst. Isuceniju Platiny*, **11** (1933) 46.
123 W. P. Griffith, J. Lewis and G. Wilkinson, *J. Chem. Soc.*, (1959) 1775.
124 W. Manchot and A. Waldmüller, *Chem. Ber.*, **59** (1926) 2363.
125 J. Smidt and R. Jira, *Chem. Ber.*, **93** (1963) 162.
126 J. Lewis, R. J. Irving and G. Wilkinson, *J. Inorg. Nucl. Chem.*, **7** (1958) 32.
127 W. Beck and K. Schmidtner, *Chem. Ber.*, **100** (1967) 3363.
128 J. A. Weaver, P. Hambright, P. T. Talbert, E. Kang and A. N. Thorpe, *Inorg. Chem.*, **9** (1970) 268.
129 M. N. Hughes and K. J. Rutt, *J. Chem. Soc.*, *(A)* (1970) 3015.
130 M. N. Hughes and K. J. Rutt, *Inorg. Chem.*, **10** (1971) 414.
131 G. Booth, *Adv. Inorg. Radiochem.*, **6** (1964) 1.
132 L. M. Venanzi, *Chem. in Britain*, **4** (1968) 162.
133 K. A. Jensen, *Z. anorg. allgem. Chem.*, **229** (1936) 237.
134 J. Chatt and R. G. Wilkins, *J. Chem. Soc.*, (1951) 2532.
135 G. B. Kauffman and L. A. Teter, *Inorg. Synth.*, **7** (1963) 245.
136 S. H. Mastin and P. Haake, *Chem. Comm.*, (1970) 202.
137 J. Chatt, *J. Chem. Soc.*, (1951) 652.
138 J. Chatt and L. M. Venanzi, *J. Chem. Soc.*, (1955) 2787.
139 R. J. Goodfellow and L. M. Venanzi, *J. Chem. Soc.*, (1965) 7533.
140 A. C. Smithies, M. Rycheck and M. Orchin, *J. Organometal. Chem.*, **12** (1968) 199.
141 J. Chatt and L. M. Venanzi, *J. Chem. Soc.*, (1957) 2351.
142 K. Issleib and E. Wenschuh, *Z. anorg. allgem. Chem.*, **305** (1960) 15.
143 R. G. Hayter and F. S. Humiec, *Inorg. Chem.*, **2** (1963) 306.
144 R. G. Hayter, *Nature* **193** (1962) 872.
145 R. G. Hayter, *Inorg. Chem.*, **3** (1964) 301.
146 R. G. Hayter, *J. Amer. Chem. Soc.*, **84** (1962) 3046.
147 F. Klanberg and E. L. Muetterties, *J. Amer. Chem. Soc.*, **90** (1968) 3296.
148 J. Chatt and B. T. Heaton, *J. Chem. Soc. (A)*, (1968) 2745.
149 J. Chatt and A. A. Williams, *J. Chem. Soc.*, (1951) 3061.
150 P. Schutzenberger and M. Fontaine, *Bull. Soc. Chim. France*, **17** (1872) 386, 482; **18** (1872) 148.
151 A. E. Arbuzov and V. H. Zoroastrova, *Izvest. Akad. Nauk. SSSR., Otdel. Khim. Nauk.*, (1952) 809; *Chem. Abs.*, **47** (1953) 9898.
152 W. Strecker and M. Schurigin, *Chem. Ber.*, **42** (1909) 1767.
153 M. J. Church and M. J. Mays, *J. Inorg. Nucl. Chem.*, **33** (1971) 253.
154 G. A. Levishina, A. D. Troitskaya and R. R. Shagidullin, *Russ. J. Inorg. Chem.*, **11** (1966) 985.
155 A. Pidcock and C. R. Waterhouse, *J. Chem. Soc. (A)*, (1970) 2080.
156 A. Pidcock and C. R. Waterhouse, *Inorg. Nucl. Chem. Lett.*, **3** (1967) 487.
157 R. D. W. Kemmitt, D. I. Nicholls and R. D. Peacock, *Chem. Comm.*, (1967) 599.
158 R. D. W. Kemmitt, D. I. Nicholls and R. D. Peacock, *J. Chem. Soc. (A)*, (1968) 2149.
159 J. Chatt and R. G. Wilkins, *J. Chem. Soc.*, (1952) 273.
160 J. Chatt and R. G. Wilkins, *J. Chem. Soc.*, (1952) 4300.
161 J. Chatt and R. G. Wilkins, *J. Chem. Soc.*, (1953) 70.
162 A. F. Wells, *Z. Phys.*, **167** (1938) 169.
163 S. F. Watkins, *J. Chem. Soc.(A)*, (1970) 168.
164 M. Black, R. H. B. Mais and P. G. Owston, *Acta Cryst.*, **B25** (1969) 1760.
165 J. Chatt and F. A. Hart, *J. Chem. Soc.*, (1953) 2363.
166 J. Chatt and F. A. Hart, *Nature*, **169** (1952) 673.
167 J. Chatt and F. A. Hart, *J. Chem. Soc.*, (1960) 2807.
168 J. Chatt and L. M. Venanzi, *J. Chem. Soc.*, (1955) 3858.
169 J. Chatt and L. M. Venanzi, *J. Chem. Soc.*, (1957) 2445.
170 J. Chatt, N. P. Johnson and B. L. Shaw, *J. Chem. Soc.*, (1964) 1662.
171 J. Chatt and B. L. Shaw, *J. Chem. Soc.*, (1962) 5075.
172 L. Malatesta and C. Cariello, *J. Chem. Soc.*, (1958) 2323.
173 T. H. Chan, *Chem. Comm.*, (1968) 895.

174   B. Bosnich and S. B. Wild, *J. Amer. Chem. Soc.*, **92** (1970) 459.
175   P. Groth, *Acta Chem. Scand.*, **24** (1970) 2785.
176   D. L. Weaver, *Inorg. Chem.*, **9** (1970) 2250.
177   R. Mason and D. R. Russell, *Chem. Comm.*, (1966) 26.
178   R. Mason, G. B. Robertson, P. O. Whimp and D. A. White, *Chem. Comm.*, (1968) 1655.
179   R. Mason and P. O. Whimp, *J. Chem. Soc. (A)*, (1969) 2709.
180   D. S. Payne, J. A. A. Mokuolu and J. C. Speakman, *Chem. Comm.*, (1965) 599.
181   N. A. Bailey, J. M. Jenkins, R. Mason and B. L. Shaw, *Chem. Comm.*, (1965) 237 and 296.
182   N. A. Bailey and R. Mason, *J. Chem. Soc. (A)*, (1968) 2594.
183   M. Black, R. H. B. Mais and P. G. Owston, *Acta Cryst.*, **B25** (1969) 1760.
184   F. Cariati, R. Mason, G. B. Robertson and R. Ugo, *Chem. Comm.*, (1967) 408.
185   E. M. Badley, J. Chatt, R. L. Richards and G. A. Sim, *Chem. Comm.*, (1969) 1322.
186   U. A. Gregory, J. A. J. Jarvis, B. T. Kilbourn and P. G. Owston, *J. Chem. Soc. (A)*, (1970) 2770.
187   G. G. Messmer, E. L. Amma and J. A. Ibers, *Inorg. Chem.*, **6** (1967) 725.
188   P. G. Owston, J. M. Partridge and J. M. Rowe, *Acta Cryst.*, **13** (1960) 246.
189   R. Ugo, F. Conti, S. Cenini, R. Mason and G. B. Robertson, *Chem. Comm.*, (1968) 1498.
190   R. Eisenberg and J. A. Ibers, *Inorg. Chem.*, **4** (1965) 773.
191   W. J. Bland, R. D. W. Kemmitt, I. W. Nowell and D. R. Russell, *Chem. Comm.*, (1968) 1065.
192   K. W. Muir and L. Manojlovic-Muir, unpublished results.
193   G. G. Messmer and E. L. Amma, *Inorg. Chem.*, **5** (1966) 1775.
194   H. C. Clark, P. W. R. Corfield, K. R. Dixon and J. A. Ibers, *J. Amer. Chem. Soc.*, **89** (1967), 3360.
195   S. F. Watkins, *Chem. Comm.*, (1968) 504.
196   S. F. Watkins, *J. Chem. Soc. (A)*, (1970) 168.
197   L. Pauling, *The Nature of The Chemical Bond*, Cornell University Press, 3rd edn., (1960), p. 246.
198   J. F. Nixon and A. Pidcock, *Ann. Rev. NMR Spectroscopy*, **2** (1969) 346.
199   K. J. Packer, *J. Chem. Soc.*, (1963) 960.
200   A. Pidcock, R. E. Richards and L. M. Venanzi, *J. Chem. Soc. (A)*, (1966) 1707.
201   R. D. Bertrand, F. B. Ogilvie and J. G. Verkade, *J. Amer. Chem. Soc.*, **92** (1970) 1908.
202   F. H. Allen, A. Pidcock and C. R. Waterhouse, *J. Chem. Soc. (A)*, (1970) 2087.
203   R. R. Dean and J. C. Green, *J. Chem. Soc. (A)*, (1968) 3047.
204   S. O. Grim, R. L. Keiter and W. McFarlane, *Inorg. Chem.*, **6** (1967) 1133.
205   A. Pidcock, R. E. Richards and L. M. Venanzi, *Proc. Chem. Soc.*, (1962) 184.
206   D. P. Craig, A. Maccoll, R. S. Nyholm, L. E. Orgel and L. E. Sutton, *J. Chem. Soc.*, (1954) 332.
207   J. M. Jenkins and B. L. Shaw, *Proc. Chem. Soc.*, (1963) 279.
208   J. M. Jenkins and B. L. Shaw, *J. Chem. Soc. (A)*, (1966) 770.
209   A. Bright, B. E. Mann, C. Masters, B. L. Shaw, R. M. Slade and R. E. Stainbank, *J. Chem. Soc. (A)*, (1971) 1826.
210   P. R. Brookes and B. L. Shaw, *J. Chem. Soc. (A)*, (1967) 1079.
211   D. G. Hendricker, R. E. McCarley, R. W. King and J. G. Verkade, *Inorg. Chem.*, **5** (1966), 639.
212   A. Pidcock, *Chem. Comm.*, (1968) 92.
213   R. J. Goodfellow, *Chem. Comm.*, (1968) 114.
214   D. A. Duddell, J. G. Evans, P. L. Goggin, R. J. Goodfellow, A. J. Rest and J. G. Smith, *J. Chem. Soc. (A)*, (1969) 2134.
215   F. B. Ogilvie, J. M. Jenkins and J. G. Verkade, *J. Amer. Chem. Soc.*, **92** (1970) 1916.
216   J. R. Mass and B. L. Shaw, unpublished results quoted in reference 209.
217   F. B. Ogilvie, R. J. Clark and J. G. Verkade, *Inorg. Chem.*, **8** (1969) 1904.
218   W. McFarlane, *Chem. Comm.*, (1968) 393.
219   A. Pidcock, R. E. Richards and L. M. Venanzi, *J. Chem. Soc. (A)*, (1968) 1970.
220   J. A. Pople and D. P. Santry, *Mol. Phys.*, **8** (1964) 1.

221 J. G. Verkade, R. W. King and C. W. Heitsch, *Inorg. Chem.*, **3** (1964) 884.
222 W. McFarlane, *J. Chem. Soc. (A)*, (1967) 1922.
223 F. H. Allen and A. Pidcock, *J. Chem. Soc. (A)*, (1968) 2700.
224 B. T. Heaton and A. Pidcock, *J. Organometal. Chem.*, **14** (1968) 235.
225 R. S. Berry, *Ann. Rev. Phys. Chem.*, **20**, (1969) 357.
226 J. R. Blackburn, R. Nordberg, F. Stevie, R. G. Albridge and M. M. Jones, *Inorg. Chem.*, **9** (1970) 2374.
227 D. T. Clark and D. B. Adams, *Chem. Comm.*, (1971) 602.
228 G. E. Coates and C. Parkin, *J. Chem. Soc.*, (1963) 421.
229 D. M. Adams, J. Chatt, J. Gerratt and A. D. Westland, *J. Chem. Soc.*, (1964) 734.
230 P. L. Goggin and R. J. Goodfellow, *J. Chem. Soc. (A)*, (1966) 1462.
231 R. J. Goodfellow, P. L. Goggin and L. M. Venanzi, *J. Chem. Soc. (A)*, (1967) 1897.
232 R. J. Goodfellow, P. L. Goggin and D. A. Duddell, *J. Chem. Soc. (A)*, (1968) 504.
233 R. J. Goodfellow, J. G. Evans, P. L. Goggin and D. A. Duddell, *J. Chem. Soc. (A)*, (1968) 1604.
234 D. A. Duddell, P. L. Goggin, R. J. Goodfellow and M. G. Norton, *Chem. Comm.*, (1968) 879.
235 P. J. D. Park and P. J. Hendra, *Spectrochim. Acta*, **25A** (1969) 227.
236 D. M. Adams and P. J. Chandler, *J. Chem. Soc. (A)*, (1969) 588.
237 P. J. D. Park and P. J. Hendra, *Spectrochim. Acta.*, **25A** (1969) 909.
238 J. Chatt, G. J. Leigh and D. M. P. Mingos, *J. Chem. Soc. (A)* (1969) 2972.
239 D. A. Duddell, P. L. Goggin, R. J. Goodfellow, M. G. Norton and J. G. Smith, *J. Chem. Soc. (A)*, (1970) 545.
240 K. Shobatake and K. Nakamoto, *J. Amer. Chem. Soc.*, **92** (1970) 3332.
241 M. A. Bennett, R. J. H. Clark and A. D. J. Goodwin, *Inorg. Chem.*, **6** (1967) 1625.
242 J. Chatt, G. A. Gamlen and L. E. Orgel, *J. Chem. Soc.* (1959) 1047.
243 C. W. Fryer, *Chem. Ind. (London)*, (1970) 115.
244 R. J. P. Williams, *Chem. in Britain*, **4** (1968) 277.
245 S. Ahrland, J. Chatt and N. R. Davies, *Quart. Rev.*, **12** (1958) 265.
246 C. K. Jørgenson, *Structure and Bonding*, **3** (1967) 106.
247 R. G. Pearson, *J. Amer. Chem. Soc.*, **85** (1963) 3533.
248 R. G. Pearson, *Science*, **151** (1966) 172.
249 R. G. Pearson, *Chem. in Britain*, **3** (1967) 103.
250 S. Ahrland, *Structure and Bonding*, **1** (1966) 207.
251 S. Ahrland, *Structure and Bonding*, **5** (1968) 118.
252 D. S. Payne, *Topics in Phosphorus Chemistry*, **4** (1967) 85.
253 A. D. Westland, *J. Chem. Soc.*, (1965) 3060.
254 C. R. Kistner, J. D. Blackman and W. C. Harris, *Inorg. Chem.*, **8** (1969) 2165.
255 A. D. Westland and M. Northcott, *Can. J. Chem.*, **48** (1970) 2907.
256 G. W. Parshall, *J. Amer. Chem. Soc.*, **86** (1964) 5367.
257 G. W. Parshall, *J. Amer. Chem. Soc.*, **88** (1966), 704.
257a A. J. Carty and A. Efraty, *Chem. Comm.*, (1968) 1559.
257b W. D. Horrocks and R. C. Taylor, *Inorg. Chem.*, **2** (1963) 723.
258 A. Tolman, *J. Amer. Chem. Soc.*, **92** (1970) 2953.
259 J. Chatt and R. G. Wilkins, *J. Chem. Soc.*, (1953) 70.
260 J. Chatt and R. G. Wilkins, *J. Chem. Soc.*, (1956) 525.
261 T. Thirunamachandran, personal communication.
262 L. D. Pettit, *Quart. Rev.*, **25** (1971) 1.
263 R. S. Berry, *J. Chem. Phys.*, **32** (1960) 933.
264 L. M. Venanzi, *Angew Chem. Int. Ed.*, **3** (1964) 453.
265 T. E. W. Howell, S. A. J. Pratt and L. M. Venanzi, *J. Chem. Soc.*, (1961) 3167.
266 J. G. Hartley, L. M. Venanzi and D. C. Goodall, *J. Chem. Soc.*, (1963) 3930.
267 C. A. Savage and L. M. Venanzi, *J. Chem. Soc.*, (1962) 1548.
268 G. A. Mair, H. M. Powell and L. M. Venanzi, *Proc. Chem. Soc.*, (1961) 170.
269 J. A. Brewster, C. A. Savage and L. M. Venanzi, *J. Chem. Soc.*, (1961) 3699.
270 M. J. Norgett, J. H. M. Thornley and L. M. Venanzi, *J. Chem. Soc. (A)*, (1967) 540.

271   J. W. Dawson, L. M. Venanzi, J. R. Preer, J. E. Hix and H. B. Gray, *J. Amer. Chem. Soc.*, **93** (1971) 778.

272   T. L. Blundell and H. M. Powell, *J. Chem. Soc. (A)*, (1967) 1650.

273   R. J. Gillespie, *J. Chem. Soc.*, (1963) 4679.

274   Reference 46, Chapter 2.

275   C. M. Harris and R. S. Nyholm, *J. Chem. Soc.*, (1956) 4375.

276   C. M. Harris, R. S. Nyholm and D. J. Phillips, *J. Chem. Soc.*, (1960) 4379.

277   C. M. Harris, R. S. Nyholm and N. C. Stephenson, *Nature*, **177** (1956) 1127.

278   J. W. Collier, F. G. Mann, D. G. Watson and H. R. Watson, *J. Chem. Soc.*, (1964) 1803.

279   A. C. Cope and E. C. Friedrich, *J. Amer. Chem. Soc.*, **90** (1968) 909.

280   S. Trofimenko, *J. Amer. Chem. Soc.*, **93** (1971) 1808.

281   S. Trofimenko, private communication quoted in reference 282.

282   G. W. Parshall, *Acc. Chem. Res.*, **3** (1970) 139.

283   H. Takahashi and J. Tsuji, *J. Organometal. Chem.*, **10** (1967) 511.

284   R. F. Heck, *J. Amer. Chem. Soc.*, **90** (1968) 313.

285   S. P. Molnar and M. Orchin, *J. Organometal. Chem.* **16** (1969) 196.

286   A. Kasahara, *Bull. Chem. Soc. Japan*, **41** (1968) 1272.

287   G. E. Hartwell, R. V. Lawrence and M. J. Smas, *Chem. Comm.* (1970) 912.

288   H. Onoue, K. Minami and K. Nakagawa, *Bull. Chem. Soc. Japan*, **43** (1970) 3480.

289   A. J. Cheney, B. E. Mann, B. L. Shaw and R. M. Slade, *Chem. Comm.*, (1970) 1176.

290   B. L. Shaw, *Joint Annual Meeting of The Chemical Society. and The Royal Institute of Chemistry*, Brighton, April 1971.

291   G. Longoni, P. Chini, F. Canziani and P. Fantucci, *Chem. Comm.*, (1971) 470.

292   E. W. Ainscough and S. D. Robinson, *Chem. Comm.*, (1971) 130.

293   D. L. Weaver, *Inorg. Chem.*, **9** (1970) 2250.

294   D. R. Fahey, *Chem. Comm.*, (1970) 417.

295   D. R. Fahey, *J. Organometal. Chem.*, **27** (1971) 283.

296   B. Crociani, T. Boschi, R. Pietropaolo and U. Belluco, *J. Chem. Soc. (A)*, (1970) 531.

297   R. Countryman and W. S. McDonald, personal communication.

298   D. I. Hall and R. S. Nyholm, *J. Chem. Soc. (A)*, (1971) 1491.

299   H. P. Fritz, I. R. Gordon, K. E. Schwarzhans and L. M. Venanzi, *J. Chem. Soc.*, (1965) 5210.

CHAPTER 8

Divalent Complexes with Group VIB Elements

Platinum(II) and palladium(II) are both class 'b' or 'soft' metal ions (pp. 13–15) and therefore form stronger complexes with sulphur than with oxygen ligands. The stabilities of complexes of sulphur, selenium and tellurium are very similar; the actual stability sequence within these three donor atoms depending on the nature of the other ligands bound to the metal. A comparison of the complexes of the simple oxygen and sulphur ligands shows the following stability order:

$$\left.\begin{array}{c} H_2O \\ ROH \end{array}\right\} \text{ form more stable complexes than } \left\{\begin{array}{c} H_2S \\ RSH \end{array}\right.$$

$$OH^- \quad \text{forms complexes of comparable stability to} \quad SH^-$$

$$\left.\begin{array}{c} R_2O \\ RO^- \end{array}\right\} \text{ form very much less stable complexes than } \left\{\begin{array}{c} R_2S \\ RS^- \end{array}\right.$$

The relative positions of the neutral ligands are those predicted on an electrostatic model from the permanent dipole moments of these ligands since the permanent dipole moments and co-ordinating abilities for the oxygen ligands decrease in the order $H_2O > ROH > R_2O$ whereas for the sulphur ligands both the permanent dipole moments and the co-ordinating abilities increase in the order $H_2S < RSH < R_2S$. This permanent dipole moment will be augmented by an induced dipole moment which will be greater for the sulphur than for the oxygen ligands due to the greater polarisability of sulphur. The presence of empty, relatively low-energy d-orbitals on sulphur could also contribute to the strength of the metal(II)–sulphur bond by accepting π-back-donation of electron density from the metal atom.

Both an electrostatic and a covalent model would suggest that the stability of the complexes of the uninegative ligands should decrease in the order $RO^- > RS^-$. Thus on an electrostatic argument the bond strength $(-\Delta H)$ will be given by:

$$-\Delta H = \frac{Ze \times e}{r_{M^{2+}} + r_{L^-}}$$

Ze = effective nuclear charge on the metal; e = electronic charge; r = ionic radius. The greater size of the sulphur atom should result in $-\Delta H_{M-OR}$

being greater than $-\Delta H_{\text{M–SR}}$ which is the exact opposite of what is observed. If, on the other hand, the M–L bond is covalent, then:

$$-\Delta H = I_{\text{M}} - E_{\text{L}} + f(\chi_{\text{M}} \times \chi_{\text{L}})$$

I_{M} = ionisation potential of the metal; E_{L} = electron affinity of the ligand; $(\chi_{\text{M}} \times \chi_{\text{L}})$ = a function of the product of the electronegativities (χ) of the metal and ligand. Since oxygen is more electronegative than sulphur, $-\Delta H_{\text{M–OR}}$ should be larger than $-\Delta H_{\text{M–SR}}$, whereas in fact the reverse is found experimentally. The fact that neither the electrostatic nor the covalent descriptions of the σ-bonding in metal(II) complexes of uninegative ligands can explain the observed stability order $\text{RS}^- > \text{RO}^-$ suggests that π-back donation of electron density from metal to ligand may be important in the mercaptide complexes. It will be absent in the alkoxide complexes due to the lack of any suitable empty low energy orbitals on oxygen.

Ligands such as sulphite ions, thiosulphate ions, thiourea and dialkyl-sulphides that bind to platinum(II) and palladium(II) through a sulphur atom generally exhibit a high *trans*-effect as deduced from preparative studies.[1] However, the *trans*-influence of these ligands as deduced from the infra-red stretching frequencies of metal–chlorine bonds *trans* to dialkyl-sulphides (see Table 60) and from the lengthening of the metal–ligand bond in the *trans*-position is negligible (see p. 177 for details on dialkylsulphide complexes, p. 180 for details on sulphite complexes and p. 183 for details on thiosulphate complexes). Thus the *trans*-effect of these ligands is probably a π-*trans*-effect resulting from π-back-donation of electron density from the metal atom to sulphur, which leaves the ligand with little or no *trans*-influence (see pp. 301–303). It must be remembered, however, that it is possible for a ligand that has no apparent *trans*-influence to exert a σ-*trans*-effect (pp. 300–301).

Since the ligands containing group VIB elements are rather diverse in type the first part of this chapter has been divided up into subsections, as indicated in the list of contents. In the second part of the chapter multidentate and ambidentate ligands containing both group VB and group VIB donor atoms are considered.

BINARY CHALCOGENIDES AND HYDROXIDES

Whilst the oxides and sulphides of platinum(II) and palladium(II) are typical compounds the selenides and tellurides lie on the borderline of compounds and alloys (Table 36). Thus whilst the preparation of palladium(II) selenide suggests that it is a typical compound prepared similarly to palladium(II) sulphide,[16] the true formula of platinum selenide is $\text{PtSe}_{0.8}$ suggesting it is an alloy.

In keeping with the ability of sulphur to form polysulphides $(\text{Pr}_4\text{N}^+)_2$-$[\text{PtS}_{10}]^{2-}$ has been prepared in which a planar platinum(II) atom is surrounded by two pentasulphide (S_5^{2-}) chelating ligands.[32]

TABLE 36

Preparation and Properties of the Binary Chalcogenides and Hydroxides

Compound	Preparation	Chemical properties	Physical properties
PdO	(i) Heat Pd metal in oxygen[2] (ii) Fuse $PdCl_2$ with $NaNO_3$ at 600°C[3]	Insoluble in acids, including aqua-regia. Strong oxidising agent which glows in contact with hydrogen[4] and oxidises CO to CO_2 above 100°C.[5,6] PdO is a powerful catalyst for the reduction of organic compounds, such as olefins (to alkanes)[7] and aldehydes (to alkanes),[8] by hydrogen	Black powder. X-ray diffraction[9] shows Pd has a distorted square planar environment and O has a distorted tetrahedral environment Pd–O bond length = 2·01(1) Å
$Pd(OH)_2$ or $PdO . H_2O$	Addition of alkali to a Pd(II) compound, especially $Pd(NO_3)_2$[4]	Soluble in acids to give Pd(II) salts and alkalies to give palladites (PdO_2^{2-}).[10] Loses H_2O slowly on heating, but does not lose all of it until heated to 500–600°C, when it begins to lose oxygen.[4] $Pd(OH)_2$ is stable in water, although easily reduced to metallic Pd by formic acid, hydrogen peroxide and hydrogen.[10] The following equilibrium data has been reported:[11] $\log_{10} k_1 = 13\cdot0 \pm 0\cdot4$ at 25°C and zero-ionic strength. $\log_{10} k_2 = 25\cdot8 \pm 0\cdot4$ at 25°C and zero-ionic strength. $\log_{10} k$ solubility product $= -2\cdot7$ ($K_n = [Pd(OH)_n^{2-n+}]/[Pd^{2+}]\cdot[OH^-]^n$)	
PdS	(i) Action of H_2S on a Pd(II) salt[12] (ii) Heat metallic Pd with elemental S[12]	Insoluble in water or dilute HCl.[12] Slight soluble in HNO_3 and aqua-regia[13]	When made in wet way it is a brown powder; whereas when made in dry way it forms bluish insoluble crystals.[12] X-ray diffraction[14] shows Pd has a distorted square-planar environment and S has a distorted tetrahedral environment. Semiconductor.[15] Diamagnetic[15]
PdSe	(i) Add a solution of $PdCl_2$ drop by drop to a saturated solution of hydrogen selenide[16] (ii) Heat metallic Pd with elemental Se	Insoluble in water, but soluble in aqua-regia[13]	Dark brown solid[16] with PdS structure[17] Semiconductor.[15] Diamagnetic[15]

TABLE 36 (contd.)

Compound	Preparation	Chemical properties	Physical properties
PdTe	Heat metallic Pd with metallic Te		NiAs structure.[17] Superconductor below 4°K[18–20]
PtO	(i) Heat Pt under 8 atm. of oxygen at 420–440°C[21] (ii) An impure material containing some free Pt can be obtained by heating Pt(OH)$_2$[22]	PtO catalyses the reduction of aldehydes to alcohols by hydrogen[8]	Grey powder. X-ray diffraction[9] shows Pt has a distorted square-planar environment and O has a distorted tetrahedral environment; Pt–O bond length = 2·02(2) Å
Pt(OH)$_2$ or PtO . H$_2$O	(i) Action of hot KOH on PtCl$_2$[23] (ii) Action of hot KOH on K$_2$PtCl$_6$ in presence of SO$_2$[23]	Very rapidly oxidised in air so that it must be prepared in an inert atmosphere. It behaves as a weak base. It is stable in water but unstable in hot alkalies and hot acids disproportionating to a mixture of the Pt(IV) compound and Pt metal.[24] Pt(OH)$_2$ is reduced by H$_2$O$_2$ and CH$_3$COOH[25] to Pt metal and oxidised by ozone or KMnO$_4$ to PtO$_2$[24]	Black powder
PtS	(i) Heat Pt sponge and S in a closed tube[26] (ii) Fuse PtCl$_2$ with Na$_2$CO$_3$ and S[24] (iii) Pass hydrogen sulphide into a solution of a [PtCl$_4$]$^{2-}$ salt[27]	It is very resistant to acids and alkalies and insoluble in aqua-regia[27]	Black precipitate as prepared by (iii) that forms grey needles. X-ray diffraction[28] shows that there are four sulphur atoms around each Pt at corners of a rectangle 3·040 Å × 3·470 Å. Shortest Pt–S = 2·312 Å. Diamagnetic.[28] Semiconductor[15]
PtSe$_{0·8}$	Heat Pt and Se[29]		Metallic and diamagnetic[28]
PtTe	Heat PtTe$_2$ formed on heating Pt and Te[30]		Structure uncertain as more recent work[28] inconsistent with earlier work[31] which showed NiAs-type structure. Diamagnetic.[28] Superconductor below 4°K[19,20]

COMPLEXES WITH UNIDENTATE GROUP VIB LIGANDS

(a) H₂E complexes

Complexes of divalent platinum and palladium containing H_2E where E is a group VIB element, are only known when E is oxygen because hydrogen sulphide and selenide both react with compounds of the divalent metals to precipitate the sulphide or selenide. Although $[Pd(H_2O)_4]^{2+}(ClO_4)_2$ has been prepared[33] aquo-complexes are rarely isolated because water is a very poor ligand for both platinum(II) and palladium(II) (Table 57, p. 239). Aquo-complexes are very important in the theory of the mechanism of the reactions of the square-planar divalent complexes as they are considered to be responsible for the first-order term in the rate law in reactions that take place in aqueous solution (p. 294).

(b) HE⁻ complexes

Hydroxide ions are very 'hard' bases and are not therefore expected to react readily with platinum(II) and palladium(II) compounds. However, several hydroxy-complexes are known. Although rarely isolated, hydroxy-complexes are responsible for the acidities of solutions of aquo-complexes[34-9] (reaction 224). Although unstable hydroxy complexes were reported earlier,[40] the

$$[PtCl_4]^{2-} + H_2O \rightleftharpoons Cl^- + [PtCl_3(H_2O)]^- \rightleftharpoons$$
$$[PtCl_3(OH)]^{2-} + H^+ \quad (224)$$

first stable crystalline one (see note p. 229) was $[(PEt_3)_2Pt(GePh_3)(OH)]$[41]

$$[(PEt_3)_2Pt(GePh_3)I] \xrightarrow{H_2O, LiOH} [(PEt_3)_2Pt(GePh_3)(OH)] + LiI \quad (225)$$

(reaction 225), which was unaffected by hydrogen or tertiary phosphines. The hydroxy-group reacts reversibly with alcohols to give alkoxy-complexes[41] (reaction 226). The stability of this complex may be due to the very

$$[(PEt_3)_2Pt(GePh_3)(OH)] + ROH \underset{}{\overset{(R = Me, Et \text{ or } Pr)}{\rightleftharpoons}}$$
$$[(PEt_3)_2Pt(GePh_3)(OR)] + H_2O \quad (226)$$

high *trans*-influence of the triphenylgermyl group (p. 82) which renders the *trans*-position more class 'a' in character than it otherwise would be.

Air stable crystalline hydroxy-complexes of dimeric phosphinato-bridged platinum(II) complexes are prepared by reaction 227 where R = Ph or Et, M′ = P or As and R′ = Me, Et, Bu or Ph.[42] The hydroxy-group is replaced

$$2\begin{bmatrix} R'_3M' & & Cl \\ & \diagdown \diagup & \\ & Pt & \\ & \diagup \diagdown & \\ ClR_2P & & Cl \end{bmatrix} \xrightarrow[+H_2O + LiOH]{in\ C_6H_6}$$

$$\begin{bmatrix} R'_3M' & & (R_2PO) & & OH \\ & \diagdown \diagup & & \diagdown \diagup & \\ & Pt & & Pt & \\ & \diagup \diagdown & & \diagup \diagdown & \\ HO & & (R_2PO) & & M'R'_3 \end{bmatrix} \quad (227)$$

by chloride on treatment with hydrochloric acid (reaction 228). The corresponding arsinato-bridged hydroxy-complexes cannot be prepared, possibly due to a greater involvement of the platinum d-orbitals in the

$$[(R'_3M')(HO)Pt(R_2PO)_2Pt(OH)(M'R'_3)] + 4HCl \rightarrow$$
$$2\ cis\text{-}[(M'R'_3)Pt(PR_2OH)Cl_2] + 4H_2O \quad (228)$$

phosphinato than in the arsinato bridge, which draws the d-electrons away from the terminal positions, so making these terminal positions more class 'a' in character in the phosphinato-bridged complexes.

The only platinum(II) complexes of SH$^-$ and SeH$^-$ are $[(PPh_3)_2PtH(EH)]$, where E = S or Se, prepared by treating $[Pt(PPh_3)_2]$ with an excess of hydrogen sulphide or hydrogen selenide[43] (reaction 229). The high n.m.r. chemical shift of the EH$^-$ proton (11·4 p.p.m. in SH$^-$ and 14·2 p.p.m. in SeH$^-$ relative

$$[Pt(PPh_3)_2] + H_2E \rightarrow \begin{bmatrix} & H & \\ & / & \\ E & & PPh_3 \\ \diagdown & / & \\ & Pt & \\ / & \diagdown & \\ Ph_3P & & H \end{bmatrix} \quad (229)$$

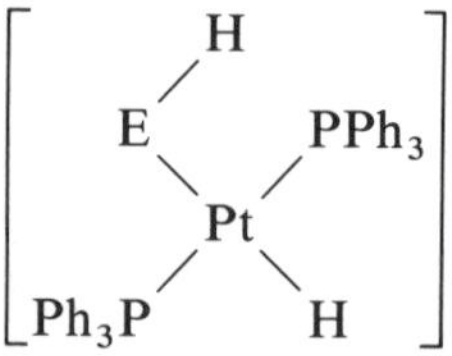

to tetramethylsilane at 10 p.p.m.) suggests that this proton interacts slightly with the platinum atom. The complexes $[(PPh_3)_2PtH(EH)]$ do not react with class 'b' or 'soft' ligands such as carbon monoxide, ethylene or acetylene. Oxygen converts $[(PPh_3)_2PtH(SH)]$ into $[(PPh_3)_2PtS]_2$ and hard electrophilic reagents such as hydrogen chloride and alkyl, acyl and aroyl halides react to give the hydrido–platinum(II) complex $[(PPh_3)_2PtHCl]$[43] (reactions 230–2). The inability of the hydrosulphide and hydroselenide com-

$$[(PPh_3)_2PtH(EH)] + HCl \rightarrow [(PPh_3)_2PtHCl] + H_2E \quad (230)$$

$$[(PPh_3)_2PtH(EH)] + MeI \rightarrow [(PPh_3)_2PtHI] + MeEH \quad (231)$$

$$[(PPh_3)_2PtH(EH)] + RCOCl \rightarrow [(PPh_3)_2PtHCl] + RCOEH \quad (232)$$

plexes to react with 'soft' ligands such as carbon monoxide and unsaturated organic compounds is of relevance to the poisoning of platinum surfaces by sulphur and selenium compounds since metallic platinum is used to catalyse the reactions of such molecules.

(c) E^{2-} complexes

The simple binary compounds of group VIB element anions have already been described (pp. 166–168). Of the various group VIB element anions only the sulphide is found in complexes where it acts as a bridging ligand. Examples of sulphide complexes are $[(PMe_2Ph)_4Pt_2S_2]$ (65) formed by treating cis-$[(PMe_2Ph)_2PtCl_2]$ with sodium sulphide in ethanol in the absence of

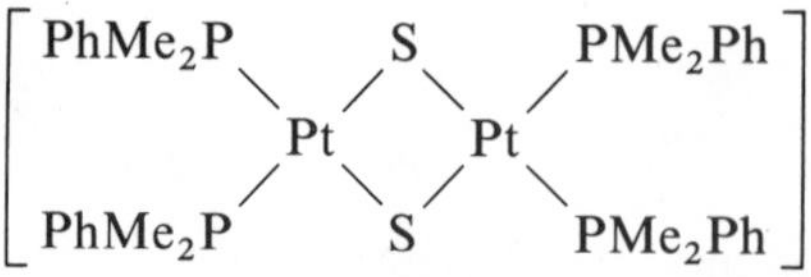

$$\begin{bmatrix} PhMe_2P & & S & & PMe_2Ph \\ \diagdown & / & \diagdown & / & \\ & Pt & & Pt & \\ / & \diagdown & / & \diagdown & \\ PhMe_2P & & S & & PMe_2Ph \end{bmatrix}$$

65

bulky anions[44] and $[(PMe_2Ph)_6Pt_3S_2]X_2$, where $X = ClO_4^-$, BPh_4^- or BF_4^- (**66**) formed in the presence of bulky anions.[44] $[(PPh_3)_4Pt_2S]$, formed

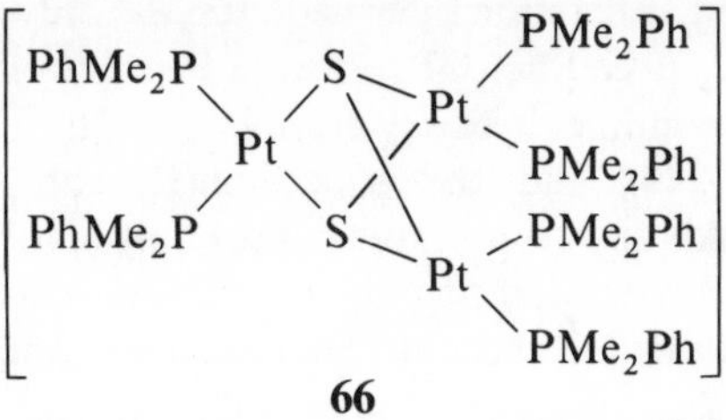

66

by treating cis-$[(PPh_3)_2PtCl_2]$ with sodium sulphide in ammoniacal ethanolic solution,[44] has a structure analogous to that of $[(PPh_3)_3Pt_2S(CO)]^{[45,46]}$ (**67**)

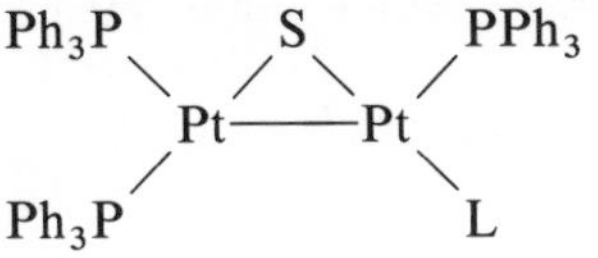

$(L = CO$ or $PPh_3)$

67

prepared by heating $[(PPh_3)_2Pt(COS)]$ in chloroform.[47]

Polysulphide complexes $[(PPh_3)_2MS_4]$, where $M = Pd$ or Pt, which probably contain five-membered MS_4 rings are prepared by heating $[M(PPh_3)_4]$ with $6\,mol$ of elemental sulphur.[44] Polymeric complexes

$[(PPh_3)_2PtS]_n$ and $[(PPh_3)_2PtSe]_n$ containing $Pt \overset{E}{\underset{E}{\diagdown\diagup}} Pt$ bridges are pre-

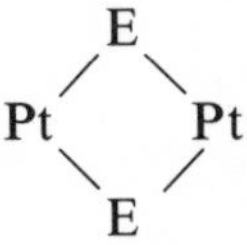

pared by treating $[Pt(PPh_3)_4]$ with smaller amounts of sulphur or selenium.[43] $[(PPh_3)_2PtS]_n$ is inert to attack by carbon monoxide, oxygen or ethylene but does give an adduct $[(PPh_3)_2PtS(SO_2)]_2$ with sulphur dioxide. Both the sulphur and the selenium complexes $[(PPh_3)_2PtE]_n$ are model compounds for studying the poisoning of platinum metal surfaces by sulphur and selenium and their inability to react with carbon monoxide, oxygen or ethylene appears to parallel that of platinum metal catalysts that have been poisoned by either sulphur or selenium.

(d) REH complexes

No alcohol complexes of platinum(II) and palladium(II) have been isolated. However, from a study of the kinetics of substitution reactions of platinum(II) complexes (pp. 295, 307–308 and Table 87) it has been established that straight chain alcohols such as methanol,[48] ethanol[49] and propanol[49] give a significant amount of alcohol–platinum(II) complex in alcoholic solution, whereas the branched alcohol t-butanol gives an insignificant amount of alcohol–platinum(II) complex.[49] No platinum(II) or palladium(II) complexes containing mercaptans are known, since mercaptans react with simple platinum(II) and palladium(II) salts to give mercaptide complexes (see below, next section).

(e) RE⁻ complexes

Although many alkoxy–complexes of both platinum(II) and palladium(II) have been postulated as reaction intermediates, the only ones that have been isolated are $[(PEt_3)_2Pt(GePh_3)(OR)]$, where R = Me, Et or Pr, formed by treating the corresponding hydroxy-complexes with alcohol[41] (reaction 226) (see note p. 229), and the palladium(II) complex trans-$[(PEt_3)_2$-$Pd(OCH_3)CN]$.[50] Alkoxy-complexes are probably involved in the oxidation of alcohols to carbonyl compounds in the presence of palladium(II) chloride[51,52] (reactions 233 and 234). They are also thought to be intermedi-

$$RCH_2OH + PdCl_2 \rightarrow [RCH_2O-PdCl] + HCl \qquad (233)$$

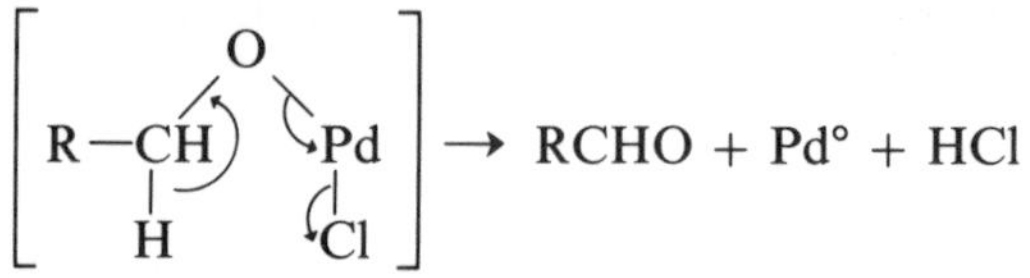

$$\left[\begin{array}{c} R-CH \overset{O}{\underset{H}{\big\langle}} \overset{\diagdown}{\underset{Cl}{\big\rangle}} Pd \end{array} \right] \rightarrow RCHO + Pd^° + HCl \qquad (234)$$

ates in the carbonylation of ethanol catalysed by palladium(II) chloride[53] (reactions 235 and 236). Sulphur dioxide appears to insert into palladium(II)–alkoxyl bonds since when sulphur dioxide is bubbled into a solution of

$$C_2H_5OH + CO \xrightarrow{PdCl_2 + LiCl}$$

$$ClCOOC_2H_5 + CH_3COOC_2H_5 + Pd \text{ metal} \qquad (235)$$

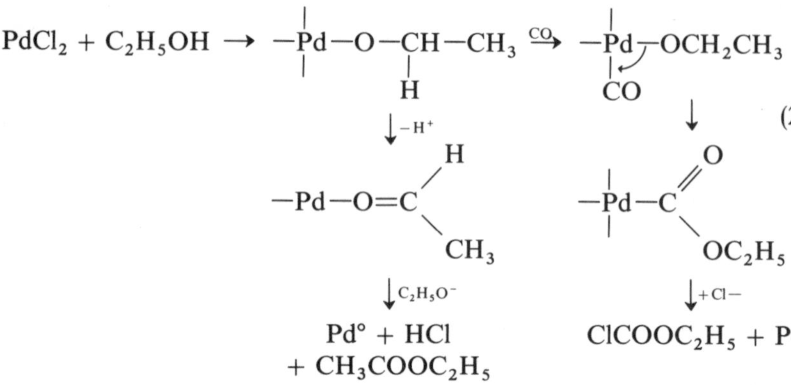

$$PdCl_2 + C_2H_5OH \rightarrow -\overset{|}{\underset{|}{Pd}}-O-\overset{|}{\underset{H}{C}H}-CH_3 \xrightarrow{CO} -\overset{|}{\underset{CO}{Pd}}-OCH_2CH_3$$

$$\downarrow{-H^+} \qquad\qquad\qquad \downarrow \qquad (236)$$

$$-Pd-O=C\overset{H}{\underset{CH_3}{\diagup}} \qquad\qquad -Pd-C\overset{O}{\underset{OC_2H_5}{\diagup\!\!\diagup}}$$

$$\downarrow{C_2H_5O^-} \qquad\qquad \downarrow{+Cl-}$$

$$Pd^° + HCl \qquad\qquad ClCOOC_2H_5 + Pd^°$$
$$+ CH_3COOC_2H_5$$

palladium(II) chloride in alcohol a sulphinato–palladium(II) complex is formed (reaction 237).[54]

$$2SO_2 + 2ROH + 2PdCl_2 \xrightarrow{(R = CH_3 \text{ or } C_2H_5)}$$

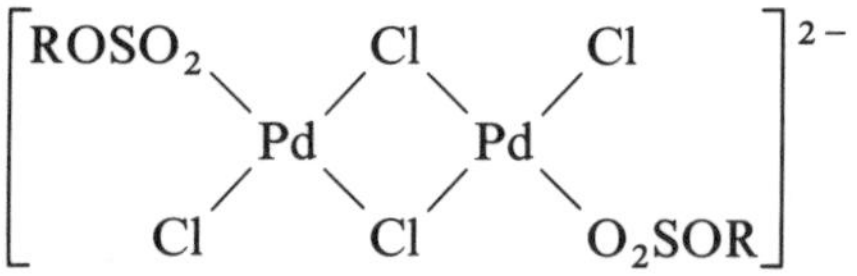

$$\left[\begin{array}{c} ROSO_2 \diagdown \quad Cl \quad \diagup Cl \\ Pd \qquad Pd \\ Cl \diagup \quad \diagdown Cl \diagup \quad \diagdown O_2SOR \end{array} \right]^{2-} + 2H^+ \qquad (237)$$

Mercaptide ions (RS⁻) are highly polarisable and therefore form strong complexes with class 'b' metal ions. However, the simple mercaptides such as $[Pd(SR)_2]$ where R = Et, Pr^n, Bu^n, Am^n,[55] Ph[55,56] and C_6F_5[57] and $[Pt(SR)_2]$ where R = Ph[56] and C_6F_5[57] are all polymeric and highly associated. The palladium(II) complexes and $[Pt(SC_6F_5)_2]$ are prepared by treating the metal(II) acetate with the appropriate mercaptan in benzene. Thiophenol does not react with platinum(II) salts under these conditions,[55] but $[Pt(SPh)_2]$ can be prepared by the self-phenylation reaction that occurs on heating the diphenyliodonium salt of bis(dithiooxalato)platinum(II) in the solid state[56] (reaction 238). Treatment of sodium tetrachloropalladate(II) with diphenyldisulphide in benzene yields a polymeric palladium(II) complex,

$$(Ph_2I^+)_2[Pt(S_2C_2O_2)]^{2-} \xrightarrow[146°C]{\text{heat at}} [Pt(SPh)_2] + 2PhI + 2SCO + 2CO \quad (238)$$

which contains alternate chloride and mercaptide bridges.[58] The chloride bridges in this complex are cleaved by neutral ligands such as pyridine or triphenylphosphine (reaction 239). If bis(acetonitrile)palladium(II) dichloride

$$nNa_2PdCl_4 + nPhSSPh \xrightarrow{\text{MeOH}}$$

$$2nNaCl + \left[\begin{array}{c} Ph \\ | \\ S \\ Cl \diagdown \diagup \diagup Pd \diagdown \diagdown Pd \diagup \\ S \diagdown Cl \\ | \\ Ph \end{array}\right] \xrightarrow{\text{pyr}} \left[\begin{array}{c} Ph \\ | \\ pyr \diagdown S \diagdown Cl \\ Pd \diagdown Pd \\ Cl \diagup S \diagdown pyr \\ | \\ Ph \end{array}\right] \quad (239)$$

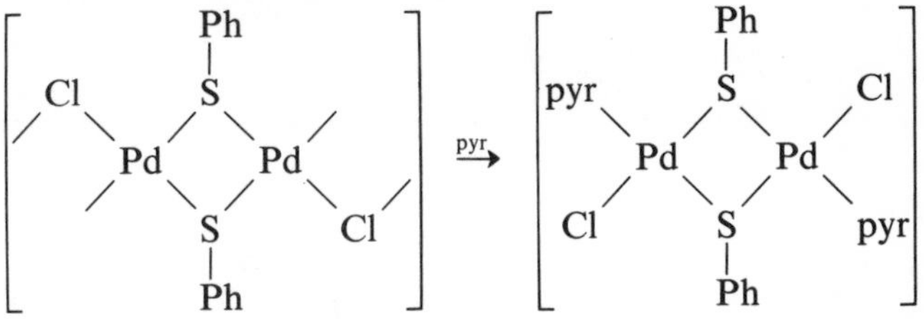

is used as the starting material a bisulphide bridged complex is formed in benzene which on treatment with methanol gives the original monosulphide bridged polymer (reaction 240).

$$2[Pd(CH_3CN)_2Cl_2] + 2PhSSPh \xrightarrow{\text{benzene}}$$

$$4CH_3CN + \left[\begin{array}{c} Ph\ Ph \\ |\ \ | \\ Cl \diagdown S-S \diagdown Cl \\ Pd \diagdown Pd \\ Cl \diagup S-S \diagdown Cl \\ |\ \ | \\ Ph\ Ph \end{array}\right] \xrightarrow[\text{MeOH}]{\text{reflux in}} \left[\begin{array}{c} Ph \\ | \\ Cl \diagdown S \diagdown \\ Pd \diagdown Pd \\ Cl \diagup S \diagdown Cl \\ | \\ Ph \end{array}\right]_n \quad (240)$$

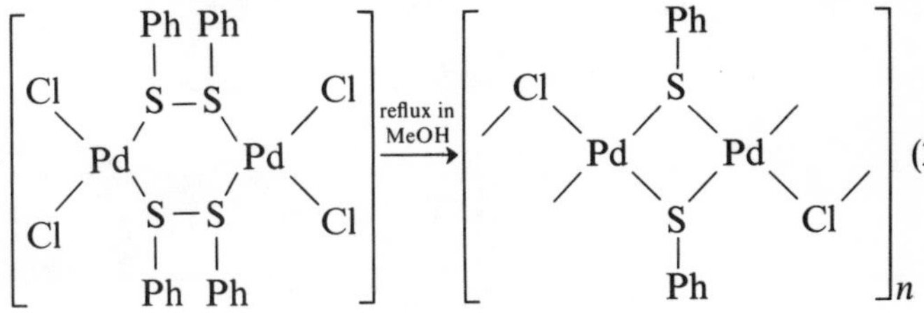

Monomeric phenylmercaptide complexes of palladium(II)[59] and perfluorophenylmercaptide complexes of both platinum(II) and palladium(II)[60] have been prepared (reactions 241–3). X-ray diffraction has shown that the

$$[(PR_3)_2PdCl_2] + 2PhSNa \xrightarrow[\substack{(PR_3 = PEt_3PEt_2Ph \\ \text{or } \frac{1}{2}Ph_2PCH_2CH_2PPh_2)}]{\text{EtOH}}$$

$$[(PR_3)_2Pd(SPh)_2] + 2NaCl \quad (241)$$

$$K_2MCl_4 + 2Bu_4NCl + 4C_6F_5SH \xrightarrow[(M = Pd, Pt)]{\text{EtOH/H}_2\text{O}}$$

$$(Bu_4N^+)_2[M(SC_6F_5)_4]^{2-} + 2KCl + 4HCl \quad (242)$$

$$[M(SC_6F_5)_2] + 2PPh_3 \xrightarrow[(M = Pd, Pt)]{\text{boil in EtOH}} [(PPh_3)_2M(SC_6F_5)_2] \quad (243)$$

$[(PPh_3)_2Pt(SC_6F_5)_2]$, formed in reaction 243, is the *trans*-isomer[61] and that with palladium reaction 243 also yields some of the mercaptide bridged dimer *trans*-[(PPh$_3$)(C$_6$F$_5$S)Pd(SC$_6$F$_5$)$_2$Pd(C$_6$F$_5$S)(PPh$_3$)][62] (see Appendix II, structure 89).

Whereas halogen-bridged dimeric complexes of platinum(II) and palladium(II) are readily cleaved by unidentate ligands such as pyridine and triphenylphosphine, the corresponding thio-bridged complexes are not.[58,63-65] Also both *cis*- and *trans*-isomers of the platinum(II) thiobridged complexes [(PPr$_3^n$)M(RS)Cl]$_2$ can be prepared,[64,65] whereas with palladium(II) only the *cis*-isomer is formed.[64] Both isomers of the platinum(II) ethylmercaptide complex are formed simultaneously in reaction 244, whereas with the phenyl-mercaptide analogues two different preparative

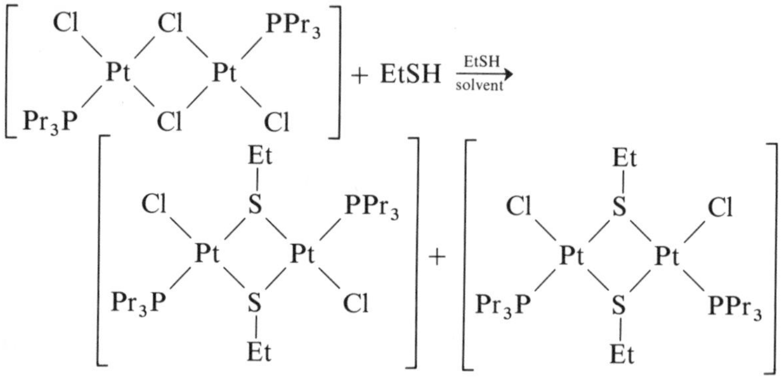

$$
\left[\begin{array}{c} Cl \diagdown \diagup Cl \diagdown \diagup PPr_3 \\ Pt \qquad Pt \\ Pr_3P \diagup \diagdown Cl \diagdown Cl \end{array}\right] + EtSH \xrightarrow[solvent]{EtSH}
$$

$$
\left[\begin{array}{c} Et \\ | \\ Cl \diagdown \diagup S \diagdown \diagup PPr_3 \\ Pt \qquad Pt \\ Pr_3P \diagup \diagdown S \diagdown Cl \\ | \\ Et \end{array}\right] + \left[\begin{array}{c} Et \\ | \\ Cl \diagdown \diagup S \diagdown \diagup Cl \\ Pt \qquad Pt \\ Pr_3P \diagup \diagdown S \diagdown PPr_3 \\ | \\ Et \end{array}\right] \quad (244)
$$

routes are required to prepare the two isomers (reactions 245 and 246). The

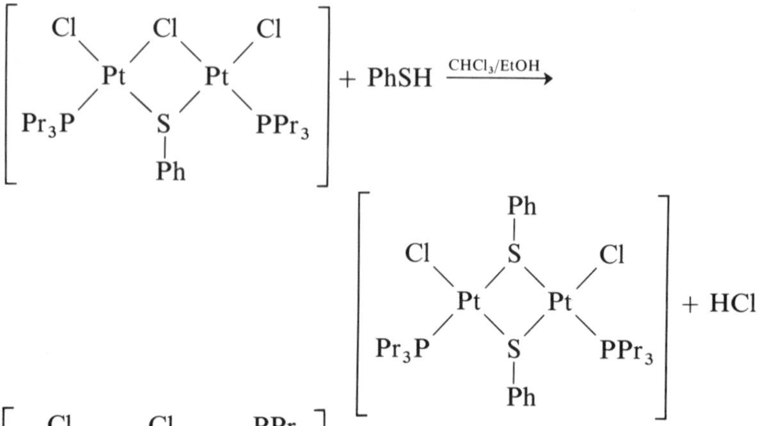

$$
\left[\begin{array}{c} Cl \diagdown \diagup Cl \diagdown \diagup Cl \\ Pt \qquad Pt \\ Pr_3P \diagup \diagdown S \diagdown PPr_3 \\ | \\ Ph \end{array}\right] + PhSH \xrightarrow{CHCl_3/EtOH}
$$

$$
\left[\begin{array}{c} Ph \\ | \\ Cl \diagdown \diagup S \diagdown \diagup Cl \\ Pt \qquad Pt \\ Pr_3P \diagup \diagdown S \diagdown PPr_3 \\ | \\ Ph \end{array}\right] + HCl \quad (245)
$$

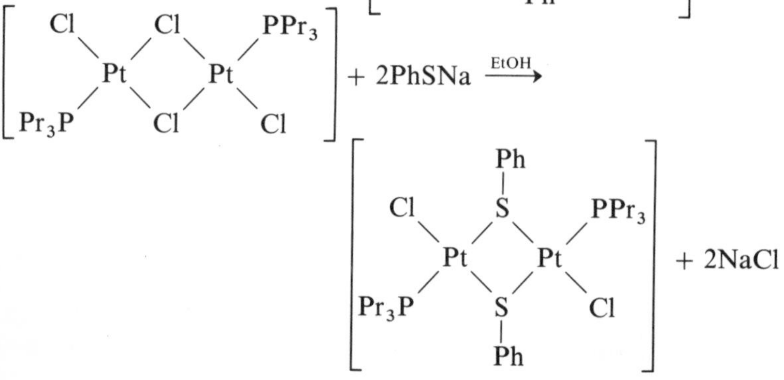

$$
\left[\begin{array}{c} Cl \diagdown \diagup Cl \diagdown \diagup PPr_3 \\ Pt \qquad Pt \\ Pr_3P \diagup \diagdown Cl \diagdown Cl \end{array}\right] + 2PhSNa \xrightarrow{EtOH}
$$

$$
\left[\begin{array}{c} Ph \\ | \\ Cl \diagdown \diagup S \diagdown \diagup PPr_3 \\ Pt \qquad Pt \\ Pr_3P \diagup \diagdown S \diagdown Cl \\ | \\ Ph \end{array}\right] + 2NaCl \quad (246)
$$

bridged complex $[(PPr_3)ClPd(Cl)(SPh)PdCl(PPr_3)]$ containing both chloride- and thio-bridging ligands is always *cis* which is consistent with the high *trans*-effect of mercaptide ligands.[65] A further type of isomerism shown in **68** and **69** involving the bridging thiol group is exhibited by the alkyl thio-bridged platinum(II) complexes containing two different alkylmercaptides.[65] The monomeric phenylmercaptide and phenylselenide hydrido-

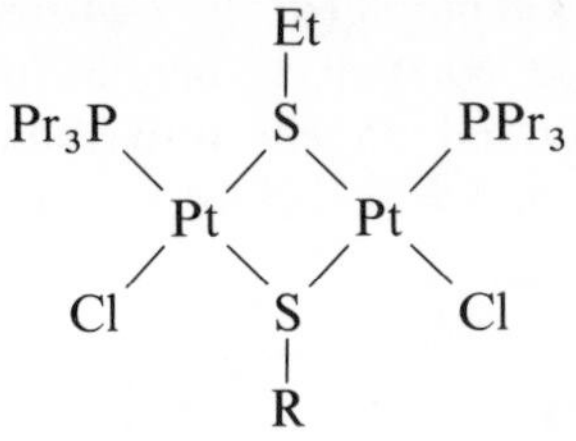

68 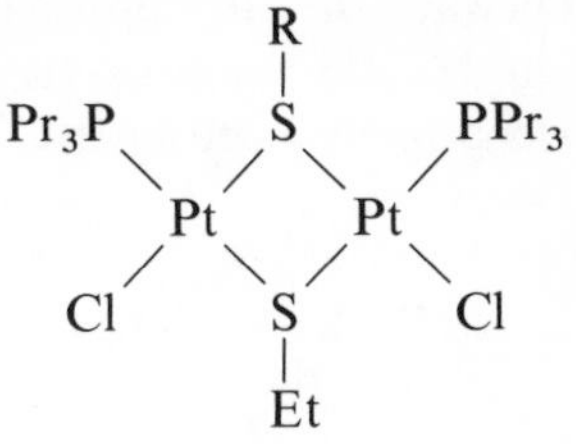   **69**

complexes of platinum(II), prepared by treating zerovalent $[Pt(PPh_3)_2]$ with phenylmercaptan or its selenium analogue[43] (reaction 247) have similar chemical properties to their HS^- and HSe^- analogues (p. 170).

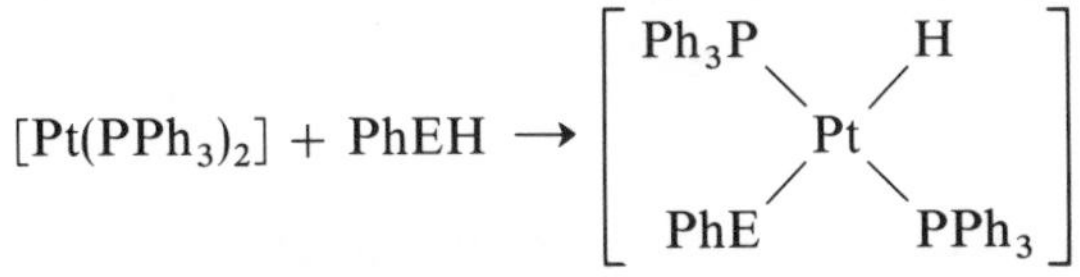

$$[Pt(PPh_3)_2] + PhEH \rightarrow \begin{bmatrix} Ph_3P \quad\quad H \\ \diagdown\quad\diagup \\ Pt \\ \diagup\quad\diagdown \\ PhE \quad\quad PPh_3 \end{bmatrix} \tag{247}$$

(f) R_2E complexes
No platinum(II) or palladium(II) complexes of ethers have been reported. Thioethers, selenoethers and telluroethers all form complexes with platinum(II) and palladium(II) which, for convenience, are divided here into monomeric and dimeric complexes.

(i) *Monomeric R_2E complexes*
Platinum(II) forms both *cis*- and *trans*-complexes with dialkylsulphides[66-8] and dialkylselenides.[69,70] The *trans*-isomers unexpectedly exhibit moderate dipole moments in solution although the *trans*-structure in the solid state has been confirmed by x-ray diffraction.[71] The origin of these finite dipole moments in the *trans*-complexes is uncertain as the crystal structure of the palladium(II) complex *trans*-$[Pd(Et_2Se)_2Cl_2]$ has shown it to be completely symmetrical in the solid state[72] (Appendix II, structure 45). Whilst the dimethyl–, diethyl– and tetramethylene–sulphide complexes *trans*-$[Pt(R_2S)_2X_2]$, where X = Cl or Br, are stable to heat, the pentamethylene–sulphide complex undergoes an unusual *trans* to *cis* rearrangement, which may be due to crystal structure effects.[73] In addition to *cis*- and *trans*-$[Pt(Me_2S)_2Cl_2]$, dimethylsulphide forms an ionic complex[66] $[Pt(Me_2S)_4]^{2+}$-$[PtCl_4]^{2-}$. The telluroether complexes of platinum(II) are much less stable than their sulphur and selenium analogues. The best characterised compound is $[Pt\{(PhCH_2)_2Te\}_2Cl_2]$ which probably has a *trans*-configuration as it is soluble in chloroform. It is stable in the solid state and does not

appear to have any isomeric forms, but it decomposes rapidly in solution to platinum, tellurium and dibenzyl.[74,75]

Dialkylsulphides[76–8] and dialkyl- and diaryl-selenides[74,75] form monomeric palladium(II) complexes [Pd(R$_2$E)$_2$X$_2$], which only exist in the *trans*-configuration, as confirmed by x-ray diffraction of *trans*-[Pd(Et$_2$Se)$_2$Cl$_2$][72] (Appendix II, structure 45). Alkylphenylsulphides form similar monomeric complexes when the alkyl group has a straight carbon chain. However, with t-butyl or Me$_2$EtC groups (R) polymeric complexes [(PhSR)Pd$_2$Cl$_4$]$_n$, which are probably tetrameric **70**, are formed.[79] The only monomeric

70

cis-dialkylselenide complexes of palladium(II) involve chelating ligands such as 1,2-bis(isopropylseleno)ethane.[80,81]

(ii) Dimeric R$_2$E complexes
The stabilities of the chloro-bridged platinum(II) complexes, [L$_2$PtCl$_4$], fall in the order[82] L = R$_3$P ~ R$_2$S > R$_3$As > amine > R$_2$Te > R$_3$Sb > R$_2$Se whereas for palladium(II) the stability order[83] is L = R$_3$P > R$_3$As ~ R$_2$S > R$_2$Se > R$_2$Te > R$_3$Sb. The platinum(II) complexes are generally more stable than their palladium(II) analogues except for the R$_2$Se complexes where the palladium(II) complexes are the more stable. This may be due to the orbitals used by selenium in forming a σ-bond to the metal being comparable in size to the palladium(II) orbitals but smaller than the platinum(II) orbitals.[83] The chloro-bridged complexes are readily cleaved by unidentate ligands such as amines to give monomeric [M(R$_2$E)(am)Cl$_2$] complexes, where M = Pd or Pt, E = S, Se or Te and am = amine.[84]

The dimeric complexes [(R$_2$S)$_2$Pt$_2$X$_4$], where R = Me or Et and X = Cl or Br, prepared by treating the monomeric [Pt(R$_2$S)$_2$X$_2$] complexes in ethanol with a solution of Na$_2$PtX$_4$ in ethanol, were found to have different solubility properties and infra-red spectra to other [(R$_2$S)$_2$M$_2$X$_4$] complexes, where M = Pt or Pd and X = halogen.[85] An X-ray diffraction study of [(Et$_2$S)$_2$Pt$_2$Br$_4$] and [(Me$_2$S)$_2$Pd$_2$Br$_4$] (Appendix II structures 148 and 25 respectively) confirmed this difference and showed that whereas the palladium(II) complex had the expected bromo-bridged structure (**71**) the platinum(II) complex had diethylsulphide bridging ligands(**72**).[85,86] The most notable feature of these two structures is the shorter M–S bond length in the platinum(II) complex (2·22(1) Å) than in the palladium(II) complex (2·30(2) Å), which suggests significantly stronger bonding when the ligand is bridging than when it is terminal. This is probably due to the fact that the sulphur atom in the terminal position in the palladium(II) complex carries a

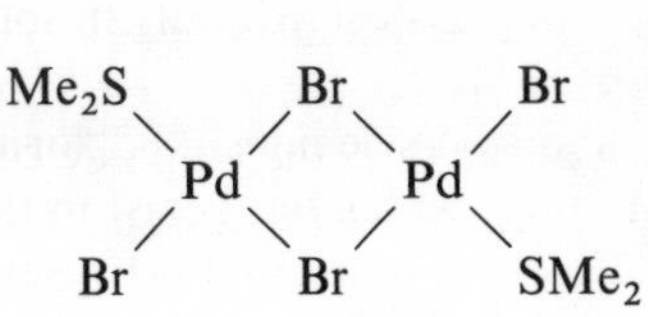

$$
\begin{array}{ccc}
\text{Me}_2\text{S} & \text{Br} & \text{Br} \\
& \diagdown\diagup\;\; \diagdown\diagup \\
& \text{Pd} \quad\quad \text{Pd} \\
& \diagup\diagdown\;\; \diagup\diagdown \\
\text{Br} & \text{Br} & \text{SMe}_2
\end{array}
\qquad\qquad
\begin{array}{ccc}
& \text{Et}_2 & \\
\text{Br} & \text{S} & \text{Br} \\
\diagdown\diagup\;\; \diagdown\diagup \\
\text{Pt} \quad\quad \text{Pt} \\
\diagup\diagdown\;\; \diagup\diagdown \\
\text{Br} & \text{S} & \text{Br} \\
& \text{Et}_2 &
\end{array}
$$

71 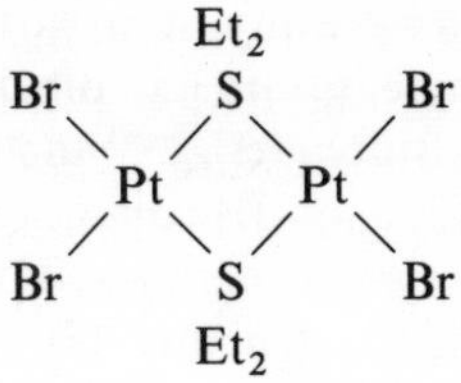   **72**

lone-pair of electrons which can act repulsively with non-bonded d-electrons on the metal whereas when the sulphur is in the bridging position in the platinum(II) complex all the sulphur outer electrons are accommodated in bonding orbitals, eliminating the repulsive interactions and allowing a strong bond.

(iii) Physical properties of R_2E complexes

An analysis of the x-ray diffraction data in Appendix II indicates that the metal(II)–sulphur bond lengths found in dialkylsulphide complexes of platinum(II) and palladium(II) fall in the range 2·28–2·45 Å, which is comparable to the sum of the covalent radii (2·35 Å),[87] and suggests that there is little or no π-bonding in the metal(II)–group VIB element bond. Metal–halogen bonds *trans* to group VIB elements show no apparent *trans* bond lengthening (Table 37) indicating that the ligands have a negligible or zero *trans*-influence.

TABLE 37

Metal–halogen bond lengths *trans* to R_2E ligands

Complex	M–X bond length trans to R_2E	Comparison M–X bond length	Reference
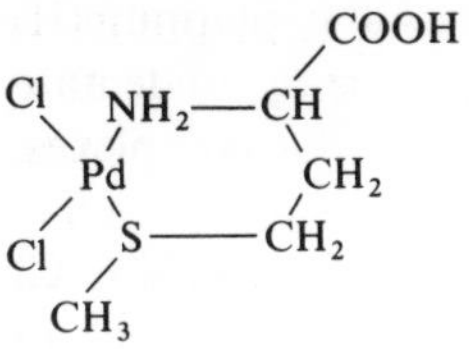 Cl, NH$_2$—CH(COOH), Pd, Cl, S—CH$_2$ (CH$_2$), CH$_3$	2·332 Å	2·308 Å for Pd–Cl *trans* to NH$_2$ in same complex	88
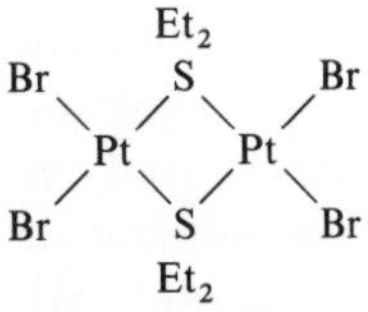 Br, S(Et$_2$), Br, Pt, Pt, Br, S(Et$_2$), Br	2·338 Å	2·32 Å for terminal Pd–Br in $(\text{NEt}_4^+)_2[\text{Pt}_2\text{Br}_6]^{2-}$	86, 89
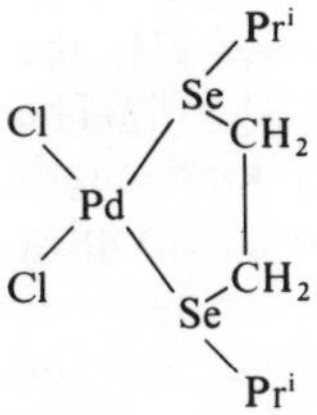 Cl, Se(Pr^i)(CH$_2$), Pd, Cl, Se(CH$_2$)(Pr^i)	2·31 Å	2·31 Å for Pd–Cl in $[\text{Pd}(\text{H}_4\text{EDTA})\text{Cl}_2]$	81, 90

The infra-red and Raman spectra of a series of *trans*-[M(EMe$_2$)X$_2$]. complexes, where M = Pt or Pd, E = S, Se or Te and X = Cl, Br and I, at atmospheric pressure[91] and in the case of the dimethylsulphide complexes at high pressures,[92] (Table 38) indicate that the metal–group VIB element

TABLE 38
Metal–Group VIB element stretching frequencies in *trans*-[M(EMe$_2$)$_2$X$_2$]
complexes (data from ref. 91)

M	*X*	*EMe$_2$*	$\nu_{M-E(sym.)}$ *(cm^{-1})*	$\nu_{M-E(asym.)}$ *(cm^{-1})*
Pt	Cl	SMe$_2$	346	311
Pt	Br	SMe$_2$	344	315
Pt	I	SMe$_2$	—	313
Pt	Cl	SeMe$_2$	176	225
Pt	Br	SeMe$_2$	173	220
Pt	I	SeMe$_2$	172	233
Pt	Br	TeMe$_2$	169	233
Pd	Cl	SMe$_2$	322	310
Pd	Br	SMe$_2$	313	316
Pd	I	SMe$_2$	—	307
Pd	I	SeMe$_2$	—	219

stretching frequencies fall in the order Pt–S > Pt–Se $\approx$ Pt–Te and consequently the approximate force constants vary in the order Pt–S > Pt–Se < Pt–Te. The higher strength of the platinum(II)–tellurium than the platinum(II)–selenium bond may be due to the orbitals used by tellurium in forming the σ-bond being comparable in size to the platinum(II) orbitals, whereas the selenium orbitals are rather smaller. The N–H stretching frequencies in *trans*-[PtL(am)Cl$_2$], where am = amine, suggest that the σ-inductive effect transmitted from the donor atom of ligand L across the platinum atom to the N–H bond increases on changing L in the order L = γ-picoline < piperidine < R$_2$S < R$_2$Se < R$_2$Te < R$_3$As < R$_3$P < R$_3$Sb < P(OR)$_3$.[84]

The electronic spectra of *trans*-[ML$_2$X$_2$] where M = Pd or Pt, X = Cl or Br and L = Bu$_2^n$S, Bu$_2^i$S, Bu$_2^s$S, Pr$_2^n$S and Et$_2$Se exhibit three mean bands: one due to a *d–d* transition ($d_{xy} \rightarrow d_{x^2-y^2}$) at 22,000–24,000 cm^{-1} (Pd) and 20,800–21,700 cm^{-1} (Pt), one due to charge transfer from the sulphur or selenium bonding orbitals to the metal $d_{x^2-y^2}$ at about 32,000 cm^{-1} (Pd) and 40,000–44,000 cm^{-1} (Pt) and a third at 40,700–42,300 cm^{-1} in [Pd(Et$_2$Se)$_2$Cl$_2$] due to charge transfer from the chloride ligands to the palladium $d_{x^2-y^2}$ orbital.[93] The ligand field splittings inferred from the energies of the lowest energy *d–d* transitions ($d_{xy} \rightarrow d_{x^2-y^2}$) in the complexes *trans*-[PtL(piperidine)-Cl$_2$] decrease across the series L = P(OMe)$_3$ > PPr$_3$ > piperidine > AsPr$_3$ > Et$_2$S > Et$_2$Se > Et$_2$Te.[94] The range of energies, however, is fairly small varying from 33,400 cm^{-1} when L is trimethylphosphite to 29,400 cm^{-1} when L is dimethyltelluride.

The n.m.r. spectra of *cis*- and *trans*-[Pt(Me$_2$Se)$_2$Cl$_2$] in dichloromethane solution exhibit positive[195]platinum–[77]selenium coupling constants.[95]

The ratio of the reduced coupling constants in the *cis* and *trans* selenium complexes was found to be less than the ratio in the phosphine analogues. Although a complete explanation for this is not possible, the result would be consistent with less π-bonding in the platinum(II)–selenium than in the platinum(II)–phosphorus bond. The proton n.m.r. spectra of thioether complexes indicate that at low temperature the sulphur atoms are pyramidal, but on warming inversion occurs and the pyramidal sulphur averages to a planar conformation.[96–9] The rate of inversion increases as the metal–sulphur bond strength decreases when the *trans*-halogen is altered in the order I < Br < Cl. The sulphur atoms appear to be co-ordinated throughout, and as one S–M bond is broken another is formed by the same sulphur atom, so that this is in effect a nucleophilic substitution reaction of one orbital of sulphur by another.[98,99]

COMPLEXES OF SMALL ANIONS THAT BIND THROUGH OXYGEN

In this section complexes of the bidentate oxygen donor anions sulphate and carbonate and the monodentate oxygen donor anion nitrate are described. Nitrite complexes are considered elsewhere (pp. 211 and 214).

(i) Sulphato-complexes

Sulphato-complexes of platinum(II) and palladium(II) are rare and must be prepared under forcing conditions because the sulphate group which binds through oxygen rather than sulphur is too poor a ligand, for these metal ions, to displace even such weakly co-ordinated ligands as water. Thus the complexes $[M(SO_4)(NH_3)_2(H_2O)]$, where M = Pd or Pt, prepared by treating aqueous solutions of $[M(NH_3)_2Cl_2]$ with silver sulphate,[100,101] dissociate instantly in water to give free sulphate ions and $[M(NH_3)_2(H_2O)_2]^{2+}$ ions. $[(PPh_3)_2M(SO_4)]$, where M = Pd or Pt, can be prepared by treating $[(PPh_3)_2MO_2]$ with sulphur dioxide[102–5] (pp. 41–43). The platinum complex $[(PPh_3)_2Pt(SO_4)]$ is stable in air and fairly resistant to hydrolysis;[102] no details about the palladium(II) complex have yet been published.

(ii) Carbonato-complexes

Palladium(II) and platinum(II) carbonato-complexes can be prepared either by treating the zerovalent tertiary phosphine complexes with a mixture of oxygen and carbon dioxide (reaction 248 see also p. 41), or, in the case of

$$M(PPh_3)_4 + \tfrac{1}{2}O_2 + CO_2 \xrightarrow[(M = Pt, Pd)]{} [(PPh_3)_2M(CO_3)] + 2PPh_3 \quad (248)$$

platinum only, by treating the dichloroplatinum(II) complex *cis*-$[(PPh_3)_2PtCl_2]$ with silver carbonate.[106,107] An x-ray diffraction study of $[(PPh_3)_2Pt(CO_3)]$ (Appendix II structure 196) showed that the platinum, phosphorus and oxygen atoms are all coplanar.[108] These complexes react with acids, such as hydrogen chloride, acetic, trifluoroacetic or benzoic acid to liberate carbon dioxide[107] (reaction 249).

$$[(PPh_3)_2Pt(CO_3)] + RCOOH \xrightarrow[\text{heat}]{\text{benzene}}$$
$$[(PPh_3)_2Pt(RCOO)_2] + \uparrow CO_2 + H_2O \quad (249)$$

(iii) Nitrato-complexes

Two main methods have been used to prepare the relatively few known nitrato-complexes. In the first method, nitrate complexes such as $[Pd(NO_3)_2(H_2O)_2]$ are prepared directly by dissolving the metal in concentrated nitric acid.[109] This bis(aquo) complex reacts with other ligands such as 2,2'-bipyridyl to give complexes such as $[Pd(bipyr)(NO_3)_2]$,[110] and treatment with N_2O_5 results in the formation of the palladium(IV) nitrate $[Pd(NO_3)_4]$.[111] The alternative method for preparing nitrato-complexes is to treat the zerovalent oxygen complexes $[(PPh_3)_2M(O_2)]$, where M = Pd or Pt, with nitrogen dioxide[102,103,105] (reaction 250). The nitrate ligand is readily liberated from these complexes by boiling in water.[102]

$$[(PPh_3)_2M(O_2)] + N_2O_4 \rightarrow [(PPh_3)_2M(NO_3)_2] \quad (250)$$

The infra-red spectra of nitrato-complexes are consistent with metal–oxygen bonding[109] which has been confirmed by x-ray diffraction[112,113] (see Appendix II structures 27 and 190). It is noteworthy that the M–O–N atoms are not colinear and that the co-ordinated N–O distance in cis-$[Pd(Me_2SO)_2(NO_3)_2]$ is 1·32(1) Å as compared to 1·217(9) Å for the non-co-ordinated N–O bonds.

COMPLEXES OF SMALL ANIONS THAT BIND THROUGH SULPHUR

(i) Sulphito-complexes

Although the sulphite ligand can occupy two co-ordination positions it is generally a unidentate ligand towards platinum(II) and palladium(II) binding through the sulphur atom as in $[Pd(SO_3)(H_2O)_3]$.[101] Metal–sulphur bonding is suggested by the infra-red spectra of these complexes,[114,115] which show an increase in the S–O stretching frequency of the sulphite ion on co-ordination to platinum(II) and palladium(II), whereas if metal–oxygen bonding were present, both higher and lower frequencies should be observed due to a reduction of the symmetry of an oxygen bonded sulphite ligand from C_{3V} to C_S. Metal–sulphur bonding was confirmed by x-ray diffraction studies of trans-$Na_2[Pd(SO_3)_2(NH_3)_2]$ and $[Pd(SO_3)(NH_3)_3]$[116,117] (Appendix II, structures 10 and 11). The appreciably shorter Pd–S bond length trans to ammonia (2·245(6) Å) than trans to another sulphite ligand (2·294(6) Å) could be due to π-bonding in the palladium–sulphur bond. The alternative explanation, namely that the sulphite ligand has a substantial σ-trans-effect and hence a substantial tr–ns-influence (see pp. 301–303) is unlikely since the palladium(II)–nitrogen bond lengths in $[Pd(SO_3)(NH_3)_3]$ cis and trans to the sulphite ligand are effectively equal (2·107(20) Å and 2·103(20) Å respectively[117]). Although it has no trans-influence the sulphite-group, in common with other sulphur-bonded ligands has a substantial trans-effect as deduced from preparative studies.[118] Its high position in the spectrochemical series[119] is indicated by the white colour of the ammine–sulphitoplatinum(II) complexes.

Palladium(II) sulphate reacts with 1,10-phenanthroline to form insoluble [Pd(SO₃)(phen)], which is dimeric with two bidentate sulphite bridging ligands (73).[120] Palladium(II) sulphite complexes such as $K_2[Pd(SO_3)_2$-

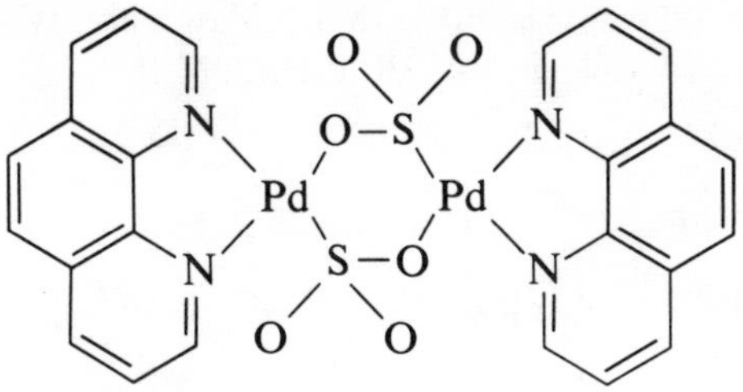

73

$(H_2O)_2]$ have been used to detect the presence of carbon monoxide in gas streams.[101] This depends on the high *trans*-effect of the sulphito-group which labilises a water ligand so that it is readily replaced by carbon monoxide to give a transient sulphito–carbonyl palladium(II) complex that readily decomposes to palladium metal, which is detected visually.

(ii) Arylsulphinate complexes

Arylsulphinate complexes[121] of both platinum(II)[122,123] and palladium(II)[124] can be prepared by treating a halide complex with the sodium salt of the arylsulphinate ion[123,124] (reaction 251) or by treating the zero-

$$[PdX_4]^{2-} + 2PhSO_2Na \xrightarrow{(X = Cl, Br)}$$
$$[PdX(PhSO_2)_2(H_2O)]^- + 2NaX + X^- \quad (251)$$

valent tertiaryphosphine complex with the arylsulphonylchloride[122] (reaction 252). Their general stability and infra-red spectra are consistent with the

$$[Pt(PPh_3)_4] + RSO_2Cl \rightarrow [(PPh_3)_2Pt(O_2SR)Cl] \quad (252)$$

presence of metal–sulphur bonding. The reactions (reactions 253 and 254) of the palladium(II) complexes led to the suggestion that the benzenesulphinate ligand, in common with other sulphur bonded ligands, had a high

$$[PdCl(PhSO_2)_2(H_2O)]^- + H_2O \rightarrow [Pd(PhSO_2)_2(H_2O)_2] + Cl^- \quad (253)$$

$$[PdCl(PhSO_2)_2(H_2O)]^- + NaSCN \rightarrow$$
$$[Pd(SCN)(PhSO_2)_2(H_2O)]^- + NaCl \quad (254)$$

trans-effect, which the original authors ascribed to a high π-acceptor ability of the benzenesulphinate ligand.[124] However, later work with platinum(II) complexes suggested that the high *trans*-effect of the arylsulphinate ligands was in fact due to the arylsulphinate group being a very strong σ-donor[123] since the platinum–chlorine stretching frequency *trans* to a *para*-toluenesulphinate ligand (304 cm⁻¹ in *trans*-[(PEt₃)₂PtCl(SO₂tol)]) was similar to that found *trans* to a pentafluorophenyl ligand (302 cm⁻¹ in *trans*-[(PEt$_3$)$_2$PtCl(C$_6$F$_5$)] and *trans* to a triethylphosphine ligand (310 cm^{-1} in *cis*-[(PEt$_3$)$_2$PtCl(C$_6$F$_5$)]).

Arylsulphinate complexes eliminate sulphur dioxide on heating to form aryl complex[122] (reaction 386 p. 329). Arylsulphinate ions are poor bridging ligands so that dimeric complexes containing arylsulphinate bridging ligands can only be prepared by treating the chloro-bridged complexes with stannous chloride (reaction 255), consistent with a bridging ability series of SnCl$_3$ < RSO$_2$ < Cl < Br < I < R$_2$PO < SR < PR$_2$.[42,123]

$$
\begin{bmatrix} \text{tol SO}_2 & \text{Cl} & \text{PEt}_3 \\ & \text{Pt} \quad \text{Pt} & \\ \text{Et}_3\text{P} & \text{Cl} \quad \text{SO}_2 \text{ tol} \end{bmatrix} + 2\text{SnCl}_2 \longrightarrow
$$

$$
\begin{bmatrix} & \text{tol} & \\ \text{Cl}_3\text{Sn} & \text{SO}_2 & \text{PEt}_3 \\ & \text{Pt} \quad \text{Pt} & \\ \text{Et}_3\text{P} & \text{SO}_2 & \text{SnCl}_3 \\ & \text{tol} & \end{bmatrix} \qquad (255)
$$

(iii) Thiosulphate complexes

Both platinum(II) and palladium(II) form stable complexes such as [Pd(en)(S$_2$O$_3$)$_2$]$^{2-}$,[125] [Pd(dien)(S$_2$O$_3$)],[126] [Pd(Et$_4$dien)S$_2$O$_3$)],[126] [Pt(S$_2$O$_3$)$_4$]$^{6-}$[127-30] and *trans*-[Pt{SC(NH$_2$)$_2$}$_2$(S$_2$O$_3$)(H$_2$O)],[131] in which the thiosulphate group is bound to the metal by a single metal–sulphur bond (see Appendix II, structure 20). The complexes are prepared by displacement of halide ligands by thiosulphate ions.

In addition to complexes in which the thiosulphate anion acts as a unidentate ligand a number of complexes such as [Pt(S$_2$O$_3$)Cl$_2$]$^{2-}$, [Pt(S$_2$O$_3$)$_2$]$^{2-}$ and [Pd(S$_2$O$_3$)$_2$]$^{2-}$ in which the thiosulphate ligand is bidentate are known[132] and one complex [Pt(S$_2$O$_3$)$_3$]$^{4-}$ is known in which both unidentate and bidentate thiosulphate groups are present.[132] In these chelate complexes the ligand is bound to the metal through one sulphur and one oxygen atom, so that *cis*- and *trans*-isomers of [Pt(S$_2$O$_3$)$_2$]$^{2-}$ can be isolated.[132] These two isomers can be distinguished by their reactions with ethylenediamine, since the more soluble *cis*-isomer gives *cis*-[Pt(S$_2$O$_3$)$_2$(en)]$^{2-}$ (reaction 256), whilst the non-electrolyte [Pt(S$_2$O$_3$)(en)] is obtained from

$$
\begin{bmatrix} \text{S} & \text{S} \\ \text{O}_2\text{S} \quad \text{Pt} \quad \text{SO}_2 \\ \text{O} & \text{O} \end{bmatrix}^{2-} + \text{en} \longrightarrow \begin{bmatrix} & \text{NH}_2 \\ \text{O}_3\text{S}_2 \quad \text{CH}_2 \\ \text{Pt} \\ \text{O}_3\text{S}_2 \quad \text{CH}_2 \\ & \text{NH}_2 \end{bmatrix}^{2-} \qquad (256)
$$

the *trans*-isomer (reaction 257).

$$
\begin{bmatrix} \text{S} & \text{O} \\ \text{O}_2\text{S} \quad \text{Pt} \quad \text{SO}_2 \\ \text{O} & \text{S} \end{bmatrix}^{2-} + \text{en} \longrightarrow \begin{bmatrix} & \text{NH}_2 \\ \text{S} \quad \text{CH}_2 \\ \text{O}_2\text{S} \quad \text{Pt} \\ \text{O} \quad \text{CH}_2 \\ & \text{NH}_2 \end{bmatrix} + \text{S}_2\text{O}_3^{2-} \qquad (257)
$$

As a consequence of its co-ordination through sulphur, the thiosulphate ligand has a high *trans*-effect as deduced from preparative studies,[1] but no *trans*-influence since the palladium–nitrogen bond lengths in $[Pd(en)_2]^{2+}[Pd(en)(S_2O_3)]^{2-}$ are 2·071(7) Å *trans* to thiosulphate in the anion and 2·080(15) Å *trans* to ethylenediamine in the cation.[125] Treatment of $K_6[Pt(S_2O_3)_4]$ with ammonia or pyridine results in the displacement of two thiosulphate ligands.[133] With ammonia the product is *cis*-$K_2[Pt(S_2O_3)_2$-$(NH_3)_2]$ whereas with pyridine, although *cis*-$K_2[Pt(S_2O_3)_2(pyr)_2]$ is undoubtedly formed, the high *trans*-effect of the thiosulphate ligands labilises the pyridine so that the product actually isolated is $K_2[Pt(S_2O_3)_2]$.

The reactions of platinum(II) complexes with thiosulphate ions can be used to determine their isomerism in an analogous way to Kurnakov's thiourea test[129,134] (see below pp. 183–184). Thus *cis*-$[Pt(NH_3)_2X_2]$, reacts with 1 mol of sodium thiosulphate to yield $[Pt(NH_3)_2(S_2O_3)]$, whereas the *trans*-isomer yields $[Pt(NH_3)_2(S_2O_3)(H_2O)]$. If excess thiosulphate is used the *cis*-isomer yields $[Pt(S_2O_3)_3]^{6-}$ and the *trans*-isomer yields $[Pt(NH_3)_2$-$(S_2O_3)_2]^{2-}$.

UREA AND THIOUREA COMPLEXES

Infra-red studies on the bis(urea) complexes $[M(NH_2CONH_2)_2Cl_2]$, where M = Pd or Pt, indicate that urea acts as a unidentate ligand bonding through one of its nitrogen atoms.[135] Thiourea[136,137] and N,N'-disubstituted selenoureas[138,139] also act as unidentate ligands towards platinum(II) and palladium(II) but co-ordinate through their sulphur or selenium atoms. X-ray diffraction[136,137] has shown that $[Pd(NH_2CSNH_2)_4]Cl_2$ is distorted slightly from a square-planar towards a tetrahedral arrangement with palladium-sulphur bond lengths of 2·33 ± 0·01 Å (See Appendix II structure 31).

Thiourea, in common with other sulphur donor ligands, has a high *trans*-effect as deduced from preparative studies,[1] which forms the basis of the Kurnakov test for distinguishing the *cis*- and *trans*-isomers of dihalogeno-diammine platinum(II).[140] This test, which relies on the relative *trans*-effects of the three ligands being thiourea > Cl^- > NH_3, involves the formation of $[Pt\{SC(NH_2)_2\}_4]^{2+}$ from *cis*-$[Pt(NH_3)_2Cl_2]$ and $[Pt(NH_3)_2$-$\{SC(NH_2)_2\}_2]^{2+}$ from *trans*-$[Pt(NH_3)_2Cl_2]$ (reactions 258 and 259). In contrast to these reactions the thiourea in $[Pd\{SC(NH_2)_2\}_4]Cl_2$ is displaced on dissolving the complex in liquid ammonia.[141]

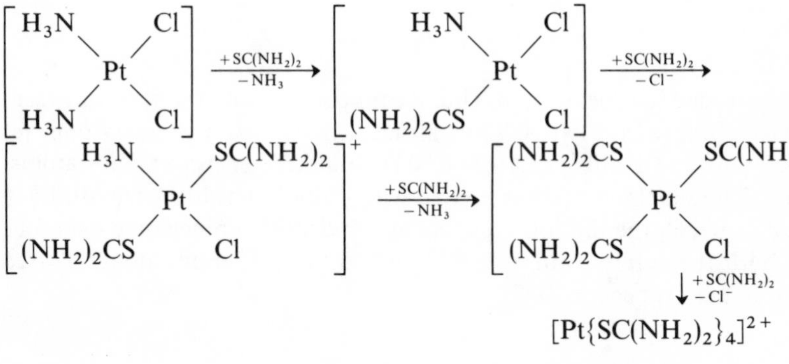

$$[Pt\{SC(NH_2)_2\}_4]^{2+} \quad (258)$$

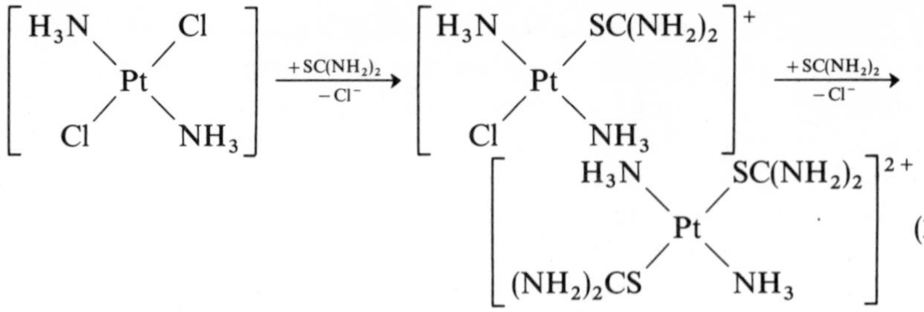

$$\left[\begin{array}{c} H_3N \diagdown \quad Cl \\ Pt \\ Cl \diagup \quad \diagdown NH_3 \end{array}\right] \xrightarrow[-Cl^-]{+SC(NH_2)_2} \left[\begin{array}{c} H_3N \diagdown \quad SC(NH_2)_2 \\ Pt \\ Cl \diagup \quad \diagdown NH_3 \end{array}\right]^+ \xrightarrow[-Cl^-]{+SC(NH_2)_2}$$

$$\left[\begin{array}{c} H_3N \diagdown \quad SC(NH_2)_2 \\ Pt \\ (NH_2)_2CS \diagup \quad \diagdown NH_3 \end{array}\right]^{2+} \qquad (259)$$

Diphenylphosphinothioyl derivatives of thiourea react with palladium(II) chloride to form complexes (**74**) in which the palladium atom is bound to the

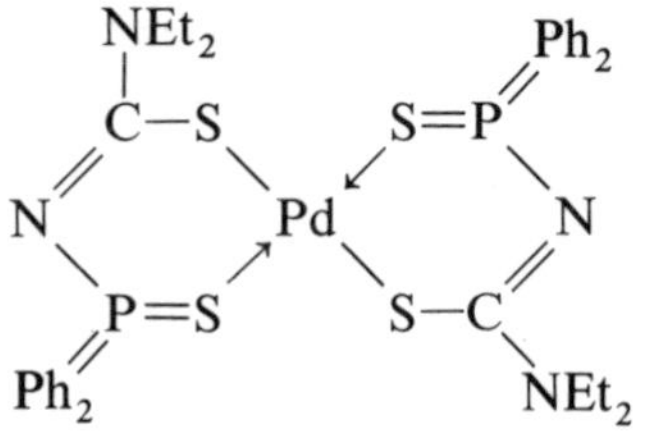

74

two sulphur atoms in the ligand and not to either of the nitrogen atoms or the phosphorus atom.[142] This is very similar to the situation in the planar, diamagnetic bis-complexes of imidodithiodiphosphinates (**75**) with platinum(II) and palladium(II).[143]

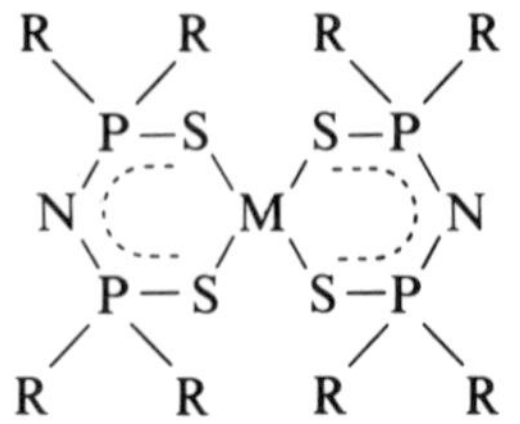

75

Dithiobuiret (**76**) has three lone-pairs of electrons all potentially capable

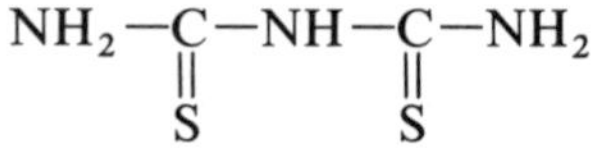

$$NH_2-\underset{\underset{S}{\|}}{C}-NH-\underset{\underset{S}{\|}}{C}-NH_2$$

76

of being donated to a metal atom. Thus it could bind to a metal (a) via nitrogen, (b) via sulphur, (c) via two sulphur atoms, (d) via two nitrogen atoms, (e) via a sulphur and a nitrogen atom or (f) using three or more donor atoms. When palladium(II) chloride is treated with dithiobiuret (dtb), [Pd(dtb)$_2$] is isolated in which the dithiobiuret acts as a bidentate uninegative ligand in which both the sulphur atoms are bound to the palladium atom[144] (see structure 28 in Appendix II).

DIALKYLSULPHOXIDE COMPLEXES

Infra-red and x-ray diffraction studies of dimethylsulphoxide complexes of platinum(II) and palladium(II), $[M(DMSO)_2X_2]$, where X is a halide ion, a nitrate ion or an amine, have shown that in the solid state the sulphoxide is bound to the metal through a metal–sulphur bond.[145–50] N.m.r. studies have shown that this metal–sulphur bonding is retained in dimethylsulphoxide solution.[148,151] In addition to the bis-complexes tetrakis (dialkylsulphoxide)palladium(II) complexes have been prepared, in which both sulphur and oxygen bonded dialkylsulphoxide ligands are present.[152–4] These complexes, prepared by mixing palladium(II) chloride and silver perchlorate or silver tetrafluoroborate in dialkylsulphoxide show interesting structural changes as the steric requirements of the alkyl group are increased. Thus dimethylsulphoxide forms cis-$[Pd(Me_2\underline{S}O)_2(Me_2S\underline{O})_2]^{2+}$ (where the donor atom is underlined), di-n-propylsulphoxide and di-n-butylsulphoxide both form $trans$-$[Pd(R_2\underline{S}O)_2(R_2S\underline{O})_2]^{2+}$ and di-iso-pentylsulphoxide forms an entirely oxygen bonded cation $[Pd\{(i\text{-pentyl})_2S\underline{O}\}_4]^{2+}$.[154] This series suggests that in the absence of severe steric effects, for example in $[M(Me_2SO)_2Cl_2]$, dialkylsulphoxides exhibit their true preference for metal–sulphur bonding. However, increasing the steric effects, for example in $[Pd(Me_2SO)_4]^{2+}$, can prevent sulphur to metal bonding but still allow oxygen to metal bonding so that a mixed sulphur and oxygen bonded dimethylsulphoxide cation is obtained.

CARBOXYLATE $\left(-C\begin{smallmatrix}\;\diagup O \\ \diagdown \\ \quad O^-\end{smallmatrix}\right)$ COMPLEXES

Palladium(II) acetate is a brown crystalline compound obtained on treating slightly acid solutions of palladium(II) nitrate with glacial acetic acid or by dissolving palladium sponge suspended in hot glacial acetic acid by the addition of a small quantity of nitric acid.[155] Palladium(II) propionate can be made similarly, whilst the benzoate, trifluoroacetate and pentafluoropropionate can be obtained from the acetate by exchange reactions. Although the fluorocarboxylates are monomeric in benzene solution at 37°C, the acetate, propionate and benzoate are trimeric in benzene at 37°C.[155] The acetate, however, is monomeric in glacial acetic acid and exists as a mixture of monomer and dimer in dimethylformamide at 65°C.[156] The crystal structure of palladium(II) acetate shows that in the solid state it is made up of trimeric units in which each palladium atom is approximately square-planar and joined to the other two palladium atoms by double acetate bridges[157] (see Appendix II, structure 21).

The palladium(II) carboxylates are indefinitely stable in air, soluble in a number of organic solvents (see Appendix I, p. 455) but decompose on warming in alcoholic solvents to give palladium metal. Bidentate ligands such as acetylacetone or salicylaldehyde displace the carboxylate ligands, but unidentate ligands (L) such as amines, triphenylphosphine and triphenylarsine give trans-[Pd(RCOO)$_2$L$_2$] in which the carboxylate groups are unidentate.[155] Treatment of palladium(II) acetate with excess chloride displaces the acetate ions to give tetrachloropalladate(II) ions.[156] On heating palladium(II) acetate with lithium nitrate in acetic acid under reflux, nitrous fumes are given off and brown polymeric [Pd(CN)(CH$_3$COO)]$_n$ is formed, which probably has a macromolecular layer structure with linear bridging cyanide ions.[158]

Platinum(II) acetate can be prepared by careful reduction of a solution of Na$_2$Pt(OH)$_6$ in acetic acid with formic acid.[155] Although it is trimeric in acetone, chloroform, benzene and chlorobenzene, x-ray powder photographs show that it is not isostructural with palladium(II) acetate. It does not undergo simple cleavage reactions with amines, triphenylphosphine and triphenylarsine, but instead gives compounds of very high molecular weight and uncertain composition.[155]

Oxalic acid readily forms mono- and bis-oxalato complexes with platinum(II)[159–61] (reactions 260 and 261). On oxidation with chlorine potassium tetraoxalatoplatinum(II) is oxidised to the platinum(IV) derivative K$_2$[Pt(OX)$_2$Cl$_2$] without affecting the co-ordinated oxalate groups.[162] Monomeric oxalate complexes of palladium(II) such as [Pd(NH$_3$)$_2$(OX)] in which the oxalate ions acts as bidentate ligands have been prepared by treating the corresponding dichloro complexes with potassium oxalate.[163] The dimeric complex [(PBu$_3$)$_2$Pd$_2$Cl$_2$(OX)], prepared by treatment of

$$K_2Pt(NO_2)_4 + H_2OX \rightarrow K_2[Pt(NO_2)_2(OX)_2] \cdot H_2O \xrightarrow[H_2OX]{+\text{excess}}$$

$$K_2[Pt(OX)_2] \cdot \text{aq.} \quad (260)$$

$$K_2PtCl_4 + 2K_2OX \underset{HCl}{\rightleftharpoons} K_2[Pt(OX)_2] + 4KCl \quad (261)$$

dimeric [(PBu$_3$)$_2$Pd$_2$Cl$_4$] with potassium oxalate was shown by x-ray diffraction to have structure 77.[164] The oxalate bridge in this complex is cleaved by p-toluidine to give a 1:1 mixture of [(PBu$_3$)Pd(p-toluidine)(OX)] and [(PBu$_3$)Pd(p-toluidine)Cl$_2$].

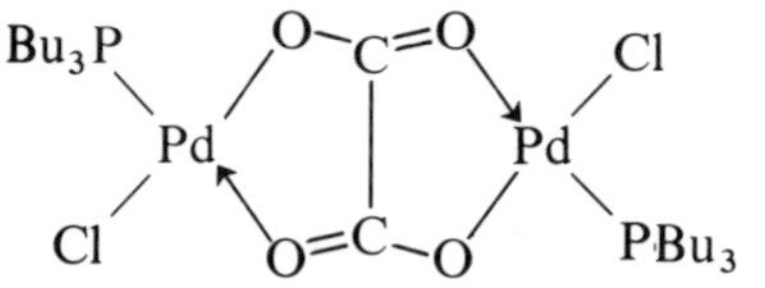

77

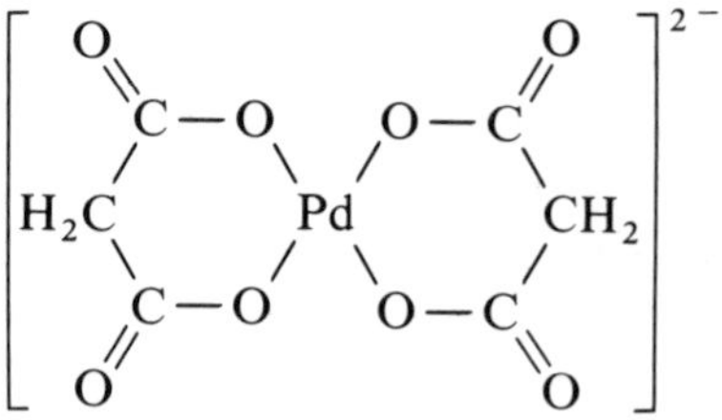

78

Potassium bis(malonato)palladium(II) (78) can be prepared by warming a suspension of palladium(II) chloride in concentrated potassium malonate solution.[165] A comparison of the infra-red spectra of the bis(malonate)[165] and bis(oxalate)[166] complexes of palladium(II) suggests that the metal–ligand bond in the malonate complex is more ionic than that in the oxalate complex.

MONOTHIOCARBOXYLATE $\left(-C\begin{smallmatrix}\diagup O\\ \diagdown S^-\end{smallmatrix}\right)$ COMPLEXES

Dark coloured, very stable dithio–oxalate complexes $K_2[M(C_2O_2S_2)_2]$, where M = Pd or Pt, are formed when platinum(II) or palladium(II) di-chloride is added to a concentrated solution of potassium dithio-oxalate.[167,168] X-ray diffraction indicates that the anions are square-planar with structure **79**.[169] The infra-red spectra of these complexes suggest that

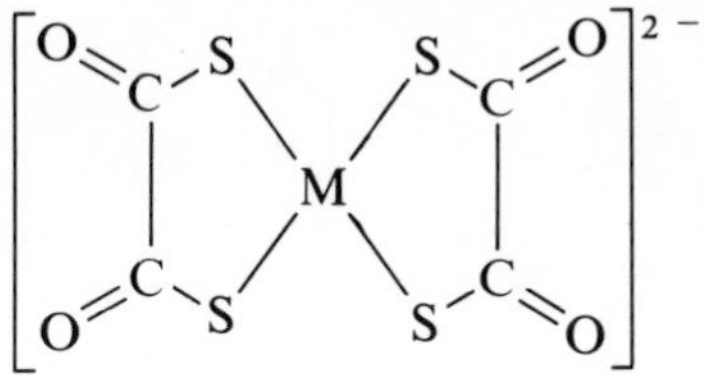

79

the metal (II)–sulphur bonds are extremely strong and may well involve partial $d_\pi - p_\pi$ or $d_\pi - d_\pi$ bonding.[170] The complexes are potentially capable of co-ordinating to a class 'a' metal via the terminal oxygen ligands, and by treating a suspension of the bis(dithio-oxalate)palladium(II) with stannic chloride either one or two tin(II) atoms can be co-ordinated to the palladium(II) complex, depending on the solvent[171] (reaction 262). Thio-

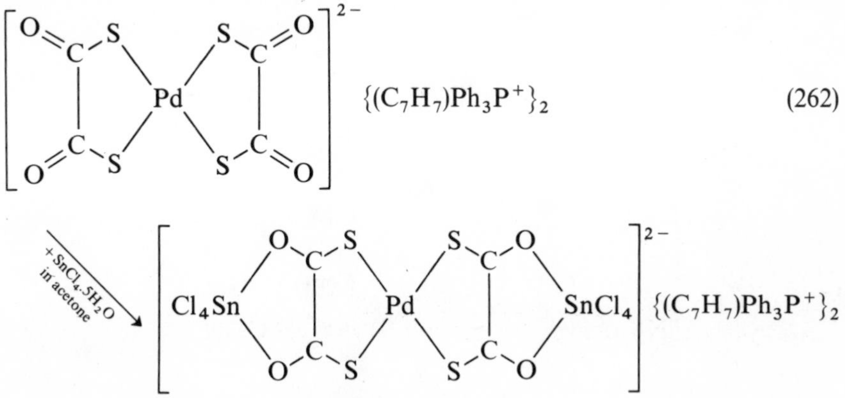

(262)

carbamic esters, $R-NH-\overset{\underset{\|}{S}}{C}-OR'$, co-ordinate to both platinum(II) and palladium(II) as unidentate ligands through sulphur.[172,173]

DITHIOCARBOXYLATE $\left(-C\begin{smallmatrix}\nearrow S\\\searrow S^-\end{smallmatrix}\right)$ **AND DITHIOPHOSPHATE**

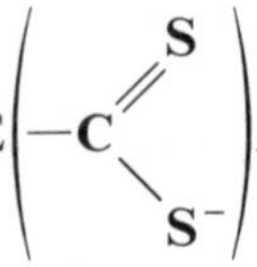

$\left(-P\begin{smallmatrix}\nearrow S\\\searrow S^-\end{smallmatrix}\right)$ **COMPLEXES**

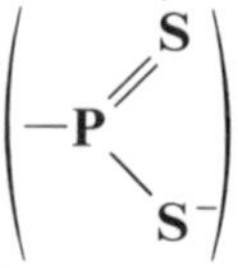

Four-membered chelate rings are formed by platinum(II) and palladium(II) with dithiobenzoates[174] (**80**), alkylxanthates[175,176] (**81**), dialkyldithio-

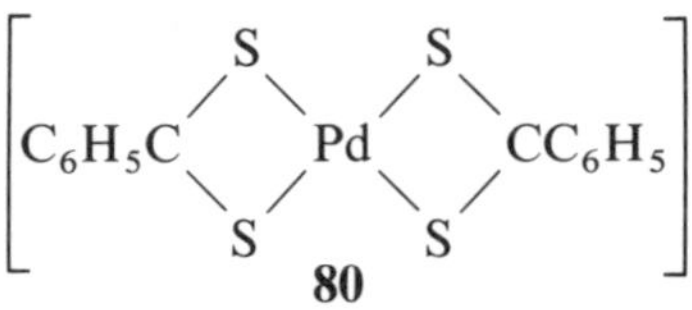

$$\left[C_6H_5C\begin{smallmatrix}S\\S\end{smallmatrix}Pd\begin{smallmatrix}S\\S\end{smallmatrix}CC_6H_5\right]$$
80

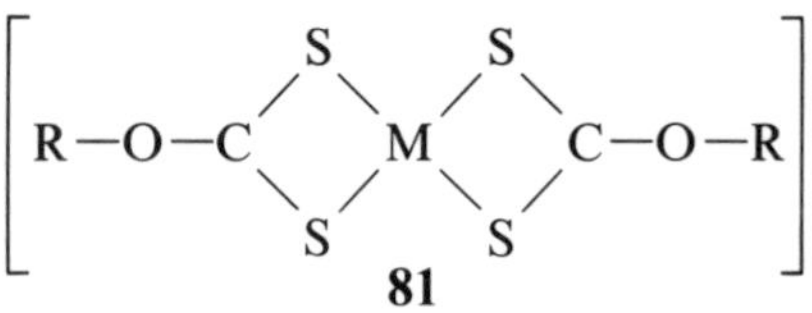

$$\left[R-O-C\begin{smallmatrix}S\\S\end{smallmatrix}M\begin{smallmatrix}S\\S\end{smallmatrix}C-O-R\right]$$
81

carbamates[175,177–80] (**82**), dithiocarbamates[179] (**82** where R = H), dithio-

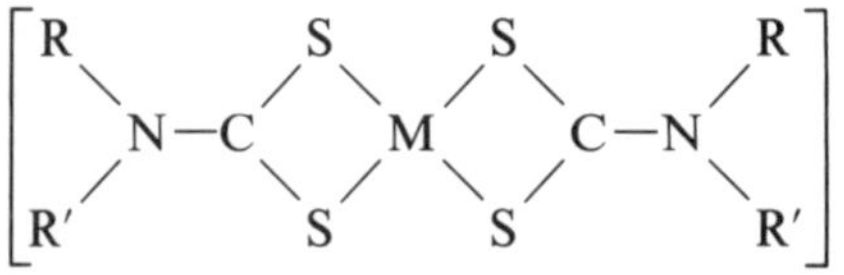

$$\left[\begin{smallmatrix}R\\R'\end{smallmatrix}N-C\begin{smallmatrix}S\\S\end{smallmatrix}M\begin{smallmatrix}S\\S\end{smallmatrix}C-N\begin{smallmatrix}R\\R'\end{smallmatrix}\right]$$
82

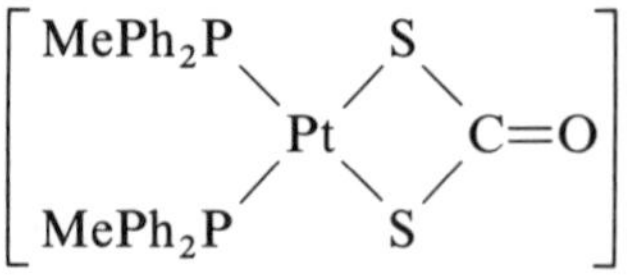

$$\left[\begin{smallmatrix}MePh_2P\\MePh_2P\end{smallmatrix}Pt\begin{smallmatrix}S\\S\end{smallmatrix}C=O\right]$$
83

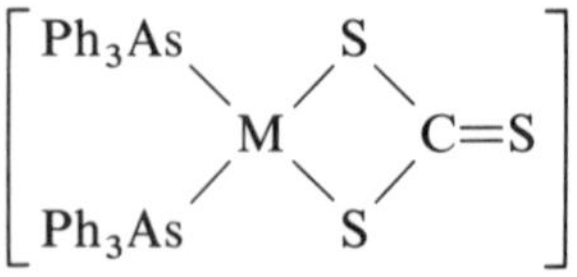

$$\left[\begin{smallmatrix}Ph_3As\\Ph_3As\end{smallmatrix}M\begin{smallmatrix}S\\S\end{smallmatrix}C=S\right]$$
84

carbonate[181] (**83**), trithiocarbamate[182] (**84**). N-cyanodithiocarbamate[182] (**85**),   1,1-dicyanoethylene-2,2'-dithiolate[182]   (**86**),   dialkyldithiophos-

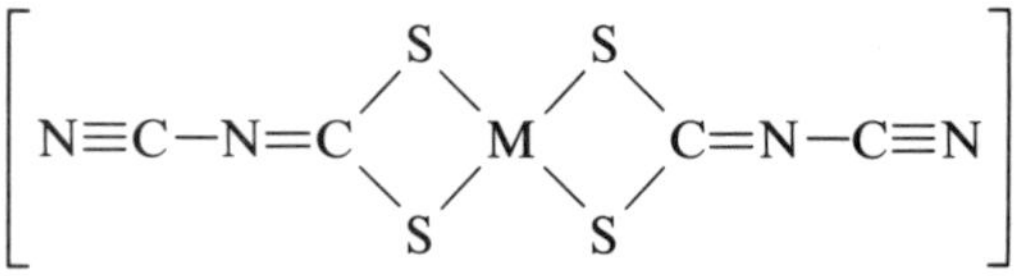

$$\left[N\equiv C-N=C\begin{smallmatrix}S\\S\end{smallmatrix}M\begin{smallmatrix}S\\S\end{smallmatrix}C=N-C\equiv N\right]$$
85

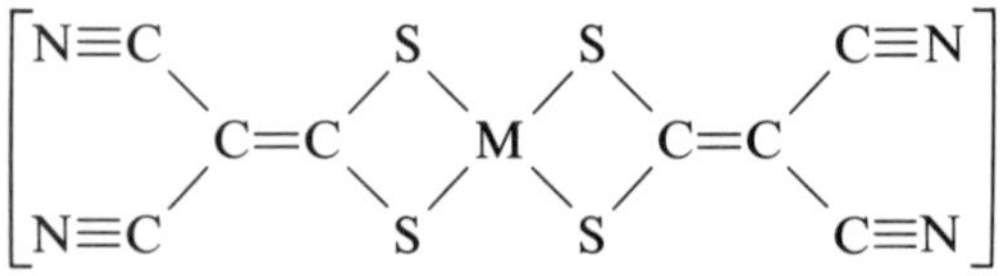

$$\left[\begin{smallmatrix}N\equiv C\\N\equiv C\end{smallmatrix}C=C\begin{smallmatrix}S\\S\end{smallmatrix}M\begin{smallmatrix}S\\S\end{smallmatrix}C=C\begin{smallmatrix}C\equiv N\\C\equiv N\end{smallmatrix}\right]$$
86

phates[178,183] (**87**), dialkyl- and diaryl-phosphinodithioates[184–186] (**88**), diethylselenothiophosphinates[184] (**89**) and diethyldiselenophosphinates[187] (**90**). The sulphur complexes are more stable than their selenium analogues, which tend to decompose in daylight and on heating.

$$
\begin{bmatrix}
\text{R}-\text{O} & \text{S} & \text{S} & \text{O}-\text{R} \\
& \text{P} \quad \text{M} \quad \text{P} & \\
\text{R}-\text{O} & \text{S} & \text{S} & \text{O}-\text{R}
\end{bmatrix}
\qquad
\begin{bmatrix}
\text{R} & \text{S} & \text{S} & \text{R} \\
& \text{P} \quad \text{M} \quad \text{P} & \\
\text{R} & \text{S} & \text{S} & \text{R}
\end{bmatrix}
$$

87   88

$$
\begin{bmatrix}
\text{Et} & \text{S} & \text{S} & \text{Et} \\
& \text{P} \quad \text{Pd} \quad \text{P} & \\
\text{Et} & \text{Se} & \text{Se} & \text{Et}
\end{bmatrix}
\qquad
\begin{bmatrix}
\text{Et} & \text{Se} & \text{Se} & \text{Et} \\
& \text{P} \quad \text{Pd} \quad \text{P} & \\
\text{Et} & \text{Se} & \text{Se} & \text{Et}
\end{bmatrix}
$$

89   90

The xanthate complexes are generally less stable than the dithiocarbamate complexes. A comparison of the infra-red spectra of these two complexes[177,179,188] has suggested that there is considerable double-bond character in the C–N bond of co-ordinated dithiocarbamates (**91**), but little double-bond character in the C–O bond of co-ordinated xanthates (**92**). This

$$
\begin{array}{c}
\text{R} \\
^{+}\!\!\diagdown \\
\text{N}{=}\text{C} \\
\diagup \\
\text{R}
\end{array}
\begin{array}{c}
\text{S} \\
\diagup \quad \diagdown \\
\quad \text{M} \\
\diagdown \quad \diagup \\
\text{S}
\end{array}
\qquad\qquad
\text{R}{-}\overset{+}{\text{O}}{=}\text{C}
\begin{array}{c}
\text{S} \\
\diagup \quad \diagdown \\
\quad \text{M} \\
\diagdown \quad \diagup \\
\text{S}
\end{array}
$$

91   92

leads to a mesomeric drift of electrons to the sulphur in the dithiocarbamate complexes so increasing the σ-donor ability of the sulphur atoms of the dithiocarbamate ligand over that of the sulphur atoms of the xanthate ligand, so that the dithiocarbamates form the more stable complexes. The electronic

spectra of complexes containing the $-\text{C}\begin{smallmatrix}\text{S}\\ \diagdown\\ \text{S}^{-}\end{smallmatrix}$ or $-\text{P}\begin{smallmatrix}\text{S}\\ \diagdown\\ \text{S}^{-}\end{smallmatrix}$ groups indicate

that the ligand field strength increases in the order $\text{Cl}^{-} < (\text{EtO})_2\text{PS}^{-} < \text{Et}_2\text{NCS}_2^{-} < \text{HOCH}_2\text{CH}_2(\text{S}^{-})\text{CH}_2\text{S}^{-} < \text{CH}_2(\text{COS})_2^{2-} \ll \text{amines.}^{[178]}$

Platinum(II) xanthate complexes undergo nucleophilic attack by amines to yield dithiocarbamate complexes[189] (reaction 263). Both xanthate and dithiocarbamate complexes of platinum(II) react with 1 mol of a tertiary

$$
\left[\text{PhCH}_2\text{O}-\text{C}\begin{smallmatrix}\text{S}\\ \diagdown\\ \text{S}\end{smallmatrix}\text{Pt}\begin{smallmatrix}\text{S}\\ \diagdown\\ \text{S}\end{smallmatrix}\text{C}-\text{OCH}_2\text{Ph}\right] + \text{Me}-\!\!\bigcirc\!\!-\text{N}-\text{H} \xrightarrow{\text{toluene}}
$$

$$
\left[\text{Me}-\!\!\bigcirc\!\!-\text{N}-\text{C}\begin{smallmatrix}\text{S}\\ \diagdown\\ \text{S}\end{smallmatrix}\text{Pt}\begin{smallmatrix}\text{S}\\ \diagdown\\ \text{S}\end{smallmatrix}\text{C}-\text{OCH}_2\text{Ph}\right] + \text{PhCH}_2\text{OH} \quad (263)
$$

phosphine to give a five-co-ordinate adduct (reaction 264);[190] with two moles the four-co-ordinate complexes [(PMePh$_2$)$_2$M(S$_2$COEt)] are

$$[Pt(S_2COEt)_2] + PMePh_2 \rightarrow [(PMePh_2)Pt(S_2COEt)_2] \qquad (264)$$

formed.[181,190] 1:1 and 1:2 adducts are formed when [Pd(S$_2$PPh$_2$)$_2$] is treated with 1 and 2 mol of a tertiary phosphine,[185] although these are not five- and six-co-ordinate palladium(II) complexes as they were first thought to be but four-co-ordinate complexes formed by reaction 265.[186]

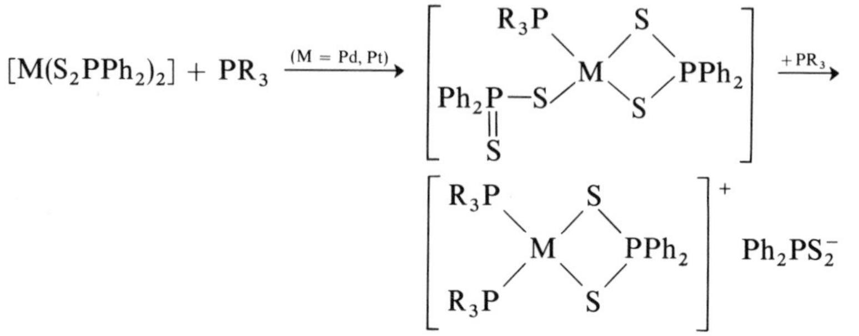

$$[M(S_2PPh_2)_2] + PR_3 \xrightarrow{(M = Pd, Pt)} \cdots \xrightarrow{+PR_3} \cdots \qquad Ph_2PS_2^- \quad (265)$$

β-DIKETONE COMPLEXES

β-Diketones form a variety of complexes with platinum(II) and palladium(II),[191,192] the simplest of which involve β-diketones acting as bidentate oxygen bonding ligands. However, in addition, products in which β-diketones bond to platinum(II) through the γ-carbon atom or through the double-bond of the enol are known.

(i) Complexes containing oxygen-bonded chelating β-diketone ligands
Palladium(II) acetylacetonate is prepared by treating palladium(II) chloride with acetylacetone and potassium hydroxide, and can be recrystallised from hot benzene solution[193,194] (reaction 266). The potassium hydroxide en-

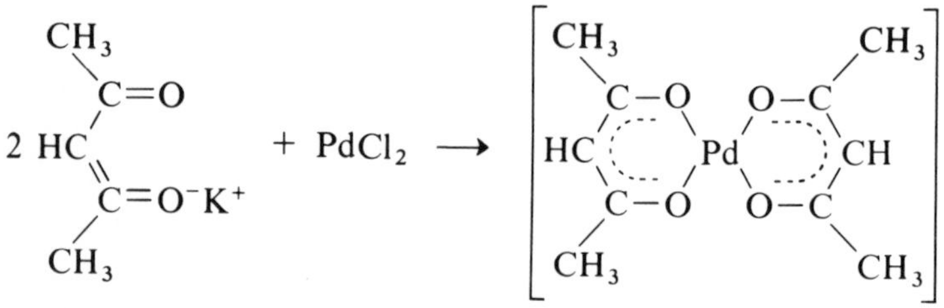

$$2\ HC \cdots + PdCl_2 \rightarrow \cdots \qquad (266)$$

sures that the keto-enol tautomerisation (reaction 267) lies largely on the enol side. Platinum(II) acetylacetonate is prepared similarly using four

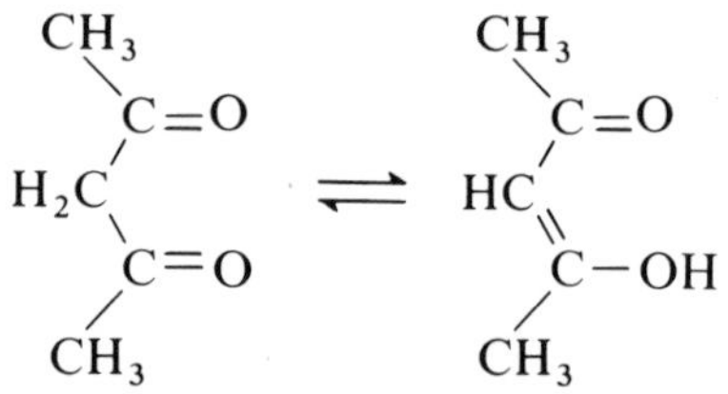

$$\cdots \rightleftharpoons \cdots \qquad (267)$$

equivalents of acetylacetone.[195] Both platinum(II) and palladium(II) acetylacetonates are yellow crystalline complexes that are slightly soluble in benzene and very soluble in chloroform. An alternative method for preparing bis–β-diketonate complexes of palladium(II) involves treating a methanolic solution of disodium hexachlorodipalladate(II) with an excess of the β-diketone.[196] With acetylacetone this reaction yields [Pd(acac)$_2$] directly but with 2,2,6,6-tetramethylheptane-3,5-dione a methoxy-bridged complex, which can be isolated, is formed first (reaction 268).

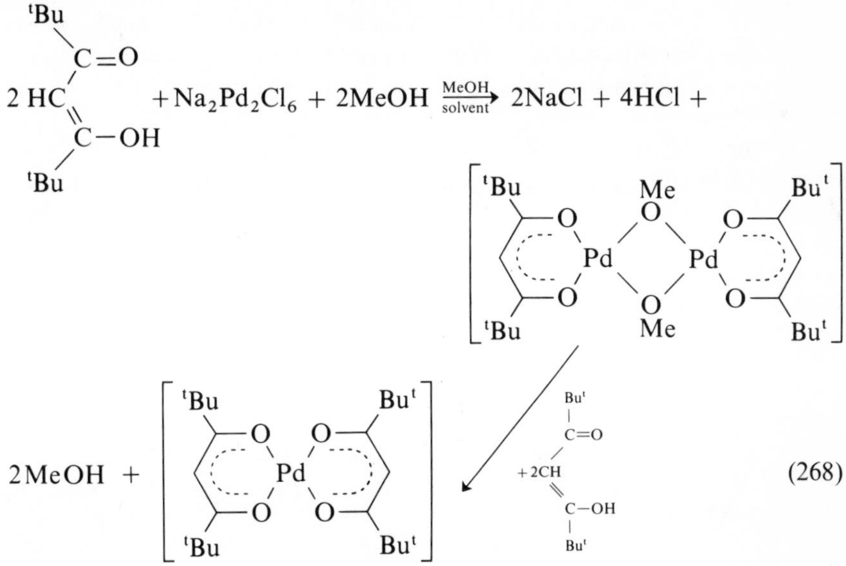

X-ray diffraction studies[197–9] have shown that the metal atom and the β-diketone ring are virtually coplanar and that the O–M–O bond is between 90° and 94° indicating that β-diketones are virtually ideal ligands for giving undistorted square-planar geometry about the metal. The palladium(II)–oxygen bond lengths vary between 1·97 and 2·085 Å, which is considerably less than the 2·70 Å calculated from the sum of the covalent radii[200] (for further details see Appendix II structures 54, 64 and 72).

The infra-red spectra of the bis–β-diketonates are complex due to interactions between the two chelate rings.[200] Accordingly K[Pt(acac)Cl$_2$] and its deuterated analogue were prepared using only one equivalent of acetylacetone and their infra-red spectra analysed by normal co-ordinate analysis.[202] The band at 478 cm^{-1} was assigned to the Pt–O stretching vibration coupled with a ring deformation, and the Urey–Bradley force constant for the Pt–O stretching mode calculated to be 2·46 mdyn/Å, which was smaller than for Pt–CN in K$_2$[Pt(CN)$_4$] (3.425 mdyn/Å[203]) but slightly larger than for Pt–NH$_3$ in trans-[Pt(NH$_3$)$_2$Cl$_2$] (2·09 mdyn/Å[204]) and for Pt–Cl in K[Pt(acac)Cl$_2$] (1·78 mdyn/Å[202]). More recently the infra-red spectra of [Pt(acac)$_2$] and [Pd(acac)$_2$] have been analysed.[205] The bands involving palladium were identified by recording the spectra of ^{104}Pd and ^{110}Pd isotopically enriched [Pd(acac)$_2$].[206] This is the only safe way to assign the spectra of these complexes because it shows which bands involve the palladium atom, since the frequencies of such bands are shifted when the atomic weight of the palladium isotope is varied.

The C–H stretching vibration of the proton attached to the γ-carbon atom of the acetylacetone ring occurs above 3000 cm^{-1},[207] which is consistent with the 'aromatic character' of the acetylacetone ring as deduced from its chemical reactions (see below). This proton resonates at 5·53 p.p.m. below tetramethylsilane in the proton n.m.r. spectrum of $[Pt(acac)_2]$.[207] Whilst consistent with the presence of 'aromatic character' this does not provide any evidence for aromaticity[208,209] because the shifts that might be expected to arise from a ring current are masked by large upfield displacements which are probably associated with changes in the structure of the β-diketone ligand on co-ordination.[210]

The chemical reactions of metal(II) acetylacetonates are generally those characteristic of an aromatic ring system.[211,212] Thus $[Pd(acac)_2]$ is brominated at the γ-carbon on treatment with N-bromo-succinimide. On treatment with nitric oxide an electrophilic substitution of the proton attached to the γ-carbon atom occurs, followed by rearrangement to give a final product which contains a palladium–nitrogen bond[213] (reaction 269).

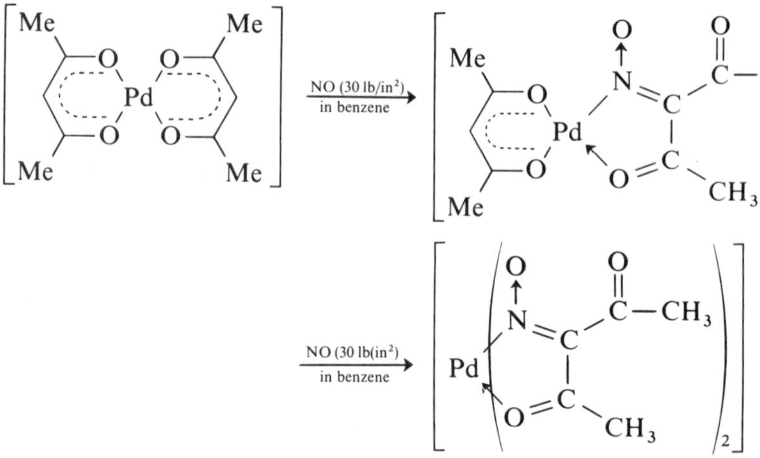

$$\text{Rate} = \{k_1 + k_2[\text{X}]\}[\text{Pd(acac)}_2]$$

(269)

The β-diketone ring can be cleaved from the metal by treatment with tetrafluoroboric acid followed by addition of triphenylphosphine.[214]

A study of the substitution reactions of $[Pd(acac)_2]$ in a medium containing 90% water and 10% methanol by water and various nucleophiles showed that the observed rate was consistent with the typical two term rate-law found for square-planar substitution reactions (see Chapter 11), where k_1 and k_2

$$\text{Rate} = \{k_1 + k_2[\text{X}]\}[\text{Pd(acac)}_2]$$

are the first and second-order rate constants respectively and both k_1 and k_2 are functions of the proton concentration.[215] The reactivity of the nucleophile, X, increased in the order $\text{X} = \text{H}_2\text{O} \approx \text{OH}^- < \text{Cl}^- < \text{Br}^- < \text{I}^- < \text{SCN}^-$. The dependence of the rate on the proton concentration arises because the leaving group is a chelate ligand which, once one end has been displaced from the metal, can either reform the chelate ring or react with a proton to form a stable unidentate ligand that is subsequently displaced by a second nucleophile.

The isotopic exchange rate between free ^{14}C-acetylacetone and [Pd(acac)$_2$] in anisole, benzene and methylene chloride again follows a two term rate-law, that is typical of square-planar substitution reactions.[216]

$$Rate = \{k_1 + k_2[Hacac]\}[Pd(acac)_2]$$

The k_1 term probably involves a dissociation mechanism since a solvent path is unlikely in these media. The k_2 term corresponds to the reagent path in which acetylacetone itself, rather than its anion, is the nucleophilic reagent. In the presence of acids a third term that is first-order in both acid and complex is added to the rate expression, although it is uncertain whether the function of the acid is to stabilise the complex formed by the unidentate acetylacetone ligand after one palladium(II)–oxygen bond has been cleaved, as suggested above for the nucleophilic substitution reactions of [Pd(acac)$_2$], or whether the acid promotes cleavage of the first palladium(II) oxygen bond.

(ii) Complexes involving β-diketone ligands that are not oxygen bonded

When sodium or potassium tetrachloroplatinate(II) are treated with acetylacetone in alkaline solution, K[Pt(acac)$_2$Cl] and Na[Pt(acac)$_2$Cl$_2$].5H$_2$O are obtained as well as K[Pt(acac)Cl$_2$] and [Pt(acac)$_2$].[195,217] In addition K[Pt(acac)$_3$] can also be prepared using high concentrations of acetylacetone.[207]

(a) [Pt(acac)$_2$Cl]$^-$

X-ray diffraction of K[Pt(acac)$_2$Cl] indicates one bidentate oxygen-bonded acetylacetone ligand and one acetylacetone ligand bound to the metal via the γ-carbon atom[218,219] (93) (see Appendix II, structure 149). The carbon-

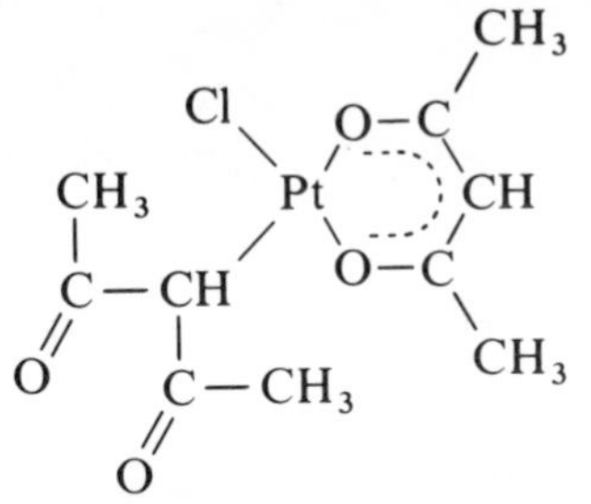

3

bonded acetylacetone ligand can be distinguished from the bidentate oxygen-bonded ligand by infra-red and proton n.m.r. spectroscopy. In the infra-red the C–H stretching frequency of the γ-proton lies above 3000 cm^{-1} in the oxygen-bonded ligand, but below 3000 cm^{-1} in the carbon-bonded ligand. The C=O stretching frequency in the carbon-bonded ligand is higher than in the bidentate oxygen-bonded ligand. Carbon-bonded acetylacetone ligands are readily detected by n.m.r. because the ^{195}platinum–

[1]hydrogen coupling constant increases from about 10 Hz in the bidentate oxygen-bonded ligand to about 120 Hz in the γ-carbon bonded ligand.[207] These criteria have been used to establish that the complexes K[Pt(diketone)$_2$X], where X = Cl or Br, formed by β-diketones such as trifluoroacetylacetone and benzoylacetone have structures similar to **93**.[220] The anion [Pt(acac)$_2$Cl]$^-$ can act as a terdentate ligand, enabling complexes of Mn^{2+}, Fe^{2+}, Co^{2+}, Ni^{2+}, Cu^{2+}, Zn^{2+} and Cd^{2+} to be prepared in which the first row transition metal is bound to two [Pt(acac)$_2$Cl]$^-$ anions via two carbonyl oxygen atoms and a carbon atom[221] (**94**).

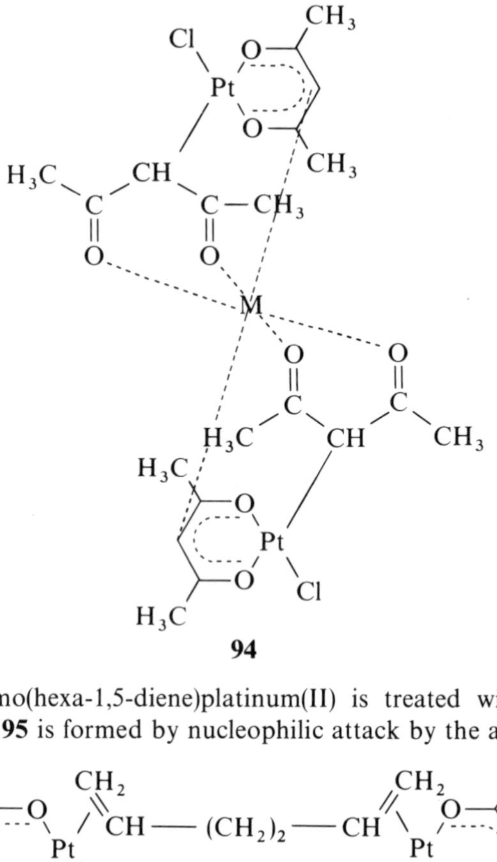

94

When dibromo(hexa-1,5-diene)platinum(II) is treated with thallium(I) acetylacetonate **95** is formed by nucleophilic attack by the acetylacetonate

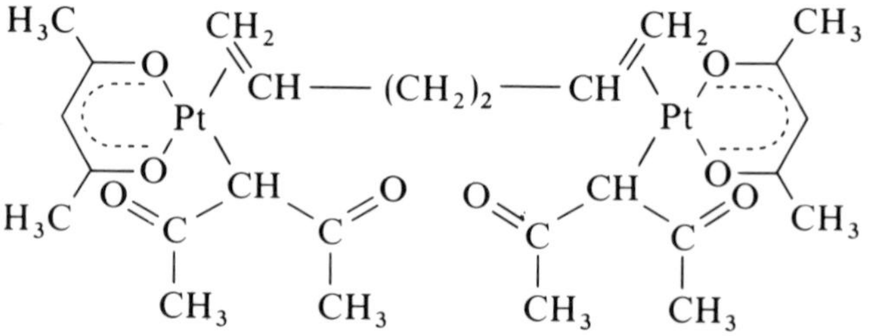

95

anion at the metal.[222] **95** is very different to **96**, obtained by nucleophilic attack by the acetylacetonate ion on the diolefin when [(diene)MX$_2$], where diene = cycloocta-1,5-diene, norbornadiene or dicyclopentadiene, M = Pd or Pt and X = halogen, is treated with thallium(I) acetylacetonate.[223,224]

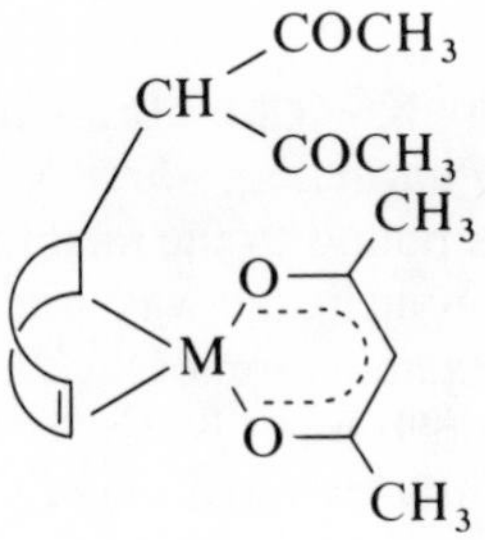

96

(b) $[Pt(acac)_2Cl_2]^{2-}$

The infra-red and n.m.r. criteria outlined above (pp. 191–192) indicate that $[Pt(acac)_2Cl_2]^{2-}$ contains two γ-carbon bonded acetylacetone ligands.[207,225] The stretching force constant of 2·50 mdyn/Å obtained for the platinum–carbon bond using a Urey–Bradley force field[226] is very similar to the 2·46 mdyn/Å found for the Pt–O stretching vibration in $K[Pt(acac)Cl_2]$.[202] This is consistent with the acetylacetone readily forming both types of co-ordinate bonds not only under the same conditions, but in the same reaction mixture.[217]

(c) $[Pt(acac)_3]^{-}$

The infra-red and n.m.r. spectra of $K[Pt(acac)_3]$ suggest that this complex contains one oxygen-bonded acetylacetone ligand and two γ-carbon bonded ligands.[207] A series of complexes $M[Pt(acac)_3]_2$ where M = Mn, Fe, Co, Cu, Cd and Pd have been prepared and shown to involve two $[Pt(acac)_3]^{-}$ anions acting as terdentate ligands by using three of the carbonyl groups of the acetylacetone ligands that are bonded to platinum through their γ-carbon atoms.[221]

When $K[Pt(acac)_3]$ is treated with hydrobromic or hydrochloric acid, the two γ-carbon bonded acetylacetone groups condense together to form a bicyclic dienyl compound which co-ordinates to platinum to give the complex $[Pt(C_{10}H_{14}O_3)X_2]$, where X = Cl or Br, (**97**)[227–9] (see Appendix II,

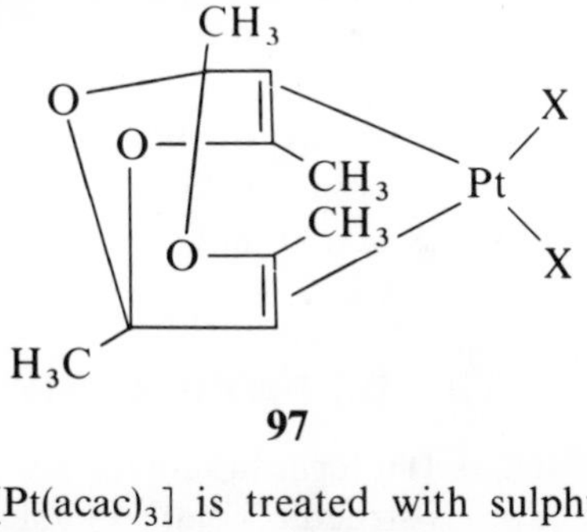

97

structure 150). If $K[Pt(acac)_3]$ is treated with sulphuric acid instead of hydrochloric, the resulting product is polymeric $[Pt(acac)_2]_n$ in which both the acetylacetone ligands are bonded to the platinum atom through their γ-carbon atoms and a co-ordination number of four is attained by some of the carbonyl groups of the acetylacetone ligands bonding to platinum.[229]

(d) H[Pt(acac)$_2$X]

When K[Pt(acac)$_2$X], where X = Cl or Br, is treated with a strong acid it yields the yellow 'acid' H[Pt(acac)$_2$X], where X = Cl or Br[229,230] in which one acetylacetone ligand is bound to the metal through two oxygen atoms, whilst the second acetylacetone ligand was originally thought to be bonded as an olefin (**98**).[229] A re-examination of the data by the same group found that the π-allylic structure (**99**) fitted better,[229] although a detailed infrared investigation subsequently favoured the original olefin structure **98**.[231]

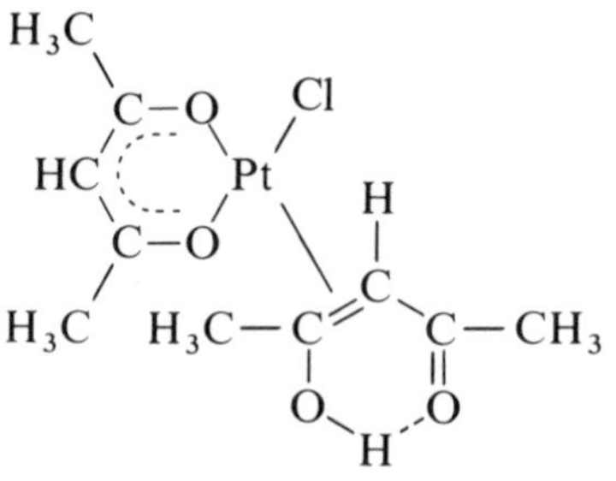

98 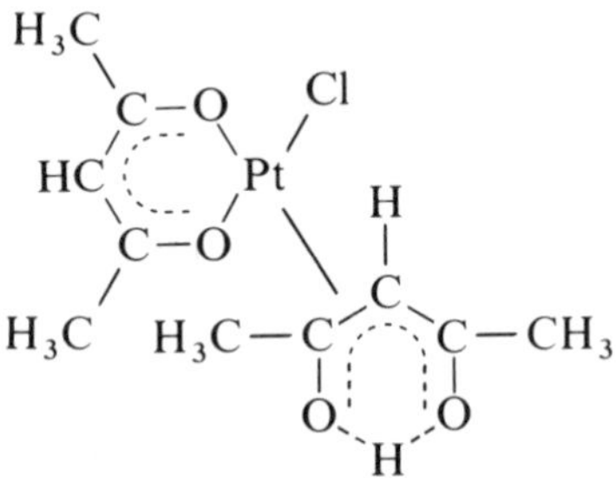   **99**

The acid is therefore more correctly formulated as [Pt(acac)(acacH)X], since the hydrogen is not a free proton as implied by H[Pt(acac)$_2$X] but bound to the enolic oxygen atom of the π-bonded acetylacetone ligand.

H[Pt(acac)$_2$X], where X = Cl or Br, undergoes three types of reaction.[232] With unidentate ligands, such as tertiary phosphines, arsines, stibines, phosphites, amines, carbon monoxide, olefins or acetylenes the π-bonded acetylacetone ligand is displaced (reaction 270). With bidentate ligands such as 2,2'-bipyridyl, 1,10-phenenthroline, ethylenediamine or 1,2-bis-diphenyl-phosphinoethane one acetylacetone ligand is displaced and the remaining ligand is bound to platinum through its γ-carbon atom reaction 271). With a bidentate diolefin such as butadiene, isoprene or cyclo-octa-1,5-diene, [(acac)ClPt(diene)PtCl(acac)], which contains bidentate oxygen-bonded acetylacetone ligands and bridging diene ligands is formed (reaction 272). The bonding of the acetylacetone ligand to the metal in these products appears to be determined by the number of potential acceptor sites on the metal. When two sites are available oxygen bonded acetylacetone derivatives are formed whereas when only one site is available the acetylacetone is bound to the metal via its γ-carbon atom.

β-MONOTHIODIKETONE COMPLEXES

Monothiodiketones such as **100** form bis-complexes with platinum(II) and palladium(II) which have *cis*-structures both in solution and in the solid state (**101**).[233-5] The x-ray diffraction structure[236] of the π-2-methyl-allylpalladium(II) complex of this ligand (**102**) is given in Appendix II (structure 69). It has recently been shown that palladium(II) and platinum(II) can be separated by gas chromatography as their bis(monothiotrifluoro-

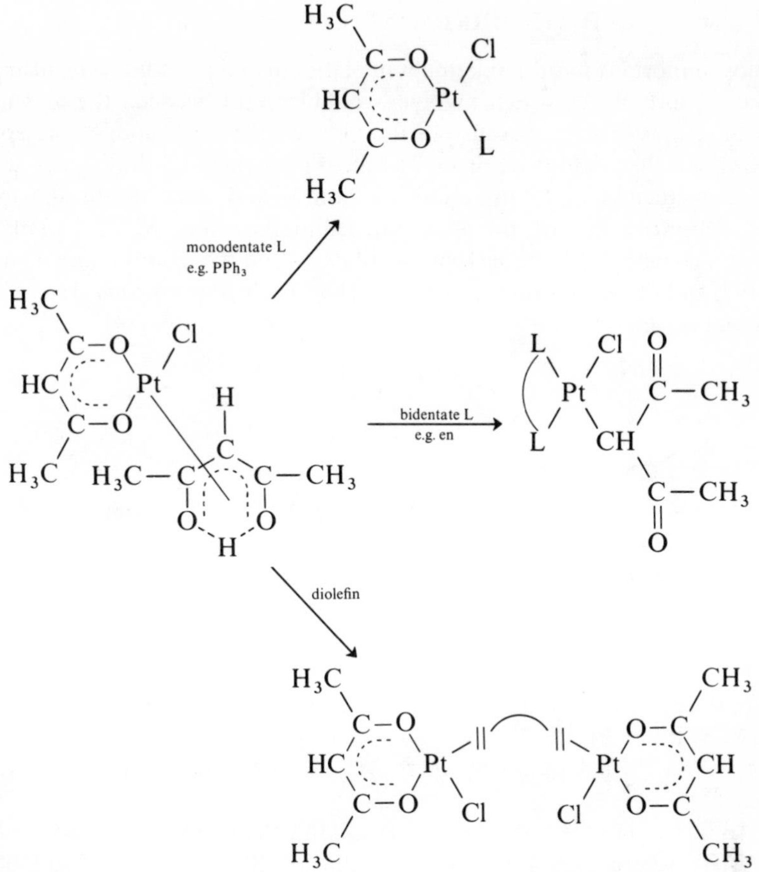

(270)

(271)

(272)

acetylacetonates) (**103**).[237] A similar separation can be achieved using the bis(acetylacetone) ethylenedi-imine Schiff's base complexes **104**.[238]

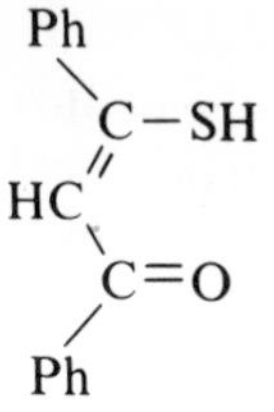

100

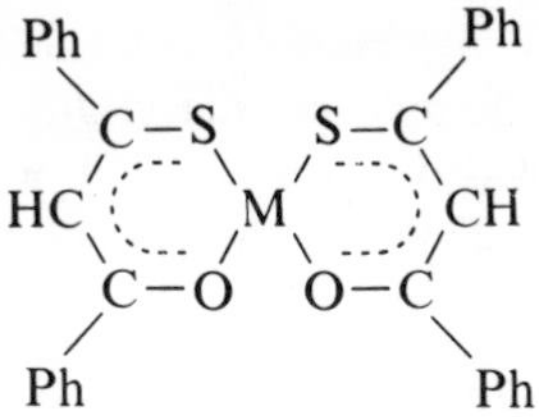

101

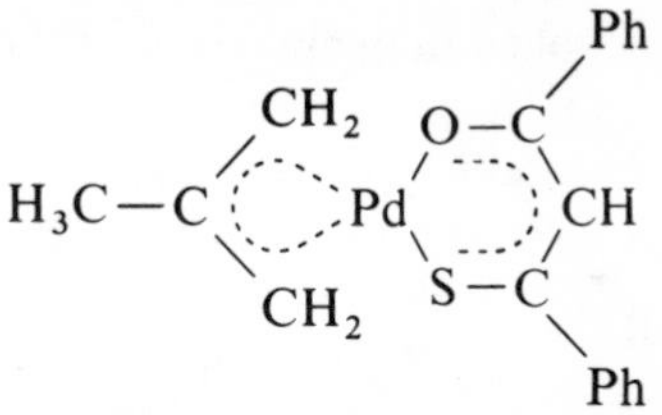

102

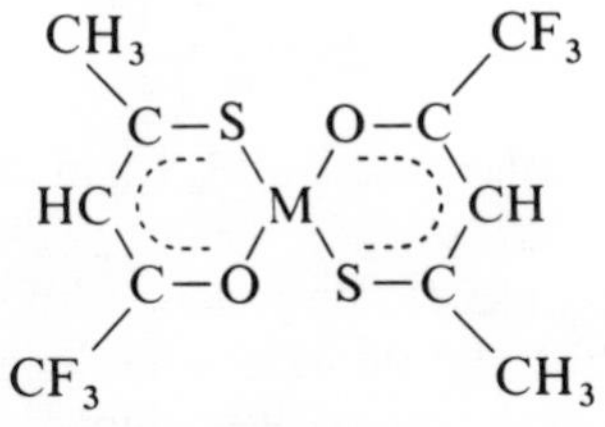

103

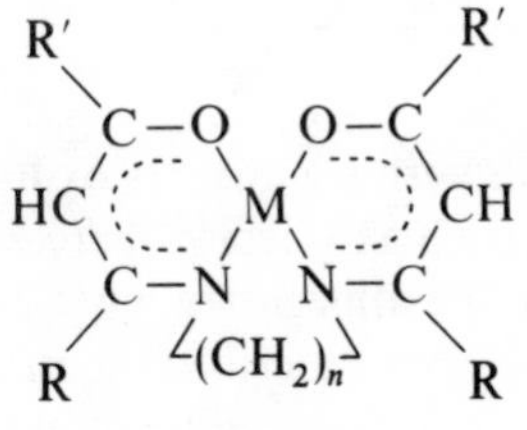

104

R, R' = CH_3 or CF_3
n = 2 or 3

β-DITHIODIKETONE COMPLEXES

The most important β-dithiodiketones in platinum and palladium chemistry are the 1,2-dithiolenes which involve a double-bond between the α- and β-carbon atoms. This is because of the ability of these ligands to accept electrons and then delocalise them. The most important 1,2-dithiolenes are cis-1,2-dicyanoethylene-1,2-dithiolate dianion (also known as the dimer-captomaleonitrile ion or the maleonitriledithiolate ion, MNT^{2-}) (105) bis-perfluoromethyldithietene (106), the dialkyl- and diaryl-dithiolate dian-ions (107) and the aryldithiolate dianions (108). Their preparations are given in a recent review.[239]

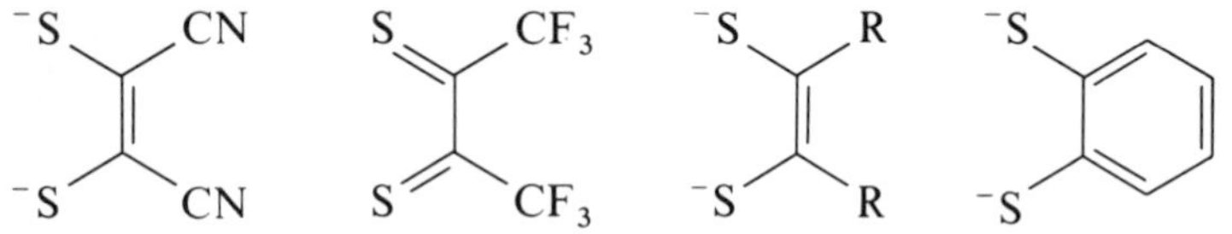

105   106   107   108

(i) Preparation of 1,2-dithiolene and related complexes
A number of procedures have been used to prepare 1,2-dithiolene and related complexes:

(a) Treatment of the metal(II) halide with the sodium salt of the dithiolate anion in the presence of a bulky cation yields $(Bu_4N^+)_2[M(S_2C_2(CN)_2)_2]^{2-}$, where M = Pd or Pt.[240] When the bis-tertiary phosphine metal(II) halide is used, $[(PR_3)_2M(S_2C_2(CN)_2)]$, where M = Pd or Pt, is obtained as the product.[241]

(b) Addition of bis(trifluoromethyl)-1,2-dithietene to a solution of $[M(PPh_3)_4]$, where M = Pd or Pt, in dichloromethane yields crystals of $[(PPh_3)_2M(S_2C_2(CF_3)_2)]$.[241]

(c) Air and light sensitive $(Ph_4As^+)_2[Pt(S_2C_6H_3Me)_2]^{2-}$ is prepared by treating K_2PtCl_4 with an alkaline solution of toluene-3,4-dithiol under nitrogen in the presence of tetraphenylarsonium chloride.[242]

(d) A series of complexes of dimethyl- and diaryl-dithiolate anions (107) have been synthesised by treating K_2PtCl_4 and K_2PdCl_4 with the appropriate thiophosphoric ester[243] (reaction 273). The thiophosphoric ester is formed

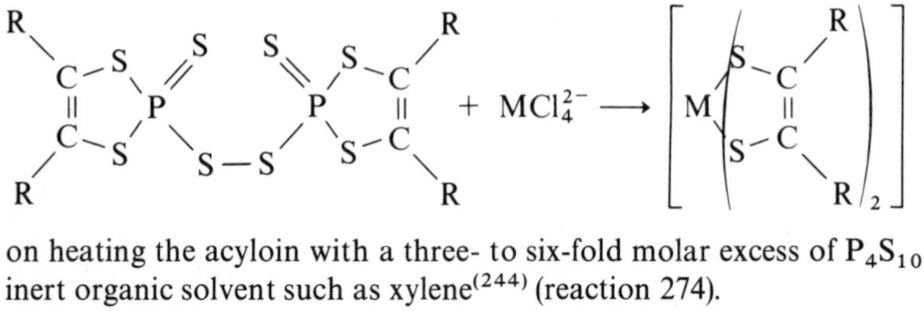

$$+ MCl_4^{2-} \longrightarrow \qquad (273)$$

on heating the acyloin with a three- to six-fold molar excess of P_4S_{10} in an inert organic solvent such as xylene[244] (reaction 274).

(e) In addition to the dinegative bis(1,2-dithiolene) complexes, whose preparations have just been described, neutral bis(1,2-dithiolene) complexes can also be prepared and these can be reduced to the dinegative complexes with the aid of suitable reducing agents. Thus for example $(Ph_4As^+)_2$-

$$\begin{matrix} R \\ | \\ C=O \\ 2 \quad | \\ H-C-OH \\ | \\ R \end{matrix} + P_4S_{10} \xrightarrow[\text{xylene}]{\text{heat in}} \begin{matrix} R \\ \diagdown \\ C-S \qquad S \qquad S \qquad S-C \\ \| \qquad P \qquad P \qquad \| \\ C-S \diagup \qquad \diagdown \qquad \diagdown \qquad S-C \\ \diagup \qquad S-S \qquad S-S \qquad \diagdown \\ R \qquad\qquad\qquad\qquad\qquad R \end{matrix} \qquad (274)$$

$[Pt(S_2C_2(CF_3)_2)_2]^{2-}$ has been prepared by reducing $[Pt(S_2C_2(CF_3)_2)_2]$ with sodium amalgam in tetrahydrofuran; the palladium analogue can be reduced by hydrazine.[241] The neutral complexes are obtained by treating the zerovalent complexes $[M(PPh_3)_4]$ with excess bis(perfluoromethyl)-1,2-dithietene (cf. method b) first in dichloromethane when $[(PPh_3)_2-M(S_2C_2(CF_3)_2)]$ is formed and then in benzene when $[(PPh_3)_2M(S_2C_2(CF_3)_2)]$ reacts with further 1,2-dithietene to give the neutral bis(1,2-dithiolene) complex $[M(S_2C_2(CF_3)_2)_2]$.

(ii) Structure of the 1,2-dithiolene complexes
Although no palladium(II) and platinum(II) 1,2-dithiolene complexes have been investigated by x-ray diffraction, they probably involve square-planar co-ordination around the metal atom since this is the commonest geometry observed for platinum(II) and palladium(II) and also for the bis(1,2-dithiolene) complexes of other metals.[239]

(iii) Oxidation states of the metal in 1,2-dithiolene complexes
One of the most important properties of 1,2-dithiolene complexes is the ease with which they undergo electron transfer reactions (eqn. 275). These electron

$$\begin{bmatrix} S \qquad S \\ \diagup \qquad \diagdown \\ (\quad M \quad) \\ \diagdown \qquad \diagup \\ S \qquad S \end{bmatrix}^{0} \underset{+\varepsilon}{\rightleftharpoons} \begin{bmatrix} S \qquad S \\ \diagup \qquad \diagdown \\ (\quad M \quad) \\ \diagdown \qquad \diagup \\ S \qquad S \end{bmatrix}^{-} \underset{+\varepsilon}{\rightleftharpoons} \begin{bmatrix} S \qquad S \\ \diagup \qquad \diagdown \\ (\quad M \quad) \\ \diagdown \qquad \diagup \\ S \qquad S \end{bmatrix}^{2-} \qquad (275)$$

transfer reactions have been systematised by a study of the half-wave potentials of each one-electron transfer step[241] (see Table 39). A comparison of experimental observations of the chemical behaviour of these complexes with the data in Table 39 has led to the following empirical observations.[245]

(a) All reduced species in couples less positive than ~ 0 V are readily oxidised in solution by air, whereas all reduced species in couples more positive than this value are stable to air oxidation.

(b) Oxidised species in couples more positive than $\sim +0.2$ V are unstable to reduction by weakly basic solvents such as ketones and alcohols, whilst those in couples within the approximate range -0.12 V to $+0.20$ V are reduced by stronger bases such as aromatic amines.

(c) Oxidised forms in couples more negative than ~ 0.12 V are reduced by stronger reducing agents such as hydrazine or sodium amalgam.

(d) Reduced forms in couples less positive than ~ 0.4 V can be oxidised by iodine.

From Table 39 it is apparent that the decreasing oxidative stability of the 1,2-dithiolene dianions $[ML_2]^{2-}$, that is the order of increasingly negative

TABLE 39
Polarographic and voltammetric data obtained for bis-1,2-dithiolene
complexes in methyl cyanide (data from ref. 239)

| | Half-wave potentials | | | |
| | $[ML_2]^- + \varepsilon \rightleftharpoons [ML_2]^{2-}$ (volts) | | $[ML_2]^0 + \varepsilon \rightleftharpoons [ML_2]^-$ (volts) | |
Ligand (L)	M = Pd	M = Pt	M = Pd	M = Pt
$S_2C_2(CN)_2$	+0·46	+0·21	not detected	not detected
$S_2C_2(CF_3)_2$	+0·081	−0·267	+0·963	+0·819
$S_2C_2Ph_2$	−0·51[a]	−0·806[a]	+0·15[a]	+0·06[a]
$S_2C_2H_2$	−0·75[a]	—	+0·14[a]	—
$S_2C_2(p\text{-MeOC}_6H_4)_2$	−0·75[a]	−0·95[a]	+0·06[a]	−0·03[a]
$S_2C_2Me_2$	−0·90[a]	−1·10[a]	−0·09[a]	−0·16[a]

[a] These $E^{1/2}$ values have been arbitrarily converted to a standard methyl-cyanide standard calomel electrode scale.

potentials for reaction 276 (L = 1,2-dithiolene), increases as the group R

$$[ML_2]^- + \varepsilon \rightleftharpoons [ML_2]^{2-} \tag{276}$$

in the ligand $R_2C_2S_2^{2-}$ is altered in the order R = CN > CF_3 > Ph > H > Me > Et. Thus the dianions are stabilised by electron-withdrawing R groups such as cyanide groups and destabilised by electron-releasing R groups such as ethyl groups.[246] It is noteworthy that it is the electron-withdrawing R groups that are also most effective in stabilising the dianionic forms of the free ligands.[247]

(iv) Infra-red spectra

In the infra-red spectra of 1,2-dithiolene complexes (Table 40) the v_1 stretching frequency, which is due principally to the C=C stretching vibration,

TABLE 40
Infra-red stretching frequencies in 1,2-dithiolene complexes (data from refs. 241 and 243)

	Infra-red frequency (cm^{-1})				
			v_3		
			$R-C\!\!\begin{smallmatrix}S\\\\S\end{smallmatrix}$		
	v_1 Perturbed	v_2 Perturbed		v_4	v_5
Complex	C=C stretch	C=S stretch	stretch	M–S stretch	M–S stretch
$[Pd(S_2C_2Ph_2)_2]^0$	1342	1136	884	401	352
$[Pd(S_2C_2(CF_3)_2)_2]^{1-}$	1513				
$[Pt(S_2C_2Me_2)_2]^0$	1324	908	563	405	310
$[Pt(S_2C_2Ph_2)_2]^0$	1351	1139	877	403	373
$[Pt(S_2C_2(CF_3)_2)_2]^0$	1422				
$[Pt(S_2C_2(CF_3)_2)_2]^{1-}$	1493				
$[Pt(S_2C_2(CF_3)_2)_2]^{2-}$	1515				

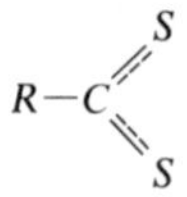

decreases slightly as R is altered in the order $CF_3 > Ph > CH_3$ and decreases markedly as the overall negative charge on the complex decreases. Two explanations are possible. The first[214] is based on the assumption that as the overall charge is decreased from -2 to 0, the metal is oxidised by two units with a consequent increase in donation of charge from ligand to metal and hence decrease in the $C=C$ stretching frequency due to the reduction of the electron density in this bond. The second[248] assumes that the oxidation which accompanies the reduction in the overall charge on the complex from $2-$ to 0 is due to the removal of electrons that were predominantly ligand in character, with the result that the ligands are progressively oxidised from dithiolates to dithiodiketones. This latter explanation is consistent with such x-ray data as is available which shows that there is a lengthening of the 'ethylenic' bond on oxidation (e.g. '$C=C$' bond length $= 1.33(2)$ Å in $[Ni(S_2C_2(CN)_2)_2]^{2-(249)}$ and $1.356(7)$ Å in $[Ni(S_2C_2(CN)_2)_2]^{1-(250)}$).

(v) E.s.r. spectra
Whilst the neutral and dianionic complexes contain no unpaired electrons, the monoanionic complexes $[M(S_2C_2R_2)_2]^-$ have a single unpaired electron which gives rise to an e.s.r. signal.[241,251] Single crystal studies have indicated that there are three anisotropic g values suggesting that the unpaired electron is not in a metal-localised orbital, but in an orbital which involves significant mixing of the metal d and ligand orbitals.

(vi) Electronic spectra
The electronic spectra of the neutral and monoanionic 1,2-dithiolene complexes exhibit a number of very intense charge transfer bands in the visible region, which give rise to the deep colours of these complexes. In the dianionic complexes (Table 41) these charge transfer bands move to higher energy

TABLE 41
Electronic spectra of $[M(S_2C_2(CN)_2)_2]^{2-}$ anions (data from ref. 252)

Complex	Colour	Band (cm^{-1})	Extinction coefficient	Assignment
$[Pd(S_2C_2(CN)_2)_2]^{2-}$	Green	15,700	64	$d_{x^2-y^2} \rightarrow d_{xy}$
		22,700	5,700	$L(\pi) \rightarrow M$
		25,800	2,840(sh)	$M \rightarrow L(\pi)$
		30,800	20,200(sh)	$L(\pi) \rightarrow L(\pi^*)$
		33,900	47,000	$L(\sigma) \rightarrow M$
		37,800	45,000	
		42,800	34,000	
$[Pt(S_2C_2(CN)_2)_2]^{2-}$	Red	14,410	49(sh)	$d \rightarrow d$ (singlet $\rightarrow$ triplet)
		15,650	56(sh)	$d \rightarrow d$ (singlet $\rightarrow$ triplet)
		18,500	1,220(sh)	$d_{x^2-y^2} \rightarrow d_{xy}$
		21,100	3,470	$M \rightarrow L(\pi)$
		29,700	15,600	$L(\pi) \rightarrow L(\pi^*)$
		32,300	13,400	$L(\pi) \rightarrow M$
		38,500	17,000(sh)	
		43,800	43,500	$L(\sigma) \rightarrow M$

enabling the less intense d–d-transitions to be observed. A comparison of the electronic spectra of a series of palladium(II) complexes[252,253] indicates that 1,2-dithiolenes occur at the low end of the spectrochemical series, i.e. $S_2C_2(CN)_2^{2-} < Br^- < Cl^- < S_2C_2(CF_3)_2^{2-} < \underline{SCN}^- < S_2P(OEt)_2^- < S_2CNR_2^- < H_2O \sim O_2C_2O_2^{2-} < NH_3 < NH_2\overline{CH_2CH_2}NH_2 < CN^-$.

(vii) Bonding in 1,2-dithiolene complexes

Three major attempts have been made to describe the bonding in metal–1,2-dithiolene complexes. Three possible structures for the neutral $[M(S_2C_2R_2)_2]^\circ$ complexes can be considered:[248] **109**, in which the metal is formally zero-valent was considered unlikely since the nickel complexes would be tetrahedral whereas $[Ni(S_2C_2Ph_2)_2]$ is square planar;[254,255] **110** was also eliminated on the grounds that the highly polarisable sulphur atoms would be unlikely to stabilise the metals in the tetravalent oxidation state; **111**,

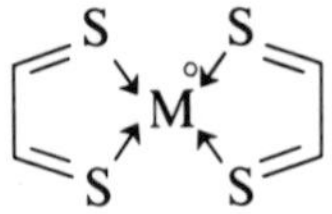

| **109** | **110** | **111** |

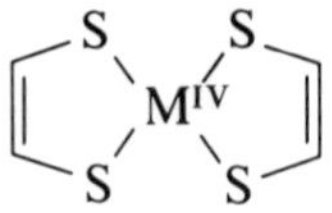

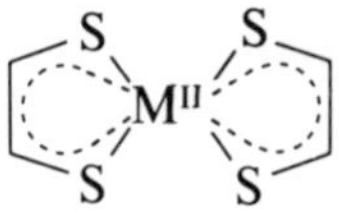

which can also be represented as a mixture of the canonical forms **112** and **113**, involves significant interaction between the ligand and metal

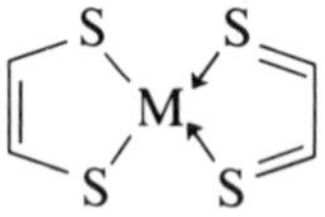

| **112** | **113** |

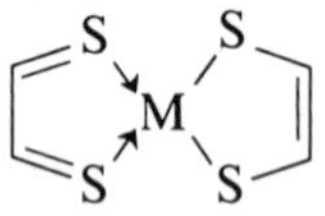

orbitals giving rise to a calculated ground state for the $[M(S_2C_2R_2)_2]^\circ$ complexes of $\cdots(2b_{1u})^2(2b_{2g})^2(3a_g)^2(2b_{3g})^2(4a_g)^2(3b_{2g})^0(3b_{1g})^0$. The nature and occupancy of the three uppermost orbitals, summarised in Table 42, indi-

TABLE 42

Three highest energy molecular orbitals in $[M(S_2C_2R_2)_2]^{n-}$ complexes
(data from ref. 248)

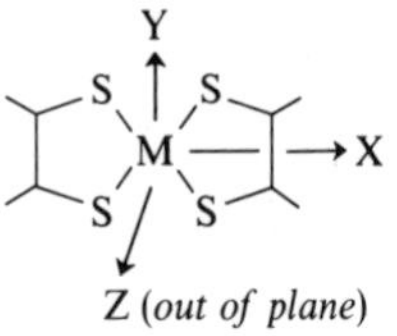

Z (*out of plane*)

Orbital	Symmetry of orbital	Description of orbital	Occupancy of orbital n = 0	n = 1	n = 2
$3b_{1g}$	xy	Mainly metal (d_{xy})	—	—	—
$3b_{2g}$	xz	82% ligand + 18% metal (d_{xz})	—	↑	↑↓
$4a_g$	$x^2 - y^2$	Mainly metal ($d_{x^2-y^2}$)	↑↓	↑↓	↑↓

cates that the mainly ligand $3b_{2g}$ orbital is empty in the neutral complexes and full in the dianionic complexes. When the $3b_{2g}$ level is full the sulphur ligands become true dithiolate ligands so that the dianionic complexes are correctly represented as containing divalent metal ions complexed by two dianionic dithiolate ligands. The monoanionic and neutral complexes can also be regarded as containing divalent metal ions and ligands that have been oxidised progressively towards dithiodiketones.

This bonding scheme is supported by several experimental observations:

(a) There is little reduction in the C=C bond length on increasing the overall negative charge (n in Table 42) on the complex (e.g. C=C bond length $= 1.356(7)$ Å in $[Ni(S_2C_2(CN)_2)_2]^{-(250)}$ and $1.33(2)$ Å in $[Ni(S_2C_2(CN)_2)_2]^{2-(249)}$). This is consistent with an increasing electron density on the 1,2-dithiolene ligand and a decrease in the contribution of the dithiodiketone form of the ligand with increasing negative charge on the complex. The increasing frequency of the v_1 infra-red band with increasing negative charge on the complex (see Table 40, p. 200) is explained similarly.

(b) The predominance of the ligand (π-molecular orbital) character in the $3b_{2g}$ level is consistent with the variation of the polarographic half-wave potentials for the oxidation and reduction of the complexes with the substituents on the 1,2-dithiolene ligands (see Table 39). In particular electron donating substituents, which would increase the absolute energy of the $3b_{2g}$ level, destabilise the dianionic complexes $[M(S_2C_2R_2)_2]^{2-}$ and electron withdrawing substituents which lower the absolute energy of the $3b_{2g}$ level stabilise the dianionic complexes.

(c) Single occupancy of the $3b_{2g}$ orbital in the monoanionic complexes $[M(S_2C_2R_2)_2]^-$ is consistent with the three-fold anisotropy of the g tensor obtained from e.s.r. studies.

(d) The absence in the dianionic complex of the very intense, low-energy, charge transfer bands that are present in the electronic spectra of the neutral and monoanionic complexes is consistent with the $3b_{2g}$ level being full in the dianionic complexes.

(e) The proton n.m.r. spectrum of $[Ni(S_2C_2H_2)_2]^0$ exhibits a proton resonance in a position normally associated with aromatic rather than aliphatic protons. Furthermore $[Ni(S_2C_2H_2)_2]^0$ can be alkylated by a Friedel–Crafts alkylation reaction. Both observations support structure **111** rather than **109** or **110**.

The second description of the bonding in 1,2-dithiolene complexes was put forward on the basis of modified Wolfsberg–Helmholtz calculations designed to interpret the electronic spectra of these complexes.[252] The ground state of the dianionic molecule was calculated to be

$$\cdots(3b_{1u})^2(2b_{1g})^2(4b_{2g})^2(4a_g)^2(3b_{1g})^0$$

in which the upper three levels were considered to be as follows:

$3b_{1g}$	xy-symmetry	55·7% metal character
$4a_g$	$x^2 - y^2$-symmetry	26·0% metal character
$4b_{2g}$	xz-symmetry	more ligand than metal character

Whilst very successful in interpreting the electronic spectra of these complexes this bonding scheme is incompatible with the voltammetric and e.s.r. data mentioned above. Subsequently it was suggested that the order of the $4a_g$ and $4b_{2g}$ levels may be opposite to that suggested originally.[256]

A third description of the bonding in 1,2-dithiolene complexes resulted from detailed calculations designed to interpret the e.s.r. results.[257] These suggested an outer-orbital configuration of

$$\cdots(b_{3g})^1(b_{1g})^0$$

for the monoanionic complexes $[M(S_2C_2R_2)_2]^{1-}$ in which the b_{3g} orbital was calculated to have 50% metal and 50% ligand character. The principal difference between this bonding scheme and that in Table 42 is that in this case the metal in the monoanionic complexes is considered to be a d^7 trivalent ion, whereas the bonding description in Table 42 considers it to be a d^8 divalent ion with the unpaired electron strongly delocalised over the 1,2-dithiolene ligands.

(viii) Reactions of 1,2-dithiolene complexes

Weak oxygen or nitrogen bases such as ketones, alcohols or aromatic amines reduce $[M(S_2C_2(CF_3)_2)_2]$ to $[M(S_2C_2(CF_3)_2)_2]^-$ [241,251] With triphenylphosphine on the other hand, one 1,2-dithiolene ligand is displaced from the neutral platinum(II) complex to give $[(PPh_3)_2Pt(S_2C_2R_2)]$ together with triphenylphosphine sulphide[258,259] (reaction 277). Other tertiary phos-

$$[Pt(S_2C_2R_2)_2] \xrightarrow[\text{(R = CH}_3, \text{CF}_3, \text{C}_6\text{H}_5)]{+PPh_3} [(PPh_3)_2Pt(S_2C_2R_2)] + Ph_3PS \quad (277)$$

phines add to the bis(1,2-dithiolene) complexes (reactions 278 and 279),

$$[Pt(S_2C_2Ph_2)_2] + Ph_2PCH_2CH_2PPh_2 \rightarrow$$
$$[Pt(S_2C_2Ph_2)_2(Ph_2PCH_2CH_2PPh_2)] \quad (278)$$

$$[Pt(S_2C_2Ph_2)_2] + 2PBu_3 \rightarrow [Pt(S_2C_2Ph_2)_2(PBu_3)_2] \quad (279)$$

although the structures of the products are unknown.[259]

Olefins such as butadiene, 2,3-dimethyl-1,3-butadiene, isoprene, 1,3-cyclohexadiene, norbornadiene and norbornene react with $[Pd(S_2C_2Ph_2)_2]$ to give 1:1 adducts, **114**, in which labile carbon–sulphur bonds have been

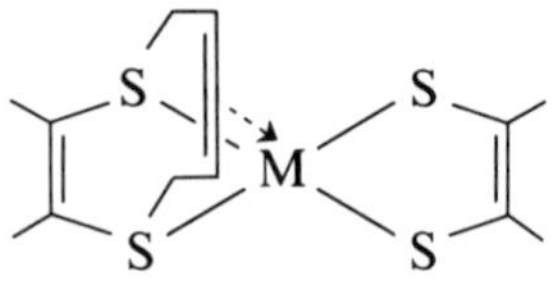

114

formed.[243] The palladium adducts are more stable than their nickel or platinum analogues. Although less reactive than their nickel analogues, neutral platinum and palladium 1,2-dithiolene complexes react with alkynes to give thiophene derivatives[243] (reaction 280).

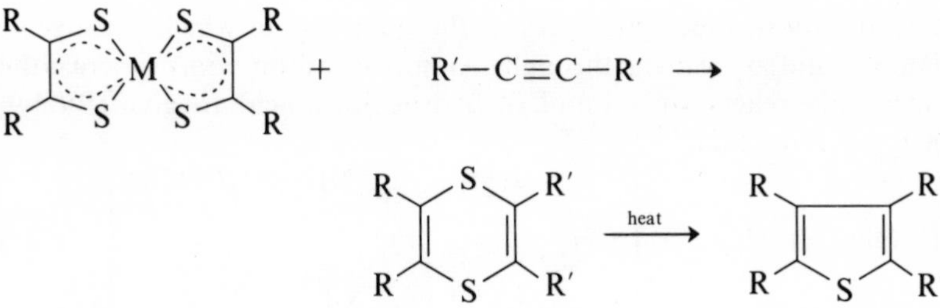

$$(280)$$

(ix) Tertiary phosphine sulphide and selenide and tertiary arsine sulphide complexes

Complexes of platinum(II) and palladium(II), $[ML_2Cl_2]$ where L is Ph_3PS, Ph_3PSe or Ph_3AsS, are prepared by treating the metal(II) chloride in dilute hydrochloric acid solution with the ligand in alcoholic or acetone solution.[260,261] The P–S or P–Se stretching frequency of the ligand in the infra-red is lowered on co-ordination, indicating metal–sulphur or –selenium co-ordination,[260] which is consistent with the charge distribution in the single bond formulation of these ligands $(Ph_3P^+ —S^-)$.[261] The rate of extraction of platinum(II) from an aqueous solution increases in the order $Ph_3PS < Ph_3PSe < Ph_3AsS$ (relative rates $1:16:112$), which is the anticipated order of increasing negative charge on the group VIB atom.

COMPLEXES WITH MULTIDENTATE LIGANDS CONTAINING BOTH A GROUP VB AND A GROUP VIB DONOR ATOM

It is convenient to consider multidentate ligands with group VB and VIB donor atoms in two groups: first the multidentate, potentially chelating ligands and secondly the ambidentate ligands.

(i) Multidentate ligands containing oxygen and nitrogen donor atoms

The majority of multidentate ligands containing both oxygen and nitrogen donor atoms that have been investigated have been either naturally occurring amino-acids such as glycine (NH_2CH_2COOH) or man-made amino-acids such as nitriloacetic and $(N\{CH_2COOH\}_3)$. Of these the most widely investigated is glycine which can form $K_2[Pt(gly)_4]^{(262)}$ if a large excess of glycine is used, $K(Pt(gly)_3]$ when only 3 mol of glycine are used,[263] *cis*- and *trans*-$[M(gly)_2]$, where M $=$ Pd or Pt, when slightly in excess of 2 mol are used,[264–70] and $K[M(gly)Cl_2]$ when only 1 mol of glycine is used.[263] In $K[M(gly)Cl_2]$ and $[M(gly)_2]$ the glycine ligands are bidentate binding through both oxygen and nitrogen; in $K[Pt(gly)_3]$ one glycine is bidentate and two are unidentate nitrogen donor ligands and in $K_2[Pt(gly)_4]$ all four ligands are unidentate bound through nitrogen.[263] *Cis*- and *trans*-$[Pt(gly)_2]$ both retain their configurations on boiling in water unless excess glycine is added when the *cis*-complex isomerises to the *trans*-product. With palladium(II) a *trans* to *cis* isomerisation occurs which is most unusual in palladium(II) chemistry where *trans*-isomers are generally the more stable.[271] The reactions of the *cis*- and *trans*-platinum(II) isomers with concentrated hydrochloric acid are different, the *trans*-isomer giving *trans*-[Pt(glyH)$_2$Cl$_2$] in which the glycine ligands are bound to the metal through their nitrogen atoms only, whereas the *cis*-isomer only reacts with 1 mol of hydrochloric acid to give [Pt(gly)-(glyH)Cl] (reaction 281).[272]

$$\text{cis-}[Pt(gly)_2] + HCl \rightarrow \left[\begin{array}{c} \underset{CH_2}{\overset{H_2N}{\diagdown}} \underset{COO}{\overset{}{\diagup}} Pt \underset{Cl}{\overset{NH_2CH_2COOH}{\diagup}} \end{array}\right] \qquad (281)$$

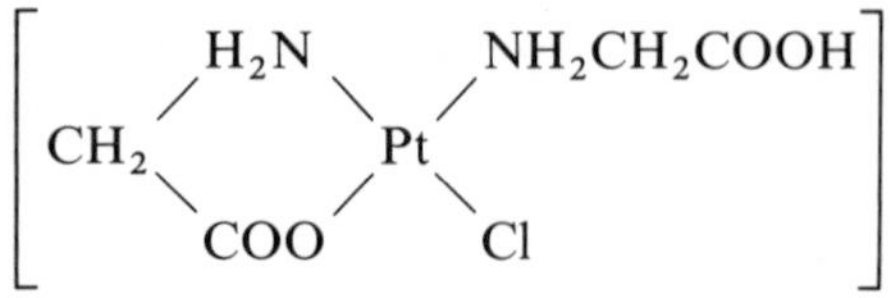

Glycine is a strongly co-ordinating ligand for these metals, as indicated by the stability constants for equilibria 282.[273] The kinetics of displacement

$$[PdCl_4]^{2-} + gly \underset{\mu = 0.1\,M;\,25°C}{\overset{k_1 = 1.3 \times 10^9}{\rightleftharpoons}} [Pd(gly)Cl_2]^- \underset{\mu = 0.1\,M;\,25°C}{\overset{k_2 = 2.7 \times 10^8}{\rightleftharpoons}} [Pd(gly_2)] \qquad (282)$$

of glycine from [Pd(gly)$_2$] by perchloric and hydrochloric acids indicate that the reaction follows eqn. 283,[267] which is similar to that found for the

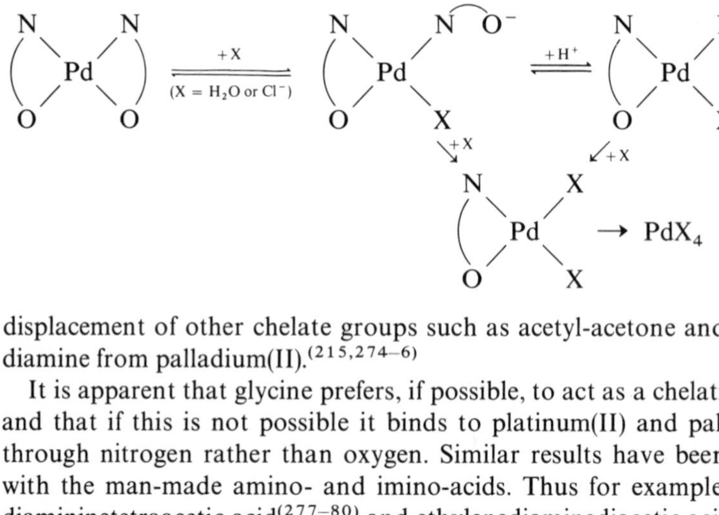

$$(283)$$

displacement of other chelate groups such as acetyl-acetone and ethylene-diamine from palladium(II).[215,274-6]

It is apparent that glycine prefers, if possible, to act as a chelating ligand, and that if this is not possible it binds to platinum(II) and palladium(II) through nitrogen rather than oxygen. Similar results have been obtained with the man-made amino- and imino-acids. Thus for example ethylene-diamininetetraacetic acid[277-80] and ethylenediaminediacetic acid[281] normally bind to platinum(II) and palladium(II) as bidentate ligands through their nitrogen atoms and only when the concentrations of all other co-ordinating ligands, including hydroxide ions, are reduced below equimolar amounts is it possible for EDTA to act as either a tridentate (two nitrogen and one oxygen donor atoms) or a quadridentate (two nitrogen and two oxygen donor atoms) ligand.[277,278] Similar results have been found for aminodiacetic acid ligands RN(CH$_2$COOH)$_2$, where R = H, CH$_3$ or CH$_2$COOH,[282,283] pyrrolidine- and piperidine-2-carboxylic acids[284] and 115[285] which all show a preference for nitrogen co-ordination.

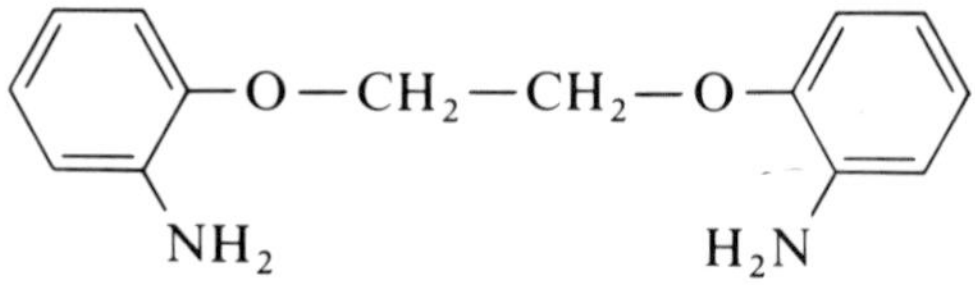

115

The infra-red spectra[266,269,286,287] of $[M(gly)_2]$, where M = Pd or Pt, exhibit asymmetric carboxylate stretching frequencies at higher wavenumbers ($v_{asym.}COO^- = 1643\ cm^{-1}$ (*trans*-$[Pd(gly)_2]$), $1642\ cm^{-1}$ (*trans*-$[Pt(gly)_2]$) and $1647\ cm^{-1}$ (*trans*-$[Pt(alanine)_2]$)) and symmetric carboxylate stretching frequencies at lower wavenumbers ($v_{sym.}COO^- = 1373\ cm^{-1}$ (*trans*-$[Pd(gly)_2]$), $1374\ cm^{-1}$ (*trans*-$[Pt(gly)_2]$) and $1382\ cm^{-1}$ (*trans*-$[Pt(alanine)_2]$)) than most other divalent cations.[286] This is probably due to some covalent

bonding in the metal–oxygen bond giving rise to $-C\overset{\displaystyle O}{\underset{\displaystyle O-M}{\diagup\diagdown}}$ in addition

to the ionic binding of $-C\overset{\displaystyle O}{\underset{\displaystyle O}{\diagup\diagdown}} - M^{2+}$. The stretching force constants of the

metal–oxygen and metal–nitrogen bonds in these complexes (Table 43) are

TABLE 43
Metal–ligand stretching vibrations in the infra-red spectra of bis(glycine)
complexes (data from ref. 287)

Complex	Stretching frequencies (cm^{-1})		Force constants ($mdyn/Å$)	
	v_{M-O}	v_{M-N}	K_{M-O}	K_{M-N}
trans-$[Pt(gly)_2]$	415	549	2·10	2·10
cis-$[Pt(gly)_2]$	406,393	554	2·10	2·10
trans-$[Pd(gly)_2]$	420	550	2·00	2·00

similar and less for the palladium than for the platinum complexes.[287] These results illustrate very well the dangers of equating bond strengths with stretching force constants, since the preparative results suggest that the metal–nitrogen bond is stronger than the metal–oxygen bond. The reason for the inequivalence of bond strengths and force constants can be understood from the potential curve for a diatomic molecule (Fig. 19). The stretching force constant is a measure of the curvature of the potential well near the equilibrium position whereas the dissociation energy (D) is a measure of the depth of that potential well. Thus a large stretching force constant means a sharp curvature of the potential well near the bottom, but does not necessarily indicate a deep potential well.

Infra-red spectroscopy can be used to distinguish between unidentate and bidentate glycine ligands since unidentate glycine exhibits a band at 1700 cm^{-1} due to the C–O antisymmetric stretching mode which is absent from the bidentate complexes.[288] In addition all the platinum(II)–glycine complexes exhibit a weak absorption at about 548 cm^{-1} whereas only those which contain a glycinato anion acting as a bidentate ligand exhibit an absorption at about 400 cm^{-1} as well.[288]

The photochemical behaviour of the bis(glycine) complexes is interesting. Irradiation of *trans*-$[M(gly)_2]$ in the *d–d* region of the spectrum has no effect,

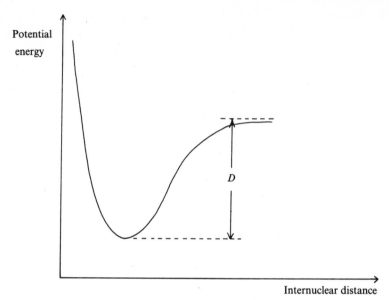

Fig. 19. Potential curve for a diatomic molecule.

whereas irradiation in the electron transfer region gives decomposition with formation of the metal oxide.[289] Irradiation of *cis*-[Pt(gly)$_2$] in the *d–d* region of the spectrum in the absence of excess glycine isomerises it to *trans*-[Pt(gly)$_2$]. Since the most likely mechanism for this isomerisation is intra-molecular it probably involves the formation of a tetrahedral transition state formed by excitation of one electron into the $d_{x^2-y^2}$ antibonding orbital. The circular dichroism spectra of palladium(II) complexes of *N*-methylamino acids and dipeptides can be understood in terms of a hexadecant rule, which is consistent with planar chelate rings.[290]

An interesting pair of isomers of [Pt(gly)(C$_2$H$_4$)Cl], which may be called the *N-trans* and the *O-trans* isomers, have been synthesised by routes that depend on the *trans-* effects of the groups present being in the order C$_2$H$_4$ > RNH$_2$ > Cl. The *N-trans* isomer, in which the amino-group of the glycine is *trans* to ethylene, can be prepared by reaction 284,[291,292] whereas the

$$K[Pt(C_2H_4)Cl_3] + NH_2CH_2COOH \rightarrow$$

$$\left[\begin{array}{c} CH_2 \overset{\displaystyle H_2N}{\underset{\displaystyle COO}{\diamondsuit}} Pt \overset{\displaystyle Cl}{\underset{\displaystyle CH_2}{\diamondsuit}} \\ \end{array} \right] + HCl + KCl \quad (284)$$

O-trans isomer must be prepared by reaction 285.[293] The two isomers have broadly similar infra-red and ultra-violet spectra but can be distinguished by a detailed examination of these spectra.

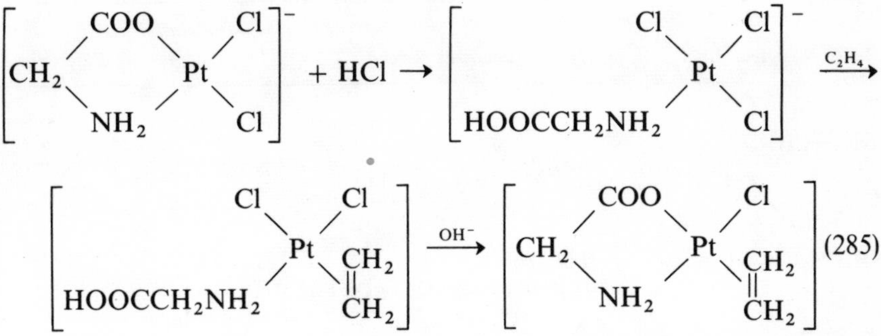

$$\left[\begin{array}{c} CH_2 \underset{NH_2}{\overset{COO}{\diagdown}} Pt \underset{Cl}{\overset{Cl}{\diagup}} \end{array} \right]^- + HCl \rightarrow \left[\begin{array}{c} \underset{HOOCCH_2NH_2}{\overset{Cl}{\diagdown}} Pt \underset{Cl}{\overset{Cl}{\diagup}} \end{array} \right]^- \xrightarrow{C_2H_4}$$

$$\left[\begin{array}{c} \underset{HOOCCH_2NH_2}{\overset{Cl}{\diagdown}} Pt \underset{\overset{\|}{CH_2}}{\overset{Cl}{\diagup}} CH_2 \end{array} \right] \xrightarrow{OH^-} \left[\begin{array}{c} CH_2 \underset{NH_2}{\overset{COO}{\diagdown}} Pt \underset{\overset{\|}{CH_2}}{\overset{Cl}{\diagup}} CH_2 \end{array} \right] \quad (285)$$

In addition to direct co-ordination, amino-acids readily form Schiff's bases and peptides, both of which are good ligands for platinum(II) and palladium(II). It is interesting that whilst glycine and sodium pyruvate react with palladium(II) chloride to give the anticipated Schiff's base (reaction 286), glyoxylic acid and α-alanine also react with palladium(II) chloride to give

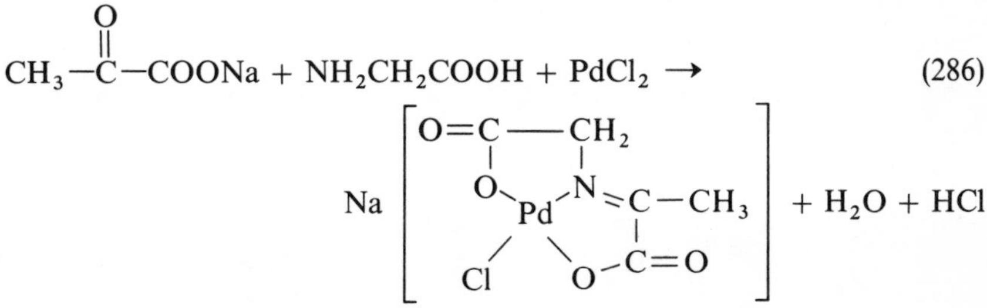

$$CH_3 - \overset{\overset{O}{\|}}{C} - COONa + NH_2CH_2COOH + PdCl_2 \rightarrow \quad (286)$$

$$Na \left[\begin{array}{c} O=C \text{---} CH_2 \\ \overset{O}{\diagdown} \underset{Pd}{} \overset{N}{\diagup} \underset{\|}{} C-CH_3 \\ Cl \overset{\diagdown}{} O^{-}C=O \end{array} \right] + H_2O + HCl$$

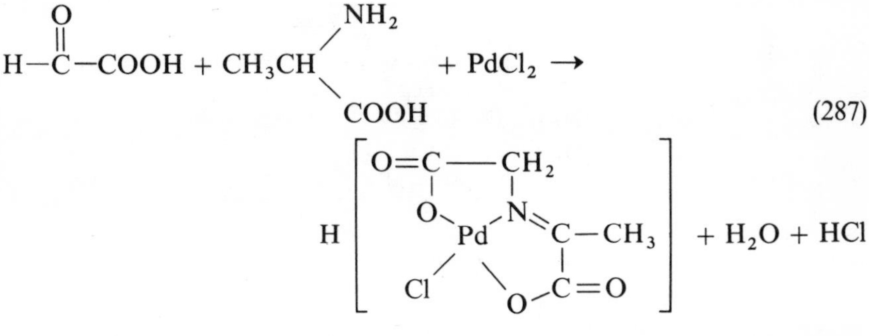

$$H - \overset{\overset{O}{\|}}{C} - COOH + CH_3CH \overset{NH_2}{\underset{COOH}{\diagup\diagdown}} + PdCl_2 \rightarrow$$
$$(287)$$

$$H \left[\begin{array}{c} O=C \text{---} CH_2 \\ \overset{O}{\diagdown} \underset{Pd}{} \overset{N}{\diagup} \underset{\|}{} C-CH_3 \\ Cl \overset{\diagdown}{} O^{-}C=O \end{array} \right] + H_2O + HCl$$

the same Schiff's base (reaction 287) rather than **116**, which might have been expected.[294] Of the divalent cations studied in detail palladium(II) is one of the most effective for ionising peptide hydrogen atoms to yield square-planar complexes with amide nitrogen donor atoms.[295]

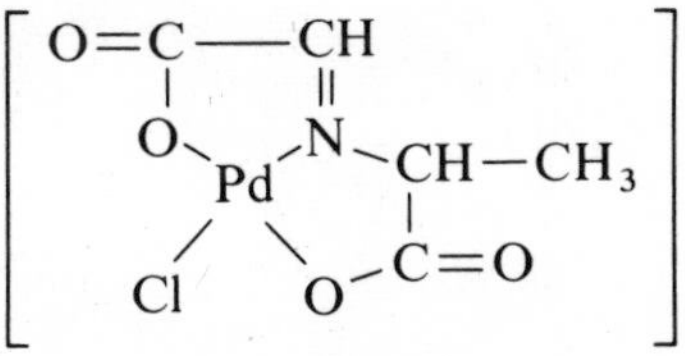

$$\left[\begin{array}{c} O=C \text{---} CH \\ \overset{O}{\diagdown} \underset{Pd}{} \overset{\|}{} \overset{N}{\diagup} \underset{}{} CH-CH_3 \\ Cl \overset{\diagdown}{} O^{-}C=O \end{array} \right]$$

116

TABLE 44
Complexes formed by polydentate ligands containing both sulphur and nitrogen donor atoms

Ligand (L)	Complexes formed	Co-ordination through	Reference
$NH_2-CH{<}^{COOH}_{CH_2}$ CH_3S-CH_2	$[M(LH)Cl_2]$	N and S	296, 297, 298
$NH_2CH_2CH_2SH$	$[PdL_2]$ $[(L_2Pd)M(PdL_2)]\,(M = Pd, Pt)$	N and S N and S to Pd, S to M	299 300
(pyridine)CH_2CH_2SH	As with $NH_2CH_2CH_2SH$		301
(pyridine)$CH_2CH_2SCH_3$	$[MLX_2]$ and $[ML_2]$ $[PtL_2X]^+$	N and S N and S	302 302
(quinoline, X at 5-position, SH at 8-position) $(X = H, Cl, Br)$	$[ML_2]\,(M = Pd, Pt)$	N and S	303–5
(quinoline, SCH_3 at 8-position)	$[MLX_2]$	N and S	306
(pyridine)CH_2SCH_3	$[MLX_2]\,(M = Pd, X = Cl, Br, I, SCN; M = Pt, X = SCN)$ $[ML_2]^{2+}\,(M = Pd, Pt)$	N and S N and S	307 307
(benzene)$<^{NH_2}_{SCH_3}$	$[MLX_2]\,(M = Pd, X = Cl, Br, I, SCN; M = Pt, X = Br, I, SCN)$ $[ML_2]^{2+}\,(M = Pd, Pt)$	N and S N and S	308 308
S_4N_4	$\left[\begin{smallmatrix} S{-}NH\;\;NH{-}S \\ \mid\quad M\quad \mid \\ N{-}S\;\;\;S{-}N \end{smallmatrix}\right]$	N and S	309–12
$NH_2-\underset{\parallel}{C}-NH-\underset{\parallel}{C}-NH_2$ (LH) (S,S)	$[Pd(LH)_2]^{2+}$ $[Pd(NH_3)_2L]^+$	N and S N and S	313, 314 313, 314
$H_2N-C{<}^{NH-NH_2}_{S}$ $\left(\leftrightarrow H_2N-C{<}^{N-NH_2}_{SH}\right)$	$[MLCl_2]$ $[ML_2]^{2+}$	N and S N and S	315 315

(ii) Multidentate ligands containing sulphur and group VB donor atoms

Some of the more important multidentate ligands containing both nitrogen and sulphur donor atoms are shown in Table 44, where it can be seen that both the nitrogen and sulphur atoms co-ordinate to the metal. The complexes are all square-planar except for the five-co-ordinate platinum(II) complexes of 2-(2-methylthioethyl)pyridine (**117**). Table 45 summarises the more important

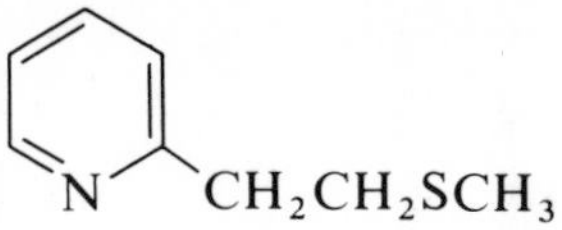

117

complexes formed by multidentate ligands containing sulphur and either phosphorus or arsenic donor atoms. The range of stereochemistries exhibited by the metal atoms in these complexes is limited, square-planar geometry being the commonest. This suggests that sulphur atoms do not support five-co-ordinate structures nearly as readily as phosphorus or arsenic atoms, and is probably due to the less covalent nature of the metal–sulphur bond and the lower ability of sulphur to delocalise the excess negative charge away from the metal by π-back acceptance.

A number of palladium(II) complexes of ligands containing thiomethyl groups have been found to undergo ready S-demethylation in boiling dimethyl formamide[321,323—6] (for example reaction 288). The demethylated products react with methyl halide to regenerate the original S-methylated palladium(II) dihalide.

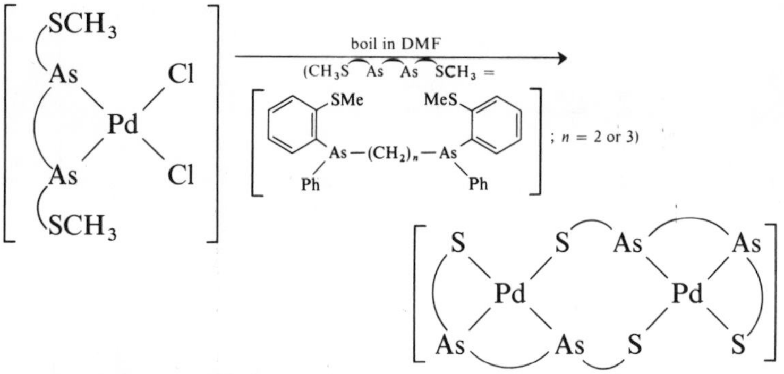

(288)

(iii) Ambidentate ligands

Here an ambidentate ligand is considered to be 'a ligand containing two or more types of potential σ-donor sites, only one of which is involved in co-ordination at any one time',[327] and we shall be concerned with the co-ordination of nitrite, cyanate, thiocyanate and selenocyanate ions together with thiomorpholine ligands to platinum(II) and palladium(II).

Nitrite complexes

Although the nitrite ion can bind to metals through both nitrogen and oxygen,[327–9] all the known platinum(II) and palladium(II) complexes involve the N-bonded nitro-ligand ([(PEt$_3$)$_2$PtH(NO$_2$)] is a possible exception to this generalisation since its proton n.m.r. spectrum does not show the broadening

TABLE 45

Types of complexes formed by multidentate sulphur and either phosphorus or arsenic ligands

Ligand (L)	Complexes formed	Co-ordination through	Stereochemistry of metal	Reference
o-C6H4(AsMe2)(SMe)	[PdLCl2]	As and S	Square-planar	316
	[PdL2]2+	As and S	Square-planar	316
	[PdL2Cl]+	As and S	five-co-ordinate; structure unknown	316
H2C(CH2—AsMe2)(CH2—SMe)	[PdL2I2]	As only	Square-planar	317
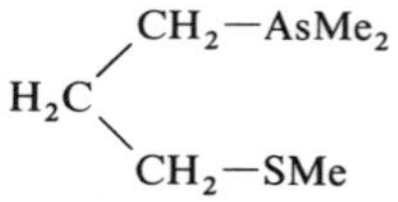	[MLX2] (M = Pd, X = Cl, Br, I; M = Pt, X = I)	As and S	Square-planar	318
	[PtL2]2+[PtX4]2- (X = Cl, Br)	As and S	Square-planar	318
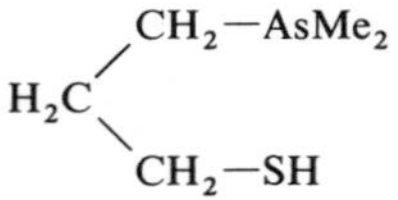 H2C(CH2—AsMe2)(CH2—SH)	[ML2] (M = Pd or Pt)	As and S	Square-planar	319
o-C6H4(PPh2)(SMe)	[PdLCl2]	P and S	Square-planar	320, 321
	[PdL2]2+	P and S	Square-planar	320, 321
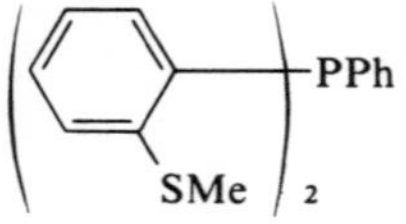 [C6H3(PPh)(SMe)]2	[PdLCl2]	?	Square-planar	320, 321
	[PdLCl]+	Both S atoms and P atom	Square-planar	320, 321
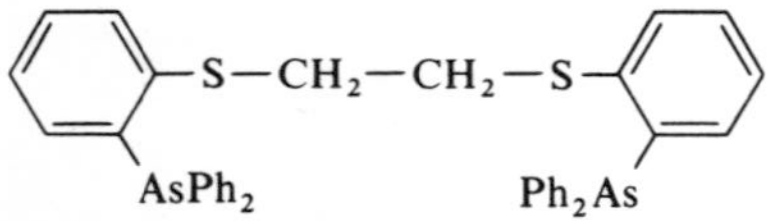 Ph2As—C6H4—S—CH2—CH2—S—C6H4—AsPh2	[PdLCl2]	One S atom and one As atom, both on same benzene ring. One S and one As are free	Square-planar	285

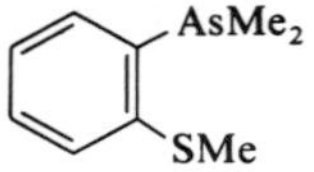

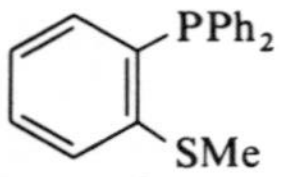

Ligand	Complex	Coordination	Geometry	Ref.
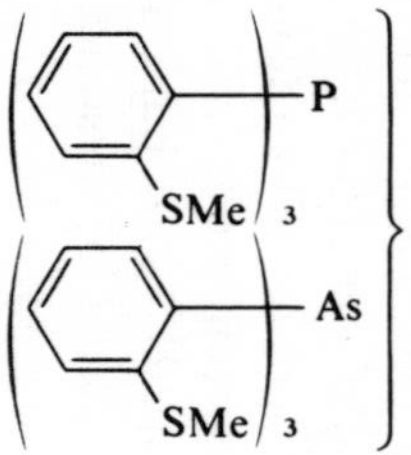 (structure: P-bridged tris ligand with SMe, ×3; and As-bridged tris ligand with SMe, ×3)	[PdLCl₂]	?	Square-planar	320
	[Pd₂LCl₄]	One S and one As to one Pd and two S to second Pd	Square-planar	320
Ph₂P(CH₂)₃S(CH₂)₃S(CH₂)₃PPh₂	[MLCl]⁺ (M = Pd, Pt)	Both S and both P	Square pyramidal	322
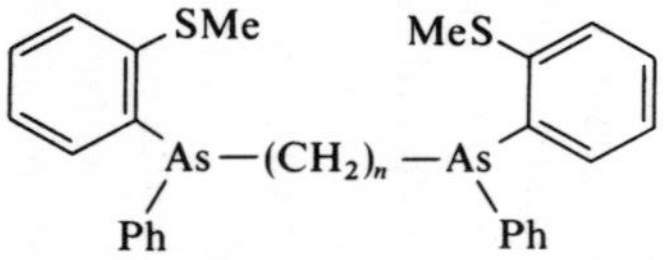 (structure: MeS/SMe aryl with As-(CH₂)ₙ-As, Ph/Ph, n = 2 or 3)	[PdLCl₂]	Both As atoms	Square-planar	323
	[Pd₂LCl₄]	One As and one S to each Pd	Either square-planar or trigonal bipyramidal	323
	[Pd₂L₂Cl₂]²⁺	Both As to one Pd. One S to other Pd. One S is free	Square-planar	323
	[Pd₂L₂]²⁺	Two As and one S of each ligand to one Pd. Remaining S to other Pd	Square-planar	323
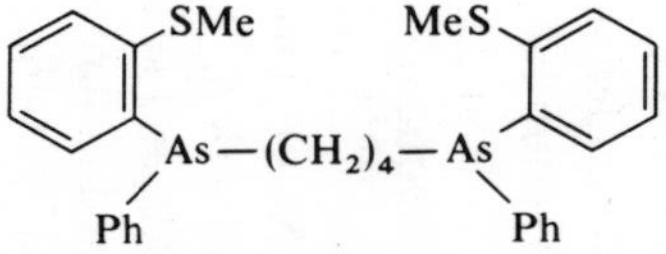 (structure: MeS/SMe aryl with As-(CH₂)₄-As, Ph/Ph)	[Pd₂LCl₄]	One As and one S to each Pd.	Square-planar	324
	[Pd₂L₂Cl₂]²⁺	Both As to one Pd. One S to other Pd. One S is free	Square-planar	324

that might be expected from an *N*-bonded nitro-group;[330] however, its infra-red spectrum is consistent with an *N*-bonded nitro-group[331]). *N*-bonded nitro-complexes can be distinguished from *O*-bonded nitrito-complexes by infra-red spectroscopy since in the nitro-complexes both $\nu_{asym.}$ (1328 cm^{-1} in $NaNO_2$) and $\nu_{sym.}$ (1261 cm^{-1} in $NaNO_2$) are raised in frequency by co-ordination whereas in the nitrito complexes $\nu_{asym.}$ is increased while $\nu_{sym.}$ is decreased in frequency.[332] Nitro-complexes are prepared by treating the metal(II) halide with silver nitrate.[333,334] In addition the zero-valent oxygen complexes [(PPh$_3$)$_2$M(O$_2$)] react with nitric oxide to form [(PPh$_3$)$_2$M(NO$_2$)$_2$].[103,105,335]

The influence of the ligand L in [L$_2$M(NO$_2$)$_2$] on the infra-red stretching frequencies of the co-ordinated nitro-group suggests that there is some π-bonding between the filled d_π orbitals of the metal and the vacant π^*-(antibonding) orbital of the nitrite ion.[333] This is consistent with the observation that the plane containing the oxygen and nitrogen atoms of the nitro-group in [Pd(NO$_2$)(NH$_3$)$_3$] is coplanar with the square-plane containing the palladium atom, although this planarity could arise from crystal packing forces.[336] The presence of π-back-donation probably accounts for the lack of platinum(II) and palladium(II) nitrito-complexes where such π-back donation would not be possible. π-Back-donation in the nitro–metal bond but not in the nitrito–metal bond accounts for the higher position of the *N*-bonded nitro-group than the *O*-bonded nitrito-group in the spectro-chemical series.[337]

Cyanate complexes

Treatment of platinum(II) and palladium(II) halide complexes with either silver or potassium cyanate yields *N*-bonded isocyanato-complexes such as (Me$_4$N)$_2$[Pd(NCO)$_4$],[338] [L$_2$M(NCO)$_2$], where M = Pd or Pt and L were ligands of varying σ-donor and π-acceptor abilities[339,340] and [(PR$_3$)$_2$-PtH(NCO)].[331,341] In addition complexes such as [(PPh$_3$)$_2$M(NCO)$_2$], where M = Pd or Pt, have been prepared either by treating [M(NCO)$_4$]$^{2-}$ with triphenylphosphine[342] or by treating the corresponding azide complexes, [(PPh$_3$)$_2$M(N$_3$)$_2$], with carbon monoxide[343] (see p. 125). So far all the platinum(II) and palladium(II) cyanate complexes that have been prepared have been shown to contain *N*-bonded cyanate ligands, although it is possible that a small proportion of the *O*-bonded cyanate is present in solutions of [(PPh$_3$)$_2$PtH(NCO)] in benzene.[330]

The *N*-bonded nature of the cyanate ligand in platinum(II) and palladium(II) complexes has been mainly established by infra-red spectroscopy since the C–N and C–O stretching frequencies both increase and the N–C–O bending frequency decreases on co-ordination to platinum(II) and palladium(II).[340] Varying the σ-donor and π-acceptor abilities of the ligands L in [L$_2$M(NCO)$_2$] from ammonia to triphenylstibine does not alter the mode of co-ordination. This is in direct contrast to the results obtained with thiocyanate and selenocyanate ligands (see below, pp. 217–219), but can be rationalised in terms of the charge distributions calculated for these ligands (Table 46),[344] since this shows that the nitrogen of the cyanate ligand carries

TABLE 46
Calculated atomic charges for NCX⁻ anions (data from ref. 344)

Anion	Charge on N	Charge on C	Charge on X	Difference between charge on N and X
NCO⁻	−0·7712	−0·0442	−0·1846	0·5876
NCS⁻	−0·4826	+0·1943	−0·7108	0·2282
NCSe⁻	−0·3941	+0·2345	−0·8404	0·4463

substantially more negative charge than oxygen. This, coupled with the enhanced possibility of π-back-donation of electron density from the metal to the cyanate π^*-(antibonding) orbitals accounts for the N-bonded nature of the cyanate ligand co-ordinated to platinum(II) and palladium(II). As expected, in view of the linear nature of the M–N–C–O group, this is not altered by steric crowding such as is present in [Pd(Et₄dien)(NCO)]BPh₄.[345] The bands in the electronic spectra of Na₂[M(NCO)₄], where M = Pd or Pt, are consistent with the cyanate ligand having a high ligand field strength, but lower than that of cyanide.[346]

Thiocyanate complexes
The most widely studied ambidentate ligand is the thiocyanate ligand, partly because of its versatility in forming both S-bonded and $\underline{N}$-bonded complexes and partly because of the ease with which these complexes can be distinguished using infra-red spectroscopy[347,348] (Table 47). This is not

TABLE 47
Typical ranges for the fundamental frequencies of co-ordinated thiocyanate ligands
(data taken from ref. 327 and modified to include the results from ref. 349)

	$v_1 \ (cm^{-1})$	$v_2 \ (cm^{-1})$	$v_3 \ (cm^{-1})$
Approximate origin of stretching frequency	C–N stretch	N–C–S bend	C–S stretch
Typical range for N-bonded thiocyanate	2040–2100	448–480	780–860
Typical range for S-bonded thiocyanate	2080–2100	410–470	690–720

entirely satisfactory since the ranges overlap to some extent, and assignment is further complicated by the first overtone of v_2 having an intensity comparable to v_3, so that a band in the 800–880 cm⁻¹ region may be due to either $2v_2$ of an $\underline{S}$-bonded or v_3 of an $\underline{N}$-bonded thiocyanate ligand.[350] However, the two possibilities can generally be distinguished by measuring the integrated intensity of v_1 since for $\underline{N}$-bonded thiocyanate ligands this is greater than the free ion value whereas for $\underline{S}$-bonded thiocyanate ligands it is less.[351-3] The far infra-red region can also help in distinguishing $\underline{N}$- and $\underline{S}$-bonded thiocyanate ligands, since Pd–SCN complexes exhibit a band between 290 and 320 cm^{-1} that has been ascribed to the Pd–S stretching mode whereas Pd–NCS complexes have a band in the region 260–270 cm^{-1} due to the Pd–N stretching mode.[349] When recording the far infra-red spectra of thiocyanate complexes it is important to avoid using a caesium iodide

TABLE 48

X-ray diffraction data for palladium(II) and platinum(II) thiocyanate complexes

Complex	M–S bond length (Å)	M–N bond length (Å)	No. of Structure in Appendix II	Reference
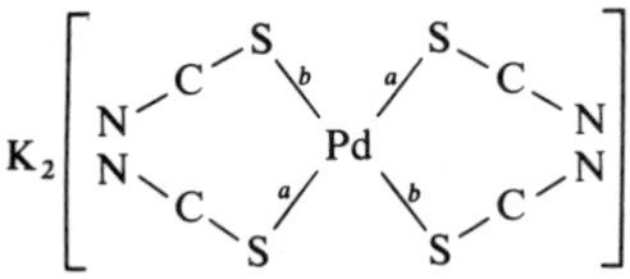	$a = 2.312(96)$ $b = 2.392(90)^a$		32	355
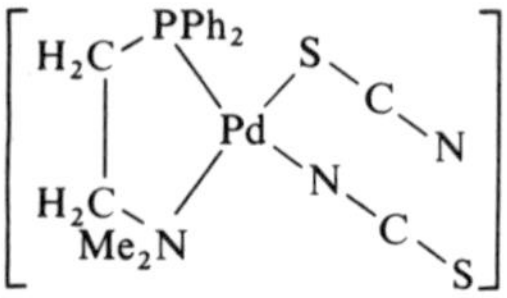	2.295	2.063	70	356
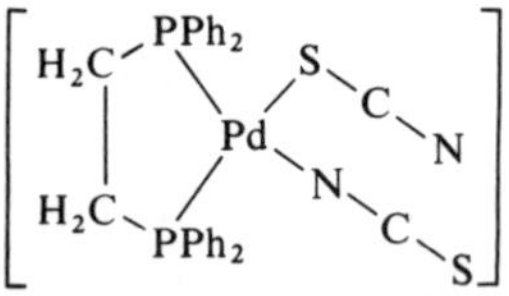	2.364	2.000	82	357
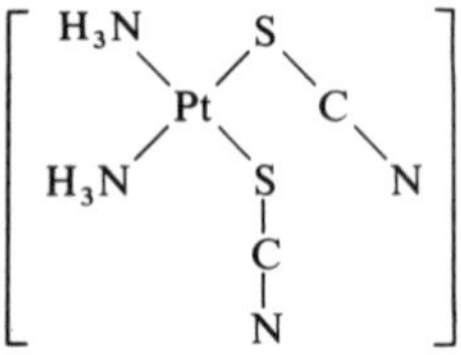	2.30		122	358
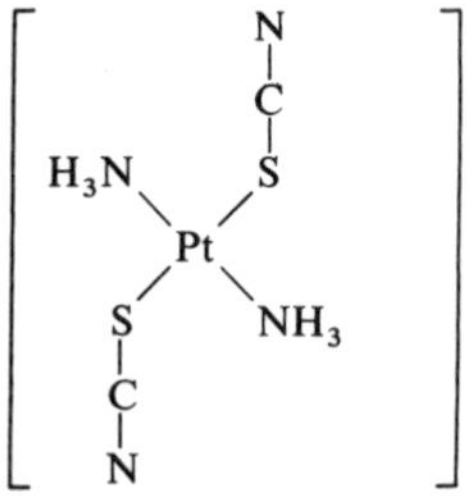	2.265		123	359
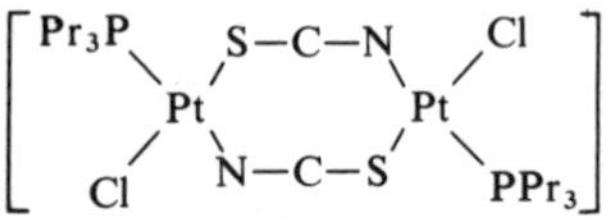	2.327(5)	2.08(1)	182	360, 361
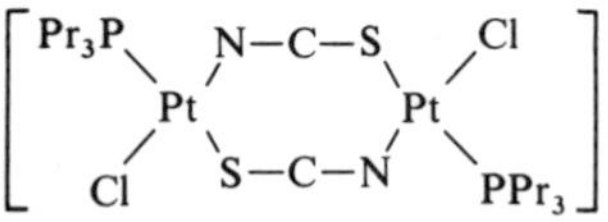	2.408(4)	1.97(1)	182	360, 361

^{a} These two sulphur atoms are associated with a second palladium atom at a distance of 3·66 Å.

pellet technique because the thiocyanate in the complex exchanges with the iodide in the matrix. Nujol mulls sandwiched between thin sheets of polythene are ideal for determining these spectra.

An alternative, less commonly used, method for distinguishing between N- and S-bonded thiocyanate complexes, that is limited to very soluble samples, is ^{14}N n.m.r. spectroscopy. It depends on the ^{14}N resonance being shifted slightly downfield in S-bonded thiocyanate complexes relative to free thiocyanate ions, whereas in N-bonded complexes it is shifted upfield by a comparatively large amount.[354]

X-ray diffraction, which is the most reliable method for determining the structure of thiocyanate complexes indicates that thiocyanate ions can bond in at least four ways (Table 48):

$$M-S-C\equiv N \qquad M-N=C=S \qquad M-S-C\equiv N-M \qquad N=C-S\overset{\displaystyle M}{\underset{\displaystyle M}{\diagup\diagdown}}$$

The metal–sulphur and metal–nitrogen bond lengths are similar to those calculated from the sum of the covalent radii (M–S, 2·35 Å; M–N, 1·95 Å) suggesting that the contribution of π-bonding to these bonds is negligible.[361]

When discussing the S- or N-bonded nature of thiocyanate ligands it is important to remember that the factors that tip the balance in favour of one mode of bonding or another must be very small since the complex [(PPr$_3$)$_2$-Pt$_2$Cl$_2$(SCN)$_2$] forms two readily isolatable isomers one containing the sulphur atom *trans* to the tertiary phosphine ligand and the other with the sulphur atom *trans* to the chloride ligand. Similarly with [(AsPh$_3$)$_2$Pd(NCS)$_2$] and [Pd(bipyr)(NCS)$_2$] both the N- and S-bonded linkage isomers can be isolated by first forming the kinetically favoured S-bonded isomer and then rearranging this to the thermodynamically more stable N-bonded isomer[362] (reaction 289). The factors that must be considered in a discussion of the mode

$$\begin{bmatrix} NCS & & SCN \\ & \diagdown \diagup & \\ & Pd & \\ & \diagup \diagdown & \\ NCS & & SCN \end{bmatrix}^{2-} + \text{AsPh}_3 \xrightarrow[\text{in EtOH}]{0°C}$$

$$\begin{bmatrix} \text{Ph}_3\text{As} & & SCN \\ & \diagdown \diagup & \\ & Pd & \\ & \diagup \diagdown & \\ NCS & & \text{AsPh}_3 \end{bmatrix} \xrightarrow[\text{solid state}]{150°C; 30 \text{ mins,}} \begin{bmatrix} \text{Ph}_3\text{As} & & NCS \\ & \diagdown \diagup & \\ & Pd & \\ & \diagup \diagdown & \\ SCN & & \text{AsPh}_3 \end{bmatrix} \qquad (289)$$

of bonding of the thiocyanate ligand to the metal include electronic and steric factors, the physical state (i.e. solid or melt) of the sample and, in solution the nature of the solvent, and in the crystalline state the nature of the anion.

(a) *Electronic factors.* A casual glance at Table 49 suggests that ligands that can accept electron density from the metal encourage the formation of

TABLE 49

The nature of the bonding in metal–thiocyanate complexes

Complex	M–SCN	M–NCS	Reference
$[M(SCN)_4]^{2-}$ (M = Pt, Pd)	✓		348
$[M(NH_3)_2(SCN)_2]$ (M = Pt, Pd)	✓		348, 358, 359
$[Pd(pyr)_2(SCN)_2]$	✓		349
$[Pd(phen)(SCN)_2]$	✓		363
$[Pd(5\text{-}NO_2phen)(NCS)_2]$		✓	363
$[Pd(4,4'\text{-}Me_2bipyr)(NCS)(SCN)]$	✓	✓	363
$[Pd(bipyr)(NCS)_2]^a$		✓	362, 364
$[Pd(PPh_3)_2(NCS)_2]$		✓	362
$[Pd(AsPh_3)_2(NCS)_2]^a$		✓	362, 364
$[Pd(SbPh_3)_2(SCN)_2]$	✓		362
$[Pd(dien)(SCN)]^+$	✓		365
$[Pd(Et_4dien)(SCN)]^+ BPh_4^-$	✓		366
$[Pd(Et_4dien)(SCN)]^+ SCN^-$		✓	367, 368

^{a} Stable linkage isomer.

Pd–NCS bonds, whereas ligands with no π-bonding ability tend to encourage Pd–SCN bonding, although there are some glaring exceptions to this, such as SbPh₃. This could be due to π-bonding in the metal–ligand bond increasing the effective positive charge on the metal making it 'harder'[369] so favouring the formation of the more ionic palladium–nitrogen bond,[348] although this explanation is probably too simple because with cobalt(III)

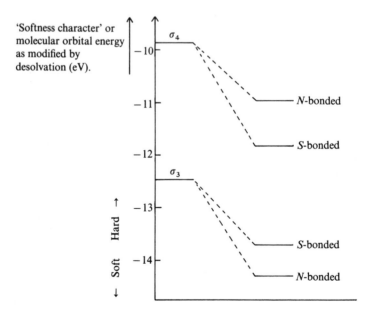

Fig. 20. Energies of the σ_3 and σ_4 molecular orbitals of thiocyanate as modified by desolvation from water in the case of both nitrogen and sulphur bonding to the metal ion (data from ref. 371).

complexes the electronic effects of ligands are exactly the reverse. This apparently contradictory result can be understood in terms of the 'softness character' of the potential donor orbitals of the thiocyanate ligand. The 'softness character' is the energy of the ligand molecular orbital as modified by the desolvation process necessary for the bond formation. Molecular orbital calculations indicate that only the highest two occupied bonding orbitals σ_3 and σ_4 of the thiocyanate ligand are available for co-ordination to metal atoms.[370] The influence of nitrogen or sulphur co-ordination on the energies of these orbitals as modified by desolvation (i.e. the 'softness character'), calculated using Klopman's polyelectronic perturbation theory,[371] is shown schematically in Fig. 20. Hard class 'a' metal ions will bond to the hardest nucleophilic centre available which will be the nitrogen atom of the σ_4 orbital whereas soft class 'b' metal ions will bond to the softest nucleophilic centre available which will be the nitrogen atom of the σ_3 orbital. The softer 'a' metal ions (e.g. cobalt(III) with soft ligands) bond to the sulphur atom of the σ_4 orbital and the harder class 'b' metal ions (e.g. palladium(II) with hard ligands) bond to the sulphur atom of σ_3.[371] Thus a class 'b' metal will bond to thiocyanate via sulphur if it is surrounded by hard ligands and via nitrogen if surrounded by soft ligands. In other words the nitrogen of the thiocyanate ligand is softer than sulphur with class 'b' metals whereas the reverse is true with class 'a' metals.

(b) *Steric factors.* The small energy difference between S̲- and N̲-bonded thiocyanate enables steric factors to modify the mode of bonding. Steric crowding around the metal might be expected to lead to the preferential formation of N̲-bonded thiocyanate complexes because the M–N–CS unit is linear whereas the M–S–CN unit has an M–S–C bond angle of about 107°.[356] Thus in $[Pd(dien)(SCN)]^+NO_3^-$, the diethylenetriamine ligand allows sufficient room for a non-linear Pd–S–CN unit to be present,[365] whereas in $[Pd(Et_4dien)(NCS)]^+SCN^-$ the four bulky ethyl groups prevent the non-linear Pd–S–CN unit being formed but allow the linear Pd–N–CS unit to be formed.[368]

(c) *The physical state.* The physical state of the complex can also affect the mode of bonding of the thiocyanate ligand to the metal. Thus in the solid state the thiocyanate anion is S̲-bonded in $[(AsBu_3^n)_2Pd(SCN)_2]$ whereas on melting this complex partially isomerises to a mixture of the N̲- and S̲-bonded linkage isomers. Resolidification of the melt yields only the S̲-bonded isomer.[350]

(d) *The nature of the solvent.* The nature of the palladium(II)–thiocyanate bonding in $[(Ph_2AsCH_2CH_2AsPh_2)Pd(SCN)_2]$ depends on whether the material is in the solid state, in which case it is entirely S̲-bonded, or whether it is dissolved in dichloromethane when a mixture of N̲- and S̲-bonded thiocyanate is present.[372] This isomerisation is reversible. As yet no palladium(II) or platinum(II) complexes analogous to $[(AsPh_3)Mn(CO)_4-(SCN)]$,[373] which shows N̲-bonded thiocyanate in one solvent (methyl cyanide) and S̲-bonded thiocyanate in another (chloroform), appear to have been reported.

(*e*) *The nature of the anion.* The nature of the anion can be very important for determining the mode of bonding in the solid state. For example the complex cation [Pd(Et$_4$dien)(SCN)]$^+$ has an N̲-bonded thiocyanate ligand in its most stable form with thiocyanate[367] counteranions, whereas with tetraphenylborate[366] and hexafluorophosphate[367] as the counteranions the stable isomer in the solid state has S̲-bonded thiocyanate ligands. Although the size of the counteranion is probably important no systematic survey of counteranions has yet been carried out.

Selenocyanate complexes

The co-ordination behaviour of the selenocyanate ion is similar to that of the thiocyanate ion. Once again the S̲e- and N̲-bonded linkage isomers can be distinguished by infra-red spectroscopy (Table 50). Again greater reliability is possible if the integrated intensity of the band is determined since for N̲-bonded selenocyanate ligands this is greater than in the free ion whereas for S̲e-bonded selenocyanate ligands it is less.[374]

TABLE 50

Approximate ranges for the fundamental frequencies of co-ordinated selenocyanate groups (data taken from refs. 374 and 375)

Approximate origin of frequency	v_1 (cm^{-1}) C–N stretch	v_2 (cm^{-1}) N–C–Se bend	v_3 (cm^{-1}) C–Se stretch
Typical range for N-bonded selenocyanate	2055–2118	424–430	617–673
Typical range for Se-bonded selenocyanate	2060–2125	363–418	515–589

The mode of bonding of the selenocyanate ligand to palladium(II) is much less sensitive to electronic effects than that of the thiocyanate ligand. Thus in (Bu$_4$N$^+$)$_2$[M(SeCN)$_4$]$^{2-}$, where M = Pd or Pt, and in [L$_2$Pd(SeCN)$_2$] when L is varied from a σ-bonding to a π-bonding ligand as in Table 49 above, the selenocyanate ligand is always S̲e-bonded, with the exception of [(PBu$_3^n$)$_2$Pd(NCSe)$_2$] and [Pd(Et$_4$dien)(SeCN)]$^+$BPh$_4^-$.[366,376,377] Both these exceptions are due to steric crowding which favours the linear M–N–CSe unit;[366] [Pd(Et$_4$dien)(SeCN)]$^+$BPh$_4^-$ only exists as the N̲-bonded isomer in solution, the S̲e-bonded isomer being the stable form in the solid. The obvious preference of the selenocyanate ligand to bind to platinum(II) and palladium(II) through selenium is consistent with the calculated atomic charges (see Table 46, p. 215).

Thiomorpholine

Thiomorpholine (**118**) complexes have been used to study the comparative co-ordinating ability of dialkylamines and dialkylsulphides. Two series of

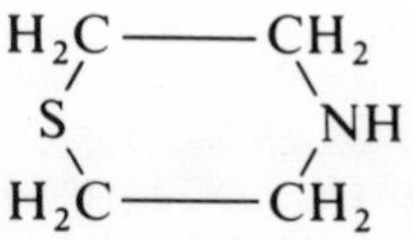

$$
\begin{array}{ccc}
H_2C & \!\!\!\!\!\text{------}\!\!\!\!\! & CH_2 \\
/ & & \backslash \\
S & & NH \\
\backslash & & / \\
H_2C & \!\!\!\!\!\text{------}\!\!\!\!\! & CH_2
\end{array}
$$

118

complexes are formed, one monomeric and the other polymeric. Although it might have been anticipated that thiomorpholine would bond to class 'b' metal ions such as platinum(II) and palladium(II) through sulphur it binds through nitrogen in the monomeric $[PdL_2Cl_2]$ complexes.[378] This is consistent with the observation that amines will displace dialkylsulphides from platinum(II) complexes whereas the reverse does not take place[379] (reactions 290 and 291). In the polymeric complexes the thiomorpholine

$$[Pt(SR_2)_2Cl_2] + pyr \longrightarrow [Pt(SR_2)(pyr)Cl_2] + R_2S \qquad (290)$$

$$[Pt(SR_2)_2Cl_2] + 4am \xrightarrow{\text{(am = amine)}} [Pt(am)_4]Cl_2 + 2R_2S \qquad (291)$$

acts as a bridging group (**119**). Thiomorpholin-3-one (**120**) only forms the monomeric nitrogen bonded *trans*-$[ML_2X_2]$ complexes,[378] where M = Pd or Pt and X = halogen.

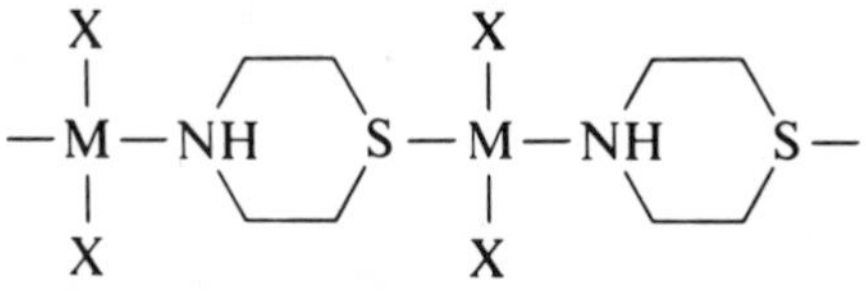

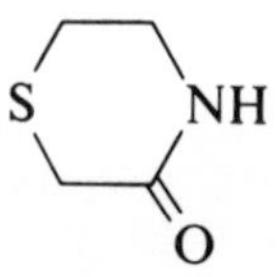

119   **120**

REFERENCES

1   J. V. Quagliano and L. Schubert, *Chem. Rev.*, **50** (1952) 201.
2   L. Wöhler, *Z. Elektrochem.*, **11** (1905) 836.
3   R. L. Shriner and R. Adams, *J. Amer. Chem. Soc.*, **46** (1924) 1683.
4   N. V. Sidgwick, *The Chemical Elements and Their Compounds*, Oxford 1950, vol. 2, p. 1558.
5   P. V. McKinney, *J. Amer. Chem. Soc.*, **54** (1932) 4498.
6   P. V. McKinney, *J. Amer. Chem. Soc.*, **55** (1933) 3626.
7   J. W. Kern, R. L. Shriner and R. Adams, *J. Amer. Chem. Soc.*, **47** (1925) 1147.
8   R. L. Shriner and R. Adams, *J. Amer. Chem. Soc.*, **46** (1924) 1683.
9   W. J. Moore and L. Pauling, *J. Amer. Chem. Soc.*, **63** (1941) 1392.
10 M. J. N. Pourbaix, J. van Muylder and N. de Zoubov, *Platinum Metals Review*, **3** (1959) 100.
11 R. M. Izatt, D. Eatough and J. J. Christensen, *J. Chem. Soc. (A)*, (1967) 1301.
12 L. Wöhler, K. Ewold and H. G. Krall, *Chem. Ber.*, **66** (1933) 1638.
13 *Handbook of Chemistry and Physics*, Ed. by R. C. Weast, The Chemical Rubber Co., Ohio, 47th edn., 1966, p. B202.
14 T. F. Gaskell, *Z. Krist.*, **96** (1937) 203; *Chem. Abs.*, **31** (1937) 6142.
15 F. Hulliger, *J. Phys. Chem. Solids*, **26** (1965) 639.
16 L. Moser and K. Atynski, *Mon.*, **54** (1925) 235; *Chem. Abs.*, **19** (1925) 1231.
17 F. Hulliger, *Structure and Bonding*, **4** (1968) 83.
18 A. Kjekshus and W. B. Pearson, *Canad. J. Phys.*, **43** (1965) 438.

19 C. J. Raub, V. B. Compton, T. H. Geballe, B. T. Matthias, J. P. Maita and G. W. Hull, *J. Phys. Chem. Solids*, **26** (1965) 205.

20 B. T. Matthias and S. Geller, *J. Phys. Chem. Solids*, **4** (1958) 318.

21 L. Wöhler and F. Martin, *Z. Elektrochem.*, **15** (1909) 791.

22 L. Wöhler, *Z. anorg. allgem. Chem.*, **40** (1904) 423.

23 L. Wöhler and F. Martin, *Z. Elektrochem.*, **15** (1909) 791.

24 Reference 4, p. 1581.

25 J. Thomsen, *Thermochemistry*, Longmans, Green & Co., London, 1908.

26 E. Davey, *Phil. Mag.*, **40** (1812) 28.

27 H. Remy, *Treatise on Inorganic Chemistry*, translated by J. S. Anderson, ed. by J. Kleinberg, Elsevier, (1956), vol. II, p. 346.

28 F. Grønvold, H. Haraldsen and A. Kjekshus, *Acta Chem. Scand.*, **14** (1960) 1879.

29 F. Roessler, *Z. anorg. allgem. Chem.*, **9** (1895) 31.

30 F. Roessler, *Z. anorg. allgem. Chem.*, **15** (1897) 405.

31 W. Hume-Rothery and G. V. Raynor, *The Structure of Metals and Alloys*, The Institute of Metals, London, 3rd edn., 1954, p. 187.

32 A. E. Wickenden and R. A. Krause, *Inorg. Chem.*, **8** (1969) 779.

33 S. E. Livingstone, *J. Chem. Soc.*, (1957) 5091.

34 L. F. Grantham, T. S. Elleman and D. S. Martin, *J. Amer. Chem. Soc.*, **77** (1955) 2965.

35 R. G. Denning, F. R. Hartley and L. M. Venanzi, *J. Chem. Soc. (A)*, (1967) 324.

36 A. A. Grinberg and D. Rjabtschikoff, *Acta Physicochim. URSS*, **3** (1935) 555; *Chem. Abs.*, **30** (1936) 4074.

37 K. A. Jensen, *Z. anorg. allgem. Chem.*, **242** (1939), 87.

38 D. S. Martin and R. J. Adams, *Proc. Sixth Int. Coord. Chem. Conf.*, (1961) 579.

39 T. S. Elleman, T. W. Reishus and D. S. Martin, *J. Amer. Chem. Soc.*, **80** (1958) 536.

40 K. A. Jensen, *Z. anorg. allgem. Chem.*, **229** (1936) 225.

41 R. J. Cross and F. Glockling, *J. Chem. Soc.*, (1965) 5422.

42 J. Chatt and B. T. Heaton, *J. Chem. Soc. (A)*, (1968) 2745.

43 R. Ugo, G. La Monica, S. Cenini, A. Segre and F. Conti, *J. Chem. Soc. (A)*, (1971) 522.

44 J. Chatt and D. M. P. Mingos, *J. Chem. Soc. (A)*, (1970) 1243.

45 A. C. Skapski and P. G. H. Troughton, *Chem. Comm.*, (1969) 170.

46 A. C. Skapski and P. G. H. Troughton, *J. Chem. Soc. (A)*, (1969) 2772.

47 M. C. Baird and G. Wilkinson, *J. Chem. Soc. (A)*, (1967) 865.

48 U. Belluco, L. Cattalini, F. Basolo, R. G. Pearson and A. Turco, *J. Amer. Chem. Soc.*, **87** (1965) 241.

49 R. G. Pearson, H. B. Gray and F. Basolo, *J. Amer. Chem. Soc.*, **82** (1960) 787.

50 F. Glockling and E. H. Brooks, *Preprints, Div. Petr. Chem.*, 14, No. 2, B135 (1969).

51 A. V. Nikiforova, I. I. Moiseev and Ya. K. Syrkin, *Zh. Obshch. Khim.*, **33** (1963) 3239; *Chem. Abs.*, **60** (1964) 3995.

52 W. G. Lloyd, *J. Org. Chem.*, **32** (1967) 2816.

53 M. Graziani, P. Uguagliati and G. Carturan, *J. Organometal Chem.*, **27** (1971) 275.

54 M. Graziani, R. Ros and G. Carturan, *J. Organometal Chem.*, **27** (1971) C19.

55 F. G. Mann and D. Purdie, *J. Chem. Soc.*, (1935) 1549.

56 G. E. Hunter and R. A. Krause, *Inorg. Chem.*, **9** (1970) 537.

57 T. A. Stephenson, S. M. Morehouse, A. R. Powell, J. P. Heffer and G. Wilkinson, *J. Chem. Soc.*, (1965) 3632.

58 T. Boschi, B. Crociani, L. Toniolo and U. Belluco, *Inorg. Chem.*, **9** (1970) 532.

59 R. G. Hayter and F. S. Humiec, *J. Inorg. Nucl. Chem.*, **26** (1964) 807.

60 R. S. Nyholm, J. F. Skinner and M. H. B. Stiddard, *J. Chem. Soc. (A)*, (1968) 38.

61 R. H. Fenn and G. R. Segrott, *J. Chem. Soc. (A)*, (1970) 2781.

62 R. H. Fenn and G. R. Segrott, *J. Chem. Soc. (A)*, (1970) 3197.

63 J. Chatt and F. G. Mann, *J. Chem. Soc.*, (1938) 1949.

64 J. Chatt and F. A. Hart, *J. Chem. Soc.*, (1953) 2363.

65 J. Chatt and F. A. Hart, *J. Chem. Soc.*, (1960) 2807.

66 L. A. Tschugaev and W. Subbotin, *Chem. Ber.*, **43** (1910) 1200.

67 K. A. Jensen, *Z. anorg. allgem. Chem.*, **225** (1935) 97; *Chem. Abs.*, **30** (1936) 919.

68 K. A. Jensen, *Z. anorg. allgem. Chem.*, **225** (1935) 115; *Chem. Abs.*, **30** (1936) 1321.
69 J. E. Fritzmann, *Z. anorg. allgem. Chem.*, **73** (1911) 239; *Chem. Abs.*, **6** (1912) 839.
70 E. Fritzmann and V. V. Krinitzkiv, *J. Appl. Chem. Russ.*, **11** (1939) 1610.
71 E. G. Cox, H. Saenger and W. Wardlaw, *J. Chem. Soc.*, (1934) 182.
72 P. E. Skakke and S. E. Rasmussen, *Acta Chem. Scand.*, **24** (1970) 2634.
73 E. A. Allen, N. P. Johnson, D. T. Rosevear and W. Wilkinson, *Chem. Comm.*, (1971) 171.
74 E. Fritzmann, *J. Russ. Chem. Soc.*, **47** (1915) 588.
75 E. Fritzmann, *Z. anorg. allgem. Chem.*, **133** (1924) 119, 133.
76 L. Tschugaev, *Z. anorg. allgem. Chem.*, **134** (1924) 277.
77 L. Tschugaev and C. Ivanov, *Z. anorg. allgem. Chem.*, **135** (1924) 153.
78 F. G. Mann and D. Purdie, *J. Chem. Soc.*, (1935) 1549.
79 V. N. Ipatieff and B. S. Friedman, *J. Amer. Chem. Soc.*, **61** (1939) 684.
80 N. N. Greenwood and G. Hunter, *J. Chem. Soc. (A)*, (1967) 1520.
81 H. J. Whitfield, *J. Chem. Soc. (A)*, (1970) 113.
82 J. Chatt and L. M. Venanzi, *J. Chem. Soc.*, (1955) 2787.
83 J. Chatt and L. M. Venanzi, *J. Chem. Soc.*, (1957) 2351.
84 J. Chatt, L. A. Duncanson and L. M. Venanzi, *J. Chem. Soc.*, (1955) 4461.
85 P. L. Goggin, R. J. Goodfellow, D. L. Sales, J. Stokes and P. Woodward, *Chem. Comm.*, (1968) 31.
86 D. L. Sales, J. Stokes and P. Woodward, *J. Chem. Soc. (A)*, (1968) 1852.
87 L. Pauling, *The Nature of the Chemical Bond*, Cornell University Press, 3rd edn., 1960.
88 R. C. Warren, J. F. McConnell and N. C. Stephenson, *Acta Cryst.*, **B26** (1970) 1402.
89 N. C. Stephenson, *Acta Cryst.*, **17** (1964) 587.
90 H. J. Whitfield, *J. Chem. Soc. (A)*, (1970) 113.
91 J. R. Allkins and P. J. Hendra, *J. Chem. Soc. (A)*, (1967) 1325.
92 J. R. Allkins, R. J. Obremski, C. W. Brown and E. R. Lippincott, *Inorg. Chem.*, **8** (1969) 1450.
93 B. E. Aires, J. E. Fergusson, D. T. Howarth and J. M. Miller, *J. Chem. Soc. (A)*, (1971) 1144.
94 J. Chatt, G. A. Gamlen, and L. E. Orgel, *J. Chem. Soc.*, (1959) 1047.
95 W. McFarlane, *Chem. Comm.*, (1968) 755.
96 E. W. Abel, R. P. Bush, F. J. Hopton and C. R. Jenkins, *Chem. Comm.*, (1966) 58.
97 P. Haake and P. C. Turley, *J. Amer. Chem. Soc.*, **89** (1967) 4611.
98 P. C. Turley and P. Haake, *J. Amer. Chem. Soc.*, **89** (1967) 4617.
99 R. J. Cross, G. J. Smith and R. Wardle, *Inorg. Nucl. Chem. Letters*, **7** (1971) 191.
100 H. J. S. King, *J. Chem. Soc.*, (1938) 1338.
101 G. A. Earwicker, *J. Chem. Soc.*, (1960) 2620.
102 C. D. Cook and G. S. Jauhal, *J. Amer. Chem. Soc.*, **89** (1967) 3066.
103 J. J. Levison and S. D. Robinson, *Inorg. Nucl. Chem. Letters*, **4** (1968) 407.
104 R. W. Horn, E. Weissberger and J. P. Collman, *Inorg. Chem.*, **9** (1970) 2367.
105 J. J. Levison and S. D. Robinson, *J. Chem. Soc. (A)*, (1971) 762.
106 C. J. Nyman, C. E. Wymore and G. Wilkinson, *Chem. Comm.*, (1967) 407.
107 C. J. Nyman, C. E. Wymore and G. Wilkinson, *J. Chem. Soc. (A)*, (1968) 561.
108 F. Cariati, R. Mason, G. B. Robertson and R. Ugo, *Chem. Comm.*, (1967) 408.
109 B. M. Gatehouse, S. E. Livingstone and R. S. Nyholm, *J. Chem. Soc.*, (1957) 4222.
110 S. E. Livingstone, *J. Proc. Roy. Soc. New South Wales*, **86** (1952) 32; *Chem. Abs.*, **47** (1953) 7932b.
111 C. C. Addison and B. G. Ward, *Chem. Comm.*, (1966) 155.
112 R. Countryman and W. S. McDonald, Personal communication.
113 D. A. Langs, C. R. Hare and R. G. Little, *Chem. Comm.*, (1967) 1080.
114 A. V. Babaeva, Yu. Ya. Kharitonov and Z. M. Novozhenyuk, *Russ. J. Inorg. Chem.*, **6** (1961) 1159.
115 G. Newman and D. B. Powell, *Spectrochim. Acta*, **19** (1963) 213.
116 M. V. Capparelli and L. N. Becka, *J. Chem. Soc. (A)*, (1969) 260.
117 M. A. Spinnler and L. N. Becka, *J. Chem. Soc. (A)*, (1967) 1194.
118 S. E. Livingstone, *Quart. Rev.*, **19** (1965) 386.

119 C. K. Jørgensen, *Absorption Spectra and Chemical Bonding in Complexes*, Pergamon, Oxford, 1962.
120 R. Eskenazi, J. Raskovan and R. Levitus, *J. Inorg. Nucl. Chem.*, **27** (1965) 371.
121 G. Vitzthum and E. Lindner, *Angew. Chem. Int. Ed.*, **10** (1971) 315.
122 C. D. Cook and G. S. Jauhal, *Canad. J. Chem.*, **45** (1967) 301.
123 J. Chatt and D. M. P. Mingos, *J. Chem. Soc. (A)*, (1969) 1770.
124 B. Chiswell and L. M. Venanzi, *J. Chem. Soc. (A)*, (1966) 1246.
125 S. Baggio, L. M. Anzel and L. N. Becka, *Acta Cryst.*, **B26** (1970) 1698.
126 J. B. Goddard and F. Basolo, *Inorg. Chem.*, **7** (1968) 936.
127 P. Shottlander, *Ann.*, **140** (1866) 200.
128 D. I. Ryabchikov, *Compt. rend. Acad. Sci. URSS*, **27** (1940) 349.
129 D. I. Ryabchikov, *Compt. rend. Acad. Sci. URSS*, **28** (1940) 231.
130 D. I. Ryabchikov, *Compt. rend. Acad. Sci. URSS*, **44** (1944) 277.
131 D. I. Ryabchikov, *Compt. rend. Acad. Sci. URSS*, **27** (1940) 685.
132 D. I. Ryabchikov, *Compt. rend. Acad. Sci. URSS*, **41** (1943) 208.
133 D. I. Ryabchikov, *Compt. rend. Acad. Sci. URSS*, **40** (1943) 229.
134 D. I. Ryabchikov, *Compt. rend. Acad. Sci. URSS*, **32** (1941) 344.
135 R. B. Penland, S. Mizushima, C. Curran and J. V. Quagliano, *J. Amer. Chem Soc.*, **79** (1957) 1575.
136 S. Oi, T. Kawase, K. Nakatsu and H. Kuroya, *Bull. Chem. Soc. Japan*, **33** (1960) 861.
137 D. A. Berta, W. A. Spofford, P. Boldrini and E. L. Amma, *Inorg. Chem.*, **9** (1970) 136.
138 T. Tarantelli and C. Furlani, *J. Chem. Soc. (A)*, (1968) 1717.
139 P. J. Hendra and Z. Jovic, *Spectrochim. Acta.*, **24A** (1968) 1713.
140 N. S. Kurnakov, *J. Prakt. Chem.*, **50** (1894) 483.
141 G. W. Watt and J. S. Thompson, *J. Inorg. Nucl. Chem.*, **33** (1971) 1319.
142 I. Ojima, T. Iwamoto, T. Onishi, N. Inamoto and K. Tamaru, *Chem. Comm.*, (1969) 1501.
143 A. Davison and E. S. Switkes, *Inorg. Chem.*, **10** (1971) 837.
144 R. L. Girling and E. L. Amma, *Chem. Comm.*, (1968) 1487.
145 F. A. Cotton and R. Francis, *J. Amer. Chem. Soc.*, **82** (1960) 2986.
146 F. A. Cotton, R. Francis and W. D. Horrocks, *J. Phys. Chem.*, **64** (1960) 1534.
147 J. Selbin, W. E. Bull and L. H. Holmes, *J. Inorg. Nucl. Chem.*, **16** (1961) 219.
148 W. Kitching, C. J. Moore and D. Doddrell, *Inorg. Chem.*, **9** (1970) 541.
149 M. J. Bennett, F. A. Cotton and D. L. Weaver, *Nature*, **212** (1966) 286.
150 M. J. Bennett, F. A. Cotton, D. L. Weaver, R. J. Williams and W. H. Watson, *Acta. Cryst.*, **23** (1967) 788.
151 S. Thomas and W. L. Reynolds, *Inorg. Chem.*, **8** (1969) 1531.
152 B. B. Wayland and R. F. Schramm, *Chem. Comm.*, (1968) 1465.
153 R. B. Wayland and R. F. Schramm, *Inorg. Chem.*, **8** (1969) 971.
154 J. H. Price, R. F. Schramm and B. B. Wayland, *Chem. Comm.*, (1970) 1377.
155 T. A. Stephenson, S. M. Morehouse, A. R. Powell, J. P. Heffer and G. Wilkinson, *J. Chem. Soc.*, (1965) 3632.
156 M. Tamura and T. Yasui, *J. Chem. Soc. Japan, Ind. Chem. Sect.*, **71** (1968) 1855. (English summary p. A116) *Chem. Abs.*, **70** (1969) 61532.
157 A. C. Skapski and M. L. Smart, *Chem. Comm.*, (1970) 658.
158 B. Braithwaite and D. Wright, *Chem. Comm.*, (1969) 1329.
159 M. Vèzes, *Compt. Rendu.*, **125** (1897) 525.
160 M. Vèzes, *Bull. Soc. Chim. France*, [3], **21** (1899) 143.
161 M. Vèzes, *Bull. Soc. Chim. France*, [3], **19** (1898) 875.
162 A. Werner and E. Grebe, *Z. anorg. allgem. Chem.*, **21** (1899) 377.
163 F. G. Mann, D. Crowfoot, D. C. Gattiker and N. Wooster, *J. Chem. Soc.*, (1935) 1642.
164 J. Chatt, F. G. Mann and A. F. Wells, *J. Chem. Soc.*, (1938) 2086.
165 M. J. Schmelz, I. Nakagawa, S. Mizushima and J. V. Quagliano, *J. Amer. Chem. Soc.*, **81** (1959) 287.
166 M. J. Schmelz, T. Miyazawa, S. Mizushima, T. J. Lane and J. V. Quagliano, *Spectrochim. Acta*, **9** (1957) 51.
167 H. O. Jones and H. S. Tasker, *J. Chem. Soc.*, **95** (1909) 1904.

168 C. S. Robinson and H. O. Jones, *J. Chem. Soc.*, **101** (1912) 62.
169 E. G. Cox, W. Wardlaw and K. C. Webster, *J. Chem. Soc.*, (1935) 1475.
170 J. Fujita and K. Nakamoto, *Bull. Chem. Soc. Japan*, **37** (1964) 528.
171 D. Coucouvanis, *J. Amer. Chem. Soc.*, **92** (1970) 707.
172 T. Tarantelli and C. Furlani, *J. Chem. Soc. (A)*, (1971) 1213.
173 P. Porta, *J. Chem. Soc. (A)*, (1971) 1217.
174 M. Bonamico and G. Dessy, *Chem. Comm.*, (1968) 483.
175 J. Chatt, L. A. Duncanson and L. M. Venanzi, *Nature*, **177** (1965) 1042.
176 G. W. Watt and B. J. McCormick, *J. Inorg. Nucl. Chem.*, **27** (1965) 898.
177 J. Chatt, L. A. Duncanson and L. M. Venanzi, *Suomen Kemi*, **B29** (1956) 75; *Chem. Abs.*, **51** (1957) 5559d.
178 C. K. Jørgenson, *J. Inorg. Nucl. Chem.*, **24** (1962) 1571.
179 K. Nakamoto, J. Fujita, R. A. Condrate and Y. Morimoto, *J. Chem. Phys.*, **39** (1963) 423.
180 B. J. McCormick, B. P. Stormer and R. I. Kaplan, *Inorg. Chem.*, **8** (1969) 2522.
181 J. P. Fackler and W. C. Seidel, *Inorg. Chem.*, **8** (1969) 1631.
182 J. P. Fackler and D. Coucouvanis, *J. Amer. Chem. Soc.*, **88** (1966) 3913.
183 S. H. H. Chaston, S. E. Livingstone, T. N. Lockyer, V. A. Pickles and J. S. Shannon, *Aust. J. Chem.*, **18** (1965) 673.
184 W. Kuchen and H. Hertel, *Angew. Chem. Int. Ed.*, **8** (1969) 89.
185 T. A. Stephenson and B. D. Faithful, *J. Chem. Soc. (A)*, (1970) 1504.
186 J. M. C. Alison and T. A. Stephenson, *Chem. Comm.*, (1970) 1092.
187 W. Kuchen and B. Knop, *Angew. Chem. Int. Ed.*, **4** (1965) 244.
188 G. W. Watt and B. J. McCormick, *Spectrochim. Acta.*, **21** (1965) 753.
189 J. P. Fackler, D. Coucouvanis, W. C. Seidel, R. C. Masek and W. Holloway, *Chem. Comm.*, (1967) 924.
190 J. P. Fackler, W. C. Seidel and J. A. Fetchin, *J. Amer. Chem. Soc.*, **90** (1968) 2707.
191 D. Gibson, *Coord. Chem. Rev.*, **4** (1969) 225.
192 D. W. Thompson, *Structure and Bonding*, **9** (1971) 27.
193 G. A. Barbieri, *Atti. Acad. Lincei. Rend.*, **23** (1914) 334; *Chem. Abs.*, **8** (1914) 2988.
194 A. A. Grinberg and L. K. Simonova, *Zh. Prikl. Khim.*, **26** (1953) 880; *Chem. Abs.*, **47** (1953) 11060g.
195 A. A. Grinberg and I. N. Chapurskii, *Russ. J. Inorg. Chem.*, **4** (1959) 137.
196 D. A. White, *J. Chem. Soc. (A)*, (1971) 143.
197 M. R. Churchill, *Inorg. Chem.*, **5** (1966) 1608.
198 P.-K. Hon, C. E. Pfluger and R. F. Belford, *Inorg. Chem.*, **6** (1967) 730.
199 J. F. Malone and W. S. McDonald, *J. Chem. Soc. (A)*, (1970) 3124.
200 E. C. Lingafelter and R. L. Braun, *J. Amer. Chem. Soc.*, **88** (1966) 2951.
201 K. Nakamoto, P. J. McCarthy and A. E. Martell, *J. Amer. Chem. Soc.*, **83** (1961) 1272.
202 G. T. Behnke and K. Nakamoto, *Inorg. Chem.*, **6** (1967) 433.
203 D. M. Sweeny, I. Nakagawa, S. Mizushima and J. V. Quagliano, *J. Amer. Chem. Soc.*, **78** (1956) 889.
204 K. Nakamoto, P. J. McCarthy, J. Fujita, R. A. Condrate and G. T. Behnke, *Inorg. Chem.*, **4** (1965) 36.
205 M. Mikami, I. Nakagawa and T. Shimanouchi, *Spectrochim. Acta.*, **23A** (1967) 1037.
206 K. Nakamoto, C. Udovich and J. Takemoto, *J. Amer. Chem. Soc.*, **92** (1970) 3973.
207 J. Lewis, R. F. Long and C. Oldham, *J. Chem. Soc.*, (1965) 6740.
208 R. H. Holm and F. A. Cotton, *J. Amer. Chem. Soc.*, **80** (1958) 5658.
209 B. Bock, K. Flatau, H. Junge, M. Kuhr and H. Musso, *Angew. Chem. Int. Ed.*, **10** (1971) 225.
210 J. A. S. Smith and J. D. Thwaites, *Disc. Faraday Soc.*, **34** (1962) 143.
211 J. P. Collman, *Adv. Chem. Ser.*, **37** (1963) 78.
212 J. P. Collman, *Angew. Chem. Int. Ed.*, **4** (1965) 132.
213 D. A. White, *J. Chem. Soc. (A)*, (1971) 233.
214 B. F. G. Johnson, J. Lewis and D. A. White, *J. Amer. Chem. Soc.*, **91** (1969) 5186.
215 R. G. Pearson and D. A. Johnson, *J. Amer. Chem. Soc.*, **86** (1964) 3983.
216 K. Saito and M. Takahashi, *Bull. Chem. Soc. Japan*, **42** (1969) 3462.

217 A. Werner, *Chem. Ber.*, **34** (1901) 2584.
218 B. N. Figgis, J. Lewis, R. Mason, R. S. Nyholm, P. J. Pauling and G. B. Robertson, *Nature*, **195** (1962) 1278.
219 R. Mason, G. B. Robertson and P. J. Pauling, *J. Chem. Soc. (A)*, (1969) 485.
220 D. Gibson, J. Lewis and C. Oldham, *J. Chem. Soc. (A)*, (1966) 1453.
221 J. Lewis and C. Oldham, *J. Chem. Soc. (A)*, (1966) 1456.
222 D. A. White, *Synth. Inorg. and Metal-Org. Chem.*, **1** (1971) 59.
223 B. F. G. Johnson, J. Lewis and M. S. Subramanian, *Chem. Comm.*, (1966) 117.
224 B. F. G. Johnson, J. Lewis and M. S. Subramanian, *J. Chem. Soc. (A)*, (1968) 1993.
225 R. A. D. Wentworth and C. H. Brubaker, *Inorg. Chem.*, **3** (1964) 1472.
226 G. T. Behnke and K. Nakamoto, *Inorg. Chem.*, **6** (1967) 440.
227 D. Gibson, C. Oldham, J. Lewis, D. Lawton, R. Mason and G. B. Robertson, *Nature*, **208** (1965) 580.
228 R. Mason and G. B. Robertson, *J. Chem. Soc. (A)*, (1969) 492.
229 D. Gibson, J. Lewis and C. Oldham, *J. Chem. Soc. (A)*, (1967) 72.
230 G. Allen, J. Lewis, R. F. Long and C. Oldham, *Nature*, **202** (1964) 589.
231 G. T. Behnke and K. Nakamoto, *Inorg. Chem.*, **7** (1968) 2030.
232 G. Hulley, B. F. G. Johnson and J. Lewis, *J. Chem. Soc. (A)*, (1970) 1732.
233 S. H. H. Chaston and S. E. Livingstone, *Aust. J. Chem.*, **20** (1967) 1065.
234 L. P. Eddy, J. W. Hayes, S. E. Livingstone, H. L. Nigam and D. V. Radford, *Aust. J. Chem.*, **24** (1971) 1071.
235 E. A. Shugam, L. M. Shkol'nikova and S. E. Livingstone, *Zh. Strukt. Khim.*, **8** (1967) 550; *Chem. Abs.*, **68** (1968) 33978.
236 J. J. Lippard and S. M. Morehouse, *J. Amer. Chem. Soc.*, **91** (1969) 2504.
237 R. Belcher, W. I. Stephen, I. J. Thomson and P. C. Uden, *Chem. Comm.*, (1970) 1019.
238 R. Belcher, M. Pravica, W. I. Stephen and P. C. Uden, *Chem. Comm.*, (1971) 41.
239 J. A. McCleverty, *Prog. Inorg. Chem.*, **10** (1968) 49.
240 E. Billig, R. Williams, I. Bernal, J. H. Waters and H. B. Gray, *Inorg. Chem.*, **3** (1964) 663.
241 A. Davison, N. Edelstein, R. H. Holm, and A. H. Maki, *Inorg. Chem.*, **3** (1964) 814.
242 R. Williams, E. Billig, J. H. Waters and H. B. Gray, *J. Amer. Chem. Soc.*, **88** (1966) 43.
243 G. N. Schrauzer and V. P. Mayweg, *J. Amer. Chem. Soc.*, **87** (1965) 1483.
244 G. N. Schrauzer, V. P. Mayweg and W. Heinrich, *Inorg. Chem.*, **4** (1965) 1615.
245 A. Davison and R. H. Holm, *Inorg. Synth.*, **10** (1967) 8.
246 D. C. Olson, V. P. Mayweg and G. N. Schrauzer, *J. Amer. Chem. Soc.*, **88** (1966) 4876.
247 A. L. Balch and R. H. Holm, *J. Amer. Chem. Soc.*, **88** (1966) 5201.
248 G. N. Schrauzer and V. P. Mayweg, *J. Amer. Chem. Soc.*, **87** (1965) 3585.
249 R. Eisenberg, J. A. Ibers, R. J. H. Clark and H. B. Gray, *J. Amer. Chem. Soc.*, **86** (1964) 113.
250 C. J. Fritchie, *Acta Cryst.*, **20** (1966) 107.
251 A. Davison, N. Edelstein, R. H. Holm and A. H. Maki, *Inorg. Chem.*, **2** (1963) 1227.
252 S. I. Shupack, E. Billig, R. J. H. Clark, R. Williams and H. B. Gray, *J. Amer. Chem. Soc.*, **86** (1964) 4594.
253 H. B. Gray, *Trans. Metal Chem.*, **1** (1965) 240.
254 D. Sartain and M. R. Truter, *Chem. Comm.*, (1966) 382.
255 D. Sartain and M. R. Truter, *J. Chem. Soc. (A)*, (1967) 1264.
256 M. J. Baker-Hawkes, E. Billig, and H. B. Gray, *J. Amer. Chem. Soc.*, **88** (1966) 4870.
257 A. H. Maki, N. Edelstein, A. Davison and R. H. Holm, *J. Amer. Chem. Soc.*, **86** (1964) 4580.
258 A. Davison and D. V. Howe, *Chem. Comm.*, (1965) 290.
259 V. P. Mayweg and G. N. Schrauzer, *Chem. Comm.*, (1966) 640.
260 P. C. Nicpon and D. W. Meek, *Chem. Comm.*, (1966) 398.
261 M. G. King and G. P. McQuillan, *J. Chem. Soc. (A)*, (1967) 898.
262 A. A. Grinberg and L. H. Volshtein, *Compt. rend. Acad. Sci. URSS*, **2** (1935) 485; *Chem. Abs.*, **29** (1935) 6860.
263 L. E. Erickson, J. W. McDonald, J. K. Howie and R. P. Clow, *J. Amer. Chem. Soc.*, **90** (1968) 6371.

264 F. W. Pinkard, E. Sharratt, W. Wardlaw and E. G. Cox, *J. Chem. Soc.*, (1934) 1012.
265 L. M. Volshtein and G. G. Motyagina, *Russ. J. Inorg. Chem.*, 5 (1960) 840.
266 T. J. Lane, J. A. Durkin and R. J. Hooper, *Spectrochim. Acta.*, 20 (1964) 1013.
267 J. S. Coe and J. R. Lyons, *J. Chem. Soc. (A)*, (1971) 829.
268 V. Balzani, V. Carassiti, L. Moggi and F. Scandola, *Inorg. Chem.*, 4 (1965) 1243.
269 A. J. Saraceno, I. Nakagawa, S. Mizushima, C. Curran and J. V. Quagliano, *J. Amer. Chem. Soc.*, 80 (1958) 5018.
270 L. M. Volshtein, M. F. Mogilevkina and G. G. Motyagina, *Russ. J. Inorg. Chem.*, 6 (1961) 564.
271 J. S. Coe and J. R. Lyons, *Inorg. Chem.*, 9 (1970) 1775.
272 L. M. Volshtein and I. O. Volodina, *Russ. J. Inorg. Chem.*, 5 (1960) 949.
273 L. E. Maley and D. P. Mellor, *J. Aust. Sci. Res.*, A2 (1949) 579; *Chem. Abs.*, 45 (1951) 3279c.
274 A. J. Poë and D. H. Vaughan, *Inorg. Chim. Acta.*, 1 (1967) 255.
275 J. S. Coe, J. R. Lyons and M. D. Hussain, *J. Chem. Soc. (A)*, (1970) 90.
276 L. Cattalini, M. Martelli and G. Marangoni, *Inorg. Chim. Acta.*, 2 (1968) 405.
277 Y. O. Aochi and D. T. Sawyer, *Inorg. Chem.*, 5 (1966) 2085.
278 B. B. Smith and D. T. Sawyer, *Inorg. Chem.*, 8 (1969) 1154.
279 D. J. Robinson and C. H. L. Kennard, *Chem. Comm.*, (1967) 1236.
280 D. J. Robinson and C. H. L. Kennard, *J. Chem. Soc. (A)*, (1970) 1008.
281 L. E. Erickson, H. L. Fritz, R. J. May and D. A. Wright, *J. Amer. Chem. Soc.*, 91 (1969) 2513.
282 B. B. Smith and D. T. Sawyer, *Inorg. Chem.*, 7 (1968) 1526.
283 B. B. Smith and D. T. Sawyer, *Inorg. Chem.*, 8 (1969) 379.
284 K. Freund and H. Frye, *Inorg. Nucl. Chem. Letters*, 7 (1971) 107.
285 R. D. Cannon, B. Chiswell and L. M. Venanzi, *J. Chem. Soc. (A)*, (1967) 1277.
286 K. Nakamoto, Y. Morimoto and A. E. Martell, *J. Amer. Chem. Soc.*, 83 (1961) 4528.
287 R. A. Condrate and K. Nakamoto, *J. Chem. Phys.*, 42 (1965) 2590.
288 J. A. Kieft and K. Nakamoto, *J. Inorg. Nucl. Chem.*, 29 (1967) 2561.
289 V. Balzani, V. Carassiti, L. Moggi and F. Scandola, *Inorg. Chem.*, 4 (1965) 1243.
290 E. W. Wilson and R. B. Martin, *Inorg. Chem.*, 10 (1971) 1197.
291 A. Panunzi, R. Palumbo, C. Pedone and G. Paiaro, *J. Organometal. Chem.*, 5 (1966) 586.
292 J. A. Kieft and K. Nakamoto, *J. Inorg. Nucl. Chem.*, 30 (1968) 3103.
293 J. Fujita, K. Konya and K. Nakamoto, *Inorg. Chem.*, 9 (1970) 2794.
294 H. Yoneda, Y. Morimoto, Y. Nakao and A. Nakahara, *Bull. Chem. Soc. Japan*, 41 (1968) 255.
295 E. W. Wilson and R. B. Martin, *Inorg. Chem.*, 9 (1970) 528.
296 L. M. Volshtein and M. F. Mogilevkina, *Russ. J. Inorg. Chem.*, 8 (1963) 304.
297 C. A. McAuliffe, *J. Chem. Soc. (A)*, (1967) 641.
298 H. C. Freeman and M. L. Golomb, *Chem. Comm.*, (1970) 1523.
299 D. C. Jicha and D. H. Busch, *Inorg. Chem.*, 1 (1962) 872.
300 D. C. Jicha and D. H. Busch, *Inorg. Chem.*, 1 (1962) 878.
301 J. W. Wrathall and D. H. Busch, *Inorg. Chem.*, 2 (1963) 1182.
302 P. S. K. Chia, S. E. Livingstone and T. N. Lockyer, *Aust. J. Chem.*, 19 (1966) 1835.
303 J. Bankovskis, G. Mezarups and A. Ievins, *Zhur. analit. Khim.*, 17 (1962) 721; *Chem. Abs.*, 58 (1963) 13123a.
304 J. Bankovskis, G. Mezarups and A. Ievins, *Latvijas P.S.R. Zinatnu Akad., Vestis, Kim. Ser.*, (1962) 323; *Chem. Abs.*, 59 (1963) 5749c.
305 J. Cirule, J. Bankovskis, A. Ievins and J. Asaks, *ibid.*, (1963) 135; *Chem. Abs.*, 60 (1964) 3702c.
306 L. F. Lindoy, S. E. Livingstone and T. N. Lockyer, *Aust. J. Chem.*, 19 (1966) 1391.
307 P. S. K. Chia, S. E. Livingstone and T. N. Lockyer, *Aust. J. Chem.*, 20 (1967) 239.
308 L. F. Lindoy, S. E. Livingstone and T. N. Lockyer, *Aust. J. Chem.*, 20 (1967) 471.
309 E. Fluck, M. Gochring and J. Weiss, *Z. anorg. allgem. Chem.*, 287 (1956) 51.
310 J. Weiss and M. Becke-Goehring, *Z. Naturforsch.*, 13B (1958) 198; *Chem. Abs.*, 52 (1958) 16113e.

311   I. Lindqvist and J. Weiss, *J. Inorg. Nucl. Chem.*, **6** (1958) 184.
312   I. Lindqvist and R. Rosenstein, *J. Inorg. Nucl. Chem.*, **7** (1958) 421.
313   S. N. Poddar and P. Ray, *J. Indian Chem. Soc.*, **29** (1952) 279.
314   P. Ray, *Chem. Rev.*, **61** (1961) 313.
315   K. A. Jensen, *Z. anorg. allgem. Chem.*, **221** (1934) 6.
316   S. E. Livingstone, *J. Chem. Soc.*, (1958) 4222.
317   J. P. Beale and N. C. Stephenson, *Acta Cryst.*, **B26** (1970) 1655.
318   B. Chiswell and S. E. Livingstone, *J. Inorg. Nucl. Chem.*, **23** (1961) 37.
319   S. E. Livingstone, *J. Chem. Soc.*, (1956) 437.
320   G. Dyer, M. O. Workman and D. W. Meek, *Inorg. Chem.*, **6** (1967) 1404.
321   S. E. Livingstone and T. N. Lockyer, *Inorg. Nucl. Chem. Lett.*, **3** (1967) 35.
322   T. D. DuBois and D. W. Meek, *Inorg. Chem.*, **8** (1969) 146.
323   R. L. Dutta, D. W. Meek and D. H. Busch, *Inorg. Chem.*, **9** (1970) 1215.
324   R. L. Dutta, D. W. Meek and D. H. Busch, *Inorg. Chem.*, **9** (1970) 2098.
325   L. F. Lindoy, S. E. Livingstone and T. N. Lockyer, *Nature*, **211** (1966) 519.
326   L. F. Lindoy, S. E. Livingstone and T. N. Lockyer, *Inorg. Chem.*, **6** (1967) 652.
327   A. H. Norbury and A. I. P. Sinha, *Quart. Rev.*, **24** (1970) 69.
328   J. L. Burmeister, *Coord. Chem. Rev.*, **1** (1966) 205.
329   J. L. Burmeister, *Coord. Chem. Rev.*, **3** (1968) 225.
330   J. Powell and B. L. Shaw, *J. Chem. Soc.*, (1965) 3879.
331   J. Chatt and B. L. Shaw, *J. Chem. Soc.*, (1962) 5075.
332   K. Nakamoto, *Infrared Spectra of Inorganic and Coordination Compounds*, John Wiley, New York and London, 1963, pp. 151–5.
333   J. L. Burmeister and R. C. Timmer, *J. Inorg. Nucl. Chem.*, **28** (1966) 1973.
334   M. Vèzes, *Bull. Soc., Chim. France*, **19** (1898) 875.
335   J. P. Collman, M. Kubota and J. W. Hosking, *J. Amer. Chem. Soc.*, **89** (1967) 4809.
336   F. P. Boer, V. B. Carter and J. W. Turley, *Inorg. Chem.*, **10** (1971) 651.
337   H. B. Gray, *Electrons and Chemical Bonding*, W. A. Benjamin, New York, 1965.
338   D. Forster and D. M. L. Goodgame, *J. Chem. Soc.*, (1965) 1286.
339   A. H. Norbury and A. I. P. Sinha, *Inorg. Nucl. Chem. Lett.*, **3** (1967) 355.
340   A. H. Norbury and A. I. P. Sinha, *J. Chem. Soc. (A)*, (1968) 1598.
341   J. C. Bailar and H. Itatani, *J. Amer. Chem. Soc.*, **89** (1967) 1592.
342   W. Beck and E. Schuierer, *Chem. Ber.*, **98** (1965) 298.
343   W. Beck and W. P. Fehlhammer, *Angew. Chem. Int. Ed.*, **6** (1967) 169.
344   E. L. Wagner, *J. Chem. Phys.*, **43** (1965) 2728.
345   J. L. Burmeister and N. J. DeStefano, *Inorg. Chem.*, **8** (1969) 1546.
346   W. Beck and K. Feldl, *Z. anorg. allgem. Chem.*, **341** (1965) 113.
347   P. C. H. Mitchell and R. J. P. Williams, *J. Chem. Soc.*, (1960) 1912.
348   A. Turco and C. Pecile, *Nature*, **191** (1961) 66.
349   R. N. Keller, N. B. Johnson and L. L. Westmoreland, *J. Amer. Chem. Soc.*, **90**, (1968) 2729.
350   A. Sabatini and I. Bertini, *Inorg. Chem.*, **4** (1965) 1665.
351   S. Fronaeus and R. Larsson, *Acta Chem. Scand.*, **16** (1962) 1447.
352   C. Pecile, *Inorg. Chem.*, **5** (1966) 210.
353   R. Larsson and A. Miezis, *Acta Chem. Scand.*, **23** (1969) 37.
354   O. W. Howarth, R. E. Richards and L. M. Venanzi, *J. Chem. Soc.*, (1964) 3335.
355   A. Mawby and G. E. Pringle, *Chem. Comm.*, (1970) 385.
356   G. R. Clark, G. J. Palenik and D. W. Meek, *J. Amer. Chem. Soc.*, **92** (1970) 1077.
357   G. Beran and G. J. Palenik, *Chem. Comm.*, (1970) 1354.
358   Ia. Ia. Bleidelis, *Kristallografiya*, **2** (1957) 278; *Chem. Abs.*, **51** (1957) 14364b.
359   Ia. Ia. Bleidelis and G. B. Bokii, *Kristallografiya*, **2** (1957) 281; *Chem. Abs.*, **51** (1957) 14363i.
360   P. G. Owston and J. M. Rowe, *Acta Cryst.*, **13** (1960) 253.
361   U. A. Gregory, J. A. J. Jarvis, B. T. Kilbourn and P. G. Owston, *J. Chem. Soc. (A)*, (1970) 2770.
362   J. L. Burmeister and F. Basolo, *Inorg. Chem.*, **3** (1964) 1587.

363 I. Bertini and A. Sabatini, *Inorg. Chem.*, **5** (1966) 1025.

364 F. Basolo, J. L. Burmeister and A. J. Poë, *J. Amer. Chem. Soc.*, **85** (1963) 1700.

365 F. Basolo, H. B. Gray and R. G. Pearson, *J. Amer. Chem. Soc.*, **82** (1960) 4200.

366 J. L. Burmeister, H. J. Gysling and J. C. Lim, *J. Amer. Chem. Soc.*, **91** (1969) 44.

367 F. Basolo, W. H. Baddley and K. Wiedenbaum, *J. Amer. Chem. Soc.*, **88** (1966) 1576.

368 F. Basolo, W. H. Baddley and J. L. Burmeister, *Inorg. Chem.*, **3** (1964) 1202.

369 R. G. Pearson, *J. Chem. Ed.*, **45** (1968) 581.

370 L. DiSipio, L. Oleari and G. DeMichelis, *Coord. Chem. Rev.*, **1** (1966) 7.

371 A. H. Norbury, *J. Chem. Soc. (A)*, (1971) 1089.

372 D. W. Meek, P. E. Nicpon and V. I. Meek, *J. Amer. Chem. Soc.*, **92** (1970) 5351.

373 M. F. Farona and A. Wojcicki, *Inorg. Chem.*, **4** (1965) 1402.

374 J. L. Burmeister and L. E. Williams, *Inorg. Chem.*, **5** (1966) 1113.

375 A. Sabatini and I. Bertini, *Inorg. Chem.*, **4** (1965) 959.

376 J. L. Burmeister and H. J. Gysling, *Inorg. Chim. Acta.*, **1** (1967) 100.

377 H.-H. Schmidtke and D. Garthoff, *Helv. Chim. Acta.*, **50** (1967) 1631.

378 E. A. Allen, N. P. Johnson, D. T. Rosevear and W. Wilkinson, *J. Chem. Soc. (A)*, (1970) 2137.

379 S. P. Derendyaev, *Zhur. Neorg. Khim.*, **3** (1958) 2295; *Chem. Abs.*, **55** (1961) 1263g.

Note added in proof. X-ray diffraction has shown that the compound formulated as $[(PEt_3)_2Pt(GePh_3)(OH)]$ is actually cis-$[(PEt_3)_2Pt(GePh_2OH)(C_6H_5)]$ formed by internal rearrangement (R. J. D. Gee and H. M. Powell, *J. Chem. Soc. (A)*, (1971) 1957).

CHAPTER 9

Divalent Metal Halides

The halides and halide complexes of platinum(II) and palladium(II) have been widely studied (Tables 51–4) for several reasons: they are easily prepared, stable both in solution and in the solid state, and the relative simplicity of the halide ions has led to a number of studies of the electronic properties of the halides. The divalent metals are class 'b' or 'soft' acids and accordingly prefer to co-ordinate to the heavier halide ligands. This has several consequences. In preparative experiments it is essential to allow for halide exchange if a number of halides are present in the mixture. In recording the infra-red spectra the use of potassium bromide or caesium iodide discs may give rise to halide exchange. This can be avoided in the near infra-red by using sodium chloride discs, and in the far infra-red by separating the caesium iodide plate from the mull by a thin film of polythene. The vibrational spectra of the metal halides are extremely useful as a first indication of the structure of the compound because the metal–halide stretching vibrations give rise to very intense bands that are readily identified by replacing one halide by another.

The most important points about the simple halides and halide complexes (Tables 51–4) are the absence of PtF_2 and $[Pt_2F_6]^{2-}$ and the fact that $[PtF_4]^{2-}$, $[PdF_4]^{2-}$ and $[Pd_2F_6]^{2-}$ have not been fully characterised. Since the enthalpy of formation of PtF_2 can be calculated to be about 30 kcal/mol[21] it is probable that the difluoride disproportionates rapidly (eqn. 292) as a result of the high enthalpy of formation of the tetrafluoride. PdF_2 has a distorted

$$2PtF_2 \longrightarrow PtF_4 + Pt \tag{292}$$

octahedral structure, which results in its being paramagnetic, and presumably if platinum formed a difluoride this would also have a distorted octahedral structure, since a square-planar structure would require fluoride bridging groups, and the absence of $[Pt_2F_6]^{2-}$ and doubt surrounding $[Pd_2F_6]^{2-}$ indicates that fluoride ions are not good bridging ligands, possibly due to the absence of any low energy empty d-orbitals on fluorine. If sufficient energy is expended to give platinum an octahedral configuration then it is energetically worthwhile promoting platinum to the +4 oxidation state and so PtF_2 does not exist. The questionable existence of the other fluorides and fluoride anions of divalent platinum and palladium probably reflects this ability of fluoride ions to stabilise the +4 oxidation state. The instability of the complex

TABLE 51
The preparation, properties and structures of the palladium(II) halides

Halide	Preparation	Properties	Structure
PdF$_2$	(1) Treatment of PdI$_2$ with BrF$_3$ at room temperature followed by refluxing with SeF$_4$.[1] $PdI_2 + BrF_3 \xrightarrow[\text{temp.}]{\text{room}} PdF_3.BrF_3$ $PdF_3.BrF_3 + SeF_4 \xrightarrow{\text{reflux}} PdF_4.2SeF_4$ $PdF_4.2SeF_4 \xrightarrow[\text{excess SeF}_4]{\text{reflux in}} PdF_2$ (2) Reflux Pd^{II}[Pd^{IV}F$_6$] with SeF$_4$.[2] (3) Heat PdS$_2$ and SF$_4$ at 150–300°C in a bomb at 5–50 atm.[3] (4) Thermal decomposition of PdGeF$_6$ at 350°C in vacuo.[4]	Violet powder,[1] which is hydrolysed in water to the hydroxide.[1,5] PdF$_2$ is the only paramagnetic salt of Pd^{II}.[1]	X-ray powder diffraction studies[1,6,7] indicate that it is tetragonal ($a = 4.956$ Å and $c = 3.389$ Å) with a rutile structure. Pd is octahedral with four fluorines at 2.155 Å and 2 at 2.171 Å.[6]
PdCl$_2$	Direct chlorination of Pd at 300°C	Red, hygroscopic solid. Soluble in water to give PdCl$_2$.2H$_2$O.[9] Very readily reduced by H$_2$[10] and C$_2$H$_4$[11] etc. in aqueous solution. Decomposes above 600°C into palladium and chlorine.[8,12–14]	Differential thermal analysis and x-ray diffraction studies indicate that there are at least three polymorphs[15] which depend on mode of preparation. Best known is α-PdCl$_2$ which is orthorhombic[16] ($a = 3.81$ Å, $b = 3.34$ Å and $c = 11.0$ Å) with two formula units in the unit cell. Structure consists of infinite chains in which each Pd is surrounded by four coplanar bridging Cl atoms at 2.31 Å; $\angle$PdClPd = 93°, $\angle$ClPdCl = 87°. A second polymorph (β-PdCl$_2$) is hexameric Pd$_6$Cl$_{12}$ with octahedral clusters of Pd atoms[17] (cf. PtCl$_2$).
PdBr$_2$	(1) Direct bromination of Pd metal.[18] (2) Treatment of Pd metal with nitric acid containing free bromine.[9]	Brown solid, insoluble in H$_2$O but soluble in hydrobromic acid.[9]	Not investigated.
PdI$_2$	Dissolve PdCl$_2$ in HCl and add potassium iodide.[19]	Black solid, insoluble in H$_2$O, alcohol and ether. Dissolves in excess KI giving K$_2$PdI$_4$. Dissociates into metal and iodine above 350°C.[19]	Not investigated.

TABLE 52
The preparation, properties and structures of the platinum(II) halides

Halide	Preparation	Properties	Structure
PtF$_2$	Claimed to be formed when Pt wire is electrically exploded in SF$_4$ vapour.[20] The analysis was unsatisfactory and the formulation PtF$_2$ was based on the fact that it differed from known Pt fluorides. Earlier work where PtF$_2$ could not be prepared suggests that it is probable that the formulation PtF$_2$ is incorrect.[21]		
PtCl$_2$	(1) Direct chlorination of Pt metal at elevated temperatures.[22] (2) Chlorination of PtO$_2$ at 550°C.[23] (3) Thermal decomposition of PtCl$_4$.[22,24,25] (4) Pass Cl$_2$ over H$_2$PtCl$_6$ at 500°C.[23] (5) Reduction of H$_2$PtCl$_6$ with hydrazine.[26]	Green-brown solid, insoluble in water.[23] On heating gives PtCl[25] and at higher temperatures Pt metal and chlorine.[12,27]	Single crystal x-ray diffraction[28,29] showed β-PtCl$_2$ (prepared by method 4) to be hexameric with discrete 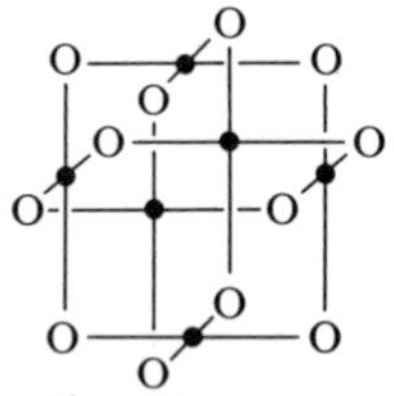 Pt$_6$Cl$_{12}$ units^a with Pt–Pt distances of 3·32 and 3·40 Å, and Pt–Cl distances of 2·34 and 2·39 Å. On heating β-PtCl$_2$ at 500°C for 1–2 days it gives α-PtCl$_2$ which is not isotypic with α-PdCl$_2$.[30]
PtBr$_2$	Direct bromination of Pt followed by thermal decomposition of the initial mixture of PtBr$_3$ and PtBr$_4$.[31]	Brown solid, insoluble in water.[32]	Not determined.
PtI$_2$	Thermal decomposition of PtI$_3$ at 270°C.[33,34] PtI$_3$ is prepared by adding KI to a solution of H$_2$PtCl$_6$ and heating the resulting precipitate to 200°C.[33,34]	Black powder, which is insoluble in H$_2$O, alcohol, ether and ethyl acetate[32] and which decomposes to Pt metal above 500°C.[33]	Investigated powder by x-ray diffraction. Data did not fit cubic, hexagonal or trigonal systems.[33]

^{a} ·PtO, Cl.

TABLE 53
The preparation, properties and structures of the halide complexes of palladium(II)

Halide	Preparation	Properties	Structure
$[PdF_4]^{2-}$	No preparative details recorded.[35]	Potassium salt is yellow.[35]	Not investigated.
$[PdCl_4]^{2-}$	(1) Interaction of $PdCl_2$ and MCl (M = alkali metal) in HCl solution gives M_2PdCl_4.[36] (2) Reduction of $[PdCl_6]^{2-}$ by (a) thermal decomposition[8,13] (b) treatment with Pd sponge in HCl solution[37] or (c) hydrazine.[38] (3) Dissolution of Pd sponge in aqua-regia and evaporation to dryness. Residue dissolved in 0·5 N HCl containing MCl.[39,40] (4) Reaction between $Pd(OAc)_2$ and LiCl or $CuCl_2$.[41]	The alkali metal salts are reddish-brown crystalline solids, soluble in water.	Potassium and ammonium salts are isostructural with the corresponding Pt salts.[42,43] Pd–Cl bond length = 2·30 Å.[43]
$[PdBr_4]^{2-}$	(1) Halogen exchange with $[PdCl_4]^{2-}$ [44] (2) Reduction of $[PdBr_6]^{2-}$ with calculated amount of oxalic acid.[39] (3) Dissolution of Pd sponge in 48% HBr containing a little HNO_3 and evaporation to dryness. Residue dissolved in 0·5 N HBr containing MBr.[39,40]	K_2PdBr_4 is a dark-red crystalline solid.	$[Pd(NH_3)_4][PdBr_4]$ shown to have the same tetragonal structure[45] as $[Pt(NH_3)_4][PtCl_4]$[46] confirming the square-planar arrangement of $[PdBr_4]^{2-}$.
$[PdI_4]^{2-}$	Add K_2PdCl_4 to an aqueous solution containing a tenfold excess of KI.[40]	Only stable in aqueous solution.	Not investigated.
$[Pd_2F_6]^{2-}$(?)	Interaction of an equimolar mixture of CsF and PdF_2 in refluxing SeF_4 gives $CsPdF_3$ on removal of solvent.[2]	Pinkish-brown solid.[2]	The structure is unknown. The compound, which has an empirical formula of $CsPdF_3$ is paramagnetic with a magnetic moment of 1·60 $B.M.^2$
$[Pd_2Cl_6]^{2-}$	(1) Addition of K_2PdCl_4 in water to $Ph_3MeAs^+Cl^-$ in water gives a precipitate of $(Ph_3MeAs)_2[Pd_2Cl_6]$.[47] (2) Treatment of $PdCl_2$ with MCl (M = Li[48] or Na[49]) in acetic acid gives $M_2[Pd_2Cl_6]$.	Pinkish-brown solid.[47] $Li_2[Pd_2Cl_6]$ reacts reversibly with excess LiCl in acetic acid to give Li_2PdCl_4. For the equilibrium $Li_2Pd_2Cl_6 + 2LiCl \xrightarrow{\text{HOAc}} 2LiPdCl_4$ $K = 0·1$ mol^{-1} litre at 25°C.[48]	Molecular weight[47] measurements and ultraviolet and visible spectra[39] are consistent with planar $[Pd_2Cl_6]^{2-}$ ions.

TABLE 53 cont.

Halide	Preparation	Properties	Structure
$[Pd_2Br_6]^{2-}$	Treatment of aqueous K_2PdBr_4 with $NR_4^+Br^-$ ($R = CH_3, C_2H_5$ etc.) gives $(NR_4)_2[Pd_2Br_6]$.[47]	Brown crystalline solid.[47]	Isostructural with the corresponding platinum(II) complex which contains planar $[Pt_2Br_6]^{2-}$ ions.[47]
$[Pd_2I_6]^{2-}$	Treat PdI_2 in aqueous alcohol with excess NaI containing $NEt_4^+I^-$ or $Me_3PhN^+I^-$.[47]	Black crystalline solid.[47]	Not determined, but assumed to be analogous to other $[M_2X_6]^{2-}$ salts where M = Pt or Pd and X = halogen.[47] The ultra-violet and visible spectra have been successfully interpreted on this assumption.[50]

TABLE 54

The preparation, properties and structures of the halide complexes of platinum(II)

Halide	Preparation	Properties	Structure
$[PtF_4]^{2-}$	No preparative details given.[35] No evidence for $[PtF_4]^{2-}$ in aqueous solution.[51]	Potassium salt is yellow.[35]	Not investigated.
$[PtCl_4]^{2-}$	(1) Dissolution of $PtCl_2$ in HCl to give H_2PtCl_4 followed by addition of alkali metal chloride or carbonate.[52] (2) Reduction of $[PtCl_6]^{2-}$ with H_2S or stoichiometric amount of hydrazine.[40,53]	Red, crystalline solids, soluble in water without decomposition.	Crystal structure of K_2PtCl_4[42,43] consists of alternate layers of potassium ions and square-planar $[PtCl_4]^{2-}$ units. Pt–Cl bond length = 2·33 Å,[42] or 2·29 Å.[43] Pt–Cl bond length in $[Pt(NH_3)_4][PtCl_4]$ = 2·34 Å.[46]
$[PtBr_4]^{2-}$	(1) Treat $[PtCl_4]^{2-}$ with NaBr in HBr solution.[44] (2) Reduction of $[PtBr_6]^{2-}$ with a stoichiometric amount of hydrazine sulphate[40,54] or an oxalate.[44]	K_2PtBr_4 is a brown crystalline solid.[54]	$[Pt(CH_3NH_2)_4][PtBr_4]$ has the same tetragonal structure[45] as $[Pt(NH_3)_4][PtCl_4]$ confirming the square-planar arrangement of $[PtBr_4]^{2-}$. K_2PtBr_4 is isostructural with K_2PtCl_4.[55]

TABLE 54 cont.

Halide	Preparation	Properties	Structure
$[PtI_4]^{2-}$	(1) Treatment of $PtCl_4$ with KI gives K_2PtI_4 and iodine.[56] (2) Addition of $Pt(OH)_2$ in conc. HCl to a hot conc. KI solution gives K_2PtI_4.[57] (3) Treatment of aqueous K_2PtCl_4 with a ten-fold excess of KI.[40]	K_2PtI_4 is a black crystalline solid[57] of limited stability.[58] It is difficult to isolate from solution, probably due to the greater insolubility of $K_2[Pt_2I_6]$.[59] In fact Pöe doubts whether $[PtI_4]^{2-}$ has ever been isolated.[59]	The Raman and infrared spectra are consistent with $[PtI_4]^{2-}$ having D_4 symmetry i.e. $[PtI_4]^{2-}$ is square-planar.[58]
$[Pt_2Cl_6]^{2-}$	Treatment of an aqueous solution of M_2PtCl_4 with $Ph_3MeAs^+Cl^-$ led to impure $(Ph_3MeAs)_2[Pt_2Cl_6]$.[47] It was originally reported to be impossible to separate this material from the $[PtCl_4]^{2-}$ contaminent due to the insolubility of both products,[47] but later reported that $(Ph_4As)_2[Pt_2Cl_6]$ could be recrystallised from dichloromethane.[60]	$(Ph_4As)_4[Pt_2Cl_6]$ is a pinkish-brown microcrystalline solid of limited solubility in water. $[Pt_2Cl_6]^{2-}$ is stable in water.[61]	The far infra-red of $(Ph_4As)_2[Pt_2Cl_6]$ is consistent with a planar structure for $[Pt_2Cl_6]^{2-}$, analogous to that of $[Pt_2Br_6]^{2-}$.
$[Pt_2Br_6]^{2-}$	Treat an aqueous solution of M_2PtBr_4 with $R_4N^+Br^-$ $(R = CH_3, C_2H_5$ etc.) to precipitate $(R_4N)_2[Pt_2Br_6]$.[47]	$(Et_4N)_2[Pt_2Br_6]$ is a yellowish-brown crystalline solid of limited solubility in water.[47] $[Pt_2Br_6]^{2-}$ is stable in water.[47]	Single crystal x-ray diffraction study[62] of $(NEt_4)_2[Pt_2Br_6]$ showed that $[Pt_2Br_6]^{2-}$ is planar with a slightly distorted square-planar co-ordination about each Pt atom (for further details see Appendix II, structure 116).
$[Pt_2I_6]^{2-}$	Treat PtI_2 in aqueous alcohol with excess NaI and $R_4N^+I^-$ $(R = CH_3, C_2H_5$ etc.) to precipitate $(R_4N)_2[Pt_2I_6]$.[47]	$(Et_4N)_2[Pt_2I_6]$ is a buff coloured crystalline solid of limited solubility in water.[47] $[Pt_2I_6]^{2-}$ is stable in water.[47]	No crystallographic data are available. The polarised crystal spectra can be satisfactorily interpreted by assuming a structure analogous to $[Pt_2Br_6]^{2-}$.[50]

iodides $[PdI_4]^{2-}$ and $[PtI_4]^{2-}$ [40,58] (see Tables 53 and 54) probably results from the high bridging ability of iodide ions which leads to the formation of $[Pd_2I_6]^{2-}$ and $[Pt_2I_6]^{2-}$.

The complexes in which platinum(II) and palladium(II) are bonded directly to chloride, bromide and iodide ligands are too numerous to describe in detail. However, fluoride complexes are very rare. The platinum(II) complexes $[(PPh_3)_2PtF_2]$ and $[\{P(OPh)_3\}_2PtF_2]$, formed by the action of hydrogen fluoride (liquid at $-78°C$ or gaseous at room temperature in benzene solution) on $[Pt(PPh_3)_4]$ and $[Pt\{P(OPh)_3\}_4]$ respectively,[63] may well be ionic[64] (e.g. $[(PPh_3)_3PtF]^+(HF_2^-)$). Both complexes appear to be co-ordinatively unsaturated and add carbon monoxide under pressure[63] (reaction 293). Although iodo-complexes are very common, tri-iodo com-

$$[(PR_3)_2PtF_2] + 2CO \xrightarrow[\text{(R = Ph or OPh)}]{\text{pressure}} [(PR_3)_2Pt(CO)_2F_2] \qquad (293)$$

plexes are rare and limited to $[(PPh_3)_2Pt(I_3)CH_3]$ prepared by reaction 294.[65]

$$[(PPh_3)_2Pt(o\text{-tolyl})_2] + MeI \rightarrow [(PPh_3)_2Pt(I_3)Me] \qquad (294)$$

The results of studies of the equilibria set up when halide complexes are dissolved in aqueous solution (see Tables 55 and 56) are incomplete partly

TABLE 55
Equilibrium constants for palladium(II)–halide complexes[a]

System	Ionic strength	Temp (°C)	$\log_{10} K$	Ref.	Remarks
$Pd^{++} + Cl \rightleftharpoons [PdCl]^+$	0 (corrected)	21	6·2	66 ⎫	$\Delta H = -8$
		29·5	6·0	66 ⎬	kcal/mol $\Delta S = 1$ cal/
		38	5·9	66 ⎭	deg/mol
	0·8	20	4·34	67	
	0 (corrected)	25	6·0	68	
	0·86	25	3·98	68	
	1·0	25	3·88	69	
	1·0	25	3·88	70	
	0·1	20	5·1	71	
	b	*b*	4·7	72	
$[PdCl]^+ + Cl^- \rightleftharpoons$ $[PdCl_2]$	0 (corrected)	21	4·7	66 ⎫	$\Delta H = -9$ kcal/
		29·5	4·5	66 ⎬	mol $\Delta S = -10$ cal/
		38	4·1	66 ⎭	deg/mol
	0·8	20	3·54	67	
	0 (corrected)	25	4·6	68	
	0·86	25	3·24	68	
	1·0	25	3·03	69	
	1·0	25	3·06	70	
	b	*b*	3·0	72	
$[PdCl_2] + Cl^- \rightleftharpoons$ $PdCl_3^-$	0 (corrected)	21	2·5	66 ⎫	$\Delta H = -8$ kcal/
		29·5	2·4	66 ⎬	mol $\Delta S = -14$ cal/
		38	2·2	66 ⎭	deg/mol
	0·8	20	2·68	67	
	0 (corrected)	25	2·5	68	
	0·86	25	2·3	68	
	1·0	25	2·18	69	
	1·0	25	2·14	70	
	b	*b*	2·6	72	

TABLE 55 contd.

System	Ionic strength	Temp (°C)	$\log_{10}$ K	Ref.	Remarks
$[PdCl_3]^- + Cl^- \rightleftharpoons$ $[PdCl_4]^{2-}$	0 (corrected)	21	2·6	66 ⎤	$\Delta H = -8$ kcal/ mol $\Delta S = -15$ cal/ deg/mol
		29·5	2·6	66 ⎬	
		38	2·5	66 ⎦	
	0·8	20	1·68	67	
	0 (corrected)	25	2·0	68	
	0·86	25	2·0	68	
	1·0	25	1·34	69	
	1·0	25	1·34	70	
	b	b	1·62	72	
$Pd^{++} + 4Cl^- \rightleftharpoons$ $[PdCl_4]^{2-}$	b	b	12·2	73	
	3·4	b	11·4	74	
	b	b	11·9	72	
$[PdCl_4]^{2-} + OH^- \rightleftharpoons$ $[Pd(OH)Cl_3]^{2-} + Cl^-$	1·0	19–50	5·7	75	stated to be 'independent of temperature'
$Pd^{++} + Br^- \rightleftharpoons [PdBr]^+$	0·1	20	6·8	71	
	0·8	20	4·37	76	
$[PdBr]^+ + Br^- \rightleftharpoons$ $[PdBr_2]$	0·8	20	4·08	76	calculated
$[PdBr_2] + Br^- \rightleftharpoons$ $[PdBr_3]^-$	0·8	20	3·79	76	calculated
$[PdBr_3]^- + Br^- \rightleftharpoons$ $[PdBr_4]^{2-}$	0·8	20	3·50	76	
$Pd^{++} + 4Br^- \rightleftharpoons$ $[PdBr_4]^{2-}$	0·8	20	15·7	76	
	b	b	16·1	73	
	0 (corrected)	25	13·1	77	
$Pd^{++} + I^- \rightleftharpoons [PdI]^+$	0·1	20	10·0	71	
$Pd^{++} + 4I^- \rightleftharpoons [PdI_4]^{2-}$	b	b	24·9	73	
$[PdCl_4]^{2-} + Br^- \rightleftharpoons$ $[PdCl_3Br]^{2-} + Cl^-$	4·5	25	1·55	78	
$[PdCl_3Br]^{2-} + Br^- \rightleftharpoons$ $[PdCl_2Br_2]^{2-} + Cl^-$	4·5	25	1·09	78	
$[PdCl_2Br_2]^{2-} + Br^- \rightleftharpoons$ $[PdClBr_3]^{2-} + Cl^-$	4·5	25	0·95	78	
$[PdClBr_3]^{2-} + Br \rightleftharpoons$ $[PdBr_4]^{2-} + Cl^-$	4·5	25	0·55	78	
$[PdBr_4]^{2-} + I^- \rightleftharpoons$ $[PdBr_3I]^{2-} + Br^-$	4·5	25	2·75	79	
$[PdBr_3I]^{2-} + I^- \rightleftharpoons$ $[PdBr_2I_2]^{2-} + Br^-$	4·5	25	3·00	79	
$[PdBr_2I_2]^{2-} + I^- \rightleftharpoons$ $[PdBrI_3]^{2-} + Br^-$	4·5	25	1·70	79	
$[PdBrI_3]^{2-} + I^- \rightleftharpoons$ $[PdI_4]^{2-} + Br^-$	4·5	25	0·80	79	

[a] All complexes are four-co-ordinate. Where less than four ligands are shown the remainder are water ligands.

[b] Not stated.

TABLE 56
Equilibrium constants for platinum(II)–halide complexes[a]

System	Ionic strength	Temp (°C)	$\log_{10} K$	Ref.	Remarks
$[PtCl_2] + Cl^- \rightleftharpoons$	0·32	25	3·27	80	
$[PtCl_3]^-$	0 (corrected)	25	3·0	80	
	0·1	35	5·52	81 ⎫	$\Delta H = -31·2$
	0·1	45	4·06	81 ⎬	kcal/mol
	0·1	55	3·13	81 ⎭	
$[PtCl_3]^- + Cl \rightleftharpoons$	0·32	25	1·82	80	
$[PtCl_4]^{2-}$	0 (corrected)	25	1·52	80	
	0·318	15	1·89	82 ⎫	$\Delta H = -5·10$
	0·318	25	1·74	82 ⎬	kcal/mol
	0·318	30	1·68	82 ⎭	
	0·1	17–18	1·72	83	
	0·1	25	5·95	81 ⎫	
	0·1	35	5·24	81 ⎪	$\Delta H = -5·35$
	0·1	45	4·77	81 ⎬	kcal/mol
	0·1	55	4·46	81 ⎭	
$Pt^{2+} + 4Cl^- \rightleftharpoons$	b	b	16·6	84	
$[PtCl_4]^{2-}$	0 (corrected)	25	16	85	
$[Pt(OH)_4]^{2-} + Cl^- \rightleftharpoons$	1	25	10·48	86	
$[Pt(OH)_3Cl]^{2-} + OH^-$					
$[Pt(OH)_3Cl]^{2-} + Cl^- \rightleftharpoons$	1	25	10·00	86	
$[Pt(OH)_2Cl_2]^{2-} + OH^-$					
$[Pt(OH)_2Cl_2]^{2-} + Cl^- \rightleftharpoons$	1	25	9·52	86	
$[Pt(OH)Cl_3]^{2-} + OH^-$					
$[Pt(OH)Cl_3]^{2-} + Cl^- \rightleftharpoons$	1	25	8·66	86	
$[PtCl_4]^{2-} + OH^-$					
$[PtBr_3]^- + Br^- \rightleftharpoons$	0·1	17–18	2·56	83, 37	
$[PtBr_4]^{2-}$	0·1	25	2·52	87	
$Pt^{2+} + 4Br^- \rightleftharpoons$	0 (corrected)	25	20·5	85	
$[PtBr_4]^{2-}$	b	b	20·4	84	
$[PtI_2] + I^- \rightleftharpoons [PtI_3]^-$	10^{-3}	25	2·3	59	
$[PtI_3]^- + I^- \rightleftharpoons [PtI_4]^{2-}$	1·0	25	1·7	59	
$Pt^{2+} + 4I^- \rightleftharpoons [PtI_4]^{2-}$	b	b	29·6	84	
$[PtCl_4]^{2-} + Br^- \rightleftharpoons$	0·318	25	1·16	88	
$[PtCl_3Br]^{2-} + Cl^-$					

[a] All complexes are four-co-ordinate. Where less than four ligands are shown, the remainder are water ligands.
[b] Not stated.

due to the fact that the number of complexes present in solution is large and partly, in the case of platinum(II) complexes, due to the fact that the equilibria are only established slowly. However, the results do support the hypothesis that both platinum(II) and palladium(II) are class 'b' or 'soft' acceptors which exhibit a preference for halide ions of $I^- > Br^- > Cl^-$. The great preference of palladium(II) to be four-co-ordinate is well illustrated by the high degree of dissociation of $[PdCl_5]^{3-}$ and $[PdCl_6]^{4-}$, which can only be detected in very concentrated chloride solutions.[89,90] Traces of $[PdBr_6]^{4-}$

have been detected in strong bromide solutions of palladium(II).[91] The equilibrium constants of a number of ligands (L) with $[Pt(C_2H_4)Cl_3]^-$ (equilibrium 295) in aqueous solution[92,93] (Table 57) show a stability constant order of L = H_2O < F^- < Cl^- < Br^- < I^- < SCN^- < Me_3N < NH_3 < Me_2NH < $MeNH_2$. This is a particularly favourable system to

$$[Pt(C_2H_4)Cl_3]^- + L \rightleftharpoons \textit{trans-}[Pt(C_2H_4)LCl_2] + Cl^- \qquad (295)$$

study because equilibrium is reached rapidly due to the high *trans*-effect of the olefin.

TABLE 57

Stability constants for the system $[Pt(C_2H_4)Cl_3]^-$ + L $\rightleftharpoons$ *trans*-$[Pt(C_2H_4)LCl_2]$ + Cl^- in 0·2 M aqueous perchloric acid at 25°C

$K = [trans-Pt(C_2H_4)LCl_2][Cl^-]/[Pt(C_2H_4)Cl_3^-][L]$

L	K	Reference
H_2O	3×10^{-3}	93
F^-	<0·03	93
Cl^-	1	93
Br^-	3·4	93
I^-	120	93
SCN^-	>120	93
NH_3	2×10^5	92
$MeNH_2$	$1·25 \times 10^6$	92
Me_2NH	$3·16 \times 10^5$	92
Me_3N	*ca.* 10^3	92

The metal(II)–halogen bond involves a σ-bond formed by overlap of a filled halogen orbital with an empty metal orbital. Whether or not there is any π-bonding in either platinum(II)– or palladium(II)–halogen bonds is at present uncertain, but it would appear that all the properties of these bonds can be explained without π-bonding, although this does not necessarily exclude its presence. The covalent character of the metal(II)–halogen bond increases with increasing halogen atomic weight corresponding to the decrease in the electronegativity of the halogen, as shown by the relative positions of the halide ions in the nephelauxetic series ($F^- \ll Br^- < I^{-[94]}$), which reflects the ability of a ligand by forming a covalent metal–ligand bond to decrease the effective nuclear charge of the metal and so allow the d-orbitals to expand.

All the halides except fluoride are good bridging ligands. Their ability to act as bridging ligands relative to their ability to act as terminal ligands increases with increasing atomic number (see Table 58), so that iodide is as effective as a bridging ligand as it is as a terminal ligand, whereas chloride is a better terminal ligand. A study of equilibrium 296 showed that for a given amine this equilibrium lies increasingly to the left as the atomic number of the

$$[(PPr_3^n)_2Pt_2X_4] + 2\text{ amine} \xrightleftharpoons{\text{acetone}} 2 \textit{ trans-}[(PPr_3^n)Pt(amine)X_2] \quad (296)$$

TABLE 58
Force constants (mdyn/Å) for $[M_2X_6]^{2-}$ anions (data taken from ref. 60)

Anion	k($M-X_{terminal}$)	k($M-X_{bridge}$)	$k_{terminal}/k_{bridge}$
$[Pt_2Cl_6]^{2-}$	2·02	1·42	0·70
$[Pt_2Br_6]^{2-}$	1·62	1·56	0·96
$[Pt_2I_6]^{2-}$	1·29	1·29	1·00
$[Pd_2Cl_6]^{2-}$	1·59	1·36	0·86
$[Pd_2Br_6]^{2-}$	1·37	1·30	0·95

halide increases indicating that the heavier halogens show a greater tendency to act as bridging ligands.[95]

VIBRATIONAL SPECTROSCOPY

The infra-red spectra of platinum(II) and palladium(II) halogen complexes have been widely investigated and the stretching frequencies assigned. These assignments may be treated with a fair degree of confidence because of the very high intensities of the absorptions and the ease with which the bands can be identified by replacing one halogen by another. However, as will be seen in what follows, the lack of correlation between the metal–halogen stretching frequencies and the *trans*-influence of the ligand *trans* to the halogen provides strong evidence that bands assigned to metal–halogen stretching vibrations do in fact involve a considerable amount of other vibrational modes and that the degree of mixing in these other modes is dependent on the nature of the other ligands present. Hence, the statement that 'a given band is due to a metal halogen stretching vibration' is necessarily only an approximate statement of the truth. Relatively few Raman studies have been reported because of the coloured nature of the complexes. However, the increasing use of laser Raman spectrometers is yielding this data.

$[MX_4]^{2-}$
The infra-red[96–104] and Raman[58,105] spectra of the square-planar $[MX_4]^{2-}$ ions are summarised in Table 59. The metal–halogen stretching force constants (Table 59), obtained using a modified generalised force field in which only five force constants were considered to be appreciable, should be regarded as relative rather than absolute values. They decrease in the order M–Cl > M–Br > M–I and Pt–X > Pd–X. Although there is no general relationship between force constant and bond strength (see pp. 207–208), because the plot of potential energy against interatomic distance is not the same for all bonds, these force constants do parallel the relative bond strengths anticipated both from chemical properties and from the heats of formation of both the crystalline metal halides and the aqueous $[MCl_4]^{2-}$ anions.[106]

$[MLCl_3]^-$
The infra-red spectra of a series of anionic $[MLCl_3]^-$ complexes[107] (Table 60) indicate that the ligand L has relatively little effect on the stretching

TABLE 59
Metal–halogen stretching frequencies in $[MX_4]^{2-}$ anions (data from ref. 58)

	$v(M-X)_{sym.} (cm^{-1})$ (Raman)	$v(M-X)_{asym.} (cm^{-1})$ (Infra-red)	M–X stretching force constant (mdyn/Å)
$K_2[PtCl_4]$	333	321	1·99[a]
$K_2[PtBr_4]$	232	205	1·71
$K_2[PtI_4]$	180	142	~0·9
$K_2[PdCl_4]$	336	310	1·44
$K_2[PdBr_4]$	260	192	1·13

[a] An alternative calculation giving a Pt–Cl stretching force constant of 1·87 mdyn/Å has been described in detail by L. H. Jones, *Coord. Chem. Rev.*, **1**, (1966) 351.

frequencies of the *cis*-chloride ligands. However, it has a profound effect on the stretching frequency of the *trans*-chloride ligand in agreement with the concept of *trans*-influence (pp. 301–303). The ligands L are listed in Table 60 in order of decreasing frequency of the *trans*-M–Cl bond and it can be seen that the two series for platinum(II) and palladium(II) whilst broadly similar, differ somewhat in detail. The discrepancies are probably due to the

TABLE 60
Infra-red spectra of $[MLCl_3]^-$ (data taken from ref. 107 except where stated)

M	L	$v(M-Cl)$ trans to L (cm^{-1})	$v(M-Cl_2)$(Cl cis to L) asymmetric (cm^{-1})	symmetric (cm^{-1})
Pd	CO	331	352	300
	C_2H_4	317	337	288
	SEt_2	311	339	288
	SMe_2	307	344	293
	$AsEt_3$	272	338	296
	PPh_3	271	348	298
	PMe_3	265	343	295
	PEt_3	265	340	293
	PPr_3^n	262	337	293
	$AsMe_3$	253	338	294
Pt	CO	322	343	343
	$Me_2C(OMe)C\equiv C-C(OMe)Me_2$[a]	314	325	335
	SMe_2	310	325	344
	C_2H_4	309	330	330
	SEt_2	307	345(?)	307
	NO_2[b]	298	322	298
	$AsEt_3$	280	350	280
	PPh_3	279	333	279
	PMe_3	275	332	332
	$AsMe_3$	272	329	329
	PEt_3	271	330	330
	PPr_3^n	270	329	329

[a] Reference 108.   [b] Reference 109.

interaction of the M–Cl stretching mode with other vibrations within the molecule. This is part of the basis for rejecting infra-red spectroscopy as a technique for determining the *trans*-influence of ligands (see p. 303).

cis- and trans-[MX₂L₂]

cis- and *trans-*$[MX_2L_2]$
The metal–halogen stretching frequency in *trans*-$[MX_2L_2]$ is almost insensitive to the nature of the ligand L (Table 61); typical values are: Pd–Cl =

TABLE 61

Infra-red active metal–halogen stretching frequencies in *trans*-$[MX_2L_2]$ complexes in the solid state

X	L	$v(Pd-X)(cm^{-1})$	Reference	$v(Pt-X)(cm^{-1})$	Reference
Cl	NH₃	332	110	331·5	111
	pyr	356	110	342·6	111
	SMe₂	359	110, 112		
	SEt₂	358	110, 112	341·9	111
	SeEt₂			337·2	111
	AsEt₃	353	110	338·8	111
	AsPh₃	360	113		
	PMe₃			$\{$326	110
				339	115
	PEt₃	355	110	340·3	111
	PPr₃ⁿ	353	110	338·8	111
	PBu₃ⁿ	355	110		
	PPh₃	357	110		
	(CH₃)₂N₂O	362	114		
Br	NH₃			227·0	111
	pyr			251·2	111
	SMe₂	260	112	215	116
	SEt₂	270	112	253·5	111
	SeEt₂			240·7	111
	AsMe₃	279	115		
	AsEt₃			250·9	111
	PEt₃			254·4	111
	PPr₃			259·0	111
	PPh₃	286	113		
	AsPh₃	282	113		
I	SMe₂	202	112		
	SEt₂	188	112		
	SPr₂ⁱ	178	112		

$357 \pm 5\,cm^{-1}$; Pt–Cl $= 340 \pm 3\,cm^{-1}$; Pd–Br $= 275 \pm 15\,cm^{-1}$; Pt–Br $= 244 \pm 20\,cm^{-1}$; Pd–I $= 190 \pm 12\,cm^{-1}$. The small *cis*-influence of the ligands L on the metal–halogen stretching frequency is consistent with the relatively small kinetic *cis*-effect of these ligands (pp. 303–305). On the other hand, the metal–halogen stretching frequencies in *cis*-$[MX_2L_2]$ are very sensitive to the nature of the ligand L (Tables 62 and 63), generally decreasing with increasing *trans*-influence of L (pp. 301–303). There are discrepancies between the *trans*-influence series on p. 302 and those deduced from Tables 62 and 63, which arise from coupling between the metal–halogen and other vibrational modes within the molecule.

TABLE 62
Infra-red active palladium–chlorine stretching frequencies in
cis-$[PdCl_2L_2]$ in the solid state

L	$v(Pd–X)(cm^{-1})$		Reference
NH_3	327	306	117
$MeNH_2$	317	308	117
$EtNH_2$	317	296	117
$H_2NCH_2CH_2NH_2$	$\begin{cases}307\\306\end{cases}$	296 / 272	117 / 110
$Me_2NCH_2CH_2NH_2$	331	321	110
$Me_2NCH_2CH_2NHMe$	331	321	110
$Me_2NCH_2CH_2NMe_2$	330	321·5	110
$Ph_2NCH_2CH_2NPh_2$	317	306	117
pyr	342	333	117
bipyr	354	342	117
1,10-phen	353	346	117
$MeSCH_2CH_2SMe$	314·5	296	110
$EtSCH_2CH_2SEt$	322·6	309·6	110
1,5-COD	335, 325	296	110
$AsMe_3$	316	298·5	110
PMe_3	284	271	115
$Ph_2PCH_2CH_2PPh_2$	310	286	110

TABLE 63
Infra-red active platinum–halogen stretching frequencies in cis-$[PtX_2L^1L^2]$ complexes

L^1	L^2	$v(Pt–Cl)(cm^{-1})$		$v(Pt–Br)(cm^{-1})$		Reference
$TeEt_2$	As L^1	302·2	282·2	217·0	208·2	111
PMe_3	As L^1	303	277			115
PPh_3	As L^1	303	280			111
PEt_3	As L^1	303·2	280·8	211·6	193·7	118
$SbPr_3$	As L^1	311·0	280·6	217·4	—	111
$AsMe_3$	As L^1	313·0	294·0	211·0	—	115
$AsEt_3$	As L^1	314·2	287·5			111
PMe_3	PPh_2Cl	316	290			118
PMe_3	$AsPh_2Cl$	325	296			118
NH_3	As L^1	330·0	323·0	224·3	220·5	111, 119
SEt_2	As L^1	330·0	318·2	254·0	226·2	111
$SeEt_2$	As L^1	333·0	316·6			111
Cyclo-octa-1,5-diene	As L^1	338·0	316·2	229·5	215·3	111
pyr	As L^1	343·0	327·9	219·4	210·6	111

MX_2

$PdCl_2$, $PtCl_2$ and $PtBr_2$ all have halogen-bridged structures in the solid
state. $PdCl_2$ consists of infinite chains of square-planar $PdCl_4$ units[16] and
the stretching vibration of the Pd—Cl bridges of the chain is observed at
340 cm^{-1}.[120] $PtCl_2$ consists of Pt_6Cl_{12} units involving bridging chlorine
atoms,[28,29] and the stretching vibration of the Pt–Cl bridges is observed
at 318 cm^{-1}.[120] The structure of $PtBr_2$ is unknown, but it absorbs at
230 cm^{-1}[120] which would be consistent with the presence of Pt–Br bridges.

$[MX_2L]_2$

Three generalisations can be made about the infra-red spectra of halogen-bridged $[MX_2L]_2$ complexes (Table 64): (i) The M–X$_{terminal}$ stretching frequency is largely independent of the nature of the ligand L, which is of course *cis* to the terminal halogen. (ii) The M–X (bridging, *trans* to X) stretching frequency is also largely independent of the nature of the ligand L. (iii) The M–X (bridging, *trans* to L) stretching frequency is strongly dependent on the nature of L but this dependence is different for the palladium(II) and platinum(II) series, reflecting the differing importance of mixing in other vibrational modes in the two series.

TABLE 64

Stretching frequencies observed in $[MX_2L]_2$ complexes (in cm^{-1})

X	L	$v(Pd-X)$ (terminal)	$v(Pd-X)$ (bridging trans to X)	$v(Pd-X)$ (bridging trans to L)	$v(Pt-X)$ (terminal)	$v(Pt-X)$ (bridging trans to X)	$v(Pt-X)$ (bridging trans to L)	Ref.
Cl	PMe$_3$	346	294	256	347	330	260	121
	PEt$_3$	356	301	249	351	327	265	121
	PPr$_3^n$	356	299	253	356	323	257	121
	AsMe$_3$	339	303	241	351	323	257	121
	AsEt$_3$	352	300	261	350	322	261	121
	SMe$_2$	360	308	283	—	—	—	121
	C$_2$H$_4$	355	302	271	359	317	287	121
	CO	—	—	—	368	331	301	121
Cl	Cl^a	340	302	—	346	315	—	60
Br	Br^b	266	192	—	237	196	—	60
I	I^b	222	144	—	179	157	—	60

^{a} For Pd used Bu$_4^n$N$^+$ salt; for Pt used Ph$_4$As$^+$ salt.
^{b} Used Et$_4$N$^+$ salt.

ELECTRONIC SPECTROSCOPY

Much effort has been expended in studying the electronic spectra of the square-planar $[MX_4]^{2-}$ complexes[122–45] (Table 65). At first there appeared to be as many assignments as there were workers in the field, but recently more agreement has been obtained and results from polarised crystal spectroscopy,[126,128–30,132,133,137] Faraday effect measurements[132] and luminescence spectroscopy[145] have all supported the assignment given in Table 65. It would be beyond the scope of this book to include a detailed discussion of the arguments that lie behind the assignment of these spectra, but the interested reader is referred to the references cited, especially that of Day and his co-workers.[126]

A detailed investigation of the electronic spectra of the dimeric $[M_2Br_6]^{2-}$ and $[M_2I_6]^{2-}$ anions, where M = Pd or Pt, has shown that they can be understood by direct extension of the assignment of the $[MX_4]^{2-}$ spectra.[123] There is no evidence for any electronic interaction between the two metal atoms in the dimeric anion, so that the *d–d* transitions in $[M_2X_6]^{2-}$ anions

TABLE 65

Solid and solution spectra of K_2MX_4 (band energies are in cm^{-1}; extinction coefficients are molar)

K₂PtCl₄ Solution[a] λ	ε	Reflectance[b] λ	Crystal[b] λ	ε	K₂PtBr₄ Solution[c] λ	ε	K₂PdCl₄ Solution[d] λ	ε	Reflectance[c] λ	Crystal[b] λ	ε	K₂PdBr₄ Solution[d] λ	ε	Reflectance[d] λ	Assignment[e]
17,700	3	17,500	17,300	5, xy											$^1A_{1g} \rightarrow {}^3A_{2g}$ Singlet → Triplet
21,000	15	20,400	20,300	17, xy; 20, z					16,700	17,500	19, xy; 7, z				$^1A_{1g} \rightarrow {}^3E_g$ Singlet → Triplet
			23,800	25, xy; 15, z	16,700	5									$^1A_{1g} \rightarrow {}^3B_{1g}$ Singlet → Triplet
25,500	59	27,000	26,000	45, xy	19,700	15	22,100	200	21,500	20,000	67, xy			16,000	$^1A_{1g} \rightarrow {}^1A_{2g}$ $d_{xy} \rightarrow d_{x^2-y^2}$
30,200	64	29,000	28,800	57, xy	24,300	100	25,000	250	23,300	22,800	125, xy; 80, z			20,000	$^1A_{1g} \rightarrow {}^1E_g$ $d_{xz}, d_{yz} \rightarrow d_{x^2-y^2}$
37,900	250	36,500			28,200	120	30,200	1,200	31,500	29,500	67, xy			26,000	$^1A_{1g} \rightarrow {}^1B_{1g}$ $d_{z^2} \rightarrow d_{x^2-y^2}$
46,000	9580	42,500			37,300	7,000	36,000	12,000	36,500			30,100	10,400		$^1A_{1g} \rightarrow {}^1A_{2u}$ $Cl_{pz} \rightarrow d_{x^2-y^2}$
							45,190	28,200[e]				40,500	30,400		$^1A_{1g} \rightarrow {}^1E_u$ $Cl_{p\sigma} \rightarrow d_{x^2-y^2}$

[a] Reference 125.　[b] Reference 126.　[c] Reference 122.　[d] Reference 39.　[e] Reference 123.　[f] Reference 124.

are analogous to those in $[MX_4]^{2-}$ anions. In the charge transfer region $[M_2X_6]^{2-}$ species show twice as many transitions as $[MX_4]^{2-}$ due to the different energies of the p_σ and p_π orbitals in the terminal and bridging halide ligands.

NUCLEAR QUADRUPOLE RESONANCE SPECTROSCOPY

Although nuclear quadrupole resonance studies of metal(II)–halide complexes might be expected to give information about the covalent character of the metal–halogen bond, they have not been very enlightening largely because of the number of assumptions that must be made before quantitative data can be obtained. However, the the n.q.r. spectra of K_2PtCl_4,[146] K_2PtBr_4,[147] $(Et_4N)_2[PdCl_4]$[148] and K_2PdBr_4[147] have been recorded and the covalent character of the metal(II)–halogen bonds calculated to be 39% (Pt–Cl), 43% (Pt–Br), 40% (Pd–Cl) and 40% (Pd–Br). In these calculations it was assumed that the halide bonding orbitals had 15% s-character, 85% p-character and 0% d-character and that there was no π-bonding in the metal–halogen bonds. Although the figures should not be considered as absolute their relative orders of magnitude are almost certainly significant, with the probable exception of $(Et_4N)_2[PdCl_4]$ where the change of cation may give rise to solid state effects.

N.q.r. resonance frequencies have also been obtained for α- and β-$PdCl_2$.[149] The magnitudes of the coupling constants indicate that the two forms have nearly identical electron distributions, the differences being due to differences in the Pd–Cl–Pd bond angle (93° in α-$PdCl_2$ and 90° in β-$PdCl_2$). Comparison of data obtained for β-$PdCl_2$ and β-$PtCl_2$ with that for $[PdCl_4]^{2-}$ and $[PtCl_4]^{2-}$ showed that the charges on the chlorine atoms of the dichlorides were considerably smaller than on the chlorine atoms of the tetrachloro complexes where the metal has to delocalise the charges of four chlorides (β-$PdCl_2$, Cl = −0·26, Pd = +0·52;[149] $[PdCl_4]^{2-}$, Cl = −0·67, Pd = +0·72;[148] β-$PtCl_2$, Cl = −0·16, Pt = +0·29;[149] $[PtCl_4]^{2-}$, Cl = −0·67, Pt = +0·68[146]). In the three cases currently available (MCl_2, $[MCl_4]^{2-}$ and $[MBr_4]^{2-}$) there is a difference in the charge on the palladium and platinum atoms (charge on Pd minus charge on Pt) that is positive and which indicates that the valence-state electronegativity of platinum(II) is about 0·1 units greater than that of palladium(II).[149]

TABLE 66

[195]Platinum n.m.r. chemical shifts in aqueous solution at 28°C relative to H_2PtCl_6 (data from ref. 150)

Compound	δ (p.p.m.)
Na_2PtCl_4	1650
Na_2PtBr_4	2690
$Na_2Pt(CN)_4$	4770

NUCLEAR MAGNETIC RESONANCE SPECTROSCOPY

Although ^{195}platinum nuclear magnetic resonance might be expected to give information about the nature of the platinum–halogen bond, the results (Table 66) are too few at present to do more than state that the order of ^{195}platinum chemical shifts is Cl < Br < CN (cf. p. 60).

REFERENCES

1   N. Bartlett and M. A. Hepworth, *Chem. Ind. (London)*, (1956) 1425.
2   N. Bartlett and J. W. Quail, *J. Chem. Soc.* (1961) 3728.
3   W. C. Smith, *U.S. Patent*, 2,952,514 (1960); *Chem. Abs.*, **55** (1961) 3939c.
4   N. Bartlett and P. R. Rao, *Proc. Chem. Soc.*, (1964) 393.
5   A. G. Sharpe, *J. Chem. Soc.*, (1950) 3444.
6   N. Bartlett and R. Maitland, *Acta Cryst.*, **11** (1958) 747.
7   F. Ebert, *Z. anorg. allgem. Chem.*, **196** (1931) 395; *Chem. Abs.*, **25** (1931) 3892.
8   F. Puche, *Ann. Chim. (Paris)*, **9** (1938) 233; *Chem. Abs.*, **32** (1938) 5322.
9   N. V. Sidgwick, *The Chemical Elements and Their Compounds*, Oxford University Press, Oxford, 1950, vol. II, p. 1560.
10 V. V. Ipatiev and V. G. Tronev, *J. Gen. Chem. Russ.*, **5** (1935) 643; *Chem. Abs.*, **29** (1935) 6819.
11 S. C. Ogburn and W. C. Brastow, *J. Amer. Chem. Soc.*, **55** (1933) 1307.
12 J. Krustinsons, *Z. Elektrochem.*, **44** (1938) 537; *Chem. Abs.*, **32** (1938) 8891.
13 F. Puche, *Compt. Rend.*, **200** (1935) 1206.
14 W. E. Bell, U. Merten and M. Tagam, *J. Phys. Chem.*, **65** (1961) 510.
15 J. R. Soulen and W. H. Chappell, *J. Phys. Chem.*, **69** (1965) 3669.
16 A. F. Wells, *Z. Krist.*, **100** (1938) 189.
17 H. Schäfer, U. Wiese, K. Rinke and K. Brendel, *Angew. Chem. Int. Ed.*, **6** (1967) 253.
18 J. H. Canterford and R. Colton, *Halides of the Second and Third Row Transition Metals*, John Wiley, New York, 1968, p. 368.
19 S. A. Shchukarev, T. A. Tolmacheva and Yu. L. Pazukhina, *Russ. J. Inorg. Chem.*, **9** (1964) 1354.
20 E. Cook and B. Siegel, *J. Inorg. Nucl. Chem.*, **29** (1967) 2739.
21 N. Bartlett and D. H. Lohmann, *J. Chem. Soc.*, (1964) 619.
22 L. Wöhler and S. Streicher, *Chem. Ber.*, **46** (1913) 1591; *Chem. Abs.*, **7** (1913) 2727.
23 J. M. Lutton and R. W. Parry, *J. Amer. Chem. Soc.*, **76** (1954) 4271.
24 V. P. Kazakov and B. I. Peshchevitskii, *Radiokhimiya*, **4** (1962) 509; *Chem. Abs.*, **59** (1963) 5781f.
25 S. A. Shchukarev, M. A. Oranskaya and T. S. Shemyakina, *Zhur. Neorg. Khim.*, **1** (1956) 17; *Chem. Abs.*, **50** (1956) 9833i.
26 W. E. Cooley and D. H. Busch, *Inorg. Synth.*, **5** (1957) 208.
27 J. Krustinsons, *Z. Elektrochem.*, **45** (1939) 83; *Chem. Abs.*, **33** (1939) 3238.
28 K. Brodersen, G. Thiele and H. G. Schnering, *Z. anorg. allgem. Chem.*, **337** (1965) 120.
29 K. Brodersen, *Angew. Chem. Int. Ed.*, **3** (1964) 519.
30 U. Wiese, H. Schäfer, H. G. V. Schnering, C. Brendel and K. Rinke, *Angew. Chem. Int. Ed.*, **9** (1970) 158.
31 L. Wöhler and F. Müller, *Z. anorg. allgem. Chem.*, **149** (1925) 377; *Chem. Abs.*, **20** (1926) 718.
32 Reference 9, p. 1582.
33 G. R. Argue and J. J. Banewicz, *J. Inorg. Nucl. Chem.*, **25** (1963) 923.
34 S. A. Shchukarev, T. A. Tolmacheva and G. M. Slavutskaya, *Russ. J. Inorg. Chem.*, **9** (1964) 1351.
35 R. Hoppe, Personal communication to R. D. Peacock, *Prog. Inorg. Chem.*, **2** (1960) 193.
36 E. Lodewijk and D. Wright, *J. Chem. Soc. (A)*, (1968) 119.
37 F. S. Clements and E. V. Nutt, *Brit. Patent*, 879,074 (1961); *Chem. Abs.*, **56** (1962) 8298c.

38 Kh. I. Gil'dengershel and G. A. Shagisultanova, *Zhur. Priklad. Khim.*, **26** (1953) 222; *Chem. Abs.*, **47** (1953) 11061b.
39 C. M. Harris, S. E. Livingstone and I. H. Reece, *J. Chem. Soc.*, (1959) 1505.
40 S. E. Livingstone, *Synthesis in Inorganic and Metal-Organic Chemistry*, **1** (1971) 1.
41 M. Tamura and T. Yasui, *J. Chem. Soc. Japan, Ind. Chem. Sect.*, **71** (1968) 1855 (English summary, p. A116); *Chem. Abs.*, **70** (1969) 61532.
42 R. G. Dickinson, *J. Amer. Chem. Soc.*, **44** (1922) 2404.
43 W. Theilacker, *Z. anorg. allgem. Chem.*, **234** (1937) 161; *Chem. Abs.*, **32** (1938) 1204.
44 E. Biilmann and A. C. Andersen, *Chem. Ber.*, **36** (1903) 1565.
45 J. R. Miller, *Proc. Chem. Soc.*, (1960) 318.
46 M. Atoji, J. W. Richardson and R. E. Rundle, *J. Amer. Chem. Soc.*, **79** (1957) 3017.
47 C. M. Harris, S. E. Livingstone and N. C. Stephenson, *J. Chem. Soc.*, (1958) 3697.
48 P. M. Henry and O. W. Marks, *Inorg. Chem.*, **10** (1971) 373.
49 N. R. Davies, *Aust. J. Chem.*, **17** (1964) 212.
50 P. Day, M. J. Smith and R. J. P. Williams, *J. Chem. Soc. (A)*, (1968) 668.
51 T. P. Perros, W. F. Sager and D. E. Icenhower, *J. Amer. Chem. Soc.*, **79** (1957) 1301.
52 A. J. Cohen, *Inorg. Synth.*, **6** (1960) 209.
53 N. G. Klyuchnikov and R. N. Savel'eva, *Zhur. Neorg. Khim.*, **1** (1956) 2764; *Chem. Abs.*, **51** (1957) 10288b.
54 G. A. Shagisultanova, *Russ. J. Inorg. Chem.*, **6** (1961) 904.
55 G. B. Bokii and G. A. Kukina, *Kristallografiya, SSSR*, **2** (1957) 400; *Structure Reports*, **21** (1957) 417.
56 Reference 9, p. 1605.
57 N. M. Nikolaeva and E. D. Pastukhova, *Russ. J. Inorg. Chem.*, **12** (1967) 794.
58 P. J. Hendra, *J. Chem. Soc. (A)*, (1967) 1298.
59 B. Corain and A. J. Poë, *J. Chem. Soc. (A)*, (1967) 1318.
60 D. M. Adams, P. J. Chandler and R. G. Churchill, *J. Chem. Soc. (A)*, (1967) 1272.
61 J. Chatt, *J. Chem. Soc.*, (1950) 2301.
62 N. C. Stephenson, *Acta Cryst.*, **17** (1964) 587.
63 J. McAvoy, K. C. Moss and D. W. A. Sharp, *J. Chem. Soc.*; (1965) 1376.
64 R. D. W. Kemmitt, R. D. Peacock and J. Stocks, *J. Chem. Soc. (A)*, (1971) 846.
65 C. R. Kistner, D. A. Drew, J. R. Doyle and G. W. Rausch, *Inorg. Chem.*, **6** (1967) 2036.
66 H. A. Droll, B. P. Block and W. C. Fernelius, *J. Phys. Chem.*, **61** (1957) 1000.
67 S. A. Shchukarev, O. A. Lobaneva, M. A. Ivanova and M. A. Kononova, *Vestnik Leningrad Gos. Univ.*, **16** (1961) 152; *Chem. Abs.*, **55** (1961) 24362c.
68 A. A. Biryukov and V. A. Shlenskaya, *Russ. J. Inorg. Chem.*, **9** (1964) 450.
69 K. Burger and D. Dyrssen, *Acta Chem. Scand.*, **17** (1963) 1489.
70 K. Burger, *Magy. Kem. Folyoirat*, **70** (1964) 179; *Chem. Abs.*, **61** (1964) 1316a.
71 A. A. Grinberg, M. I. Gel'fman and N. V. Kiseleva, *Russ. J. Inorg. Chem.*, **12** (1967) 620.
72 M. I. Gel'fman and N. V. Kiseleva, *Russ. J. Inorg. Chem.*, **14** (1969) 258.
73 A. A. Grinberg, N. V. Kiseleva and M. I. Gel'fman, *Dokl. Akad. Nauk. SSSR*, **153** (1963) 1327; *Chem. Abs.*, **60** (1964) 8697g.
74 O. G. Levanda, I. I. Moiseev and M. N. Vargaftik, *Izv. Akad. Nauk. SSSR, Ser. Khim.*, (1968) 2368; *Chem. Abs.*, **70** (1969) 23588.
75 V. I. Kazakova and B. V. Ptitsyn, *Russ. J. Inorg. Chem.*, **12** (1967) 323.
76 S. A. Shchukarev, O. A. Lobaneva, M. A. Ivanova and M. A. Kononova, *Russ. J. Inorg. Chem.*, **9** (1964) 1503.
77 W. M. Latimer, *The Oxidation States of the Elements and Their Potentials in Aqueous Solutions*, Prentice-Hall, New York, 2nd edn., 1952, p. 203.
78 S. C. Srivastava and L. Newman, *Inorg. Chem.*, **5** (1966) 1506.
79 S. C. Srivastava and L. Newman, *Inorg. Chem.*, **6** (1967) 762.
80 C. I. Saunders and D. S. Martin, *J. Amer. Chem. Soc.*, **83** (1961) 807.
81 N. M. Nikolaeva, B. V. Ptitsyn and I. I. Gorbacheva, *Russ. J. Inorg. Chem.*, **10** (1965) 570.
82 L. F. Grantham, T. S. Elleman and D. S. Martin, *J. Amer. Chem. Soc.*, **77** (1955) 2965.
83 A. A. Grinberg and G. A. Shagisultanova, *Russ. J. Inorg. Chem.*, **5** (1960) 134.

84　A. A. Grinberg and M. I. Gel'fman, *Dokl. Akad. Nauk. SSSR.*, **133** (1960) 1081; *Chem. Abs.*, **54** (1960) 23632b.

85　Reference 77, p. 205.

86　B. I. Pishchevitskii, B. V. Ptitsyn and N. M. Leskova, *Izv. Sibirsk. Otd. Akad. Nauk. SSSR,* (1962) 143; *Chem. Abs.*, **58** (1963) 13418d.

87　A. A. Grinberg and G. A. Shagisultanova, *Russ. J. Inorg. Chem.*, **5** (1960) 920.

88　W. W. Dunning and D. S. Martin, *J. Amer. Chem. Soc.*, **81** (1959) 5566.

89　A. K. Sundaram and E. B. Sandell, *J. Amer. Chem. Soc.*, **77** (1955) 855.

90　A. A. Grinberg and N. V. Kiseleva, *Zhur. Neorg. Khim.*, **3** (1958) 1804; *Chem. Abs.*, **53** (1959) 21344i.

91　C. M. Harris, S. E. Livingstone and I. H. Reece, *Aust. J. Chem.*, **10** (1957) 282.

92　J. Chatt and G. A. Gamlen, *J. Chem. Soc.*, (1956) 2371.

93　I. Leden and J. Chatt, *J. Chem. Soc.*, (1955) 2936.

94　C. K. Jørgensen, *Absorption Spectra and Chemical Bonding in Complexes*, Pergamon, Oxford, 1962.

95　J. Chatt and L. M. Venanzi, *J. Chem. Soc.*, (1955) 3858.

96　H. Stammreich and R. Forneris, *Spectrochim. Acta.*, **16** (1960) 363.

97　D. M. Adams, *Proc. Chem. Soc.*, (1961) 335.

98　C. H. Perry, D. P. Athans, E. F. Young, J. R. Durig and B. R. Mitchell, *Spectrochim. Acta.*, **23A**, (1967) 1137.

99　D. M. Adams and H. A. Gebbie, *Spectrochim. Acta.*, **19** (1963) 925.

100　J. H. Fertel and C. H. Perry, *J. Phys. Chem. Solids*, **26** (1965) 1773.

101　A. Sabatini, L. Sacconi and V. Schettino, *Inorg. Chem.*, **3** (1964) 1775.

102　H. Poulet, P. Delorme and J. P. Mathieu, *Spectrochim. Acta*, **20** (1964) 1855.

103　J. Hiraishi and T. Shimanouchi, *Spectrochim. Acta*, **22** (1966) 1483.

104　D. M. Adams and D. C. Newton, *J. Chem. Soc. (A)*, (1969) 2998.

105　P. J. Hendra, *Nature*, **212** (1966) 179.

106　R. N. Goldberg and L. G. Hepler, *Chem. Rev.*, **68** (1968) 229.

107　R. J. Goodfellow, P. L. Goggin and D. A. Duddell, *J. Chem. Soc. (A)*, (1968) 504.

108　A. D. Allen and T. Theophanides, *Can. J. Chem.*, **42** (1964) 1551.

109　M. Le Postollec, J. P. Mathieu and H. Poulet, *J. Chim. Phys.*, **60** (1963) 1319; *Chem. Abs.*, **60** (1964) 10066d.

110　G. E. Coates and C. Parkin, *J. Chem. Soc.*, (1963) 421.

111　D. M. Adams, J. Chatt, J. Gerratt and A. D. Westland, *J. Chem. Soc.*, (1964) 734.

112　R. J. H. Clark, G. Natile, U. Belluco, L. Cattalini and C. Filippin, *J. Chem. Soc. (A)*, (1970) 659.

113　K. Shobatake and K. Nakamoto, *J. Amer. Chem. Soc.*, **92** (1970) 3332.

114　R. D. Brown and G. E. Coates, *J. Chem. Soc.*, (1962) 4723.

115　R. J. Goodfellow, J. G. Evans, P. L. Goggin, and D. A. Duddell, *J. Chem. Soc. (A)*, (1968) 1604.

116　P. J. Hendra and E. R. Lippincott, *Nature*, **212** (1966) 1448.

117　J. R. Durig, R. Layton, D. W. Sink and B. R. Mitchell, *Spectrochim. Acta.*, **21** (1965) 1367.

118　J. Chatt and B. T. Heaton, *J. Chem. Soc. (A)*, (1968) 2745.

119　H. Poulet, P. Delorme and J. P. Mathieu, *Spectrochim. Acta.*, **20** (1964) 1855.

120　D. M. Adams, M. Goldstein and E. F. Mooney, *Trans. Faraday Soc.*, **59** (1963) 2228.

121　R. J. Goodfellow, P. L. Goggin and L. M. Venanzi, *J. Chem. Soc. (A)*, (1967) 1897.

122　H. B. Gray and C. J. Ballhausen, *J. Amer. Chem. Soc.*, **85** (1963) 260.

123　P. Day, M. J. Smith and R. J. P. Williams, *J. Chem. Soc. (A)*, (1968) 668.

124　H. Ito, J. Fujita and K. Saito, *Bull. Chem. Soc. Japan*, **40** (1967) 2584.

125　J. Chatt, G. A. Gamlen and L. E. Orgel, *J. Chem. Soc.*, (1958) 486 .

126　P. Day, A. F. Orchard, A. J. Thompson and R. J. P. Williams, *J. Chem. Phys.*, **42** (1965) 1973.

127　R. F. Fenske, D. S. Martin and K. Ruedenberg, *Inorg. Chem.*, **1** (1962) 441.

128　D. S. Martin and C. A. Lenhardt, *Inorg. Chem.*, **3** (1964) 1368.

129   D. S. Martin, M. A. Tucker and A. J. Kassman, *Inorg. Chem.*, **4** (1965) 1682.
130   O. S. Mortensen, *Acta Chem. Scand.*, **19** (1965) 1500.
131   H. B. Gray, *Transition Metal Chem.*, **1** (1965) 239.
132   D. S. Martin, J. G. Foss, M. E. McCarville, M. A. Tucker and A. J. Kassman, *Inorg. Chem.*, **5** (1966) 491.
133   D. S. Martin, M. A. Tucker and A. J. Kassman, *Inorg. Chem.*, **5** (1966) 1298.
134   D. S. Martin, *Coord. Chem. Rev.*, **1** (1966) 39.
135   D. S. Martin, M. A. Tucker and A. J. Kassman, *Coord. Chem. Rev.*, **1** (1966) 44.
136   R. A. Bailey and J. A. McIntyre, *Inorg. Chem.*, **5** (1966) 1824.
137   B. G. Anex, M. E. Ross and M. W. Hedgcock, *J. Chem. Phys.*, **46** (1967) 1090.
138   H. Basch and H. B. Gray, *Inorg. Chem.*, **6** (1967) 365.
139   F. A. Cotton and C. B. Harris, *Inorg. Chem.*, **6** (1967) 369.
140   B. P. Kovarskaya, *Teor. Eksp. Khim.*, **3** (1967) 687; *Chem. Abs.*, **68** (1968) 72326.
141   W. R. Mason and H. B. Gray, *J. Amer. Chem. Soc.*, **90** (1968) 5721.
142   A. J. McCaffery, P. N. Schatz and P. J. Stephens, *J. Amer. Chem. Soc.*, **90** (1968) 5730.
143   O. Lobaneva and M. A. Kononova, *Probl. Sovrem. Khim. Koord. Soedin, Leningrad Gos. Univ.*, (1968) 180; *Chem. Abs.*, **70** (1969) 15682.
144   Q. Looney and B. E. Douglas, *Inorg. Chem.*, **9** (1970) 1955.
145   D. L. Webb and L. A. Rossiello, *Inorg. Chem.*, **9** (1970) 2622.
146   E. P. Marram, E. J. McNiff and J. L. Ragle, *J. Phys. Chem.*, **67** (1963) 1719.
147   K. Ito, D. Nakamura, Y. Kurita, K. Ito and M. Kubo, *J. Amer. Chem. Soc.*, **83** (1961) 4526.
148   C. W. Fryer and J. A. S. Smith, *J. Chem. Soc. (A)*, (1970) 1029.
149   W. van Bronswijk and R. S. Nyholm, *J. Chem. Soc. (A)*, (1968) 2084.
150   A. v. Zelewsky, *Helv. Chim. Acta*, **51** (1968) 803.

CHAPTER 10

Compounds and Complexes of the Tetravalent Metals

During this chapter we shall become increasingly aware that platinum readily forms platinum(IV) compounds and complexes, whereas palladium is reluctant to form palladium(IV) compounds. This is ascribed (p. 13) largely to the fact that the sum of the first four ionisation potentials of palladium (109·5 eV) is greater than that of platinum (97·16 eV). Moreover, the relative covalent bond strengths in platinum(IV) and palladium(IV) compounds also influence the overall energy of these compounds, but this contribution is difficult to assess. At a first glance the covalent bond energy of a platinum(IV)-element bond might well be expected to be similar to that of a palladium(IV)-element bond owing to the similarity in the octahedral covalent radii of the elements in the tetravalent state (Table 5) which should enable these metals to each give similar overlaps with the orbitals of a given atom. Approximate calculations of metal(IV)–halogen bond energies (Table 67) support this conclusion.

Another point that will recur throughout the chapter is the considerable similarity between the square-planar platinum(II) and octahedral platinum(IV) complexes. Many of the properties of platinum(IV) complexes can be understood by looking upon them as platinum(II) complexes with two extra ligands bound one above and one below the square-plane. These extra ligands have little effect on the metal–ligand bonds in the square-plane. The justification for this approach is that in both platinum(IV) and platinum(II) complexes the residual charge on the platinum atom is very small,[1,2] in agreement with Pauling's Electroneutrality principle.[3] Thus in their bond

TABLE 67

Mean bond energy of the metal–halogen bond in $[MX_4]^{2-}$ and $[MX_6]^{2-}$ anions (data from ref. 4)

Compound	Mean M–X bond energy (kcal/mol)
K_2PdCl_6	70
K_2PtCl_6	77
K_2PtBr_6	70
K_2PdCl_4	83
K_2PtCl_4	89
K_2PtBr_4	80

energies, for which the residual charge on the metal is important, we should expect divalent and tetravalent complexes to behave similarly. This is borne out by Table 67, where the mean bond energies of the metal–halogen bonds are similar in the divalent $[MX_4]^{2-}$ and tetravalent $[MX_6]^{2-}$ anions.

The layout of the present chapter follows that adopted in Chapters 4 to 9; the material is divided into sections according to the position of the principal donor atom in the periodic table. The reader is also referred to Chapter 11 (pp. 315–320) which deals with substitution reactions of the tetravalent complexes, and to Chapter 12 (pp. 351–356) where organometallic complexes of the tetravalent metal ions are considered.

HYDRIDE COMPLEXES

Neither isolatable hydride complexes of palladium(IV) nor likely reaction intermediates are known. Only a few hydride complexes of platinum(IV) are well established. These include $[(PEt_3)_2PtH_2Cl_2]$, prepared by treating trans-$[(PEt_3)_2PtHCl]$ in ether with dry hydrogen chloride,[5] $[(Ph_2PCH_2CH_2-PPh_2)Pt(SnMe_3)_3H]$ and $[(Ph_2PCH_2CH_2PPh_2)Pt(SnMe_3)_2HCl]$, prepared by treating the platinum(II) complexes $[(Ph_2PCH_2CH_2PPh_2)Pt(SnMe_3)_2]$ and $[(Ph_2PCH_2CH_2PPh_2)Pt(SnMe_3)Cl]$ with trimethylstannane[6,7] (reactions 305 and 306 on p. 258), and $[(PEt_3)_2Pt(EH_2X)HI_2]$, where E = Si or Ge and X = halogen, prepared by oxidative addition of EH_3X to trans-$[(PEt_3)_2PtI_2]$.[8,9] These platinum(IV)–hydrido complexes are all unstable and readily revert to the corresponding platinum(II) complexes. $[(PEt_3)_2-PtH_2Cl_2]$ loses HCl on storage, except under hydrogen chloride, and also on treatment with water to give the original hydrido–platinum(II) complex trans-$[(PEt_3)_2PtHCl]$. The decomposition of the other hydrido–platinum(IV) complexes to platinum(II) complexes is discussed below (pp. 258–259). This ready reduction to a platinum(II) complex is typical of hydrido–platinum(IV) complexes, and thus they are widely implicated as unstable intermediates in those reactions of hydrido–platinum(II) complexes which appear to involve an initial oxidative-addition reaction (see pp. 62, 83–87, 89–91).

COMPLEXES WITH ELEMENTS FROM GROUPS I, II AND III

No compounds or complexes containing ligands with donor atoms from groups I, II or III have been reported.

COMPLEXES WITH TRANSITION METALS

There are two types of complex formed by platinum(IV) and palladium(IV) with transition metals. Both involve vertical stacks of these metal atoms in two different oxidation states. In the first type, which have bridging halide ions ($-Pt^{II}-X-Pt^{IV}-X-Pt^{II}-$), the metal atoms in the two different oxidation states are in crystallographically non-equivalent environments; whereas in the second type, in which there is direct metal–metal interaction ($-Pt^{II}-Pt^{IV}-$

Pt^{II}–Pt^{IV}–), the metal atoms in the two different oxidation states are in crystallographically equivalent environments and are indistinguishable.

(i) Complexes with metal atoms separated by a halide bridging ligand

The first of the complexes with vertical stacks of alternate platinum(II) and platinum(IV) ions separated by halide bridging ligands to be reported was Wolffram's red salt[10] $[Pt^{IV}(EtNH_2)_4Cl_2][Pt^{II}(EtNH_2)_4]Cl_4 \cdot 4H_2O$ and it is typical of a whole range of mixed valence complexes with atoms in both the +2 and +4 oxidation states (Table 68). It is a bright-red precipitate obtained

TABLE 68

References to the preparation and properties of mixed valence complexes containing palladium(IV) and platinum(IV)

Complex	Preparation	Electronic spectrum	x-ray diffraction investigation
$[M^{IV}a_4X_2]^{2+}[M^{II}a_4]^{2+}$			
$[Pt_2(EtNH_2)_8Cl_2]^{4+}$	10, 11	12	13
$[Pt_2(EtNH_2)_8Br_2]^{4+}$	14		14
$[Pt_2(PrNH_2)_8Cl_2]^{4+}$	10		
$[Pt_2(en)_4Cl_2]^{4+}$	15		
$[M^{II}a_2X_2][M^{IV}a_2X_4]$			
$[Pd_2(NH_3)_4Cl_6]$	16^a	18	
$[PdPt(NH_3)_4Cl_6]$	17	18	17
$[Pt_2(NH_3)_4Cl_6]$	19, 20	18	17
$[Pt_2(NH_3)_4Br_6]$	19, 20		17, 21
$[Pt_2(NH_3)_4I_6]$	19, 20		
$[Pt_2(en)_2Cl_6]$	22, 23	24, 25	
$[Pt_2(en)_2Br_6]$	24	24	26
$[Pt_2(en)_2I_6]$	24	24	
$[Pt^{IV}(NH_3)_2Cl_2(OH)_2][Pt^{II}(NH_3)_2Cl_2]$	23	27	

^{a} The electrical conductivity of the complex is 500 times greater in the direction of the Cl–Pd^{II}–Cl–Pd^{IV}–Cl chains than in the direction perpendicular to these chains (refs. 28, 28a).

by mixing solutions of $[Pt^{IV}(EtNH_2)_4Cl_2]Cl_2$ and $[Pt(EtNH_2)_4]Cl_2$, or by oxidising $[Pt(EtNH_2)_4]Cl_2$ with hydrogen peroxide in the presence of hydrochloric acid.[11] X-ray diffraction has shown[13] that Wolffram's red salt has chains of platinum atoms, alternately in the +2 and +4 oxidation states, separated by chloride ligands (Fig. 21). The chloride ions are 2·26 Å from the platinum(IV) ions and 3·31 Å from the platinum(II) ions so that the +2 and +4 ions are in structurally different environments. However, with neither this compound nor the corresponding bromide[14] is it possible to resolve the structure completely because there are stacking faults due to the chains slipping past one another. The structures of $[Pt^{IV}enBr_4][Pt^{II}enBr_2]$[26] and $[Pt^{IV}(NH_3)_2Br_4][Pt^{II}(NH_3)_2Br_2]$[17,21] (Appendix II, structures 124 and 117) proved to be similar to that of Wolffram's red salt, with platinum(II) and platinum(IV) ions in structurally distinguishable environments. It is interesting that in spite of the variety of the mixed valence halides of palladium and

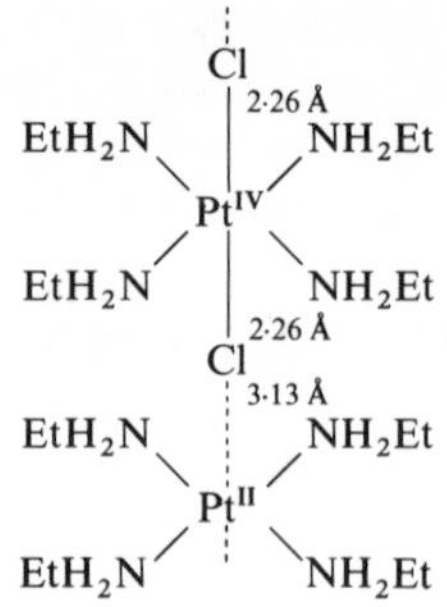

$$
\begin{array}{c}
\vdots \\
\mathrm{Cl} \\
\mathrm{EtH_2N} \quad \Big|\, {}^{2\cdot26\,\text{Å}} \quad \mathrm{NH_2Et} \\
\diagdown \quad \diagup \\
\mathrm{Pt^{IV}} \\
\diagup \quad \Big| \quad \diagdown \\
\mathrm{EtH_2N} \qquad \qquad \mathrm{NH_2Et} \\
\Big|\, {}^{2\cdot26\,\text{Å}} \\
\mathrm{Cl} \\
\vdots\, {}^{3\cdot13\,\text{Å}} \\
\mathrm{EtH_2N} \qquad \qquad \mathrm{NH_2Et} \\
\diagdown \quad \diagup \\
\mathrm{Pt^{II}} \\
\diagup \quad \vdots \quad \diagdown \\
\mathrm{EtH_2N} \qquad \qquad \mathrm{NH_2Et}
\end{array}
$$

Fig. 21. Structure of Wolffram's red salt (from ref. 13).

platinum known in the solid state there is no evidence for such compounds in solution.[18]

The electronic spectra of these complexes (Table 68) are generally similar to that of Wolffram's red salt in having an extra, very intense band in the visible region that is not present in the spectra of the individual platinum(II) and platinum(IV) complexes and that is strongly polarised in the direction of the metal atom chain.[12] This band is almost certainly due to the transfer of an electron from the highest filled d-orbital of a platinum(II) ion (probably d_{z^2} since in these complexes the ligands are purely σ-bonding) to the empty d_{z^2} orbital of the adjacent platinum(IV) ion. Because these two d_{z^2} orbitals point at one another and can overlap via the $3p_\sigma$ orbital of the intervening chloride ion, such transitions are electronically allowed, with a polarisation along the metal atom chain.[29] That the electrical conductivity of $[Pd^{II}(NH_3)_2Cl_2][Pd^{IV}(NH_3)_2Cl_4]$ is 500 times greater in the direction of the $-Cl-Pd^{II}-Cl-Pd^{IV}-$ chains than perpendicular to them[28] is consistent with this model.

(ii) Complexes with direct metal–metal interaction

A series of mixed valence platinum oxalate complexes were first prepared by Werner by treating sodium hexachloroplatinate(IV) with oxalic acid.[30] They have bright copper colours in contrast to the yellow of the normal platinum(II) oxalate. X-ray diffraction of $K_{1\cdot6}[Pt(C_2O_4)_2].2\cdot5H_2O$ showed the complex ions to be stacked directly above one another with a platinum–platinum distance of only 2·75 Å.[31] Furthermore, it was impossible to distinguish between the platinum(II) and platinum(IV) atoms in this structure. This structure is quite different from that of the platinum(II) complex $K_2[Pt(C_2O_4)_2].2H_2O$ which contains isolated planar $[Pt(C_2O_4)_2]^{2-}$ anions[32] (Appendix II, structure 126). It is suggested that, whereas in $K_2[Pt(C_2O_4)_2].2H_2O$ the d_{z^2} band is filled, on oxidation electrons are drawn off from that band and an overall bonding results, and the colour deepens because of the new allowed electronic transitions.

Partial oxidation of $K_2[Pt(CN)_4]$ and also partial reduction of $K_2[Pt(CN)_4X_2]$, where $X = Cl$ or Br, leads to the isolation of

$K_2[Pt(CN)_4X_{0.32}].2.6H_2O$, which can also be obtained by the evaporation of aqueous solutions of $K_2[Pt(CN)_4]$ and $K_2[Pt(CN)_4X_2]$ mixed in a $5:1$ molar ratio.[33,34] The products have a coppery sheen and absorb or reflect all light polarised parallel to the needle axis and travelling perpendicular to this axis, whereas light polarised perpendicular to this direction is absorbed only very slightly in the visible region. X-ray diffraction of $K_2[Pt(CN)_4Cl_{0.32}]$. $2.6H_2O$ indicates that it contains square-planar $[Pt(CN)_4]$ groups stacked in columns along the needle axis of the crystal with a platinum–platinum distance of 2·88 Å.[33,34] The chloride ions are not bound directly to any of the platinum atoms, which are all in crystallographically identical environments, so that it is impossible to distinguish the platinum(II) from the platinum(IV) ions. A d.c. potential of more than 150 V/cm applied along the needle axis of crystals of $K_2[Pt(CN)_4Br_{0.32}].2.5H_2O$ causes an oxidation-reduction reaction which commences at the anode and spreads through the crystal to the cathode. This reaction, which produces polycrystalline $K_2[Pt(CN)_4].2H_2O$ and bromine, is unusual in that it is not directly related to electrolysis and takes place away from the electrodes.[35]

Nitric acid oxidation of platinum metal in the presence of cyanide ions yields $K_2[Pt(CN)_5].3H_2O$ in which platinum is formally in the trivalent oxidation state.[36] X-ray diffraction has shown vertical stacks of planar $[Pt(CN)_4]$ units separated by a platinum–platinum distance of 2.92 Å. Each $[Pt(CN)_4]$ unit is twisted by 45° about the Pt–Pt axis relative to its nearest neighbours. The platinum–carbon bond lengths in all the $[Pt(CN)_4]$ units are 2.02 Å so that it is impossible to distinguish the platinum(II) from the platinum(IV) ions.[36] The fifth cyanide ion is not co-ordinated to the platinum but bound as a free ion within the lattice (Appendix II, structure 139).

COMPLEXES WITH GROUP IVB ELEMENTS

None of the group IVB elements form complexes with palladium(IV). The carbon ligands that bond to platinum(IV) by a σ-bond are considered in Chapter 12 (pp. 351–356), so that in this section only cyanide and π-cyclopentadienyl complexes together with complexes of silicon, germanium and tin ligands are considered. No complexes of the tetravalent metals with either isocyanide or carbon monoxide ligands have been reported.

(i) Cyanide complexes

Potassium hexacyanoplatinate(IV) is prepared by grinding a mixture of K_2PtI_6 and potassium cyanide and recrystallising the product from water;[37] when K_2PtCl_6 is used reduction of the platinum(IV) occurs. Similarly K_2PdCl_6 and K_2PdBr_6 are reduced by potassium cyanide in the solid state yielding $[Pd(CN)_4]^{2-}$ and palladium metal.[38] However, $K_2Pd(CN)_6$ can be prepared by treating K_2PdCl_6 with potassium cyanide in the presence of a strong oxidising agent such as potassium peroxydisulphate.[38] Both $K_2Pd(CN)_6$ and $K_2Pt(CN)_6$ are stable to water and dilute acids at room temperature, although in boiling water $K_2Pd(CN)_6$ is slowly decomposed to $K_2Pd(CN)_4$ and palladium(II) cyanide. Treatment of a solution of

$Ag_2[Pt(CN)_6]$ with hydriodic acid yields $H_2Pt(CN)_6$,[39] which is a colourless crystalline air stable complex that readily dissolves in water to give a strongly acidic solution that dissolves metallic zinc with the liberation of hydrogen.

Attempts to prepare platinum(IV) cyano-complexes by treating $H_2[Pt(CN)_4]$ or $K_2[Pt(CN)_4]$ with hydrogen peroxide yield the 'platinum(III)' complexes $H[Pt(CN)_4]$ and $K[Pt(CN)_4]$.[40,41] On heating hydrated $H[Pt(CN)_4]$, $Pt(CN)_3$ is formed. All three 'platinum(III)' products are probably mixtures of platinum(II) and platinum(IV) derivatives.

Although there is no evidence for the formation of mixed halogeno–cyano-platinum(IV) complexes in solution,[42] oxidation of $K_2Pt(CN)_4$ by chlorine, bromine or iodine yields $K_2[Pt(CN)_4X_2]$, where X = Cl, Br or I.[43–5] The '$Pt(CN)_4$' group in these complexes is inert to hydrolysis[45] and ammoniation[46] (reactions 297 and 298). Similarly the cyano-groups in the bis–cyano complexes $[PtL_2(CN)_2X_2]$, where L = NH_3 or $\frac{1}{2}$en and X =

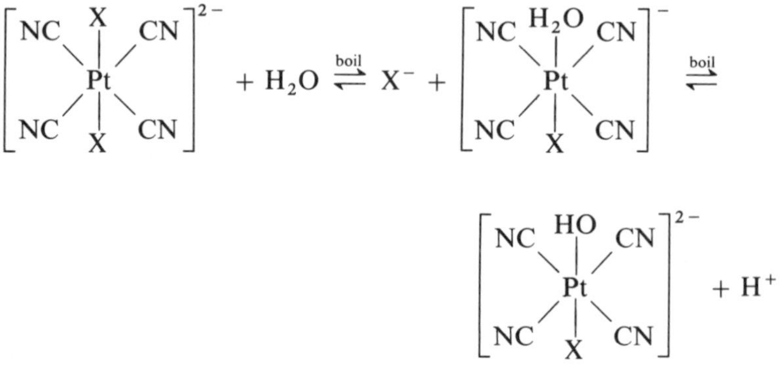

$$K_2[Pt(CN)_4X_2] + 2NH_3 \xrightarrow{\text{boil}} [Pt(CN)_4(NH_3)_2] + 2KX \quad (298)$$

halide ion, formed by oxidation of $[PtL_2(CN)_2]$ by halogen in the presence of hydrogen peroxide, are also inert to substitution.[47,48]

The hydroxy complex, $[Pt(en)(OH)_2(CN)_2]$, prepared by treating $[Pt(en)(CN)_2]$ with hydrogen peroxide does not react with solutions containing chloride, bromide, iodide and cyanide ions.[47] One of the difficulties in preparing cyano-complexes of the tetravalent metals is that when the corresponding chloro-complexes of these metals are treated with potassium cyanide they very often yield hydroxy-complexes as in reaction 299[49] or occasionally

$$[Pt(NH_3)_2Cl_2(NO_3)_2] \xrightarrow{KCN} [Pt(NH_3)_2Cl_2(OH)_2] \quad (299)$$

give reduction to a platinum(II) complex as in reaction 300.[48] However, when the iodo-complex *trans*-$[Pt(NH_3)_2I_2(CN)_2]$ is treated with potassium cyanide the *tris*-cyano complex $[Pt(NH_3)_2I(CN)_3]$ is formed.[48] This reaction

$$K_2[Pt(CN)_4X_2] + 2KCN \rightarrow K_2Pt(CN)_4 + C_2N_2 + 2KX \quad (300)$$

occurs because the halide ligand displaced lies *trans* to an iodide ligand which has a high *trans*-effect (p. 299). Reaction 301 exploits this ability of

$$[PtI_4(MeNH_2)_2] + 2KCN \rightarrow cis\text{-}[PtI_2(CN)_2(MeNH_2)_2] + 2KI \text{ (301)}$$

iodide ligands to encourage the co-ordination of cyanide ligands in the *trans*-position[50] and the *cis*-nature of the product suggests that the *trans*-effect of the cyanide ligand in platinum(IV) complexes is less than that of the iodide ligand.[48]

X-ray diffraction has shown that the $[Pt(CN)_6]^{2-}$ ion has a regular octahedral structure with linear Pt–C–N bonds.[51] The infrared and Raman spectra[38,52] of $K_2[M(CN)_6]$, where M = Pt or Pd, have C–N stretching vibrations at higher frequencies and C–N stretching constants which are larger than the corresponding divalent complexes ($K_2Pt(CN)_6$,[52] $v_{C-N} = 2216 \text{ cm}^{-1}$, $K_{C-N} = 17\cdot5 \text{ mdyn/Å}$; $K_2Pt(CN)_4$,[53] $v_{C-N} = 2150 \text{ cm}^{-1}$, $K_{C-N} = 16\cdot8 \text{ mdyn/Å}$). This is consistent with less π-back-donation from the metal to the empty π^*-(antibonding) orbital of the cyanide ion in the complexes of the tetravalent metals. The stretching force constants of the metal–carbon bonds in $K_2[Pt(CN)_4X_2]$, where X = Cl, Br or I, are about $2\cdot80 \text{ mdyn/Å}$ compared with metal–halogen stretching force constants in these complexes of $2\cdot09$ (X = Cl), $1\cdot73$ (X = Br) and $1\cdot37 \text{ mdyn/Å}$ (X = I).[54]

(ii) π-Cyclopentadienyl complexes

The only π-cyclopentadienyl platinum(IV) complex, $[(\pi\text{-}C_5H_5)Pt(CH_3)_3]$, is prepared by treating trimethyl platinum(IV) iodide with sodium cyclopentadienide in tetrahydrofuran.[55,56] The two peaks in the proton n.m.r. spectrum (relative intensities 9 : 5) are consistent with π-bonding of the cyclopentadienyl ring to the metal, but since σ-cyclopentadienyl ligands can also yield a single 'time-averaged' proton n.m.r. signal this does not rigorously exclude a σ-bonded structure.[57] However, x-ray diffraction investigation[58,59] has confirmed the π-bonding description of the platinum(IV)–cyclopentadienyl bond, which is also indicated by the infra-red and Raman spectra.[61] $(\pi\text{-}C_5H_5)Pt(CH_3)_3$ is thermally very stable, possibly owing to its monomeric, approximately octahedral, structure which allows the platinum atom to have 18 electrons in its outer bonding orbitals.[60] The fragmentation pattern observed in the mass spectrometer is consistent with the dominant fragmentation path involving the elimination of methyl radicals.[60,62]

(iii) Complexes containing silicon, germanium and tin ligands

Many of the reactions of the silyl, germyl and stannyl complexes of platinum(II) appear to take place by an oxidative addition-elimination mechanism involving the intermediate formation of a platinum(IV) complex (pp. 83–91). Although no such platinum(IV) complexes had been isolated until recently it has now been found that they can be prepared by at least three routes. No palladium(IV) complexes of these ligands have been reported.

(a) Action of HX on a platinum(II) complex

Stable platinum(IV) complexes with group IVB ligands have been isolated by the addition of hydrogen halides to the platinum(II) complexes[8,63] (reactions 302 and 303).

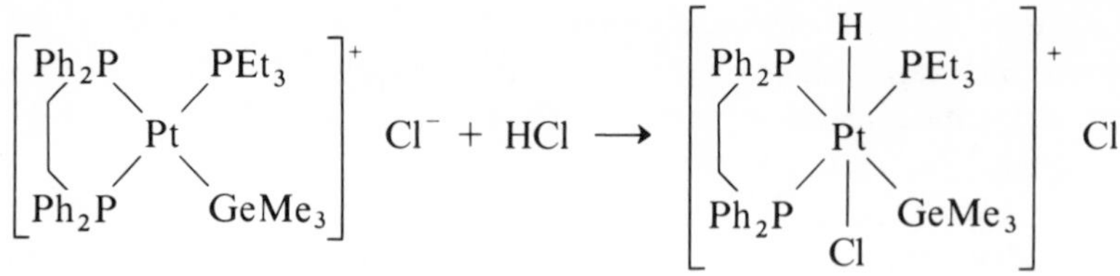

$$\begin{bmatrix} Ph_2P & PEt_3 \\ & Pt \\ Ph_2P & GeMe_3 \end{bmatrix}^+ Cl^- + HCl \longrightarrow \begin{bmatrix} & H & \\ Ph_2P & | & PEt_3 \\ & Pt & \\ Ph_2P & | & GeMe_3 \\ & Cl & \end{bmatrix}^+ Cl^- \qquad (302)$$

$$trans\text{-}[(PEt_3)_2Pt(SiH_2I)I] + HI \longrightarrow \begin{bmatrix} & H & PEt_3 \\ & | & / \\ I-Pt-SiH_2I \\ & / & | \\ Et_3P & & I \end{bmatrix} \qquad (303)$$

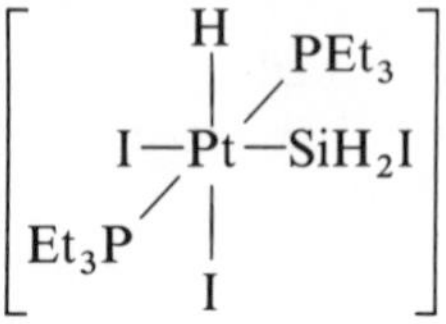

(b) Addition of R_3EH to platinum(O) and platinum(II) complexes
Trimethylstannyl platinum(IV) complexes can be prepared by treating both platinum(O) and platinum(II) complexes with trimethylstannane[7] (reactions 304–6).

$$[(Ph_2PCH_2CH_2PPh_2)_2Pt] + 3Me_3SnH \longrightarrow$$

$$[(Ph_2PCH_2CH_2PPh_2)Pt(SnMe_3)_3H] + Ph_2PCH_2CH_2PPh_2 + \uparrow H_2 \quad (304)$$

$$[(Ph_2PCH_2CH_2PPh_2)Pt(SnMe_3)_2] + Me_3SnH \longrightarrow$$

$$[(Ph_2PCH_2CH_2PPh_2)Pt(SnMe_3)_3H] \quad (305)$$

$$[(Ph_2PCH_2CH_2PPh_2)Pt(SnMe_3)Cl] + Me_3SnH \longrightarrow$$

$$[(Ph_2PCH_2CH_2PPh_2)Pt(SnMe_3)_2HCl] \quad (306)$$

(c) Addition of H_3ECl to platinum(II) complexes
Monochlorosilyl and -germyl complexes of platinum(IV) are formed when platinum(II) complexes are treated with monochlorosilane or -germane[8,9,64] (reaction 307).

$$trans\text{-}[(PEt_3)_2PtI_2] + SiH_3Cl \longrightarrow [(PEt_3)_2Pt(SiH_2Cl)HI_2] \quad (307)$$

These platinum(IV) complexes containing silyl, germyl and stannyl ligands are relatively unstable and, as predicted in Chapter 6, from the reactions in which they are thought to be intermediates, they readily eliminate two ligands to form platinum(II) complexes, either slowly at room temperature, as in the elimination of hydrogen in reaction 308[9] or rapidly on attempting to recrystallise the platinum(IV) complexes as in reactions 309 and 310.[7,63]

$$trans\text{-}[(PEt_3)_2Pt(EH_2Cl)HI_2] \xrightarrow{(E\ =\ Si\ or\ Ge)} trans\text{-}[(PEt_3)_2Pt(EHClI)I] + I_2$$

$$(308)$$

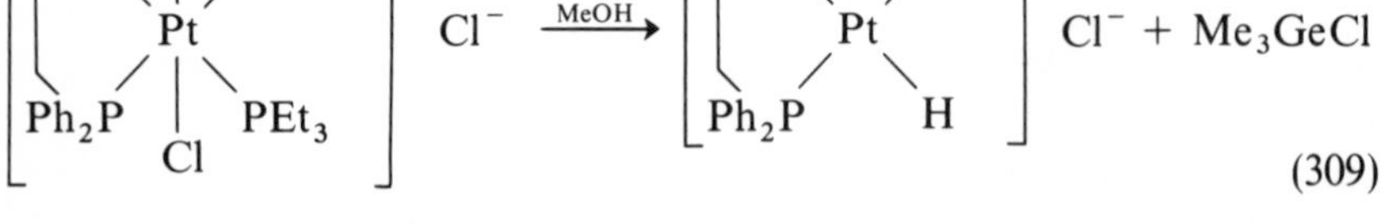

$$\begin{bmatrix} & H & \\ Ph_2P & | & GeMe_3 \\ & Pt & \\ Ph_2P & | & PEt_3 \\ & Cl & \end{bmatrix}^+ Cl^- \xrightarrow{MeOH} \begin{bmatrix} Ph_2P & PEt_3 \\ & Pt & \\ Ph_2P & H \end{bmatrix}^+ Cl^- + Me_3GeCl$$

$$(309)$$

$$
\begin{bmatrix}
& \text{SnMe}_3 & \\
\text{Ph}_2\text{P} & | \quad \text{SnMe}_3 & \\
& \diagdown \diagup & \\
& \text{Pt} & \\
& \diagup \diagdown & \\
\text{Ph}_2\text{P} & | \quad \text{SnMe}_3 & \\
& \text{H} &
\end{bmatrix}
\xrightarrow{\ C_6H_6\ }
\begin{bmatrix}
\text{Ph}_2\text{P} & \quad \text{SnMe}_3 \\
\diagdown & \diagup \\
& \text{Pt} \\
\diagup & \diagdown \\
\text{Ph}_2\text{P} & \quad \text{SnMe}_3
\end{bmatrix}
+ \text{Me}_3\text{SnH} \quad (310)
$$

These reactions all involve loss of a group IVB ligand plus the remaining ligand that lies *trans* to the group of highest *trans*-influence. Thus in reaction 309 the hydride group has the highest *trans*-influence and so Cl and GeMe$_3$ are eliminated. In reaction 310 the trimethylstannyl ligand has a slightly higher *trans*-influence than the hydride ligand (pp. 301–303) and so H and SnMe$_3$ are eliminated. Reaction 308 is more complex and probably involves elimination of I and EH$_2$Cl as the first step. This, however, is followed by reaction of the EH$_2$ClI so formed with the hydrido-complex *trans*-[(PEt$_3$)$_2$PtHI] to eliminate hydrogen and yield *trans*-[(PEt$_3$)$_2$Pt(EHClI)I] (cf. reaction 92, p. 78).

COMPLEXES WITH GROUP VB ELEMENTS

Whilst platinum(IV) forms a wide range of complexes with nitrogen, phosphorus and arsenic donor atoms, palladium(IV) is reluctant to form such complexes. Thus with ammonia the platinum(IV) complexes [(Pt(NH$_3$)$_n$-Cl$_{6-n}$]$^{(n-2)+}$ are known for $n = 0$ to 6 whereas with palladium(IV) only ammine complexes with $n = 0$ and 2 are known. The complexes of the tetravalent metals with the group IVB elements are considered under the following headings: (i) amine, tertiary phosphine and tertiary arsine complexes, (ii) alkyl and aryl cyanide complexes, (iii) azide complexes and (iv) azobenzene complexes.

(i) Amine, tertiary phosphine and tertiary arsine complexes
These complexes are all considered together in the present chapter rather than splitting off the nitrogen complexes as was done in Chapter 7 when considering the divalent metals. This is feasible here because of the greater similarities between the complexes of the nitrogen, phosphorus and arsenic ligands due to the reduced possibility of π-back-donation of electron density from the metal, which probably plays an important role in differentiating the complexes of nitrogen from those of the other group VB elements with the divalent metals.

Preparation
(a) *Amine complexes.* There are two main methods for preparing amine complexes. In the first, a platinum(IV) complex is treated with an amine as in the preparation of [Pt(NH$_3$)$_6$]Cl$_4$[65] and [Pt(NH$_3$)$_5$Cl]$^{3+}$(PO$_4$$^{3-}$)[66,67] which are prepared by treating ammonium hexachloroplatinate(IV) with liquid ammonia. The tripositive cation is preferentially precipitated in the presence of sodium phosphate. The second method used for platinum(IV) complexes containing four or less amine groups, involves oxidation of the

corresponding amine–platinum(II) complexes with halogens[68] (reaction 311) or hydrogen peroxide[69] (reaction 312). With bis–amino complexes of platinum(II) the *cis-* or *trans* nature of the platinum(II) complex is preserved

$$[Pt(NH_3)_4X_2] + Br_2 \rightarrow [Pt(NH_3)_4Br_2]X_2 \qquad (311)$$

$$[Pt(en)_2]^{2+} + H_2O_2 \rightarrow [Pt(en)_2(OH)_2]^{2+} \qquad (312)$$

during the oxidation.[70] Concentrated nitric acid can also be used as the oxidising agent[71] (reaction 313). The halide, but not the amine groups, in these amino–platinum(IV) complexes are readily replaced by other ligands

$$[Pt(en)Cl_2] + 2HNO_3 \rightarrow H_2O + [Pt(en)Cl_2(NO_2)(NO_3)] \xrightarrow[(X\,=\,Cl,\,Br,\,I)]{KX}$$

$$[Pt(en)Cl_2(NO_2)X] \quad (313)$$

such as NO_3^-, NO_2^-, SCN^- or OH^- as well as by other halide ions. These substitution reactions, which are discussed in detail in Chapter 11 (pp. 315–320) are catalysed by platinum(II) complexes.

The few amino–palladium(IV) complexes known, such as [Pd(NH_3)_2-Cl_4],[72] [Pd(en)Cl_4][72] and [Pd(pyr)_2X_2Cl_2],[72,73] where X = Cl, Br or I, have been prepared by oxidation of the corresponding dichloro–palladium(II) complexes with halogen in chloroform solution.

(b) *Tertiary phosphine and tertiary arsine complexes.* Tertiary phosphine and tertiary arsine complexes of the tetravalent metals have been prepared by oxidation of the corresponding platinum(II) and palladium(II) complexes by halogens.[74,75] No substitution reactions on halide complexes of the tetravalent metals appear to have been reported and consequently complexes containing five or six tertiary phosphine or tertiary arsine ligands have not been described. *Cis-trans*-isomerisation of the bis-complexes is much easier than for the complexes of divalent platinum so that whilst chlorination of *trans*-[(PPr_3)_2PtCl_2] gives the *trans*-isomer of the corresponding platinum(IV) complex, chlorination of *cis*-[(PPr_3)_2PtCl_2] yields a mixture of 70% *cis*- and 30% *trans*-[(PPr_3)_2PtCl_4].[75] Although palladium(IV) complexes of tertiary arsines such as [(Ph_2MeAs)_2PdBr_4] are not particularly stable, palladium(IV) complexes of the ditertiary arsine **121** (diars) are stable and can be prepared

121

by nitric acid oxidation of the palladium(II) complexes [Pd(diars)_2X_2], where X = Cl or Br, followed by addition of perchloric acid to precipitate the perchlorate salts [Pd(diars)_2X_2]^{2+}(ClO_4^-)_2.[76] In addition to the use of halogens and nitric acid as oxidising agents alkyl halides can undergo oxidative addition reactions with platinum(II) complexes to give platinum(IV) products[77–9] (reaction 447, p. 355).

Properties

(*a*) *Chemical.* The amine, tertiary phosphine and tertiary arsine complexes of platinum(IV) are stable crystalline complexes which are generally more deeply coloured than their planar platinum(II) analogues. They are very resistant to mild reducing agents such as alcohols. The tertiary phosphine and some of the tertiary arsine complexes are resistant to reduction by sulphur dioxide, although the complexes of the diarsine **121** are reduced cleanly to platinum(II) complexes by sulphur dioxide[76] (reaction 314). The tertiary phosphine complexes can be reduced cleanly to the corresponding

$$[Pt(diars)_2X_2]^{2+}(ClO_4^-)_2 \xrightarrow[X = Cl, Br, I]{SO_2} [Pt(diars)_2]^{2+}(ClO_4^-)_2 \quad (314)$$

platinum(II) complexes with stannous chloride.[75] The tertiary arsine complexes melt without decomposition, whereas, on heating the tertiary phosphine complexes *cis*-elimination of chlorine occurs[75] (reactions 315 and 316). Where *cis*- and *trans*-isomers are formed these isomerise more readily

$$trans\text{-}[(PPr_3)_2PtCl_4] \xrightarrow[\text{for 15 min.}]{\text{heat at 195°C}} cis\text{-}[(PPr_3)_2PtCl_2]$$
$$9\%$$
$$+ \; trans\text{-}[(PPr_3)_2PtCl_2] + \text{black oil} \quad (315)$$
$$14\% \qquad\qquad 77\%$$

$$cis\text{-}[(PPr_3)_2PtCl_4] \xrightarrow[\text{for 10 min.}]{135\text{-}140°C} trans\text{-}[(PPr_3)_2PtCl_4] +$$
$$43.5\%$$
$$(PPr_3Cl^+)_2[PtCl_6]^{2-} + trans\text{-}[(PPr_3)_2PtCl_2] + cis\text{-}[(PPr_3)_2PtCl_2] \quad (316)$$
$$33\% \qquad\qquad 15\% \qquad\qquad 8.5\%$$

than the corresponding platinum(II) isomers, their relative stability depending very much on the other ligands present, the *cis*-isomer increasing in stability with increasing electro-negativity of the halogen. As with the corresponding platinum(II) complexes the reactivities of the *cis*- and *trans*-isomers are very different. Thus the *cis*-complexes *cis*-[(ER$_3$)$_2$PtCl$_4$], where E = P or As, react instantaneously with silver nitrate to give a precipitate of silver chloride, whereas the *trans*-isomers take some hours to react.[74,75] This has been used as a test for distinguishing between the isomers. It depends on the high *trans*-effect of the tertiary phosphine and arsine ligands, which being largely a σ-*trans*-effect is transmitted across an octahedral complex.

The palladium(IV) complexes of ligands containing group VB donor atoms are less stable than their platinum(IV) analogues; the arsine complexes decrease in stability in the order diarsine **121** > Ph$_2$MeAs $\ggg$ MeAs(CH$_2$-CH$_2$CH$_2$AsMe$_3$)$_2$.[76,80] Even with the most stable, it is noteworthy that, whereas the order of stability of the [MCl$_6$]$^{2-}$ anions is Pt > Pd $\gg$ Ni, the order of stability of the diarsine complexes [M(diars)$_2$X$_2$]$^{2+}$ is Pt $\gg$ Ni > Pd.[81] This very low stability of tertiary arsine complexes of palladium(IV) relative to halide complexes of palladium(IV) arises from changes that occur in the relative orders of the promotion energies necessary to form palladium(IV) ions consequent to changes of the residual charge on the metal which results from changes in the nature of the ligands surrounding that metal ion.

(b) *X-ray diffraction.* X-ray diffraction, which has only been applied to the platinum(IV) complexes of group VB ligands (Table 69) indicates that the platinum(IV)–ligand bond lengths are almost identical to platinum(II)–ligand bond lengths. The complexes have regular octahedral structures, except where chelating or bulky ligands cause minor distortions.

TABLE 69

X-ray diffraction results for platinum(IV) complexes of group VB ligands together with some comparison platinum(II) complexes

	Pt–E bond length (Å)	Pt–Cl bond length (Å)	Structure No. in Appendix II	Reference
$[Pt(NH_3)_5Cl]^{3+}(Cl^-)_3$	2·00	2·30	92	82
trans-$[Pt(NH_3)_4Cl_2]^{2+}(Cl^-)_2$	2·03	2·25	95	83
cis-$[Pt(en)_2Cl_2]^{2+}(Cl^-)_2$	2·06	2·30	134	84
trans-$[(PEt_3)_2PtCl_4]$	2·39	2·33	158	85
(structure, see diagram)	2·44	2·47 (Pt–Br) trans to Br	187	86
		2·55 (Pt–Br) trans to CH₂		

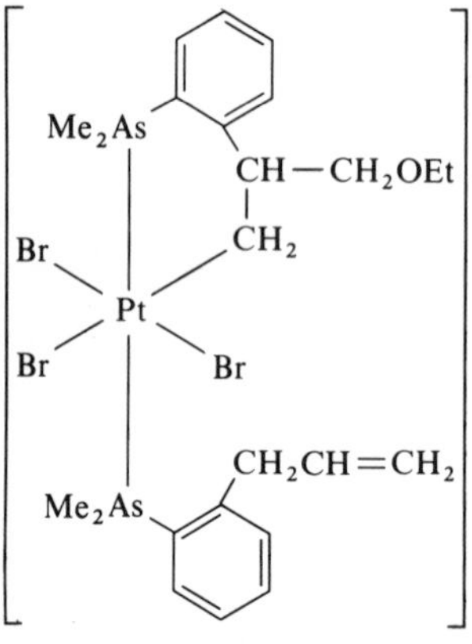

Equivalent platinum(II) complexes

cis- and trans-$[Pt(NH_3)_2Cl_2]$	2·03 (mean)	2·32 (mean)	93, 94	87
trans-$[(PEt_3)_2PtCl_2]$	2·30	2·30	157	88
trans-$[(PEt_3)_2PtBr_2]$	2·32	2·43 (Pt–Br)	156	88
$[Pt(QAs)I]^+BPh_4^{-a}$	2·46 (mean of Pt–As bonds in the trigonal plane)	—	208	89

a QAs = **57** (p. 147).

(c) *Infra-red spectroscopy.* The infra-red spectra of the cis- and trans-complexes $[L_2PtX_4]$, where X = Cl, Br or I and L = NH₃, pyr, SMe₂, PEt₃ or AsEt₃ have confirmed the stereochemistries of the complexes.[90–3] The low frequency spectra of the *trans*-isomers can be interpreted as a superimposition of the expected spectra of the square-planar $[PtX_4]^{2-}$ complexes and the linear L–Pt–L unit.[90] In the *cis*-complexes the spectra are more complicated and difficult to assign. Although the platinum–arsenic and platinum–chlorine stretching frequencies cannot be distinguished the platinum–phosphorus stretching frequency can be identified and it is almost identical to that in the corresponding platinum(II) complex (*cis*-[(PEt$_3$)$_2$PtCl$_4$], v(Pt–P) = 444, 428 cm^{-1}; *cis*-[(PEt$_3$)$_2$PtCl$_2$], v(Pt–P) = 443, 427 cm^{-1}; *cis*-[(PEt$_3$)$_2$PtBr$_4$], v(Pt–P) = 444, 425 cm^{-1}; *cis*-[(PEt$_3$)$_2$PtBr$_2$], v(Pt–P) = 440, 424 cm^{-1}[90,94,95]). Thus the oxidation of a divalent complex to a tetravalent complex results in very little change in the physical properties of the original square-planar unit.

(d) Electronic spectroscopy. Platinum(IV) complexes of amines, tertiary phosphines and tertiary arsines are generally deeper in colour than the corresponding platinum(II) complexes, due to both the high extinction coefficients of the bands in the visible and near ultra-violet regions of the spectrum and the presence of bands at lower energies in the platinum(IV) complexes than in the platinum(II) complexes. Thus the lowest energy band in [Pt(en)Br$_2$] at 30,500 cm^{-1} has an extinction coefficient of about 450 l.mol^{-1} whereas [Pt(en)Br$_4$] has two bands at lower energies (24,000 cm^{-1} (shoulder) and 27,000 cm^{-1} with an extinction coefficient of about 2300 l.mol^{-1}[24]). Very high extinction coefficients have been observed for all the bands in [Pt(NH$_3$)$_6$]Cl$_4$,[96] [Pt(NH$_3$)$_5$Cl]Cl$_3$[96] [Pt(en)Cl$_2$XY], where X and Y = Cl, Br, OH, NO$_2$ and NO$_3$,[97] and in *trans*-[L$_2$PtX$_4$], where L = PEt$_3$ or AsEt$_3$ and X = Cl or Br,[98] suggesting that these are due to charge transfer rather than *d–d*-transitions. The only spectra that have been analysed in detail are those of the bis-tertiary phosphine and bis-tertiary arsine complexes *trans*-[L$_2$PtX$_4$] where the first charge transfer bands were observed at lower energies than in the corresponding [PtX$_6$]$^{2-}$ anions, suggesting that these lowest energy charge transfer bands arise largely from the transfer of charge from orbitals on the tertiary phosphine or arsine ligand to the empty platinum(IV) e_g orbitals.

(e) N.m.r. spectroscopy. ^{31}P n.m.r. spectroscopy can distinguish between the *cis* and *trans*-isomers of [(PBu$_3$)$_2$PtCl$_4$] since the *cis*-isomer shows a much greater ^{31}P–^{195}Pt coupling constant than the *trans*-isomer.[99] The ratio of these coupling constants for the *cis* and *trans*-platinum(IV) complexes (1.41) is very similar to the value of 1·47 observed for the ratio of these coupling constants for the corresponding platinum(II) complexes, again supporting the hypothesis that the oxidation of a divalent to a tetravalent platinum complex results in very little change in the bonding in the original square-planar unit.

(f) Conformational aspects of platinum(IV) complexes containing bidentate ligands. Since the octahedral geometry of the platinum(IV) ion is well established and its complexes are stable with respect to hydrolysis,

platinum(IV) complexes of bidentate ligands such as ethylenediamine are potentially useful for the study of stereospecific reactions analogous to the studies that have been reported for cobalt(III) complexes.[100,101] However, cis-[Pt(en)$_2$Cl$_2$] has been synthesised and resolved into its two optical isomers only recently.[102] The two isomers are exceedingly stable towards racemisation[103] and much more stable than the corresponding cobalt(III) complexes. Treatment of the $(+)_{450}$-[Pt(en)$_2$Cl$_2$]Cl$_2$ isomer with a stoichiometric amount of ethylenediamine in aqueous solution at room temperature yields optically pure d-[Pt(en)$_3$]Cl$_4$[102] originally prepared by separation of racemic [Pt(en)$_3$]Cl$_4$ with d-tartrate.[104] Again the optical isomers of this tris-complex are stable towards recemisation.[105] The high optical stability of the isomers of cis-[Pt(en)$_2$Cl$_2$]Cl$_2$ and the stereoselectivity of their reactions, for example with ethylenediamine, make this platinum(IV) complex potentially more useful for mechanistic studies than the more easily racemised cobalt(III) complexes.[100] Accordingly the crystal structure of $(+)_{450}$-[Pt(en)$_2$Cl$_2$]Cl$_2$ has been investigated and the absolute configuration of the two chelate rings determined by anomalous dispersion studies[103] (Appendix II, structure 134). A considerable amount of work employing optically pure cis-[Pt(en)$_2$Cl$_2$]Cl$_2$ in the study of reaction mechanisms will probably be reported in the near future. Investigations of [Pt(en)$_3$]$^{4+}$ cations have shown that the chemical shift differences of the protons in the two conformations of the ligands are too small to be distinguished by n.m.r..[106,107] However, the two conformations of propylenediamine in [Pt(propylenediamine)$_3$]$^{4+}$ can be distinguished because the chemical shift difference between the axial and equatorial methyl protons is 0·13 p.p.m.[108]

(ii) Alkyl and aryl cyanide complexes

Although no palladium(IV) complexes containing alkyl or aryl cyanides appear to have been reported, platinum(IV) complexes [Pt(RCN)$_2$X$_4$], where R = Et, Pr, Bu and Ph and X = Cl or Br, can be prepared by treating [Pt(RCN)$_2$X$_2$] with the halogen in chloroform or bromoform.[109,110] If the bromination of [Pt(RCN)$_2$Cl$_2$] is carried out in ethanol rather than chloroform as a solvent, saponification of the nitrile occurs and the platinum(II) is oxidised to yield (NH$_4$)$_2$[PtBr$_6$]. A mono(methyl cyanide) complex of platinum(IV) is formed on refluxing a solution of sodium hexachloroplatinate(IV) in methyl cyanide by a complex reaction yielding initially [Pt(NH$_3$)$_3$Cl$_3$]$^+$[Pt(MeCN)Cl$_5$]$^-$ from which Cs$^+$[Pt(MeCN)Cl$_5$]$^-$ can be precipitated.[111]

(iii) Azide complexes

An azide complex of platinum(IV), [Pt(N$_3$)$_6$]$^{2-}$, is prepared by treating chloroplatinic acid with excess sodium azide.[112,113] The symmetric N$_3$ stretching vibration at 1270 cm^{-1} and the asymmetric stretching vibration at 2030 cm^{-1},[113] in the infra-red are consistent with a non-linear Pt–N–N$_2$ group.[114,115] A ligand field band at about 20,000 cm^{-1} and a charge transfer band at 32,600 cm^{-1} in the electronic spectrum are consistent with the azide ion being between S-bonded thiocyanate and diethyldithiophos-

phate in the spectrochemical series and between bromide and diethyldi-
thiophosphate in the nephelauxetic series.[116] These positions suggest that
the platinum(IV)–nitrogen σ-bond is augmented by π-donation of electron
density from the azide ligand to an empty orbital of π-symmetry on the
platinum[114] (122). No azide complexes of palladium(IV) appear to have been
reported.

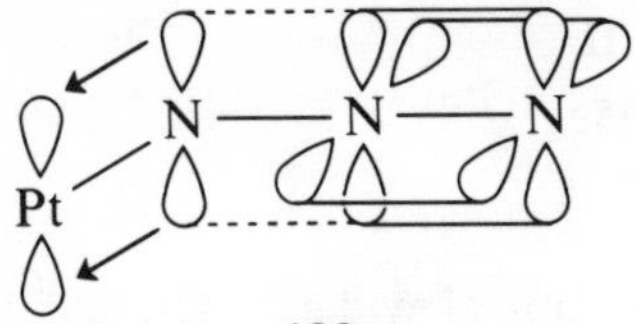

122

(iv) Azobenzene complexes

No azobenzene complexes of either platinum(IV) or palladium(IV) have
been prepared. Re-investigation of the product obtained by treating platinum
tetrachloride with azobenzene in glacial acetic acid has shown that it is a
salt $(PhNH{=}NPh^+)_2[PtCl_6]^{2-}$ in which the azobenzene is not co-ordinated
to the platinum atom,[117] and not $[Pt(PhN{=}NPh)_2Cl_4]$ as first claimed.[118]

COMPLEXES WITH LIGANDS CONTAINING A GROUP VB AND A GROUP IVB DONOR ATOM

Two series of complexes, one monomeric (e.g. **123**) and the other dimeric (e.g.
124) containing both a group VB and a carbon donor atom have been

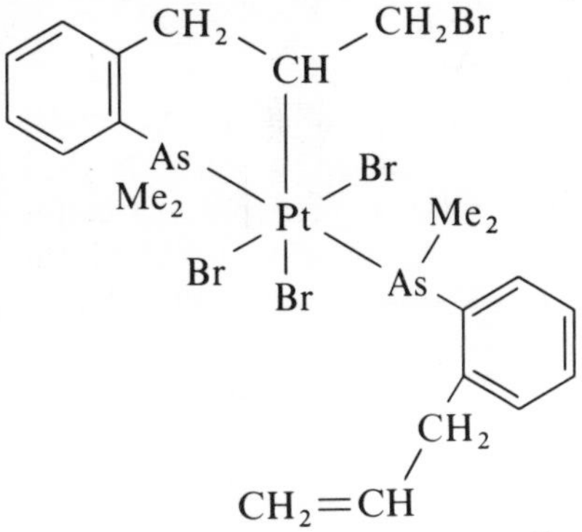

123

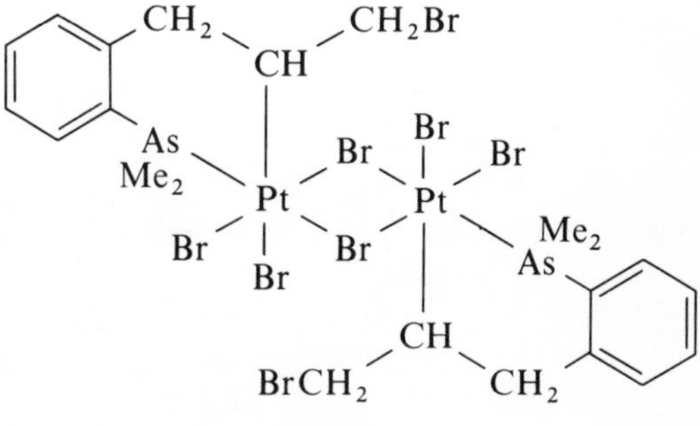

124

prepared by bromine oxidation of the bis– ([Pt(ligand)$_2$X$_2$]) and mono–platinum(II) complexes ([Pt(ligand)X$_2$]) of the ligands **125** and **126** respectively (eqns. 317 and 318).[119,120] The structures of **123** and **124** are based

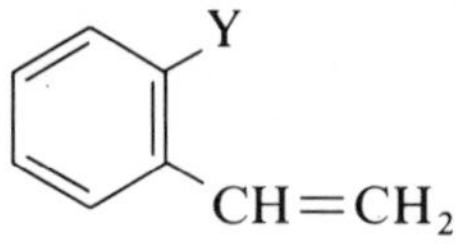

Y = PPh$_2$ or AsMe$_2$

125

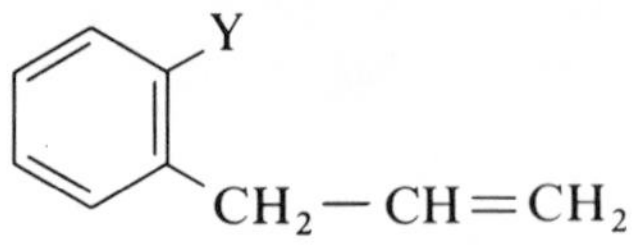

Y = PPh$_2$ or AsMe$_2$

126

on infra-red, n.m.r. and chemical evidence. On boiling in ethanol the mono-meric complexes (**123**) give platinum(IV) complexes in which the platinum(IV)

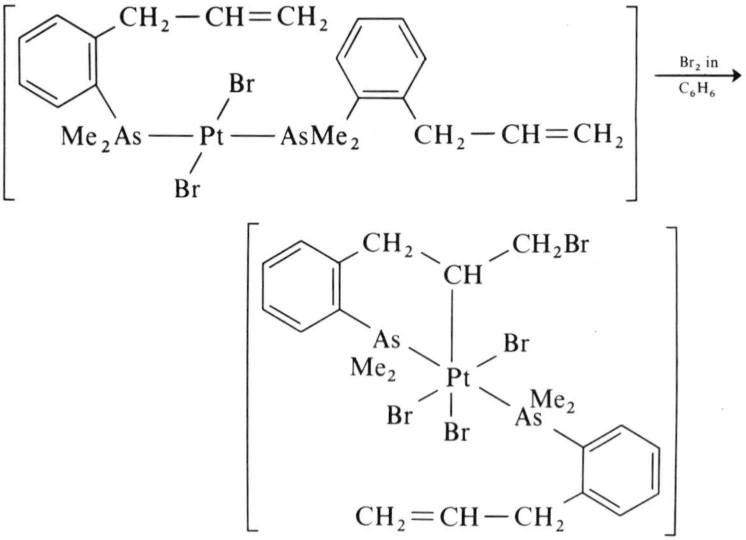

$$\xrightarrow{\text{Br}_2 \text{ in } \text{C}_6\text{H}_6}$$

(317)

$$\xrightarrow{\text{Br}_2 \text{ in } \text{C}_6\text{H}_6}$$

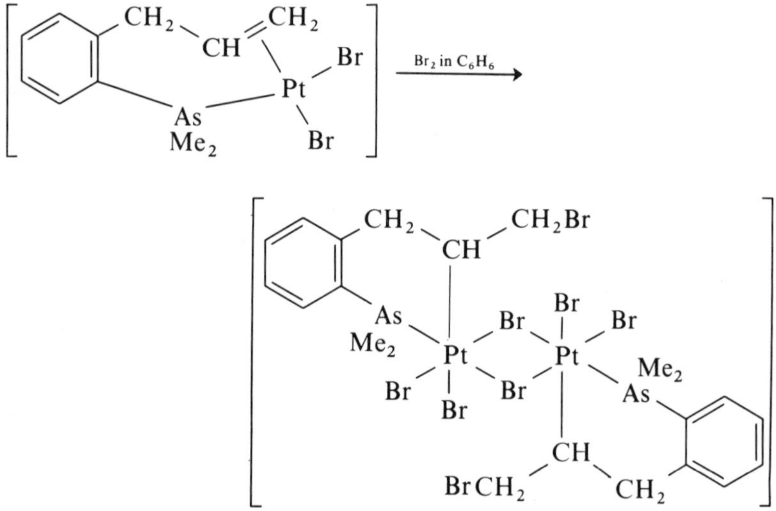

(318)

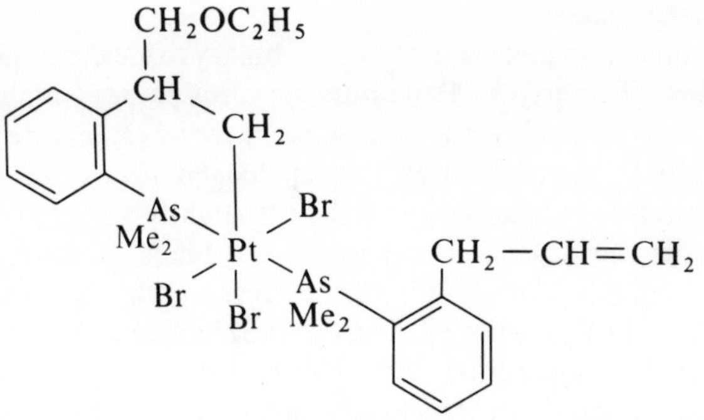

$$CH_2OC_2H_5$$

[Structure showing aromatic ring with CH(CH₂OC₂H₅)CH₂ substituent, AsMe₂ groups, Pt center with Br ligands, and CH₂—CH=CH₂ group]

127

–carbon σ-bond is retained[121] (**127**) (see Appendix II, structure 187). By contrast the dimeric platinum(IV) complexes (**124**) react with alcohols to eliminate bromide and yield platinum(II)–olefin complexes[120,122] (reaction 319). The platinum(IV)–carbon σ-bond in the monomeric platinum(IV)

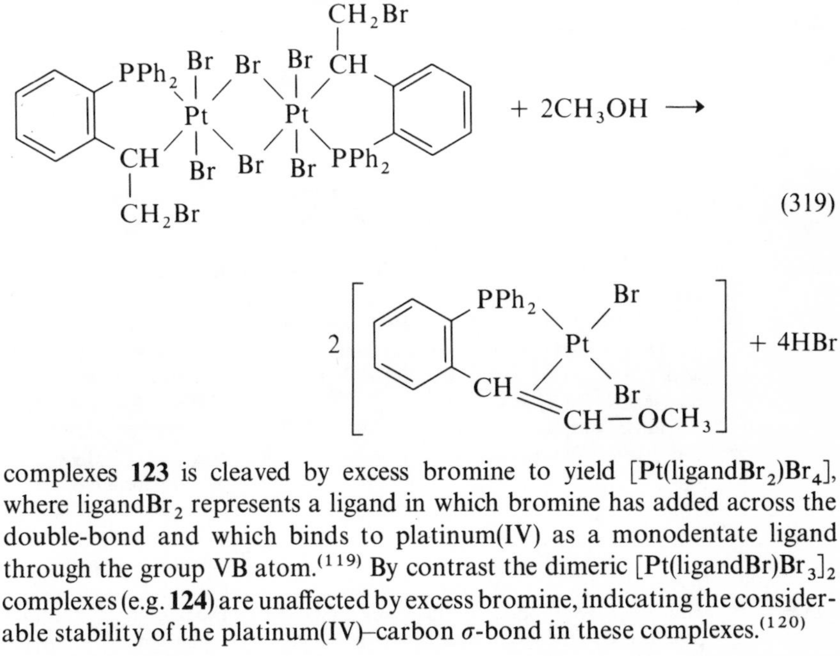

[Reaction structure showing dimeric platinum complex with PPh₂, Br ligands, CH₂Br groups]

$$+ \; 2CH_3OH \; \longrightarrow$$

(319)

$$2 \left[\text{[platinum complex with PPh}_2\text{, Pt, Br, CH=CH—OCH}_3\text{]} \right] \; + \; 4HBr$$

complexes **123** is cleaved by excess bromine to yield [Pt(ligandBr₂)Br₄], where ligandBr₂ represents a ligand in which bromine has added across the double-bond and which binds to platinum(IV) as a monodentate ligand through the group VB atom.[119] By contrast the dimeric [Pt(ligandBr)Br₃]₂ complexes (e.g. **124**) are unaffected by excess bromine, indicating the considerable stability of the platinum(IV)–carbon σ-bond in these complexes.[120]

COMPLEXES WITH GROUP VIB ELEMENTS

Palladium(IV) compounds and complexes of the group VIB elements are rare and limited to the binary oxide, hydroxide and chalcogenides. Platinum(IV) forms a much wider range of complexes which are considered under the headings: (i) binary chalcogenides, (ii) hydroxides and hydroxy-complexes, (iii) dialkyl sulphide and selenide complexes, (iv) nitrate complexes, (v) carbonate complexes and (vi) sulphite complexes.

(i) Binary chalcogenides

Both platinum(IV) and palladium(IV) form binary oxides, sulphides, selenides and tellurides (Table 70). Palladium(II) disulphide, sulphide selenide (PdSSe) and diselenide all form an elongated pyrites structure in which the palladium atom is surrounded by four chalcogen atoms in a square-plane with two more axial chalcogen atoms at a slightly greater distance. This distortion splits the degeneracy of the two orbitals in the e_g level giving diamagnetic semiconductors.[126] On increasing the pressure the axial sulphur atoms in PdS_2 are forced closer to the metal and the semiconduction becomes a metallic conduction.[129] The other dichalcogenides in Table 70 all have cadmium–iodide structures in which the metal atoms occupy perfect octahedral co-ordination sites. PtS_2 and $PtSe_2$ are semiconductors, $PtTe_2$ is a metallic conductor and $PdTe_2$ is a superconductor. Thus whilst the oxides

TABLE 70

Preparation, properties and structures of the oxides, sulphides, selenides and tellurides of palladium(IV) and platinum(IV)

Compound	Preparation	Chemical properties	Structure
$PdO_2.nH_2O$	Formed as a dark red precipitate by treating M_2PdCl_6 with alkali.[123]	Loses oxygen slowly at room temp. and more rapidly on heating to 200°C to give PdO. It is a strong oxidising agent.[123]	
PdS_2	(1) Heat $PdCl_2$ with excess sulphur at 400–500°C[124] (2) Treat Na_2PdS_3 with acid.[125] (3) Heat K_2PdCl_6 or Rb_2PdBr_6 with S at 200–300°C.[125]	Blackish-grey crystalline solid, soluble in carbon disulphide and aqua-regia but not other mineral acids.[125]	Diamagnetic, semiconductor.[126] Elongated pyrites structure in which each Pd is surrounded by four S in a square-plane (Pd–S = 2·30 Å) and two axial S slightly further away.[127,128] High pressure form has less elongation of pyrites structure and metallic conduction properties.[129]
$PdSe_2$	Heat $PdCl_2$ with excess Se at 600°C.[124]	Olive-grey crystalline compound, soluble in conc. HNO_3 and aqua-regia.[124]	Diamagnetic,[126] semiconductor.[126] Elongated pyrites structure in which each Pd is surrounded by four Se in a square-plane (Pd–Se = 2·44 Å) and two axial Se are slightly further away.[127,128]
$PdTe_2$	Heat $PdCl_2$ with excess Te at 750°C.[124]	Silver, crystalline compound, soluble in dilute nitric acid.[124]	Superconductor below 1·69°K.[130,131] Has cadmium iodide structure.[132]

TABLE 70 contd.

Compound	Preparation	Chemical properties	Structure
PtO_2	(1) Boil $PtCl_4$ with excess NaOH, add HOAc and boil white precipitate until it is yellow to give $PtO_2.3H_2O$.[133] (2) Heating $PtO_2.3H_2O$ drives off most of the water, although the last traces cannot be removed without some loss of oxygen.[133] (3) Heating Pt powder in oxygen at 450°C, 150 atm. gives a mixture of PtO and PtO_2. PtO and unreacted Pt can be removed using aqua-regia.[134] (4) Fuse H_2PtCl_6 with $NaNO_3$ at 500–550°C and leach the cooled melt with H_2O.[135,136]	$PtO_2.3H_2O$ is yellow and PtO_2 is a black powder. On heating it decomposes to Pt metal and oxygen.[135] $PtO_2.3H_2O$ is ampho-teric—it dissolves in alkalies to give $M_2[Pt(OH)_6]$ and in hydrochloric acid (easily), nitric acid (with difficulty) and sulphuric acid (with difficulty).[133] $PtO_2.3H_2O$ is an excel-lent catalyst for the hydrogenation of c com-ds.[135-7]	Unit cell is 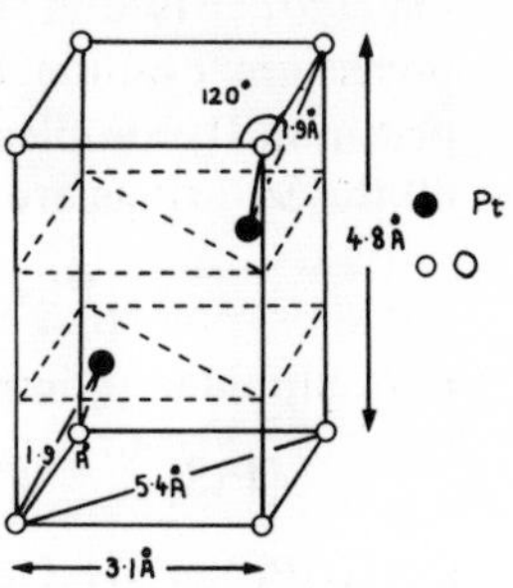 Pt–O = 1·9 Å.[138]
PtS_2	(1) Heat Pt and S at 650°C.[139] (2) Heat $(NH_4)_2PtCl_6$ and S.[140] (3) Treat M_2PtCl_6 solution with H_2S.[140]	When prepared in the dry way it is a steel-grey powder which is quite stable in air.[140] When made in wet way it is black, readily oxi-dised by air and must be filtered and dried in an atmosphere of H_2S.[140]	Diamagnetic,[141] semi-conductor,[142] with CdI_2-type struc-ture[141,143] in which each Pt is octahedrally surrounded by six S at 2·407 Å.[161]
$PtSe_2$	(1) Heat Pt and Se.[144] (2) Heat $PtCl_4$ with excess Se.[124] (3) Treat M_2PtCl_6 solution with H_2Se.[145]	When prepared in wet way $PtSe_2$ is a black amorphous powder, that is less stable than PtS_2 and readily oxi-dised by air.[145]	Diamagnetic,[141] semi-conductor,[142] with CdI_2-type struc-ture[141,143] in which each Pt is surrounded octahedrally by six Se at 2·499 Å.[143]
$PtTe_2$	(1) Heat Pt and Te.[146] (2) Heat $PtCl_4$ with excess Te.[124] (3) Treat M_2PtCl_6 solution with H_2Te.[147]	When prepared in wet way it is black, less stable than $PtSe_2$ and insoluble in dilute acids.[147]	Diamagnetic,[141] metallic conductor,[142] which has CdI_2-type structure[141,143,148] in which each Pt is sur-rounded octahedrally by six Te at 2·666 Å.[141]

of platinum(IV) and palladium(IV) are typical compounds there is a transi-tion in the properties of the ME_2 compounds, as the chalcogens (E) increase in atomic number, towards those of alloys as exemplified by the metallic conduction properties of $PtTe_2$.

Sulphur readily forms polysulphide anions such as S_5^{2-} which form complexes with platinum(IV). $(NEt_4)_2[PtS_{15}].2H_2O$, formed from chloroplatinic acid and ammonium polysulphide,[149] has an approximately octahedral arrangement of three bidentate S_5^{2-} groups around the central platinum(IV) ion[150,151] (see Appendix II, structure 114). $(NH_4)_2[PtS_{15}]$ reacts with potassium cyanide by electron transfer and desulphuration to form the platinum(II) complex anion $[PtS_{10}]^2$ (reaction 320).[152] With excess cyanide all the S_5^{2-} rings are displaced (reaction 321). No unambiguously identified

$$[PtS_{15}]^2 + 5CN^- \rightarrow [PtS_{10}]^{2-} + 5SCN^- \qquad (320)$$

polysulphide   complexes   of   palladium(IV)   have   been   reported;

$$[PtS_{15}]^{2-} + 17CN^- \rightarrow [Pt(CN)_4]^{2-} + 13SCN^- + 2S^{2-} \qquad (321)$$

$(NH_4)_2PdS_{11}.\tfrac{1}{2}H_2O$, prepared from sodium tetrachloropalladate(II) and ammonium pentasulphide[153] is probably a palladium(II) complex, although its structure has not been investigated.

(ii) Hydroxides and hydroxy-complexes
The preparations and properties of the more important hydroxides and hydroxy-complexes of the tetravalent metals are summarised in Table 71 ($[(CH_3)_3Pt(OH)]_4$ is considered with the other trimethylplatinum(IV) complexes in Chapter 12 (pp. 351–354)). The simple hydroxides of platinum(IV) and palladium(IV), $[M(OH)_4]$, cannot be isolated as anhydrous materials, since decomposition occurs before all the water has been removed. Palladium(IV) hydroxide is less stable than its platinum(IV) analogue and is unstable in air. In addition to the simple hydroxide, platinum(IV) forms a wide range of hydroxy-complexes containing halide (see Table 71), ammonia,[161,172,173] ethylenediamine[172] and nitrite[174] ligands. These are prepared either by treating a chloroplatinum(IV) complex with sodium hydroxide (for further references see ref. 175) or by treating the corresponding platinum(II) complex with hydrogen peroxide.[161] The latter method is limited to the preparation of the dihydroxy–platinum(IV) complexes, whereas titration with alkali can be used to prepare any hydroxy-complex of platinum(IV).

X-ray diffraction has shown $K_2[Pt(OH)_6]$ to be an octahedral complex, isomorphous with $K_2Sn(OH)_6$[176] having a platinum–oxygen bond length of 1·96 Å.[160] The electronic spectra of *trans*-$[PtCl_4(OH)_2]^{2-}$ salts exhibit two ligand field bands at lower energy than in the hexachloroplatinate(IV) anions,[167] in direct contrast to the prediction of the spectrochemical series.[177] Extended Hückel calculations indicate that this arises because the π-antibonding influence of hydroxide and chloride ions is larger than their σ-antibonding influence so that hydroxide ions give a weaker ligand field in platinum(IV) complexes than chloride ions.

(iii) Dialkyl sulphide and selenide complexes
Dialkyl sulphide[178,179] and dialkylselenide[180] complexes of platinum(IV) of the type *cis*- and *trans*-$[Pt(ER_2)_2X_4]$, where E = S or Se and X = Cl, Br or

TABLE 71

Preparations and properties of palladium(IV) and platinum(IV) hydroxides and hydroxy-complexes

Compound or complex	Preparation	Properties
$Pd(OH)_4$ $(PdO_2.2H_2O)$	Treat M_2PdCl_6 with alkali to give a dark-red precipitate which turns black on attempting to dry it.[154]	A strong oxidising agent that is unstable in air and decomposes with liberation of oxygen. Readily reduced by hydrogen, organic acids and hydrogen peroxide. It is amphoteric, dissolving in HCl, HNO$_3$ and H$_2$SO$_4$ to give Pd^{2+} ions, and although insoluble in dilute NaOH it dissolves in 10 N KOH to give palladate (PdO$_3^{2-}$) ions.[155]
$H_2Pt(OH)_6$ $(Pt(OH)_4.2H_2O)$	Heat H_2PtCl_6 with boiling alkali and precipitate with acetic acid.[156]	Pale-yellow powder, which loses two molecules of water at 100°C and two more at higher temperatures. Soluble in water and alcohol.[156] It is amphoteric, dissolving in alkalis to give $M_2Pt(OH)_6$ and acids to give $[PtX_6]^{2-}$ salts.[157] It is a moderately energetic oxidising agent.[158]
$M_2Pt(OH)_6$	Dissolve H_2PtCl_6 in alkali.[156]	Does not lose water at 100°C, but decomposes with loss of water at higher temperatures.[156,159] Very stable. X-ray diffraction study of $K_2Pt(OH)_6$ shows Pt surrounded by six O in an octahedral arrangement with Pt–O = 1·96 Å.[160]
$M_2[PtCl_4(OH)_2]$	(1) Treat M_2PtCl_4 with hydrogen peroxide[161,162] (this preparation applies to $[PtL_4(OH)_2]^{n-}$ where L = Cl, Br, NH$_3$ etc.). (2) Dissolve PtCl$_4$ in water when $H_2[PtCl_4(OH)_2]$.aq. crystallises out.[163] ($M_2[PtX_4(OH)_2]$ where X = Br or I are prepared similarly from PtX$_4$).[164,165] (3) Treat M_2PtCl_4 with hydrogen persulphide (H$_2$S$_2$).[166]	The hydroxo-ligands in these complexes can be replaced by acid residues by using the calculated amount of acid.[161] The electronic and vibrational spectra of *trans*-$[PtCl_4(OH)_2]^{2-}$ have been reported[167] and it was found that the ligand field of OH$^-$ is lower than that of Cl$^-$ in Pt(IV) complexes.
$[PtX_n(OH)_{6-n}]^{2-}$ (X = Cl or Br)	Prepared by successive replacement of X$^-$ in $[PtX_6]^{2-}$ by OH$^-$.[168–171]	

I, are prepared by oxidising the corresponding *cis*- or *trans*-platinum(II) complexes with halogens. The infra-red spectra[90] of *trans*-[Pt(SMe$_2$)$_2$X$_4$], where X = Cl, Br or I, can be interpreted as a superposition of the spectra of square-planar [PtX$_4$]$^{2-}$ and linear Me$_2$S–Pt–SMe$_2$ units. The Pt–(SMe$_2$)$_2$ stretching frequencies in *trans*-[Pt(SMe$_2$)$_2$X$_4$] complexes occur at 311, 321 and 319 cm^{-1} in the chloride, bromide and iodide respectively, which is very close to the 311, 315 and 313 cm^{-1} respectively found for the chloride, bromide and iodide *trans*-[Pt(SMe$_2$)$_2$X$_2$] complexes of platinum(II).[181] The infra-red spectrum of *cis*-[Pt(SMe$_2$)$_2$Cl$_4$] is complex due to the relatively low symmetry and the fact that v(Pt–Cl) and v(Pt–S) have similar frequencies so that substantial mixing of the fundamental vibrational modes occurs.

(iv) Nitrate complexes

Nitrato-complexes of palladium(IV) have not been reported. The nitrato-complexes of platinum(IV) can be prepared by two main routes, one of which involves the oxidation of nitrato–platinum(II) complexes, either by concentrated nitric acid[182] (reaction 322) or by hydrogen peroxide[183] (reaction 323), and the other of which involves treating a platinum(IV) halide complex

$$\textit{cis or trans-}[Pt(NH_3)_2Cl_2] + 2AgNO_3 \xrightarrow{\text{dil. HNO}_3}$$

$$2AgCl + \textit{cis- or trans-}[Pt(NH_3)_2(NO_3)_2]$$

$$\downarrow \substack{\text{conc. HNO}_3\text{, evaporate} \\ \text{to dryness.}}$$

$$\textit{cis- or trans-}[Pt(NH_3)_2(NO_3)_4] \qquad (322)$$

with a stoichiometric amount of silver nitrate[184] (reaction 324). In addition

$$[Pt(NH_3)_2(NO_3)_2] + H_2O_2 \rightarrow [Pt(NH_3)_2(NO_3)_2(OH)_2] \qquad (323)$$

$$[Pt(NH_3)_2Br_4] + 2AgNO_3 \rightarrow [Pt(NH_3)_2Br_2(NO_3)_2] + 2AgBr \quad (324)$$

K$_2$[Pt(NO$_3$)$_4$(OH)$_2$] has been prepared by addition of a stoichiometric amount of potassium nitrate to a solution of freshly prepared platinum(IV) oxide (PtO$_2$.H$_2$O) in nitric acid.[185]

No platinum(IV) nitrate complexes have been studied by x-ray diffraction so that the exact mode of co-ordination of the nitrato-group to the platinum atom is uncertain.[186] However, the infra-red spectra of *cis*- and *trans*-[Pt(NH$_3$)$_2$(NO$_3$)$_4$] suggest that co-ordination occurs through only one oxygen atom.[182] The nitrato-group binds fairly weakly to platinum(IV) so that the complexes are readily hydrolysed in water leading to hydroxo-complexes[184] (reaction 325). The nitrato-group can also be replaced by halide ions.[184]

$$\textit{trans-}[Pt(NH_3)_2(NO_3)_4] \xrightarrow{H_2O}$$

$$\textit{trans-}[Pt(NH_3)_2(NO_3)_2(OH)_2] + 2H^+ + 2NO_3^- \quad (325)$$

(v) Carbonate complexes

A white *trans*-dicarbonate-platinum(IV) complex [Pt(en)$_2$(CO$_3$)$_2$], in which the carbonate ligands appear to bind to the metal as unidentate ligands through oxygen, has been prepared[69] by reaction 326. The carbonato–ligands are readily displaced either by heating in water when $[Pt(en)_2(OH)_2]CO_3$ is regenerated or by treating with hydrochloric acid when $[Pt(en)_2Cl_2]Cl_2$ is formed.[69]

$$[Pt(en)_2Cl_2] + Ag_2CO_3 \rightarrow 2AgCl + [Pt(en)_2]CO_3 \xrightarrow{H_2O_2}$$

$$[Pt(en)_2(OH)_2]CO_3 \underset{\substack{\text{heat in} \\ \text{water}}}{\overset{CO_2}{\rightleftharpoons}} [Pt(en)_2(CO_3)_2] \quad (326)$$

(vi) Sulphite complexes

The product formed when $[Pt(NH_3)_4Cl_2](NO_3)_2$ is treated with sulphurous acid was originally thought to be the platinum(II) complex $[Pt(NH_3)_4]$-$(HSO_3)_2$.[187] However, later work suggested that it is the platinum(IV) complex $[Pt(NH_3)_4(SO_3)_2].2H_2O$,[188,189] which is also formed when $[Pt(NH_3)_4]Cl_2$ is treated with sulphur dioxide and potassium nitrite.[189] On heating $[Pt(NH_3)_4(SO_3)_2].2H_2O$ at 100°C, sulphur dioxide is eliminated and the platinum(II) complex $[Pt(NH_3)_4]SO_4$ is formed;[189] with strong hydrochloric acid $[Pt(NH_3)_4(SO_3)_2].2H_2O$ yields $[Pt(NH_3)_4Cl_2]Cl_2$. It is, as yet, uncertain whether the sulphite ligands are bound to the metal through sulphur or oxygen.

COMPLEXES WITH MULTIDENTATE LIGANDS CONTAINING BOTH A GROUP VB AND A GROUP VIB DONOR ATOM

As in Chapter 8, this section is conveniently divided into two parts, the first covering multidentate potentially chelating ligands and the second covering ambidentate ligands such as thiocyanate ions.

(i) Multidentate ligands

The only multidentate ligand containing both a group VB donor atom that has been used to prepare complexes of the tetravalent metals is di(2-aminoethyl)-sulphide, which forms optically active complexes of platinum(IV) in which the sulphur and one of the nitrogen atoms are co-ordinated to the metal[190] (reaction 327). On boiling in water the second nitrogen atom

$$S(CH_2CH_2NH_3^+Cl^-)_2 + [PtCl_4]^2 \rightarrow 2Cl^- +$$

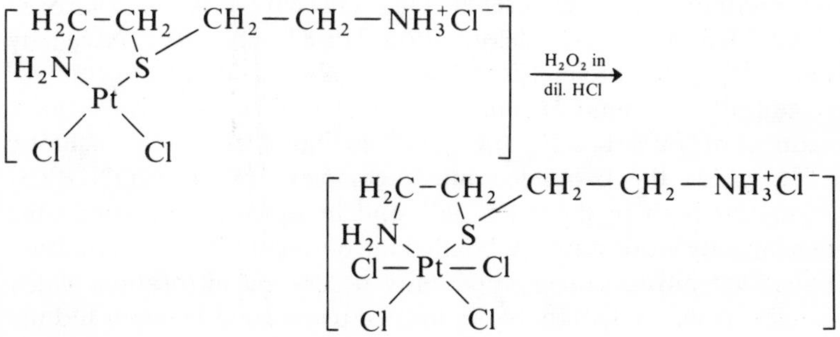

co-ordinates to the metal (reaction 328), but the overall stereochemistry

$$\begin{bmatrix} \text{H}_2\text{C}-\text{CH}_2 & \text{CH}_2-\text{CH}_2-\text{NH}_3^+\text{Cl}^- \\ \text{H}_2\text{N} \quad \text{S} \\ \text{Cl}-\text{Pt}-\text{Cl} \\ \text{Cl} \quad \text{Cl} \end{bmatrix} \xrightarrow[\text{water}]{\text{boil in}}$$

$$\begin{bmatrix} \text{H}_2\text{C}-\text{CH}_2 \\ \text{H}_2\text{N} \quad \text{S}-\text{CH}_2 \\ \text{Cl}-\text{Pt} \quad | \\ \text{Cl} \quad \text{Cl} \quad \text{NH}_2 \quad \text{CH}_2 \end{bmatrix}^+ + \text{Cl}^- + \text{HCl} \quad (328)$$

(i.e. fac or mer) of the product is unknown.

(ii) Ambidentate ligands
Of the various possible ambidentate ligands containing donor atoms from group VB and group VIB only nitrite, thiocyanate and selenocyanate ions form complexes with platinum(IV), and none form complexes with palladium(IV).

Nitro- and nitro-complexes
As with the divalent metals the majority of the nitrite complexes of platinum(IV) involve N-bonded nitro rather than O-bonded nitrito ligands. By treating sodium hexachloroplatinate(IV) with the required quantity of sodium nitrite chloronitro complexes of platinum(IV), $Na_2[Pt(NO_2)_nCl_{6-n}]$, where $n = 0$ to 5, have been prepared.[191] This reaction is not a simple substitution reaction but involves intermediate reduction of platinum(IV) to platinum(II), and reaction of nitrite ion with the intermediate platinum(II) complex. The hexa-nitro compound $K_2[Pt(NO_2)_6]$ cannot be prepared by substitution of the final chloride ion in $K_2[Pt(NO_2)_5Cl]$ possibly because of the very low *trans*-effect of the nitro-group.[191,192] However, oxidation of the platinum(II) complex $K_2[Pt(NO_2)_4]$ with concentrated nitric acid saturated with the oxides of nitrogen yields blue $K_2[Pt(NO_2)_4(NO)(NO_3)]$ which readily isomerises through a red intermediate to the colourless stable $K_2[Pt(NO_2)_6]$.[193] The infra-red spectra of nitrite complexes have nitrite vibrations (asymmetric N–O stretch 1443–1497 cm^{-1}, symmetric N–O stretch 1300–1373 cm^{-1}, NO_2 deformation 816–836 cm^{-1} and NO_2 wag 585–621 cm^{-1}[194]) that are consistent with an N-bonded nitro-ligand.[195] However, not all platinum(IV) complexes involve N-bonded nitro-ligands since treatment of $[Pt(NH_3)_5(H_2O)]Cl_4$ with sodium nitrite in hydrochloric acid at 0°C yields the O-bonded nitrito-complex $[Pt(NH_3)_5(ONO)]Cl_3$ which isomerises both in the solid state and in aqueous solution to the thermodynamically more stable N-bonded nitro-isomer.[196] The formation of the O-bonded nitrito-isomer apparently occurs via nitrosation of the aquo-complex (reaction 330) by N_2O_3 formed by reaction between sodium nitrite and hydrochloric acid (reaction 329).

$$2NaNO_2 + 2HCl \rightarrow 2NaCl + 2HNO_2 \rightarrow N_2O_3 + H_2O \quad (329)$$

$$[(NH_3)_5Pt–OH_2]^{4+} \rightleftharpoons H^+ + [(NH_3)_5Pt–OH]3 + \xrightarrow{N_2O_3}$$

$$\begin{bmatrix} (NH_3)_5Pt–O\cdots H \\ | \\ O–N\cdots NO_2 \end{bmatrix}^{3+} \rightarrow HNO_2 + [(NH_3)_5Pt–ONO]^{3+} \quad (330)$$

Thiocyanate and selenocyanate complexes

The known platinum(IV) complexes of thiocyanate and selenocyanate, which include $[Pt(SCN)_6]^{2-}$, $[Pt(SeCN)_6]^{2-}$ [197–202] and *cis* and *trans*-$[Pt(NH_3)_2$-$(NO_2)_2(SCN)_2]$,[203] all involve co-ordination through the group VIB atom (details of the structure of $K_2[Pt(SeCN)_6]$ are given in Appendix II, structure 141).

COMPLEXES WITH GROUP VIIB ELEMENTS

The preparation and properties of the halides and halide complex anions of the tetravalent metals are summarised in Tables 72 and 73. It is apparent that

TABLE 72
Preparations, properties and structures of the halides of palladium(IV) and platinum(IV)

Halide	Preparation	Properties	Structure
PdF_4	Action of fluorine on $Pd^{II}[Pd^{IV}F_6]$ at 150–300°C.[204] The product gives a sharper x-ray diffraction pattern if the fluorination is carried out at 100 lb/in².[204]	Brick-red solid easily reduced chemically or thermally to $Pd^{II}[Pd^{IV}F_6]$.[204]	X-ray diffraction investigation of the powder shows a tetragonal unit cell ($a = 6.585$ Å and $c = 5.835$ Å) with four formula units in the cell. Structurally similar to PtF_4.[204]
PtF_4	(1) Thermal decomposition of $(BrF_3)_2PtF_4$ obtained by interaction BrF_3 with $PtCl_4$ or $PtBr_4$.[205] (2) Fluorination of $PtCl_2$ below 200°C.[206]	Yellow-brown crystalline solid, only slowly hydrolysed by water.[207] Reduced to Pt metal by hydrogen at 100°C.[207]	Investigated by x-ray powder diffraction.[207] Monoclinic ($a = 6.68$ Å, $b = 6.68$ Å, $c = 5.71$ Å, $\gamma = 92.02°$). Four molecules to the unit cell; similar to UCl_4 in which metal is eight-co-dinate.
$PtCl_4$	(1) Chlorinate Pt metal at elevated temperature.[208] (2) Chlorinate $PtCl_2$ at 275°C.[209] (3) Heat H_2PtCl_6 alone or in Cl_2.[210]	Reddish-brown crystals soluble in H_2O, acetone and alcohol.[211] Thermally unstable gives $PtCl_2$ above 350°C.[209,212,213]	Structure unknown—powder x-ray diffraction study[210] has been interpreted on basis of tetrahedral $PtCl_4$ units. This is most unlikely to be correct[211] and a much more likely structure that is consistent with the x-ray diffraction pattern involves tetrameric Pt_4Cl_{16} units analogous to the tetrameric units in $[Me_3PtI]_4$.[214]

TABLE 72 cont.

Halide	Preparation	Properties	Structure
$PtBr_4$	(1) Brominate Pt metal at 150°C.[215] (2) Pass Br_2 over H_2PtBr_6 at 285–310°C.[215] (3) Dissolve metallic Pt in concentrated HBr containing excess Br_2 in a sealed tube at 150°C.[211]	Dark red solid only slightly soluble in water.[211] Thermally less stable than $PtCl_4$ giving $PtBr_2$ above 180°C.[215,216]	X-ray powder diffraction pattern recorded,[216] but not interpreted in terms of a structure.
PtI_4	Heat Pt metal in a sealed tube with either a stoichiometric amount of iodine at 150°C[217] or excess iodine at 240°C.[218]	Black crystalline solid,[217] which decomposes slightly at room temperature and quantitatively at 200°C.[218]	Investigated by powder x-ray diffraction.[217,218] Data did not fit cubic, hexagonal or tetragonal systems.

TABLE 73

Preparations, properties and structures of the complex halides of palladium(IV) and platinum(IV)

Complex	Preparation	Properties	Structure
$Pd^{II}[Pd^{IV}F_6]$ (PdF_3)	(1) Fluorination of PdX_2 (X = halogen) at 200–250°C in a flow system.[219] (2) BrF_3 treatment of PdX_2 gives $PdF_3.BrF_3$ which is decomposed at 180°C in vacuo.[204,205,220]	Black crystalline solid, rapidly hydrolysed by water.[205]	Isostructural with $Pd^{2+}[MF_6]^{2-}$ where M = Ge, Pt, Sn.[204] Room temperature magnetic moment shows two unpaired electrons associated with Pd^{2+} in an octahedral environment.[204] X-ray diffraction study of powder shows Pd–F = 2·04 Å in $[PdF_6]^{2-}$.[221]
PdF_6^{2-}	(1) Fluorination of M_2PdCl_6 (M = K, Rb and Cs) at 200–300°C gives M_2PdF_6.[222] (2) Dissolution of M_2PdCl_6 in BrF_3 followed by heating to 150°C gives M_2PdF_6.[223]	Pale-yellow salts, immediately hydrolysed by water[223] (contrast PtF_6^{2-}).	X-ray powder diffraction studies[221-6] indicate that the salts can form a number of phases according to the method of preparation. Pd–F bond lengths vary between 1·98 Å (Mg^{2+} cation) and 2·22 Å (Ca^{2+} cation).[221]
$PdCl_6^{2-}$	(1) Dissolve Pd metal in aqua regia or HCl saturated with Cl_2. Isolate by addition of alkali metal chloride.[227] (2) Dissolve K_2PdF_6 in HCl.[223] (3) Dissolve $(SeF_3)_2PdF_6$ in a solution of KCl in HCl.[228]	Dark-red crystalline solids, which lose Cl_2 to give M_2PdCl_4 above about 200°C.[227,229,230]	Isostructural with $(NH_4)_2PtCl_6$.[231]

TABLE 73 cont.

Complex	Preparation	Properties	Structure
$PdBr_6^{2-}$	Action of bromine on M_2PdBr_4 in HBr solution.[232]	Black solids giving dark red solutions in HBr.[233] Stable in air, slightly soluble in cold water and decomposed by hot water with liberation of bromine.[233]	X-ray powder diffraction studies show that M_2PdBr_6 are cubic with $(NH_4)_2PtCl_6$ structure.[231,234]
PtF_6^{2-}	(1) Treatment of $PtCl_4^{2-}$ with F_2[235] or BrF_3[223] (2) Neutralisation of H_2PtF_6, prepared from $La_2(PtF_6)_3$ using an ion-exchange resin, with MOH.[236] $La_2(PtF_6)_3$ is prepared by fluorinating Pt foil and LaF_3 at 525°C.[237] (3) Treatment of Pt metal with a mixture of Br_2 and BrF_5 gives $(BrF_2)_2PtF_6$, which on addition of KF yields K_2PtF_6.[238]	Pale-yellow, diamagnetic solids which are stable to H_2O (cf. PdF_6^{2-}). Na^+ salt decomposed in air just above room temperature,[235] whilst Mg^{2+}, Ca^{2+} and Sr^{2+} salts decompose at 600°C to Pt metal and alkaline earth difluoride.[236]	PtF_6^{2-} in K_2PtF_6 is octahedral with Pt–F = 1·91 Å.[239]
$PtF_3Cl_3^{2-}$	(1) Interaction of BrF_3 and M_2PtCl_6 (where M = alkali metal) under controlled conditions.[240] (2) BrF_3 + K_2PtCl_4 at room temperature gives $K_2[PtF_3Cl_3]$.[240]	Stable, non-hygroscopic orange solids which disproportionate only slowly in aqueous solution to PtF_6^{2-} and $PtCl_6^{2-}$ [240]	X-ray powder diffraction data showed that $PtF_3Cl_3^{2-}$ is a species distinct from PtF_6^{2-} and $PtCl_6^{2-}$.[240] No details are available as to fac or mer nature of the complex anion.
$PtCl_6^{2-}$	(1) Dissolve metal in aqua regia or HCl saturated with Cl_2.[241] Isolate by addition of alkali metal chloride. (2) Dissolve $PtCl_4$ in HCl and add alkali metal chloride.[242] (3) Pass Cl_2 or CCl_4 over a heated mixture of Pt metal and alkali metal chloride.[243]	Dark-red crystalline solids.	Alkali metal salts are isostructural with $(NH_4)_2PtCl_6$, which has been shown by single crystal x-ray diffraction to contain Pt octahedrally surrounded by six Cl at 2·44 Å.[244] In H_2PtCl_6 solution Pt–Cl $\approx$ 2·39 Å[245] and in K_2PtCl_6 solid Pt–Cl = 2·33 Å.[246]
$PtBr_6^{2-}$	Salts are generally prepared from H_2PtBr_6,[247,248] which is prepared by dissolving Pt metal in HBr containing free bromine[248] or by dissolving $PtCl_4$ in HBr.[249]	H_2PtBr_6 is a carmine crystalline solid. Deliquescent and very soluble in H_2O, EtOH, HOAc, $CHCl_3$ and Et_2O.[249] Salts of K^+, Ag^+, NH_4^+, Rb^+, Cs^+ and Tl^+ are insoluble in water; Pb^{2+} salt is soluble in water.[249]	X-ray powder diffraction studies show that most salts of $PtBr_6^{2-}$ adopt the cubic K_2PtCl_6 structure.[250] In solution $PtBr_6^{2-}$ is a regular octahedron with a Pt–Br bond length of 2·43 Å.[245]

TABLE 73 cont.

Complex	Preparation	Properties	Structure
PtI_6^{2-}	(1) Treat H_2PtCl_6 with excess KI.[251] (2) Treat PtI_4 in HI with alkali metal iodide.[252] (3) Treat H_2PtCl_4 in HI with alkali metal iodide.[253]	Brown crystalline compounds with a metallic lustre.[252] In aqueous solution they are rapidly decomposed by heat and light.[252]	X-ray powder diffraction studies have been carried out which only give information on the symmetry of the unit cell.[240] ^{127}I n.q.r. results suggest that there are three crystallographically non-equivalent iodine atoms.[254]

whilst platinum(IV) forms halides PtX_4 and halide complexes $[PtX_6]^{2-}$ with all the halogens, palladium(IV) only forms a halide with fluorine and halide complexes with fluorine, chlorine and bromine. This result is typical of what has already been found on many occasions in this chapter, namely that whereas platinum readily forms compounds and complexes in the tetravalent state, palladium is reluctant to form tetravalent palladium (p. 13). The concentration of the lighter halide needed to displace a heavier halide from its $[PtX_6]^{2-}$ anion[255] (Table 74) indicates that platinum(IV) has strong class 'b' characteristics in solution and exhibits a much greater preference for the heavier than the lighter halides. In spite of this result, which is in accordance with evidence from preparative studies, preliminary results for equilibrium 330a have suggested that the ratios of successive equilibrium constants (K_n/K_{n+1}) are virtually theoretical[256] (Table 75).

$$[PtCl_{7-n}Br_{n-1}]^{2-} + Br^- \rightleftharpoons [PtCl_{6-n}Br_n]^{2-} + Cl^- \qquad (330a)$$

TABLE 74
Concentrations of Y^- necessary to force the reaction $[PtX_6]^{2-} + 6Y^- \rightleftharpoons [PtY_6]^{2-} + 6X^-$ to go to 99% completion (data taken from ref. 255)

X	Y	Concentration of Y needed relative to concentration of $[PtX_6]^{2-}$
Br	Cl	660
I	Br	25,000
I	Cl	16,000,000

TABLE 75
Ratios of successive equilibrium constants for equilibrium 330a in aqueous solution at a total platinum(IV) concentration of 2 M (data from ref. 256)

K_n/K_{n+1}	K_1/K_2	K_2/K_3	K_3/K_4	K_4/K_5	K_5/K_6
Experimental	2·48	1·76	1·68	1·80	2·01
Statistical	2·40	1·875	1·78	1·875·	2·40

TABLE 76
Fundamental bands in the vibrational spectra of K_2MX_6 complexes

Band	Symmetry	Raman or infra-red activity	Approximate description of vibrational mode	K_2PdF_6	K_2PdCl_6	K_2PtCl_6	K_2PtBr_6	K_2PtI_6
ν_1	A_{1g}	R	M–X stretch	317[a]	600[b]	344[a]	207[a]	
ν_2	E_g	R	M–X stretch	292[a]	576[b]	320[a]	190[a]	
ν_3	F_{1u}	IR	M–X stretch	340[c]	571[b]	343[c]	244[c]	186[d]
ν_4	F_{1u}	IR	X–M–X def.	175[c]	281[b]	182[c]	90[c]	46[d]
ν_5	F_{2g}	R	X–M–X def.	164[a]	210[b]	162[a]	97[a]	

[a] Reference 258. [b] Reference 235. [c] Reference 262. [d] Reference 259.

(i) Vibrational spectroscopy

A comparison of the fundamental bands in the infra-red and Raman spectra of the $[MX_6]^{2-}$ ions[235,257—62] (Table 76) as well as the calculated force constants (Table 77) with those for the divalent metal ions (see Table 59, p.

TABLE 77
Force constants for the M–X stretching vibration in
$[MX_6]^{2-}$ anions

Anion	$K(M–X)(mdyn/Å)$	Reference
$[PdCl_6]^{2-}$	1·40	262
$[PtCl_6]^{2-}$	$\begin{cases} 1·87 \\ 1·89 \end{cases}$	262 / 263
$[PtBr_6]^{2-}$	1·54	262

241) indicates that the increase in the metal–halogen stretching frequency on going from the divalent to the tetravalent state (e.g. $\nu_{asym.}$(Pt–Cl) in $[PtCl_4]^{2-}$ = 321 cm^{-1} and $\nu_{asym.}$(Pt–Cl) in $[PtCl_6]^{2-}$ = 343 cm^{-1} and $\nu_{sym.}$(Pt–Cl) in $[PtCl_4]^{2-}$ = 333 cm^{-1} and $\nu_{sym.}$(Pt–Cl) in $[PtCl_6]^{2-}$ = 344 cm^{-1}), is accompanied by a slight decrease in the platinum–chlorine stretching force constants.

(ii) Electronic spectroscopy

The electronic spectra of $[MX_6]^{2-}$ anions[264—6] (see Table 78) exhibit both weak bands, which have been assigned to *d–d* transitions and strong bands

TABLE 78
Electronic spectra of $[MX_6]^{2-}$ anions

Complex anion	$\lambda(cm^{-1})$	Extinction coefficient	Assignment[a]	Reference
PtF$_6^{2-}$	28,750	~50	$d \rightarrow d$ (singlet $\rightarrow$ singlet)	264, 267
(colourless)	36,350	~50	$d \rightarrow d$ (singlet $\rightarrow$ singlet)	264, 267
PtCl$_6^{2-}$	22,100	50	$d \rightarrow d$ (singlet $\rightarrow$ triplet)	264
(yellow)	28,300	490	$\begin{cases} d \rightarrow d \text{ (singlet } \rightarrow \text{ singlet)} \\ \text{ligand } \pi \ (t_{1u}) \rightarrow \text{metal } d \text{ (singlet } \rightarrow \text{ triplet)} \end{cases}$	264 / 265
	38,200	24,500	ligand π (t_{1u}) $\rightarrow$ metal d (singlet $\rightarrow$ singlet)	265
	49,500		ligand σ (t_{1u}) $\rightarrow$ metal d (singlet $\rightarrow$ singlet)	265
PtBr$_6^{2-}$	19,100	140	$d \rightarrow d$ (singlet $\rightarrow$ triplet)	264
(dark-red)	23,000	1,500	$d \rightarrow d$ (singlet $\rightarrow$ singlet)	264
	27,500	7,400	ligand π (t_{1u}) $\rightarrow$ metal d (singlet $\rightarrow$ triplet)	265
	31,900	v.wk. shoulder	ligand π (t_{2u}) $\rightarrow$ metal d (singlet $\rightarrow$ triplet)	265
	32,900	shoulder	ligand π (t_{1u}) $\rightarrow$ metal d (singlet $\rightarrow$ singlet)	265
	40,500	shoulder	ligand σ (t_{1u}) $\rightarrow$ metal d (singlet $\rightarrow$ triplet)	265
	44,500		ligand σ (t_{1u}) $\rightarrow$ metal d (singlet $\rightarrow$ singlet)	265
PtI$_6^{2-}$	20,250	12,800	ligand π (t_{1u}) $\rightarrow$ metal d (singlet $\rightarrow$ triplet)	264, 265
(brown)	22,400	8,300	ligand π (t_{2u}) $\rightarrow$ metal d (singlet $\rightarrow$ triplet)	264, 265
	29,150	17,700	ligand π (t_{1u}) $\rightarrow$ metal d (singlet $\rightarrow$ singlet)	264, 265
PdCl$_6^{2-}$	29,400		ligand π (t_{1u}) $\rightarrow$ metal d (singlet $\rightarrow$ singlet)	265
(red)	40,800		ligand σ (t_{1u}) $\rightarrow$ metal d (singlet $\rightarrow$ singlet)	265

[a] Based on the qualitative molecular orbital scheme in Fig. 22. The assignment of the charge-transfer bands was aided by magnetic circular dichroism studies.[265]

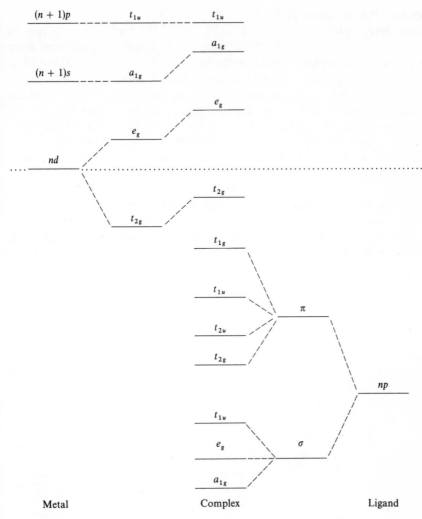

Fig. 22. Schematic energy level diagram for $[MX_6]^{2-}$. All levels below the dotted line are filled. All charge transfer transitions to metal e_g orbitals can originate only from the ligand t_{1u} and t_{2u} levels. Reproduced by permission from reference 265.

due to charge transfer from the halide orbitals to the metal. These charge transfer bands lie at lower energies than in the corresponding platinum(II) and palladium(II) complexes due to the d-orbitals being more stable as a result of the higher effective nuclear charges of the tetravalent metal ions. The ligand(π) to metal charge transfer band is less sensitive to the nature of the ligand in the platinum(IV) complexes than in the platinum(II) complexes (for example this band varies between 31,700 and 37,100 cm^{-1} for platinum(IV) complexes and between 36,000 and 47,000 cm^{-1} for platinum(II) for the series of ligands Cl$^-$, Br$^-$, N$_3^-$, SeCN$^-$, SCN$^-$ and NO$_2^-$). This could be due to either less π-bonding or to a greater involvement of the metal acceptor orbital in σ-bonding in the octahedral platinum(IV) complexes.[266] The spectrochemical series of ligands for platinum(IV), obtained from the spectra of the $[PtX_6]^{2-}$ ions, is $SeCN^- < Br^- < SCN^- < N_3^- < Cl^- < NH_3 \sim$ en $< NO_2^- < CN^-$.[266]

(iii) Nuclear quadrupole resonance studies

N.q.r. is potentially a very useful technique for investigating the nature of the bonding in metal–halogen bonds, although at present the interpretation of the data is empirical, because the classical interpretation of n.q.r. data by the Townes and Dailey method[268] (eqn. 331) describes the chemical bonding in terms of three parameters, whilst at most only two are available experimentally. Thus n.q.r. frequencies can sometimes give relative information about a

$$eQq = \{(1 - i)(1 - S) - \pi\}eQq_{at.} \tag{331}$$

eQq = N.q.r. coupling constant of the halogen in the complex; $eQq_{at.}$ = coupling constant in atomic halogen; i = ionic character of the M–X bond; S = percentage of s-character in the bonding orbital of halogen; π = degree of π-bonding.

series of metal–halogen bonds, but rarely give absolute information. The n.q.r. frequencies for A_2MX_6 complexes (Table 79) show a negligible depend-

TABLE 79
Nuclear quadrupole resonance frequencies for $[MX_6]^{2-}$ anions at room temperature

Anion	Cation	Nucleus	Resonance frequency (MHz)	Reference
$[PdCl_6]^{2-}$	K^+	^{35}Cl	26.55	269
$[PdBr_6]^{2-}$	K^+	^{81}Br	201·81	269
$[PtCl_6]^{2-}$	$H^+(+6H_2O)$	^{35}Cl	26.55	270
	$Na^+(+6H_2O)$	^{35}Cl	25·73, 26·47, 27·04	270
	K^+	^{35}Cl	25·81	270
	NH_4^+	^{35}Cl	26·07	271
	Rb^+	^{35}Cl	26·29	271
	Cs^+	^{35}Cl	26·60	271
$[PtBr_6]^{2-}$	K^+	^{81}Br	200·2	271
	NH_4^+	^{81}Br	202·5	271
	Rb^+	^{81}Br	204·4	271
	Cs^+	^{81}Br	207·2	271
$[PtI_6]^{2-}$	K^+	^{127}I	v_1, 202·20, 201·99, 198·90	254
			v_2, 403·80, 403·56, 397·68	254

ence on the cation. The covalent character of the metal–halogen bonds in these complexes (Table 80) is calculated from eqn. 331, assuming that there is no π-bonding. This is reasonable since the t_{2g} metal orbitals are completely filled, and is consistent with the negative temperature coefficients of the resonance frequencies of these complexes (Table 81), since π-bonding is generally accompanied by positive temperature coefficients.[272] Whilst the values in Table 80 are not absolute their relative magnitudes are probably correct and indicate that the covalent characters of the metal–halogen bonds in the $[MX_6]^{2-}$ anions are virtually independent of the metal. This is not

TABLE 80

The covalent character of the metal–halogen bonds and the net charge on palladium and platinum in $[MX_6]^{2-}$ anions

Compound	Covalent Character of M–X bond		Net Charge on Pd or Pt		Reference
	Assuming $S = 0$	Assuming $S = 0.15$	Assuming $S = 0$	Assuming $S = 0.15$	
K_2PdCl_6	0·48	0·57	1·12	0·58	269
K_2PdBr_6	0·54	0·63	0·76	0·22	269
K_2PtCl_6	0·47	0·56	1·18	0·64	254
K_2PtBr_6	0·53	0·62	0·82	0·28	254
K_2PtI_6	0·59	0·70	0·46	−0·20	254

TABLE 81

Temperature dependence of the nuclear quadrupole resonance frequencies for $[MX_6]^{2-}$ anions in the temperature range dry ice temperature to room temperature

Compound	$\Delta v/\Delta T(kHz/\deg)$	Reference
K_2PdCl_6	−1·00	269
K_2PdBr_6	−6·2	269
K_2PtCl_6	−0·98	271
$(NH_4)_2PtCl_6$	−0·94	271
Rb_2PtCl_6	−0·78	271
Cs_2PtCl_6	−0·61	271
$(NH_4)_2PtBr_6$	−6·7	271
Rb_2PtBr_6	−6·4	271
Cs_2PtBr_6	−5·9	271
K_2PtI_6	v_1 −15(average)	254
	v_2 −30(average)	254

surprising since platinum(IV) and palladium(IV) have identical electronic configurations and almost the same electronegativity (pp. 9–10).

It is significant that the net charges on the platinum(IV) and palladium(IV) atoms are very similar to those found from n.q.r. data on the divalent halide complexes (see p. 246). This is consistent with the x-ray absorption spectra of divalent and tetravalent platinum complexes which suggest that in both series of complexes the net charge on the platinum atom is very small.[1,2]

(iv) Nuclear magnetic resonance studies

An attempt to interpret the ^{195}Pt n.m.r. chemical shifts in platinum(IV) complexes (Table 82) using the simplifications of the Ramsey theory[273] suggested by Griffith and Orgel[274] to interpret ^{59}Co n.m.r. chemical shifts in d^6 cobalt(III) complexes has failed.[256] This is similar to the situation found for platinum(II) complexes (see p. 138) and indicates that whereas in the case of cobalt(III) the chemical shifts are related largely to the σ-covalency of the

TABLE 82

^{195}Platinum n.m.r. chemical shifts in aqueous solution at 28°C (data from ref. 256)

Compound	$\delta(p.p.m.)$
H_2PtCl_6	0
H_2PtCl_5Br	284
$H_2PtCl_4Br_2$	579
$H_2PtCl_3Br_3$	882
$H_2PtCl_2Br_4$	1190
$H_2PtClBr_5$	1522
H_2PtBr_6	1860
H_2PtI_6	6300

metal–ligand bond as reflected by the position of the ligand in the spectrochemical series, in the platinum(IV)–halide complexes this term is dominated by the π-covalency of the metal–ligand bond as reflected by the position of the ligand in the nephelauxetic series,[177] so that the ^{195}Pt chemical shifts increase in the order $Cl^- < Br^- < I^-$. Much more data will be required before ^{195}Pt n.m.r. chemical shifts can be used to give information about the nature of the metal–ligand bonding in platinum(IV) complexes.

The ^{19}F n.m.r. spectra of K_2PtF_6 and K_2PdF_6 in anhydrous hydrogen fluoride indicate that the exchange of fluoride ligands in these complexes is slow relative to the n.m.r. time scale.[275] The complexes exhibit very large ^{19}F chemical shifts which are independent of temperature.

(v) Reactions

The mechanisms of the substitution reactions of platinum(IV) complexes are described in Chapter 11 (pp. 315–320). In addition to simple substitution reactions, halide and pseudo–halide ions can react with the halogenoplatinum(IV) complexes to give either substitution or reduction depending on the nature of (a) the other ligands present, (b) the attacking anion and (c) the presence or absence of halogen scavengers. Similarly, although the first stage in the hydrolysis of hexaiodoplatinate(IV) anions involves hydrolysis (i.e. substitution by water ligands) to give $[PtI_5(H_2O)]^-$, subsequent steps involve both further hydrolysis and reduction of the platinum(IV) complex to platinum(II).[276] The equilibrium constants for the first stage of this hydrolysis (i.e. reaction 332) and for the corresponding chloride and bromide, where no reduction occurs in the later steps are 4×10^{-4} (X = I at 25°C[276]),

$$[PtX_6]^{2-} + H_2O \rightleftharpoons [PtX_5(H_2O)]^- + X^- \tag{332}$$

$1·4 \times 10^{-3}$ (X = Br at 50°C[277]) and $3·25 \times 10^{-2}$ (X = Cl at 50°C[277]). Unless the temperature dependence of these constants reverses their relative values, the greater stability of the hexaiodo-complex relative to the other hexahalo-complexes is in contrast with the slightly lower stability of the $[PtI_4]^{2-}$ anion compared with the $[PtBr_4]^{2-}$ and $[PtCl_4]^{2-}$ anions.[278]

The first hydrolysis reaction (reaction 332), which is relatively slow, is followed by a very rapid deprotonation of the aquo-complex to yield $[PtX_5(OH)]^{2-}$.[279,280]

REFERENCES

1   V. Collet, Thesis of the University of Paris, 1959.
2   V. Collet, *Compt. Rend.*, **248** (1959) 1314; *Chem. Abs.*, **53** (1959) 12820i.
3   L. Pauling, *J. Chem. Soc.*, (1948) 1461.
4   F. R. Hartley, *Nature Physical Science*, **236** (1972) 75.
5   J. Chatt and B. L. Shaw, *J. Chem. Soc.*, (1962) 5075.
6   A. F. Clemmit and F. Glockling, *Chem. Comm.*, (1970) 705.
7   A. F. Clemmit and F. Glockling, *J. Chem. Soc. (A)*, (1971) 1164.
8   J. E. Bentham and E. A. V. Ebsworth, *Inorg. Nucl. Chem. Lett.*, **6** (1970) 145.
9   J. E. Bentham, S. Cradock and E. A. V. Ebsworth, *J. Chem. Soc. (A)*, (1971) 587
10   H. Wollfram, *Dissertation*, Königsberg, 1900.
11   H. Reihlen and E. Flohr, *Chem. Ber.*, **67** (1934) 2010.
12   S. Yamada and R. Tsuchida, *Bull. Chem. Soc. Japan*, **29** (1956) 894; *Chem. Abs.*, **51** (1957) 6269b.
13   B. M. Craven and D. Hall, *Acta. Cryst.*, **14** (1961) 475.
14   B. M. Craven and D. Hall, *Acta. Cryst.*, **21** (1966) 177.
15   S. Kida, *Bull. Chem. Soc. Japan*, **38** (1965) 1804.
16   H. S.-C. Deville and H. Debray, *Compt. Rend.*, **86** (1878) 927.
17   J. Wallen, C. Brosset and N. G. Vannerburg, *Arkiv. Kemi.*, **18** (1962) 541; *Chem. Abs.*, **58** (1963) 1968b.
18   A. J. Cohen and N. Davidson, *J. Amer. Chem. Soc.*, **73** (1951) 1955.
19   A. A. Grinberg and F. M. Filinov, *Bull. Acad. Sci. URSS. Classe. Sci. Math., Ser. Chim.*, (1937) 1245; *Chem. Abs.*, **32** (1938) 4098.
20   A. A. Grinberg and F. M. Filinov, *Bull. Acad. Sci. URSS. Classe. Sci. Chem.*, (1941) 361; *Chem. Abs.*, **35** (1941) 7863.
21   D. Hall and P. P. Williams, *Acta. Cryst.*, **11** (1958) 624.
22   H. D. K. Drew and H. J. Tress, *J. Chem. Soc.*, (1933) 1335.
23   L. Chugaev and I. Chernyaev, *Z. anorg. allgem. Chem.*, **182** (1929) 159; *Chem. Abs.*, **23** (1929) 5427.
24   G. W. Watt and R. E. McCarley, *J. Amer. Chem. Soc.*, **79** (1957) 4585.
25   S. Yamada and R. Tsuchida, *Bull. Chem. Soc. Japan*, **29** (1956) 421; *Chem. Abs.*, **50** (1956) 12582f.
26   T. D. Ryan and R. E. Rundle, *J. Amer. Chem. Soc.*, **83** (1961) 2814.
27   G. B. Bokii and G. I. Distler, *Dokl. Akad. Nauk. SSSR*, **56** (1967) 923.
28   T. W. Thomas and A. E. Underhill, *Chem. Comm.*, (1969) 1344.
28a T. W. Thomas and A. E. Underhill, *J. Chem. Soc. (A)*, (1971) 512.
29   M. B. Robin and P. Day, *Adv. Inorg. Radiochem.*, **10** (1967) 247.
30   A. Werner, *Z. anorg. allgem. Chem.*, **12** (1896) 46.
31   K. Krogmann, *Angew. Chem., Int. Ed.*, **3** (1964) 147.
32   R. Mattes and K. Krogmann, *Z. anorg. allgem. Chem.*, **332** (1964) 247.
33   K. Krogmann and H.-D. Hausen, *Z. anorg. allgem. Chem.*, **358** (1968) 67.
34   K. Krogman, *Angew. Chem. Int. Ed.*, **8** (1969) 35.
35   P. S. Gomm and A. E. Underhill, *Chem. Comm.*, (1971) 511.
36   A. Piccinin and J. Toussaint, *Bull. Soc. Roy. Sci. Liège*, **36** (1967) 122; *Chem. Abs.* **68** (1968) 73213.
37   I. I. Chernyaev and A. V. Babkov, *Dokl. Akad. Nauk. SSSR*, **152** (1963) 882; *Chem. Abs.*, **60** (1964) 6510f.
38   H. Siebert and A. Siebert, *Angew. Chem. Int. Ed.*, **8** (1969) 600.
39   I. I. Chernyaev and A. V. Babkov, *Russ. J. Inorg. Chem.*, **9** (1964) 1217.
40   L. A. Levy, *J. Chem. Soc.*, **101** (1912) 1081.
41   H. Terrey, *J. Chem. Soc.*, (1928) 202.

42 S. Kida, *Bull. Chem. Soc. Japan*, **33** (1960) 587; *Chem. Abs.*, **55** (1961) 204f.
43 C. W. Blomstrand, *J. Prakt. Chem.*, **3** (1871) 208.
44 A. Miolati and I. Bellucci, *Gazz. Chim. Ital.*, **30** (part 2) (1900) 588.
45 I. I. Chernyaev, A. V. Babkov and N. N. Zheligovskaya, *Russ. J. Inorg. Chem.*, **8** (1963) 1279.
46 F. Vil'm, *Zhur. Russ. Fiz.-Khim. Obshch.*, **21** (1889) 436.
47 I. I. Chernyaev, A. V. Babkov and N. N. Zheligovskaya, *Russ. J. Inorg. Chem.*, **9** (1964) 319.
48 I. I. Chernyaev and A. V. Babkov, *Russ. J. Inorg. Chem.*, **10** (1965) 433.
49 I. I. Chernyaev, N. N. Zheligovskaya and A. V. Babkov, *Russ. J. Inorg. Chem.*, **6** (1961) 26.
50 I. I. Chernyaev and T. N. Leonova, *Russ. J. Inorg. Chem.*, **9** (1964) 1372.
51 A. Piccinin, *Bull. Soc. Roy. Sci. Liège*, (1967) 476; *Chem. Abs.*, **69** (1968) 71313.
52 H. Siebert and A. Siebert, *Z. Naturforsch.*, **B22** (1967) 674.
53 D. M. Sweeny, I. Nakagawa, S.-I. Mizushima and J. V. Quagliano, *J. Amer. Chem. Soc.*, **78** (1956) 889.
54 L. H. Jones and J. M. Smith, *Inorg. Chem.*, **4** (1965) 1677.
55 S. D. Robinson and B. L. Shaw, *Z. Naturforsch.*, **B18** (1963) 507.
56 S. D. Robinson and B. L. Shaw, *J. Chem. Soc.*, (1965) 1529.
57 H. P. Fritz and K. E. Schwarzhans, *J. Organometal. Chem.*, **5** (1966) 181.
58 V. A. Semion, A. Z. Rubezhov, Yu. T. Struchkov and S. P. Gubin, *Russ. J. Struct. Chem.*, **10** (1969) 144; *Chem. Abs.*, **70** (1969) 100682.
59 G. W. Adamson and J. C. J. Bart, personal communication to ref. 60.
60 K. W. Egger, *J. Organometal. Chem.*, **24** (1970) 501.
61 J. R. Hall and B. E. Smith, *Aust. J. Chem.*, **24** (1971) 911.
62 R. B. King, *Appl. Spectry.*, **23** (1969) 148; *Chem. Abs.*, **70** (1969) 114431.
63 F. Glockling and K. A. Hooton, *J. Chem. Soc.* (*A*), (1968) 826.
64 J. E. Bentham and E. A. V. Ebsworth, *Inorg. Nucl. Chem. Lett.*, **6** (1970) 671.
65 L. Chugaev, *Z. anorg. allgem. Chem.*, **137** (1924) 1.
66 L. Chugaev and N. Vladimirov, *Compt. Rend.*, **160** (1915) 840.
67 L. Chugaev and I. Chernyaev, *Compt. Rend.*, **161** (1915) 637.
68 A. Werner, *Ber.*, **40** (1907) 4093.
69 F. Basolo, J. C. Bailar and B. R. Tarr, *J. Amer. Chem. Soc.*, **72** (1950) 2433.
70 E. G. Cox and G. H. Preston, *J. Chem. Soc.*, (1933) 1089.
71 I. I. Chernyaev, N. N. Zheligovskaya, Leti-k'en and D. N. Kurganovich, *Russ. J. Inorg. Chem.*, **9** (1964) 312.
72 H. D. K. Drew, F. W. Pinkard, G. H. Preston and W. Wardlaw, *J. Chem. Soc.*, (1932) 1895.
73 A. Rosenheim and T. A. Maass, *Z. anorg. allgem. Chem.*, **18** (1898) 334.
74 R. S. Nyholm, *J. Chem. Soc.*, (1950) 843.
75 J. Chatt, *J. Chem. Soc.*, (1950) 2301.
76 C. M. Harris, R. S. Nyholm and D. J. Phillips, *J. Chem. Soc.*, (1960) 4379.
77 J. Chatt and B. L. Shaw, *J. Chem. Soc.*, (1959) 705.
78 J. Chatt and B. L. Shaw, *J. Chem. Soc.*, (1959) 4020.
79 J. D. Ruddick and B. L. Shaw, *J. Chem. Soc.* (*A*), (1969) 2964.
80 G. A. Barclay, R. S. Nyholm and R. V. Parish, *J. Chem. Soc.*, (1961) 4433.
81 R. S. Nyholm and M. L. Tobe, *Adv. Inorg. Radiochem.*, **5** (1963) 1.
82 G. B. Bokii and L. A. Popova, *Dokl. Akad. Nauk. SSSR*, **67** (1949) 73; *Chem. Abs.*, **43** (1949) 7287b.
83 G. B. Bokii and M. A. Porai-Koshits, *Dokl. Akad. Nauk. SSSR*, **64** (1949) 337; *Chem. Abs.*, **43** (1949) 4535i.
84 C. F. Liu and J. A. Ibers, *Inorg. Chem.*, **9** (1970) 773.
85 L. Aslanov, R. Mason, A. G. Wheeler and P. O. Whimp, *Chem. Comm.*, (1970) 30.
86 M. A. Bennett, G. J. Erskine, J. Lewis, R. Mason, R. S. Nyholm, G. B. Robertson and A. D. C. Towl, *Chem. Comm.*, (1966) 395.
87 G. H. W. Milburn and M. R. Truter, *J. Chem. Soc.* (*A*), (1966) 1609.

88   G. G. Messmer and E. L. Amma, *Inorg. Chem.*, **5** (1966) 1775.
89   G. A. Mair, *D. Phil. Thesis*, Oxford, 1961.
90   D. M. Adams and P. J. Chandler, *J. Chem. Soc. (A)*, (1967) 1009.
91   M. Le Postelloc, *J. Chim. Phys.* **62** (1965) 67; *Chem. Abs.*, **63** (1965) 1342h.
92   T. C. Ray and A. D. Westland, *Inorg. Chem.*, **4** (1965) 1501.
93   J. Chatt, G. J. Leigh and D. M. P. Mingos, *J. Chem. Soc. (A)*, (1969) 1674.
94   D. M. Adams and P. J. Chandler, *Chem. Comm.*, (1966) 69.
95   D. M. Adams, J. Chatt, J. Gerratt and A. D. Westland, *J. Chem. Soc.*, (1964) 734.
96   C. K. Jørgensen, *Acta. Chem. Scand.*, **10** (1956) 518.
97   I. I. Chernyaev, N. N. Zheligovskaya and Leti-k'en, *Russ. J. Inorg. Chem.*, **9** (1964) 315.
98   G. J. Leigh and D. M. P. Mingos, *J. Chem. Soc. (A)*, (1970) 587.
99   A. Pidcock, R. E. Richards and L. M. Venanzi, *J. Chem. Soc. (A)*, 1966, 1707.
100 R. D. Gillard and H. M. Irving, *Chem. Rev.*, **65** (1965) 603.
101 A. M. Sargeson, *Trans. Metal. Chem.*, **3** (1966) 303.
102 C. F. Liu and J. Doyle, *Chem. Comm.*, (1967) 412.
103 C. F. Liu and J. A. Ibers, *Inorg. Chem.*, **9** (1970) 773.
104 J. P. Mathieu, *Bull. Soc. Chim. France*, **6** (1939) 1258.
105 F. P. Dwyer and A. M. Sargeson, *J. Amer. Chem. Soc.*, **81** (1959) 5272.
106 T. G. Appleton, J. R. Hall and C. J. Hawkins, *Inorg. Chem.*, **9** (1970) 1299.
107 J. R. Collogly, C. J. Hawkins and J. K. Beattie, *Inorg. Chem.*, **10** (1971) 317.
108 J. K. Beattie and L. H. Novak, *J. Amer. Chem. Soc.*, **93** (1971) 620.
109 L. Ramberg, *Chem. Ber.*, **40** (1907) 2578.
110 V. A. Golovnya and C.-C. Ni, *Zhur. Neorg. Khim.*, **3** (1959) 1954; *Chem. Abs.*, **55** (1966), 7134d.
111 V. V. Lebedinskii and V. A. Golovyna, *Izvest. Sekt. Platiny*, **20** (1947) 84; *Chem. Abs.*, **44** (1950) 5309.
112 P. Senise and L. R. M. Pitombo, *Anais. Assoc. Brasil. Quim.*, **20** (1961) 93; *Chem. Abs.*, **58** (1963) 1487c.
113 W. Beck, E. Schuierer and K. Feldl, *Angew. Chem. Int. Ed.*, **5** (1966) 249.
114 W. Beck, W. P. Fehlhammer, P. Pöllmann, E. Schuierer and K. Feldl, *Chem. Ber.*, **100** (1967) 2335.
115 T. Theophanides and Y. Lafortune, *Can. Spectrosc.*, **13** (1968) 40; *Chem. Abs.*, **69** (1968) 31669.
116 H.-H. Schmidtke and D. Garthoff, *J. Amer. Chem. Soc.*, **89** (1967) 1317.
117 R. G. Denning and J. Thatcher, *J. Amer. Chem. Soc.*, **90** (1968) 5917.
118 M. S. Kharasch and T. A. Ashford, *J. Amer. Chem. Soc.*, **58** (1936) 1733.
119 M. A. Bennett, J. Chatt, G. J. Erskine, J. Lewis, R. F. Long and R. S. Nyholm, *J. Chem. Soc. (A)*, (1967) 501.
120 M. A. Bennett, G. J. Erskine and R. S. Nyholm, *J. Chem. Soc. (A)*, (1967) 1260.
121 M. A. Bennett, G. J. Erskine, J. Lewis, R. Mason, R. S. Nyholm, G. B. Robertson and A. D. C. Towl, *Chem. Comm.*, (1966) 395.
122 M. A. Bennett, W. R. Kneen and R. S. Nyholm, *J. Organometal. Chem.*, **26** (1971) 293.
123 Gmelin, *Handbuch Der Anorganischen Chemie*, No. 65 (Palladium), Verlag Chemie, Berlin, 1942, p. 267.
124 L. Wöhler, K. Ewald and H. G. Krall, *Ber.*, **66** (1933) 1638.
125 N. V. Sidgwick, *The Chemical Elements and Their Compounds*, Oxford University Press, Oxford, (1950) vol. 2, p. 1575.
126 F. Hulliger, *J. Phys. Chem. Solids*, **26** (1965) 639.
127 F. Grønvold and E. Røst, *Acta Chem. Scand.*, **10** (1956) 1620.
128 F. Grønvold and E. Røst, *Acta Cryst.*, **10** (1957) 329.
129 R. A. Munson and J. S. Kasper, *Inorg. Chem.*, **8** (1969) 1198.
130 C. J. Raub, V. B. Compton, T. H. Geballe, B. T. Matthias, J. P. Maita and G. W. Hull, *J. Phys. Chem. Solids.*, **26** (1965) 205.
131 B. T. Matthias and S. Geller, *J. Phys. Chem. Solids.*, **4** (1958) 318.
132 F. Hulliger, *Structure and Bonding*, **4** (1968) 83.
133 L. Wöhler, *Z. anorg. allgem. Chem.*, **40** (1904) 423.

134 P. Lafitte and P. Grandadam, *Compt. Rend.*, **198** (1934) 1925; *Chem. Abs.*, **28** (1934) 4328.
135 R. Adams, V. Voorhees and R. L. Shriner, *Org. Synth. Coll. Vol.*, **1** (1941) 463.
136 V. L. Frampton, J. D. Edwards and H. R. Henze, *J. Amer. Chem. Soc.*, **73** (1951) 4432.
137 F. A. Vandenheuval, *Anal. Chem.*, **28** (1956) 362.
138 N. A. Shishakov, *Kristallografiya*, **2** (1957) 689; *Chem. Abs.*, **53** (1959) 7712.
139 W. Biltz and R. Juza, *Z. anorg. allgem. Chem.*, **190** (1930) 161.
140 E. Davey, *Phil. Mag.*, **40** (1812) 28.
141 F. Grønvold, H. Haraldsen and A. Kjekshus, *Acta. Chem. Scand.*, **15** (1960) 1879.
142 F. Hulliger, *Helv. Phys. Acta.* **33** (1960) 959.
143 L. Thomassen, *Z. physik. Chem.*, **B2** (1929) 349.
144 A. Minozzi, *Atti. reale. accad. Lincei*, **18** (1909) 150.
145 L. Moser and K. Atynski, *Monatsch*, **54** (1925) 235, *Chem. Abs.*, **19** (1925) 1231.
146 C. Roessler, *Z. anorg. allgem. Chem.*, **15** (1897) 405.
147 A. Brukl, *Monatsch.*, **45** (1925) 471; *Chem. Abs.*, **20** (1926) 881.
148 W. O. Groeneveld Meijer, *J. Am. Mineralogist*, **40** (1955) 647.
149 K. A. Hofmann and F. Höchtlen, *Chem. Ber.*, **36** (1903) 3090.
150 P. E. Jones and L. Katz, *Chem. Comm.*, (1967) 842.
151 P. E. Jones and L. Katz, *Acta Cryst.*, **B25** (1969) 745.
152 A. E. Wickenden and R. A. Krause, *Inorg. Chem.*, **8** (1969) 779.
153 K. A. Hofmann and F. Höchtlen, *Chem. Ber.*, **37** (1904) 245.
154 H. Remy, *Treatise on Inorganic Chemistry* translated by J. S. Anderson, Ed. by J. Kleinberg, Elsevier, Amsterdam, (1956) vol. II, p. 341.
155 M. J. N. Pourbaix, J. Van Muylder and N. de Zoubov, *Platinum Metals Review*, **3** (1959) 100.
156 I. Bellucci, *Atti R.*, **12** (1900) [5], ii, 635.
157 Reference 125, p. 1613.
158 M. J. N. Pourbaix, J. Van Muylder and N. de Zoubov, *Platinum Metals Review*, **3** (1959) 47.
159 M. Blondel, *Ann. Chim. Phys.*, **6** (1905) [8] 81.
160 C. O. Björling, *Arkiv. Kemi. Min. Geol.*, **B15** (1941) No. 2: *Chem. Abs.*, **36** (1942) 697.
161 L. Chugaev and V. Klopin, *Z. anorg. allgem. Chem.*, **151** (1926) 253; *Chem. Abs.*, **20** (1926) 1765.
162 A. A. Grinberg and F. M. Filinov, *Compt. Rend. Acad. Sci. URSS.*, **17** (1937) 23; *Chem. Abs.*, **32** (1938) 2449.
163 A. Miolati, *Z. anorg. allgem. Chem.*, **33** (1903) 251.
164 A. Miolati and I. Bellucci, *Atti. R.*, **9** (1900) [5], ii, 51.
165 I. Bellucci, *Atti. R.*, **11** (1902) [5], i, 8.
166 A. A. Grinberg and F. M. Filinov, *Bull. Acad. Sci. URSS.*, (1937) 907; *Chem. Abs.*, **32** (1938) 2045.
167 L. E. Cox and D. G. Peters, *Inorg. Chem.*, **9** (1970) 1927.
168 E. H. Archibald, *J. Chem. Soc.*, **117** (1920) 1104.
169 E. H. Archibald and W. A. Gale, *J. Chem. Soc.*, **121** (1922) 2849.
170 W. Halberstadt, *Chem. Ber.*, **17** (1884) 2962.
171 A. Miolati, *Z. anorg. allgem. Chem.*, **22** (1899) 445.
172 R. G. Johnson, F. Basolo and R. G. Pearson, *J. Inorg. Nucl. Chem.*, **24** (1962) 59.
173 A. A. Grinberg and Yu. N. Kukushkin, *Dokl. Akad. Nauk. SSSR.*, **132** (1960) 1071; *Chem. Abs.*, **55** (1961) 13008e.
174 A. A. Grinberg and Kh. I. Gil'dengershel, *Izvest. Akad. Nauk. SSSR. Otdel. Khim. Nauk*, (1948) 479; *Chem. Abs.*, **43** (1949) 1673f.
175 L. G. Sillen and A. E. Martell, *Stability Constants of Metal-Ion Complexes*, Chem. Soc. Special Publ. No. 17, 1964.
176 A. Hantzsch, *Z. Physikal. Chem.*, **72** (1910) 362.
177 C. K. Jørgensen, *Absorption Spectra and Chemical Bonding in Complexes*, Pergamon, Oxford, 1962.
178 F. G. Angell, H. D. K. Drew and W. Wardlaw, *J. Chem. Soc.*, (1930) 349.
179 E. G. Cox, H. Saenger and W. Wardlaw, *J. Chem. Soc.*, (1934) 182.

180 Gmelin, *Handbuch Der Anorganischen Chemie*, Verlag Chemie, Berlin, (1957) *Platin*, part D, pp. 594–6.
181 J. R. Allkins and P. J. Hendra, *J. Chem. Soc. (A)*, (1967) 1325.
182 B. M. Gatehouse, S. E. Livingstone and R. S. Nyholm, *J. Chem. Soc.*, (1957) 4222.
183 I. I. Chernyaev and N. N. Krasovskaya, *Zhur. Neorg. Khim.*, **3** (1958) 2025.
184 I. I. Chernyaev, N. N. Zheligovskaya and A. V. Babkov, *Russ. J. Inorg. Chem.*, **6** (1961) 26.
185 R. Eskenazi, J. Raskovan and R. Levitus, *Chem. Ind. (London)*, (1962) 1327.
186 C. C. Addison, N. Logan, S. C. Wallwork and C. D. Garner, *Quart. Rev.*, **25** (1971) 289.
187 P. T. Cleve, *Nova Acta. Sci. Upsaliensis*, **6** (1866) [3], 1.
188 D. Strömholm, *Z. anorg. allgem. Chem.*, **108** (1919) 184; *Chem. Abs.*, **14** (1920) 2310.
189 S. A. Borisov-Pototskii, *Ann. Secteur Platine, Inst. Chim. Gen. USSR*, **16** (1939) 41; *Chem. Abs.* **34** (1940) 4685.
190 F. G. Mann, *J. Chem. Soc.*, (1930) 1745.
191 I. I. Chernyaev, L. A. Nazarova and A. S. Mironova, *Russ. J. Inorg. Chem.*, **4** (1959) 340.
192 I. I. Chernyaev, L. A. Nazarova and A. S. Mironova, *Russ. J. Inorg. Chem.*, **6** (1961) 144.
193 I. I. Chernyaev, L. A. Nazarova and A. S. Mironova, *Russ. J. Inorg. Chem.*, **6** (1961) 1238.
194 A. V. Babaeva, O. N. Evstaf'eva and Yu. Ya. Kharitonov, *Russ. J. Inorg. Chem.*, **7** (1962) 15.
195 J. L. Burmeister, *Coord. Chem. Rev.*, **3** (1968) 225.
196 F. Basolo and G. S. Hammaker, *Inorg. Chem.*, **1** (1962) 1.
197 J. Lewis, R. S. Nyholm and P. W. Smith, *J. Chem. Soc.*, (1961) 4590.
198 D. Forster and D. M. L. Goodgame, *Inorg. Chem.*, **4** (1965) 715.
199 A. Sabatini and I. Bertini, *Inorg. Chem.*, **4** (1965) 959.
200 H.-H. Schmidtke, *Ber. Bunsengesellschaft Phys. Chem.*, **71** (1967) 1138; *Chem. Abs.*, **68** (1968) 34451.
201 G. V. Tsintsadze, V. V. Skopenko and A. E. Shvelashvili, *Tr. Gruz. Politekh. Inst.*, (1968) 19; *Chem. Abs.*, **70** (1969) 100619.
202 G. S. Cavicchi and M. P. Quaglio, *Boll. Chim. Farm.*, **103** (1964) 660; *Chem. Abs.*, **62** (1965) 3887c.
203 A. V. Babaeva, O. N. Evstaf'eva and Yu. Ya. Kharitonov, *Fiz. Probl. Spektroskopii*, *Akad. Nauk. SSSR*, **1** (1960) 415; *Chem. Abs.*, **59** (1963) 14745b.
204 N. Bartlett and P. R. Rao, *Proc. Chem. Soc.*, (1964) 393.
205 A. G. Sharpe, *J. Chem. Soc.*, (1950) 3444.
206 N. Bartlett and D. Lohmann, *Proc. Chem. Soc.*, (1960) 14.
207 N. Bartlett and D. Lohmann, *J. Chem. Soc.*, (1964) 619.
208 L. Wöhler and S. Streicher, *Chem. Ber.*, **46** (1913) 1591; *Chem. Abs.*, **7** (1913) 2727.
209 V. P. Kazakov and B. I. Peshchevitskii, *Radiokhimiya*, **4** (1962) 509; *Chem. Abs.*, **59** (1963) 5781f.
210 M. T. Falqui, *Ann. Chim. (Rome)*, **48** (1958) 1160; *Chem. Abs.*, **53** (1959) 12784f.
211 J. H. Canterford and R. Colton, *Halides of the Second and Third Row Transition Metals*, John Wiley, New York, 1968, p. 364.
212 J. Krustinsons, *Z. Elektrochem.*, **44** (1938) 537; *Chem. Abs.*, **32** (1938) 8891.
213 S. A. Shchukarev, M. A. Oranskaya and T. S. Shemyakina, *Zhur. Neorg. Khim.*, **1** (1956) 17; *Chem. Abs.*, **50** (1956) 9833i.
214 J. R. Miller, *Adv. Inorg. and Radiochem.*, **4** (1962) 133.
215 L. Wöhler and F. Müller, *Z. anorg. allgem. Chem.*, **149** (1925) 377; *Chem. Abs.*, **20** (1926) 718.
216 S. A. Shchukarev, T. A. Tolmacheva, M. A. Oranskaya and L. V. Komandrovskaya, *Zhur. Neorg. Khim.*, **1** (1956) 8; *Chem. Abs.*, **50** (1956) 9833g.
217 G. R. Argue and J. J. Banewicz, *J. Inorg. Nucl. Chem.*, **25** (1963) 923.
218 S. A. Shchukarev, T. A. Tolmacheva and G. M. Slavutskaya, *Russ. J. Inorg. Chem.*, **9** (1964) 1351.
219 O. Ruff and E. Ascher, *Z. anorg. allgem. Chem.*, **183** (1929) 204.
220 N. Bartlett and M. A. Hepworth, *Chem. Ind. (London)*, (1956) 1425.

221   H. Henkel and R. Hoppe, *Z. anorg. allgem. Chem.*, **359** (1968) 160.
222   R. Hoppe and W. Klemm, *Z. anorg. allgem. Chem.*, **268** (1952) 364.
223   A. G. Sharpe, *J. Chem. Soc.*, (1953) 197.
224   D. H. Brown, K. R. Dixon, R. D. W. Kemmitt and D. W. A. Sharp, *J. Chem. Soc.*, (1965) 1559.
225   B. Cox, *J. Chem. Soc.*, (1956) 876.
226   N. Bartlett and J. W. Quail, *J. Chem. Soc.*, (1961) 3728.
227   F. Puche, *Ann. Chim. (Paris)*, **9** (1938) 233; *Chem. Abs.*, **32** (1938) 5322.
228   B. Weinstock, J. G. Malm and G. L. Goodman, *Proc. 8th Intern. Conf. Low Temp. Physics*, London, (1962), p. 405.
229   F. Puche, *Compt. Rend.*, **200** (1935) 1206.
230   K. Watanabe, *Nippon Kagaku Zasshi*, **77** (1956) 1675; *Chem. Abs.*, **52** (1958) 2636e.
231   A. G. Sharpe, *J. Chem. Soc.*, (1953) 4177.
232   A. Gutbier and A. Krell, *Chem. Ber.*, **38** (1905) 2385.
233   Reference 125, p. 1576.
234   J. A. A. Ketelaar and J. F. van Walsem, *Rec. Trav. Chim.*, **57** (1938) 964.
235   L. A. Woodward and M. J. Ware, *Spectrochim. Acta.*, **19** (1963) 775.
236   M. K. Norr, T. P. Perros and C. R. Naeser, *J. Amer. Chem. Soc.*, **80** (1958) 5035.
237   T. P. Perros and C. R. Naesser, *J. Amer. Chem. Soc.*, **75** (1953) 2516.
238   I. Chernyaev, N. S. Nikolaev and E. G. Ippolitov, *Dokl. Akad. Nauk. SSSR*, **130**, (1960) 1041.
239   D. P. Mellor and N. C. Stephenson, *Aust. J. Sci. Res.*, **A4** (1951) 406.
240   D. H. Brown, K. R. Dixon and D. W. A. Sharp, *J. Chem. Soc. (A)*, (1966) 1244.
241   Reference 125, p. 1623.
242   N. S. Kurnakov and E. A. Nikitina, *J. Gen. Chem. USSR.*, **10** (1940) 577; *Chem. Abs.*, **34** (1940) 7718.
243   M. Delépine, *Bull. Soc. Chim. France*, (1956) 282.
244   R. W. G. Wyckoff and E. Posnjak, *J. Amer. Chem. Soc.*, **43** (1921) 2292.
245   P. A. Vaughan, J. H. Sturdivant and L. Pauling, *J. Amer. Chem. Soc.*, **72** (1950) 5477.
246   F. J. Ewing and L. Pauling, *Z. Krist.*, **68** (1928) 223; *Chem. Abs.*, **23** (1929) 2083.
247   A. Gutbier and A. Krell, *Chem. Ber.*, **38** (1905) 2385.
248   A. Gutbier and F. Bauriedel, *Chem. Ber.*, **42** (1909) 4243.
249   C. Duval in *Nouveau Traité de Chimie Minerale*, ed. by P. Pascal, Masson, Paris, 1958, vol. XIX, p. 753.
250   R. W. G. Wyckoff, *Crystal Structures*, Interscience, New York, 1965, vol. 3, p. 341.
251   R. L. Datta, *J. Amer. Chem. Soc.*, **35** (1913) 1185.
252   Reference 249, p. 755.
253   G. A. Shagisultanova, *Russ. J. Inorg. Chem.*, **6** (1961) 904.
254   D. Nakamura, Y. Kurita, K. Ito and M. Kubo, *J. Amer. Chem. Soc.*, **82** (1960) 5783.
255   H. I. Schlesinger and R. E. Palmateer, *J. Amer. Chem. Soc.*, **52** (1930) 4316.
256   A. V. Zelewsky, *Helv. Chim. Acta.*, **51** (1968) 803.
257   D. M. Adams, *Proc. Chem. Soc.*, 1961, 335.
258   L. A. Woodward and J. A. Creighton, *Spectrochim. Acta.*, **17** (1961) 594.
259   D. M. Adams and H. A. Gebbie, *Spectrochim. Acta.*, **19** (1963) 925.
260   D. M. Adams, J. Chatt, J. M. Davidson and J. Gerratt, *J. Chem. Soc.*, (1963) 2189.
261   M. Le Postelloc, J. P. Mathieu and H. Poulet, *J. Chim. Phys.* **60** (1963) 1319; *Chem. Abs.*, **60** (1964) 10066e.
262   J. Hiraishi, I. Nakagawa and T. Shimanouchi, *Spectrochim. Acta.*, **20** (1964) 819.
263   J. N. Murrell, *J. Chem. Soc. (A)*, (1969) 297.
264   C. K. Jørgensen, *Acta Chem. Scand.*, **10** (1956) 518.
265   G. N. Henning, P. A. Dobosh, A. J. McCaffery and P. N. Schatz, *J. Amer. Chem. Soc.*, **92** (1970) 5377.
266   D. L. Swihart and W. R. Mason, *Inorg. Chem.*, **9** (1970) 1749.
267   T. E. Wheeler, T. P. Perros and C. R. Naeser, *J. Amer. Chem. Soc.*, **77** (1955) 3488.
268   C. H. Townes and B. P. Dailey, *J. Chem. Phys.*, **17** (1949) 782.
269   K. Ito, D. Nakamura, Y. Kurita, Ko. Ito and M. Kubo, *J. Amer. Chem. Soc.*, **83** (1961) 4526.

270   Ko. Ito, K. Ito, D. Nakamura and M. Kubo, *Bull. Chem. Soc. Japan*, **35** (1962) 518.
271   D. Nakamura and M. Kubo, *J. Phys. Chem.*, **68** (1964) 2986.
272   W. van Bronswijk, *Structure and Bonding*, **7** (1970) 87.
273   N. F. Ramsey, *Phys. Rev.*, **78** (1950) 699.
274   J. S. Griffith and L. E. Orgel, *Trans. Faraday Soc.*, **53** (1957) 601.
275   N. A. Matwiyoff, L. B. Asprey, W. E. Wageman, M. J. Reisfeld and E. Fukushima, *Inorg. Chem.*, **8** (1969) 750.
276   B. Corain and A. J. Poë, *J. Chem. Soc. (A)*, (1967) 1633.
277   C. M. Davidson and R. F. Jameson, *Trans. Faraday Soc.*, **61** (1965) 2462.
278   B. Corain and A. J. Poë, *J. Chem. Soc. (A)*, (1967) 1318.
279   R. Dreyer, I. Dreyer and D. Rettig, *Z. Phys. Chem. (Leipzig)*, **224** (1963) 199.
280   A. T. Hubbard and F. C. Anson, *Anal. Chem.*, **38** (1966) 1887.

CHAPTER 11

Substitution Reactions

In the other chapters in this book the reactions of platinum and palladium have been considered with the main emphasis being on the nature of the reactants and products and, except in occasional instances very little importance being attached to the detailed mechanism of the reaction. This is because the detailed mechanisms of relatively few reactions of platinum and palladium complexes are understood. However, some reactions are fairly well understood, particularly the substitution reactions of complexes of both the divalent and tetravalent metals, and these form the subject matter of the bulk of the material presented in this chapter. The first part of the chapter is devoted to the substitution reactions of the complexes of the divalent metals, and although exceptions will crop up most of these substitution reactions occur by an associative mechanism. The evidence on which this assertion is based is presented and the reactions are considered from the point of view of the entering ligand, the leaving ligand, the substrate, the non-labile ligands with particular reference to their *cis*- and *trans*-effects, and the solvent. Having established the usual mechanism for the substitution reactions of the square-planar complexes of the divalent metals the exceptions to this typical behaviour are considered, before considering the substitution reactions of trigonal bipyramidal complexes of the divalent metals and mechanisms for the *cis-trans* isomerisation of the square-planar complexes.

A consideration of the reactions of the octahedral complexes of platinum(IV) leads to the conclusion that they are not as simple as those of octahedral cobalt(III) complexes, which have been widely investigated and generally found to occur by a dissociative mechanism, because many reactions of platinum(IV) complexes have been found to be catalysed by platinum(II) complexes. It will become apparent during the course of this part of the chapter that a great deal less is understood about the detailed mechanisms of the substitution reactions of the octahedral complexes of platinum(IV) than of the square-planar complexes of platinum(II). The chapter concludes with a summary of the present knowledge of the mechanism of the oxidation-reduction reactions that occur between platinum(II) and platinum(IV). It will be apparent in this section that this is one of the least understood areas of the chemistry considered in this present chapter.

As is customary in discussing kinetics, the square brackets normally used to enclose Werner complexes have been omitted throughout this chapter to avoid confusion with square brackets used as a symbol for concentration.

SUBSTITUTION REACTIONS OF COMPLEXES OF THE DIVALENT METALS

The substitution reactions of platinum(II) complexes generally occur at a fairly slow rate, so that they are ideal for kinetic studies made on simple, relatively inexpensive equipment. As a result they have been widely investigated and are at least thought to be fairly well understood. However, a note of uncertainty has recently been sounded following the observation that reaction 333 did not go to completion in spite of the fact that iodide is

$$Pt(dien)NO_2^+ + I^- \rightleftharpoons Pt(dien)I^+ + NO_2^- \tag{333}$$

a very good displacing ligand. The authors are concerned that much of the data on the kinetics of platinum(II) substitution reactions with weakly displacing ligands may be in error due to equilibration between appreciable amounts of initial and product material.[1]

TABLE 83
Reviews of substitution reactions of platinum(II) and palladium(II) complexes

Title	Reference
Mechanisms of substitution reactions of metal complexes	2
The *trans* effect in metal complexes	3
Inorganic reaction mechanism	4
Substitution reactions of square-planar complexes	5
Square-planar substitutions	6
Substitution reactions of square-planar complexes	7
Anomalies in ligand exchange reactions for platinum(II) complexes	8
The application of reaction mechanisms to the synthesis of co-ordination compounds	9
The intimate mechanism of replacements in d^8 square-planar complexes	10

The subject has been reviewed in detail by a number of workers (Table 83) and here, accordingly, only such material is presented as is essential to an understanding of the chemistry of platinum(II) and palladium(II) complexes. Incidentally, palladium(II) complexes react about 10^5–10^6 times faster than platinum(II) complexes giving reaction rates that are usually too fast for the simple equipment necessary to study the platinum(II) complexes. Hence most of the information discussed in this section has been derived from work on platinum(II) complexes.

(a) General mechanism

Ligand substitution reactions of square-planar platinum(II) complexes occur with retention of configuration, that is *cis*-reactants give *cis*-products and similarly *trans*-reactants give *trans*-products. The kinetics of ligand substitution reactions of square-planar complexes (reaction 334) follow a

$$PtA_3X + Y^- \rightarrow PtA_3Y + X^- \tag{334}$$

two-term rate law:

$$\text{rate} = \{k_1 + k_2[Y^-]\}[PtA_3X]$$

where k_1 is a first-order rate constant, k_2 is a second-order rate constant and $[Y^-]$ represents the concentration of the entering ligand. A two-term rate law implies, of course, two parallel mechanisms and these are now considered to be a solvent-dependent and a solvent-independent path. Both routes are thought to involve an associative mechanism, although the evidence is indirect rather than direct.[1] It includes the following:

(i) Variation of the charge on the complex has only a slight effect on the rate.[11-13] Some typical results, shown in Table 84, indicate that on going

TABLE 84
The effect of the charge on the complex on the rate of
hydrolysis of platinum(II) complexes in water at 25°C
(from ref. 11)

Complex	First-order rate constant $(sec^{-1} \times 10^5)$
$PtCl_4^{2-}$	3·9
$Pt(NH_3)Cl_3^-$ *(trans-Cl$^-$)*	0·62
$Pt(NH_3)Cl_3^-$ *(cis-Cl$^-$)*	5·6
cis-$Pt(NH_3)_2Cl_2$	2·5
trans-$Pt(NH_3)_2Cl_2$	9·8
$Pt(NH_3)_3Cl^+$	2·6

from the unipositive complex $Pt(NH_3)_3Cl^+$ to the binegative complex $PtCl_4^{2-}$ the first-order rate constant remains roughly constant. This is inconsistent with a dissociative mechanism, where a small change in the charge on the complex has a dramatic effect on the rate of reaction (for example, *trans*-$Co(NH_3)_4Cl_2^+$ hydrolyses approximately 3700 times faster[14] than $Co(NH_3)_5Cl^{++}$), but is consistent with an associative pathway.

(ii) Steric hindrance causes a sharp decrease in the rate of reaction of the *cis*- and *trans*-$(PEt_3)_2PtRCl$ complexes with pyridine as the R group is changed from phenyl to *o*-tolyl to mesityl (Table 85). The *ortho*-methyl groups in these complexes respectively block zero, one and two potential co-ordination sites on the metal. That an increase in steric hindrance is accompanied by a decrease in reactivity is strong evidence for an associative mechanism, as is shown by comparison with cobalt(III) complexes, which react by a dissociative mechanism and exhibit steric acceleration.[16]

(iii) It is well established that the rates of reaction of square-planar platinum(II) complexes are strongly dependent on the reactivity of the entering ligand (Table 86). This result is entirely consistent with an associative mechanism.

TABLE 85

The effect of steric blocking of the potential co-ordination
sites on platinum(II) available for association with the
incoming ligand on the rate of reaction of *cis*- and *trans*-
$(PEt_3)_2PtRCl$ with pyridine in ethanol (from ref. 15)

R	Rate $(sec^{-1}) \times 10^6$	
	R trans *to Cl (at 25°C)*	R cis *to Cl (at 0°C)*
Mesityl	3·42	1·0
o-Tolyl	17·4	200
Phenyl	124	80,000

TABLE 86

Second-order rate constants for the reaction
$Pt(dien)Br^+ + Y^- \rightarrow Pt(dien)Y^+ + Br^-$ in water at 25°C
(from ref. 17)

Y^-	Second-order rate constant $(l.mol^{-1} sec^{-1}) \times 10^4$
OH^-	0
Cl^-	8·8
pyr	33
NO_2^-	37
N_3^-	77
I^-	2300
SCN^-	4300
$SC(NH_2)_2$	8300

(iv) The first-order rate constants of platinum(II) complexes depend on
the solvent (Table 87). This suggests that the k_1 term also involves an associative mechanism. Moreover, the very high rate of reaction in dimethylsulphoxide (DMSO), which is known to co-ordinate strongly to plati-
num(II),[19] as opposed to that in hydroxylic solvents, which are known to
co-ordinate more weakly, is inconsistent with a dissociative mechanism.

TABLE 87

Dependence of the first-order rate constant of the
reaction *trans*-$Pt(pyr)_2Cl_2 + 2R_4N^{36}Cl \rightleftharpoons$
trans-$Pt(pyr)_2{}^{36}Cl_2 + 2R_4NCl$ on the solvent at
25°C (from ref. 18; R_4N = *n*-octadecylbenzyl-
dimethylammonium)

Solvent	First-order rate constant $(sec^{-1}) \times 10^5$
DMSO	38
H_2O	3·5[a]
CH_3NO_2	3·2
C_2H_5OH	1·4
$n-C_3H_7OH$	0·42

[a] Value refers to *trans*-$Pt(NH_3)_2Cl_2$.

(v) The rate of reaction 335 is decreased by a factor of 1.4

$$\text{trans-}(PEt_3)_2PtHCl + pyr \longrightarrow \text{trans-}(PEt_3)_2PtH(pyr)^+ + Cl^- \quad (335)$$

on replacing the hydride by a deuteride ligand.[20] Since the trans-effects of the hydride and deuteride groups are unlikely to be sufficiently different to account for the change in rate, it is probable that there is some weakening of the Pt–H (or Pt–D) bond in the transition state relative to that in the initial reactant. The weakening could arise from electron donation to the platinum by the incoming pyridine ligand; this would accord with a five-co-ordinate transition state and an associative mechanism for this sub-stitution reaction.

(vi) Activation volumes for the acid hydrolyses of $PtCl_4^{2-}$ (-17 ml/mol) and $Pt(NH_3)Cl_3^-$ (-14 ml/mol)[21] and for the bromide substitution of $(PEt_3)_2PtCl_2$ (-28 ml/mol)[22] are indicative of an associative mechanism.

(b) Effect of entering ligand

Table 86 shows that the nature of the entering group has a profound effect on the rate of substitution reactions of square-planar complexes. All attempts to identify those properties of a nucleophile make it effective as an entering group have been based on linear free-energy relationships.[23,24] The first important and qualitatively successful generalisation was embodied in the *principle of hard and soft acids and bases*;[25] this states that soft nucleophiles are most effective towards soft substrates such as platinum(II) and palla-dium(II) complexes. It has not been possible, however, to establish a quanti-tative scale of softness, because the softness of a nucleophile depends upon a number of factors some of which also depend on the substrate. The more important factors include the ionic or covalent character of the nucleophile-substrate bond, the possibility of π-bonding between the nucleophile and the substrate, and finally, the polarisability of the nucleophile.[26]

A rather more successful approach to those properties of a nucleophile, which make it suitable for substitution at platinum(II), has been the nucleo-philicity scale, developed by studying the reaction of nucleophiles with *trans-*$Pt(pyr)_2Cl_2$ in methanol.[27] The rates of these reactions are given by:

$$\text{Rate} = \{k_1 + k_2[Y]\}[\text{trans-}Pt(pyr)_2Cl_2]$$

The nucleophilicity of a ligand (n_{Pt}) is defined as the ratio of the logarithm of the second-order rate constant for the ligand (k_2) to the logarithm of the rate constant for the standard reagent methanol (k_1^{MeOH}).

Hence

$$n_{Pt} = \log_{10}(k_2^Y/k_1^{MeOH}).$$

Nucleophilicity constants can be made dimensionless by dividing through by the concentration of the methanol (26 mol/litre), which converts the first-order rate constant for methanol to a second-order rate constant, to give constants known as n_{Pt}^0.[28,29] When the substrate is altered, the nucleo-philicity constants are related to the second-order rate constants by the linear free-energy relationship

$$S \times n^0_{Pt} = \log_{10}(k^Y_2/k^0_2)$$

where S is a function of the substrate known as the nucleophilic discrimination factor and $\log_{10} k^0_2$, which is a measure of the intrinsic reactivity of
each substrate, refers to the reaction in which the entering group (Y) is
methanol.[30]

Nucleophilicity constants for a wide range of ligands have been reported
(Table 88). However, the nucleophilicity constant has not been meaningfully

TABLE 88
Nucleophilicity constants (n^0_{Pt}) for typical nucleophiles (from ref. 29)

Nucleophile	n^0_{Pt}	Nucleophile	n^0_{Pt}	Nucleophile	n^0_{Pt}
CH_3OH	0·00	N_2H_4	3·86	$C_6H_{11}NC$	6·34
CH_3COO^-	<2·0	C_6H_5SH	4·15	$Sb(C_6H_5)_3$	6·79
CO	<2·0	Br^-	4·18	$As(C_6H_5)_3$	6·89
CH_3O^-	<2·4	$S(C_2H_5)_2$	4·52	$SeCN^-$	7·11
F^-	<2·2	$P(N\{C_2H_5\}_2)_3$	4·54	CN^-	7·14
Cl^-	3·04	$S(CH_3)_2$	4·87	$C_6H_5S^-$	7·17
NH_3	3·07	$(CH_3O)_2PO^-$	5·01	$SC(NH_2)_2$	7·17
Imidazole	3·08	$S(CH_2)_5$	5·02	$P(OCH_3)_3$	7·23
Piperidine	3·13	$S(CH_2)_4$	5·14	$S_2O_3^{2-}$	7·34
Aniline	3·16	$SnCl_3^-$	5·44	$As(C_2H_5)_3$	7·68
Pyridine	3·19	I^-	5·46	$P(C_6H_5)_3$	8·93
NO_2^-	3·22	$Se(CH_2C_6H_5)_2$	5·53	$P(n-C_4H_9)_3$	8·96
$S(CH_2C_6H_5)_2$	3·43	$Se(CH_3)_2$	5·70	$P(C_2H_5)_3$	8·99
N_3^-	3·58	SCN^-	5·75		
NH_2OH	3·85	SO_3^{2-}	5·79		

related to other properties of the ligand such as basicity or redox potential,
nor has a means been found of calculating the constant other than from
kinetic measurements. The most important factor controlling its value appears to be the softness of the ligand. It is interesting however, that the
nucleophilicity order in Table 88 is remarkably similar to the *trans*-effect
order (see below p. 299), indicating that a good *trans*-labiliser is also a good
entering group in square-planar substitutions. This is understandable in
terms of the reaction mechanism discussed below (section (g) pp. 308–311)
since both the *trans*-group and the entering group occupy similar positions
with respect to the leaving group in the trigonal plane of the trigonal bipyramidal reaction intermediate.

Whilst the nucleophilicity constants of many of the ligands recorded in
Table 88 are independent of the substrate there are a number of π-bonding
nucleophiles such as $SeCN^-$, thiourea and NO_2^- which exhibit a positive
deviation from the calculated rate when they react with negatively charged
complexes such as $PtCl_4^{2-}$ and a negative deviation with positively charged
complexes such as $Pt(dien)Cl^+$. Such nucleophiles are termed biphilic,[31] to
emphasise that it is the electrophilic properties of these ligands resulting
from their ability to accept electron density back-donated from the metal

to their empty π-orbitals, that leads to an enhanced rate with negatively charged substrates and a reduced rate with positively charged substrates.[31-33]

Relatively little is known about the effect of the entering group on the rate of palladium(II) substitution reactions although the kinetics of reaction 336 showed that the order of nucleophile reactivity was[34] $H_2O \approx OH^-$

$$Pd(acac)_2 + 2Y^- + H^+ \longrightarrow Pd(acac)Y_2^- + Hacac \qquad (336)$$

$< Cl^- < Br^- < I^- < SCN^-$ which was analogous to that found for platinum(II) complexes (see Table 88).

A comparison of unidentate and bidentate amines as entering ligands has been made by studying the reaction of (bipyr)PtCl$_2$ with amines. The activation enthalpy and entropy both decrease on replacing a unidentate amine by a bidentate ligand suggesting that not only is the bonding stronger in the transition state but also the transition state is more compact in the case of the bidentate entering ligands.[35] This could be due to hydrogen-bonding between the unco-ordinated end of the amine and the chloride ligands in the transition state as in **128** or **129**

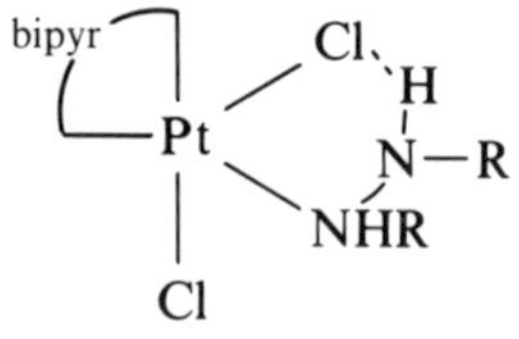

128 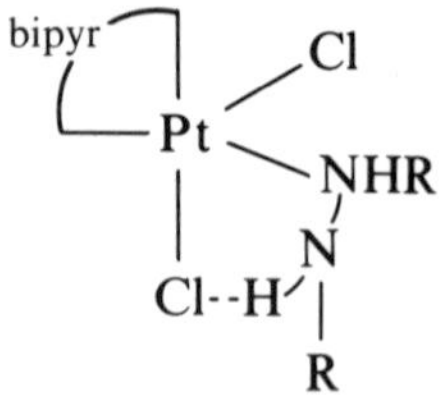   **129**

(c) Effect of substrate

As mentioned above (Table 84, p. 294), variation of the charge of the substrate has only a limited effect on the rate of substitution reactions and provides strong evidence for an associative mechanism. A way to evaluate the effect of the substrate on this rate is to examine the nucleophilic discrimination factors (S) and intrinsic reactivities ($\log_{10} k_2^0$) for a number of substrates. The experimental values (Table 89) indicate that as the intrinsic

TABLE 89

Nucleophilic discrimination factors (S) and intrinsic reactivities ($\log_{10} k_2^\circ$) for several platinum(II) complexes (from ref. 7, p. 401)

Complex	Solvent	S	$\log_{10} K_S^0$
trans-(PEt$_3$)$_2$PtCl$_2$	CH$_3$OH	1·43	−8·82
trans-(AsEt$_3$)$_2$PtCl$_2$	CH$_3$OH	1·25	−7·49
trans-(SeEt$_2$)$_2$PtCl$_2$	CH$_3$OH	1·05	−6·13
trans-Pt(pip)$_2$Cl$_2$	CH$_3$OH	0·91	−5·83
Pt(en)Cl$_2$	H$_2$O	0·64	−5·26
Pt(dien)Br$^+$	H$_2$O	0·75	−5·16
Pt(dien)Cl$^+$	H$_2$O	0·65	−4·56
Pt(dien)H$_2$O^{2+}	H$_2$O	0·44	−1·08

reactivity of the complex increases, its ability to discriminate between nucleophiles decreases. This is to be expected since a complex of low reactivity will be one that does not readily reach the transition state and therefore one for which the rate of reaction will depend more on the properties of the nucleophile.

The tertiary-phosphine and -arsine complexes exhibit the greatest nucleophilic discrimination factors, probably because they withdraw charge from the platinum atom through π-back-donation, thus facilitating the donation of electron density from the incoming nucleophile to platinum. This in turn allows the incoming nucleophile to make a greater contribution to the transition state, so that the reactions of tertiary–phosphine and –arsine complexes of platinum(II) exhibit greater sensitivity to changes in the nucleophile than complexes of other ligands less capable of π-bonding.

(d) Effect of the non-labile ligand
The non-labile ligands in square-planar complexes influence the rate of reaction in a number of ways.

(i) The trans-effect
One of the first ways to be recognised[36] in which a non-labile ligand can influence the rate and mode of substitution in a square-planar complex was by its *trans*-effect. The *trans*-effect has been defined[3] as 'the effect of a coordinated group on the rate of the substitution reactions of ligands opposite to it in a metal complex'. It is apparent in both preparative and kinetic studies. An approximate order of the decreasing labilising ability of ligands[5,37] is:

$$C_2H_4 \sim Me_2C(OH)C\equiv C-C(OH)Me_2 \sim NO \sim CO \sim CN^-$$
$$> R_3Sb > R_3P > R_3As \sim H^- \sim SC(NH_2)_2 > CH_3^- > C_6H_5^-$$
$$> SCN^- > NO_2^- > I^- > Br^- > Cl^- > NH_3 > OH^- > H_2O.$$

Preparatively the *trans*-effect has been used on many occasions of which the preparation of the three possible geometric isomers of $Pt(pyr)(NH_3)BrCl$ is a classic example[38] (reactions 337–9). Note that in addition to the *trans*-

$$
\begin{array}{ccccc}
\text{Cl} & & \text{Br} & & \text{Br} \\
| & & | & & | \\
\text{Cl}-\text{Pt}-\text{NH}_3 & \xrightarrow{\text{Br}^-} & \text{Cl}-\text{Pt}-\text{NH}_3 & \xrightarrow{\text{pyr}} & \text{Cl}-\text{Pt}-\text{NH}_3 \\
| & & | & & | \\
\text{Cl} & & \text{Cl} & & \text{pyr}
\end{array}
\qquad (337)
$$

$$
\begin{array}{ccccc}
\text{Cl} & & \text{Br} & & \text{Br} \\
| & & | & & | \\
\text{Cl}-\text{Pt}-\text{pyr} & \xrightarrow{\text{Br}^-} & \text{Cl}-\text{Pt}-\text{pyr} & \xrightarrow{\text{NH}_3} & \text{Cl}-\text{Pt}-\text{pyr} \\
| & & | & & | \\
\text{Cl} & & \text{Cl} & & \text{NH}_3
\end{array}
\qquad (338)
$$

$$
\begin{array}{ccccc}
\text{Cl} & & \text{NH}_3 & & \text{NH}_3 \\
| & & | & & | \\
\text{Cl}-\text{Pt}-\text{pyr} & \xrightarrow{\text{NH}_3} & \text{Cl}-\text{Pt}-\text{pyr} & \xrightarrow{\text{Br}^-} & \text{Cl}-\text{Pt}-\text{Br} \\
| & & | & & | \\
\text{pyr} & & \text{pyr} & & \text{pyr}
\end{array}
\qquad (339)
$$

effect this synthesis depends on the relative strengths of the metal–ligand bonds, the platinum–nitrogen bond being stronger than the platinum–chlorine bond. Kinetically, the *trans*-effect is well illustrated by the results in Table 90 which indicate that the *trans*-ligand modifies the rate of a reaction by several orders of magnitude.

TABLE 90
Effect of the *trans*-ligand on the rate of replacement of chloride by pyridine in ethanol
(from ref. 15)

			Rate constants	
Complex	*Temp (°C)*	*Trans-ligand*	*First-order* (sec^{-1})	*Second-order* $(M^{-1}\,sec^{-1})$
cis-$(PMe_3)_2PtCl_2$	0	PMe_3	$8{\cdot}3 \times 10^{-2}$	—
cis-$(PEt_3)_2PtCl_2$	0	PEt_3	$1{\cdot}7 \times 10^{-2}$	$3{\cdot}8$
trans-$(PEt_3)_2PtHCl$	25	H^-	$1{\cdot}8 \times 10^{-2}$	$4{\cdot}2$
trans-$(PEt_3)_2Pt(CH_3)Cl$	25	CH_3^-	$1{\cdot}7 \times 10^{-4}$	$6{\cdot}7 \times 10^{-2}$
trans-$(PEt_3)_2Pt(C_6H_5)Cl$	25	$C_6H_5^-$	$3{\cdot}3 \times 10^{-5}$	$1{\cdot}6 \times 10^{-2}$
trans-$(PEt_3)_2PtCl_2$	25	Cl^-	1×10^{-6}	4×10^{-4}

Many papers have been written on the *trans*-effect and the theories advanced to explain it. Before the discovery of the high *trans*-effect of pure σ-donors such as the hydride and methyl groups, the π-bonding theory[39,40] was considered the most satisfactory, but to account for the high *trans*-effect of σ-donors the earlier electrostatic theory of Grinberg must also be considered.[41]

The polarisation or σ-*trans*-effect theory considers the *trans*-effect to be principally electrostatic in origin and transmitted through σ-bonds. Thus in a complex with four identical ligands the polarisation of each of the ligands by the metal ion will be the same and a dipole will not result. However, if one ligand (L) is more polarisable than the others, then an induced dipole will result and the distribution of electron density will move through the σ-bond towards the ligand (X) *trans* to L. It has often been stated that as a result of this the Pt–X bond becomes weakened and lengthened (see ref. 42). Although sometimes true, this is not always so, since the kinetic *trans*-effect may arise simply because the *trans*-group (L) owns rather more of the empty p_σ orbital of the metal in the transition state than in the ground state, thus reducing the energy difference between the ground state and the transition state. The stronger the σ-donor ability of the *trans*-group the lower becomes the energy difference between these two states and, hence, the greater the *trans*-effect of that group.[6,43] Thus a very polarisable *trans*-ligand (L), of which the hydride ion is a good example, should exhibit a strong *trans*-effect. Also, since the polarisability of the metal ion is very important, the *trans*-effect of a ligand should be less in palladium(II) than platinum(II) because the covalency of palladium(II) complexes is less than that of platinum(II) complexes. This prediction is born out in practice. This theory has been put on a semi-quantitative basis by calculating the overlap of the valence orbitals of a number of ligands with the platinum $6p_\sigma$ orbital,[6] from this a σ-*trans*-effect order of $H^- > PR_3 > -SCN^- > I^-, CH_3^-, CO,$ $CN^- > Br^- > Cl^- > NH_3 > OH^-$ emerges.

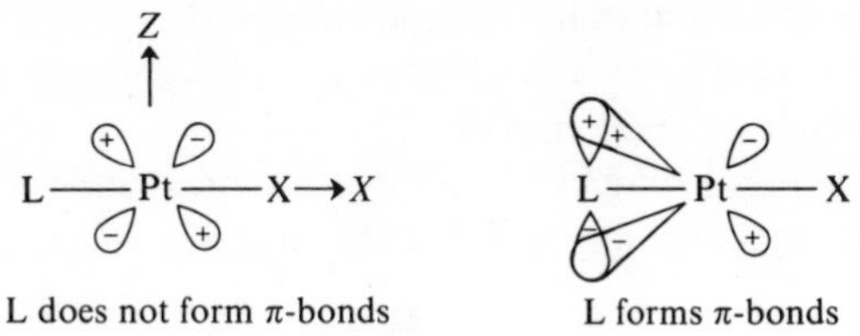

L——Pt——X→X L does not form π-bonds

L——Pt——X L forms π-bonds

Fig. 23. Schematic representation of the π-bonding mechanism for the *trans*-effect.

In the π-*trans*-effect theory a ligand L (Fig. 23) which can form π-bonds will withdraw some of the electron density of the metal d_{xz} orbital away from the ligand (X) in the *trans*-position. This will[6,44] first direct an incoming nucleophile toward X, which is the region of lowest electron density and, second, stabilise the trigonal bipyramidal transition state since L,X and the incoming ligand are all in the trigonal plane where L can be most effective at delocalising the extra electronic charge from the metal. This theory has been put on a semi-quantitative basis by calculating the overlap of the valence orbitals of a number of ligands with the platinum $6p_\pi$ orbital[6] to give a π-*trans*-effect order of $CH_2{=}CH_2$, $CO > CN^- > -NO_2^- >$ $-SCN^- > I^- > Br^- > Cl^- > NH_3 > OH^-$.

A combination of the order of ligands from the σ- and π-*trans*-effects gives rise to the overall series cited earlier (p. 299). It should be noted that although the σ-*trans*-effect can be manifest in the initial molecule (as a bond-weakening) as well as in the transition state, the π-*trans*-effect is not apparent in the ground state but exerts its influence by its effect on the transition state. Although it has not been observed, it is of course possible for a ligand which exerts a π-*trans*-effect to affect the ground state, since if the ligand in the *trans*-position also π-bonds to the metal then bond-weakening might be expected, whereas if there is only weak π-bonding in the *trans*-bond, that bond will not be weakened and may even be strengthened.[45]

The selective deprotonation of the nitrogen *trans* to X in Pt(dien)X^+ (where $X = I^-$, SCN^- and NO_2^-) indicates a major change in the bonding of diethylenetriamine on co-ordination, since deprotonation of the central nitrogen of free $NH_2(CH_2)_2NH(CH_2)_2NH_2$ has statistically only a 20% chance of occurring and this is reduced still further because the acidity of secondary amines is 5 to 10 times less than that of primary amines. Furthermore the rate of the reaction of the deprotonated complex with CH_3^+ decreases in the order $I^- > SCN^- > NO_2^-$.[46] Both these observations have been ascribed to a secondary *trans*-effect which, since it should only be a σ-*trans*-effect, may be of use in separating the σ- and π-*trans*-effects.

(ii) Trans-*influence*

So far we have discussed the *trans*-effect as observed kinetically, which depends on the energies of both the original complex and the transition state. In this book, the term 'trans-effect' is reserved strictly for the interpretation of kinetic and preparative data. However, the term trans-effect has also been applied to data obtained from x-ray diffraction and infra-red studies in which a ligand (L) is found to weaken the bond trans to itself. This effect, which is present in the ground state of the complex, is much more conveniently described by the recently introduced term 'trans-influence',[47,48] which we shall employ.

The value of distinguishing between the trans-effect and the trans-influence is well illustrated by considering Tables 91 and 92. The results given show

TABLE 91

Platinum–chlorine bond lengths in some platinum(II) complexes

Molecule	Trans-atom (or ligand)	Pt–Cl bond length $(\text{Å})^a$	No. of structure in Appendix II	Ref.
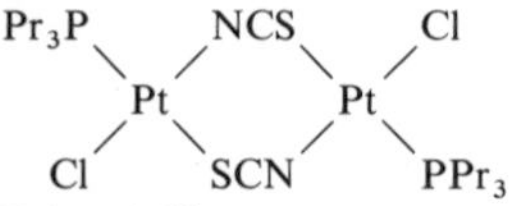 Pr_3P, NCS, Cl / Pt–Pt / Cl, SCN, PPr_3	—NCS	2·277(4)	182	49
$Pt(acac)_2Cl^-$	O	2.28(1)	149	50
trans-$(PEt_3)_2PtCl_2$	Cl	2·30(1)	157	51
trans-$(PEt_3)_2Pt(CO)Cl$	CO	2·30	169	52
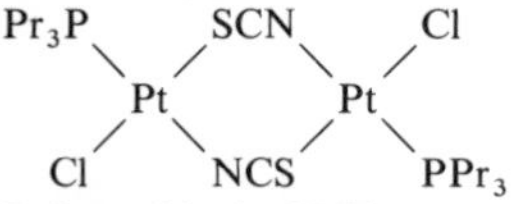 Pr_3P, SCN, Cl / Pt–Pt / Cl, NCS, PPr_3	—SCN	2·304(4)	182	49
$Pt(L\text{-methionineH})Cl_2$	S	2·32	140	53
$K^+Pt(NH_3)Cl_3^-.H_2O$	N	2·321(7)	96	54
$K^+Pt(C_2H_4)Cl_3^-.H_2O$	$CH_2{=}CH_2$	2·327(7)	121	55
cis-$PtCl_2\{C(OEt)NHC_6H_5\}(PEt_3)$	C of carbene	2·365(5)	172	56
cis-$(PMe_3)_2PtCl_2$	P	2·37(1)	143	57
trans-$(PPh_2Et)_2PtHCl$	H	2·42(1)	191	58
trans-$(PPhMe_2)_2(SiPh_2Me)PtCl$	Si	2·45(1)	192	59
Pt–Cl bond lengths in complexes where Cl bridges two Pt atoms				
trans-$(Me_3As)_2Pt_2Cl_4$	Cl	2·312(5)	142	60
trans-$(Me_3As)_2Pt_2Cl_4$	As	2·394(6)	142	60
trans-$(Pr_3P)_2Pt_2Cl_4$	P	2·425(8)	178	61

^{a} The figure in brackets is the estimated error on the last figure.

TABLE 92

Palladium–chlorine bond lengths in some palladium(II) complexes

Complex	Trans-atom (or ligand)	Pd–Cl bond length $(\text{Å})^a$	No. of structure in Appendix II	Ref.
K_2PdCl_4	Cl	2·30	3	62
$Pd(H_4EDTA)Cl_2$	N	2·31(1)	46	63
$Pd(^iPrSeCH_2CH_2Se^iPr)Cl_2$	Se	2·31(1)	44	64
$Pd(norbornadiene)Cl_2$	C=C	2·32(1)	40	65
trans-$(PPh_3)_2PdCl(o\text{-}C_6H_4{-}N{=}N{-}C_6H_5)$	C	2·382(5)	88	66

^{a} The figure in brackets is the estimated error on the last figure.

that the only ligands that lengthen the *trans*-metal–chlorine bond are those containing C, H, P, As and Si as the donor atoms. These are just those ligands which exert their *trans*-effect by a σ-mechanism. By contrast ligands such as carbon monoxide and olefins, which exert their *trans*-effect by a π-mechanism, exhibit little or no *trans*-influence.

Besides x-ray diffraction, infrared and n.m.r. data can also be used to determine the *trans*-influence of ligands. However, the interpretation of this data is not as straightforward as that from x-ray diffraction. For example the use of infra-red data really requires a knowledge of the force constants, relatively few of which have been evaluated, because the use of frequencies alone is liable to give incorrect results due to the interaction of one stretching mode with others in the complex (see pp. 240–242). Furthermore, recent work with isotopically labelled complexes has shown that many earlier assignments are in error.[67] Nor does n.m.r. data always lead simply to a disclosure of the *trans*-influence of ligands. An analysis of the ^{195}platinum–^{31}phosphorus coupling in the n.m.r. spectra of tertiaryphosphine-platinum(II) complexes indicated that the tertiary phosphine ligands are much more powerful σ-donor ligands than previously thought.[47] This is, of course, also indicated by the x-ray data in Table 91. However, whilst in this case the n.m.r. results are not open to too much criticism, except that the authors may have overemphasised their rejection of the π-bonding capacity of tertiary phosphines,[68] a further application of n.m.r., in which the contribution of the π-acceptor capacity of a ligand to its *trans*-influence was determined, has been criticised. This is the oft cited determination of the n.m.r. chemical shifts of a series of complexes *trans*-$(PEt_3)_2PtXY$ where

Y = F—⟨benzene ring⟩—   or   ⟨benzene ring with F⟩—   and X = '*trans*-ligand'; it is discussed

in detail in Chapter 12 (pp. 335–336).

A new technique used for studying the *trans*-influence of ligands is nuclear quadrupole resonance spectroscopy. Recent work with platinum(II) and palladium(II) complexes[69–71] has confirmed the indication from n.m.r. that both *cis*- and *trans*-influences operate exclusively through the σ-component of the metal-ligand bond. This work, coupled with the results of x-ray diffraction, has shown that a given reduction in the platinum–chlorine σ-bond order by a particular ligand causes a much greater increase in the length of the platinum–chlorine bond *trans* to the ligand than in that of the platinum–chlorine bond *cis* to the ligand.

(iii) Cis-*Effect*

Whereas ligands *trans* to the leaving group have a profound effect on its rate of substitution, ligands *cis* to it generally have only a minor effect. As Table 93 indicates, the rate of substitution of a group *cis* to a methyl group is $3\frac{1}{2}$ times faster than that *cis* to a chloride group, whereas Table 90 indicates a factor of 170 between the corresponding groups when in the *trans*-position. However, the relative unimportance of the *cis*-effect should

TABLE 93

Effect of the *cis*-ligand on the rate of substitution of the
chloride in *cis*-$(PEt_3)_2Pt(L)Cl$ by pyridine in methanol
at $0°C$ (data from ref. 15)

cis-ligand (L)	First-order rate constant (sec^{-1})
Cl^-	1.7×10^{-2}
C_6H_5	3.8×10^{-2}
CH_3	6.0×10^{-2}

not be over-stressed, since *cis*-effects can be more important than *trans*-effects when groups of nearly equal *trans*-effects are compared. This is illustrated in Table 94 where formation rates for a series of consecutive complexes have been analysed in respect of the *cis*- and *trans*-effects of the non-halide ligand. The *cis*-effects here are between 5 and 20,000 times as important as the *trans*-effects.

TABLE 94

Coefficients for the *cis*- and *trans*-effects of ligands (L) on the rate of replacement of halide ligands (X) by L ligands according to the equation $MX_{4-n}L_n + L \longrightarrow$ $MX_{3-n}L_{n+1} + X$ (where M = Pt or Pd) analysed according to the formula

$$\frac{Rate}{no.\ of\ equivalent\ X\ in\ complex} = A \times (T)^t \times (C)^c$$

where t = no. of L ligands *trans* and c = no. of L ligands *cis* to the halide to be replaced.

M	L	X	T	C	Reference
Pt	H_2O	Cl	3×10^{-3}	3	72
Pt	H_2O	Br	3×10^{-4}	6	73
Pt	NH_3	Cl	0.5	2.4	74
Pd	NH_3	Cl	0.1	10	75

Early studies[76,77] suggested that in addition to being smaller the effect of *cis*-groups on rates of substitution was of an opposite sign to that of *trans*-groups. This, whilst sometimes so is not always true as Tables 90 and 93 show, since both the *cis* and *trans*-effects of chloride, phenyl and methyl groups are in the order $CH_3 > C_6H_5 > Cl$. A further complication any theory of the *cis*-effect must be able to accommodate is that several examples are known in which inversion of the apparent *cis*-labilising effect may occur, depending on the nature of the entering ligand. This is illustrated by reaction 340; here, when $Y^- = Cl^-$, the rate

$$trans\text{-}PtL_2Cl_2 + Y^- \longrightarrow trans\text{-}PtL_2ClY + Cl^- \tag{340}$$

increases from L = PEt_3 $(2.9 \times 10^{-5} M^{-1} sec^{-1})$ to L = pyridine $(4.5 \times 10^{-4} M^{-1} sec^{-1})$, but, when $Y^- = I^-$, the rate decreases from L = PEt_3 $(0.236 M^{-1} sec^{-1})$ to L = pyridine $(0.107 M^{-1} sec^{-1})$.[27] This is a consequence of a combination of the different abilities (i.e. different S factors—

see Table 89) of the two substrates to discriminate between the two nucleophiles and the effect of the absolute values of the rate constants. In certain cases the nature of the entering ligand can modify the *cis*-effect of a given ligand. Thus the *cis*-effect of halide ligands is normally $I^- > Br^- > Cl^-$ [78] but with sulphide entering ligands this order is reversed.[78,79] This is due to the fact that the lone-pair of electrons on the sulphur atom not used for bonding may interact with the *cis*-halide ligands in the transition state hindering bond formation.[78,80] This interaction and hence hindrance increases with increasing size of the *cis*-ligand so that a *cis*-effect order of $Cl^- > Br^- > I^-$ is observed with sulphide entering ligands.

The most satisfactory explanation of the *cis*-effect of a ligand considers the role of the *cis*-ligand in modifying the nucleophilic discrimination factors (S). In these terms, the *cis*-effect order for ligands in reaction 340, as measured by the S value is $PEt_3(1.62) > AsEt_3(1.3) > SeEt_2(1.1) > pyridine(1.00) >$ piperidine(0.95)[27] and can be accounted for on the assumption that the presence of *cis*-ligands capable of delocalising negative charge away from the metal by means of π-bonding enhances the electrophilic character of the metal leading to a relatively easier attack of the nucleophile. By contrast the *cis*-effect order of L ligands for reaction 341 where $Y = MeOH$, NO_2^-, N_3^-, I^-, CNS^- or thiourea is

$$Pt(bipyr)LCl + Y^- \rightarrow Pt(bipyr)LY + Cl^- \tag{341}$$

$NCS^-(1.3) > N_3^-(0.95) > NO_2^-(0.87) > Cl^-(0.75)$[81] which is the order of decreasing micropolarisability of these ligands,[82] corresponding to the decreasing ability of the substrate to delocalise away from the metal the extra charge donated by the incoming nucleophile. Thus, as in the *trans*-effect, there are two ways in which a ligand can exert a *cis*-effect. In the first, which we may call the π-*cis*-effect mechanism, the ligand alters the ground state distribution of charge so as to favour attack by the reagent, whereas in the second, which we may call the σ-*cis*-effect mechanism, the ligand has no effect on the ground state charge distribution, but, by virtue of its micropolarisability, it delocalises the extra charge donated to the metal by the incoming nucleophile, so stabilising the transition state.

(iv) Steric effect

The steric effects of the non-labile ligands can be of two types which really differ in degree rather than kind.

Lesser steric effects result simply in a decreased rate of substitution and are well illustrated by the effect of *ortho*-substituents on the phenyl ring in complexes of the type *cis*- and *trans*-$(PEt_3)_2Pt(aryl)Cl$ on the rate of substitution of the chloride ligand by pyridine (see Table 85). The much greater effect of the *ortho*-substituents when the aryl group is *cis* than when it is *trans* to the chloride ligand (Table 85) is a consequence of the trigonal bipyramidal transition state involved in the substitution reaction since the *cis*-ligands in the substrate are found in the apical position in the transition state where the *ortho*-methyl groups crowd the incoming and outgoing ligands far more than when the aryl group is in the equatorial position in the

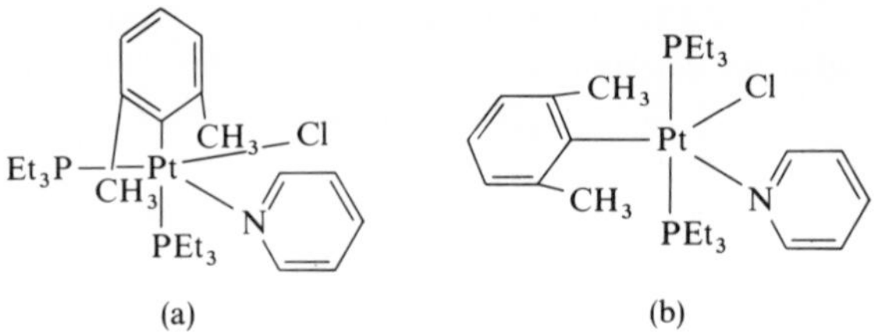

(a)   (b)

Fig. 24. Effect of *ortho*-methyl groups present on aryl ligands on the crowding in the transition state formed when the chloride ligand in $(PEt_3)_2Pt(aryl)Cl$ is substituted by pyridine. Case (a) *cis*-complex; case (b) *trans*-complex.

transition state as occurs with *trans*-complexes (see Fig. 24). This crowding results in simple nucleophiles reacting with *trans*-$(PEt_3)_2Pt(aryl)Cl$ exclusively by a dissociative mechanism (i.e. with a zero second-order rate constant).[83,84] However, with biphilic entering reagents such as cyanide and thiourea a second-order rate constant was observed probably because the very strong σ-donor ability of these aryl ligands facilitates the formation of a five-co-ordinate transition state by donation of charge from the metal to the biphilic reagent.[84]

The consequence of very large steric effects can be seen by comparing the substitution reactions of $M(dien)Cl^+$ and $M(Et_4dien)Cl^+$ where $M =$ $Pt^{[85]}$ or $Pd.^{[86]}$ In water and methanol the four ethyl groups in the Et_4dien complexes prevent the incoming nucleophile from gaining access to the metal. This has two effects. First it slows down the rate of substitution reactions by several orders of magnitude, and secondly it reduces the second-order rate constants to zero for all the nucleophiles so far examined,[86] apart from thiosulphate,[87] indicating that reaction occurs by a dissociative mechanism as for the substitution reactions of octahedral cobalt(III) complexes. For this reason the complexes $M(Et_4dien)X^+$, where X is a unidentate ligand, have been described as 'pseudo-octahedral' complexes.[86] However, when the substitution reactions were carried out in methyl cyanide the normal two-term rate law was observed[88] due to solvation of the ethyl groups of Et_4dien by methyl cyanide enabling them to spread away from the metal facilitating entry of the nucleophile.

(e) Effect of leaving group

A useful study of the role of the leaving group in the kinetics of substitution reactions has been made with $M(dien)X^+$ complexes (Table 95). The dien ligand renders the three non-labile co-ordination positions inert so that the only variable is the leaving group X. As Table 95 shows, a change of leaving group can cause a large change in the rate, indicating that a considerable perturbation of the M–X bond occurs when the transition state is formed. This is further indicated by the rate order $Cl^- > Br^- > I^-$, which is the reverse of the stability order $Cl^- < Br^- < I^-$. Another relevant point is that groups that are high in the *trans*-effect order are replaced slowly. This is to be expected since the σ- and π-*trans*-effect theories suggest that a ligand high in the *trans*-effect series will be strongly bonded to the metal.

TABLE 95

Rates of the Reaction M(dien)X$^+$ + pyr $\rightarrow$ M(dien)pyr^{2+} + X$^-$ (where M = Pt or Pd) in water at 25°C

Leaving group	Rate constant^a for Pt (sec^{-1})	Reference	Rate constant^b for Pd (sec^{-1})	Reference
NO$_3^-$	Very fast	89		
H$_2$O	1900×10^{-6c}	32		
Cl$^-$	35×10^{-6}	89	Very fast	90
Br$^-$	23×10^{-6}	89	Very fast	90
I$^-$	10×10^{-6}	89	2·10	90
N$_3^-$	$0·83 \times 10^{-6}$	89	0·78	90
SCN$^-$	$0·30 \times 10^{-6}$	89	0·20	90
NO$_2^-$	$0·050 \times 10^{-6}$	89	0·05	90
CN$^-$	$0·017 \times 10^{-6}$	89		

^{a} [pyr] = 0·0059 M.
^{b} Calculated for [pyr] = 0·0059 M.
^{c} [pyr] = 0·005 M.

However, ligands low in the *trans*-effect series are not always easy to replace, since OH$^-$ and NH$_3$ both lie low in the *trans*-effect series but are hard to replace. This must indicate that a trigonal bipyramidal transition state with these ligands in the trigonal plane has a relatively high energy.

Although the general trends in Table 95 are observed, these are not followed in detail by all reactions. For example Table 95 suggests that the rate of substitution decreases as the leaving group is altered in the order NO$_3^-$ > H$_2$O > Cl$^-$ > Br$^-$ > I$^-$ > N$_3^-$ > SCN$^-$ > NO$_2^-$ > CN$^-$, whereas for reaction 342 where Y = Cl$^-$, Br$^-$, I$^-$, NO$_2^-$ and N$_3^-$ in methanol at 25°C the

$$\text{Pt(bipyr)(NO}_2\text{)X} + \text{Y}^- \rightarrow \text{Pt(bipyr)(NO}_2\text{)Y} + \text{X}^- \tag{342}$$

ease of substitution of the leaving groups relative to N$_3^-$ is I$^-$(900) > Br$^-$(240) > Cl$^-$(140) > NO$_2^-$(7) > N$_3^-$(1).[91] Thus the effect of the leaving group on the rate of the reaction appears to depend in some way on the non-labile groups present in the substrate.

The leaving group can also have a considerable influence on the mechanism of the reaction. Thus bromide ion substitution reactions of Pd(Et$_4$dien)-(XCN)$^+$ in dimethylformamide show that when X is sulphur only the first-order solvent path is operative (a pseudo-octahedral mechanism), whereas when X is selenium both the first-order solvent path and the second-order ligand path operate concurrently.[92]

(f) Effect of the solvent

Solvents can be divided into two groups—those that can co-ordinate to the metal (the alcohols, water, nitromethane and dimethylsulphoxide) and those that either cannot (benzene and carbon tetrachloride) or can only co-ordinate very weakly (*t*-butanol, acetone and ethyl acetate). Solvents in the first group can play at least two different roles in substitution reactions. First, they can act as nucleophiles in the first-order path of the reaction and then their effectiveness decreases (see Table 87) in the order DMSO > H_2O > CH_3NO_2 > C_2H_5OH > n-C_3H_7OH. However, with the exception of DMSO, the first-order rate constant varies very little over this series of solvents. The relatively high reactivity of DMSO is in accord with its known ability to act as a ligand towards platinum(II).[19]

The second way in which a solvent can affect the rate of a substitution reaction is by its ability to solvate species involved in the reaction. However, although something is known about the solvation of square-planar complexes by 'specific' solvation arising from the weak co-ordination of two solvent molecules at the axial positions,[93] little is known about the 'non-specific' general solvation of square-planar complexes or of the trigonal bipyramidal transition intermediates. The observation that for *trans*-$(PEt_3)_2PtCl_2$ as substrate the entering group rate order of $I^- > Br^- > Cl^-$ is maintained in acetone, DMSO and methanol,[28] in spite of the fact that chloride becomes 60 times more reactive than iodide on going from methanol to acetone, suggests that the change in solvation of the entering group from the ground to the transition state is of less importance than the nature of the entering group itself. A study of solvent effects on the tertiary phosphine catalysed *cis*- *trans*-isomerisation of $(PBu_3)_2PtCl_2$ in cyclohexane has suggested that the addition of solvating solvents such as CH_3NO_2, CH_3CN, Et_2O or $CHCl_3$ results in the solvation of the original four-co-ordinate complex $(PBu_3)_2PtCl_2$ by only one solvent molecule, whereas the only hydrogen-bonding solvent investigated, methanol, solvated the complex with two molecules. The results also suggested that the actual transition intermediate was not solvated at all.[94] Leaving group solvation has been suggested as being kinetically important by studies of the replacement of an anionic ligand (chloride) by a neutral ligand (thiourea) in *trans*-$Pt(pyr)_2Cl_2$ in mixtures of water (H_2O and D_2O) and dimethylsulphoxide.[95]

A further indirect effect of the medium that has not been noted as carefully as it should is the effect of ionic strength on the rates of reactions and it has been emphasised that all rate constants that are to be compared should be recorded at constant ionic strength.[96]

In solvents that cannot co-ordinate to the metal or can only co-ordinate very weakly, such as benzene, carbon tetrachloride, *m*-cresol, *t*-butanol, ethyl acetate, acetone and dimethylformamide, the first-order path for substitution reactions is not available.[18] The second-order rate constants for reaction 343 vary over seven orders of magnitude from $10^4 \, M^{-1} \, sec^{-1}$ in carbon tetrachloride where the free chloride ion is relatively unstable to

$$trans\text{-}Pt(pyr)_2Cl_2 + 2R_4N^{36}Cl \rightleftharpoons trans\text{-}Pt(pyr)_2^{36}Cl_2 + 2R_4NCl \qquad (343)$$

$10^{-3} \, M^{-1} \, sec^{-1}$ in dimethylformamide where the free chloride ion is fairly strongly solvated and consequently quite stable.[18]

(g) The intimate mechanism

All the evidence that has been presented so far is consistent with the two-path reaction mechanism shown in Fig. 25, in which a trigonal bipyramidal transition state is formed. This mechanism is consistent with the fact that

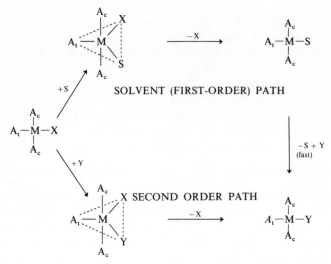

Fig. 25. Two-path reaction mechanism for the substitution reaction $MA_3X + Y \rightarrow MA_3Y + X$
(S = solvent).

the stereochemistry about the metal is maintained during substitution reactions (*cis*-starting materials give *cis*-products) because the initial *cis*- and *trans*-ligands occupy stereochemically different positions (axial and equatorial respectively) in the trigonal bipyramidal reaction intermediate. However no direct evidence, such as the isolation of a trigonal bipyramidal reaction intermediate, has been obtained.

At least three potential reaction profiles, shown in Fig. 26, are consistent with the mechanism represented in Fig. 25. A reaction which unequivocally

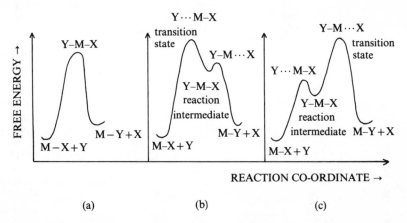

Fig. 26. Reaction profiles for substitution reactions of square-planar complexes. (a) Activation energy involves M—Y bond formation; I_A or S_N2; (b) Activation energy involves M—Y bond formation; A or $S_N2(lim)$; (c) Activation energy involves M—X bond rupture; A or S_N2. The classification I_A or A is from reference 6, p. 14 and S_N2 or $S_N2(lim)$ is from reference 7, p. 128.

follows the first reaction profile (case a) has not been found. All known instances suggest that the five-co-ordinate reaction intermediate represents a minimum between cases (b) and (c) in the reaction profile. Cases (b) and (c) themselves represent the two extremes of this situation and in practice most reactions lie somewhere between these two extremes. In case (b) bond-making is the rate controlling step for the reaction, whereas in case (c) bond-breaking is the rate controlling step. Obviously a reaction which has profile (b) in its forward direction must have profile (c) in its reverse direction.

The rate of reaction 344 was found to vary considerably with the nature

$$\text{Pt(bipyr)(NO}_2)X + RSR' \rightarrow \text{Pt(bipyr)(NO}_2)(RSR') + X^- \quad (344)$$

of the unsymmetrical thioether but relatively little with the leaving group X.[97] This is consistent with a type (b) profile. A recent study of the reverse of reaction 344 has shown that, as expected, this has a type (c) profile and is sensitive to the leaving group (the thioether) but relatively insensitive to the entering group.[98] Langford and Gray[6] have generalised this result and postulated that type (b) behaviour occurs for substitutions in which the entering group is higher in the *trans*-effect series than the leaving group and type (c) behaviour occurs when the reverse is true.

A possible alternative second-order pathway to the trigonal bipyramidal path shown in Fig. 25 is a square-pyramidal path[99] (Fig. 27). This mechanism

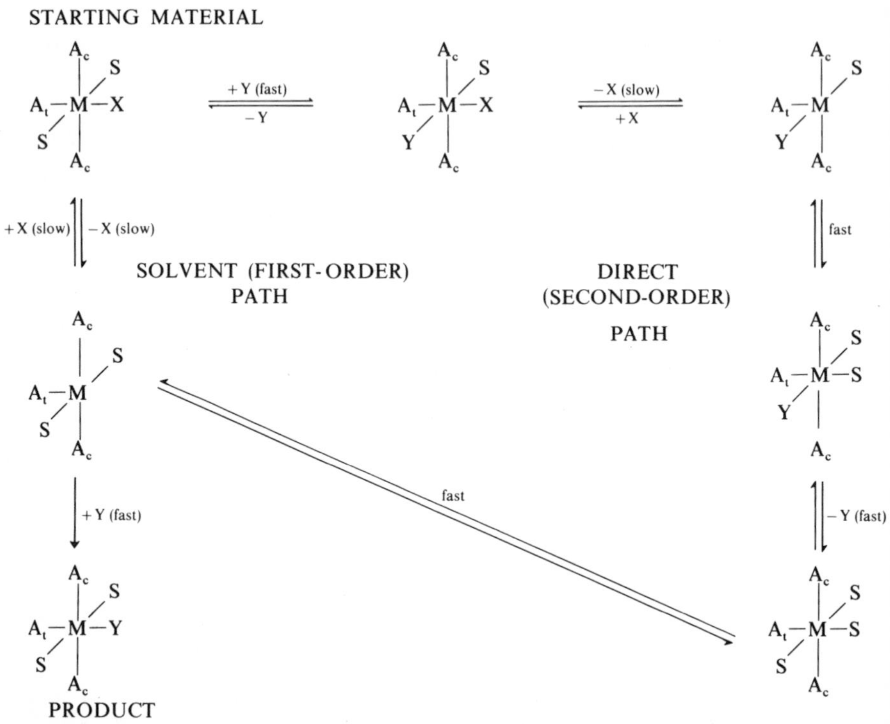

Fig. 27. Reaction mechanism of a square-planar complex involving a square-pyramidal intermediate (S = solvent).

was eliminated by a study of reaction 345 in the presence of hydroxide

$$Pt(dien)Br^+ + Y^- \rightarrow Pt(dien)Y^+ + Br^- \qquad (345)$$

ions,[32] for, had it been followed $Pt(dien)(H_2O)^{2+}$ would have been formed, but this reacts rapidly with hydroxide ions to give $Pt(dien)(OH)^+$, which is stable and inert in the presence of most nucleophilic ligands. The cation $Pt(dien)(OH)^+$ was not observed.

(h) Electrophilic catalysis
The rates of $^{36}Cl^-$ exchange with *trans*-PtL_2Cl_2 where L = pyridine[18] and piperidine[100] are accelerated by certain acids such as acetic, nitrous and boric acids. With the piperidine complex a three-term rate law was followed,

$$\text{Rate} = \{k_1 + k_2[^{36}Cl^-] + k_3[^{36}Cl^-][HA]\}\,[\textit{trans}\text{-}Pt(pip)_2Cl_2]$$

where $HA = HNO_2$ or H_3BO_3. This suggests that, in addition to the solvent (k_1) and reagent (k_2) paths, there is an acid-catalysed path for which the mechanism in Fig. 28 has been suggested. The acid catalyst removes electron

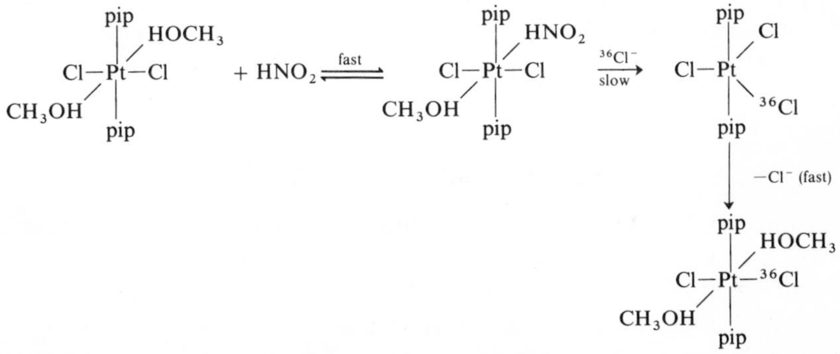

Fig. 28. Proposed mechanism for the acid-catalysed chloride exchange of *trans*-$Pt(pip)_2Cl_2$ in methanol (from ref. 100).

density from the valence d-orbitals of the metal enabling it to more readily accept electron donation from an incoming nucleophile. This interpretation can also explain the absence of a catalytic effect by nitrous acid in the reactions of *trans*-$(PEt_3)_2PtCl_2$ and *trans*-$(AsEt_3)_2PtCl_2$ since these complexes contain ligands which, by π-bonding, can themselves reduce the build-up of negative charge on the metal introduced by the incoming nucleophile.

(i) Anomalous ligand exchange reactions
The replacement of a ligand by an identical ligand, observed with the aid of radioactive isotopes, shows that the pattern of reactivity discussed in sections (a) to (g) is not universal for platinum(II) substitution reactions.[8] Isotope exchange studies enable the excess of nucleophile, usually used to provide first-order conditions, to be avoided. Under these reaction conditions several exceptions to the usual ligand replacement processes have been disclosed. Thus, although the chloride exchange reaction of $PtCl_4^{2-}$ appears

to follow the usual pattern[101,102] the rate expression for bromide exchange in solutions of $PtBr_4^{2-}$ is given by[103]

$$\text{Rate} = k_1\{[PtBr_4^{2-}] + [PtBr_3(H_2O)^-]\} + k_d[PtBr_4^{2-}][PtBr_3(H_2O)^-].$$

The presence of a term that is first-order in both $PtBr_4^{2-}$ and $PtBr_3(H_2O)^-$ is thought to indicate that a single-bromide-bridged, labile intermediate such as $(H_2O)Br_2PtBrPtBr_3^{2-}$ is involved in the reaction.

Another unusual reaction is the isotope exchange of $^{36}Cl^-$ with the *cis*-chloride ligands in $Pt(C_2H_4)Cl_3^-$. The rate expression for this reaction,

$$\text{Rate} = k_1[Pt(C_2H_4)Cl_3^-] + k_1'[\textit{trans}\text{-}Pt(C_2H_4)Cl_2(H_2O)]$$
$$+ k_d[Pt(C_2H_4)Cl_3^-][\textit{trans}\text{-}Pt(C_2H_4)Cl_2(H_2O)]$$

also contains a term that is first-order in two platinum(II) species.[104] In this case a double-chloride-bridge labile intermediate such as (C_2H_4)-$ClPtCl_2PtCl(C_2H_4)$ was postulated.

A further unusual reaction occurs in the $PtBr_4^{2-}$ catalysed bromide exchange of $Pt(dien)Br^+$ in aqueous solution. In the absence of added $PtBr_4^{2-}$, the bromide exchange of $Pt(dien)Br^+$ follows the usual two-term rate-law. However, addition of $PtBr_4^{2-}$ to the $Pt(dien)Br^+$/bromide solution results in an increase in the bromide exchange rate of both $Pt(dien)Br^+$ and $PtBr_4^{2-}$ by equal amounts.[105] This mutually catalysed exchange process follows the rate law,

$$\text{Rate} = k[Pt(dien)Br^+][PtBr_4^{2-}]$$

and a mechanism involving a single-bromide-bridged complex is suggested.

Yet another reaction that may eventually be found to take an unusual course is that of olefins with $PtCl_4^{2-}$. The olefins $CH_2{=}CH{-}CH_2X$ where $X = NH_3^+$, OH and SO_3^- react with $PtCl_4^{2-}$ in aqueous solution by an exclusively second-order path.[106] The reaction with ethylene as the olefin confirmed this,[107,108] and a preliminary observation suggested that $Pt(C_2H_4)Cl_3^-$ catalysed the aquation of $PtCl_4^{2-}$ through the formation of a double-chloride-bridged intermediate $(C_2H_4)ClPtCl_2PtCl_2^-$. This makes the aquation reaction very much faster than the rate of absorption of ethylene so that only a second-order rate is observed.[109]

(j) Substitution reactions of trigonal bipyramidal complexes

The substitution reactions of trigonal bipyramidal complexes are of interest for two reasons. Firstly these complexes are structurally intermediate between square-planar and octahedral complexes and secondly they are thought to be involved as intermediates in the reactions of square-planar complexes.

An investigation of reaction 346, where QAs is the tetraarsine *tris-(o-*

$$Pt(QAs)X^+ + Y^- \rightarrow Pt(QAs)Y^+ + X^- \qquad (346)$$

diphenylarsinophenyl)arsine and $Pt(QAs)X^+$ has the structure shown in Fig. 29, showed that the rate followed the two-term rate law[111] in eqn. 347.

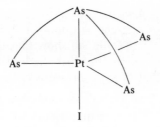

Fig. 29. Schematic diagram of the structure of Pt(QAs)I$^+$ (from ref. 110).

However, this rate law differs from that for square-planar complexes in that

$$\text{Rate} = \{k_1 + k_2[Y^-]\}\,[\text{Pt(QAs)X}^+] \qquad (347)$$

both the k_1 and k_2 terms are strongly dependent on the nature of the entering ligand. The observed rate is a result of the considerable ion-pair formation by the reactants in methanol. The overall mechanism, shown in eqn. 348–50

$$\text{Pt(QAs)X}^+ + Y^- \xrightleftharpoons{K} \text{Pt(QAs)X}^+, Y^- \qquad (348)$$

$$\text{Pt(QAs)X}^+ + Y^- \xrightleftharpoons{k_c} \text{Pt(QAs)Y}^+ + X^- \qquad (349)$$

$$\text{Pt(QAs)X}^+, Y^- + Y^- \xrightleftharpoons{k_{\text{I.P.}}} \text{Pt(QAs)Y}^+, Y^- + X^- \qquad (350)$$

gives a rate-law for excess Y^-,

$$k_{\text{obs.}} = \frac{k_c[Y^-] + k_{\text{I.P.}}[Y^-]^2}{1 + K[Y^-]}$$

Under the experimental conditions $K[Y^-]$ was much greater than unity so that the observed rate-law reduced to that of eqn. 347.

The kinetics of this reaction and the fact that the rates were about 10^4 times less than for the square-planar tritertiary arsine complexes Pt(TAs)X$^+$ suggested that the trigonal bipyramidal Pt(QAs)X$^+$ complexes are intermediate in behaviour between planar complexes, which are very susceptible to nucleophilic attack, and octahedral complexes, which are not attacked by nucleophiles.

ISOMERISATION OF SQUARE-PLANAR COMPLEXES

Square-planar complexes of the type MA$_2$B$_2$ or MA$_2$BC exhibit *cis-* *trans-*isomerisation and many pairs of *cis-* and *trans-*isomers of platinum(II)-complexes have been isolated. Rather fewer pairs of *cis-* and *trans-*isomers of palladium(II) complexes have been isolated because such isomers are sufficiently labile to ensure that a given reaction yields only the isomer which is the more stable under the particular conditions. *Cis-* and *trans-*platinum(II) complexes can be isomerised in solution in the presence of a nucleophile,[112] thermally,[113] photochemically[114] and spontaneously.[115]

Much of the current knowledge of *cis-* *trans*-isomerisation is based on a thermodynamic study of benzene solutions of $Pt(ER_3)_2X_2$, where E = P, As and Sb. The results[112] (Table 96) show that the *trans*-isomer is thermodynamically the more stable, but that the heat of formation of the *cis*-isomer is greater than that of the *trans*-isomer. The greater stability of the

TABLE 96

Thermodynamic data for the *cis-* *trans*-isomerisation of platinum(II) complexes in benzene at 25°C (from ref. 112). (Thermodynamic data refer to the equilibrium *cis* ⇌ *trans*)

Complex	Cis-*isomer at* equilibrium (%)	Conc. of trans-isomer / Conc. of cis-isomer	$-\Delta G^0$ (kcal/mol)	$+\Delta H^0$ (kcal/mol)	$+T\Delta S^0$ (kcal/mol at 25°C)
$(PEt_3)_2PtCl_2$	7·5	12·2	1·48	2·47	3·95
$(PPr_3)_2PtCl_2$	3·3	29·5	2·00	1·98	3·98
$(AsEt_3)_2PtCl_2$	0·57	176	3·07	1·18	4·25
$(SbEt_3)_2PtCl_2$	34·4	1·90	0·38	2·41	2·79

trans-isomer is due to the increase in entropy which accompanies the isomerisation of the *cis*-isomer. This arises from the greater solvation of the highly dipolar *cis*-isomer so that benzene is released when the *trans*-isomer is formed. The greater heat of formation of the *cis*-isomer has been ascribed to the more efficient π-bonding present in this isomer where the two phosphines do not have to share π-back donation of electron density from the same *d*-orbital.

Studies of the kinetics of the *cis-* *trans*-isomerisation of square-planar complexes in solution have shown:

(i) The isomerisation of nickel(II) and palladium(II) complexes is generally rapid, whereas that of platinum(II) complexes is extremely slow except in the presence of a catalyst. Suitable catalysts are good nucleophiles.

(ii) Kinetic studies of the isomerisation of the palladium(II) complexes $Pd(amine)_2X_2$ in the presence of excess amine have shown that the rate is first-order in both complex and free amine.[116]

(iii) A study of the isomerisation of *cis*-$(PR_3)_2PtCl_2$ in the presence of an excess of a different tertiary phosphine (PR'_3) again showed that the rate was first-order in both complex and added ligand (PR'_3) but also showed that no exchange of the phosphines PR_3 and PR'_3 occurred.[117] In addition the rate of isomerisation decreased as the polarity of the solvent increased.[94,118]

The observation that no phosphine exchange occurs eliminates all mechanisms in which the added phosphine and the original phosphines become equivalent. For the general reaction (351) such mechanisms include:

$$cis\text{-}MA_2X_2 \quad \overset{A' \text{ catalyst}}{\rightleftharpoons} \quad trans\text{-}MA_2X_2 \qquad (351)$$

(a) The double displacement mechanism[7] in which A' displaces X and then X displaces A.

(b) The pseudo-rotation mechanism[118] in which a trigonal bipyramidal intermediate MA_3X_2 is formed. The X–M–X angle of 90° in the *cis*-isomer could expand by pseudorotation so that dissociation of MA_3X_2 could produce *trans*-MA_2X_2.[119]

(c) The ionic-intermediate mechanism[116] (reaction 352), which is only a refinement of the double-displacement mechanism, is unsatisfactory because in addition to allowing all the A ligands to become equivalent it involves an ionic intermediate which implies that the rate of isomerisation should increase with increasing solvent polarity, whereas the reverse is found experimentally.

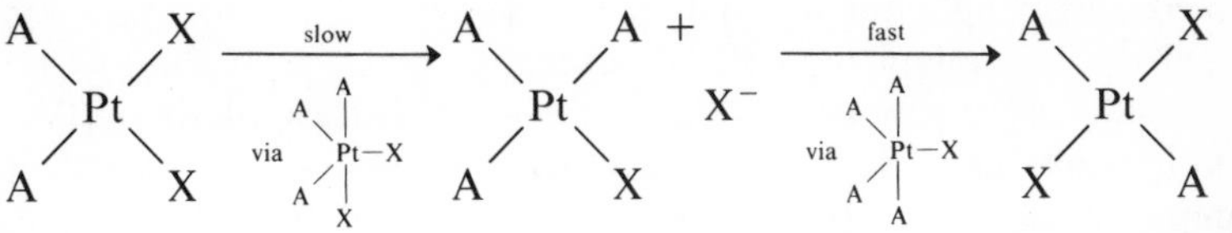

$$(352)$$

So far, the most satisfactory mechanism suggested involves a five-co-ordinate transition intermediate, which can undergo fluxional change to interconvert the positions of the original ligands, in which the catalytic nucleophile occupies a unique position.[117] Such an intermediate could have either a square pyramidal structure or a structure intermediate between square pyramidal and trigonal bipyramidal. The complex $(PR_3)_3PdBr_2$,

where PR_3 =

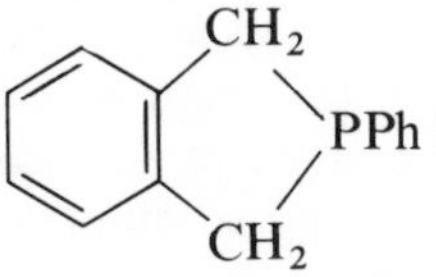

PPh is an example of a palladium(II) complex,

which contains monodentate tertiary phosphine ligands, that has a unique phosphine ligand and two further equivalent phosphine ligands.[120]

The kinetics of the uncatalysed *cis*- *trans*-isomerisation of $(PEt_3)_2Pt(o$-tolyl)Cl in alcoholic solvents have been studied and compared with the rate at which the chloride ligand is replaced by iodide and cyanide.[115] The results were consistent with an ionic dissociative mechanism involving two intermediates, one '*cis*-like' and the other '*trans*-like' (reaction 353).

$$cis\text{-}(PEt_3)_2Pt(o\text{-tolyl})Cl \rightleftharpoons \text{'}cis\text{-'}(PEt_3)_2Pt(o\text{-tolyl})^+ + Cl^- \rightleftharpoons$$

$$(353)$$

$$\text{'}trans\text{'-}(PEt_3)_2Pt(o\text{-tolyl})^+ + Cl^- \rightleftharpoons trans\text{-}(PEt_3)_2Pt(o\text{-tolyl})Cl$$

SUBSTITUTION REACTIONS OF TETRAVALENT COMPLEXES

The substitution reactions of platinum(IV) complexes are frequently cata-lysed by platinum(II) complexes and in the absence of catalyst are either very slow or do not occur at all, as has been shown both by adding an excess of platinum(II) and by removing all the platinum(II) by oxidation with cerium(IV). Frequently platinum(II) complexes have been intentionally added, but on occasions, when no platinum(II) catalysis has been reported, there is some doubt about the validity of result where it is uncertain whether there is a possibility of platinum(II) being accidentally present and affecting the rate.

The rates of these substitution reactions are first-order in respect of the platinum(IV) complex, the platinum(II) complex and the entering nucleophile.[121] A mechanism for the chloride exchange of $PtCl_6^{2-}$ put forward in 1954 suggested that platinum(III) species are formed as intermediates during the exchange,[122] although a platinum(II)–platinum(IV) bridged mechanism was suggested later.[123] In 1958 a mechanism, which is now generally accepted, involving a platinum(II)–platinum(IV) bridged complex was put forward[121] to explain the chloride exchange reaction of $Pt(en)_2Cl_2^{2+}$. In this mechanism (reactions 354–7), which is typical for platinum(IV) substitution reactions, PtL_4^{2+} is the platinum(II) catalyst, PtL_4XZ^{2+} is the platinum(IV) complex, Y is the entering ligand, X is the leaving ligand and Z is the bridging ligand. The first stage involves the addition of a fifth

$$PtL_4^{2+} + Y^- \underset{}{\overset{\text{fast}}{\rightleftharpoons}} PtL_4Y^+ \tag{354}$$

$$XPtL_4Z^{2+} + PtL_4Y^+ \underset{}{\overset{\text{slow}}{\rightleftharpoons}} XL_4PtZPtL_4Y^{3+} \tag{355}$$

$$XL_4PtZPtL_4Y^{3+} \rightleftharpoons XPtL_4^+ + ZPtL_4Y^{2+} \tag{356}$$

$$XPtL_4^+ \rightleftharpoons X^- + PtL_4^{2+} \tag{357}$$

ligand to a square-planar complex—a reaction that is well documented.[124–7] This is followed by a slow step which involves the formation of a platinum(II)–platinum(IV) halogen-bridged complex. A number of such complexes have been isolated and studied (see pp. 253–254). A platinum(IV)–platinum(II) halogen bridged complex mechanism is strongly supported by the observation that the substitution of the chloride ions in *trans*-$Pt(NH_3)_4Cl_2^{2+}$ cations by iodide in the presence of excess $Pt(en)_2^{2+}$ ions as catalyst yields $Pt(en)_2I_2^{2+}$ as the final product.[128] Recently a number of aspects of these reactions have been investigated.

(a) Effect of bridging ligand

An investigation of the effect of the bridging ligand on the rate of replacement of an ammonia molecule in halopenta-amineplatinum(IV) cations by a halide ion has shown that the effectiveness of the halide bridging ligands in promoting the reaction is $I^- > Br^- > SCN^- > Cl^-$ with relative rates of $470,000:400:32:1$ in aqueous solution at 25°C.[129,130] The differences in the entropies of activation of these reactions are small and favour a more rapid rate with iodine as the bridging atom, but the differences in the enthalpies of activation are much larger and again make iodine the most effective bridging atom. This probably arises from two effects; first, both platinum(II) and platinum(IV) form strong complexes with iodide ions and, secondly, iodine can expand its co-ordination sphere more readily than chlorine and bromine. Note, the bridging ligand (Z) in the mechanism given above is also the ligand *trans* to the leaving group (X), so that the more effective as a bridging group a ligand is, the greater its *trans*-directing ability in platinum(IV) substitution reactions.[131]

When the simple σ-donor ligand, ammonia, is replaced by a ligand such as cyanide that is also a π-acceptor, the effectiveness of the bridging ligand becomes even more important. Thus with four ammonia groups the bromide ion is about 350 times more effective than chloride as a bridging ligand, whereas with four cyanides as the *cis*-ligands this factor is increased to about 1800.[132] This may be ascribed to a stronger binding of the bridging halogen in the complexes with π-acceptor ligands placing more stringent requirements on the nature of the bridging halogen.

(b) Effect of entering group

The results, summarised in Table 97, indicate that the rate of substitution of platinum(IV) complexes decreases as the entering group is altered in the order $SCN^- > Br^- > Cl^- \sim I^- > NH_3$. Although the rates vary irregularly with halogen, the activation enthalpy decreases in the order $Cl^- > Br^- > I^- \sim SCN^-$ which parallels the increase in the platinum(IV)–

TABLE 97

Effect of entering group (Y) on the rate constants and activation parameters for the reaction
$$Pt(NH_3)_4XZ + Y \rightarrow Pt(NH_3)_4YZ + X$$

Z	Leaving group (X)	Entering group (Y)	Rate (M^{-2} sec^{-1})	ΔH* (kcal/mol)	ΔS* (entropy units)	Reference
NH_3	I^-	I^-	3.9×10^2 (25°C)	6	-29	129
NH_3	I^-	Br^-	1.2×10^4 (25°C)	8	-15	129
NH_3	I^-	Cl^-	5.6×10^2 (25°C)	11	-10	129
Cl^-	Cl^-	NH_3	1.21 (22.5°C)	6	-37	133
Cl^-	Cl^-	SCN^-	1.5×10^3 (35°C)	—	—	130
Cl^-	Cl^-	Br^-	1.7×10^2 (35°C)	—	—	130

halogen bond energies $Pt-Cl < Pt-Br < Pt-I$.[134] The activation entropy increases in the order $Cl^- < Br^- < I^- < SCN^-$ which may well be due to an increasingly rigid geometry across this series corresponding to the increase in the platinum–halogen bond strength. Although substitution reactions of platinum(IV) complexes normally take place via the bridged $Pt^{(II)}-X-Pt^{(IV)}$ intermediate mechanism shown in eqn. 354 to 357, certain examples are known in which iodide substitution of chloride or bromide in platinum(IV) complexes occurs by an initial two-electron reduction to form a platinum(II) complex and ClI or BrI in the rate-determining step.[135–40] Subsequent rapid reoxidation of this platinum(II) intermediate by ClI or BrI with incorporation of iodine into the complex can result in substitution.[137,139]

The hydroxide ion is a very powerful nucleophile which might be expected to react directly with platinum(IV) complexes by an associative mechanism. However, studies of reaction 358 have shown that it is first-order in complex

$$trans\text{-}Pt(en)_2Cl_2^{2+} + 2OH^- \rightarrow Pt(en)_2(OH)_2^{2+} + 2Cl^- \qquad (358)$$

but zero-order in hydroxide ion.[121] The kinetics are consistent with the conjugate base path (S_N1CB) shown in eqn. 359–61.

$$Pt(en)_2Cl_2^{2+} + OH^- \xrightleftharpoons{fast} Pt(en)(en-H)Cl_2^+ + H_2O \quad (359)$$

$$Pt(en)(en-H)Cl_2^+ \xrightarrow{slow} Pt(en)(en-H)Cl^{2+} + Cl^- \quad (360)$$

$$Pt(en)(en-H)Cl^{2+} + H_2O \xrightarrow{fast} Pt(en)_2(OH)Cl^{2+} \quad (361)$$

(c) Effect of leaving group

There has been little investigation into the effect of the leaving group on the rate of substitution of platinum(IV) complexes. Studies of reaction 362 show that bromide[131] and thiocyanate[130] ions are equally effective as leaving

$$Pt(NH_3)_4XCl^{2+} + Cl^- \rightarrow Pt(NH_3)_4Cl_2^{2+} + X^- \quad (362)$$

groups.

(d) Effect of substrate

There has also been little investigation into the effect of the substrate on the rate of substitution. As expected from the mechanism given in eqns. 354 to 357 the corresponding ammonia and ethylenediamine complexes react very similarly.[141,142] Again, as predicted by this mechanism, tetra-N-methylethylenediamine platinum(IV) complexes do not undergo substitution reactions because of the steric effect of the methyl groups which prevents the formation of the bridged intermediate. Instead these complexes either do not react (for instance with chloride as the potential entering group[141]) or are reduced to the platinum(II) complex when the entering group is also a reducing reagent (for instance with iodide,[140] bromide[140] and nitrite ions[142]). Complexes of the tetramethyldiarsine, o-phenylene-bis(dimethylarsine) are intermediate between ethylenediamine and tetra-N-methylethylenediamine complexes in that like ethylenediamine complexes they will undergo substitution reactions, but when the mechanism is investigated in detail it is found to occur by an oxidation-reduction route.[143]

The effect of the cis-ligand has not been investigated in any detail, but an investigation[144] of reaction 363 showed that the rate varied by only a

$$trans\text{-}Pt(dien)CCl_2^{n+} + Pt(dien)Br^+ + 2Br^- \rightarrow$$

$$trans\text{-}Pt(dien)Br_3^+ + Pt(dien)C^{n+} + 2Cl^- \quad (363)$$

factor of 5·4 at 24·2°C over the cis-ligand (C) series $NO_2^-(5·4) > Br^-(1·2) > NH_3(1·0)$.

The effect of the substrate charge on the rate is small, a decrease in the positive charge leading to an increase in the rate, e.g. trans-$Pt(NH_3)_3Cl_3^+ >$ trans-$Pt(NH_3)_4Cl_2^{2+}$ (ca. 3:1); cis-$Pt(NH_3)_4Cl_2^{2+} > Pt(NH_3)_5Cl^{3+}$ (ca. 4:1).[141] However, when the rate of substitution of the bromide ion in $Pt(CN)_4Br_2^{2-}$ by chloride ion was investigated it was found that in addition to the normal third-order rate term a chloride independent second-order term is present. This was assigned to the reaction of a neutral water molecule which is able to effectively compete with the chloride ion as a nucleophile because of the charge repulsion that would be experienced by the chloride ion. A solvent path for cationic complexes is absent because with these complexes a negative entering ligand would be strongly favoured by the positive charge. A study of reaction 364 has shown that reactivity varies with L in the order $NH_3 > CN^- > NO_2^-$ with relative rates of $28,000:70:1$

$$\text{trans-PtL}_4\text{Br}_2 + \text{Cl}^- \rightarrow \text{trans-PtL}_4\text{BrCl} + \text{Br}^- \tag{364}$$

at 25°C.[145] The activation enthalpy alone would lead to a reactivity order $NO_2^- > CN^- > NH_3$. However, the dominant term is the activation entropy which arises first from solvation changes, which are less for the ammine complex than for the others owing to the lower charge on the activated species ($3+$ for $L = NH_3$ as opposed to $5-$ for $L = CN^-$ or NO_2^-), and, secondly, from the bonding in the activated complex. Since NO_2^- is a weaker σ-donor than CN^- and since both are strong π-acceptor ligands it is likely that the strength of the bonding between the centres in the activated complexes decreases in the order $NO_2^- > CN^- > NH_3$. This gives rise both to the activation entropy order of $NO_2^- > CN^- > NH_3$ and the reverse order for the activation enthalpy of $NO_2^- < CN^- < NH_3$. These results strongly suggest that bond-making is an important feature of the activation process for anionic complexes.

(e) Oxidation–reduction reactions between platinum(IV) and platinum(II)

Recent studies of the reduction of platinum(IV) complexes of the type *cis*- or *trans*-PtL_2X_4, where L = uncharged ligand such as pyridine, piperidine, methylamine, dimethylsulphide, triethylphosphine or triethylarsine and $X = Cl$ or Br, by such anions as I^-, $S_2O_3^{2-}$, SCN^- and $SeCN^-$ to give platinum(II) complexes have been carried out to shed further light on the substitution reactions of platinum(IV) catalysed by platinum(II).[146-9] However, although it has been found that the reactions in methanol are first-order in respect of the platinum(IV) complex and first-order in respect of the reducing anion, little is known about their detailed mechanism. A study of the reduction of *trans*-$(PEt_3)_2PtX_4$, where $X = Cl^-$ or Br^- to platinum(II) by $Pt(diars)_2^{2+}$, $2NO_3^-$ in the presence of sodium halide, where halide $= Cl^-$, Br^- or I^-, in methanol, indicated that the reaction was first order in respect of platinum(IV), platinum(II) and halide ion.[150] Detailed analysis of the rate data showed that it fitted a reaction mechanism (eqns. 365 and 366) very similar to that proposed for platinum(II) catalysed substitution reactions at platinum(IV) (reactions 354–7).

$$\text{Pt(diars)}_2^{2+} + \text{Y}^- \xrightarrow{\text{fast}} \text{Pt(diars)}_2\text{Y}^+ \tag{365}$$

$$\text{trans-(PEt}_3)_2\text{PtX}_4 + \text{Pt(diars)}_2\text{Y}^+ \xrightarrow{\text{slow}} \text{X}_3(\text{PEt}_3)_2\text{Pt}-\text{X}-\text{Pt(diars)}_2\text{Y}^+$$

$$\xrightarrow{\text{slow}} \text{trans-(PEt}_3)_2\text{PtX}_2 + \text{Pt(diars)}_2\text{XY}^{2+} + \text{X}^- \tag{366}$$

Although none of the reductions mentioned above involves the formation of a platinum(III) intermediate, there is some evidence that the reduction of

320

The Chemistry of Platinum and Palladium

platinum(IV) salts by iron(II)[151] and the oxidation of platinum(II) salts by iron(III),[151] cerium(IV)[152] and hexachloroiridium anions[153] occurs via an initial rate-determining one-electron transfer reaction giving a platinum(III) intermediate. However, an earlier suggestion that the reduction of platinum(IV) salts by chromium(II) complexes also involved an initial one-electron reduction giving a platinum(III) intermediate[154] is now thought to be incorrect as further work is consistent with a rate-determining two-electron reduction of platinum(IV) to platinum(II).[155] Similarly a chronopotentiometric study of the reduction of $PtCl_6^{2-}$ in 1 M hydrochloric acid at a platinum black electrode suggested that the initial reduction product was a platinum(III) complex.[156] However, more recent work using thin layer electrochemical techniques found no evidence for the formation of a stable platinum(III) intermediate. In 1 M hydrochloric acid $PtCl_6^{2-}$ is reduced to $PtCl_4^{2-}$ and in 1 M hydrobromic acid $PtBr_6^{2-}$ is reduced to $PtBr_4^{2-}$.[157]

REFERENCES

1 P. Haake, S. C. Chan and V. Jonas, *Inorg. Chem.*, **9** (1970) 1925.
2 F. Basolo and R. G. Pearson, *Adv. Inorg. and Radiochem.*, **3** (1961) 1.
3 F. Basolo and R. G. Pearson, *Prog. Inorg. Chem.*, **4** (1962) 381.
4 J. O. Edwards, *Inorganic Reaction Mechanisms*, W. A. Benjamin, New York, 1964.
5 F. Basolo, *Adv. Chem. Ser.*, **49** (1965) 81.
6 C. H. Langford and H. B. Gray, *Ligand Substitution Processes*, W. A. Benjamin, New York, 1965, chapter 2.
7 F. Basolo and R. G. Pearson, *Mechanism of Inorganic Reactions*, Wiley, New York, 2nd edn., 1967, chapter 5.
8 D. S. Martin, *Inorg. Chim. Acta. Rev.*, **1** (1967) 87.
9 J. L. Burmeister and F. Basolo, *Prep. Inorg. Reaction*, **5** (1968) 1.
10 L. Cattalini in *Inorganic Reaction Mechanisms*, ed. by J. O. Edwards (vol. 13 of *Progress in Inorganic Chemistry*), John Wiley, 1971, p. 263.
11 M. A. Tucker, C. B. Colvin and D. S. Martin, *Inorg. Chem.*, **3** (1964) 1373.
12 L. F. Grantham, T. S. Elleman and D. S. Martin, *J. Amer. Chem. Soc.*, **77** (1955) 2965.
13 T. S. Elleman, J. W. Reishus and D. S. Martin, *J. Amer. Chem. Soc.*, **81** (1959) 10.
14 R. G. Pearson, C. R. Boston and F. Basolo, *J. Phys. Chem.*, **59** (1955) 304.
15 F. Basolo, J. Chatt, H. B. Gray, R. G. Pearson and B. L. Shaw, *J. Chem. Soc.*, (1961) 2207.
16 R. G. Pearson, C. R. Boston and F. Basolo, *J. Amer. Chem. Soc.*, **75** (1953) 3089.
17 H. B. Gray, *J. Amer. Chem. Soc.*, **84** (1962) 1548.
18 R. G. Pearson, H. B. Gray and F. Basolo, *J. Amer. Chem. Soc.*, **82** (1960) 787.
19 S. Thomas and W. L. Reynolds, *Inorg. Chem.*, **8** (1969) 1531.
20 C. D. Falk and J. Halpern, *J. Amer. Chem. Soc.*, **87** (1965) 3003.
21 H. E. Brower, L. Hathaway and K. R. Brower, *Inorg. Chem.*, **5** (1966) 1899.
22 T. Taylor and L. R. Hathaway, *Inorg. Chem.*, **8** (1969) 2135.
23 A. A. Frost and R. G. Pearson, *Kinetics and Mechanism*, Wiley, New York, 2nd edn., 1961, p. 223.
24 Reference 4, chapter 3.
25 R. G. Pearson, *J. Amer. Chem. Soc.*, **85** (1963) 3533.
26 R. G. Pearson, *Chem. Brit.*, **3** (1967) 103.
27 U. Belluco, L. Cattalini, F. Basolo, R. G. Pearson and A. Turco, *J. Amer. Chem. Soc.*, **87** (1965) 241.
28 U. Belluco, M. Martelli and A. Orio, *Inorg. Chem.*, **5** (1966) 582.
29 R. G. Pearson, H. Sobel and J. Songstad, *J. Amer. Chem. Soc.*, **90** (1968) 319.
30 U. Belluco, *Coord. Chem. Rev.*, **1** (1966) 111.

31 U. Belluco, L. Cattalini and A. Turco, *J. Amer. Chem. Soc.*, **86** (1964) 3257.
32 H. B. Gray and R. J. Olcott, *Inorg. Chem.*, **1** (1962) 481.
33 U. Belluco, L. Cattalini and A. Turco, *J. Amer. Chem. Soc.*, **86** (1964) 226.
34 R. G. Pearson and D. A. Johnson, *J. Amer. Chem. Soc.*, **86** (1964) 3983.
35 L. Baracco, L. Cattalini, J. S. Coe and E. Rotundo, *J. Chem. Soc. (A)*, (1971) 1800.
36 I. I. Chernyaev, *Ann. Inst. Platine (USSR)*, **4** (1926) 243; *Chem. Abs.*, **21** (1927) 2620.
37 T. P. Cheeseman, A. L. Odell and H. A. Raethel, *Chem. Comm.*, (1968) 1496.
38 A. D. Gel'man, E. F. Karandashova and L. N. Essen, *Dokl. Akad. Nauk. SSSR*, **63** (1948) 37; *Chem. Abs.*, **43** (1949) 1678.
39 J. Chatt, L. A. Duncanson and L. M. Venanzi, *J. Chem. Soc.*, (1955) 4456.
40 L. E. Orgel, *J. Inorg. Nucl. Chem.*, **2** (1956) 137.
41 A. A. Grinberg, *Ann. Inst. Platine (USSR)*, **5** (1927) 109.
42 D. Benson, *Mechanisms of Inorganic Reactions*, McGraw-Hill, London, 1968, p. 70.
43 S. S. Zumdahl and R. S. Drago, *J. Amer. Chem. Soc.*, **90** (1968) 6669.
44 L. Oleari, L. Di Sipio and G. De Michelis, *Ric. Sci., Rend. Sez. A.*, **8** (1965) 413; *Chem. Abs.*, **64** (1966) 4567d.
45 D. M. Adams, J. Chatt, J. Gerratt and A. D. Westland, *J. Chem. Soc.*, (1964) 734.
46 G. W. Watt and W. A. Cude, *J. Amer. Chem. Soc.*, **90** (1968) 6382.
47 A. Pidcock, R. E. Richards and L. M. Venanzi, *J. Chem. Soc. (A)*, (1966) 1707.
48 L. M. Venanzi, *Chem. Brit.*, **4** (1968) 162.
49 U. A. Gregory, J. A. J. Jarvis, B. T. Kilbourn and P. G. Owston, *J. Chem. Soc. (A)*, (1970) 2770.
50 R. Mason, G. B. Robertson and P. J. Pauling, *J. Chem. Soc. (A)*, (1969) 485.
51 G. G. Messmer and E. L. Amma, *Inorg. Chem.*, **5** (1966) 1775.
52 H. C. Clark, P. W. R. Corfield, K. R. Dixon and J. A. Ibers, *J. Amer. Chem. Soc.*, **89** (1967) 3360.
53 H. C. Freeman and M. L. Golomb, *Chem. Comm.*, (1970) 1523.
54 Y. P. Jeannin and D. R. Russell, *Inorg. Chem.*, **9** (1970) 778.
55 J. A. J. Jarvis, B. T. Kilbourn and P. G. Owston, *Acta. Cryst.*, **B27** (1971) 366.
56 E. M. Badley, J. Chatt, R. L. Richards and G. A. Sim, *Chem. Comm.*, (1969) 1322.
57 G. G. Messmer, E. L. Amma and J. A. Ibers, *Inorg. Chem.*, **6** (1967) 725.
58 R. Eisenberg and J. A. Ibers, *Inorg. Chem.*, **4** (1965) 773.
59 R. McWeeny, R. Mason and A. D. C. Towl, *Disc. Faraday Soc.*, **47** (1969) 20.
60 S. F. Watkins, *J. Chem. Soc. (A)*, (1970) 168.
61 M. Black, R. H. B. Mais and P. G. Owston, *Acta. Cryst.*, **B25** (1969) 1760.
62 W. Theilacker, *Z. anorg. allgem. Chem.*, **234** (1937) 161.
63 D. J. Robinson and C. H. L. Kennard, *J. Chem. Soc. (A)*, (1970) 1008.
64 H. J. Whitfield, *J. Chem. Soc. (A)*, (1970) 113.
65 N. C. Baenziger, G. F. Richards and J. R. Doyle, *Acta Cryst.*, **18** (1965) 924.
66 D. L. Weaver, *Inorg. Chem.*, **9** (1970) 2250.
67 K. Shobatake and K. Nakamoto, *J. Amer. Chem. Soc.*, **92** (1970) 3332.
68 R. J. P. Williams, *Chem. Brit.*, **4** (1968) 277.
69 C. W. Fryer and J. A. S. Smith, *J. Organometal. Chem.*, **18** (1969) P35.
70 C. W. Fryer and J. A. S. Smith, *J. Chem. Soc. (A)*, (1970) 1029.
71 C. W. Fryer, *Chem. Comm.*, (1970) 902.
72 L. I. Elding, *Acta Chem. Scand.*, **24** (1970) 1527.
73 L. I. Elding, *Acta Chem. Scand.*, **24** (1970) 2557.
74 M. A. Tucker, C. B. Colvin and D. S. Martin, *Inorg. Chem.*, **3** (1964) 1373.
75 R. A. Reinhart and W. W. Monk, *Inorg. Chem.*, **9** (1970) 2026.
76 A. A. Grinberg, *Russ. J. Inorg. Chem.*, **4** (1959) 683.
77 I. B. Bersuker, *Russ. J. Struct. Chem.*, **4** (1963) 419.
78 L. Cattalini, M. Martelli and G. Marangoni, *Proc. 3rd Coord. Chem. Symp.*, Decrecen, Hungary, 1970.
79 L. Cattalini, M. Martelli and G. Kirschner, *Inorg. Chem.*, **7** (1968) 1488.
80 L. Cattalini, G. Marangoni, J. S. Coe, M. Vidali and M. Martelli, *J. Chem. Soc. (A)*, (1971) 593.

81   L. Cattalini and M. Martelli, *Inorg. Chim. Acta.*, **1** (1967) 189.
82   R. G. Pearson, *J. Chem. Ed.*, **45** (1968) 581.
83   G. Faraone, V. Ricevuto, R. Romeo and M. Trozzi, *Inorg. Chem.*, **8** (1969) 2207.
84   G. Faraone, V. Ricevuto, R. Romeo and M. Trozzi, *Inorg. Chem.*, **9** (1970) 1525.
85   R. Wanguo, Master's Thesis, Northwestern University, 1965.
86   W. H. Baddley and F. Basolo, *J. Amer. Chem. Soc.*, **86** (1964) 2075.
87   J. B. Goddard and F. Basolo, *Inorg. Chem.*, **7** (1968) 936.
88   J. B. Goddard and F. Basolo, *Inorg. Chem.*, **7** (1968) 2456.
89   F. Basolo, H. B. Gray and R. G. Pearson, *J. Amer. Chem. Soc.*, **82** (1960) 4200.
90   A. Orio, V. Ricevuto and L. Cattalini, *Chim. Ind. (Milan)*, **49** (1967) 1339; *Chem. Abs.*, **68** (1968) 95044.
91   L. Cattalini and M. Martelli, *Gazz. Chim. Ital.*, **97** (1967) 498.
92   J. L. Burmeister and J. C. Lim, *Chem. Comm.*, (1969) 1154.
93   C. M. Harris, S. E. Livingstone and I. H. Reece, *J. Chem. Soc.*, (1959) 1505.
94   P. Haake and R. M. Pfeiffer, *J. Amer. Chem. Soc.*, **92** (1970) 5243.
95   R. N. Collins and D. A. Johnson, *J. Inorg. Nucl. Chem.*, **33** (1971) 1861.
96   S. C. Chan, *J. Chem. Soc. (A)*, (1966) 1000.
97   L. Cattalini, M. Martelli and G. Kirschner, *Inorg. Chem.*, **7** (1968) 1488.
98   L. Cattalini, M. Brunelli and M. Martelli, unpublished results quoted in reference 10.
99   D. Banerjea, F. Basolo and R. G. Pearson, *J. Amer. Chem. Soc.*, **79** (1957) 4055.
100   U. Belluco, L. Cattalini, F. Basolo, R. G. Pearson and A. Turco, *Inorg. Chem.*, **4** (1965) 925.
101   L. F. Grantham, T. S. Elleman and D. S. Martin, *J. Amer. Chem. Soc.*, **77** (1955) 2965.
102   C. I. Saunders and D. S. Martin, *J. Amer. Chem. Soc.*, **83** (1961) 807.
103   J. E. Teggins, D. R. Gano, M. A. Tucker and D. S. Martin, *Inorg. Chem.*, **6** (1967) 69.
104   S. J. Lokken and D. S. Martin, *Inorg. Chem.*, **2** (1963) 562.
105   J. E. Teggins and D. S. Martin, *Inorg. Chem.*, **6** (1967) 1003.
106   R. M. Milburn and L. M. Venanzi, *Inorg. Chim. Acta*, **2** (1968) 97.
107   M. Green and C. J. Wilson, *Joint Meeting of the Chemical Society and The Royal Institute of Chemistry*, Nottingham, April 1969, paper 4.11.
108   M. Green and C. J. Wilson, *Disc. Faraday Soc.*, **47** (1969) 110.
109   M. Green and C. J. Wilson, *Proc. Third Int. Inorg. Chim. Acta Symp.*, Venice, (1970) paper B2.
110   G. A. Mair, H. M. Powell and L. M. Venanzi, *Proc. Chem. Soc.*, (1961) 170.
111   R. G. Pearson, M. M. Muir and L. M. Venanzi, *J. Chem. Soc.*, (1965) 5521.
112   J. Chatt and R. G. Wilkins, *J. Chem. Soc.*, (1952) 273, 4300; (1953) 70; (1956) 525.
113   J. Chatt, *J. Chem. Soc.*, (1950) 2301.
114   P. Haake and T. A. Hylton, *J. Amer. Chem. Soc.*, **84** (1962) 3774.
115   G. Faraone, V. Ricevuto, R. Romeo and M. Trozzi, *J. Chem. Soc. (A)*, (1971) 1877.
116   L. Cattalini and M. Martelli, *J. Amer. Chem. Soc.*, **91** (1969) 312.
117   P. Haake and R. M. Pfeiffer, *J. Amer. Chem. Soc.*, **92** (1970) 4996.
118   P. Haake and R. M. Pfeiffer, *Chem. Comm.*, (1969) 1330.
119   D. R. Eaton, *J. Amer. Chem. Soc.*, **90** (1968) 4272.
120   J. W. Collier, F. G. Mann, D. G. Watson and H. R. Watson, *J. Chem. Soc.*, (1964) 1803.
121   F. Basolo, A. F. Messing, P. H. Wilks, R. G. Wilkins and R. G. Pearson, *J. Inorg. Nucl. Chem.*, **8** (1958) 203.
122   R. L. Rich and H. Taube, *J. Amer. Chem. Soc.*, **76** (1954) 2608.
123   R. Dreyer, *Kernenergie*, **5** (1962) 618; *Chem. Abs.*, **61** (1964) 15613e.
124   A. K. Sundaram and E. B. Sandell, *J. Amer. Chem. Soc.*, **77** (1956) 855.
125   L. Malatesta and L. M. Vallarino, *J. Chem. Soc.*, (1956) 1867.
126   C. M. Harris and R. S. Nyholm, *J. Chem. Soc.*, (1956) 4375.
127   C. M. Harris and N. C. Stephenson, *Chem. Ind. (London)*, (1957) 426.
128   B. I. Peshchevitskii and G. D. Mal'chikov, *Dokl. Akad, Nauk. SSSR.*, **178** (1968) 108; *Chem. Abs.*, **68** (1968) 110926.
129   W. R. Mason and R. C. Johnson, *Inorg. Chem.*, **4** (1965) 1258.
130   W. R. Mason, E. R. Berger and R. C. Johnson, *Inorg. Chem.*, **6** (1967) 248.

131   R. R. Rettew and R. C. Johnson, *Inorg. Chem.*, **4** (1965) 1565.
132   W. R. Mason, *Inorg. Chem.*, **9** (1970) 1528.
133   R. C. Johnson and E. R. Berger, *Inorg. Chem.*, **4** (1965) 1262.
134   A. J. Poë and M. S. Vaidya, *J. Chem. Soc.*, (1961) 1023.
135   A. J. Poë and M. S. Vaidya, *J. Chem. Soc.*, (1961) 2981.
136   A. J. Poë and D. H. Vaughan, *Inorg. Chim. Acta*, **2** (1968) 159.
137   E. J. Bounsall, D. J. Hewkin, D. Hopgood and A. J. Poë, *Inorg. Chim. Acta*, **1** (1967) 281.
138   B. Corain and A. J. Poë, *J. Chem. Soc. (A)*, (1967) 1633.
139   A. J. Poë and D. H. Vaughan, *J. Chem. Soc. (A)*, (1969) 2844.
140   A. J. Poë and D. H. Vaughan, *J. Amer. Chem. Soc.*, **92** (1970) 7537.
141   F. Basolo, M. L. Morris and R. G. Pearson, *Disc. Faraday Soc.*, **29** (1960) 80.
142   H. R. Ellison, F. Basolo and R. G. Pearson, *J. Amer. Chem. Soc.*, **83** (1961) 3943.
143   A. Peloso and G. Dolcetti, *Coord. Chem. Rev.*, **1** (1966) 100.
144   S. G. Bailey and R. C. Johnson, *Inorg. Chem.*, **8** (1969) 2596.
145   W. R. Mason, *Inorg. Chem.*, **8** (1969) 1756.
146   A. Peloso, R. Ettorre and G. Dolcetti, *Inorg. Chim. Acta*, **1** (1967) 307.
147   A. Peloso, G. Dolcetti and R. Ettorre, *Inorg. Chim. Acta*, **1** (1967) 403.
148   A. Peloso, G. Dolcetti and R. Ettorre, *Gazz. Chim. Ital.*, **97** (1967) 1507.
149   A. Peloso and G. Dolcetti, *J. Chem. Soc. (A)*, (1967) 1944.
150   A. Peloso and R. Ettorre, *J. Chem. Soc. (A)*, (1968) 2253.
151   A. Peloso and M. Basato, *J. Chem. Soc. (A)*, (1971) 725.
152   S. V. Zemskov, B. V. Ptitsyn, V. N. Lyubimov and V. F. Malakhov, *Russ. J. Inorg. Chem.*, **12** (1967) 648.
153   J. Halpern and M. Pribanic, *J. Amer. Chem. Soc.*, **90** (1968) 5942.
154   J. K. Beattie and F. Basolo, *Inorg. Chem.*, **6** (1967) 2069.
155   J. K. Beattie and F. Basolo, *Inorg. Chem.*, **10** (1971) 486.
156   J. J. Lingane, *J. Electroanal. Chem.*, **7** (1964) 94.
157   A. T. Hubbard and F. C. Anson, *Anal. Chem.*, **38** (1966) 1887.

CHAPTER 12

Organometallic Complexes Involving Metal–Carbon σ-Bonds

The work described in this and the remaining chapters of this book represents a field of research that is expanding very rapidly both in a pure and in an applied direction. For useful summaries of the latest developments the Annual Reports of The Chemical Society and Annual Reviews of Organometallic Chemistry may be consulted. In this chapter the more important results are discussed, but, to avoid losing the significance of these a comprehensive coverage has been avoided. For recent, more comprehensive, treatments of the subject the reader is referred to the reviews listed in Table 98.

TABLE 98
Reviews on complexes containing metal–carbon σ-bonds

Title	Reference
σ-Complexes of platinum(II) with hydrogen, carbon and other elements of group IV	1
σ-Alkyl and -aryl derivatives of transition metals	2
Organometallic chemistry of 'one-electron' ligands	3
Preparation of fluorocarbon complexes of transition metals	4
Perfluoro-complexes of transition metals	5

Platinum forms complexes containing metal–carbon σ-bonds in both the +2 and +4 oxidation states whereas, with the exception of the pentafluorophenyl complex $[(PPh_3)_2Pd(C_6F_5)_2Cl_2]$,[6] palladium forms such complexes only in the +2 oxidation state. Although complexes with platinum(IV)–carbon σ-bonds were first reported over 60 years ago,[7] it was not until 1957 that the first complexes containing platinum(II)– and palladium(II)–carbon σ-bonds were prepared.[8] However, these first complexes of the divalent metals were restricted to those with metal–carbon bonds from bidentate ligands, and it was not until 1959 that complexes containing platinum(II)–methyl σ-bonds were made.[9] Since there are substantial differences between the complexes in the +2 and +4 oxidation states we shall consider them separately.

COMPLEXES CONTAINING METAL(II)–CARBON σ-BONDS

The first complexes in which monodentate groups were bound to platinum(II) and palladium(II) by a single metal(II)–carbon σ-bond also contained phosphine and arsine ligands and for some time it was thought that the strong σ-donor and π-acceptor capacity of these ligands was essential to modify the metal d-orbitals in such a way that formation of a metal–carbon σ-bond was possible. Whilst this is still largely true it is possible to prepare complexes with organic ligands bound monofunctionally to platinum(II) that are not stabilised by phosphines or arsines. In addition to the examples shown below (130,[10] 131,[11,12] 132[13] and 133[13]), complexes stabilised by

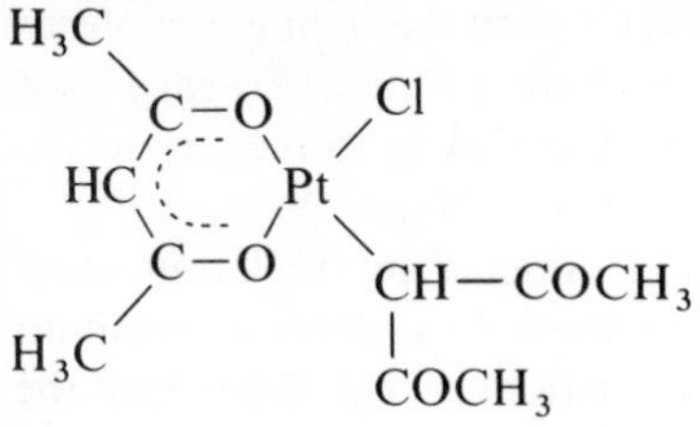

130

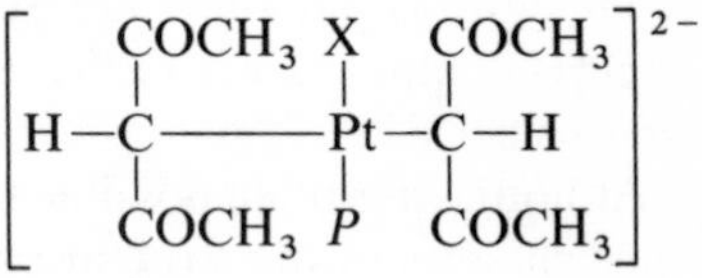

131 (X = Cl, Br)

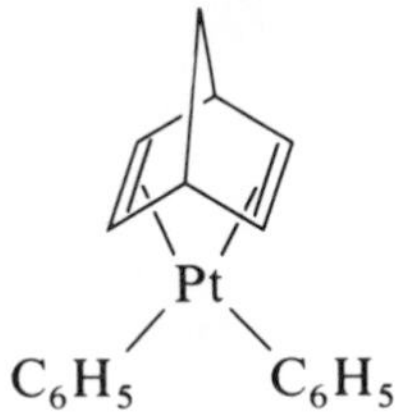

132

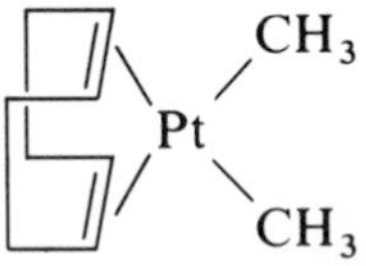

133

bidentate thioethers,[14,15] selenoethers,[15] bipyridyl,[14] pyridine,[13,16] triethylstibine[17] and triphenylstibine[18] have been described.

Preparation

Complexes containing metal(II)–carbon σ-bonds can be prepared by several reactions:

(i) Grignard and organolithium reagents

Treatment of a halometal complex with an anionic alkylating agent such as a Grignard or organolithium reagent is the most widely used method for preparing both platinum(II)– and palladium(II)–carbon σ-bonds. However, the reactions are not particularly clean and an excess of the organo reagent is generally necessary. Organolithium reagents are more powerful than Grignard reagents as exemplified by reaction 367.[14] A further difference

$$[(PEt_3)_2PdBr_2] \xrightarrow{\text{excess } CH_3MgBr} [(PEt_3)_2Pd(CH_3)Br]$$

$$\xrightarrow[\quad CH_3Li \quad]{\text{excess}}$$

$$[(PEt_3)_2Pd(CH_3)_2]$$

(367)

between Grignard and organolithium reactions is that the latter give irreversible alkylations whereas the former involve a series of equilibria[9] so that an excess of the Grignard reagent is necessary (reactions 368 and 369).

$$[(PEt_3)_2PtI_2] + CH_3MgI \rightleftharpoons [(PEt_3)_2Pt(CH_3)I] + MgI_2 \qquad (368)$$

$$[(PEt_3)_2Pt(CH_3)I] + CH_3MgI \rightleftharpoons [(PEt_3)_2Pt(CH_3)_2] + MgI_2 \qquad (369)$$

Although with platinum(II) the *cis*-halides are the more reactive starting materials, the products prefer to avoid the two groups with the highest *trans*-effects lying opposite one another, so that isomerisation frequently occurs during the reaction. With palladium(II), where the halide complexes are far more labile, only one isomer appears in the product, except in the dimethyl complexes where the initially *cis*-product slowly isomerises to the *trans*-isomer.[14]

Although it is often possible to predict the product of an alkylation, quite minor changes in the structure of the halide complex can have a profound effect on the alkylation, as is illustrated[13] in reactions 370–2. When bridged

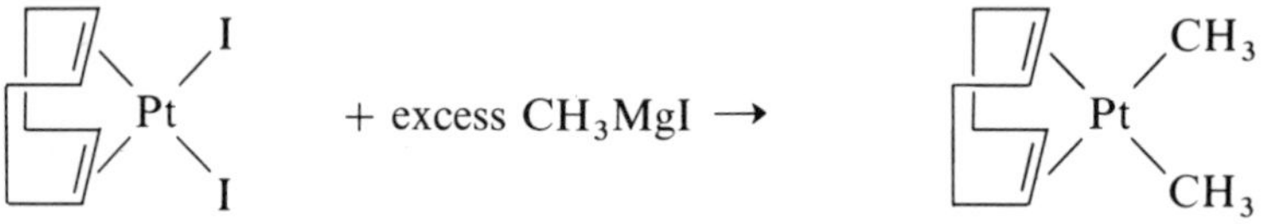

$$+ \text{ excess } CH_3MgI \rightarrow \qquad (370)$$

$$+ \text{ an excess of } CH_3MgI \rightarrow \qquad (371)$$

$$+ \text{ an excess of } CH_3MgI \rightarrow \text{ No reaction} \qquad (372)$$

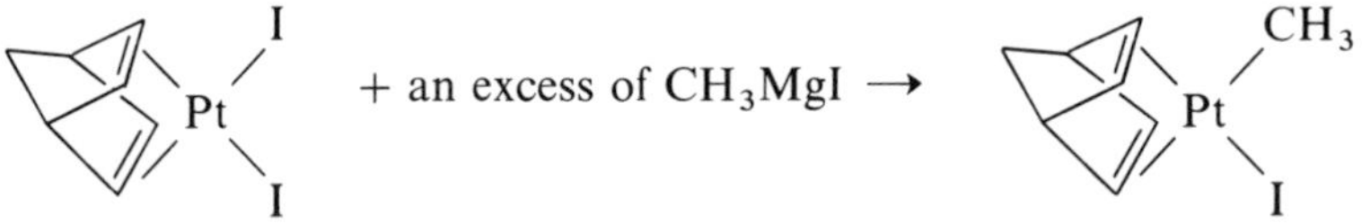

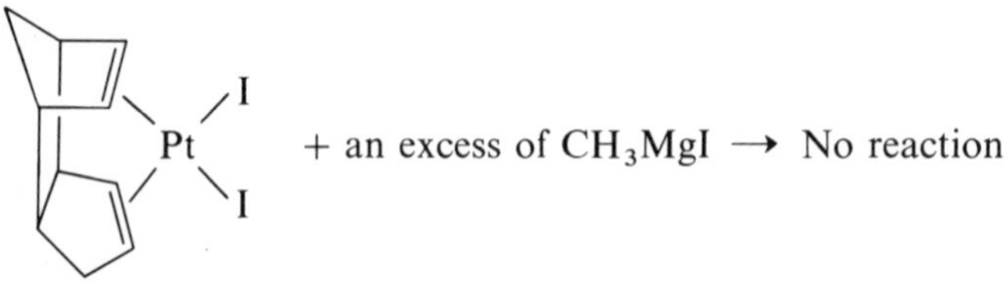

binuclear complexes are treated with phenyllithium the bridge is split when X = Cl to give *trans*-$[(P^nPr_3)_2Pt(C_6H_5)_2]$ (reaction 373), but retained intact when X = Br, I, PR$_2$, AsR$_2$ and SR.[19]

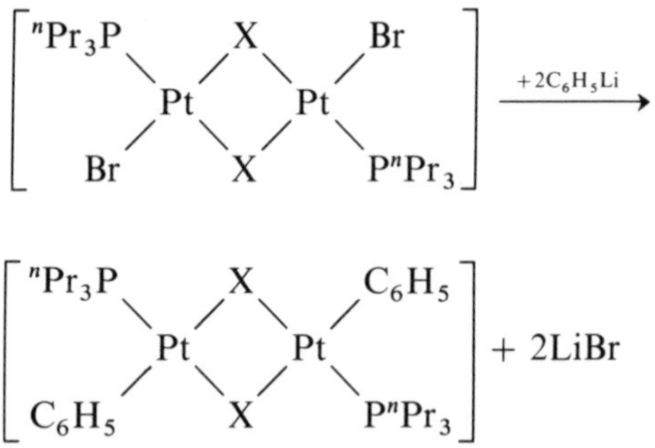

$$+ 2C_6H_5Li$$

$$+ 2LiBr \qquad (373)$$

$$(X = Br, I, PR_2, AsR_2 \text{ or } SR)$$

Although σ-metal(II)–carbon bonds in which acetylenes are bonded to platinum(II) and palladium(II) can be prepared from Grignard reagents, generally sodium acetylide in liquid ammonia is far more convenient[14,20,21] (reaction 374).

$$trans\text{-}[(AsEt_3)_2PtCl_2] + 2C_6H_5C\equiv CNa \rightarrow$$

$$trans\text{-}[(AsEt_3)_2Pt(C\equiv CC_6H_5)_2] + 2NaCl \qquad (374)$$

In certain cases Grignard reagents lead to the formation of hydrides rather than alkyl derivatives. This has been observed for cyclohexylmagnesium bromide[22] and styrylmagnesium bromide[16] with cis-[(PEt_3)_2PtCl_2] and for methylmagnesium chloride with [QAsPtCl]⁺Cl⁻ (QAs = tris(o-diphenylarsinophenyl)-arsine).[23] This reaction has recently been examined and shown to occur through a platinum–magnesium intermediate[24] (eqn. 375). In the phosphine complex, assistance

$$cis\text{-}[(PEt_3)_2PtBr_2] + C_6H_{11}MgBr \rightarrow$$

$$[(PEt_3)_2Pt(MgBr)Br] \xrightarrow{\text{H}_2\text{O}} trans\text{-}[(PEt_3)_2PtHBr] \qquad (375)$$
$$\downarrow \text{D}_2\text{O}$$
$$trans\text{-}[(PEt_3)_2PtDBr]$$

in this direction is probably provided by the relative instability of the cyclohexyl– and styryl–platinum(II) bonds (since alkylation by the Grignard reagent is reversible (see above) instability of the alkyl complex will not lead to overall decomposition but regeneration of the starting material). In the case of the QAs complex, the driving force is probably provided by the steric hindrance exercised by the phenyl groups of the QAs ligand; these prevent the more bulky methyl group binding to platinum whilst permitting the linear —MgCl moiety to react.

(ii) Cleavage reactions
Monoalkyl complexes can be prepared by selective cleavage of one of the alkyl groups from a dialkyl complex, which is possible under very carefully controlled conditions with hydrogen chloride in an anhydrous medium.[9,20] In addition cationic monomethyl complexes can be prepared by treating the chloromethyl complexes with strongly co-ordinating ligands such as tertiary phosphines[25,26] (reaction 376). With less strongly co-ordinating ligands such as methyl cyanide the reaction can be forced by the addition

$$trans\text{-}[(PMe_2Ph)_2Pt(CH_3)Cl] + PPh_3 \xrightarrow{\text{MeOH or}}_{\text{Me}_2\text{CO}}$$

$$trans\text{-}[(PMe_2Ph)_2(PPh_3)Pt(CH_3)]^+Cl^- \qquad (376)$$

of silver tetrafluoroborate.[25]

(iii) Oxidative addition
Monoalkyl complexes can be prepared by oxidative addition of alkyl halides to zerovalent complexes of platinum and palladium (reactions

378–81). The last reaction involves an initial loss of ethylene to give

$$[Pt(PPh_3)_3] + CH_3I \xrightarrow{\text{ref. 9}}$$

$$\textit{trans}\text{-}[(PPh_3)_2Pt(CH_3)I] + PPh_3 \qquad (377)$$

$$[Pt(PPh_3)_3] + C_6H_5CH{=}CHBr \xrightarrow{\text{ref. 27}}$$

$$[(PPh_3)_2Pt(CH{=}CHC_6H_5)Br] + PPh_3 \qquad (378)$$

$$[Pt(PPh_3)_3] + HC{\equiv}C{-}CR_1R_2Cl \xrightarrow{\text{ref. 28}}$$

$$[(PPh_3)_2Pt(CH{=}C{=}CR_1R_2)Cl] + PPh_3 \qquad (379)$$

$$[Pt(PPh_3)_3] + ClCH{=}C{=}CR_1R_2 \xrightarrow{\text{ref. 28}}$$

$$[(PPh_3)_2Pt(CH{=}C{=}CR_1R_2)Cl] + PPh_3 \qquad (380)$$

$$[(PPh_3)_2Pt(C_2H_4)] + CH_3I \xrightarrow{\text{ref. 29}}$$

$$[(PPh_3)_2Pt(CH_3)I] + C_2H_4 \qquad (381)$$

$Pt(PPh_3)_2$.[29]

(iv) Insertion reactions

Olefins can be inserted into metal–hydrogen bonds to yield monoakyl complexes[22] (reaction 382). Although this reaction is of limited importance

$$\textit{trans}\text{-}[(PEt_3)_2PtHCl] + C_2H_4 \rightarrow \textit{trans}\text{-}[(PEt_3)_2Pt(C_2H_5)Cl] \quad (382)$$

for the preparation of metal–carbon σ-bonds, it has great relevance to theories of the mechanism of homogeneous hydrogenation of olefins[30] and has accordingly been systematically investigated. Deuterium labelled species indicate that in the reverse reaction the hydrogen atoms attached to either of the carbon atoms of the ethyl group can provide the hydrido-ligand.[31]

When diazomethane reacts with $\textit{trans}\text{-}[(PEt_3)_2PtHBr]$ a methylene group is inserted into the platinum–hydrogen bond to give the corresponding methyl complex in low yield.[32] Three reactions (reactions 382a and 383a) in which metal–carbon σ-bonds are formed by insertion into metal–chlorine bonds have been reported. However, at present they are of specific rather than of general application.

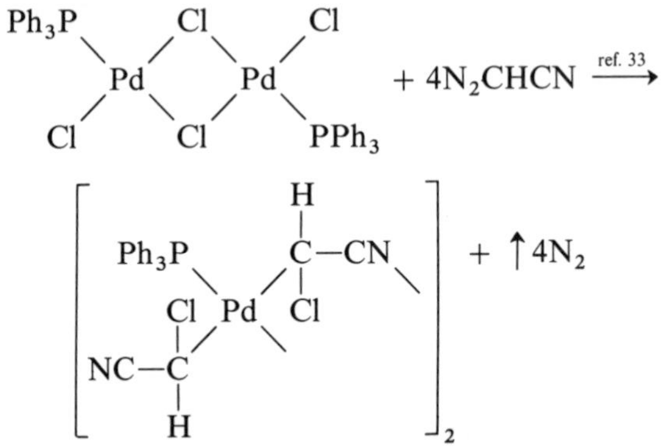

$$+ 4N_2CHCN \xrightarrow{\text{ref. 33}}$$

$$\qquad + \uparrow 4N_2 \qquad (382a)$$

$$PdCl_2 + R_2N-CH_2-C\equiv CH \xrightarrow{\text{ref. 34}} [Pd(R_2NCH_2CCl=CH)Cl]_2 \quad (383)$$

$$[(PhCN)_2PdCl_2] + 4(CF_3)_2CN_2 \xrightarrow{\text{ref. 35}}$$

$$\begin{bmatrix} PhCN & Cl & C(CF_3)_2Cl \\ & \diagdown \ / \quad \diagdown \ / & \\ & Pd \quad\quad Pd & \\ & \diagup \ \diagdown \quad \diagup \ \diagdown & \\ Cl(CF_3)_2C & Cl & NCPh \end{bmatrix}$$

$$+ [(PhCN)_2Pd\{C(CF_3)_2Cl\}_2] + 2PhCN + \uparrow 4N_2 \quad (383a)$$

(v) Elimination Reactions

Another rather specific method for forming monoalkyl and monoaryl complexes involves expulsion of a moiety originally bound between the organic group and the metal. The elimination of N_2,[36] CO,[37] SO_2,[27] $(C_6H_5)_2Sn$[38] and $(C_6H_5)_2Pb$[38] moieties have all been reported (reactions 384–8). These reactions are particularly useful for preparing aryl complexes

$$trans\text{-}[(PEt_3)_2Pt(N=N-C_6H_4NO_2-p)Cl] \xrightarrow[\substack{\text{or treated with} \\ \text{activated alumina}}]{\text{heat at 120°C}}$$

$$trans\text{-}[(PEt_3)_2Pt(p-NO_2C_6H_4)Cl] + \uparrow N_2 \quad (384)$$

$$trans\text{-}[(PEt_3)_2Pt(COCH_3)I] \xrightarrow[0.005\ mm]{\text{heat at 140°C/}} trans\text{-}[(PEt_3)_2Pt(CH_3)I] + \uparrow CO$$

$$(385)$$

$$cis\text{-}[(PPh_3)_2Pt(SO_2C_6H_5)Cl] \xrightarrow[\text{atm. pressure}]{\text{heat at 300°C/}}$$

$$cis\text{-}[(PPh_3)_2Pt(C_6H_5)Cl] + \uparrow SO_2 \quad (386)$$

$$trans\text{-}[(PPh_3)_2Pt(Sn(C_6H_5)_3)Cl] \xrightarrow{\text{reflux in acetone}}$$

$$trans\text{-}[(PPh_3)_2Pt(C_6H_5)Cl] + (C_6H_5)_2Sn \quad (387)$$

$$trans\text{-}[(PPh_3)_2Pt(Pb(C_6H_5)_3)Cl] \xrightarrow{\text{room temp.}}$$

$$trans\text{-}[(PPh_3)_2Pt(C_6H_5)Cl] + (C_6H_5)_2Pb \quad (388)$$

which have substituents on the aromatic ring that prevent the preparation of a Grignard or lithio reagent.

(vi) Use of olefins and acetylenes

The product of treating *cis*-$[(PR_3)_2PtCl_2]$ complexes with olefins or acetylenes in the presence of hydrazine is critically dependent on the phosphine employed.[39] When the phosphine is aliphatic, platinum(II)–carbon σ-complexes are formed by elimination of hydrogen chloride (reaction 389) whereas when it is aromatic platinum–carbon π-complexes are formed[40] (reaction 390). *Cis*-$[(PPh_3)_2PtCl_2]$ reacts with certain acetylenes such as

$$[(PPr_3)_2PtCl_2] + C_6H_5C\equiv CH \xrightarrow[\text{agent}]{\text{reducing}} [(PPr_3)_2Pt(C\equiv CC_6H_5)_2] \quad (389)$$

$$[(PPh_3)_2PtCl_2] + C_6H_5\equiv CH \xrightarrow[\text{agent}]{\text{reducing}} \begin{bmatrix} Ph_3P & H \\ & \diagdown \quad C \\ & Pt- \overset{\|}{\underset{\|}{\text{|||}}} \\ & \diagup \quad C \\ Ph_3P & C_6H_5 \end{bmatrix} \quad (390)$$

HC≡CC(CH$_3$)$_2$OH in the absence of a base such as hydrazine to eliminate hydrogen chloride and give the monoacetylide complex *trans*-[(PPh$_3$)$_2$-PtCl{C≡CC(CH$_3$)$_2$OH}], although if a base is added the diacetylide can be formed.[41]

(vii) Metathesis

Organometallic complexes [(R$_3$P)$_2$MR′X] can be converted into other complexes by metathetical replacement of X. The reaction is greatly facilitated by the large *trans*-effect of the organic group. The chloride is converted into the bromide or iodide by an excess of lithium bromide or sodium iodide in acetone. The thiocyanate is prepared by refluxing the iodide with potassium thiocyanate and the nitrate by treating the iodide with an equivalent of silver nitrate in aqueous methanol.[9] In some instances this exchange is accompanied by isomerisation.[42] When the neutral ligands are olefins they can be replaced by other neutral ligands such as pyridine[13] or triphenylphosphine.[16]

Properties

The platinum(II)–alkyl and –aryl complexes are colourless crystalline materials that are not hydrolysed by dilute acids or oxidised by moist air. The palladium(II) complexes are a little less stable. Both metals form a range of organometallic complexes that are considerably more stable and varied in the nature of the organic group than any of the other transition metals. With platinum both *cis*- and *trans*-isomers of the dialkyls and diaryls are known although the *trans*-isomer is more difficult to prepare. With palladium the reverse is true in that the diaryls are always *trans* and the dialkyls are usually *trans*, although the *cis*-dimethyl–palladium(II) complexes can be prepared especially when stabilised by a bidentate phosphine ligand. The monoalkyls and monoaryls of palladium(II) are always *trans*, whereas although the *trans*-complexes of platinum(II) are easier to prepare, it is possible using care to prepare the *cis*-complexes as well.

Bonding in metal(II)–carbon σ-bonds

Green[3] has conveniently summarised all the possible combinations of metal and carbon orbitals that can be involved in the formation of a metal–carbon σ-bond (see Table 99). In metal–alkyl σ-bonds the σ-bond from the alkyl group to the metal is formed by overlap of a filled sp^3 hybrid orbital on the alkyl carbon atom with an empty $sp_x d_{x^2-y^2}$ hybrid orbital on the metal (the metal–carbon bond is taken as the x-axis). Although it was originally thought that there was little π-back-donation of charge from the metal to the alkyl group because the p_y and p_z orbitals of the carbon atom were believed to be fully involved in bonding to the other atoms bound to this carbon atom, recent n.m.r. evidence has suggested that even in methyl complexes there is some π-back-donation from platinum(II) to the methyl group. 43

In metal–aryl complexes the σ-bonds are formed by overlap of a filled sp^2 hybrid orbital on the carbon with an empty $sp_x d_{x^2-y^2}$ hybrid orbital on

TABLE 99

Possible combinations of metal and carbon orbitals in the symmetry group $C_{\infty v}$ (the M–C bond is taken as the x-axis)

Symmetry	Metal orbitals	Ligand orbitals	Bond type
A_1	$s, p_x, d_{x^2 - y^2}$	s, p_x	σ
E_1	p_y, p_z, d_{xz}, d_{xy}	p_y, p_z	π
E_2	d_{z^2}, d_{yz}	—	—

the metal. In addition there is the possibility of forming π-bonds by overlap of the filled p_π-orbitals of the aryl ligand with empty hybrid orbitals on the metal, as well as π-back bonds from filled metal hybrid orbitals to the empty p_{π^*}-orbitals of the aryl ligand. It is thus not surprising to find that aryl complexes are more easily prepared and purified and are more stable than alkyl complexes.[9,14]

In metal–alkynyl complexes the σ-bonds are formed by overlap of a filled sp hybrid orbital on the carbon with an empty $sp_x d_{x^2 - y^2}$ hybrid orbital on the metal. There is also back-donation of electron density from filled $p_y d_{xy}$ and $p_z d_{xz}$ hybrid orbitals on the metal to the empty π^*-orbitals on the ligand through two π-'back-bonds', one in the xy and one in the xz plane.

Stability of metal(II)–carbon σ-bonds

One of the greater problems in determining the stability of a complex is to decide what the stability is in relation to. The phrase is often loosely applied in respect of one or all of the thermal, oxidative, hydrolytic or solvolytic factors. The platinum(II) complexes are fairly stable in respect of all of these terms, whereas the palladium(II) complexes are less so. Thus methylplatinum(II) complexes are not hydrolysed by dilute acids or oxidised by moist air, and cis-$[(PEt_3)_2Pt(CH_3)_2]$ can be distilled without decomposition at 85°C (at 10^{-4} mm/Hg). By contrast $[(PPh_3)_2Pd(CH_3)_2]$ is much less stable and decomposes in solution at 35–40°C.[14] Again in these terms, the stability order of the organic ligands is o-subst. phenyl > $C_6H_5C\equiv C-$ > p-subst. phenyl ~ phenyl > alkyl > $HC\equiv C-$.[20]

Discussion of stability becomes more meaningful when the stability is divided into its thermodynamic and kinetic components. On heating, the complexes appear to undergo homolytic fission with the formation of organic radicals, so that when $[(PEt_3)_2Pd(CH_3)_2]$ is heated at 100°C in a sealed tube a mixture of ethane and ethylene is obtained together with a trace of methane.[14] Thus the thermodynamic stability depends on the relative energies of the metal–carbon and either the carbon–carbon or carbon–hydrogen bonds. Although there is very little quantitative evidence on this, it is known that the Pt–C_6H_5 σ-bond has an energy of about 60 kcal/mol.[44] This is rather less than that of the carbon–carbon single bond (83 kcal/mol[45]) or the carbon–hydrogen bond (87 kcal/mole[37]), which are the main products of thermal decomposition. Hence it appears that the stability of platinum(II)–carbon σ-bonds must be of kinetic rather than thermodynamic origin.

An important difference between the behaviour of alkyls and aryls as ligands and that of the more familiar ligands of transition metal chemistry is that the latter complexes decompose with the formation of ions (e.g. Cl^-, NO_3^-) or of neutral molecules (e.g. H_2O, NH_3, PEt_3) all of which are stable species. Dissociation of these ligands from a complex is therefore quite likely to be reversible. By contrast dissociation of a metal–carbon σ-bond yields a very reactive product such as a free radical, or less likely a carbanion. This is an irreversible process, so that for stability a metal–carbon σ-bond must have a high activation energy barrier towards decomposition.

The kinetic stability of a compound decreases with increasing temperature, which accounts for the low temperature necessary in the preparation of most complexes containing metal–carbon σ-bonds. *Ortho*-substituted aryl ligands provide a good example of a series of very stable complexes whose extra stability, relative to complexes involving *m*-, *p*- or un-substituted aryl ligands, is largely of kinetic origin. The bulky *ortho*-substituents prevent the aryl rings from rotating about the metal–carbon σ-bond. This ensures that the *ortho*-group remains in a position where it can most effectively hinder the attack of reagents at the platinum atom.[46] Examples of such complexes are illustrated in **134**[20] and **135**.[14]

134   135

The kinetic stability of metal–carbon σ-bonds has been discussed in detail on the assumption that these bonds are split homolytically to give free radicals.[9] Homolytic dissociation can proceed either by promotion of an electron from the filled metal–carbon bonding orbital to the lowest unfilled orbital or by promotion of an electron from the highest filled orbital to the metal–carbon antibonding orbital, whichever energy difference is the lesser. Thus for kinetic stability these energy differences should be as large as possible. For the square-planar complexes *cis*-$[L_2MR_2]$ (M = Pt or Pd; R = alkyl or aryl group) we can envisage two separate cases depending on whether L is capable of giving σ-bonding only or a combination of σ- and π-bonding (Fig. 30). The value of ΔE, the energy difference between the highest filled and lowest unfilled orbitals, is considerably greater in the case in which L is capable of giving π-bonding than when it is not. Thus in $[PtCl_4]^{2-}$, ΔE has a value of $25,500\ cm^{-1}$,[47] whereas in *trans*-$[(PBu_3)_2PtCl_2]$ it is $38,000\ cm^{-1}$.[48]

With the aid of Fig. 30 we can understand not only why phosphine and arsine ligands stabilise metal–carbon σ-bonds but also why in general the stability of these bonds decreases in the order $Pt^{II} > Pd^{II} > Ni^{II}$; it is

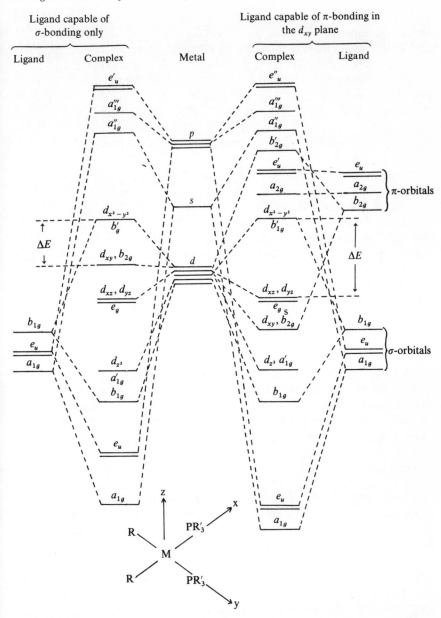

Fig. 30. Energy level diagram for *cis*-[L$_2$MR$_2$] (M = Pd, Pt; R = alkyl or aryl group).

because the value of ΔE decreases as the atomic number decreases (e.g. for [PtCl$_4$]$^{2-}$, $\Delta E = 25{,}500\ \text{cm}^{-1}$;[47] for [PdCl$_4$]$^{2-}$, $\Delta E = 22{,}100\ \text{cm}^{-1}$[49]). Furthermore, it is also possible to understand why the stability increases as the organic moiety is altered from alkyl to aryl to alkynyl, since this is the order of increasing π-acceptor ability of the respective organic groups. Thus π-bonding from the filled π-orbitals of both the aryl and alkynyl groups to empty orbitals on the metal increases the value of ΔE, Fig. 30. With nickel(II) *ortho*-substituents on the aryl groups considerably increase the stability of the aryl complex[50] because the substituents prevent free rotation of the aryl group and compel it to interact with the d_{xy} orbital, so increasing ΔE. However, this is not important in the case of the palladium(II) and platinum(II) complexes because the extra ligand field stabilisation in these complexes causes the crucial energy difference (ΔE of Fig. 30) to be between the degenerate d_{xz} and d_{yz} orbitals and the $d_{x^2-y^2}$ orbital rather than between the d_{xy} and $d_{x^2-y^2}$ orbitals. Thus the effect of the *ortho*-substituent in stabilising the platinum(II) and palladium(II) complexes is steric (see above) rather than electronic.

Although the Chatt theory is very successful for many complexes it does not realy explain how olefins, which give much lower ligand fields than phosphine (the first allowed d–d transition in $[Pt(C_2H_4)Cl_3]^-$ occurs at $23,900 \text{ cm}^{-1}$[51]), acetylacetone and even metal–carbon σ-bonds themselves can stabilise metal–carbon σ-bonds. However, Wilkinson[52] has recently suggested that this can be understood in terms of the mode of decomposition of alkyl complexes which, if possible, occurs by elimination of an olefin and formation of a metal–hydride (reaction 391). The formation of a stable metal–alkyl involves preventing reaction 391 occurring, which can be achieved

$$\begin{array}{c} CH_2R \\ | \\ M-CH_2 \end{array} \rightarrow \left[\begin{array}{c} H\cdots CHR \\ \vdots \quad | \\ M-CH_2 \end{array} \right] \rightarrow M-H + RCH = CH_2 \qquad (391)$$

either by having alkyl groups which cannot form olefins (e.g. CH_3 or CH_2Ph) or by blocking all the co-ordination sites around the metal so preventing the hydrogen atom bound to the β-carbon atom from associating with the metal, a role that is fulfilled by phosphine, olefin, acetylacetone and many other ligands. A combination of the Chatt and Wilkinson theories can explain the relative stabilities of all the known metal–carbon σ-bonded complexes.

Spectroscopic properties of metal(II)–carbon σ-bonds
A number of physical techniques have been used to investigate metal–alkyl and –aryl complexes. N.m.r. and infra-red spectroscopy together have been the most useful for characterising the organic group. However the electronic spectra and x-ray diffraction structures have been valuable in indicating the strong σ-donor properties of the ligands as shown by their high position in the spectrochemical series[55] and their ability to cause a lengthening of *trans* metal-to-ligand bonds (see pp. 302–303).

(i) Nuclear magnetic resonance
When an organic group is bound to a metal the chemical shift of the protons attached to the α-carbon atom is displaced to higher field.[56] However, the chemical shifts of protons attached to the β- and more remote carbon atoms are generally unaffected by co-ordination. For platinum, it is possible to confirm that a particular alkyl group is bound to the metal by observing the platinum–hydrogen coupling which is generally about 50–85 Hz (see Table 100).

TABLE 100

Chemical shifts and ^{195}Pt–^{1}H coupling constants for
methyl–platinum(II) complexes

Complex	τ_H (p.p.m.)	J_{Pt-H} (Hz)	Reference
cis-[(PEt₃)₂Pt(CH₃)X]			
X = Cl	9·51	52·2	57
Br	9·39	54·0	57
N₃	9·58	53·7	57
NO₂	9·64	59·4	57
CH₃	9·68	67·6	57
cis-[(AsMe₂Ph)₂Pt(CH₃)₂]	8·65	79·0	58
trans-[(PEt₃)₂Pt(CH₂D)X]			
X = Br		83·4	43
SCN		78·2	43
NO₃		85·2	43
trans-[(PMe₂Ph)₂Pt(CH₃)X]			
X = Cl	9·82	85	59
Br	9·79	83	59
I	9·73	80	59
trans-[(PMe Ph₂)₂Pt(CH₃)X]			
X = Cl	10·04	81	59
I	9·98	78	59
trans-[(AsMe₂Ph)₂Pt(CH₃)X]			
X = Cl	9·35		58
Br	9·33		58
I	9·28		58
[(C₈H₈)Pt₂(CH₃)₄]	9·22	81	13
[(C₈H₁₂)Pt(CH₃)₂]	9·19	86	13

In addition to structural evidence, n.m.r. is potentially capable of yielding
information about electronic interactions between the metal and the organic
group. Thus, as discussed above (p. 330) proton coupling constants in methyl
groups bound to platinum(II) indicate some π-back-donation of charge
from platinum(II) to the methyl groups.[43] The ^{195}Pt—^{1}H coupling constants
in a series of complexes $[(PMe_2Ph)_2PtCH_3L]^+$ follows roughly the *trans*-
influence of the ligands L, increasing with decreasing *trans*-influence of L
(L = SbPh₃ (55 Hz) > P(OPh)₃ (58 Hz) > PPh₃ (60 Hz) > CO (63 Hz) >
AsPh₃ (67 Hz) > pyr (74 Hz) > PhCN (79 Hz) ~ CH₃CN (80 Hz)).[25] A
study of the ^{195}Pt–^{31}P coupling suggests that the platinum(II)-phosphorus
bond is weakened by a methyl group in the *trans*-position;[57] the reduction
in the ^{195}Pt–^{13}C coupling constant on replacing arsenic by phosphorus in
cis-[(EMe₂Ph)₂PtMe₂], where E = P or As, may indicate that the platinum–
carbon bond is weakened on replacing the tertiary arsine by a tertiary phosphine.[60] However, some of the earlier conclusions have been questioned:
for example in interpreting the ^{19}F n.m.r. chemical shifts [61,62] of a series

of complexes trans-[PEt₃)₂PtXY] where Y = F—⟨benzene ring⟩— or ⟨benzene ring⟩—F and X = 'trans-ligand', it was assumed that the shift in the meta-compound arose from the inductive effect only, whereas that in the para-compound arose from a combination of inductive and resonance effects. Hence the difference between the chemical shifts of the para- and meta-compounds, which is proportional to the π-electron density in the aromatic ring, and which occurs as the trans-ligand X is altered reflects the π-acceptor or π-donor capacity of X. This interpretation has been criticised however[63] on the grounds that the trans-ligand X will affect the π-electron density in the aromatic ring in two ways:

(a) An increase in the π-acceptor ability of X will reduce the π-electron density available to the aryl group.

(b) A change in the σ-donor strength of X will alter the ability of the aryl group to accept electron density (thus interfering with (a)) by two mechanisms. First, an increase in the σ-donor strength of X will decrease the effective nuclear charge on the metal and thus cause an expansion of the d-orbitals thereby changing the degree of π-overlap of these orbitals with the aryl group. Secondly an increase in the trans-influence of X will weaken and lengthen the platinum(II)–aryl bond thus altering the platinum(II)–aryl interaction.

(ii) Infra-red

The infra-red spectra of metal–alkyl complexes complement rather than duplicate the evidence from n.m.r., since the infra-red spectra give information about the complex as a whole whereas n.m.r. gives more specific information about the organic ligand. The stereochemistry of the complex can be deduced from the infra-red spectrum, because there are two infra-red active metal–ligand stretching frequencies in cis-complexes and only one in trans-complexes. However, for phosphine complexes, the cis or trans nature can be established more easily from their n.m.r. spectra, since the trans-isomer exhibits very strong P–P coupling, whereas that in the cis-isomer is much weaker.[64]

The platinum(II)–carbon stretching frequencies are in the range 500–550 cm^{-1} and the palladium(II)–carbon stretching frequencies in the range 455–535 cm^{-1} (Table 101). The lower values for the M–C stretching frequencies in $[(PEt_3)_2M(CH_3)_2]$ when M = Pd (491, 457 cm^{-1}) than when M = Pt (523, 506 cm^{-1}) suggest that the stretching force constant of the palladium(II)–carbon bond is about 10 % lower than that of the platinum(II)–carbon bond.[14] The metal–carbon stretching frequency in a series of complexes $[(PEt_3)_2M(C{\equiv}CR)_2]$ where M = Pd or Pt, increases with increasing electron-withdrawing power of the substituent R,[66] because electron-withdrawing substituents decrease the energy of the π^*-(antibonding) orbitals of the alkyl group facilitating metal d to alkynyl π-back-donation so increasing the strength of the metal–carbon bond. Although the C–H stretching frequencies are generally obscured by vibrations due to the other ligands, the methyl deformation frequency at about 1200 cm^{-1} is observed. Both this and the platinum(II)–carbon stretching frequency decrease as the trans ligand (X) in the complex trans-$[(PEt_3)_2Pt(CH_3)X]$ is

TABLE 101

Infra-red data for platinum(II)– and palladium(II)–alkyl complexes

Complex	ν_{M-C} (cm^{-1})	$\delta_{Me(sym.)}$ (cm^{-1})	Reference
trans-[(PEt$_3$)$_2$Pt(CH$_3$)X]			
X = NO$_3$	566		65
Cl	551	1224	65
Br	548	1220	65
I	540	1215	65
NO$_2$	544 (or 527)		65
SCN	556		
CN	516 (or 453)	1199	65
cis-[(PMe$_3$)$_2$Pt(CH$_3$)$_2$]	525, 508	1205, 1183	65
cis-[(PMe$_3$)$_2$Pt(C$_2$H$_5$)$_2$]	511, 500	1190, 1174	65
cis-[(PEt$_3$)$_2$Pt(CH$_3$)$_2$]	526, 506	1202, 1179	65
cis-[(PEt$_3$)$_2$Pt(C$_2$H$_5$)$_2$]	516, 496	1190, 1179	65
cis-[(PMe$_2$Ph)$_2$Pt(CH$_3$)$_2$]	536, 523	1202, 1175	59
cis-[(AsMe$_2$Ph)$_2$Pt(CH$_3$)$_2$]	545, 535	1208, 1182	58
[(Et$_2$P(CH$_2$)$_2$PEt$_2$)Pt(CH$_3$)$_2$]	521, 512	1188, 1171	65
[(EtS(CH$_2$)$_2$SEt)Pt(CH$_3$)$_2$]	555, 548	1218, 1199	65
[(PEt$_3$)$_2$Pd(CH$_3$)X]			
X = Br	510	1162	14
SCN	526	1180	14
CN	502	1161	14
[(PEt$_3$)$_2$Pd(CH$_3$)$_2$]	491, 457	1164	14
[(AsEt$_3$)$_2$Pd(CH$_3$)$_2$]	498, 479	1152, 1124	14
[(PPh$_3$)$_2$Pd(CH$_3$)$_2$]	529, 482	1129	14
[(bipyr)Pd(CH$_3$)$_2$]	534, 522		14
[(MeS(CH$_2$)$_2$SMe)Pd(CH$_3$)$_2$]	525, 512	1168	14

altered[65] in the order NO$_3^-$ > Cl$^-$ > Br$^-$ > I$^-$ > NO$_2^-$ > SCN$^-$ > CN$^-$, which is the order of increasing '*trans*-effect' of these ligands.[67]

(iii) Electronic spectra

The electronic spectra of these complexes have hardly been studied at all, but their pale-yellow colour suggests that alkyl and aryl groups occur high in the spectrochemical series in agreement with the quantitative results obtained for chromium(III)[68] and ruthenium(II)[69] complexes.

(iv) X-ray crystal structures

The limited x-ray diffraction information on platinum(II)– and palladium(II)–carbon σ-bonded complexes (Table 102) indicates that the metal–carbon bond lengths are close to those predicted from the sum of the covalent radii (2·05 Å),[45] and that the metal–ligand bond *trans* to the σ-bonded carbon ligand is significantly longer than usual, indicating that these ligands have a high *trans*-influence.

Chemical properties of metal(II)–carbon σ-bonds

Most of the chemical reactions of these complexes fall into one of three categories: (a) cleavage reactions, in which the metal(II)–carbon bond is

TABLE 102
X-ray diffraction data for metal–carbon σ-bonded complexes

Complex	M–C bond length (Å)	M–X bond length trans to M–C (Å)	Comparison M–X bond length (Å)	No. of structure in Appendix II	Reference
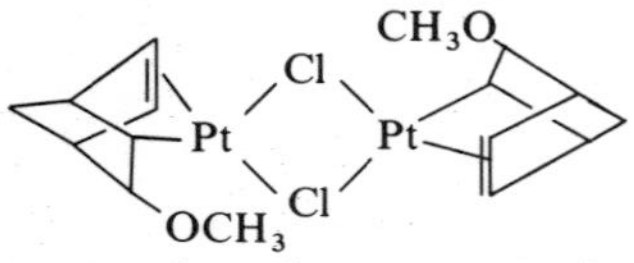	2·07	2·57 (X = Cl)	2·34 (trans to double-bond in same complex)	184	70
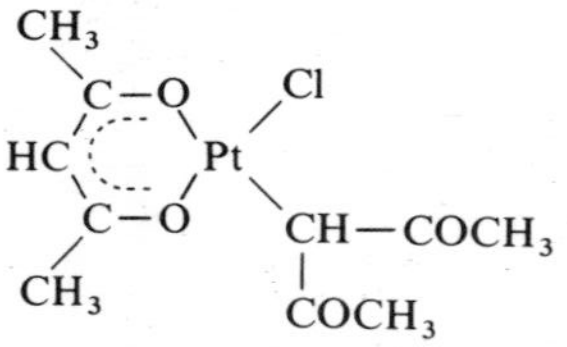	2·13	2·09 (X = O)	1·97 (trans to Cl in same complex)	149	10, 71
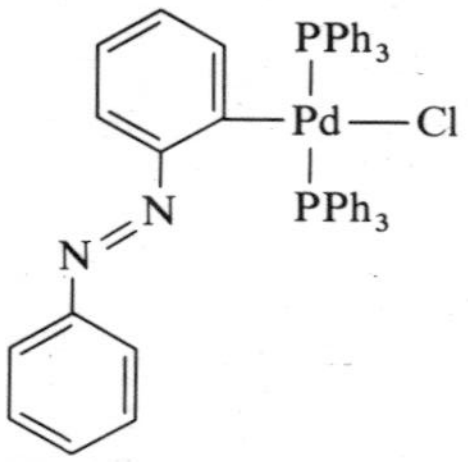	2·00	2·38 (X = Cl)	2·30 in K_2PdCl_4[a] and [(dimethylpiperazine)$PdCl_2$][b]	88	72, 73

[a] Reference 74.   [b] Reference 75.

split; (b) insertion reactions, in which a reagent is 'inserted' into the metal(II)–carbon bond; (c) replacement reactions, in which the organic moiety is replaced by another ligand.

(a) Cleavage reactions

(i) *Acids and protonic solvents.* Although platinum(II) and palladium(II) complexes are not hydrolysed by water or dilute aqueous acids, they are attacked by hydrogen chloride in ethanol or benzene solution (reaction 392). This reaction follows a two-term rate law.[76–8] A rapid pre-equilibrium is

$$trans\text{-}[(PEt_3)_2Pt(C_6H_5)_2] + HCl \rightarrow trans\text{-}[(PEt_3)_2Pt(C_6H_5)Cl] + C_6H_6$$

$$(392)$$

$$Rate = \{k_2[H^+] + k_3[H^+][Cl^-]\}[complex]$$

set up in which the platinum(IV) complexes **136** and **137**, first formed, collapse to platinum(II) derivatives with loss of benzene in the rate-determining step. Reaction 392 is an interesting example of the dependence of the

$$
\begin{array}{cc}
\begin{array}{c}
\quad\ \ S\ \ \ H \\
\quad\ \ | \ \ / \\
Et_3P—Pt—PEt_3 \\
\quad / \ | \\
C_6H_5\ \ C_6H_5
\end{array}
&
\begin{array}{c}
\quad\ \ Cl\ \ H \\
\quad\ \ | \ \ / \\
Et_3P—Pt—PEt_3 \\
\quad / \ | \\
C_6H_5\ \ C_6H_5
\end{array}
\\
\mathbf{136} & \mathbf{137}
\end{array}
$$

(S = solvent)

reactivity of metal–carbon σ-bonds on the other ligands present, since this reaction occurs sufficiently fast at 0°C for the monoaryl complex to be isolated even in the presence of an excess of hydrogen chloride. Cleavage of the second aryl group is, however, slow. By contrast the two phenyl groups in *cis*-[(PEt$_3$)$_2$Pt(C$_6$H$_5$)$_2$] are cleaved at comparable rates.[20]

Although the platinum(II) complexes are unattacked by ethanol, the palladium(II)–methyl complexes are attacked to give methane and [(PEt$_3$)$_2$Pd(OC$_2$H$_5$)CH$_3$]; the latter slowly decomposes to give palladium metal, methane, acetaldehyde and ethanol.[14] Other protonic acids that cleave palladium(II)–carbon bonds are thiophenol and *p*-nitrophenylacetylene[14] (reaction 393).

$$
[(PEt_3)_2Pd(CH_3)_2]
\begin{array}{c}
\xrightarrow[\text{fast}]{C_6H_5SH} [(PEt_3)_2Pd(SC_6H_5)_2] + 2CH_4 \\
\xrightarrow[\text{slow}]{p\text{-NO}_2C_6H_4C\equiv CH} [(PEt_3)_2Pd(p\text{-NO}_2C_6H_4C\equiv C)_2] + 2CH_4
\end{array}
$$

$$(393)$$

(ii) *Hydrogen and reducing agents.* Both platinum(II)– and palladium(II)–carbon σ-bonds are readily cleaved both by molecular hydrogen[22] (reaction 394) and by reducing agents such as LiAlH$_4$ or NaBH$_4$.[79,80] The reaction with molecular hydrogen is of potential industrial importance since it

$$cis\text{-}[(PEt_3)_2Pt(C_6H_5)Cl] + H_2 \xrightarrow[\text{1 atm.}]{20°C} trans\text{-}[(PEt_3)_2PtHCl] + C_6H_6 \quad (394)$$

may occur in the homogeneous catalytic hydrogenation of olefins after the olefin has been inserted into the platinum–hydrogen bond to form an alkyl complex. Research in this field has recently been summarised by the present author.[30]

(*iii*) *Halogens*. The cleavage of metal(II)–carbon σ-bonds by halogens (reaction 395) is often used as a diagnostic test for these bonds. With platinum,

$$cis\text{-}[(PEt_3)_2MR_2] + X_2 \xrightarrow[\text{solvent}]{C_6H_6}$$

$$[(PEt_3)_2MR_2X_2] \longrightarrow trans\text{-}[(PEt_3)_2MRX] + RX$$

$$\downarrow \begin{array}{l}\text{Excess } X_2 \text{ in}\\ C_6H_6\end{array}$$

$$trans\text{-}[(PEt_3)_2MX_2] \qquad\qquad (395)$$

when R = methyl and X = Cl the platinum(IV) intermediate $[(PEt_3)_2\text{-}Pt(CH_3)_2Cl_2]$ can be isolated but when X = I the intermediate, being unstable, decomposes to the platinum(II) derivative.[9] The diaryl platinum(IV) adducts $[(PEt_3)_2Pt(C_6H_5)_2X_2]$ can be isolated when X is either chlorine or iodine.[20] With palladium, the tetravalent complexes are unstable and only the palladium(II) products are observed.

(*iv*) *Alkyl halides*. Alkyl halides react with the platinum(II) complexes to give initially platinum(IV) adducts,[9,13,81] which may decompose on heating to give platinum(II) complexes.[9]

(*v*) *Metal halides*. Although alkali metal halides are used to substitute one halogen for another they do not react with the metal(II)–carbon σ-bond. However, magnesium iodide can cleave platinum(II)–carbon σ-bonds to give Grignard reagents[9] (reaction 396).

$$cis\text{-}[(PEt_3)_2Pt(CH_3)_2] + MgI_2 \longrightarrow$$

$$trans\text{-}[(PEt_3)_2PtI_2] + trans\text{-}[(PEt_3)_2Pt(CH_3)I] + CH_3MgI \quad (396)$$

(*vi*) *Thermal decomposition*. Methyl complexes appear to decompose thermally by a homolytic process giving free radicals which subsequently combine to give coupled products[14] (reaction 397). By contrast higher alkyl complexes decompose on heating to eliminate olefins and yield hydrido-

$$[(PEt_3)_2Pd(CH_3)_2] \xrightarrow{100°C} C_2H_6(92\%) + C_2H_4(8\%) + CH_4(\text{trace}) \quad (397)$$

complexes[22] (see p. 55).

(*b*) *Insertion reactions*
(*i*) *Carbon monoxide*. Alkyl and aryl complexes of both platinum(II) and palladium(II) are converted to acyl complexes by carbon monoxide[37,82] (reaction 398). Because square-planar palladium(II) complexes are more

$$[(PEt_3)_2M(CH_3)Cl] + CO \longrightarrow [(PEt_3)_2M(COCH_3)Cl] \quad (M = Pd, Pt) \quad (398)$$

labile than their platinum(II) analogues, the platinum complex carbonylates less readily than the palladium complex and a temperature of 90°C and pressure of 80 atm. is required for it, although the reaction of the palladium complex is complete in 24 hours at room temperature and atmospheric pressure. The latter may involve initial co-ordination of carbon monoxide along the z-axis to give a five-co-ordinate intermediate, which breaks down to give a four-co-ordinate acyl complex by migration of the alkyl group.[3] The reaction is reversible and on heating at about 140°C, which is above the melting-point, the acyl complexes lose carbon monoxide and revert to the original alkyl complexes. This mechanism is consistent with the kinetics of the carbonyl insertion reaction (reaction 399) in a range of solvents which indicate a two-step mechanism in which the initial rate-

$$[(AsPh_3)Pt(CO)Cl(C_2H_5)] + AsPh_3 \rightarrow [(AsPh_3)_2Pt(COC_2H_5)Cl] \quad (399)$$

determining step involves combination of the ethyl and carbonyl ligands to form a propionyl group. This step is not assisted by either the solvent or the incoming nucleophile[83] (reaction 400).

$$[(AsPh_3)Pt(CO)Cl(C_2H_5)] \rightleftharpoons [(AsPh_3)Pt(COC_2H_5)Cl] \xrightarrow{AsPh_3}$$

$$[(AsPh_3)_2Pt(COC_2H_5)Cl] \quad (400)$$

(*ii*) *Sulphur dioxide.* Although insertion of sulphur dioxide into platinum(II)– and palladium(II)–carbon bonds is unknown, the reverse of this hypothetical reaction[27] has been described (p. 329) as a method of preparing such metal–carbon σ-bonded species.

(*iii*) *Olefins and acetylenes.* Olefins and acetylenes such as C_2F_4 and $CF_3C\equiv CCF_3$ react with alkyl complexes to give products in which the unsaturated compound has been inserted into the metal–alkyl bond[84] (reaction 401).

$$trans\text{-}[(PMe_2Ph)_2Pt(CH_3)Cl] + C_2F_4 \rightarrow$$

$$trans\text{-}[(PMe_2Ph)_2Pt(CF_2CF_2CH_3)Cl] \quad (401)$$

The unsaturated ligand first adds on to the platinum(II)–alkyl complex to give a five-co-ordinate complex. The thermal stability of these five-co-ordinate complexes $[L_2Pt(CH_3)Cl(CF_2=CF_2)]$ varies with the nature of L in the order $L = AsMe_3 \sim AsMe_2Ph \gg SbMe_3 > PMe_3 > PEt_3 \sim PMe_2Ph$, and with trimethylarsine the five-co-ordinate adduct is indefinitely stable at room temperature whereas with trimethylphosphine it decomposes appreciably within a week to give the insertion product $[(PMe_3)_2Pt(CF_2-CF_2CH_3)Cl]$.[85] Although the five-co-ordinate adducts of fluoro-olefins and fluoroalkynes decompose to their constituents in solution and give insertion reactions on heating, the adduct with tetracyanoethylene is stable under both conditions. The n.m.r. spectrum of this complex is consistent with the tetracyanoethylene being co-ordinated in a trigonal plane with two tertiary phosphine ligands (i.e. a type T olefin complex[86]) as in **138**.[87]

$$\begin{array}{c} Me_3P \quad CH_3 \\ \diagdown \ | \quad \ C(CN)_2 \\ Pt-\| \\ \diagup \ | \quad \ C(CN)_2 \\ Me_3P \quad Cl \end{array}$$

138

Reaction 402 is an unusual example of the internal insertion of an olefin into a metal–alkyl bond[88] (for details of the structure of the product see Appendix II, structure 67).

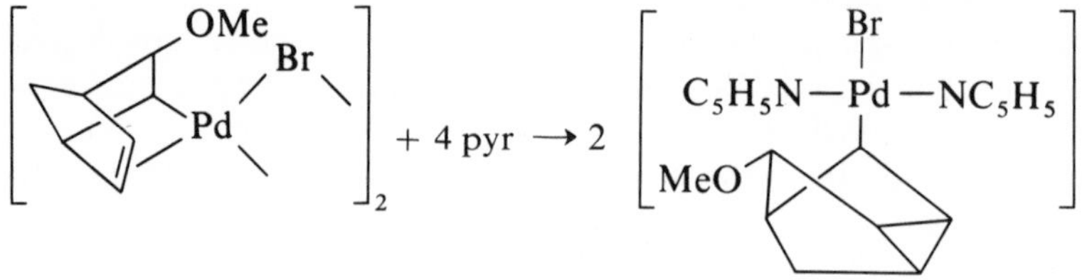

$$\left[\begin{array}{c} \text{OMe} \\ \text{Br} \\ \diagdown \\ \text{Pd} \\ \diagdown \end{array}\right]_2 + 4\,\text{pyr} \rightarrow 2 \left[\begin{array}{c} \text{Br} \\ | \\ C_5H_5N-\text{Pd}-NC_5H_5 \\ | \\ \text{MeO} \end{array}\right] \tag{402}$$

(iv) Isocyanides. Cyclohexylisocyanide reacts with palladium(II)–methyl complexes to give products in which one, two and three isocyanide ligands have inserted into the metal–carbon bond[89] (reaction 403). The n.m.r. spectra of these insertion products suggest that the single and double

$$trans\text{-}[(PMe_2Ph)_2PdCH_3I] + \text{cyclohexylNC} \rightarrow$$

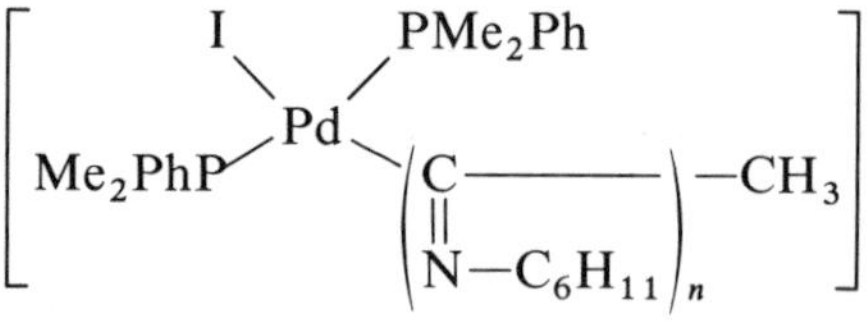

$$\left[\begin{array}{c} I \qquad PMe_2Ph \\ \diagdown \quad / \\ \text{Pd} \\ \diagup \quad \diagdown \\ Me_2PhP \qquad \begin{pmatrix} C-\!\!\!-\!\!\!-\!\!\!-\!\!\!- \\ \| \\ N-C_6H_{11} \end{pmatrix}_n \end{array}\!\!-CH_3\right] \tag{403}$$

$$(n = 1 \text{ or } 2)$$

insertion products are five-co-ordinate adducts formulated as in **139** and **140** respectively.[90] The insertion of three isocyanide ligands yields $[(PPhMe_2)PdI(C_6H_{11}N=C-C(=NC_6H_{11})-C(CH_3)=NC_6H_{11})]$ in which a five-membered chelate ring has been formed (**141**).[89]

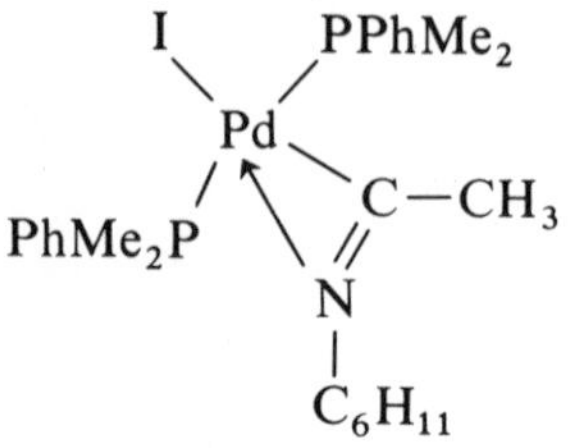

$$\begin{array}{cc}
\begin{array}{c}
I \qquad PPhMe_2 \\
\diagdown \quad / \\
\text{Pd} \\
\diagup \quad \diagdown \\
PhMe_2P \qquad C-CH_3 \\
\qquad \qquad \| \\
\qquad \qquad N \\
\qquad \qquad | \\
\qquad \qquad C_6H_{11} \\
\textbf{139}
\end{array}
&
\begin{array}{c}
I \qquad PPhMe_2 \\
\diagdown \quad / \\
\text{Pd} \\
\diagup \quad \diagdown \\
PhMe_2P \qquad C=NC_6H_{11} \\
\qquad | \qquad / \\
\qquad N=C \\
\qquad | \qquad \diagdown \\
\qquad C_6H_{11} \qquad CH_3 \\
\textbf{140}
\end{array}
\end{array}$$

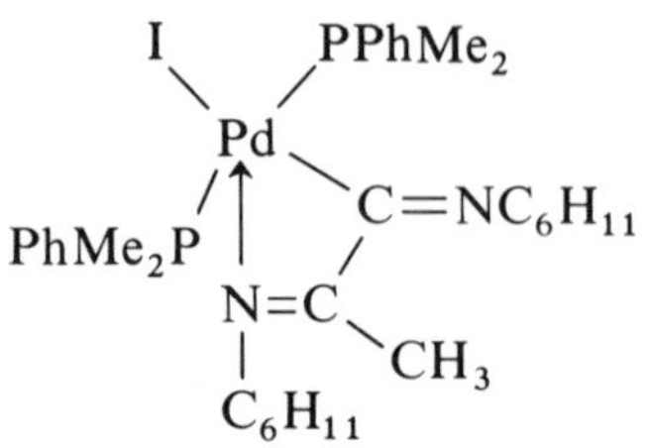

$$\begin{array}{c}
C_6H_{11} \\
| \\
N \\
\| \\
C_6H_{11}-N=C-C \qquad PMe_2Ph \\
\qquad \qquad | \quad \diagdown \quad / \\
\qquad \qquad | \quad \text{Pd} \\
\qquad \qquad | \quad \diagup \quad \diagdown \\
H_3C-C-N \qquad I \\
\qquad | \\
\qquad C_6H_{11} \\
\textbf{141}
\end{array}$$

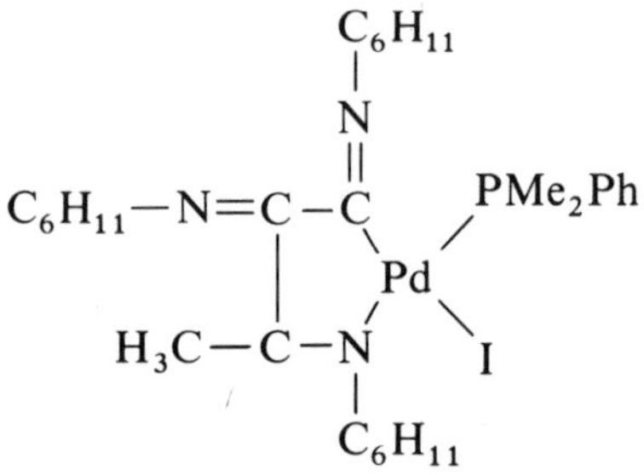

(c) Replacement reactions

(i) *Alkyl groups.* The replacement of one alkyl group by another has been used to prepare fluoroalkyl complexes[91] (reaction 404). Although this

$$[(bipyr)Pd(CH_3)_2] + C_3F_7I \rightarrow$$

$$[(bipyr)Pd(CH_3)(C_3F_7)] + [(bipyr)Pd(C_3F_7)_2] \quad (404)$$

reaction involves the replacement of a less strongly held alkyl group by one that gives a more stable complex, this is not always so, since an *o*-tolyl group can be displaced by a methyl group[13] (reaction 405).

$$[(PPh_3)_2Pt(o\text{-}CH_3C_6H_4)_2] + CH_3I \xrightarrow{115°} [(PPh_3)_2PtCH_3(I_3)] \quad (405)$$

(ii) *Other ligands.* Although ligands such as phosphines and cyanide that bind strongly to transition metals displace alkyl groups from some metals, they do not do so with platinum(II) and palladium(II) complexes. Thus treatment of $[(PEt_3)_2Pd(CH_3)Br]$ with an excess of potassium cyanide yields[14] $[(PEt_3)_2Pd(CH_3)CN]$. Similarly pyridine causes a reversible displacement of the halogen[42] (reaction 406). The equilibrium, which is

$$[(PEt_3)_2MRCl] + pyr \rightleftharpoons [(PEt_3)_2MR\ pyr]^+Cl^- \quad (406)$$

attained 100,000 times more rapidly for palladium(II) than platinum(II), normally lies well to the left; but when sodium perchlorate or sodium tetrephenylborate in acetone is added, the ionic platinum(II) species can be isolated.[63]

METAL(II)–CARBON σ-BONDED FLUOROCARBON COMPLEXES

Although closely related, metal(II)–fluorocarbon complexes are generally more robust than their hydrocarbon analogues.

Preparation

(i) From metal hydrides

The preparation of fluorocarbon complexes from metal hydrides, which has been used only with platinum(II), can involve either insertion into a platinum(II)–hydride bond[92,93] (reactions 407 and 408) or elimination of hydrogen fluoride[92,94] (reaction 409). Detailed investigation of this reaction

$$trans\text{-}[(PEt_3)_2PtHCl] + CF_3C{\equiv}CCF_3 \xrightarrow[90°C,\ 25\ hr]{cyclohexane}$$

$$trans\text{-}[(PEt_3)_2Pt(cis\text{-}CF_3C{=}CHCF_3)Cl] \quad (407)$$

$$trans\text{-}[(PEt_3)_2PtHCl] + (CF_3)_2CN_2 \xrightarrow[120°C]{hexane}$$

$$cis\text{-}[(PEt_3)_2Pt\{CH(CF_3)_2\}Cl] + N_2 \quad (408)$$

showed that $trans\text{-}[(PEt_3)_2Pt\{C(CF_2H){=}CF_2\}Cl]$ and $trans\text{-}[(PEt_3)_2Pt(CO)Cl]^+SiF_5^-$ were also formed, the latter being identified by x-ray

$$trans\text{-}[(PEt_3)_2PtHCl] + CF_2{=}CF_2 \xrightarrow[120°C,\ 50\ hr]{cyclohexane}$$

$$trans\text{-}[(PEt_3)_2Pt(CF{=}CF_2)Cl] + HF \quad (409)$$

diffraction[95] (structure 169 in Appendix II), and that the overall reaction occurred as in eqn. 410 to 412.[96]

$$[(PEt_3)_2PtHCl] + C_2F_4 \rightarrow$$

$$[(PEt_3)_2Pt(CF_2CF_2H)Cl] + [(PEt_3)_2Pt(CF{=}CF_2)Cl] + HF \quad (410)$$

$$HF + glass \rightarrow H_2O + BF_3 + SiF_4 + BF_4^- + SiF_6^{2-} \quad (411)$$

$$2[(PEt_3)_2Pt(CF{=}CH_2)Cl] + H_2O + SiF_4 \rightarrow$$

$$[(PEt_3)_2Pt(CO)Cl]^+ SiF_5^- + [(PEt_3)_2Pt\{C(CH_2F){=}CF_2\}Cl] + HF \quad (412)$$

(ii) From metal-alkyls

Perfluoroalkyl complexes can be prepared from metal–alkyl complexes either by replacing an alkyl group by a perfluoroalkyl group[91,97] (see above p. 343) or by insertion into a metal–alkyl bond. Treatment of the dimethylpalladium(II) complex $[(bipyr)Pd(CH_3)_2]$ with perfluoropropyl iodide yields ethane but not methyl iodide (reaction 413), which is quite different from the reaction of iodine with this complex which yields methyl

$$[(bipyr)Pd(CH_3)_2 + C_3F_7I \rightarrow$$

$$[(bipyr)Pd(C_3F_7)CH_3] + [(bipyr)Pd(C_3F_7)_2] \quad (413)$$

iodide and $[(bipyr)PdI_2]$, not ethane.[91] When *trans*-$[(AsMe_2Ph)_2Pt(CH_3)Cl]$ is treated with perfluorobut-2-yne, an addition reaction occurs first, but the product rearranges on standing to give an insertion compound[98] (reaction 414).

$$trans\text{-}[(AsMe_2Ph)_2Pt(CH_3)Cl] + CF_3C{\equiv}CCF_3 \rightarrow$$

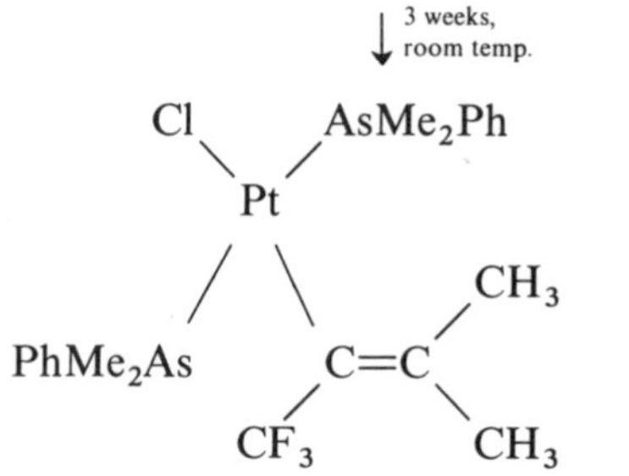

$$\downarrow \text{3 weeks, room temp.}$$

$$(414)$$

(iii) From lithium or Grignard reagent

This method, which is of wide applicability in the preparation of hydrocarbon–metal σ-bonds, is limited in the case of the fluorocarbon analogues to those reagents that are stable (e.g. C_6F_5Li and $CF_2{=}CFMgBr$). For instance CF_3MgI has only a transient existence and C_3F_7Li decomposes well below

0°C. Within this limitation, the method has been used to prepare both platinum(II) and palladium(II) complexes starting from the bisphosphine dichlorides by means of such reagents as $LiCF=CF_2$,[99] $LiCCl=CF_2$,[99] $CF_2=CFMgBr$,[99] $LiC\equiv CCF_3$,[100] LiC_6F_5,[101–103] C_6F_5MgBr[103] and *m*- and *p*-FC_6H_4MgBr[36,62] (reactions 415–417).

$$cis\text{-}[(PEt_3)_2PtCl_2] \xrightarrow{+LiC_6F_5} cis\text{-}[(PEt_3)_2Pt(C_6F_5)Cl] + LiCl$$
$$\xrightarrow{+2LiC_6F_5} cis\text{-}[(PEt_3)_2Pt(C_6F_5)_2] + 2LiCl \quad (415)$$

$$trans\text{-}[(PEt_3)_2PtCl_2] \xrightarrow{+LiC_6F_5} trans\text{-}[(PEt_3)_2Pt(C_6F_5)Cl] + LiCl$$
$$\xrightarrow{+2LiC_6F_5} trans\text{-}[(PEt_3)_2Pt(C_6F_5)_2] + 2LiCl \quad (416)$$

$$trans\text{-}[(PEt_3)_2PdCl_2] + 2LiC_6F_5 \rightarrow$$
$$trans\text{-}[(PEt_3)_2Pd(C_6F_5)_2] + 2LiCl \quad (417)$$

(iv) Oxidative addition

The oxidative addition of perfluoroalkyl iodides[104] (e.g. CF_3I, C_2F_5I and C_3F_7I) and CF_3NO[105] to $[M(PPh_3)_4]$ and $[M(diphos)_2]$ (M = Pd, Pt; diphos = $Ph_2PCH_2CH_2PPh_2$) has been used to prepare a number of perfluoroalkyl complexes (reaction 418). The reagent $(C_6F_5)_2TlBr$ also

$$[M(PPh_3)_4] + CF_3I \rightarrow cis\text{-}[(PPh_3)_2M(CF_3)I] \quad (418)$$

gives perfluorophenyl complexes by an oxidative addition.[6] With $[Pt(PPh_3)_4]$ a conventional platinum(II) complex is formed (reaction 419). However, $(C_6F_5)_2TlBr$ reacts with palladium(II) complexes to give the

$$[Pt(PPh_3)_4] + (C_6F_5)_2TlBr \rightarrow cis\text{-}[(PPh_3)_2Pt(C_6F_5)_2] \quad (419)$$

only known example of a palladium(IV) complex containing a palladium–carbon σ-bond (reaction 420).

$$trans\text{-}[(PPh_3)_2PdCl_2] + (C_6F_5)_2TlBr \rightarrow [(PPh_3)_2Pd(C_6F_5)_2Cl_2]$$
$$(420)$$

(v) Reactions of fluorocarbon complexes to give new complexes

In addition to standard reactions in which the chloride ligand is replaced by bromide, iodide, cyanide, cyanate and thiocyanate,[62] the chloride ligand in the monoperfluorohydrocarbon complexes can be replaced by a methyl group by means of methyl lithium[101] or methylmagnesium bromide[62] (reaction 421). The halogen-bridged complex $trans\text{-}[(PEt_3)_2Pt_2(CCl=$

$$trans\text{-}[(PEt_3)_2Pt(C_6F_5)Cl] + LiMe \rightarrow$$
$$trans\text{-}[(PEt_3)_2Pt(C_6F_5)CH_3] + LiCl \quad (421)$$

$CF_2)_2Br_2]$ is split on treatment with pyridine[99] to yield the monomeric complex $[(PEt_3)(pyr)Pt(CCl=CF_2)Br]$. The pentafluorophenylpalladium(II) complex $[(PPh_3)_2Pd(C_6F_5)Cl]$ formed by hydrazine reduction of the palladium(IV) complex $[(PPh_3)_2Pd(C_6F_5)_2Cl_2]$ can be oxidised by chlorine[5] to give $[(PPh_3)_2Pd^{IV}(C_6F_5)Cl_3]$.

Properties

Perfluoralkyl and perfluoroaryl complexes are generally more stable than their hydrocarbon analogues. This is exemplified by $[(bipyr)Pd(C_3F_7)_2]$ which is a stable, high-melting (180°C) crystalline product,[91] in contrast to its hydrocarbon analogue which has not been isolated.[14] Perfluorocarbon complexes are less susceptible to cleavage by acids than their hydrocarbon analogues and do not undergo insertion reactions when treated with carbon monoxide. Although methyllithium and methylamine both displace fluoride from C_6F_6 and C_6F_5H, neither reagent reacts with pentafluoro-phenyl co-ordinated to platinum(II).[4] Similarly, treatment of *cis*-$[(PPh_3)_2Pt-(C_6F_5)Cl]$ and *cis*-$[(PPh_3)_2Pt(C_6F_5)_2]$ with $NaSCH_3$ gives[106] *cis*-$[(PPh_3)_2Pt(C_6F_5)(SCH_3)]$. Such reactions suggest that perfluorocarbon ligands react in a similar way to inert halide ligands, which is consistent with approximate estimates of their electronegativities which fall between chlorine and bromine:[5] $F(4\cdot0) > Cl(3\cdot2) \sim CF_3(3\cdot2) \sim C_2F_5(3\cdot2) > C_3F_7(3\cdot1) > C_6F_5(3\cdot0) \sim Br(3\cdot0) > C_6H_5(2\cdot8) > I(2\cdot7) > CH_3(2\cdot6)$.

Bonding in metal(II)–perfluorocarbon complexes

The scheme used to describe the bonding in metal(II)–perfluorocarbon complexes is essentially that shown in Fig. 30 (p. 333) for metal(II)–hydrocarbon complexes. The extra stability of the fluorocarbon complexes, which, in view of the high lattice energies of the metal fluorides, is probably of kinetic rather than thermodynamic origin, has been suggested to arise from:

(i) The metal(II)–perfluorocarbon bond is more ionic than the metal(II)–hydrocarbon bond owing to the greater electronegativity of fluorocarbons (R_F) than hydrocarbons (R_H).[107]

(ii) The high electronegativity of R_F relative to R_H reduces the electron density on the metal, thereby contracting the d-orbitals and so decreasing the repulsions between the filled d-orbitals on the metal and the group attached to the α-carbon atom.[3]

(iii) The decrease of electron density on the metal should lead to a contraction of the metal orbital forming the metal–carbon σ-bond so that it approaches the size of the carbon orbitals more closely, thus increasing the strength of the metal–carbon σ-bond in perfluorocarbon as opposed to hydrocarbon complexes.

(iv) A further source of extra stability in fluorocarbon complexes is the possibility of back-donation of electron density from filled metal d to low energy σ^*-orbitals on the alkyl and π^*-orbitals on the aryl ligands. Although information is not available for platinum or palladium complexes, a shortening of the metal–carbon bond length in the fluorocarbon relative to the hydrocarbon complexes,[108] taken with infra-red and n.m.r. data[109] that are consistent with π-back-donation of electron density from metal to carbon, have been observed for fluorocarbon complexes of other metals.

PLATINUM–CYCLOPROPANE COMPLEXES

Platinum–cyclopropane complexes may be prepared either by heating chloroplatinic acid in acetic anhydride with cyclopropane[110] or by treating

Zeise's dimer $[(C_2H_4)_2PtCl_2]_2$ with either cyclopropane or a substituted cyclopropane.[111] All these reactions give a polymeric brown solid with the empirical formula $[(C_3H_6)PtCl_2]$; its structure has been the subject of much investigation. A platinum(IV) complex containing a four-membered ring (142) is suggested by:

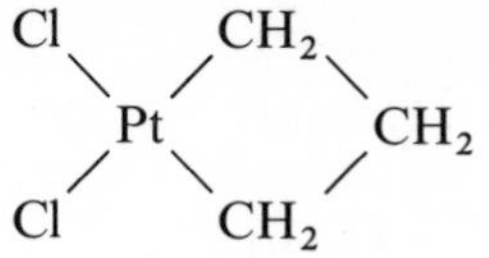

142

(i) On treatment with pyridine $[(C_3H_6)PtCl_2]_n$ gives $[(pyr)_2Pt(C_3H_6)Cl_2]$ which has the octahedral structure **143** (see structure 166 in Appendix II).[112]

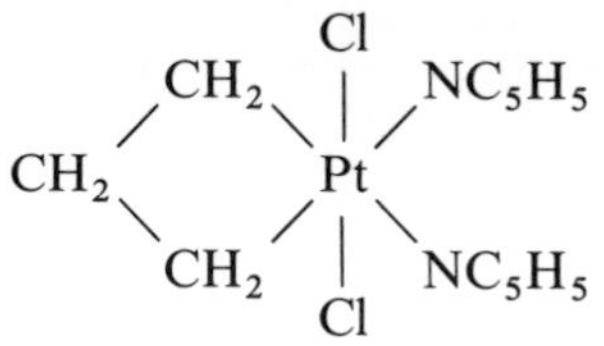

143

(ii) The infra-red spectrum of $[(C_3H_6)PtCl_2]_n$ is consistent with a four-membered ring but not with the three-membered ring of cyclopropane remaining intact in the complex.[113]

Nevertheless a structure with a three-membered ring such as **144** is suggested by the following:

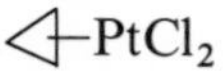

144

(i) When complexes of substituted cyclopropanes are treated with potassium cyanide,[111,113] phosphines[110] or propenylbenzene[111] the initial cyclopropane is recovered quantitatively, pure and unisomerised.

(ii) On boiling the phenylcyclopropane complex in water phenylcyclopropane together with traces of propiophenone, phenylethylcarbinol and platinum metal are produced.[110]

(iii) Competition between substituted cyclopropanes for Zeise's dimer shows that complexing ability parallels electron density in the cyclopropane ring.[110]

(iv) On heating $[(phenylcyclopropane)PtCl_2]_n$ in carbon tetrachloride the more soluble π-allylic complex $[π-(PhCH-CH-CH_2)PtCl]_2$ is formed slowly.[85].

Recently it has been suggested on the basis of both chemical and spectroscopic work that both $[(C_3H_6)PtCl_2]_n$ and the related bromide complex are tetrameric and have structure **145**, which is related to the structure of the trimethylplatinum(IV) halides[114] (see p. 352). All attempts to confirm the structure of this compound by x-ray diffraction have been frustrated by the impossibility of preparing suitable crystals.

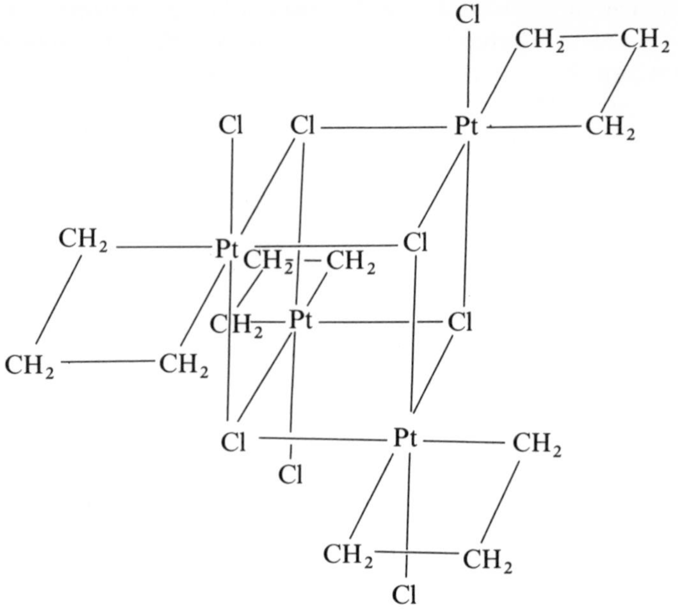

145

PLATINUM–CARBENE COMPLEXES

Although carbene complexes have only recently been reported they have attracted considerable attention so that a number of preparative routes have been devised. The most widely employed preparation involves treating an isocyanide complex with an alcohol, amine, amide ion, hydroxide ion or hydrosulphide ion[115,116] (reactions 421–3). An alternative method of preparation is to use a reactive intermediate such as an electron-rich olefin

$$[(PEt_3)Pt(CNPh)Cl_2] + C_2H_5OH \rightarrow \left[(PEt_3)Pt\left(-C\begin{matrix} OC_2H_5 \\ NHPh \end{matrix}\right)Cl_2\right] \quad (421)$$

$$trans\text{-}[(PPh_3)_2Pt(CNMe)_2]^{2+}(BF_4^-)_2 + OH^- \xrightarrow[25°C]{MeCN/H_2O}$$

$$trans\text{-}[(PPh_3)_2Pt(CNMe)(C(=O){-}NHMe)]^+BF_4^- + BF_4^- \quad (422)$$

$$trans\text{-}[(PPh_3)_2Pt(CNMe)_2]^{2+}(BF_4^-)_2 + MeC_6H_4NH^-Na^+ \rightarrow$$

$$\left[(PPh_3)_2Pt(CNMe)\left(C\begin{matrix} {=}NC_6H_4Me \\ NHMe \end{matrix}\right)\right]^+ BF_4^- + NaBF_4 \quad (423)$$

(reaction 424) or a sterically crowded imidazolidine (reaction 425) both of which are readily cleaved to yield carbene fragments.[117] However, these reactions are limited in application since other electrophilic carbene

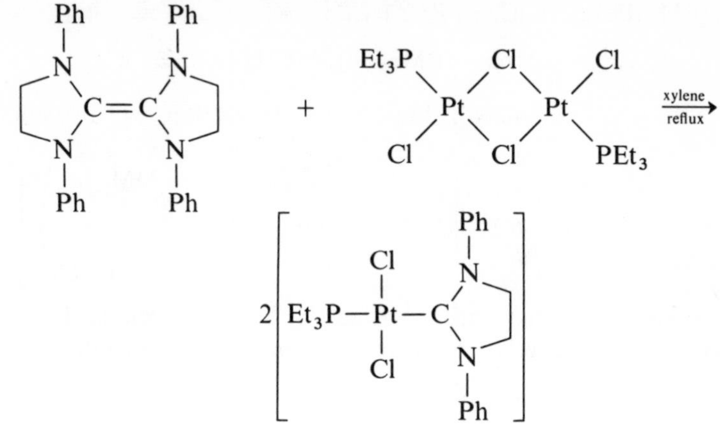

$$(424)$$

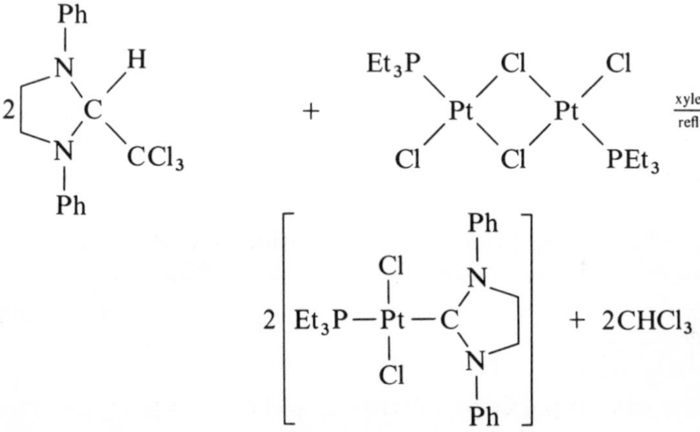

$$+ \ 2CHCl_3 \qquad (425)$$

percursors such as $(CF_3)_2CN_2$ do not give carbene complexes. The third route by which carbene complexes have been prepared is by treating the methyl complex *trans*-$[(PMe_2Ph)_2PtCH_3Cl]$ with an acetylene and silver hexafluorophosphate in methanol[118] (reaction 426). This reaction is thought to occur by initial formation of an acetylene complex followed by

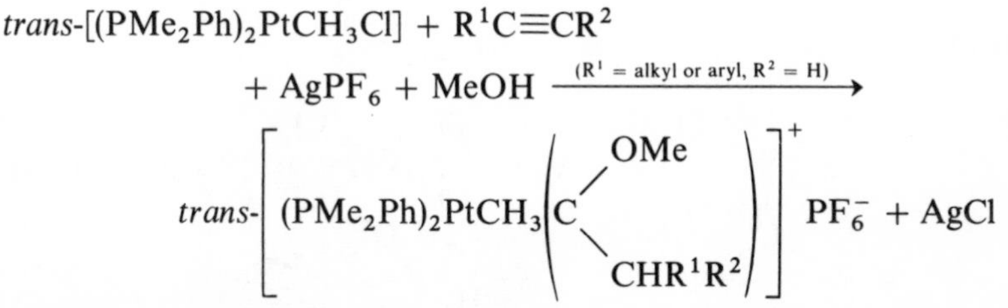

trans-$[(PMe_2Ph)_2PtCH_3Cl] + R^1C{\equiv}CR^2$

$+ \ AgPF_6 + MeOH \xrightarrow{\text{(R}^1\text{ = alkyl or aryl, R}^2\text{ = H)}}$

$$PF_6^- + AgCl \quad (426)$$

reaction of this with methanol (reaction 427) since if R^1 and R^2 in reaction 426 are phenyl groups the cationic acetylene complex *trans*-$[(PMe_2Ph)_2$-$Pt(CH_3)(PhC{\equiv}CPh)]^+PF_6^-$ is formed.

X-ray diffraction studies of the carbene complexes formed in reactions 421 and 424 have shown that the PtPCl$_2$, PtCNO and PtCN$_2$ units are planar and that the carbene moiety is approximately perpendicular to the

$$trans\text{-}[(PMe_2Ph)_2(CH_3)Cl] + R^1C{\equiv}CR^2 \rightarrow$$

$$trans\text{-}[(PMe_2Ph)_2Pt(CH_3)(R^1C{\equiv}CR^2)]^+$$
$$\downarrow \text{MeOH}$$
$$trans\text{-}[(PMe_2Ph)_2Pt(CH_3)(R^1R^2C{=}CHOMe)]^+$$
$$\downarrow \begin{array}{l}\text{rearrangement}\\\text{of olefin}\\\text{complex}\end{array}$$

$$trans\text{-}\left[(PMe_2Ph)_2Pt(CH_3)\left(-C{\overset{\displaystyle OMe}{\underset{\displaystyle CHR^1R^2}{\diagup\!\!\diagdown}}}\right)\right]^+ \qquad (427)$$

square-plane around the platinum atom[115,117] (see Appendix II structures 172 and 183 respectively). Nuclear magnetic resonance studies of *trans*-

$$\left[(PEt_3)_2PtX\left(-C{\overset{\displaystyle NHR}{\underset{\displaystyle Q}{\diagup\!\!\diagdown}}}\right)\right]^+ ClO_4^-, \text{ where } R = Ph \text{ or } Me, Q = PhNH,$$

EtNH or EtO and X = Cl or Br, indicate that there is restricted rotation about the C–N and C–O bonds of the carbene.[119] Although the reactions of these carbene complexes have not been investigated in detail it has been found that when the carboxamido complex formed in reaction 422 is treated with acid (fluoroboric acid) carbon monoxide is evolved and a mixed nitrile-isonitrile complex is formed[116] (reaction 428).

$$[(PPh_3)_2Pt(CNMe)(\overset{\displaystyle O}{\overset{\|}{C}}{-}NHMe)]^+BF_4^- + HBF_4 \xrightarrow{CH_3CN/H_2O}$$

$$[(PPh_3)_2Pt(CNMe)(NCMe)]^{2+}(BF_4^-)_2 + {\uparrow}CO + MeNH_3^+BF_4^- \quad (428)$$

When [(pyr)$_2$Pt(C$_3$H$_6$)Cl$_2$] (see above p. 347) is refluxed in a mixture of chloroform and carbon tetrachloride, the product may be formulated either as a carbene complex (146) or an ylide (147).[112] The n.m.r. evidence favours the ylide structure.[87]

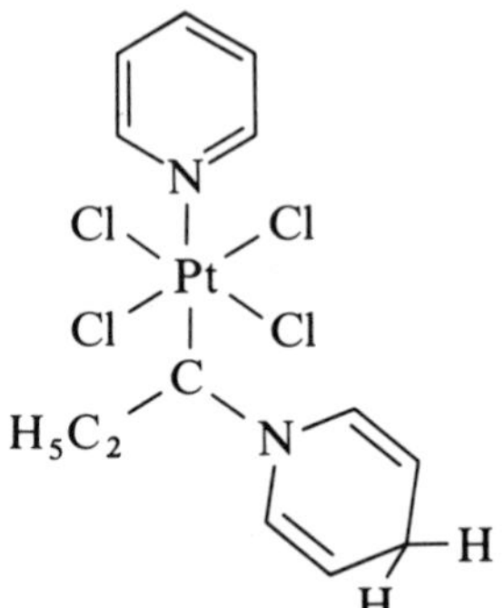

146 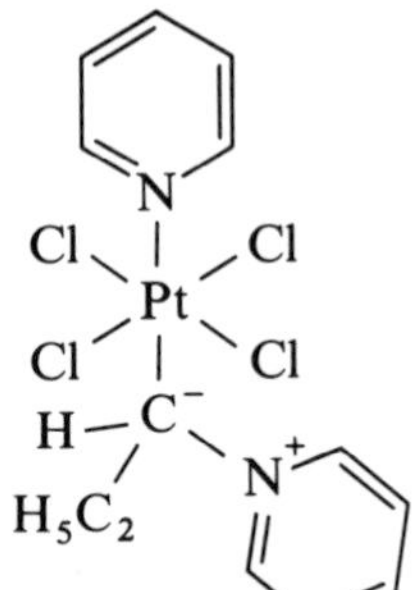   147

COMPLEXES CONTAINING METAL(IV)–CARBON σ-BONDS

Except for $[Pt_2(CH_3)_6]$ whose structure is unknown all the complexes with metal(IV)–carbon σ-bonds have octahedral co-ordination around the metal. Although only one such complex is known for palladium,[6] namely $[(PPh_3)_2Pd(C_6F_5)_2Cl_2]$, a number are known for platinum. The platinum complexes may be divided into three groups, those with the $(CH_3)_3Pt$ unit those with the $(CH_3)_2Pt$ unit and those with phosphine or arsine ligands.

Complexes containing the $(CH_3)_3Pt$ unit

The parent complex in this series $[(CH_3)_3PtI]_4$ is prepared by treating $PtCl_4^{(120,121)}$ or $K_2PtCl_6^{(122)}$ with CH_3MgI in benzene, or alternatively, *cis*-$[(pyr)_2PtCl_4]$ with CH_3MgI; the latter reaction gives initially $[(pyr)Pt-(CH_3)_3I]_2$ which by reaction with ethylenediamine followed by hydyriodic acid gives[123] $[(CH_3)_3PtI]_4$. Use of CH_3MgCl yields[124] $[(CH_3)_3PtCl]_4$. A further method that has been used to prepare $[R_3PtCl]_4$ where $R = CH_3$ and C_2H_5 involves treating $PtCl_4$ with dialkyl mercury.[125] Most $(CH_3)_3Pt^{IV}$ complexes are prepared from the iodide and β-diketone complexes as summarised in reactions 429–41, although the sulphydryl complex is prepared from the aquo-complex (reaction 442). It is noteworthy that despite a number of claims, no authentic samples of $[(CH_3)_4Pt]$ have been reported.

$$\xrightarrow[\text{ref. 126}]{\text{K in } C_6H_6} [(CH_3)_6Pt_2]_n \quad (n \geqslant 6) \tag{429}$$

$$\xrightarrow[\text{ref. 127, 128}]{\text{NaC}_5H_5} [(C_5H_5)Pt(CH_3)_3] \tag{430}$$

$$\xrightarrow[\text{ref. 129}]{\text{bipyr}} [(bipyr)Pt(CH_3)_3I] \tag{431}$$

$$\xrightarrow[\text{ref. 130}]{\text{en}} [(en)Pt(CH_3)_3I] \tag{432}$$

$$\xrightarrow[\text{ref. 131}]{\text{NH}_3} [(NH_3)_2Pt(CH_3)_3I] \tag{433}$$

$$\xrightarrow[\text{ref. 132}]{\text{pyr}} [(pyr)Pt(CH_3)_3I]_2 \xrightarrow[\text{pyr}]{\text{excess}} [(pyr)_2Pt(CH_3)_3I] \tag{434}$$

$[(CH_3)_3PtI]_4$

$$\xrightarrow[\text{ref. 133}]{\text{NaCH}_3} [(CH_3)_3Pt(OH)] \text{ (not } [(CH_3)_4Pt] \text{ as originally suggested}^{(132,133)}) \tag{435}$$

$$\xrightarrow[\text{ref. 134}]{\begin{array}{c}\text{8-quinolato-}\\\text{thallium(I)}\end{array}} \tag{436}$$

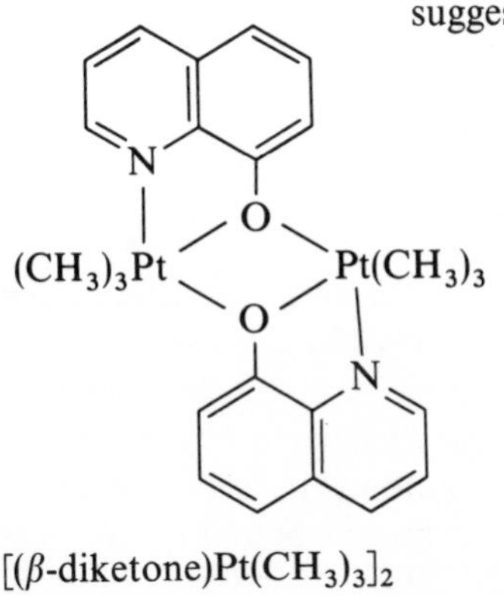

$$\xrightarrow[\text{refs. 135-7}]{\beta\text{-diketone}} [(\beta\text{-diketone})Pt(CH_3)_3]_2 \tag{437}$$

$$\xrightarrow[\text{in HX, ref. 138}]{\text{digest acac salt}} [(CH_3)_3Pt(H_2O)_3]^+X^- \quad (X^- = NO_3^-,$$

$$SO_4^{2-}, ClO_4^-, HCOO^-, CH_3COO^-, PO_4^{3-}) \quad (438)$$

$$\xrightarrow[\text{ref. 139}]{\text{L in boiling } C_6H_6} [(\beta\text{-diketone})Pt(CH_3)_3L]$$

$(L = NH_3, \text{pyr}, C_6H_5NH_2 \text{ or quinoline})$

(diketone is bonded to platinum via two oxygen atoms)   (439)

$[(\beta\text{-diketone})Pt(CH_3)_3]_2$

$$\xrightarrow[\text{ref. 139}]{\text{NN in } C_6H_6} [(\beta\text{-diketone})Pt(CH_3)_3(NN)] \quad (440)$$

(NN = 2,2'-bipyridyl or 1,10-phenanthroline) (diketone is bonded to platinum via γ-carbon atom)

$$\xrightarrow[\text{ref. 139}]{0.5 \text{ mol en}} [(\beta\text{-diketone})Pt(CH_3)_3]_2\text{en} \quad (441)$$

(diketone is bonded to platinum via two oxygen atoms)

$$2[(CH_3)_3Pt(H_2O)_3]_2(SO_4) + H_2S \xrightarrow[\text{ref. 140}]{H_2O} [(CH_3)_3PtSH]_4 . \quad (442)$$

X-ray investigations of these complexes have shown that the platinum is always octahedrally co-ordinated. The complexes $[(CH_3)_3PtX]$ where X = iodide,[141,142] chloride[124] and hydroxide[143,144] and $[(C_2H_5)_3PtCl]_4$[145] are all tetrameric with the platinum atoms located at opposite corners of a cube **148** (see Appendix II structures 162, 161, 163 and 189 respectively). The investigated β-diketone complexes, all have structure **149** in which

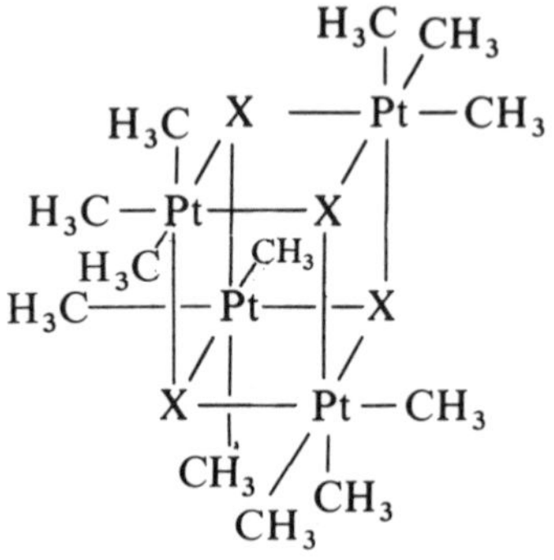

148 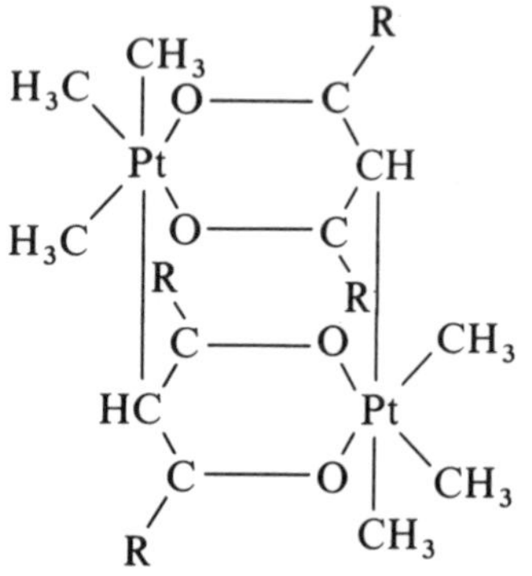   **149**

the platinum atom is bonded to the γ-carbon atom of the β-diketone[146,147] (see Appendix II, structures 176 and 188). A neutron diffraction study of the nonane-4,6-dionato complex (**149** where R = *n*-propyl) showed that the hydrogen bound to the γ-carbon atom[148] occupies a site that is sterically suitable for much larger atoms. This is surprising since attempts to synthesise complexes with bulkier groups bound to the γ-carbon atom have all failed. Two complexes $[LPt(CH_3)_3\text{-en-}Pt(CH_3)_3L]$ with ethylenediamine bridges have been investigated in which L = bidentate oxygen bonded acetylac-tone[149] and ethylenediamine[150] (see Appendix II, structures 177 and 165 respectively). Two complexes $[(CH_3)_3PtL]_2$ in which L = salicylalde-hyde[151,152] and 8-quinolinol[151,153] have been found to involve

$$Pt \diagdown \begin{matrix} O \\ \\ \\ O \end{matrix} \diagup Pt$$

Pt bridges (see Appendix II structures 179 and 185 respectively).

The only monomeric complex so far investigated, $[(CH_3)_3Pt(acac)(bipyr)]$, is also octahedral with the acetylacetone ligand bound to platinum through the γ-carbon atom[154] (see Appendix II, structure 175).

Dissolving the dimeric acetylacetone complex of trimethylplatinum(IV) in dilute mineral acid with a trace of hydrogen peroxide produces[138] the aquo ion $[(CH_3)_3Pt(H_2O)_3]^+$. This ion has C_{3V} symmetry,[155] which is consistent with a 'fac' octahedral structure. Incidentally the aquo ligands exchange very rapidly with solvent water molecules,[156] and polarography shows four well-separated waves at such negative potentials as to indicate that it has considerable stability towards electrochemical reduction.[157]

Trimethylplatinum(IV) complexes show a well resolved spin–spin splitting in their n.m.r. spectra due to coupling between the ^{195}Pt isotope and the methyl protons.[158] The coupling constant, which varies between 60 and 82 Hz, is very sensitive to the nature of the *trans*-ligand and provides a useful indication of the stereochemistry of the complex[138] (Table 103).

TABLE 103

$J_{195Pt-^1H(CH_3)}$ for $(CH_3)_3Pt$ complexes

Complex	Ligand atoms trans to the three methyl groups	J_{195Pt-^1H} (Hz)	Reference
$[(CH_3)_3PtCl]_4$	$3 \times Cl$	81·7	138
$[(CH_3)_3PtBr]_4$	$3 \times Br$	80·1	138
$[(CH_3)_3Pt(H_2O)_3]^+$	$3 \times O$	79·7	159
$[(CH_3)_3PtI]_4$	$3 \times I$	77·5	160
$[(CH_3)_3Pt(acac)]_2$	$2 \times O + 1 \times C$	75·7	138
$[\{(CH_3)_3Pt(acac)\}_2en]$	$2 \times O + 1 \times N$	75·0	138
$[(CH_3)_3Pt(SCN)_3]^{2-}$	$3 \times S$	73·4	159
$[(CH_3)_3Pt(oxime)pyr]$	$1 \times 0 + 2 \times N$	71·4	138
$[(CH_3)_3Pt(NH_3)_3]^+$	$3 \times N$	71·0	159
$[(CH_3)_3Pt(MeNH_2)_3]^+$	$3 \times N$	68·4	159
$[(CH_3)_3Pt(pyr)_3]^+$	$3 \times N$	67·5	159
$[(CH_3)_3Pt(CN)_3]^{2-}$	$3 \times C$	60·8	159

The temperature dependence of the n.m.r. spectra of the β-diketone com-plexes $[(CH_3)_3Pt(\beta\text{-diketone})]_2$ in deuteriochloroform suggests that above room temperature there is a rapid dissociation of the platinum–γ-carbon bond[161] (see **149**). Two independent detailed analyses of the infra-red and Raman spectra of $[(CH_3)_3Pt(OH)]_4$ have suggested that the band at 724 cm^{-1} previously assigned[162] to the platinum–oxygen stretching frequency is more likely to be due to an O–H wag.[163,164] This assignment suggests that the hydroxyl group is somewhat ionic in nature. A similar conclusion has been reached for the halide ligands from the infra-red and Raman spectra of $[(CH_3)_3PtCl]_4$ and $[(CH_3)_3PtI]_4$.[165] The platinum–oxygen stretching modes in $[(CH_3)_3Pt(OH)]_4$ have been assigned to bands at 369 and 396 cm^{-1} in the infra-red and 340, 369, 396 and 430 cm^{-1} in the Raman spectrum.[163]

Complexes containing the $(CH_3)_2Pt$ unit

Although halogens do not react with $[Me_3PtI]_4$ in the cold,[121,166] $[Me_3PtI]_4$ reacts with a refluxing solution of bromine in hydrobromic acid to yield $[Me_2PtBr_2]_n$.[167] The corresponding bromide and iodide can be prepared[167] following the schemes in reactions 443 and 444. The products are insoluble in non-co-ordinating solvents and water. However, the chloride

$$K_2PtCl_6 + 3MeMgI \longrightarrow [Me_3PtI] + [Me_2PtI_2] \xrightarrow[\text{CHCl}_3]{\text{en in}}$$

$$\downarrow [\{Me_3Pt(en)\}_2en]I_2 + [Me_2PtI_2(en)]$$

$$\downarrow \begin{array}{c} +\text{HI or} \\ \text{HClO}_4 \end{array}$$

$$[Me_2PtI_2]_n \qquad (443)$$

$$[Me_2PtBr_2]_n + Cl_2 + 2\,pyr \xrightarrow{\text{CHCl}_3} (pyrH^+)_2[PtCl_4Me_2]^{2-} \xrightarrow{\text{H}_2\text{O}}$$

$$[Me_2PtCl_2(pyr)_2]$$

$$\downarrow +\text{conc. HClO}_4$$

$$[Me_2PtCl_2]_n \qquad (444)$$

and bromide, but not the iodide, dissolve in chloride or bromide solutions possibly forming $[Me_2PtX_4]^{2-}$. Although the structure is uncertain the infra-red spectra are consistent with the tetrameric structure **150**.[167]

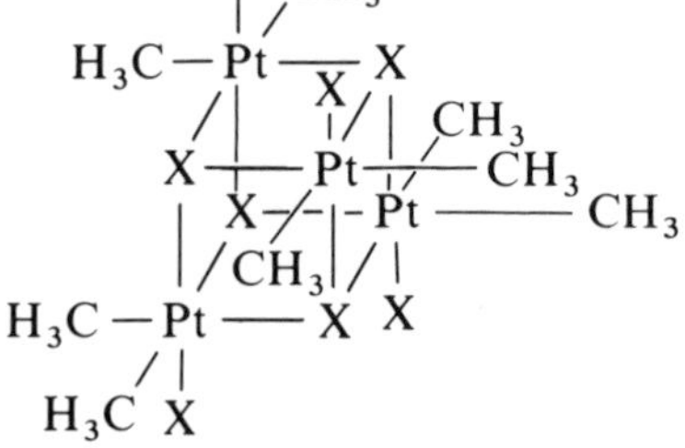

150

Complexes containing platinum(IV)–carbon σ-bonds stabilised by phosphines and arsines

Preparation

Platinum(IV)–alkyl complexes containing phosphine or arsine ligands can be prepared from the corresponding platinum(II) complexes by halogen oxidation, by the addition of alkyl halide, or by treating the platinum(IV) halo-complexes with methyl lithium.

(i) *Halogen oxidation.* Treatment of a platinum(II)–alkyl complex with 1 mol of chlorine or bromine yields the platinum(IV) complex[9,20,58,59] (reaction 445). Although *cis*-[(PEt$_3$)$_2$Pt(CH$_3$)$_2$] reacts with iodine to give a

$$cis\text{-}[(PEt_3')_2Pt(CH_3)_2] + Cl_2 \rightarrow [(PEt_3)_2Pt(CH_3)_2Cl_2] \qquad (445)$$

platinum(II)–iodo complex (reaction 446), both the *cis*- and *trans*-diphenyl-

$$cis\text{-}[(PEt_3)_2Pt(CH_3)_2] + I_2 \rightarrow trans\text{-}[(PEt_3)_2\underset{\downarrow I_2}{Pt}(CH_3)I] + CH_3I$$

$$trans\text{-}[(PEt_3)_2PtI_2] + CH_3I \qquad (446)$$

platinum(II) complexes react with iodine to give a platinum(IV)–diphenyl complex.[20] The mono-methylplatinum(II) complex stabilised by dimethyl-phenylarsine is similarly oxidised by iodine to the platinum(IV) complex[58] [(AsMe$_2$Ph)$_2$Pt(CH$_3$)I$_3$], although the corresponding phosphine complex [(PPh$_3$)$_2$Pt(CH$_3$)I$_3$], prepared by treating [(PPh$_3$)$_2$Pt(o-totyl)$_2$] with methyl iodide, proved to be a platinum(II) complex with a triodo ligand.[81]

(ii) *Addition of alkyl halide.* Methyl bromide and methyl iodide add on to platinum(II) complexes to give platinum(IV) complexes[9,20,58,168] (reaction 447).

$$trans\text{-}[(PEt_3)_2Pt(CH_3)I] + CH_3I \rightarrow [(PEt_3)_2Pt(CH_3)_2I_2] \qquad (447)$$

(iii) *Methyl lithium.* Methyl lithium reacts with platinum(IV) halide complexes to give tetramethyl platinum(IV) complexes[58,59] (reaction 448).

$$[(PMe_2Ph)_2Pt(CH_3)_nCl_{4-n}] + LiCH_3 \xrightarrow[(n=0,1)]{} [(PMe_2Ph)_2Pt(CH_3)_4]$$

$$(448)$$

Properties

The complexes are all stable crystalline compounds with fairly high melting-points (above 100°C). When heated in a vacuum above their melting-points they decompose cleanly into platinum(II) complexes together with either methyl halide or ethane.[169] The monomethyl complexes and dimethyl complexes of configuration **152** (see below) both give methyl halide whereas the dimethyl complexes of configuration **153**, the trimethyl and tetramethyl derivatives all give ethane. The dimethyl complex of configuration **154** gives 60% of methyl halide and 40% of ethane, possibly owing to rapid isomerisation before the elimination.

The ^{195}Pt–^{1}H(CH$_3$) n.m.r. coupling constant[58,59] is about 77% lower in these platinum(IV) complexes than in the corresponding platinum(II) complexes, and is very dependent on the *trans*-ligand; for example in platinum(IV)–arsine complexes $J_{^{195}Pt-^1H(CH_3)}$ has values 69–72 Hz (*trans*-halogen), 60–66 Hz (*trans*-arsenic) and 44 Hz (*trans*-methyl); in platinum(IV)–phosphine complexes the values are 67–73 Hz (*trans*-halogen), 56–59 Hz (*trans*-phosphorus) and 43·5–44 Hz (*trans*-methyl).

The infra-red spectra exhibit platinum–carbon stretching frequencies in the range 515–540 cm^{-1} for methyl groups *trans* to halogen or a group VB donor atom.[58,59] For methyl groups *trans* to methyl, which only occurs in the tetra-methyl complexes, the stretching frequency is substantially lower (*ca.* 470 cm^{-1}), which is consistent with the strong *trans*-influence of methyl groups that also gives rise to very low metal–chlorine stretching frequencies.[170,171]

The stereochemistries of the complexes have been investigated by dipole moment measurements and n.m.r.[58,59] The monomethyl complexes have structure **151** where L = PMe$_2$Ph or AsMe$_2$Ph and X = Cl or Br. The

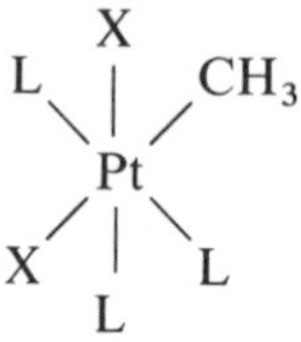

$$\begin{array}{c} X \\ L \diagdown \mid \diagup CH_3 \\ Pt \\ \diagup \mid \diagdown \\ X \quad \mid \quad L \\ L \end{array}$$

151

structures of the dimethyl derivative depend on the mode of preparation (reactions 449–53). Although the trimethyl derivatives could have three

trans-[L$_2$Pt(CH$_3$)X] + CH$_3$X → **152**

$$(L = PMe_2Ph, AsMe_2Ph; X = Br, I) \tag{449}$$

$$cis\text{-}[(PMe_2Ph)_2Pt(CH_3)_2] + Cl_2 \rightarrow \mathbf{152} + \mathbf{153} \tag{450}$$

$$cis\text{-}[(PMe_2Ph)_2Pt(CH_3)_2] + Br_2 \rightarrow \mathbf{153} \tag{451}$$

$$cis\text{-}[(AsMe_2Ph)_2Pt(CH_3)_2] \xrightarrow{\text{Cl}_2 \text{ in C}_6\text{H}_6 <10°\text{C}} \mathbf{153} \tag{452}$$

$$\xrightarrow[\text{C}_6\text{H}_6 \text{ at 20°C}]{\text{Cl}_2 \text{ or Br}_2 \text{ in}} \mathbf{154} \tag{453}$$

$$\begin{array}{ccc}
\begin{array}{c} CH_3 \\ L \diagdown \mid \diagup CH_3 \\ Pt \\ \diagup \mid \diagdown \\ X \quad \mid \quad L \\ X \end{array}
&
\begin{array}{c} CH_3 \\ X \diagdown \mid \diagup CH_3 \\ Pt \\ \diagup \mid \diagdown \\ L \quad \mid \quad X \\ L \end{array}
&
\begin{array}{c} CH_3 \\ CH_3 \diagdown \mid \diagup L \\ Pt \\ \diagup \mid \diagdown \\ X \quad \mid \quad L \\ X \end{array}
\\
\mathbf{152} & \mathbf{153} & \mathbf{154}
\end{array}$$

possible configurations only one (**155**) is formed either by addition of CH$_3$X to *cis*-[L$_2$Pt(CH$_3$)$_2$] (L = PMe$_2$Ph or AsMe$_2$Ph; X = Cl, Br or I) or by treating [(CH$_3$)$_3$PtI]$_4$ with a slight excess of phosphine.[58,59] The tetra-methyl derivatives always have the *cis* configuration **156**.

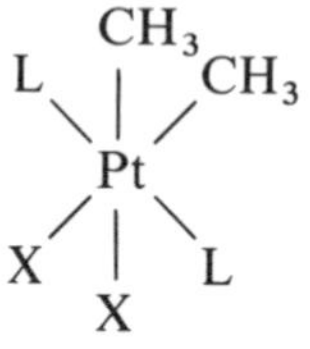

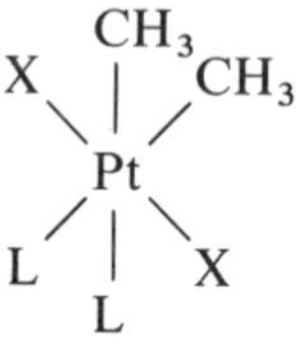

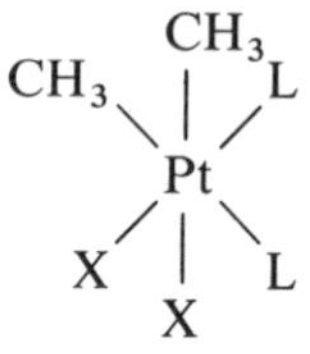

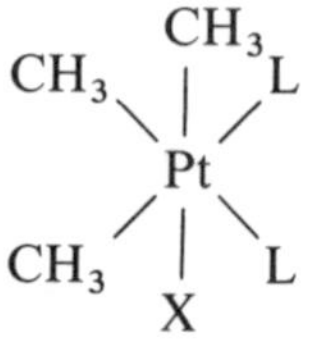

$$\begin{array}{cc}
\begin{array}{c} CH_3 \\ CH_3 \diagdown \mid \diagup L \\ Pt \\ \diagup \mid \diagdown \\ CH_3 \quad \mid \quad L \\ X \end{array}
&
\begin{array}{c} CH_3 \\ CH_3 \diagdown \mid \diagup L \\ Pt \\ \diagup \mid \diagdown \\ CH_3 \quad \mid \quad L \\ CH_3 \end{array}
\\
\mathbf{155} & \mathbf{156}
\end{array}$$

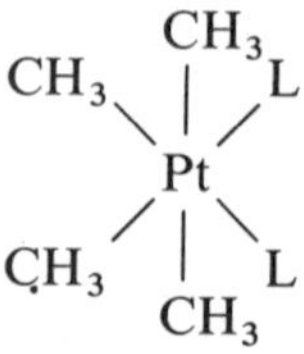

REFERENCES

1 R. J. Cross, *Organometal. Chem. Rev.*, **2** (1967) 97.
2 G. W. Parshall and J. J. Mrowca, *Adv. Organometal Chem.*, **7** (1968) 157.
3 M. L. H. Green, *Organometallic Compounds*, by G. E. Coates, M. L. H. Green and K. Wade, Methuen and Co., Ltd., London, 3rd edn., 1968, vol. 2, chapter 7.
4 M. I. Bruce and F. G. A. Stone, *Preparative Inorganic Reactions*, **4** (1968) 177.
5 R. S. Nyholm, *Quart. Rev.*, **24** (1970) 1.
6 R. S. Nyholm and P. Royo, *Chem. Comm.*, (1969) 421.
7 W. J. Pope and S. J. Peachey, *Proc. Chem. Soc.*, **23** (1907) 86.
8 J. Chatt, L. M. Vallarino and L. M. Venanzi, *J. Chem. Soc.*, (1957) 2496.
9 J. Chatt and B. L. Shaw, *J. Chem. Soc.*, (1959) 705.
10 B. N. Figgis, J. Lewis, R. F. Long, R. Mason, R. S. Nyholm, P. J. Pauling and G. B. Robertson, *Nature*, **195** (1962) 1278.
11 R. A. D. Wentworth and C. H. Brubaker, *Inorg. Chem.*, **3** (1964) 1472.
12 J. Lewis, R. F. Long and C. Oldham, *J. Chem. Soc.*, (1965) 6740.
13 C. R. Kistner, J. H. Hutchinson, J. R. Doyle and J. C. Storlie, *Inorg. Chem.*, **2** (1963) 1255.
14 G. Calvin and G. E. Coates, *J. Chem. Soc.*, (1960) 2008.
15 S. Sergi, V. Marsala, R. Pietropaolo and F. Faraone, *J. Organometal. Chem.*, **23** (1970) 281.
16 J. R. Doyle, J. H. Hutchinson, N. C. Baenziger and L. W. Tresselt, *J. Amer. Chem. Soc.*, **83** (1961) 2768.
17 A. D. Westland and M. Northcott, *Can. J. Chem.*, **48** (1970) 2907.
18 C. R. Kistner, J. D. Blackman and W. C. Harris, *Inorg. Chem.*, **8** (1969) 2165.
19 J. Chatt and J. M. Davidson, *J. Chem. Soc.*, (1964) 2433.
20 J. Chatt and B. L. Shaw, *J. Chem. Soc.*, (1959) 4020.
21 I. Collamati and A. Furlani, *J. Organometal. Chem.*, **17** (1969) 457.
22 J. Chatt and B. L. Shaw, *J. Chem. Soc.*, (1962) 5075.
23 F. R. Hartley and L. M. Venanzi, unpublished results.
24 R. J. Cross and F. Glockling, *J. Organometal. Chem.*, **3** (1965) 253.
25 H. C. Clark and J. D. Ruddick, *Inorg. Chem.*, **9** (1970) 1226.
26 K. A. Hooton, *J. Chem. Soc. (A)*, (1970) 1896.
27 C. D. Cook and G. S. Jauhal, *Can. J. Chem.*, **45** (1967) 301.
28 J. P. Collman, J. N. Cawse and J. W. Kang, *Inorg. Chem.*, **8** (1969) 2574.
29 J. P. Birk, J. Halpern and A. L. Pickard, *J. Amer. Chem. Soc.*, **90** (1968) 4491.
30 F. R. Hartley, *Chem. Rev.*, **69** (1969) 799.
31 J. Chatt, R. S. Coffey, A. Gough and D. T. Thompson, *J. Chem. Soc. (A)*, (1968) 190.
32 D. Seyferth and R. J. Cross, reported in reference 1.
33 K. Matsumoto, Y. Odaira and S. Tsutsumi, *Chem. Comm.*, (1968) 832.
34 T. Yukawa and S. Tsutsumi, *Inorg. Chem.*, **7** (1968) 1458.
35 J. Ashley-Smith, J. Clemens, M. Green and F. G. A. Stone, *J. Organometal. Chem.*, **17** (1969) P23.
36 G. W. Parshall, *J. Amer. Chem. Soc.*, **87** (1965) 2133.
37 G. Booth and J. Chatt, *J. Chem. Soc. (A)*, (1966) 634.
38 M. C. Baird, *J. Inorg. Nucl. Chem.*, **29** (1967) 367.
39 J. Chatt and G. A. Rowe, *Proc. First Int. Conf. on Coord. Chem.*, London, 1959, *Chem. Soc. Special Publ.*, no. 13, p. 117.
40 J. Chatt, G. A. Rowe and A. A. Williams, *Proc. Chem. Soc.*, (1957) 208.
41 A. Furlani, P. Bicev, M. V. Russo and P. Carusi, *J. Organometal. Chem.*, **29** (1971) 321.
42 F. Basolo, J. Chatt, H. B. Gray, R. G. Pearson and B. L. Shaw, *J. Chem. Soc.*, (1961) 2207.
43 J. D. Duncan, J. C. Green, M. L. H. Green and K. A. McLauchlan, *Disc. Faraday Soc.*, **47** (1969) 178.
44 S. J. Ashcroft and C. T. Mortimer, *J. Chem. Soc. (A)*, (1967) 930.
45 L. Pauling, *The Nature of The Chemical Bond*, Cornell University Press, 3rd edn., 1960.
46 G. Faraone, V. Ricevuto, R. Romeo and M. Trozzi, *Inorg. Chem.*, **8** (1969) 2207.
47 J. Chatt, G. A. Gamlen and L. E. Orgel, *J. Chem. Soc.*, (1958) 486.
48 A. Pidcock, R. E. Richards and L. M. Venanzi, *J. Chem. Soc. (A)*, (1968) 1970.

49   C. M. Harris, S. E. Livingstone and I. H. Reece, *J. Chem. Soc.*, (1959) 1505.
50   J. Chatt and B. L. Shaw, *J. Chem. Soc.*, (1960) 1718.
51   J. W. Moore, *Acta Chem. Scand.*, **20** (1966) 1154.
52   G. Wilkinson, Plenary lecture, *5th Int. Conf. on Organometallic Chem.*, Moscow, 1971 (for a brief report see reference 53).
53   F. R. Hartley, *Platinum Metals Review*, **16** (1972) 22.
54   D. Gibson, J. Lewis and C. Oldham, *J. Chem. Soc. (A)*, (1966) 1453.
55   C. K. Jørgensen, *Absorption Spectra and Chemical Bonding in Complexes*, Pergamon, Oxford, 1962, p. 109.
56   M. L. Maddox, S. L. Stafford and H. D. Kaesz, *Adv. Organometal. Chem.*, **3** (1965) 1.
57   F. H. Allen and A. Pidcock, *J. Chem. Soc. (A)*, (1968) 2700.
58   J. D. Ruddick and B. L. Shaw, *J. Chem. Soc. (A)*, (1969) 2964.
59   J. D. Ruddick and B. L. Shaw, *J. Chem. Soc. (A)*, (1969) 2801.
60   A. J. Cheney, B. E. Mann and B. L. Shaw, *Chem. Comm.*, (1971) 431.
61   G. W. Parshall, *J. Amer. Chem. Soc.*, **86** (1964) 5367.
62   G. W. Parshall, *J. Amer. Chem. Soc.*, **88** (1966) 704.
63   M. J. Church and M. J. Mays, *J. Chem. Soc. (A)*, (1968) 3074.
64   J. M. Jenkins and B. L. Shaw, *Proc. Chem. Soc.*, (1963) 279.
65   D. M. Adams, J. Chatt and B. L. Shaw, *J. Chem. Soc.*, (1960) 2047.
66   H. Masai, K. Sonogashira and N. Hagihara, *J. Organometal. Chem.*, **26** (1971) 271.
67   F. Basolo and R. G. Pearson, *Adv. Inorg. and Radiochem.*, **3** (1961) 1.
68   H. Dunken and G. Marx, *Z. Chem.*, **6** (1966) 436; *Chem. Abs.*, **66** (1967) 60426.
69   J. Chatt and R. G. Hayter, *J. Chem. Soc.*, (1961) 772.
70   W. A. Whitla, H. M. Powell and L. M. Venanzi, *Chem. Comm.*, (1966) 310.
71   R. Mason and G. B. Robertson, *J. Chem. Soc. (A)*, (1969) 492.
72   R. W. Siekman and D. L. Weaver, *Chem. Comm.*, (1968) 1021.
73   D. L. Weaver, *Inorg. Chem.*, **9** (1970) 2250.
74   W. Theilacker, *Z. anorg. allgem. Chem.*, **234** (1937) 161.
75   O. Hassel and B. F. Pedersen, *Proc. Chem. Soc.*, (1959) 394.
76   U. Belluco, U. Croatto, P. Uguagliati and R. Pietropaolo, *Inorg. Chem.*, **6** (1967) 718.
77   U. Belluco, M. Giustiniani and M. Graziani, *Abstr. Autumn Meeting Chem. Soc.*, Durham, 1967.
78   U. Belluco, M. Giustiniani and M. Graziani, *J. Amer. Chem. Soc.*, **89** (1967) 6494.
79   A. C. Cope and R. W. Siekman, *J. Amer. Chem. Soc.*, **87** (1965) 3272.
80   A. Kasahara, *Bull. Chem. Soc. Japan*, **41** (1968) 1272.
81   C. R. Kistner, D. A. Drew, J. R. Doyle and G. W. Rausch, *Inorg. Chem.*, **6** (1967) 2036.
82   G. Booth and J. Chatt, *Proc. Chem. Soc.*, (1961) 67.
83   R. W. Glyde and R. J. Mawby, *Inorg. Chem.*, **10** (1971) 854.
84   H. C. Clark and R. J. Puddephat, *Inorg. Chem.*, **9** (1970) 2670.
85   H. C. Clark and R. J. Puddephat, *Inorg. Chem.*, **10** (1971) 18.
86   F. R. Hartley, *Angew. Chem., Int. Ed.*, **11** (1972) 596.
87   H. C. Clark and R. J. Puddephat, *Inorg. Chem.*, **10** (1971) 416.
88   E. Forsellini, G. Bombieri, B. Crociani and T. Boschi, *Chem. Comm.*, (1970) 1203.
89   Y. Yamamoto and H. Yamazaki, *Bull. Chem. Soc. Japan*, **43** (1970) 2653.
90   Y. Yamamoto and H. Yamazaki, *Bull. Chem. Soc. Japan*, **43** (1970) 3634.
91   P. M. Maitlis and F. G. A. Stone, *Chem. Ind. (London)*, (1962) 1865.
92   H. C. Clark and W. S. Tsang, *J. Amer. Chem. Soc.*, **89** (1967) 529.
93   J. Cooke, W. R. Cullen, M. Green and F. G. A. Stone, *Chem. Comm.*, (1968) 170.
94   W. J. Cherwinski and H. C. Clark, *J. Organometal. Chem.*, **29** (1971) 451.
95   H. C. Clark, P. W. R. Corfield, K. R. Dixon and J. A. Ibers, *J. Amer. Chem. Soc.*, **89** (1967) 3360.
96   H. C. Clark, K. R. Dixon and W. J. Jacobs, *J. Amer. Chem. Soc.*, **90** (1968) 2259.
97   H. C. Clark and J. D. Ruddick, *Inorg. Chem.*, **9** (1970) 2556.
98   H. C. Clark and R. J. Puddephat, *Chem. Comm.*, (1970) 92.
99   A. J. Rest, D. T. Rosevear and F. G. A. Stone, *J. Chem. Soc. (A)*, (1967) 66.

100 M. I. Bruce, D. A. Harbourne, F. Waugh and F. G. A. Stone, *J. Chem. Soc. (A)*, (1968) 356.

101 F. J. Hopton, A. J. Rest, D. T. Rosevear and F. G. A. Stone, *J. Chem. Soc. (A)*, (1966) 1326.

102 D. T. Rosevear and F. G. A. Stone, *J. Chem. Soc.*, (1965) 5275.

103 M. D. Rausch and F. E. Tibbetts, *J. Organometal. Chem.*, **21** (1970) 487.

104 D. T. Rosevear and F. G. A. Stone, *J. Chem. Soc. (A)*, (1968) 164.

105 M. Green, R. B. L. Osborn, A. J. Rest and F. G. A. Stone, *J. Chem. Soc. (A)*, (1968) 2525.

106 M. A. Chaudhari, P. M. Treichel and F. G. A. Stone, *J. Organometal. Chem.*, **2** (1964) 206.

107 T. A. Manuel, S. L. Stafford and F. G. A. Stone, *J. Amer. Chem. Soc.*, **83** (1961) 249.

108 M. R. Churchill, *Chem. Comm.*, (1965) 86.

109 H. C. Clark and J. H. Tsai, *J. Organometal. Chem.*, **7** (1967) 515 and references therein.

110 C. F. H. Tipper, *J. Chem. Soc.*, (1955) 2045.

111 W. J. Irwin and F. J. McQuillin, *Tetra. Letters*, (1968) 1937.

112 N. A. Bailey, R. D. Gillard, M. Keeton, R. Mason and D. R. Russell, *Chem. Comm.*, (1966) 396.

113 D. M. Adams, J. Chatt, R. G. Guy and N. Sheppard, *J. Chem. Soc.*, (1961) 738.

114 S. E. Binns, R. H. Cragg, R. D. Gillard, B. T. Heaton and M. F. Pilbrow, *J. Chem. Soc. (A)*, (1969) 1227.

115 E. M. Badley, J. Chatt, R. L. Richards and G. A. Sim, *Chem. Comm.*, (1969) 1322.

116 W. J. Knebel and P. M. Treichel, *Chem. Comm.*, (1971) 516.

117 D. J. Cardin, B. Cetinkaya, M. F. Lappert, Lj. Manojlovic-Muir and K. W. Muir, *Chem. Comm.*, (1971) 400.

118 M. H. Chisholm and H. C. Clark, *Chem. Comm.*, (1970) 763.

119 E. M. Badley, B. J. L. Kilby and R. L. Richards, *J. Organometal. Chem.*, **27** (1971) C37.

120 W. J. Pope and S. J. Peachey, *Proc. Chem. Soc.*, **23** (1907) 86.

121 W. J. Pope and S. J. Peachey, *J. Chem. Soc.*, (1909) 571.

122 D. E. Clegg and J. R. Hall, *Inorg. Synth.*, **10** (1967) 71.

123 M. E. Foss and C. S. Gibson, *J. Chem. Soc.*, (1951) 299.

124 R. E. Rundle and J. H. Sturdivant, *J. Amer. Chem. Soc.*, **69** (1947) 1561.

125 S. F. A. Kettle, *J. Chem. Soc.*, (1965) 5737.

126 G. Illuminati and R. E. Rundle, *J. Amer. Chem. Soc.*, **71** (1949) 3575.

127 S. D. Robinson and B. L. Shaw, *Z. Naturforsch.*, **B18** (1963) 507; *Chem. Abs.*, **59** (1963) 8787e.

128 H. P. Fritz and K. E. Schwarzhans, *Angew. Chem. Int. Ed.*, **4** (1965) 700.

129 W. J. Lile and R. C. Menzies, *J. Chem. Soc.*, (1949) 1168.

130 A. D. Gel'man and E. A. Goruskina, *Dokl. Akad. Nauk. SSSR*, **57** (1947) 249.

131 D. O. Cowan, N. G. Krieghoff and G. Donnay, *Acta Cryst.*, **B24** (1968) 287.

132 H. Gilman and M. Lichtenwalter, *J. Amer. Chem. Soc.*, **60** (1938) 3085.

133 H. Gilman, M. Lichtenwalter and R. A. Benkeser, *J. Amer. Chem. Soc.*, **75** (1953) 2063.

134 K. Kite and M. R. Truter, *J. Chem. Soc. (A)*, (1966) 207.

135 R. C. Menzies and E. R. Wiltshire, *J. Chem. Soc.*, (1933) 21.

136 R. C. Menzies and H. Overton, *J. Chem. Soc.*, (1933) 1290.

137 A. K. Chatterjee, R. C. Menzies, J. R. Steel and F. N. Youdale, *J. Chem. Soc.*, (1958) 1706.

138 K. Kite, J. A. S. Smith and E. J. Wilkins, *J. Chem. Soc. (A)*, (1966) 1744.

139 K. Kite and M. R. Truter, *J. Chem. Soc. (A)*, (1968) 934.

140 R. Graves, J. M. Homan and G. L. Morgan, *Inorg. Chem.*, **9** (1970) 1592.

141 G. Donnay, L. B. Coleman, N. G. Krieghoff and D. O. Cowan, *Acta Cryst.*, **B24** (1968) 157.

142 G. R. Hoff and C. H. Brubaker, *Inorg. Chem.*, **7** (1968) 1655.

143 T. G. Spiro, D. H. Templeton and A. Zalkin, *Inorg. Chem.*, **7** (1968) 2165.

144 H. S. Preston, J. C. Mills and C. H. L. Kennard, *J. Organometal. Chem.*, **14** (1968) 447.

145 R. N. Hargreaves and M. R. Truter, *J. Chem. Soc. (A)*, (1971) 90.

146 A. G. Swallow and M. R. Truter, *Proc. Roy. Soc.*, **A254** (1960) 205.

147 A. C. Hazell and M. R. Truter, *Proc. Roy. Soc.*, **A254** (1960) 218.

148 R. N. Hargreaves and M. R. Truter, *J. Chem. Soc. (A)*, (1969) 2282.
149 A. Robson and M. R. Truter, *J. Chem. Soc.*, (1965) 630.
150 M. R. Truter and E. G. Cox, *J. Chem. Soc.*, (1956) 948.
151 J. E. Lydon, M. R. Truter and R. C. Watling, *Proc. Chem. Soc.*, (1964) 193.
152 M. R. Truter and R. C. Watling, *J. Chem. Soc. (A)*, (1967) 1955.
153 J. E. Lydon and M. R. Truter, *J. Chem. Soc.*, (1965) 6899.
154 A. G. Swallow and M. R. Truter, *Proc. Roy. Soc.*, **A266** (1962) 527.
155 D. E. Clegg and J. R. Hall, *Spectrochim. Acta.*, **21** (1965) 357.
156 G. E. Glass and R. S. Tobias, *J. Amer. Chem. Soc.*, **89** (1967) 6371.
157 K. Kite and D. R. Rosseinsky, *Chem. Comm.*, (1971) 205.
158 J. A. S. Smith, *J. Chem. Soc.*, (1962) 4736.
159 D. E. Clegg and J. R. Hall, *Aust. J. Chem.*, **20** (1967) 2025.
160 G. L. Morgan, R. D. Rennick and C. C. Soong, *Inorg. Chem.*, **5** (1966) 372.
161 J. R. Hall and G. A. Swile, *J. Organometal. Chem.*, **21** (1970) 237.
162 M. N. Hoeschstetter, *J. Mol. Spectry.*, **13** (1964) 407.
163 P. A. Bulliner and T. G. Spiro, *Inorg. Chem.*, **8** (1969) 1023.
164 D. E. Clegg and J. R. Hall, *J. Organometal. Chem.*, **17** (1969) 175.
165 P. A. Bulliner, V. A. Maroni and T. G. Spiro, *Inorg. Chem.*, **9** (1970) 1887.
166 O. M. Ivanova and A. D. Gel'man, *Zhur. Neorg. Khim.*, **3** (1958) 1334; *Chem. Abs.*, **53** (1959) 18724h.
167 J. R. Hall and G. A. Swile, *Aust. J. Chem.*, **24** (1971) 423.
168 J. D. Ruddick and B. L. Shaw, *Chem. Comm.*, (1967) 1135.
169 J. D. Ruddick and B. L. Shaw, *J. Chem. Soc. (A)*, (1969) 2969.
170 D. M. Adams, J. Chatt, J. Gerratt and A. D. Westland, *J. Chem. Soc.*, (1964) 734.
171 J. M. Jenkins and B. L. Shaw, *J. Chem. Soc.*, (1965) 6789.

CHAPTER 13

Olefin and Acetylene Complexes

Olefin complexes are the oldest class of organometallic complexes known; the first complex, $K^+[Pt(C_2H_4)Cl_3]^-.H_2O$, known as Zeise's salt, was prepared in 1830.[1] They have been very extensively investigated and recently a number of their reactions have been shown to be of commercial interest. For example the Wacker process for the oxidation of ethylene to acetaldehyde with a palladium(II) chloride catalyst is a full-scale industrial process[2] and the specificity of mixtures of platinum(II) and tin(II) chlorides in catalysing the hydrogenation of olefinic double-bonds has engaged attention.

The number of reviews on olefin and acetylene complexes of platinum and palladium (see Table 104), and particularly one by the present author,[4]

TABLE 104
Reviews on olefin and acetylene complexes

Title	Reference
Olefin and acetylene complexes of platinum and palladium	4
Complexes of platinum(II) with unsaturated hydrocarbons	5
Metal π-complexes with substituted olefins	6
Two-electron ligands	7
Olefin, acetylene and π-allylic complexes of transition metals	8
Mono-olefin and -acetylene complexes of nickel, palladium and platinum	9
Bonding in metal–olefin complexes	10
Olefin complexes of the transition metals	10a

which is comprehensive up to July 1968, has made a comprehensive coverage of the literature unnecessary here. Accordingly, the emphasis has been directed towards discussing the bonding in these complexes and in presenting the experimental work on which the present views are based.

Complexes in which the metal is formally in a divalent oxidation state differ substantially from those with the metal in a zerovalent state, and for convenience the two oxidation states are considered separately. Nevertheless, it will be found that the overall bonding descriptions are really very similar. It should be noted that the olefin complexes of transition metals generally fall into two main classes, class S and class T, and that the divalent and zerovalent olefin complexes of platinum are model complexes for class S and class T olefin complexes respectively.[10,11]

COMPLEXES OF DIVALENT PLATINUM AND PALLADIUM

Preparation

There are essentially three general methods for preparing these complexes.

(i) Treatment of a metal(II) salt with olefin in aqueous solution

Platinum(II) and palladium(II) salts react with an olefin in water to give olefin complexes; these may be with water soluble olefins (e.g. allylamine[12]), with water-insoluble olefins (e.g. cyclo-octatetraene[13]) or with gaseous olefins (e.g. ethylene) with[14] or without[15] pressure. The reaction, which usually takes a number of days to go to completion,[15] can be accelerated by a trace of stannous chloride.[14] The present author[15a] has described a very convenient synthesis of the parent ethylene complex $K^+[Pt(C_2H_4)Cl_3]^-.H_2O$ at room temperature and pressure (see Appendix I, pp. 462, 464).

(ii) Treatment of a metal(II) salt with olefin in non-aqueous solution

Reaction of palladium(II) dichloride with ethylene in benzene under high pressure has been used to prepare the dimer $[Pd(C_2H_4)Cl_2]_2$[16] and $[Pd(norbornadiene)Cl_2]$ has been made by treating Na_2PdCl_4 with norbornadiene in methanol.[17] Higher terminal alcohols are unsuitable solvents since they themselves react with Na_2PtCl_4 to form terminal olefin complexes[18,19] (reaction 454 where R = H, CH_3 or C_2H_5).

$$Na_2PtCl_4 + RCH_2CH_2OH \xrightarrow[Bu_4N^+Cl^-]{\text{room temp.,}} Bu_4N^+[Pt(RCH=CH_2)Cl_3]^-$$

$$(454)$$

(iii) Displacement of weakly co-ordinated ligands

Olefin complexes can be prepared by displacement of weakly co-ordinated ligands such as benzonitrile.[20] Since an excess of one olefin will displace another,[21] ethylene complexes have been widely used as starting materials for the preparation of olefin complexes. The volatility of ethylene aids its displacement.

Kinetics of formation of olefin complexes

The formation of platinum(II) complexes according to reaction 455 normally follows the two-term rate law,

$$\text{Rate} = \{k_1 + k_2[L]\}[PtX_4^{2-}]$$

where k_1, the first-order rate constant, corresponds to the rate for the

$$[PtX_4]^{2-} + L \rightarrow [PtLX_3]^- + X^- \qquad (455)$$

solvent dependent path and k_2, the second-order rate constant, corresponds to the rate for the solvent independent path.[22] However, k_1 in the reaction of olefins with the tetrachloroplatinate(II) ion is zero,[23-5] indicating that the aquo species $[PtCl_3(H_2O)]^-$ is not involved in the rate-determining step of the reaction. Two suggestions have been put forward to account for this. One is that it may arise because the d-orbitals are more compact in the less negatively charged aquo complex, giving less overlap with the olefin orbitals. This would lead to large differences in rate for olefins with $[PtCl_4]^{2-}$ and $[PtCl_3(H_2O)]^-$. An alternative suggestion,[26] put forward as a result of studies with the gaseous olefin ethylene, is that the presence of $[(C_2H_4)PtCl_3]^-$ catalyses the formation of $[PtCl_3(H_2O)]^-$ by forming a chloro-bridged complex so that the limiting step in the traditionally solvent dependent path is diffusion of the ethylene to the $[PtCl_3(H_2O)]^-$ complex. This gives rise to a dependence on the olefin concentration in the k_1 term so that this term effectively disappears and the k_2 term involves olefin interaction with both $[PtCl_4]^{2-}$ and $[PtCl_3(H_2O)]^-$. The mechanism suggested for the formation of platinum(II)–olefin complexes involves an initial co-ordination of the olefin in the axial position of a five-co-ordinate intermediate in a rapidly established pre-equilibrium followed by the displacement of an equatorial chloride ligand from this intermediate in the rate-determining step (reaction 456).

$$
\begin{bmatrix} \begin{array}{c} Cl \\ | \\ Cl-Pt-Cl \\ | \\ Cl \end{array} \end{bmatrix}^{2-} + \text{ olefin } \underset{}{\overset{fast}{\rightleftharpoons}} \begin{bmatrix} \begin{array}{c} Cl \quad olefin \\ \backslash \; | \\ Pt-Cl \\ / \; | \\ Cl \quad Cl \end{array} \end{bmatrix}^{2-} \underset{}{\overset{slow}{\rightleftharpoons}}
$$

$$
\begin{bmatrix} \begin{array}{c} Cl \\ | \\ Cl-Pt-olefin \\ | \\ Cl \end{array} \end{bmatrix}^- + Cl^- \quad (456)
$$

The stannous chloride catalysed formation of platinum(II)–olefin complexes also follows a two-term rate law; the first term is dependent on the olefin and platinum(II) salt concentrations, and the second, on the platinum(II) salt concentration and the square of the stannous chloride concentration.[27] This is consistent with two mechanisms, the first following eqn. 456 and the second involving the rate-determining formation of *cis*-$[PtCl_2(SnCl_3)_2]^{2-}$ which then reacts rapidly with the free olefin. It has recently been claimed that one olefin can catalyse the formation of the platinum(II) complex of another,[28] but the experimental evidence is inadequate to support this suggestion if Ostwald's definition[29] of a catalyst is accepted, i.e. 'a catalyst is any substance which alters the velocity of a chemical reaction without appearing in the end product of the reaction'.

Physical properties of olefin and acetylene complexes of the divalent metals
(i) X-ray structural studies
X-ray diffraction investigations of the divalent metal–olefin complexes have been fraught with more problems than almost any other x-ray investigations. The problems arise from two main sources. First, the accurate location of light atoms such as carbon in the presence of strong x-ray scatterers such as

platinum or palladium is inherently difficult and, secondly, the positions of the light atoms in a number of these structures are particularly sensitive to the refinement conditions and especially to the statistical weighting system used in the least-squares analysis. The sad tales of two molecules, the parent platinum(II)–olefin complex Zeise's salt and dichloro(norbornadiene) palladium(II), will suffice to illustrate the problems. A number of x-ray diffraction studies of Zeise's salt made in the 15 years between 1954 and 1969[30-2], have recently been shown, with the discovery of some previously undetected weak diffraction spots, to have the wrong space group.[33] Use of the new space group alters the positions of light atoms as deduced from x-ray diffraction studies[34] and will probably also alter their positions as deduced from neutron diffraction studies.[35] Three studies of the x-ray diffraction pattern obtained from dichloro(norbornadiene)palladium(II) have given different results. Initially the carbon–carbon double-bond lengths were found to be 1.52 ± 0.036 Å ;[36] re-measurement gave a value of 1.46 ± 0.037 Å[37] and a further determination at the temperature of liquid nitrogen gave a value of 1.366 ± 0.010 Å.[38]

Having emphasised how cautiously the results of x-ray diffraction studies must be treated, what can confidently be deduced from the structures that have been reported so far? First it would appear that the plane of the olefin double-bond is approximately perpendicular to the square-plane surrounding the divalent metal. The angle between these planes is not exactly 90°, as the early structures showed, but slightly different as would be expected since n.m.r. studies indicate a low energy barrier to rotation about the metal–olefin bond[39] so that the angle between the plane of the carbon–carbon double-bond and the square-plane of the metal should be dependent on the crystal forces present. Angles of 84°, 77°, 79° and 85° have been reported for $K^+[Pt(C_2H_4)Cl_3]^-.H_2O$,[34] *cis*-[Pt(*trans*CH$_3$CH=CHCH$_3$)-{S-α-(PhCHCH$_3$)NH$_2$}Cl$_2$],[40] *cis*-[Pt{(S)-1-butene}{S-α-(PhCHCH$_3$)NH$_2$}-Cl$_2$][41] and the acetylenic complex *trans*-[Pt(di-*t*-butylacetylene)(*p*-tolui-dine)Cl$_2$][37] respectively, and in the last complex the mid-point of the acetylene bond lies 0.10 Å above the platinum co-ordination plane. In diolefin complexes such as dichloro(norbornadiene)palladium(II),[38] dichloro(1,5-hexadiene)palladium(II)[43] and dichloro(cyclo-octatetraene)palladium(II)[44] where the double-bonds in the free ligands are parallel, these double-bonds lie perpendicular to and symmetrically about the palladium(II)–olefin bond. However, such an arrangement is not essential for

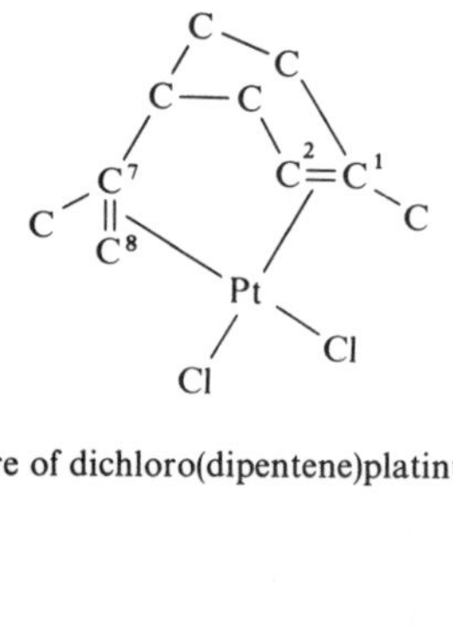

Fig. 31. The structure of dichloro(dipentene)platinum(II) (from ref. 45).

the formation of a stable complex as is illustrated by dichloro(dipentene)-platinum(II) (Fig. 31) which has one double-bond (C_1–C_2) perpendicular to the plane containing the $PtCl_2$ group, whereas the other double-bond (C_7–C_8) is at an angle of 62·1°.[45] Similarly in styrene–palladium chloride, the double-bond is neither symmetrically disposed about the palladium–

olefin bond nor perpendicular to the Pd Pd plane[46] (Fig. 32).

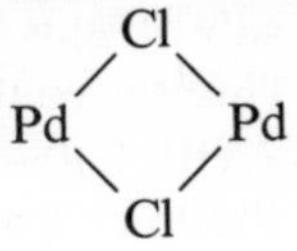

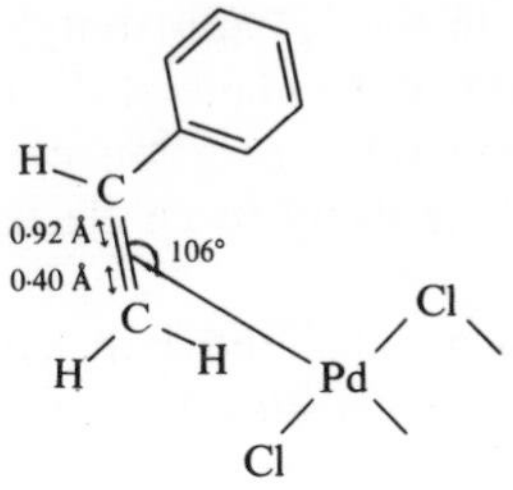

Fig. 32. The structure of $[Pd(PhCh=CH_2)Cl_2]_2$ (from ref. 46).

TABLE 105
Carbon–carbon bond lengths in co-ordinated multiple bonds

Complex	Carbon–carbon bond length (Å)		No. of structure in Appendix II	Reference
	Co-ordinated	Free		
$K^+[Pt(C_2H_4)Cl_3^-.H_2O$	1·37	1·335	121	34
$[Pd(norbornadiene)Cl_2]$	1·37	1·33	40	38
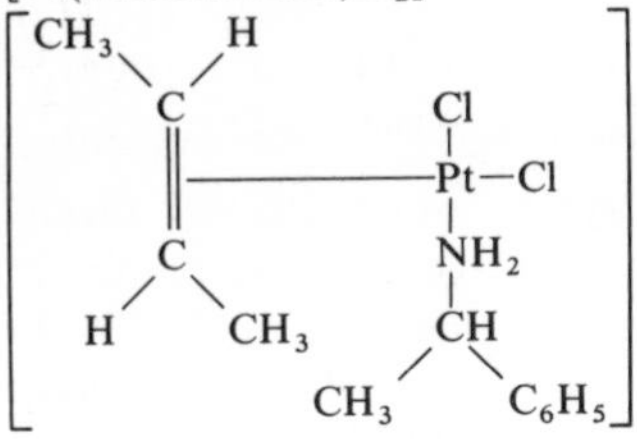 *(structure)*	1·36	1·33	153	40
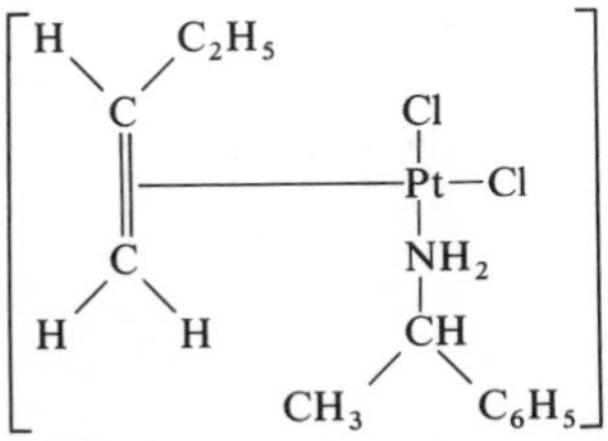 *(structure)*	1·35		154	41
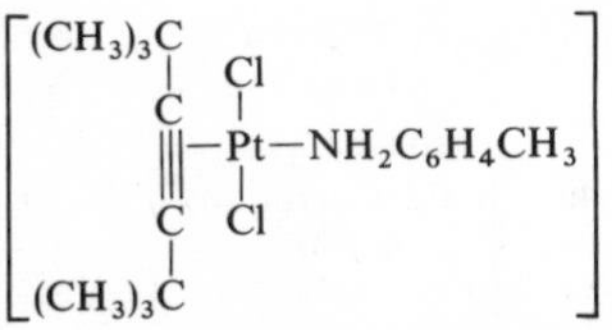 *(structure)*	1·24	1·20	174	42

A considerable amount of evidence suggests that the groups attached to the multiple-bond are bent out of the plane of that bond away from the metal, although where these groups are hydrogen atoms, as in $K^+[Pt(C_2H_4)Cl_3]^-.H_2O$, the results must be treated with caution. However, two structures, one of an olefin and the other of an acetylene complex, in which there are carbon substituents bound to the multiple bond, have confirmed this distortion. Thus the methyl groups bound to the double-bond in *cis*-[Pt(*trans*-CH$_3$CH=CHCH$_3$)(S-α-phenethylamine)Cl$_2$] are displaced $13 \pm 4°$ out of the plane of the double-bond away from the platinum atom.[40] Similarly the *t*-butyl groups in *trans*-[Pt(di-*t*-butylacetylene)(*p*-toluidine)Cl$_2$] are also bent $16.5 \pm 1.5°$ out of the plane of the triple bond away from the platinum atom.[42] Although this distortion could be the result of non-bonded interactions it is quite likely to be electronic rather than steric in origin (see p. 384 below).

Much interest has been focussed on the change in the length of the multiple bond on co-ordination. For the reasons outlined above many of the results

TABLE 106

Metal–chlorine bond lengths *trans* to olefinic groups

Complex	M–Cl bond length trans to olefin (Å)	No. of structure in Appendix II	Comparison bond lengths (Å)	Refer-ence
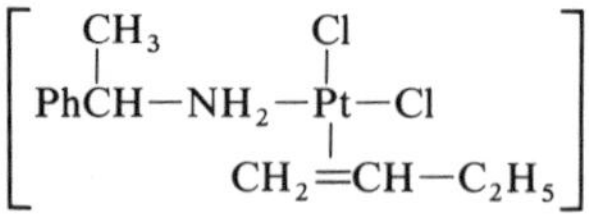 [PhCH–NH$_2$–Pt–Cl with CH$_3$, Cl, and CH$_2$=CH–C$_2$H$_5$]	2·36	154	Pt–Cl *cis* to olefin = 2·29	41
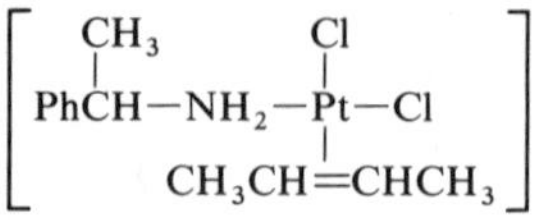 [PhCH–NH$_2$–Pt–Cl with CH$_3$, Cl, and CH$_3$CH=CHCH$_3$]	2·35	153	Pt–Cl *cis* to olefin = 2·29	40
[Pt(C$_2$H$_4$)Cl$_3$]$^-$	2·33	121	Pt–Cl *cis* to olefin = 2·30 (mean)	34
[Pt(dipentene)Cl$_2$]	2·33	151	Pt–Cl in cis-[Pt(NH$_3$)$_2$Cl$_2$] = 2·33	45, 47
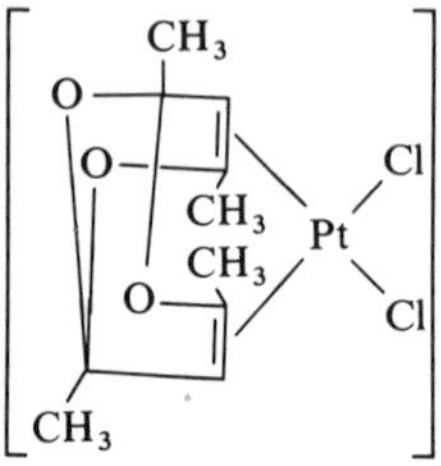 [Pt(acac-type) complex with CH$_3$, O, CH$_3$, CH$_3$, Cl, Cl, CH$_3$]	2·30, 2·31	150	Pt–Cl in trans-[Pt(NH$_3$)$_2$Cl$_2$] = 2·32	48, 47
[Cl$_3$PtCH$_2$=CH–CH=CH$_2$PtCl$_3$]$^{2-}$	2·29	127	Pt–Cl *cis* to olefin = 2·29 and 2·32	49
[Pd(norbornadiene)Cl$_2$]	2·32 (mean)	40	Pd–Cl typically 2·30–2·31	38, 50
[Pd(1,5-hexadiene)Cl$_2$]	2·30, 2·31	35		43, 50

are of doubtful value, but the most reliable, shown in Table 105, indicate that there may be a very slight lengthening of the multiple bond on co-ordination.

There has been considerable controversy as to whether or not the metal–ligand bond *trans* to an olefin or acetylene is lengthened or not. When examining the data it must be born in mind that crystal forces are known to be capable of altering platinum–chlorine bond lengths by at least 0·02 Å since the two *cis*-Pt–Cl bond lengths in $K^+[Pt(C_2H_4)Cl_3]^-.H_2O$ were found to differ by this amount.[34] With the data available at present (Tables 106 and 107) it would appear that little or no lengthening or shortening of a metal–chlorine or platinum–nitrogen bond *trans* to an unsaturated group occurs.

TABLE 107

Platinum-nitrogen bond lengths *trans* to olefins or acetylenes

Complex	Pt–N bond length (Å)	No. of structure in Appendix II	Reference
$\begin{bmatrix} CH_2 \\ \| \quad Pt-NHMe_2 \\ CH_2 \end{bmatrix}$ with Cl, Cl	2·02	132	51
$\begin{bmatrix} C(CH_3)_3 \\ C \\ \|\|\| \quad Pt-NH_2C_6H_4CH_3 \\ C \\ C(CH_3)_3 \end{bmatrix}$ with Cl, Cl	2·10 ± 0·01	174	42
Compare with			
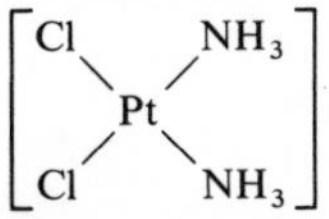 $\begin{bmatrix} Cl \quad NH_3 \\ \quad Pt \\ Cl \quad NH_3 \end{bmatrix}$	2·01 ± 0·04	93	47
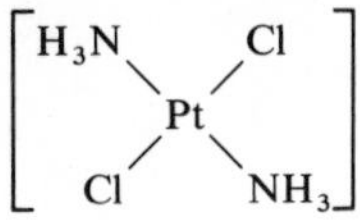 $\begin{bmatrix} H_3N \quad Cl \\ \quad Pt \\ Cl \quad NH_3 \end{bmatrix}$	2·05 ± 0·04	94	47

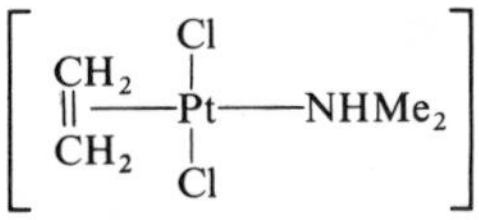

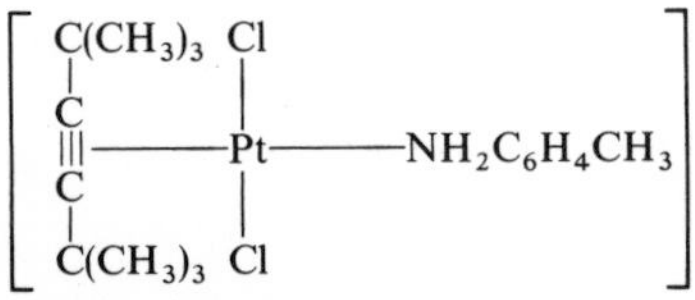

(ii) Infra-red studies

The infra-red and Raman spectra of platinum(II)– and palladium(II)–olefin complexes have been extensively studied over recent years,[18,52–64] but unfortunately the considerable amount of work that has been done has only served to emphasise how much more complicated than was first thought these spectra really are. Before going into details about what cannot be obtained from these spectra let us state that virtually the only fact that can be deduced unambiguously from the vibrational spectra is that 'the carbon–carbon double-bond stretching frequency is lowered on co-ordination to platinum(II) and palladium(II)'. It is not possible, for the reasons discussed below, to put a quantitative value to this lowering.

The vibrational spectra of metal–olefin complexes are very difficult to interpret because the symmetry of these complexes is fairly low ($[Pt(C_2H_4)Cl_3]^-$

has C_{2v} symmetry) so that many of the fundamental modes have the same symmetry properties and can therefore interact (see Table 108). Of particular importance in this connection is the fact that the $C=C$ double-bond stretching mode and the in-plane CH_2 deformation mode both have a_1 symmetry and similar frequencies. They therefore mix. Consequently they give rise

TABLE 108

Symmetry of the normal vibrational modes for the ion
$[M(C_2H_4)Cl_3]^-$ (symmetry group C_{2v})

Symmetry species	Band	Approximate vibrational mode
a_1	v_1	CH_2 symmetric stretching
	v_2	CH_2 deformation
	v_3	$C=C$ stretching
	v_4	CH_2 wagging
	v_5	PtC_2 symmetric stretching
	v_6	$PtCl_2$ symmetric stretching
	v_7	$Pt-Cl_{trans}$ stretching
	v_8	$PtCl_2$ bending
b_1	v_9	CH_2 symmetric stretching
	v_{10}	CH_2 deformation
	v_{11}	CH_2 wagging
	v_{12}	PtC_2 antisymmetric stretching
	v_{13}	PtC_2 rocking
	v_{14}	$Pt-Cl_{trans}$ bending
b_2	v_{15}	CH_2 antisymmetric stretching
	v_{16}	CH_2 twisting
	v_{17}	CH_2 rocking
	v_{18}	$PtCl_2$ antisymmetric stretching
	v_{19}	PtC_2 wagging
	v_{20}	$Pt-Cl_{trans}$ bending
a_2	v_{21}	CH_2 antisymmetric stretching
	v_{22}	CH_2 twisting
	v_{23}	CH_2 rocking
	v_{24}	PtC_2 twisting

to two bands (one at $1515 \, cm^{-1}$ and the other at $1243 \, cm^{-1}$ in Zeise's salt) both of which result from a mixture of the two fundamental modes. To assign either of these frequencies to a single vibrational mode is obviously meaningless. However, in the older literature the $1515 \, cm^{-1}$ band was assigned to the $C=C$ stretching mode and this led to an evaluation of the lowering of the $C=C$ stretching frequency on co-ordination. Not content with stopping here some authors have used this (false) lowering of the $C=C$ stretching frequency on co-ordination as a measure of the strength of the metal–olefin bond! It is hardly surprising, as the present author has demonstrated, that no correlation exists between the lowering of the $C=C$ stretching frequency on co-ordination and the metal–olefin bond strength.[65]

The PtC$_2$ symmetric and asymmetric stretching frequencies have generally been assigned to bands at about 405 and 493 cm^{-1}, but, as can be seen from Table 108, both frequencies may contain substantial contributions from other vibrational modes of the same symmetry and of similar frequency. There is fairly general agreement that the bands at 333 (strong), 324 (weak shoulder), 306·5 (strong) and 298 cm^{-1} (weak shoulder) in K$^+$[Pt(C$_2$H$_4$)Cl$_3$]$^-$ all arise largely from platinum–chlorine stretching modes, though there is little agreement about which band arises from which vibrational mode.

It would appear to the present author[66] that, until it becomes possible to determine accurately (i.e. within 1 or 2%) the magnitude of the contribution of a particular vibrational mode to a particular band in the spectrum, the statement that 'the carbon–carbon double-bond stretching frequency is lowered on co-ordination to both platinum(II) and palladium(II)' is all that can meaningfully be deduced from vibrational spectroscopy to add to our understanding of the nature of the bonding in these complexes.

Having been very destructive of the quantitative conclusions that have been deduced from vibrational spectroscopy it is important to add a further qualitative conclusion that can usefully be drawn. Thus the observation of a new band in a potential platinum(II)–olefin complex at about 140 cm^{-1} [67] below the carbon–carbon double-bond stretching frequency of the free olefin is good preliminary evidence that such a complex has been formed, and this test is widely used in preparative work. For olefins with palladium(II) this new band will be about 120 cm^{-1} [68] below that of the free olefin and for acetylenes with platinum(II) the corresponding lowering will be about 250 cm^{-1}.[69]

(iii) Nuclear magnetic resonance studies

Nuclear magnetic resonance studies have been widely used in conjunction with infra-red spectroscopy to provide preliminary evidence for the structure of olefin complexes. In particular n.m.r. has been successful in distinguishing between olefinic and π-allylic complexes. It has also given valuable information on first the rotation of the olefin groups within the molecule, which in turn has led to a modification of the early ideas on the nature of the bonding in these complexes, secondly, possible changes in the hybridisation of the carbon atoms of the double-bond on co-ordination and, thirdly, the equilibria that are set up when olefins are dissolved in certain solvents.

(a) Conformation of olefins in metal–olefin complexes

Solid-state studies. The n.m.r. spectrum of polycrystalline [PtCl$_2$(C$_2$H$_4$)]$_2$ provides evidence for a rocking motion perpendicular to the platinum(II)–olefin bond and 'wagging' about the carbon–carbon double-bond axis.[70] The data also suggest that on co-ordination the olefin bond is lengthened and the hydrogen atoms bent out of the plane of the double-bond away from the platinum atom, a result that is consistent with the x-ray data discussed above.

The results obtained when the n.m.r. spectra of single crystals of K$^+$[Pt(C$_2$H$_4$)Cl$_3$]$^-$.H$_2$O and its deuterium oxide analogue were recorded at a number of orientations relative to the magnetic field suggested that the ethylene molecule was undergoing large amplitude rotational oscillations about both the platinum–olefin bond and the carbon–carbon double-bond axis.[71]

Solution studies. Nuclear magnetic resonance solution spectra of a number of complexes have suggested that two main movements occur, namely twisting about the axis of the olefinic double-bond and rotation of the olefin about the metal–olefin bond.

Twisting about the axis of the olefinic double-bond has been deduced from the platinum–olefinic proton coupling constants for a number of complexes containing unsymmetrical olefins.[72-7] A typical result[77] is illustrated in Table 109. The lower coupling constant for H_β than either H_α or H_γ suggests

TABLE 109

J_{195Pt-^1H} coupling constants for the olefinic
protons of $Bu_4N^+[Pt(CH_3CH{=}CH_2)Cl_3]^-$ in
$(CD_3)_2CO$ solution (ref. 77)

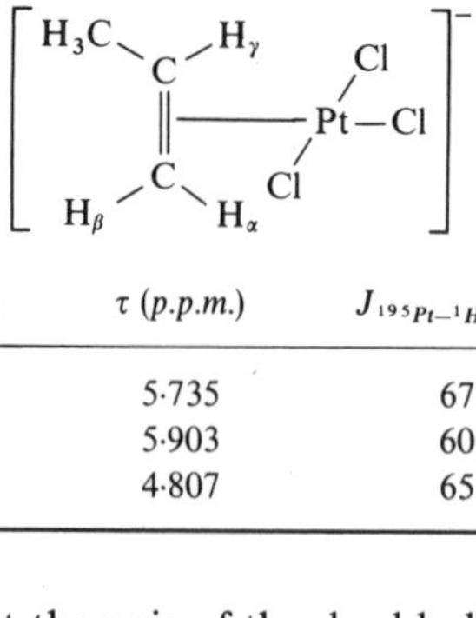

Proton	τ (p.p.m.)	J_{195Pt-^1H} (Hz)
H_α	5·735	67
H_β	5·903	60
H_γ	4·807	65

that the olefin twists about the axis of the double-bond in such a way as to minimise the non-bonded interactions between the methyl group and the central metal group. The small difference in the H_α and H_β coupling constants of about 10% suggests that the twisting is small, which is consistent with the description of the bonding given below.

Rotation of the olefin about the metal–olefin bond has been suggested to account for the temperature dependence of the n.m.r. spectra of *trans*-[PtCl$_2$(ol)(2,4,6-trimethylpyridine)],[78] where ol = styrene or *t*-butyl-ethylene and [PtCl(ol)acac],[76] where ol = propylene or *cis*-but-2-ene. In the complex [PtCl(*cis*-but-2-ene)acac], which has structure **157**, there are two proton resonances at low temperature which have been assigned to the

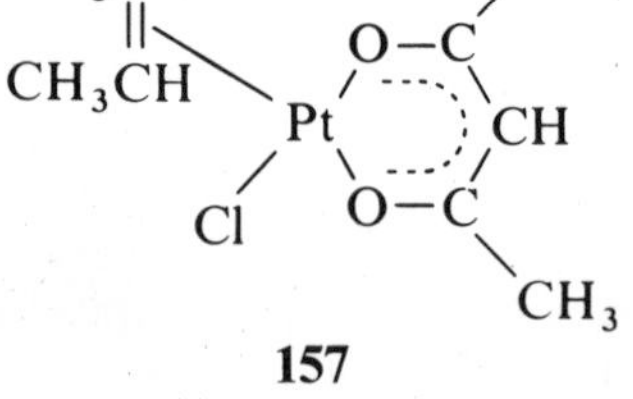

157

two isomers **158** and **159** (both drawn looking down the olefin–platinum bond). On warming, the two resonance signals coalesce, which was interpreted as being due to rotation of the olefin about the platinum–olefin

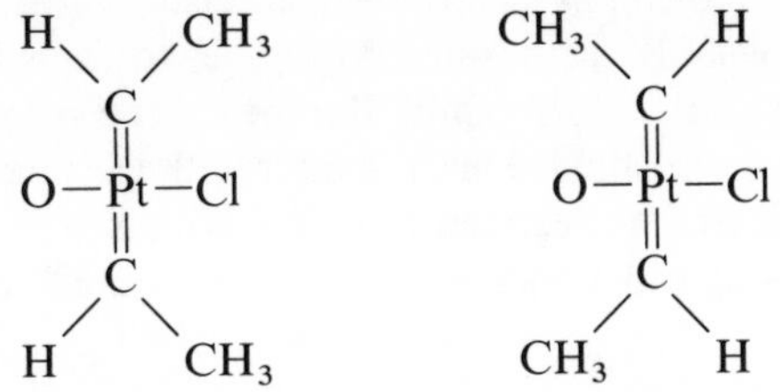

bond. A number of other possible interpretations were considered but none were consistent with all the experimental data. From the coalescence temperature the energy barrier to rotation was found to be between 10 and 15 kcal/mol and this was almost independent of the olefin, the halogen (chlorine or bromine) and whether or not the CH_3 groups in the acetylacetone ligand were changed to CF_3.[39]

(b) Nuclear magnetic resonance data relevant to bonding studies

(i) Chemical shifts. The chemical shifts of the olefinic protons in a series of *trans*-[$PtCl_2$(*p*-substituted pyridine-*N*-oxide)(C_2H_4)] complexes move downfield as the electron withdrawing character of the *para*-substituent increases.[79] This has been ascribed to a reduction in the π-back-donation of charge from platinum to ethylene which weakens the ethylene–platinum interaction. The observation that the chemical shifts of the olefinic protons of a series of diolefin complexes [MX_2(diolefin)], where M = Pd or Pt and X = halogen, move downfield on replacing platinum by palladium[80] is consistent with the lower olefin–metal interaction present in the palladium complexes.

(ii) Coupling constants. The ^{195}platinum–olefin proton coupling in the *trans*-[$PtCl_2$(*para*-substituted pyridine-*N*-oxide)(C_2H_4)] complexes increases with the increasing electron-withdrawing character of the *para*-substituent which reduces the pyridine-*N*-oxide to platinum σ-bonding so making platinum a better acceptor in the ethylene to platinum σ-bond. This increases the ^{195}Pt-^{1}H coupling by increasing the platinum s character in the platinum–ethylene bond.[81]

Both the *cis* and the *trans* proton–proton coupling constants of olefins decrease on co-ordination to platinum suggesting that the sp^2 hybridisation of the carbon atoms of the free olefin has altered slightly towards sp^3 hybridisation.[39,82] This suggests, in valence bond terms, some contribution from a structure in which the principle bonding between the olefin and platinum is two platinum–carbon σ-bonds.

(c) Nuclear magnetic resonance evidence for equilibria in solution

Nuclear magnetic resonance studies suggest that the rate of exchange of free and co-ordinated ethylene in Zeise's salt is greater than 70 sec^{-1} at

75°.[83] Studies of the mutarotation of diastereoisomers of platinum(II)–olefin complexes and of the kinetics of exchange of free and co-ordinated ethylene in $[Pt(acac)Cl(C_2H_4)]$ in chloroform and benzene[85] suggest that this exchange occurs by an associative mechanism.

The addition of fully deuterated pyridine to a solution of $[PtCl_2(C_5H_5N)(C_2H_4)]$ in chloroform shifts the methylene resonance peak in the n.m.r. spectrum to higher field. This, together with the fact that this peak has been split into a triplet, has been tentatively ascribed to the conversion of the platinum–olefin bond into a platinum–carbon σ-bond[86] (eqn. 457).

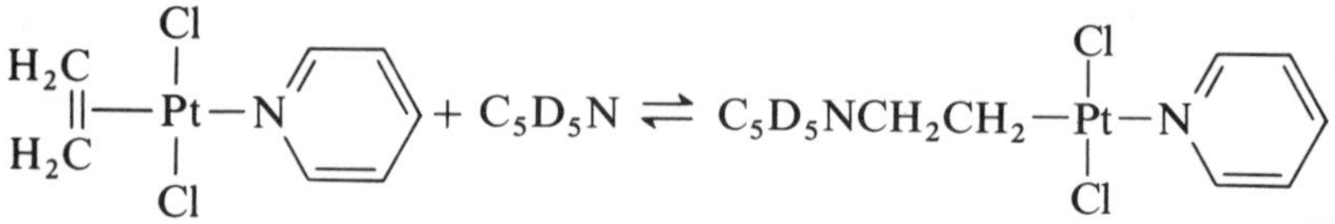

$$\begin{array}{c} Cl \\ | \\ H_2C \\ \| - Pt - N \bigcirc \\ H_2C \quad | \\ Cl \end{array} + C_5D_5N \rightleftharpoons C_5D_5NCH_2CH_2 - \begin{array}{c} Cl \\ | \\ Pt - N \bigcirc \\ | \\ Cl \end{array}$$

(457)

The n.m.r. spectra of **160** show that the coupling of the H_α protons to the platinum atom depends on the temperature, the solvent and the nature of

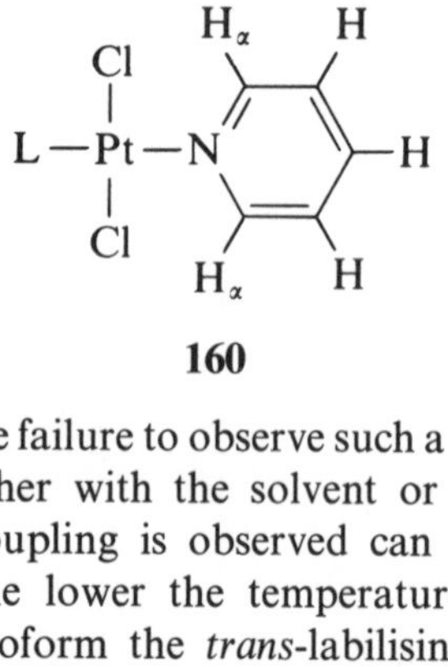

160

L.[87] By assuming that the failure to observe such a coupling is due to a rapid exchange of pyridine either with the solvent or between complexes, the temperature at which coupling is observed can be related to the *trans*-labilising effect of L; the lower the temperature the greater the *trans*-labilising effect. In chloroform the *trans*-labilising effect of a number of ligands was found to be in the order $C_2H_4 \gg PhCH{=}CH_2 \sim$ *cis*- and *trans*-$CH_3CH{=}CHCH_3 \sim CH_3CH_2CH_2CH{=}CH_2 >$ CO.

(iv) Nuclear quadrupole resonance studies
[35]Chlorine n.q.r. studies of olefin complexes have shown that on replacing one chloride ligand in K_2PtCl_4 by a mono-olefin the single resonance signal is split into three signals.[88] One of these new signals is at a lower frequency than in K_2PtCl_4 and this is assigned to the chlorine *trans* to the olefin The frequency of this resonance in the anions $[LPtCl_3]^-$ increases as L is altered in the order L = ethylene < *cis*-but-2-ene < styrene which suggests that the charge on the chloride ligand *trans* to L becomes more negative as L is changed from styrene to ethylene, which in turn suggests that the σ-donor ability of the ligands increases in the order styrene < *cis*-but-2-ene < ethylene. The other two new n.q.r. signals that appear when a chloride ion in K_2PtCl_4 is replaced by an olefin occur at higher frequency and the observation of two signals is consistent with such x-ray crystallographic data as exists, since the two *cis*–chloride ligands in $K[Pt(C_2H_4)Cl_3]$ are crystallo-

graphically inequivalent[34] (see Appendix II, structure 121). The observation of these two resonances at higher frequency than the *trans*-chloride frequency is consistent with Syrkin's polarisation model of the *trans*-effect[89] which suggests that a strong covalent bond induces extra covalency in the *cis*-metal–ligand bonds, but reduces the covalency of the *trans*-metal–ligand bond. Thus the observed splitting of the ^{35}Cl resonances in $[PtCl_4]^{2-}$ on replacing one chloride ligand by an olefin implies that the σ-bonding in the metal–olefin bond must be considerable. Although many of the results in the present chapter suggest that the metal–olefin σ-bond is weak, molecular orbital calculations have suggested that the σ-overlap integral in platinum(II) complexes increases in the order $Cl \sim N < P < $ olefin.[90] Thus it is quite possible, as was noted in Chapter 7 (pp. 139–140) in connection with tertiary phosphine complexes, that in paying attention to the π-acceptor ability of olefins in metal–olefin complexes the σ-donor ability of these ligands has been underestimated. There is a decrease in the *trans*-chloride resonance frequency and hence an increase in the negative charge carried by these chloride ligands in *cis*-$[L_2PtCl_2]$ complexes as L is altered in the order $L = \frac{1}{2}(1,5\text{-cyclo-octadiene}) > PBu_2Ph > Me_2NH,$[88,91] which is consistent with an increase in back-donation of charge from the metal in the order $N < P < $ olefin.

(v) Electronic spectra

The ultra-violet and visible spectra of Zeise's salt and a number of its analogues containing different olefins have been recorded,[12,92–5] and are all very similar to that in Fig. 33. A semi-empirical self-consistent charge and configuration molecular orbital calculation indicated that, as expected, there was a very considerable mixing of the component atomic orbitals in these complexes. However, this approach gives only qualitative agreement between the observed and calculated bands in the electronic spectrum.[94] An independent study of the electronic spectra of platinum–olefin complexes based on comparisons with similar complexes, assigned the ultra-violet bands to transitions between platinum $5d$ and olefin π^*-(antibonding) orbitals.[95] By observing the spectral shifts that occurred both on changing the solvent and on altering the ligand *trans* to the olefin, the relative energies of the d-orbitals were shown to be $d_{z^2} > d_{xz} > d_{x^2-y^2} > d_{xy} > d_{yz}$ using the axes of Fig. 34.[95]

The circular dichroism of optically active olefins is instrumentally difficult to observe since olefins absorb at about $50,000 \text{ cm}^{-1}$. However, when the olefin is co-ordinated to platinum(II), the disymmetric olefin perturbs the d–d transitions which occur between $20,000$ and $30,000 \text{ cm}^{-1}$, giving rise to a circular dichroism band in a region that is instrumentally convenient. This technique, which allows the relative and absolute configurations of the olefins to be determined, has been applied to a number of terpenoid and steroid olefins.[96,97] The sign of the circular dichroism band depends solely on the optical activity of the olefin and is independent of both the optical activity of the amine and the relative position (*cis* or *trans*) of the olefin and the amine in $[PtCl_2(RNH_2)(\text{olefin})]$.[98]

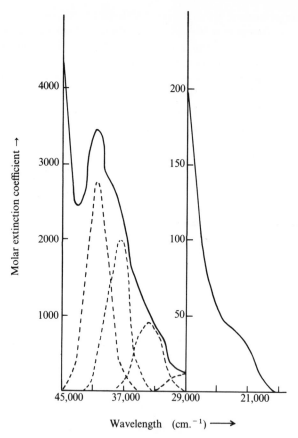

Fig. 33. The electronic spectrum of $[Pt(CH_2=CH-CH_2OH)PtCl_3]^-$ in ethanol in the region 17,000–45,000 cm^{-1} (reproduced with permission from ref. 95).

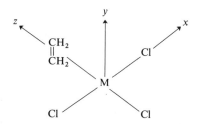

Fig. 34. Axes used in discussing metal(II)–olefin complexes.

(vi) *Stability constant studies*

The first qualitative information on the stability of platinum(II)–olefin complexes, obtained in 1936,[21] showed that the co-ordinating ability of olefins decreased with increasing substitution of the olefinic double-bond in the order $C_2H_4 > PhCH=CH_2 > Ph_2C=CH_2 \sim Ph(Me)C=CH_2$.

Since then the results of a number of workers have shown that the stability of platinum(II)–olefin complexes decreases as the overall charge on the complex ion becomes more positive,[99] indicating that when π-bonding is greatly reduced, as in cationic complexes, the stability of the platinum(II)–olefin bond is greatly decreased, despite the enhancement of the σ-bond between the olefin and platinum(II).

Systematic quantitative stability constants have only recently become available. A spectrophotometric study of equilibrium 458 indicates that

$$[PtX_4]^{2-} + \text{olefin} \rightleftharpoons [PtX_3(\text{olefin})]^- + X^- \qquad (458)$$

olefins give higher stability constants than water, lower than amines and comparable with the heavier halides.[100] The stability constants for a series of olefins[24,101-3] (Table 110) indicate that the stabilities of the complexes increase as the structure of the olefin is modified in such a way as to decrease

TABLE 110
Stability constants of platinum–olefin complexes[a]

$$PtX_4^{2-} + L \overset{K}{\rightleftharpoons} PtX_3L^- + X^-$$

$$K = \frac{[PtX_3L^-][X^-]}{[PtX_4^{2-}][L^-]}$$

Ligand X	Olefin (L)[b]	Temp. (°C)	Stability constant	ΔH (kcal/mol)	ΔS (e.u.)	Reference
Cl	all.NH$_3^+$	59·0	1022	−7·1	−7·6	100
		44·0	1737			
		30·2	2829			
Cl	all NH$_2$Et$^+$	59·0	806	−5·9	−4·4	100
		44·0	1233			
		24·0	2348			
Cl	all.NHEt$_2^+$	59·0	386	−5·6	−5·0	100
		45·3	551			
		30·0	865			
Cl	all.NEt$_3^+$	59·0	112	−4·9	−4·6	100
		45·0	153			
		25·0	260			
Cl	all.PEt$_3^+$	59·0	505			100
Cl	all.AsEt$_3^+$	58·0	917	−5·9	−4·3	100
		45·0	1320			
Cl	all.NMe$_3^+$	60·0	117	−5·3(5)	−6·6	100
		44·5	173			
		30·0	259			
Cl	all.NPr$_3^{n+}$	60·0	131			100
Cl	all.NBu$_3^{n+}$	60·0	308			100
Cl	all.NH(CH$_2$)$_4^+$	60·0	649			100
Cl	all.NH(CH$_2$)$_5^+$	60·0	437			100
Cl	but.NH$_3^+$	60·0	2038	−5·1	−0·2	100
		44·5	3017			
		30·0	4396			
Cl	but.NEt$_3^+$	60·0	4449	−3·8	+5·4	100
		44·8	5911			
		30·0	7778			

TABLE 110 (contd.)

$$PtX_4^{2-} + L \overset{K_\lambda}{\rightleftharpoons} PtX_3L^- + X^-$$

$$K = \frac{[PtX_3L^-][X^-]}{[PtX_4^{2-}][L]}$$

Ligand X	Olefin (L)[b]	Temp (°C)	Stability constant	ΔH (kcal/mol)	ΔS (e.u.)	Reference
Cl	but.AsEt$_3^+$	60·0	5459	−3·3	+7·2	100
		44·8	7014			
		30·0	8962			
Cl	pent.NH$_3^+$	60·0	1124			102
Cl	*trans*−CH$_3$CH= CH−CH$_2$−NH$_3^+$	60·2	209	−5·1	−4·6(5)	101
		44·5	304			
		30·0	450			
Cl	CH$_2$=CH−CH(CH$_3$)−NH$_3^+$	60·0	804	−6·7	−6·9	101
		45·3	1282			
		30·0	2194			
Cl	CH$_2$=C(CH$_3$)−CH$_2$−NH$_3^+$	60·0	3·2			101
Cl	(CH$_3$)$_2$C=CH−CH$_2$−NH$_3^+$	60·0	2·6			101
Cl	all.OH	60·0	3890	−8·1	−7·6	102
		44·5	7250			
		30·0	13,011			
Cl	pent.OH	60·0	2525			102
Cl	*trans*-CH$_3$−CH=CH−CH$_2$OH	60·0	3008			102
Cl	all.SO$_3^-$	55·6	1539	−6·1	−4·1	24
		45·0	2123			
		35·0	2884			
Br	all.NH$_3^+$	24·5	306			103
Br	all.NH$_2$Et$^+$	35·0	184	−4·8	−5·3	103
		25·0	241			
		0·0	506			
Br	all.NHEt$_2^+$	24·5	127			103
Br	all.NEt$_3^+$	24·5	44·1			103
Br	but.NEt$_3^+$	24·5	1190			103

[a] Determined in a medium of 1·9 M KX + 0·1 M HX using a spectrophotometric technique.
[b] all = CH$_2$=CH−CH$_2$−; but = CH$_2$=CH−(CH$_2$)$_2$−; pent = CH$_2$=CH−(CH$_2$)$_3$−.

the electron density at the olefin bond. Thus the strength of the π-back-bond from platinum(II) to olefin would appear to be more important than that of the σ-bond from olefin to platinum(II) for the formation of stable platinum(II)–olefin bonds. The platinum(II)–olefin bond strength is greater in the bromo- than the chloro-complexes, probably due to the fact that the PtBr$_3^-$ group is a poorer π-acceptor and a better σ-donor than the PtCl$_3^-$ group.[103] It is also found that in addition to electronic factors, steric factors are very important in determining the strength of the platinum(II)–olefin bond.

A number of stability constants for palladium(II)–olefin systems with simple olefins have been determined using solubility methods[104–9] (Table 111). As with the corresponding platinum(II) complexes both [Pd(olefin)Cl$_3$]$^-$ and *trans*-[Pd(olefin)Cl$_2$(H$_2$O)] are present in solution. The enthalpies and

TABLE 111

Stability constants of palladium–olefin complexes[a]

$$PdCl_4^{2-} + L \overset{K}{\rightleftharpoons} PdCl_3L^- + Cl^-$$

$$K = \frac{[PdCl_3L^-][Cl^-]}{[PdCl_4^{2-}][L]}$$

Olefin (L)	Temp. (°C)	Ionic strength	Stability constant	ΔH (kcal/mol)	ΔS (e.u.)	Reference
C_2H_4	15·0	2 M	18·7			104
	25·0		17·4			
	35·0		9·7			
C_2H_4	8·0	4 M	15·6	−1·5	0	105, 107
	13·4		16·3			
	20·0		15·2			
	25·0		13·1			
	20·0		16·9			108
C_2H_4	14·8	2 M	18·7			109
		2·1–2·2 M	15·5			
		3·1–3·2 M	15·9			
		4·5 M	16·3			
C_3H_6	10·3	4 M	8·4	0	4	107
	14·9		8·6			
	20·1		7·9			
	20·0		7·6			108
1-butene	5·0	4 M	13·9	0	5	106, 107
	10·0		12·6			
	14·8		13·6			
	20·0		12·4			
	20·0		14·3			108
1-butene	14·8		13·8			109
		2·1–2·2 M	13·8			
		3·1–3·2 M	13·6			
		4·1–4·2 M	13·6			
		5·1–5·2 M	11·3			

[a] All constants were obtained by measuring the solubility of olefins in aqueous palladium chloride solutions.

entropies of formation of $[Pd(olefin)Cl_3]^-$ are very close to zero. Although a strict comparison of palladium(II)–olefin stability constants with those obtained for palladium(II) is not possible because of the use of different olefins, the stability constants of the palladium(II)–olefin complexes are about a factor of 100 less than those of the platinum(II)–olefin complexes. Thus when comparable data are obtained, they will probably support other qualitative observations in showing that platinum(II)–olefin complexes are thermodynamically more stable than palladium(II)–olefin complexes.

The observation that *cis* olefins form complexes with higher stability constants than *trans*-olefins[110] is probably the result of two effects.

(a) Free *cis*-olefins are more strained than *trans*-olefins, as indicated by the higher heats of hydrogenation of *cis*- than *trans*-olefins,[111] and this strain is reduced on complex formation. The lowering of the double-bond stretching frequency in the infra-red and the slight lengthening of the olefin bond on complex formation both indicate that bond weakening does occur on complex formation, so relieving some of the strain in the free olefin.

(b) The actual bonds formed between *cis*-olefins and metals are stronger than between *trans*-olefins and metals. This arises from the different distortions necessary to minimise the non-bonded repulsions with the two isomeric olefins. In *cis*-olefin complexes non-bonded repulsions between the substituents present on the double-bond and the central metal group are minimised, as shown by n.m.r. spectral results,[72–7] by twisting the olefin about the double-bond axis is shown in Fig. 35. This twisting causes a small

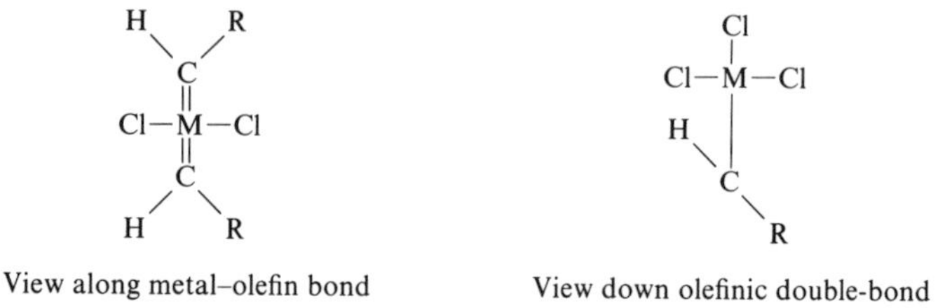

View along metal–olefin bond   View down olefinic double-bond

Fig. 35. The rotation necessary to remove steric interference in a *cis*-disubstituted olefin complex.

reduction in the orbital overlap integrals and hence gives a small reduction in the strength of the metal–olefin bond relative to that formed by ethylene. The non-bonded interaction between the substituents in a *trans*-disubstituted olefin and the *cis*-chloride ligands are minimised by a rotation of the olefin about the metal–olefin bond axis (Fig. 36). Such a rotation, which has been

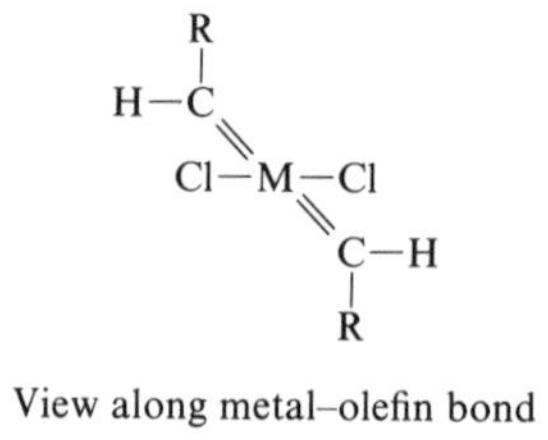

View along metal–olefin bond

Fig. 36. The rotation necessary to remove steric interference in a *trans*-disubstituted olefin complex.

found for *trans*-but-2-ene bound to platinum(II),[40] while not affecting the σ-(olefin to metal) bond, will markedly reduce the π-(metal to olefin)-bond strength by reducing the overlap between the filled *d*-orbitals and the empty olefin π*-(antibonding) orbitals. Since the π-component of the platinum(II)–olefin bond is of great importance in determining the overall strength of that bond, the rotation necessary to relieve the non-bonded interactions for *trans*-olefins will considerably reduce the overall platinum(II)–olefin bond strength and result in *cis*-olefins forming stronger bonds with metals than *trans*-olefins.

(vii) Optical activity in metal–olefin complexes

Olefins which have no symmetry planes perpendicular to the plane of the double-bond (e.g. *trans*-but-2-ene or propene) can co-ordinate to a metal in two enantiomorphous ways, **161** and **162**. If an optically active ligand

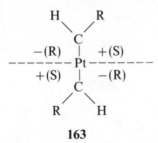

161   **162**

(e.g. 1-phenylethylamine) is also incorporated into the complex, then two diastereoisomers are formed, which can be separated by fractional re-crystallisation.[112–3] Both *cis*- and *trans*-isomers of optically active [Pt-(olefin)(amine)Cl₂] complexes have been prepared.[113–7] It is apparent that epimerisation cannot occur with a simple rotation of the olefin about its bond axis, but only by a mechanism involving cleavage of the metal–olefin bond. In the presence of traces of free olefins, epimerisation of the olefins in the *trans*-complexes takes place very rapidly due to the high rate of ex-change of free and co-ordinated olefin,[84] whereas the *cis*-complexes epimerise much more slowly.[115–6] The absolute configurations of two *cis*-complexes *cis*-[Pt(olefin)(S-α-phenylethylamine)Cl₂], where the olefin is *trans*-but-2-ene[40] or but-1-eme,[41] have been determined by x-ray diffrac-tion and preliminary results for *trans*-[Pt(olefin)(S-α-phenylethylamine)Cl₂], where the olefin is *cis*- or *trans*-but-2-ene, have been reported.[118–9] The C.D. and O.R.D. spectra of optically active olefin complexes have been studied extensively with a view to using them to determine the absolute configura-tions of olefins[96–8,116] and the correlation law for the complexes of mono-olefins with platinum(II) found to be given by the 'quadrant' rule **(163)**.

$$
\begin{array}{ccc}
\text{H} & & \text{R} \\
\diagdown & & \diagup \\
& \text{C} & \\
-(\text{R}) & | & +(\text{S}) \\
\text{------} & \text{Pt} & \text{------} \\
+(\text{S}) & | & -(\text{R}) \\
& \text{C} & \\
\diagup & & \diagdown \\
\text{R} & & \text{H}
\end{array}
$$

163

Optically active metal–olefin complexes have three main applications: (a) the resolution of optically active amines, (b) the investigation of the stereochemical course of chemical reactions and (c) the study of asymmetric induction reactions.

(a) Resolution of optically active olefins. Platinum(II) complexes containing optically active 1-phenylethylamine have been used to resolve a number of 1-olefins in which the asymmetric carbon atom is α or β to the double-bond. When *cis*-platinum(II) complexes are used, the 1-olefin whose chirality is

the same as that of the amine is preferentially complexed, whereas the reverse is true when the *trans*-complexes are used.[120] *Cis,trans*-1,5-cyclo-octadiene was partially resolved into its two enantiomorphs using 1-methyl-phenylethylamine as the optically active ligand[121] and completely resolved using 1-methylbenzylamine,[122] which was also used to resolve *trans*-cyclo-octene.[112] The olefins *trans*-cyclononene, *trans*-cyclodecene,[123] *trans*-6,7,10,11-tetrahydro-5-*H*-benzocyclononene[124] and *trans*-bicyclo[8,2,2]-tetradeca-5,10,11,13,tetraene[125] could be resolved as their diastereo-isomeric *trans*-[Pt(ol)(amine)Cl$_2$] complexes but the pure optical isomers of the free olefins could not be obtained as these undergo easy racemisation on separation from their complexes. Equilibrium constants for reaction 459 were found to be unity for *trans*-complexes,[115] but noticeably different from unity for *cis*-complexes, because of steric interaction between the amine and

$$[Pt(R\text{-olefin})(S\text{-amine})Cl_2] \rightleftharpoons [Pt(S\text{-olefin})(S\text{-amine})Cl_2] \quad (459)$$

the *cis*-co-ordinated olefin, although this effect was only observable for 1,2-disubstituted olefins.[117] As a result when *cis*-[Pt(C$_2$H$_4$)(1-phenyl-ethylamine)Cl$_2$] was equilibrated with (R,S)-*trans*-cyclo-octene, the un-reacted olefin contained an excess of the (−)-isomer, whereas treatment of the complex with potassium cyanide and examination of the olefin liberated indicated an excess of the (+)-isomer to be present.[116]

(b) *Investigation of the stereochemical course of chemical reactions.* By study-ing the reactions of complexes of known absolute configuration it is possible to determine the stereochemical course of these reactions. Thus nucleophilic attack of amines on mono-olefins co-ordinated to platinum(II) was shown to occur *trans* to the metal since co-ordinated S-butene gave only the S-tertiary amine in reaction 460.[126]

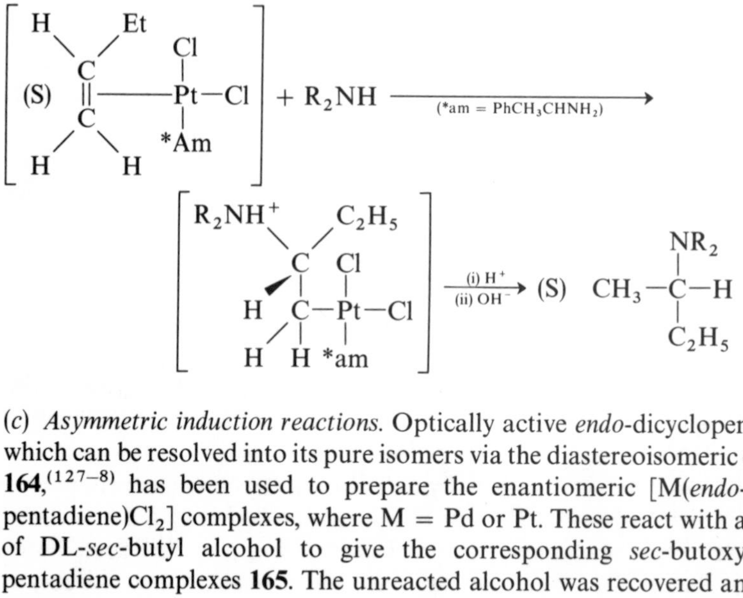

(c) *Asymmetric induction reactions.* Optically active *endo*-dicyclopentadiene, which can be resolved into its pure isomers via the diastereoisomeric complex **164**,[127–8] has been used to prepare the enantiomeric [M(*endo*-dicyclo-pentadiene)Cl$_2$] complexes, where M = Pd or Pt. These react with an excess of DL-*sec*-butyl alcohol to give the corresponding *sec*-butoxydicyclo-pentadiene complexes **165**. The unreacted alcohol was recovered and found

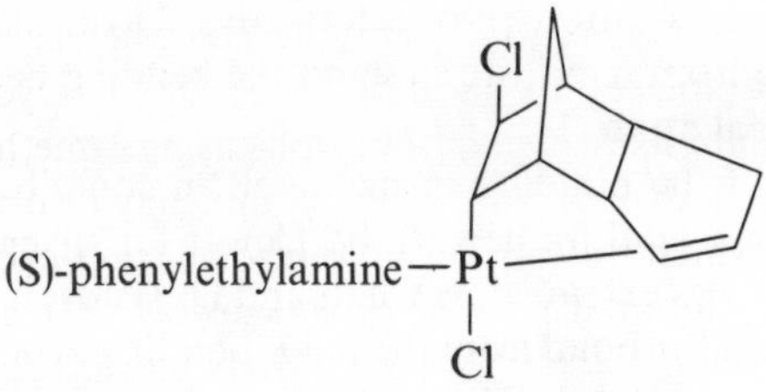

164 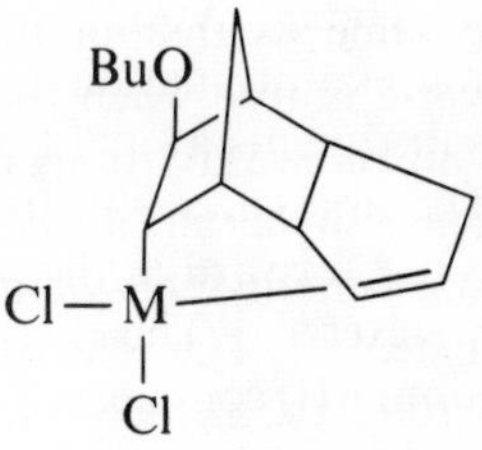   **165**

to be optically active indicating an asymmetric induction in the butoxylation reaction.[128] A slightly different type of asymmetric induction was observed in the formation of olefin complexes of platinum(II) with olefins containing asymmetric carbon atoms adjacent to the double-bond. Such asymmetric carbon atoms induce the opposite asymmetry in the co-ordinated olefin so that, for example, *trans*-[Pt(EtMeCH—CH=CH$_2$)(PhCH$_2$NH$_2$)Cl$_2$] instead of being a 1:1 mixture of the R,S and S,S complex in solution (equilibrium 461) exists as a 2:1 mixture of R,S and S,S.[129]

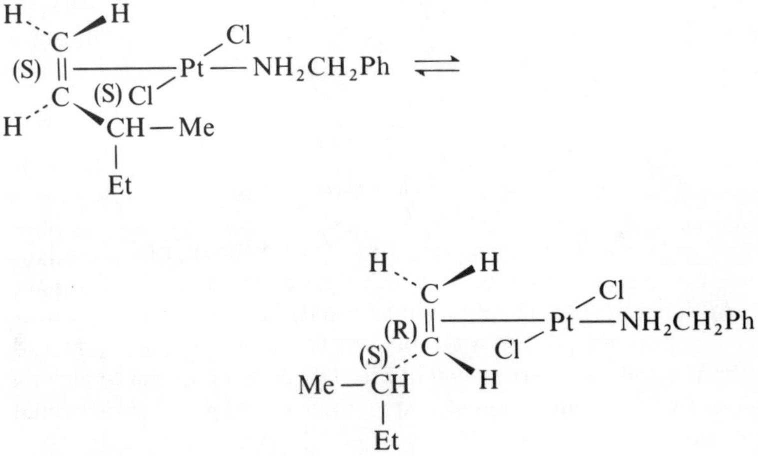

(461)

Bonding in metal(II)–olefin complexes

Any description of the bonding in metal(II)–olefin and –acetylene complexes must account for the following experimental observations:

(i) The structure of the olefin or acetylene is only modified slightly on co-ordination. The modifications include a very small lengthening of the multiple-bond and the substituents being bent out of the plane of the multiple-bond away from the metal.

(ii) The energy barrier to rotation of the olefin about the metal–olefin bond is low, but there is a preference for the olefin to be bound perpendicular to the square-plane containing the metal ion.

(iii) The olefin can twist slightly about the C=C double-bond axis.

(iv) The olefins exert a strong *trans*-effect as deduced from their influence on the substitution of ligands in the *trans*-position,[13] but little or no *trans*-influence as deduced from the lengths of the *trans*-metal–ligand bonds.

(v) The sp^2 hybridisation of the carbon atoms moves towards sp^3 hybridisation on co-ordination as deduced from n.m.r. studies and the bending back of substituents away from the metal atom.

The first successful treatment of the bonding in metal–olefin complexes was put forward in the molecular orbital framework by Dewar for silver(I) complexes.[131] This theory, which was extended by Chatt and Duncanson to platinum(II) complexes,[52] involves a σ-bond from the full π-bonding orbital of the olefin to an empty hybrid orbital on the metal complemented by π-back-donation of charge from a filled hybrid orbital on the metal to the empty π^*-(antibonding) orbital on the olefin (Fig. 37). The symmetries of

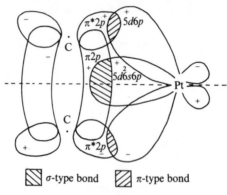

Fig. 37. Bonding in metal(II)–olefin complexes.

the relevant metal and ligand orbitals in the C_{2v} symmetry of $[Pt(C_2H_4)Cl_3]^-$ are shown in Table 112 for the olefin both perpendicular to and in the plane of the metal square-plane. The σ-(olefin to metal)–bond is formed by donation of charge from the full olefin π-orbital to the empty $sp_z d_{z^2} d_{x^2-y^2}$ hybrid orbital (the x, y and z axes are shown in Fig. 34) and the π-(metal to olefin)–bond is formed by back-donation of charge from either a $p_y d_{yz}$ hybrid orbital

TABLE 112

**Symmetries of the metal and ligand orbitals in the
C_{2v} symmetry of $[Pt(C_2H_4)Cl_3]^-$ (x, y and z axes
are those of Fig. 34)**

Symmetry	Olefin orbitals	Metal orbitals
Olefin perpendicular to metal square-plane		
A_1	π	$s, p_z, d_{z^2}, d_{x^2-y^2}$
A_2		d_{xy}
B_1	π^*	p_y, d_{yz}
B_2		p_x, d_{xz}
Olefin in the metal square-plane		
A_1	π	$s, p_z, d_{z^2}, d_{x^2-y^2}$
A_2		d_{xy}
B_1		p_y, d_{yz}
B_2	π^*	p_x, d_{xz}

(olefin perpendicular to the square-plane) or the d_{xz} orbital on the metal (olefin in the square-plane) to the olefin π^*-(antibonding) orbital. The d_{xz} orbital will give less overlap with the olefin π^*-(antibonding) orbital than the $p_y d_{yz}$ hybrid orbital because of the absence of an empty p-orbital to hybridise with it (the p_x orbital is involved in bonding to the chloride ligands in $[Pt(C_2H_4)Cl_3]^-$), since calculations have shown[132] that d_π-p_π hybridisation involving an unoccupied p-orbital can lead to a strengthened metal–ligand π-bond by increasing the overlap power of the metal d_π-orbital, and that only a small amount of p_y character would need to be added to the d_{yz} orbital in order for this effect to be noticed. This accounts for rotation of the olefin about the metal–olefin bond and explains why the perpendicular arrangement is the most stable configuration. It also accounts for the slight weakening of the olefinic double-bond on co-ordination, as required by the x-ray and infra-red studies, since both loss of electron density from the olefin π-(bonding) orbital and gain of electron density in the olefin π^*-(antibonding) orbital will result in a weakening of the olefinic double-bond. Recent molecular orbital calculations[133,134] have suggested that the platinum $6s$ orbital may not be significantly involved in the bonding. This has necessitated modification of the empty acceptor orbital on the metal from d^2sp to d^2p^2.

The high *trans*-effect of olefins (as deduced from their influence on the substitution of ligands in the *trans*-position—see p. 299) is readily explained by the above bonding scheme in terms of the π-bonding mechanism for the *trans*-effect (see Chapter 11, p. 301). The negligible *trans*-influence of olefins (see pp. 301–303) arises naturally from the bonding scheme described above, since ligands which exert their *trans*-effect by a π-acceptor mechanism will only cause *trans*-bond weakening if the *trans*-bond contains π-bonding. If there is no π-bonding (e.g. Pt–N) or only weak π-bonding (e.g. Pt–Cl) in the *trans*-bond then that bond may either not be weakened or may even be strengthened.[135]

So far the bonding scheme put forward has accounted for all the experimental observations listed at the start of this section except the slight change in the hybridisation of the olefinic carbon atoms from sp^2 to something approaching sp^3. To include this it is necessary to consider the bonding scheme of a completely sp^3 hybridised two-carbon unit. In this case the olefin is considered to be a dicarbanion which donates two electrons to platinum which is now formally in the $+4$ oxidation state. This at first sight appears very different to the σ-donation/π-back-donation scheme discussed above (see Fig. 37) but in reality the two approaches merge into one another.[11] This can be seen from Fig. 38 which shows the effect on the distribution of electron density in the region between the metal and the olefin when the relative importance of the σ- and π-components of the metal–olefin bonding scheme of Fig. 37 are altered. The two extremes, shown in Figs. 38(a) and

38(c), involve a migration of electron density from inside the $Pt\begin{smallmatrix} \diagup C \\ \diagdown C \end{smallmatrix}$ triangle,

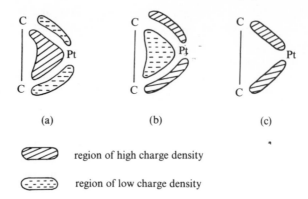

(a)   (b)   (c)

region of high charge density

region of low charge density

Fig. 38. The effect of varying the relative importance of the σ- and π-components of the metal(II)–olefin bond on the electron density in the region between the olefin and the metal. (a) sp^2-hybridised carbon atoms with a strong σ-(olefin to metal)-component and a weak π-(metal to olefin)-component in the metal–olefin bond. (b) sp^2-hybridised carbon atoms with a weak σ-component and a strong π-component in the metal–olefin bond. (c) sp^3-hybridised carbon atoms.

where it is located for sp^2 hybridised carbon atoms which give strong σ-donation and weak π-acceptance of electron density, to the edges of the

$$Pt \overset{\diagup C}{\underset{\diagdown C}{\Big|}}$$

triangle, where it is located for sp^3 hybridised carbon atoms. A carbon hybridisation scheme mid-way between sp^2 and sp^3 should give an electron density distribution mid-way between these two extremes. This mid-way point coincides with that for an sp^2 hybridisation scheme in which the π-back donation of charge (from metal to olefin) is as or more important than the σ-donation of charge (from olefin to metal). There is some evidence, such as that from stability constant studies for platinum(II)–olefin complexes,[100] that the π-back-donation of charge is more important than the σ-donation from the olefin to platinum for the formation of a stable platinum(II)–olefin bond (i.e. the situation shown in Fig. 38(b)). The electron density distribution around the olefin in Fig. 38(b) accounts for the reduced proton–proton coupling constants of co-ordinated olefins in the n.m.r. spectra, which were quoted above (p. 371) as evidence for some sp^3 hybridisation on the olefinic carbon atoms.

The bonding scheme in Fig. 38(b) can also account for the substituents R on the multiple-bond being bent out of the plane of the multiple-bond away from the metal, since on going from Fig. 38(a) to 38(b) the electron density is shifted towards the C–R bonds (where R = H or a more bulky group). This will repel the C–R bond by bond-pair/bond-pair repulsion so forcing R away from the metal.[136] When this purely qualitative argument was examined quantitatively using the Complete Neglect of Differential Overlap Molecular Orbital Theory it was shown that for acetylenes addition of electron density to the π^*-(antibonding) orbital must cause either *cis*- or *trans*-bending of the substituents. *Cis*-bending is favoured by both non-bonded interactions between the substituents and the central metal group and the fact that it enables the *s*-orbitals of the acetylene to assume the correct symmetry to contribute to the bonding.[137] Other explanations for the *cis*-bending of the substituents, that are in effect equivalent to the present one, consider the co-ordinated olefin or acetylene to be in an excited state.[90,138] However, there is little agreement as to the description of the excited state.[90,138,139]

In conclusion, the σ-donation/π-back-donation description of the bonding is metal(II)–olefin complexes, shown schematically in Fig. 37, can account for all the experimental observations so far obtained for platinum(II)– and palladium(II)–olefin and –acetylene complexes.

The greater stability of the platinum(II)– relative to palladium(II)–olefin complexes (see above p. 377) can be understood from the present molecular orbital theory of bonding. Using the co-ordinate axes of Fig. 34, the relative energies of the *d*-orbitals in platinum(II)–olefin complexes[95] are $d_{z^2} > d_{xz} > d_{x^2-y^2} > d_{xy} > d_{yz}$. If it is assumed that the same order applies in palladium(II)–olefin complexes, then the molecular orbital scheme shown in Fig. 39 is obtained. The stability of the complexes will increase with an

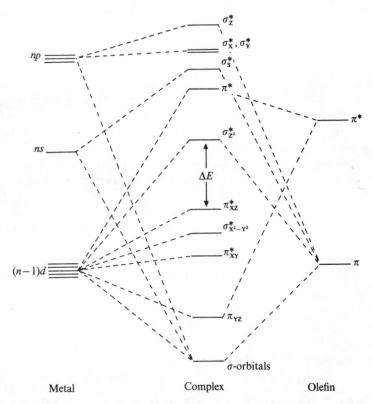

Fig. 39. Molecular orbital scheme for platinum(II)– and palladium(II)–olefin complexes. For simplicity the ligand orbitals associated with the ligands other than the olefin have been omitted.

increase in the energy difference between the highest filled and lowest unfilled molecular orbital (ΔE), and since ΔE increases on going from palladium(II) to platinum(II) (for example ΔE for $[PdCl_4]^{2-} = 19,150\ cm^{-1}$ and for $[PtCl_4]^{2-}$ $\Delta E = 23,450\ cm^{-1}$[(140)]), the stability of platinum(II)–olefin complexes will be greater than that of the palladium(II)–olefin complexes.

The greater spacial extension of the d-orbitals of platinum(II) relative to palladium(II) will confer greater stability on the platinum(II)– relative to the palladium(II)–olefin bond by giving greater overlap between the metal and olefin orbitals if both metals form metal–olefin bonds of equal length. Although no exactly comparable bond lengths have been determined the mean palladium(II)–olefin bond lengths in dichloro(norbornadiene)-palladium(II) (2·08 Å)[(38)] are, within experimental error, equal to the 2·02 Å found for the platinum(II)–olefin bond length in Zeise's salt.[(34)]

Chemical properties of metal(II)–olefin complexes

Platinum(II)–olefin complexes are a great deal more stable than palladium(II)–olefin complexes both to displacement of the olefin by a halogen (i.e. equilibrium 458, p. 375) and to reaction of the complex with a variety of reagents including water and most nucleophiles. In the following account only a brief discussion of each of the reactions of olefin complexes is given. For a fuller account the reader is referred to the author's recent comprehensive review of this subject.[(4)]

(i) Nucleophilic attack on co-ordinated olefins

Since the bonding scheme discussed above suggests that the electron density in the π-(bonding) orbital of the olefin is substantially reduced on complex formation, it is not surprising to find that olefins co-ordinated to metal ions are considerably more susceptible to nucleophilic attack than free olefins and conversely that co-ordinated olefins are more resistant to electrophilic attack than free olefins. Nucleophiles that have been shown to attack olefin complexes include OH^-, CH_3COO^-, CH_3O^-, Cl^-, alcohols, amines and amides. With the sole exception of hydroxide attack on co-ordinated mono-olefins all nucleophilic attacks that have been investigated in detail, which include attack of methoxide ions on co-ordinated diolefins[(141–5)] and attack of amines on co-ordinated mono-olefins,[(126)] occur by attack of the nucleophile on the side of the double-bond remote from the metal (eqn. 462). Furthermore,

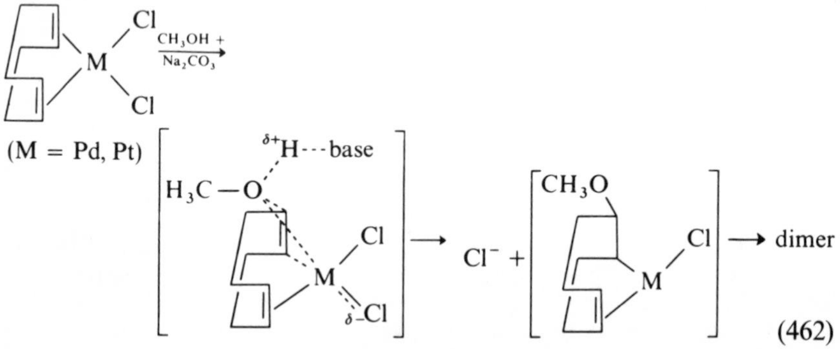

(462)

it appears likely that the nucleophile attacks between the carbon atoms of the double-bond rather than around the end of the double-bond,[146] since when the olefin is co-ordinated the regions of low and high electron density are between and around the end of the carbon atoms respectively, in direct contrast to the situation in the uncoordinated olefin. The products of nucleophilic attack on diolefin complexes are potentially susceptible to electrophilic attack, which occurs with regeneration of the original diolefin complex[147] (reaction 463).

$$
\left[\begin{array}{c} OH \\ H \\ Pd\,(\pi\text{-}C_5H_5) \end{array} \right] + HBF_4 \longrightarrow \left[\begin{array}{c} Pd\,(\pi\text{-}C_5H_5) \end{array} \right]^{+} BF_4^{-} + H_2O
$$

(463)

Nucleophilic attack on co-ordinated mono-olefins is characterised by reduction of the divalent metal to the zerovalent state and only one special case (reaction 464) in which nucleophilic attack does not lead to reduction of the metal has been reported.[148] In the case of palladium the free metal

$$
\begin{array}{ccc} Ph_3P & & PPh_3 \\ \diagdown & \diagup \\ Pt & Pt \\ \diagup & \diagdown \\ Cl & Cl\;Cl & Cl \end{array} + MeO^{-} \quad \underset{\substack{\text{pentadiene}}}{\overset{||\text{--}|| \,=\, \text{dicyclo-}}{\longrightarrow}}
$$

$$
\begin{array}{ccc} Ph_3P & Cl & Cl \\ \diagdown & \diagup\;\diagdown & \diagup \\ Pt & Pt \\ \diagup & \diagdown\;\diagup & \diagdown \\ Cl & Cl & PPh_3 \end{array}
$$

(464)

can be immediately re-oxidised to the divalent state if oxidising agents such as benzoquinone or copper(II) salts are added.[149] The copper(II) salts are reduced to copper(I) (eqn. 465) and these can be re-oxidised to copper(II) by blowing air through the solution (eqn. 466).

$$
Pd^{\circ} + 2CuCl_2 \rightarrow PdCl_2 + 2CuCl \tag{465}
$$

$$
2CuCl + 2HCl + \tfrac{1}{2}O_2 \rightarrow 2CuCl_2 + H_2O \tag{466}
$$

The metal(II)–olefin complexes can be prepared *in situ* so that a pseudo-catalytic attack of nucleophiles on olefins can occur. In general palladium(II) salts catalyse nucleophilic attack on olefins more readily than platinum(II) salts due to a number of factors:

(i) Palladium(II)–olefin complexes are formed more rapidly than platinum(II)–olefin complexes. For example, equilibria between palladium(II) salts and olefins are established in 30–40 minutes at 14·8°C.[109] whereas equilibria between platinum(II) salts and olefins require 14 days at 30°C.[100]

(ii) While the π-back donation of charge from metal to olefin is known to be considerable in the platinum(II)–olefin complexes,[100] it is possible that it is less in the palladium(II)–olefin complexes since the ionisation potentials of palladium are greater than those of platinum[150] indicating that palladium will less readily donate electrons back to the olefin than platinum. This will result in the electron density around the olefin double-bond being less when it is co-ordinated to palladium than platinum, and hence nucleophiles would more readily attack olefins co-ordinated to palladium than platinum.

(iii) The palladium–olefin bond is weaker than the platinum–olefin bond so that the activation energy of any rearrangement process that may be necessary during the course of the reaction will be lower for palladium– than platinum–olefin complexes.

(iv) Palladium(II) can more readily expand its co-ordination sphere to accept a fifth and sixth ligand than can platinum(II).[151-3] This allows the incoming nucleophile to co-ordinate to palladium(II) before attacking the olefin, thus lowering the activation energy of the nucleophilic attack.

A few important reactions involving nucleophilic attack on co-ordinated olefins are discussed briefly below:

(a) *Oxidative hydrolysis of olefins.* The oxidation of olefins co-ordinated to palladium(II) to carbonyl compounds has been developed commercially into the Wacker process.[2] The overall reaction (eqn. 467) has been shown to occur by a series of steps (eqn. 468–74) which are thought to involve the formation of an intermediate in which the olefin and attacking hydroxyl ion are co-ordinated *cis* with respect to each other.[104,154] As mentioned above,

$$C_2H_4 + PdCl_4^{2-} + H_2O \rightarrow CH_3CHO + Pd^\circ + 2HCl + 2Cl^- \quad (467)$$

$$PdCl_4^{2-} + C_2H_4 \rightleftharpoons [Pd(C_2H_4)Cl_3]^- + Cl^- \quad (468)$$

$$[Pd(C_2H_4)Cl_3]^- + H_2O \rightleftharpoons [Pd(C_2H_4)(H_2O)Cl_2] + Cl^- \quad (469)$$

$$[Pd(C_2H_4)(H_2O)Cl_2] + H_2O \rightleftharpoons [Pd(C_2H_4)(OH)Cl_2] + H_3O^+ \quad (470)$$

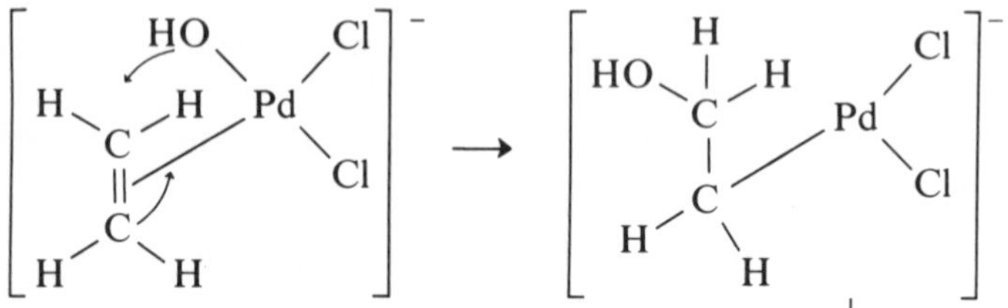

(471)

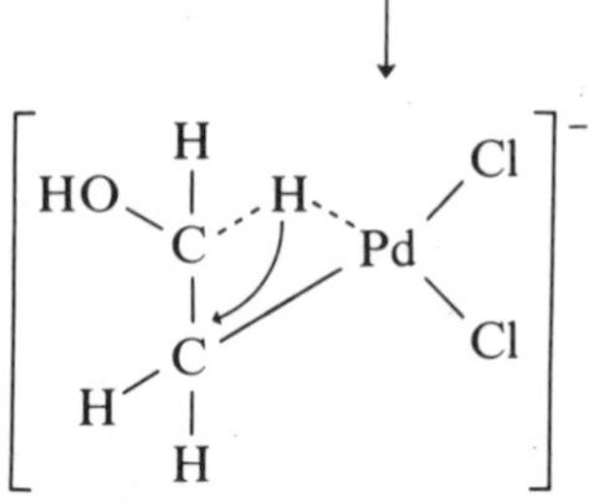

(472)

$$Pd^\circ + 2Cl^- + HO\!-\!\overset{\overset{\displaystyle H}{|}}{\underset{\underset{\displaystyle CH_3}{|}}{C}}{}^+ \quad (473)$$

$$CH_3CHOH^+ \rightarrow CH_3CHO + H^+ \qquad (474)$$

this is the only example known where nucleophilic attack on a co-ordinated olefin occurs *cis* with respect to the metal (the intramolecular attack of co-ordinated olefin by acetate ions may also be a further example of *cis*-attack, although the detailed mechanism of acetylation is less certain than that of oxidative hydrolysis[4]). It is quite possible to re-formulate the mechanism of this reaction so that it occurs by a *trans*-attack of the hydroxide ion on the co-ordinated olefin whilst still giving a mechanism that is consistent with the observed kinetic data.[155] Such a scheme involves reactions 470 and 471 in the above scheme being replaced by reaction 475. This mechanism has

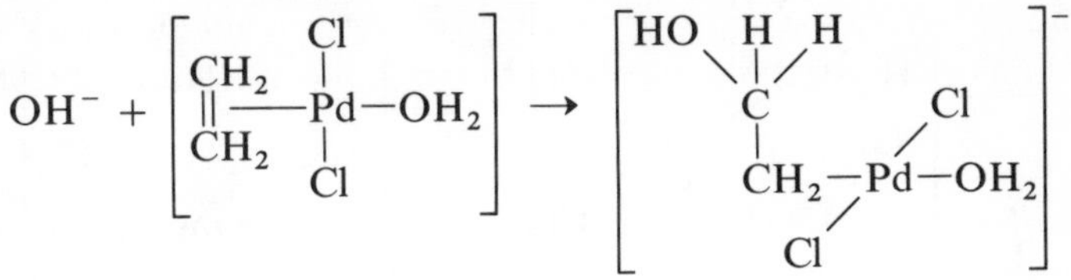

$$OH^- + \begin{bmatrix} \begin{matrix} & Cl \\ CH_2 & | \\ \| \quad -Pd-OH_2 \\ CH_2 & | \\ & Cl \end{matrix} \end{bmatrix} \rightarrow \begin{bmatrix} \begin{matrix} HO \quad H \quad H \\ \diagdown \; | \; \diagup \\ C \\ | \qquad\qquad Cl \\ CH_2-Pd-OH_2 \\ \diagup \\ Cl \end{matrix} \end{bmatrix}^- \qquad (475)$$

the advantage that the nucleophile can attack the olefin between the carbon atoms, where the electron density is depleted by co-ordination[146] rather than around the end of the carbon atoms where the π^*-(antibonding) orbitals, whose electron density is increased on co-ordination, are located, as is implied by a mechanism in which both the olefin and the hydroxide are co-ordinated to the metal before the attack occurs.[156] The co-ordinated aquo ligand in reaction 475, whose presence is indicated by the dependence of the rate of oxidation of ethylene in the presence of tetrachloropalladate(II) ions on the inverse square of the chloride concentration,[104,154] reduces the net negative charge on the complex (relative to $[Pd(C_2H_4)Cl_3]^-$), which in turn reduces the π-back donation of charge from the metal to the π^*-(antibonding) orbital of the olefin. Thus the olefin in the neutral complex $[Pd(C_2H_4)(H_2O)Cl_2]$ will be more susceptible to nucleophilic attack than the olefin in the anionic $[Pd(C_2H_4)Cl_3]^-$ complex.

Although the oxidative hydrolysis of olefins is not catalytic it can be made pseudo-catalytic by introducing copper(II) salts, which re-oxidise the palladium(0) to palladium(II) (eqn. 465), and blowing air in with the ethylene to re-oxidize the copper(I) to copper(II) (eqn. 466).

(b) Acetylation of olefins. The acetylation of olefins co-ordinated to palladium(II) is similar to the oxidative hydrolysis discussed above and has been widely investigated as a commercial method for the preparation of unsaturated esters. As with the oxidative hydrolysis of olefins the palladium(II)–olefin complex can be prepared *in situ*, so that when ethylene is passed into a solution of palladium(II) chloride in acetic acid containing an excess of sodium acetate the ethylene is oxidised to vinyl acetate, ethylidene diacetate and acetaldehyde. The mechanism of this reaction is still not completely certain, but at high acetate and low chloride concentrations it appears to proceed analogously to the oxidative hydrolysis of olefins to form *cis*-$[Pd(C_2H_4)(OAc)Cl_2]^-$ which can then either give vinyl acetate or

ethylidene diacetate depending on whether this product reacts to give a labile intermediate with a primary or a secondary carbon atom bound to

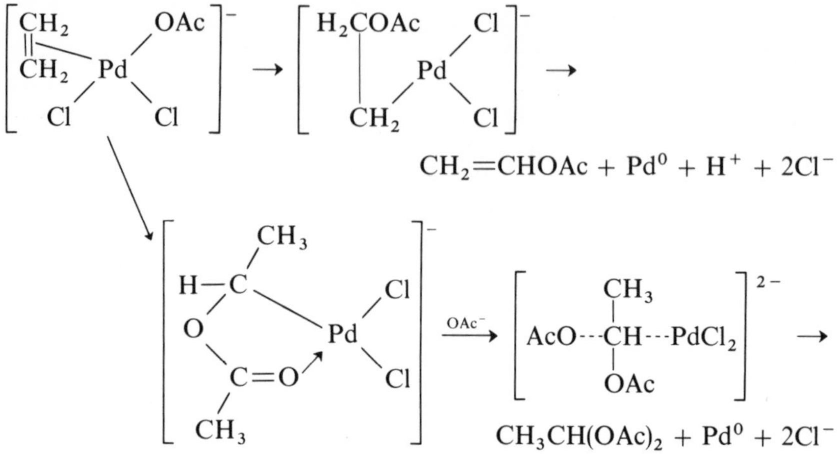

$$CH_2=CHOAc + Pd^0 + H^+ + 2Cl^-$$

$$CH_3CH(OAc)_2 + Pd^0 + 2Cl^-$$

SCHEME 4

palladium (Scheme 4). At low acetate and high chloride concentrations an intermolecular nucleophilic attack of co-ordinated olefin occurs (Scheme 5).

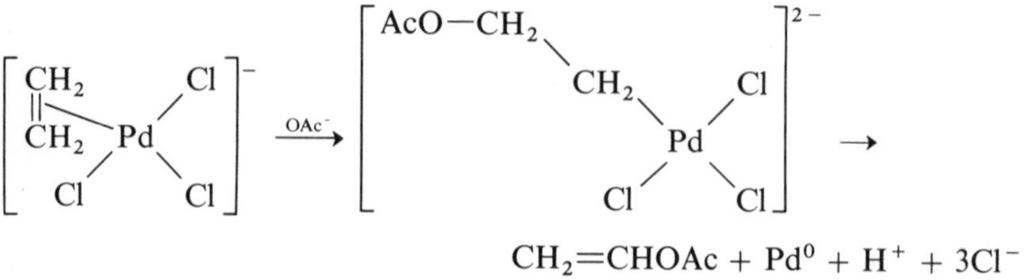

$$CH_2=CHOAc + Pd^0 + H^+ + 3Cl^-$$

SCHEME 5

(c) *Carbonylation of olefins.* The carbonylation of olefins co-ordinated to palladium(II) has been investigated as a potential commercial method for the synthesis of acid chlorides.[157] The overall reaction (eqn. 476) can be effected by passing a mixture of the olefin and carbon monoxide into a suspension of palladium(II) chloride in a non-aqueous solvent such as benzene in the presence of hydrogen chloride. By altering the solvent it is

$$2CH_2=CHR + 2CO + PdCl_2 \rightarrow 2RCHClCH_2COCl + Pd^0 \quad (476)$$

possible to obtain carboxylic acids, esters, lactones or oxo acids rather than acid chlorides. The mechanism (eqns. 477–9) involves nucleophilic attack on a co-ordinated olefin by chloride ion followed by carbon monoxide insertion into the palladium carbon σ-bond.

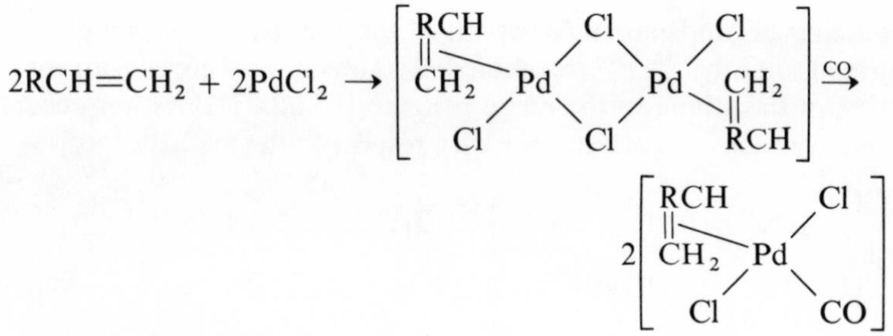

$$2RCH{=}CH_2 + 2PdCl_2 \rightarrow \left[\begin{array}{c} \underset{\overset{\|}{CH_2}}{RCH} \diagdown \\ \diagdown Pd \diagup \end{array} \begin{array}{c} Cl \\ \diagup \diagdown \diagup \\ \diagdown \diagup \diagdown \\ Cl \end{array} \begin{array}{c} Cl \\ \diagup \diagdown \\ Pd \diagdown \\ \diagup \end{array} \begin{array}{c} \\ CH_2 \\ \| \\ RCH \end{array} \right] \xrightarrow{CO}$$

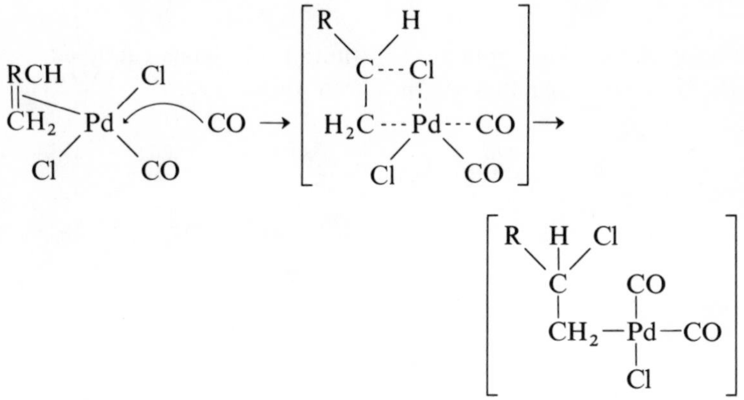

$$2 \left[\begin{array}{c} \underset{\overset{\|}{CH_2}}{RCH} \diagdown \\ \diagdown Pd \diagup \\ Cl \end{array} \begin{array}{c} Cl \\ \diagup \\ \diagdown \\ CO \end{array} \right] \qquad (477)$$

$$\begin{array}{c} \underset{\overset{\|}{CH_2}}{RCH} \diagdown \\ \diagdown Pd \diagdown \\ Cl \diagup \quad CO \end{array} \cdot CO \rightarrow \left[\begin{array}{c} R \quad H \\ \diagdown \diagup \\ C{\cdots}Cl \\ | \quad \vdots \\ H_2C{\cdots}Pd{\cdots}CO \\ \diagup \quad \diagdown \\ Cl \quad CO \end{array} \right] \rightarrow$$

$$\left[\begin{array}{c} R \quad H \quad Cl \\ \diagdown | \diagup \\ C \qquad CO \\ | \qquad | \\ CH_2{-}Pd{-}CO \\ | \\ Cl \end{array} \right] \qquad (478)$$

$$\begin{array}{c} H \\ | \\ R{-}C{-}Cl \quad CO \; CO \\ | \qquad \diagup \; \diagup \\ CH_2{-}\!\!\!-Pd\!\!-CO \\ | \\ Cl \end{array} \rightarrow \begin{array}{c} H \\ | \\ R{-}C{-}Cl \qquad CO \\ | \qquad\qquad | \\ CH_2{-}CO{-}Pd{-}CO \rightarrow \\ \diagdown \; \diagdown \\ Cl \quad CO \end{array}$$

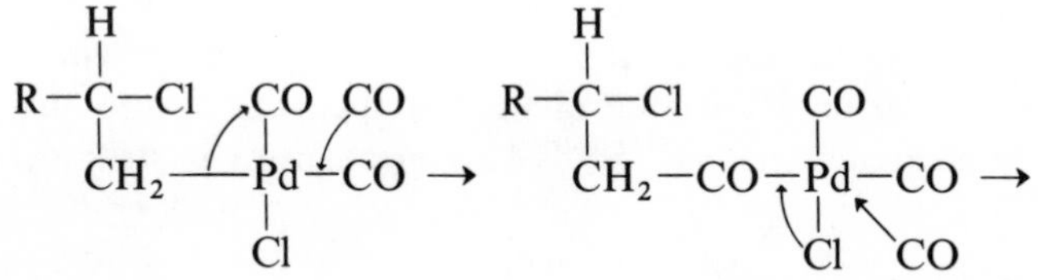

$$RCHClCH_2COCl + Pd(CO)_x \qquad (479)$$

(d) *Amination of olefins.* Studies of the nucleophilic attack of olefins by amines in the presence of palladium(II) salts have been hindered by the ability of the amine to co-ordinate to palladium(II),[158,159] although it has been established that the main organic product is a Schiff's base (reaction 480). The reaction can be studied more readily using diolefins, when a

$$C_2H_4 + RNH_2 + PdCl_2 \rightarrow CH_3CH{=}NR + Pd \text{ metal} + 2HCl \quad (480)$$

stable complex containing a palladium(II)–carbon σ-bond is formed[160-3] (reaction 481), or using tertiary phosphine or amine complexes of mono-

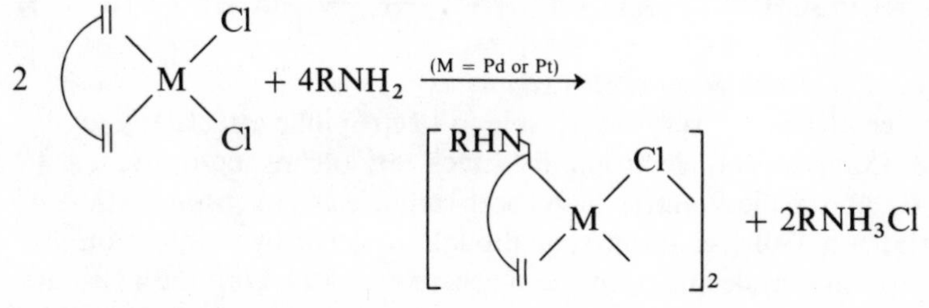

$$2 \left(\begin{array}{c} \| \\ \\ \| \end{array} \diagdown M \diagup \begin{array}{c} Cl \\ \diagdown \\ Cl \end{array} \right) + 4RNH_2 \xrightarrow{(M = Pd \text{ or } Pt)}$$

$$\left[\begin{array}{c} RHN \\ \Big(\quad \diagdown M \diagup \begin{array}{c} Cl \\ \diagdown \\ \| \quad \diagdown \end{array} \end{array} \right]_2 + 2RNH_3Cl \qquad (481)$$

olefins since these ligands stabilise the metal–carbon σ-bonded complexes that are formed initially[124,126] (reaction 482). Thus it has been shown that co-ordination of the amine to the metal prior to the attack does not occur since the amine attacks the olefin *trans* with respect to the metal.[126]

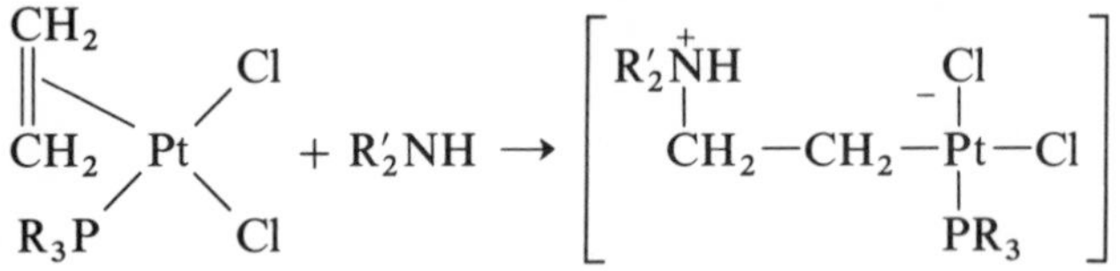

$$
\begin{array}{c}
CH_2 \\
\| \\
CH_2 \\
\diagdown Pt \diagup Cl \\
\diagup \diagdown \\
R_3P \quad Cl
\end{array}
+ R_2'NH \rightarrow
\left[
\begin{array}{c}
R_2'\overset{+}{N}H \\
| \\
CH_2-CH_2-\overset{\underset{|}{Cl}}{Pt}-Cl \\
| \\
PR_3
\end{array}
\right]
\quad (482)
$$

(e) *Arylation of olefins.* The aromatic substitution of olefins catalysed by palladium(II) complexes (reaction 483) has been widely studied.[165,166] The

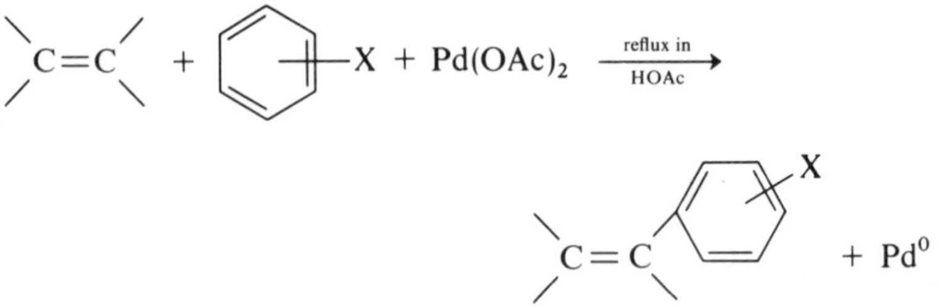

$$
\overset{\diagdown}{\diagup}C=C\overset{\diagup}{\diagdown} + \quad \text{[benzene]} + -X + Pd(OAc)_2 \xrightarrow[\text{HOAc}]{\text{reflux in}}
$$

$$
\overset{\diagdown}{\diagup}C=C \quad \text{[aryl]} -X \quad + Pd^0 \quad (483)
$$

mechanism is thought to involve the formation of a phenyl–palladium complex, followed by co-ordination of the olefin to the palladium via a σ-bond, to give a complex in which the olefin can then insert into the palladium–phenyl bond (reaction 484).[166–8] This mechanism is strongly

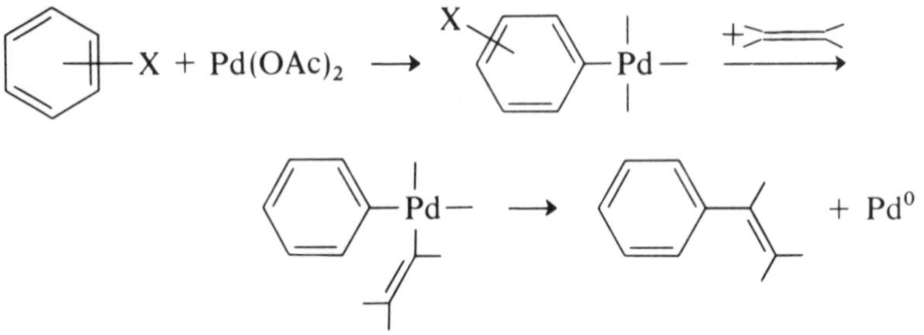

$$
\text{[benzene]} + -X + Pd(OAc)_2 \rightarrow \text{X} \text{[aryl]} -Pd- \quad \overset{+}{=\!=\!=} \rightarrow
$$

$$
\text{[aryl]} -Pd- \rightarrow \text{[aryl]} \quad + Pd^0 \quad (484)
$$

supported by the observation that the σ-bonded palladium(II)–olefin complex *trans*-[(PPh$_3$)$_2$Pd(CH=CCl$_2$)Cl] reacts with benzene to give β-β-dichlorostyrene (reaction 485).[169]

$$
trans\text{-}[(PPh_3)_2Pd(CH=CCl_2)Cl] + PhH \xrightarrow[\text{reflux}]{\text{HOAc, AgOAc}} PhCH=CCl_2 \quad (485)
$$

(ii) *Electrophilic attack on co-ordinated olefins*

Although free olefins are very susceptible to electrophilic attack,[170] no un-challenged examples of electrophilic attack on olefins co-ordinated to platinum(II) and palladium(II) have been reported. The bromination of **166**[171] (reaction 486) was at one time thought to occur by *cis*-addition of a polarised bromine molecule with the negative end attacking the platinum

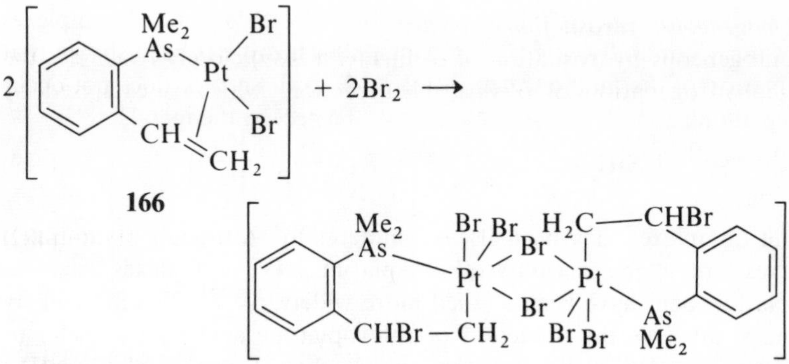

$$2 \left[\begin{array}{c} \text{Me}_2 \quad \text{Br} \\ \text{As} \diagdown \diagup \\ \quad \text{Pt} \\ \text{CH} \diagdown \text{Br} \\ \quad \diagdown \text{CH}_2 \end{array} \right] + 2\text{Br}_2 \longrightarrow$$

166

$$\left[\begin{array}{c} \text{Me}_2 \quad \text{Br Br} \quad \text{H}_2\text{C}\text{---CHBr} \\ \text{As} \diagdown \quad | \diagup \diagup \text{Br} \quad | \\ \qquad \text{Pt} \diagdown \qquad \text{Pt} \\ \quad | \qquad \text{Br} \diagup | \quad \text{As} \\ \text{CHBr---CH}_2 \quad \text{Br Br} \quad \text{Me}_2 \end{array} \right] \qquad (486)$$

and the positive end attacking the double-bond,[172] but it is now considered more likely that this and related reactions occur by a free-radical mechanism.[173]

(iii) Homogeneous hydrogenation of olefins

Both platinum(II) and palladium(II) complexes catalyse the homogeneous hydrogenation of olefins but by far the best and most widely investigated system involves platinum(II) complexes containing $SnCl_3^-$ ligands, which readily catalyse the homogeneous hydrogenation of ethylene and acetylene,[14] but have found their greatest use in the selective reduction of polyolefins (e.g. in soybean oil) to mono-olefins.[3,174] The reaction involves the formation of a complex in which both the hydrogen and the di-olefin are bound to the metal, followed by isomerisation of the double-bonds to give a conjugated di-olefin. This then reacts through an alkyl complex, as shown in eqn. 487, where M = Pd or Pt, to give a mono-olefin hydrido complex from which the mono-olefin is displaced by free diolefin. In support

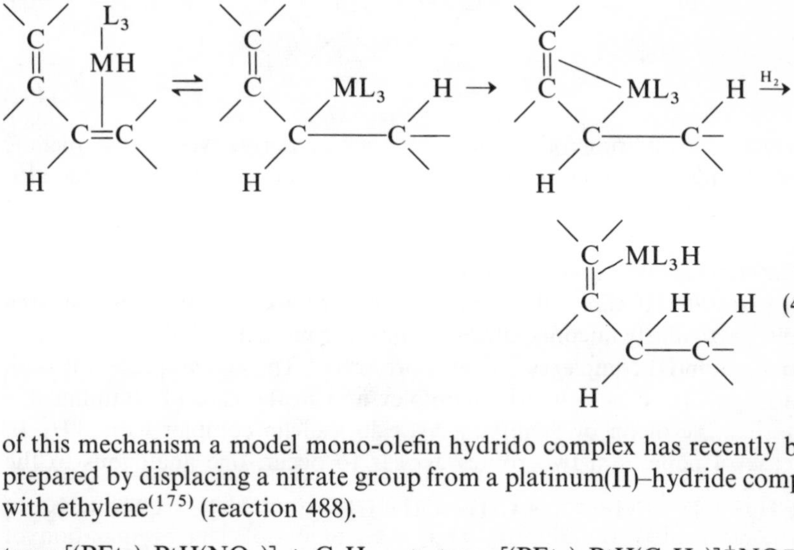

$$(487)$$

of this mechanism a model mono-olefin hydrido complex has recently been prepared by displacing a nitrate group from a platinum(II)–hydride complex with ethylene[175] (reaction 488).

$$trans\text{-}[(\text{PEt}_3)_2\text{PtH}(\text{NO}_3)] + \text{C}_2\text{H}_4 \longrightarrow trans\text{-}[(\text{PEt}_3)_2\text{PtH}(\text{C}_2\text{H}_4)]^+\text{NO}_3^-$$

$$(488)$$

(iv) *Homogeneous hydrosilation of olefins*

The homogeneous hydrosilation of olefins (eqn. 489) closely resembles the catalytic hydrogenation of olefins. The reaction, which is catalysed by

$$R_3SiH + \quad C{=}C \quad \rightarrow R_3Si{-}\overset{|}{C}{-}\overset{|}{C}{-}H \qquad (489)$$

platinum complexes, has been used commercially. Although platinum(II) complexes are effective catalysts,[176] platinum(IV) complexes such as chloroplatinic acid have been studied more widely.[177–80] The most likely mechanism involves the reduction of chloroplatinic acid by the olefin to give a platinum(IV)–olefin complex which then forms a platinum(IV) complex by addition of the silane (eqn. 490). This adduct then rearranges to give an alkyl complex (eqn. 491) from which the alkylsilicon product is displaced by reaction with further olefin (eqn. 492). An analogous reaction

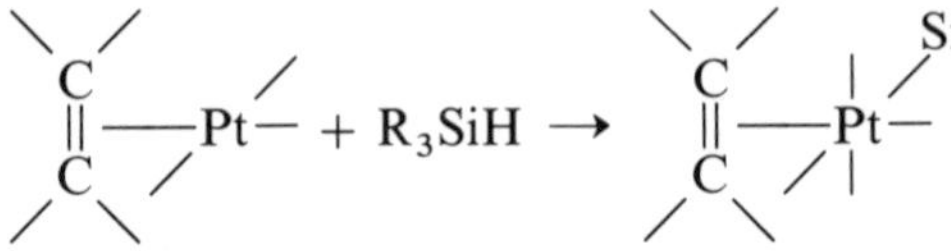

$$\qquad (490)$$

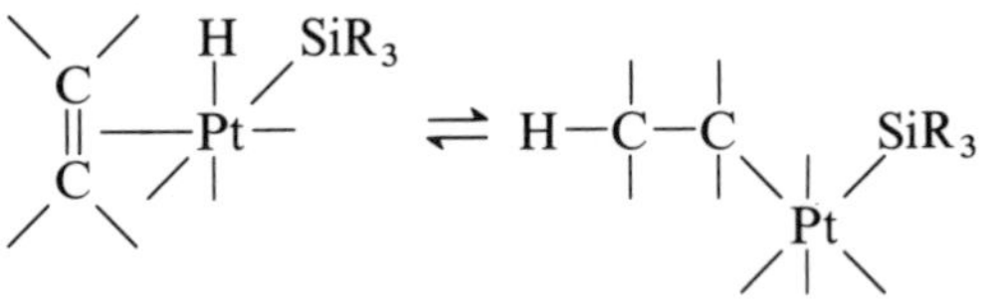

$$\qquad (491)$$

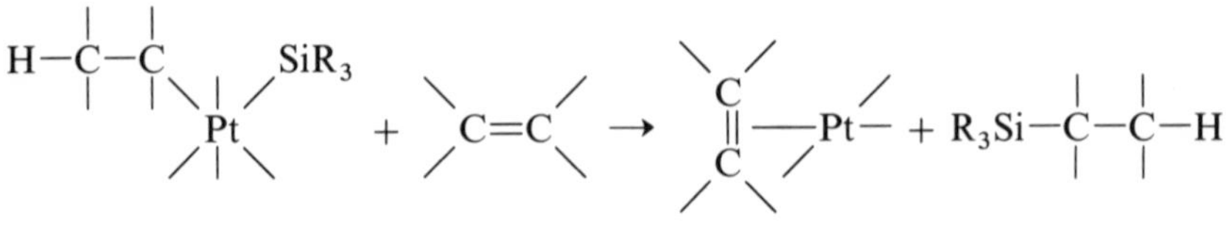

$$\qquad (492)$$

in which the addition of germanium hydride derivatives across olefinic double-bonds is homogeneously catalysed by chloroplatinic acid has also been reported.[181]

(v) *Catalytic isomerisation of olefins*

Both platinum(II) and palladium(II) complexes are capable of isomerising olefins, particularly unconjugated diolefins to conjugated diolefins, although the palladium(II) complexes are far more active. The active species for both metals appears to be a hydrido-complex and in the case of platinum this isomerises the olefin by forming a hydrido-π-olefin complex (eqn. 493). In the case of palladium two routes appear to occur, one analogous to the

$$-CH_2{-}CH{=}CH{-} \rightleftarrows -CH_2{-}\underset{M}{CH}{-}CH_2{-} \rightleftarrows -CH{=}CH{-}CH_2{-}$$
$$\underset{MH}{} \qquad\qquad \underset{M}{} \qquad\qquad\qquad \underset{MH}{}$$

$$(M = Pd^{II}, Pt^{II}) \qquad\qquad\qquad\qquad\qquad\qquad\qquad (493)$$

platinum route (eqn. 493) and the other involving a hydrido-π-allyl complex (eqn. 494). This reflects the strong tendency for palladium to form π-allyl

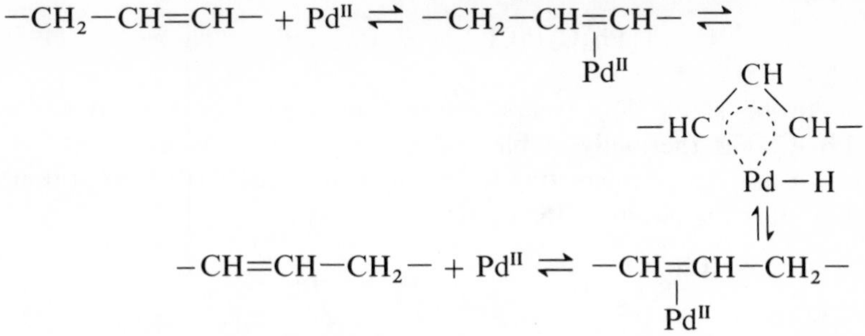

$$-CH_2-CH=CH- + Pd^{II} \rightleftharpoons -CH_2-CH=CH- \rightleftharpoons$$
$$\underset{Pd^{II}}{|}$$

$$-HC \underset{Pd-H}{\overset{CH}{\diamond}} CH-$$
$$\updownarrow$$

$$-CH=CH-CH_2- + Pd^{II} \rightleftharpoons -CH=CH-CH_2- \quad (494)$$
$$\underset{Pd^{II}}{|}$$

complexes in preference to olefin complexes (see Chapter 14, p. 441).

(vi) Polymerisation of olefins

Although palladium(II) complexes are effective homogeneous catalysts for the polymerisation of olefins only a few examples of homogeneous polymerisation catalysed by platinum(II) complexes have been reported. The palladium(II) catalysed reaction is dependent on both the nature of the solvent and the other ligands present. During the polymerisation palladium(II) is reduced to the metal unless oxidising agents such as cupric or ferric salts are present. Although a number of studies of the mechanisms of this reaction have been made, the results are far from complete; most of the mechanisms put forward involve the formation of alkyl–palladium complexes either by insertion of the olefin into the palladium–chlorine bond or by incorporation of a proton from the solvent.

(vii) Polymerisation of acetylenes

Although palladium(II) complexes do polymerise acetylenes to give polyacetylenes, by far the most widely investigated polymerisation is the dimerisation of disubstituted acetylenes to give cyclobutadienes. Treatment of Na_2PdCl_4 or $PdCl_2$ with diphenylacetylene in ethanol,[182,183] or [(PhCN)$_2$-PdCl$_2$] with diphenylacetylene or di(*para*-substituted phenyl)acetylene in ethanol–chloroform yields the dimeric complex [Pd{C$_4$Ph$_4$(OC$_2$H$_5$)}Cl]$_2$, which has been shown by x-ray diffraction to be a mixture of two π-allyl complexes one of which has the ethoxy group in an *endo* position with respect to palladium and the other in an *exo* position[188] (Fig. 40). [Pd{C$_4$Ph$_4$(OEt)}Cl]$_2$ reacts with hydrochloric acid to give [Pd(C$_4$Ph$_4$)-Cl$_2$][182] (167) in which cyclobutadiene is co-ordinated to palladium(II). The tetraphenylcyclobutadiene platinum(II) complex [(Ph$_4$C$_4$)PtCl$_2$]$_n$, where *n* is probably 2, is also prepared from diphenylacetylene following reaction

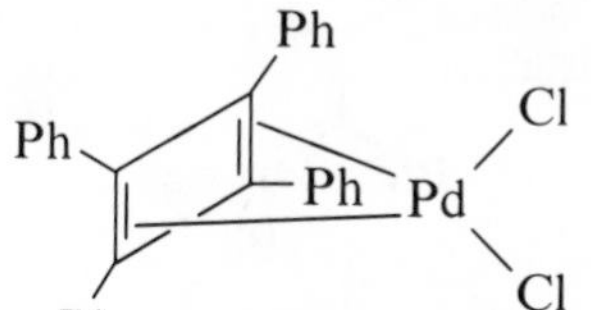

167

495.[189] Cyclobutadiene complexes such as **167** were predicted, on theoretical

$$[Pt(CO)_2Cl_2] + PhC \equiv CPh \xrightarrow{\text{ether}} [Ph_4C_4PtCl_2]_n + C_6Ph_6 + Ph\!-\!\underset{O}{\diagdown}\!-\!Ph \quad (495)$$

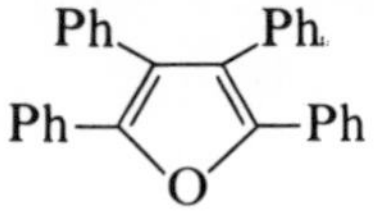

grounds, to be stable four years before they were first prepared.[190] $[Pd(C_4Ph_4)Cl_2]$ is thermally stable and resistant to oxidising agents. It undergoes a number of interesting reactions (eqns. 496–501) which appear to involve either the diradical **168** or **169**.

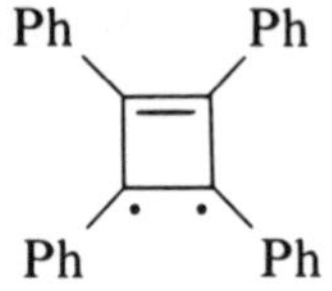

168 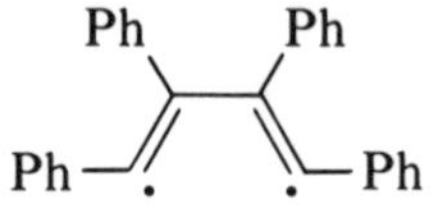   **169**

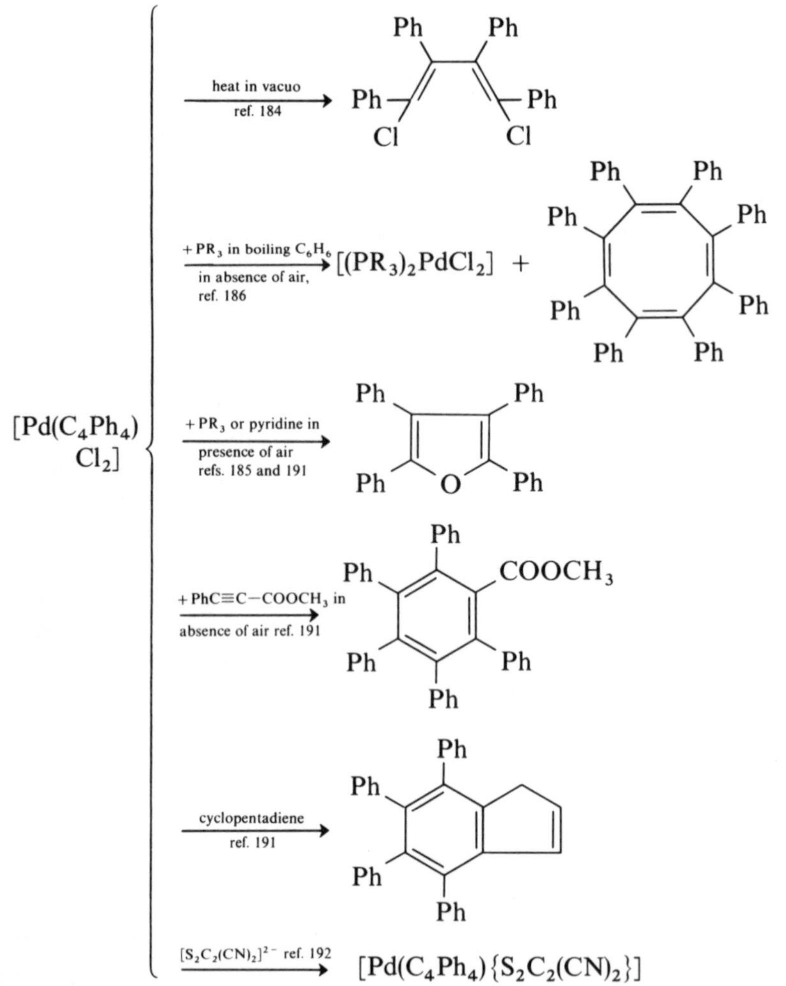

$[Pd(C_4Ph_4)Cl_2]$

$\xrightarrow[\text{ref. 184}]{\text{heat in vacuo}}$ (496)

$\xrightarrow[\substack{\text{in absence of air,}\\ \text{ref. 186}}]{+ PR_3 \text{ in boiling } C_6H_6} [(PR_3)_2PdCl_2] +$ (497)

$\xrightarrow[\substack{\text{presence of air}\\ \text{refs. 185 and 191}}]{+ PR_3 \text{ or pyridine in}}$ (498)

$\xrightarrow[\text{absence of air ref. 191}]{+ PhC \equiv C-COOCH_3 \text{ in}}$ (499)

$\xrightarrow[\text{ref. 191}]{\text{cyclopentadiene}}$ (500)

$\xrightarrow[]{[S_2C_2(CN)_2]^{2-} \text{ ref. 192}} [Pd(C_4Ph_4)\{S_2C_2(CN)_2\}]$ (501)

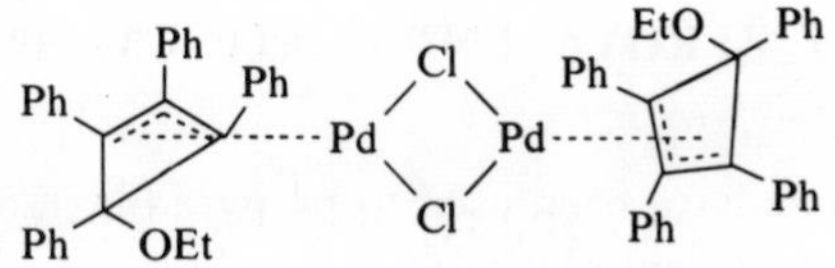

Endo

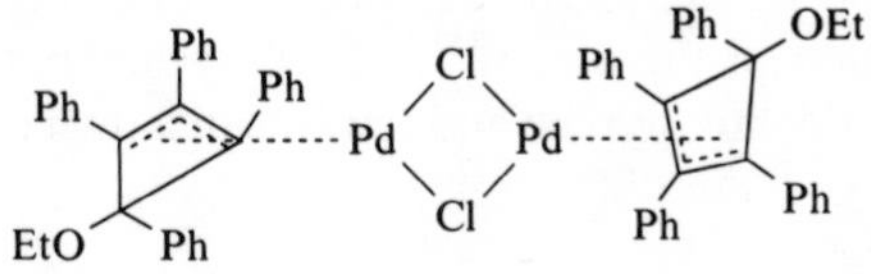

Exo

Fig. 40. Structures of *endo-* and *exo-*[Pd{C₄Ph₄(OEt)}Cl]₂.

Structure and bonding in hydroxy–acetylene complexes of the divalent metals

Hydroxyacetylenes, such as **170** and **171**, co-ordinate more strongly to

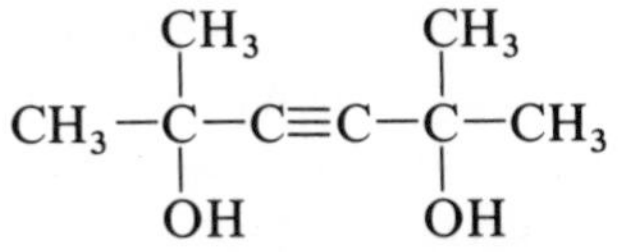

       170 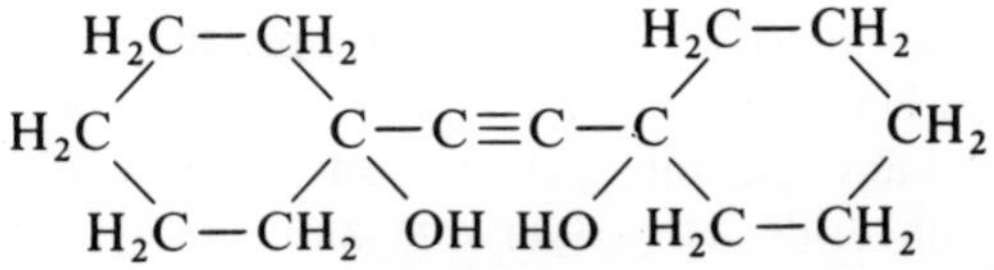   **171**

platinum(II) than ethylene to form stable complexes of the type $[Pt(ac)Cl_3]^-$ where ac = acetylene.[193–6] On co-ordination to platinum(II) the triple-bond stretching frequency in the infra-red is lowered by about $200 \, cm^{-1}$, which is consistent with the formation of a platinum(II)–acetylene π-bond,[194] and in addition the O–H stretching frequency is lowered by about $125 \, cm^{-1}$ on co-ordination.[194,197] Two possible explanations have been put forward to account for this: (i) Co-ordination of the hydroxyl oxygen to platinum(II),[197,198] (ii) Hydrogen-bonding between the hydroxyl groups and the *cis*-chloride ligands. A third possibility, namely hydrogen bonding between the hydroxyl group and the platinum atom, was eliminated by the fact that the hydroxyl proton resonance in the n.m.r. spectrum occurred at lower field than in the free ligand.[194] A recent x-ray determination of the structure of $K^+[Pt\{(C_2H_5)_2-\underset{\underset{OH}{|}}{C}-C\equiv C-\underset{\underset{OH}{|}}{C}-(C_2H_5)_2\}Cl_3]^-$ has shown that neither of the oxygen atoms are directly above or below the platinum atom thus eliminating the possibility of co-ordination of the hydroxyl oxygen to platinum(II).[199] The crystal data suggest that the acetylenic bond is slightly tilted away from the perpendicular direction and that one hydroxyl group is hydrogen bonded to the *cis*-chloride ligand whilst the other is completely free.

COMPLEXES OF ZEROVALENT PLATINUM AND PALLADIUM

Preparation

A number of methods have been used to prepare olefin and acetylene complexes of formally zerovalent platinum and palladium.

(i) From cis-$[(PPh_3)_2PtCl_2]$

Olefin complexes of *trans*-stilbene, *trans*-4,4'-dinitrostilbene, acenaphthylene and allene can be prepared by reduction of *cis*-$[(PPh_3)_2PtCl_2]$ with hydrazine followed by addition of the olefin;[200,201] ethylene, butadiene, hex-1-ene, cyclohexene, allyl alcohol, styrene and tetraphenylethylene do not react.[146]

(ii) From $[Pt(PPh_3)_4]$

Refluxing $[Pt(PPh_3)_4]$ in benzene with olefins gives olefin complexes with chloro-olefins,[202] fluoro-olefins,[203] tetracyanoethylene,[204] fumaronitrile[205] and allene[206] (reaction 502). If, however, the benzene is replaced by ethanol the chloro-olefin–platinum(0) complex is isomerised to a

$$[Pt(PPh_3)_4] + C_2(CN)_4 \rightleftharpoons [(PPh_3)_2Pt\{C_2(CN)_4\}] + 2PPh_3 \qquad (502)$$

$$[(PPh_3)_2Pt(CCl_2{=}CCl_2)] \xrightarrow{C_2H_5OH} [(PPh_3)_2PtCl(CCl{=}CCl_2)] \qquad (503)$$

platinum(II)–carbon σ-bonded product[206a] (reaction 503). By contrast, even in benzene, $[Pd(PPh_3)_4]$ reacts with chloro-olefins to give complexes containing palladium(II)–carbon σ-bonds, although olefins such as maleic anhydride, ethyl fumarate and tetracyanoethylene react to give normal olefin complexes of zerovalent palladium.[207–9] The reaction of chloro-olefins is typical of organic halogen compounds, such as methyl iodide, which react with $[Pd(PPh_3)_4]$ to give complexes which contain a palladium(II)–carbon σ-bond.[210]

(iii) From $[Pt(PPh_3)_2]$

A number of olefins and acetylenes which either co-ordinate too weakly to displace triphenylphosphine from $[Pt(PPh_3)_4]$ (reaction 502 is reversible) or which react with hydrazine, can form complexes by reaction with $[Pt(PPh_3)_2]$.[211] $[Pt(PPh_3)_2]$ is not always isolated but prepared *in situ* by reducing $[(PPh_3)_2PtO_2]$ with sodium borohydride or hydrazine.[212] If this solution is treated with an olefin (such as ethylene) the olefin complex is isolated. For olefins and acetylenes that are susceptible to sodium borohydride or hydrazine the ethylene complex $[(PPh_3)_2Pt(C_2H_4)]$ can be used as a starting material in which case the reaction occurs by liberation of ethylene and formation of $[Pt(PPh_3)_2]$, which then reacts with the new olefin or acetylene.[213] The advantages of this method are that the reaction conditions are mild and the only by-product is ethylene, which is gaseous, and therefore easily removed. Purification is therefore simple and yields are good. However, in spite of these advantages of the ethylene complex the bis-triphenylphosphine platinum(0) complexes of phenylacetylene,[204] fumaronitrile[205] and *trans*-stilbene[213] have also been used as starting materials. Recently a new method of preparing $[M(PPh_3)_2]$ *in situ* in the absence of competing ligands by the ultra-violet irradiation of $[(PPh_3)_2-M(C_2O_4)]$ (M = Pd, Pt) has been described.[214,215] Although the application of this reaction has not yet been fully investigated it is known that after the oxalate ion has been discharged as two molecules of carbon dioxide $[Pt(PPh_3)_2]$ dimerises unless suitable ligands, such as $PhC{\equiv}CPh$, are present.

(iv) From hydrido–platinum(II) complexes
Tetracyanoethylene (TCNE) reacts with *trans*-$[(PPh_3)_2PtHCl]$ to form a labile complex $[(PPh_3)_2PtHCl(TCNE)]$ which then undergoes slow loss of hydrogen chloride to give $[(PPh_3)_2Pt(TCNE)]$.[204,216] The hydrido–olefin complex $[(PPh_3)PtH(CN)(TCNE)]$ was isolated during the reaction of the corresponding cyano-complex.[216]

(v) From palladium(II) complexes
Air sensitive white crystalline palladium(0)–ethylene complexes have been prepared by treating a solution of palladium(II) acetylacetonate with (ethoxy)diethylaluminium in the presence of ethylene and a tertiary phosphine[217] (reaction 504).

$$[Pd(acac)_2] + EtOAlEt_2 + C_2H_4 + 2PR_3 \xrightarrow[\substack{R = Ph, \text{ cyclohexyl} \\ \text{or } o\text{-}CH_3C_6H_4O-}]{Et_2O, \text{ room temp.}}$$

$$[(PR_3)_2Pd(C_2H_4)] \qquad (504)$$

X-ray structural studies on zerovalent complexes
A number of olefin and acetylene complexes of formally zerovalent platinum have been reported (Table 113). In contrast to the olefin and acetylene complexes of the divalent metals, the plane of the double or triple bond lies close to the PtP_2 plane with a dihedral angle α (Fig. 41) of between 2° and 14°. In all the complexes studied the substituents bound to the multiple bond are bent out of the plane of this bond away from the metal (40° for the phenyl groups in $PhC{\equiv}CPh$,[224] 32.5° for the cyano-groups in TCNE[219] and 40.6° for the chloro-groups in tetrachloroethylene[221]).

TABLE 113
X-ray structural results for platinum(0) complexes[a]

Complex	α	β	Multiple bond length (Å)		No. of structure in Appendix II	Reference
			Complex	Free		
$[(PPh_3)_2Pt(TCNE)]$	8·3°	41·5°	1·49	1·31	206	218, 219
$[(PPh_3)_2Ni(C_2H_4)]^b$	12°	42°	1·46	1·33	—	220
$[(PPh_3)_2Pt(Cl_2C{=}CCl_2)]$	12·3°	47°	1·62		199	221
$[(PPh_3)_2Pt(Cl_2C{=}C(CN)_2)]$	1·9°	40·6°	1·42		202	222
$[(PPh_3)_2Pt(NC{-}C{\equiv}C{-}CN)]$	8°		1·40	1·19	203	223
$[(PPh_3)_2Pt(PhC{\equiv}CPh)]$	14°	39°	1·32	1·19	207	224

[a] See Fig. 41 for definition α and β.
[b] Isomorphous with the platinum(0) analogue (ref. 225).

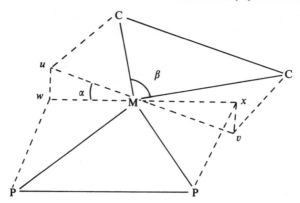

Fig. 41. The structure of olefin and acetylene complexes of platinum(0). The plane containing the points u, v, w and x together with the atom M is perpendicular to both the MCC and MPP planes.

Bonding in zerovalent metal–olefin and –acetylene complexes

The bonding in olefin and acetylene complexes of zerovalent platinum can be described in either molecular-orbital or valence-bond terms. Both descriptions are given below in an attempt to show why the molecular-orbital approach is more useful. The two descriptions are, of course, equivalent.

(i) Molecular-orbital treatment

Percyano-olefins and -acetylenes are very good π-acceptor ligands but poor σ-donors. Thus if the bonding is described in terms of the Dewar–Chatt–Duncanson model[52,131] (see above p. 382 and Fig. 42), it would be expected that the π-bond (from platinum(0) to olefin or acetylene) would be fairly strong whereas the σ-bond (from olefin or acetylene to platinum(0)) would be almost non-existent, giving a large build-up of electron density in the ligand π^*-(antibonding)-orbitals, which would weaken and lengthen the multiple bond as observed in Table 113. Furthermore, molecular orbital calculations have shown that addition of electron density to the antibonding orbitals of TCNE increases the bond order of the C–CN bond.[226] The decrease of the C–CN bond from 1·449 Å in free TCNE to 1·41 $\pm$ 0·04 Å on co-ordination to platinum[218,219] is thus consistent with the Dewar–Chatt–Duncanson model. Similarly, the bending back of the substituents attached to the multiple bond away from the platinum atom is also consistent with a substantial electron density in the π^*-(antibonding) ligand orbitals[137] (see p. 384).

So far the Dewar–Chatt–Duncanson model has been very successful in explaining the observed structures of these zerovalent complexes. How does it explain the near planarity of these complexes as opposed to the almost perpendicular arrangement of the olefin in relation to the metal square-plane in divalent complexes? Zerovalent platinum complexes are tetrahedral (e.g. $[(PPh_3)_3PtCO]$[227,228] and $[Pt(PF_3)_4]$[229,230]) because for a d^{10} atom there is no difference in the ligand field stabilisation energies

(i) *Olefin*

(ii) *Acetylene*

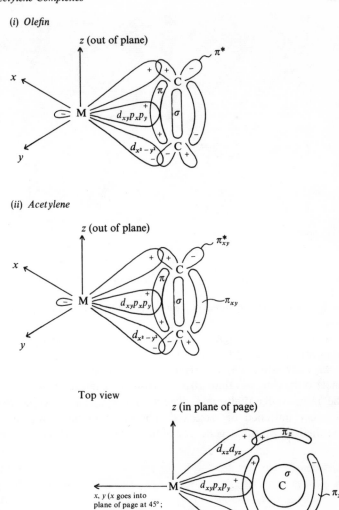

Top view

Side view

Fig. 42. Schematic diagram of the bonding in zerovalent metal–olefin and -acetylene complexes.

between a square-planar and a tetrahedral arrangement so that the ligands arrange themselves as far apart as possible (i.e. tetrahedral) to minimise bond-pair/bond-pair repulsion energies.[136] However, in an olefin (or acetylene) complex where the π-back-donation of charge from metal to olefin is large and the σ-donation from olefin to metal is small the number of d-electrons will be rather less than 10 so that if only the ligand field stabilisation energies are considered the square-planar arrangement is more stable than the tetrahedral arrangement (Fig. 43). Bond-pair/bond-pair repulsions will still favour a tetrahedral arrangement and the balance of these opposing

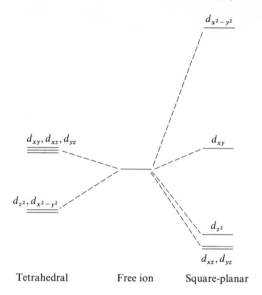

| Tetrahedral | Free ion | Square-planar |

Fig. 43. Ligand field splittings of the d-orbitals in tetrahedral and square-planar complexes.

forces results in a slight distortion from planarity with an angle α (Fig. 41) of between 2° and 14° (Table 113).

An alternative way of arriving at the same conclusion is to consider the effect of having less than 10 d-electrons in a tetrahedral environment. From the Jahn–Teller theorem a distortion will occur in which the tetrahedron is either flattened or stretched. The effect of flattening or stretching the tetrahedron on the degeneracy of the t_2 orbitals[231] is shown in Fig. 44. The

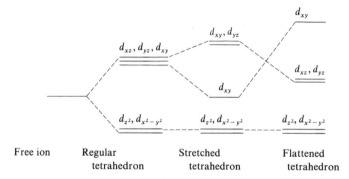

| Free ion | Regular tetrahedron | Stretched tetrahedron | Flattened tetrahedron |

Fig. 44. The effect on the degeneracy of the t_2 orbitals of distorting a regular tetrahedral environment.

ligand field stabilisation energy will clearly be greatest if the tetrahedron is flattened, since if two electrons are donated from the metal to the ligand π^*-(antibonding)-orbital the highest energy d-orbital can be left unoccupied. The observed geometry of these complexes is clearly in support of a flattened tetrahedral structure since the P–Pt–P bond angles in $[(PPh_3)_2Pt(TCNE)]$

and $[(PPh_3)_2Pt(PhC\equiv CPh)]$ are $101°^{(218,219)}$ and $102°^{(224)}$ respectively compared with 109° expected for a regular tetrahedron or 90° for a regular square-plane. A bonding scheme in which these complexes were considered to have a distorted tetrahedral structure has been suggested previously[133] on the basis of semi-empirical one-electron molecular orbital calculations, although in this case the authors concluded that there was a low energy barrier to rotation of the organic moiety, whereas in fact the results of n.m.r. spectroscopy suggest that this energy barrier is large (see pp. 407–408).

Having shown how the Dewar–Chatt–Duncanson model accounts for the almost planar arrangement in the zerovalent complexes it remains to be explained why the olefin axis lies approximately perpendicular to the square-plane around the metal in the divalent complexes. The reason is that for the divalent metal complexes the π-back donation of electron density from the metal to the olefin is greater when the olefin is perpendicular to the square-plane due to the possibility of d_π–p_π hybridisation, which gives stronger π-back-donation than an unhybridised d_π-orbital, only when the olefin is in this position. In the zerovalent complexes p_z and d_{z^2} orbitals are involved in π-back-donation to the tertiary phosphine ligands so that they are not available for π-back-donation to the olefin. As a result in the zerovalent complexes π-back-donation from the metal to the olefin is only possible when the olefin lies almost in the square-plane about the metal.

This lack of availability of the metal p_z and d_{z^2} orbitals for back-donation to the unsaturated ligand also accounts for the absence of free rotation of this ligand about the metal(0)–olefin or acetylene bond because the flattened tetrahedral structure, in which back-donation occurs from the $d_{x^2-y^2}$ orbital,[133,134] is considerably more stable than any other arrangement. This argument has recently[232,233] been put on a semiquantitative basis by calculating the total energy of the complexes $[(PH_3)_2Pt(CH_3C\equiv CCH_3)]$ and $[(PH_3)_2Pt(C_2H_4)]$ as a function of the angle between the multiple bond axis and the PtP_2 plane. The results show a single minimum when the multiple bond lies in the PtP_2 plane. The fact that these calculations predict a minimum energy when the dihedral angle α (see Fig. 41) is zero is probably an indication of the approximate nature of the calculations rather than an indication of the importance of crystal packing effects in the solid state, since crystal packing effects would be expected to vary markedly over the range of complexes shown in Table 113 whereas the complexes show a remarkably constant value of α.

(ii) Valence bond treatment

In the valence bond description the olefin or acetylene is considered to be a dicarbanion which forms two σ-bonds to platinum, which is itself in the +2 oxidation state. This description can readily account for the following experimental observations:

(a) The substituents at the multiple bond are bent away from the metal on co-ordination. This would be an immediate consequence of rehybridisation of the carbon atoms to the sp^3 arrangement necessary to form two metal–carbon σ-bonds.

(b) The geminal fluorine–fluorine coupling constant in the n.m.r. spectrum of $[(PPh_3)_2Pt(CF_3-CF=CF_2)]$ is of the same order as in a saturated fluorocarbon, which suggests sp^3 hybridisation of the 'olefinic' carbon atoms and hence platinum–carbon σ-bonds.[234]

(c) There is no free rotation of the olefin or acetylene about the metal–olefin or metal–acetylene bond.

(d) The valence bond treatment is useful in accounting for the analogies between the acetylene complexes and cyclopropenes as shown in the preparation, bromination and addition of hydrogen halide reactions shown below (reactions 505–10).[235] However, whilst the valence bond conception provides a useful way of looking at these reactions, the mechanisms of the

Preparation

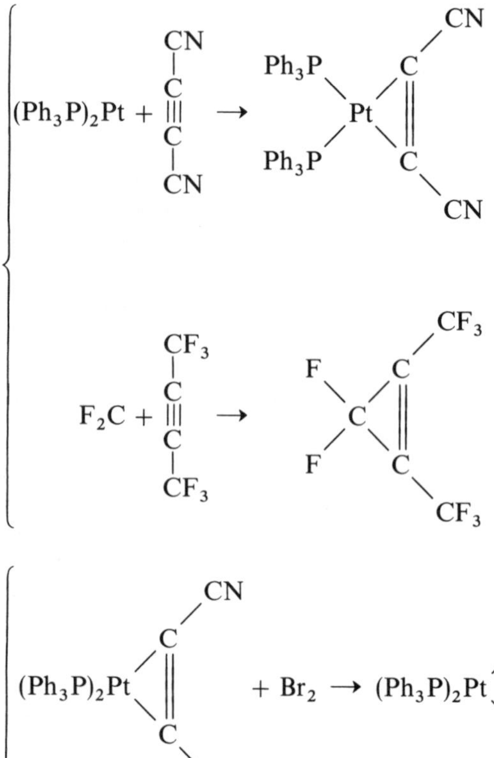

$$(Ph_3P)_2Pt + \underset{\underset{CN}{\overset{\overset{CN}{|}}{\underset{|}{\overset{C}{\underset{C}{\parallel\parallel\parallel}}}}}{} \longrightarrow \quad \text{(complex)} \qquad (505)$$

$$F_2C + \underset{\underset{CF_3}{\overset{\overset{CF_3}{|}}{\underset{|}{\overset{C}{\underset{C}{\parallel\parallel\parallel}}}}}{} \longrightarrow \quad \text{(complex)} \qquad (506)$$

Bromination

$$(Ph_3P)_2Pt\!\!\left(\!\!\begin{array}{c} C-CN \\ \parallel \\ C-CN \end{array}\!\!\right) + Br_2 \longrightarrow (Ph_3P)_2Pt\!\!\left(\!\!\begin{array}{c} C(CN)(Br) \\ | \\ C(CN)(Br) \end{array}\!\!\right) \qquad (507)$$

$$H_2C\!\!\left(\!\!\begin{array}{c} C-H \\ \parallel \\ C-H \end{array}\!\!\right) + Br_2 \longrightarrow H_2C\!\!\left(\!\!\begin{array}{c} C(H)(Br) \\ | \\ C(H)(Br) \end{array}\!\!\right) \qquad (508)$$

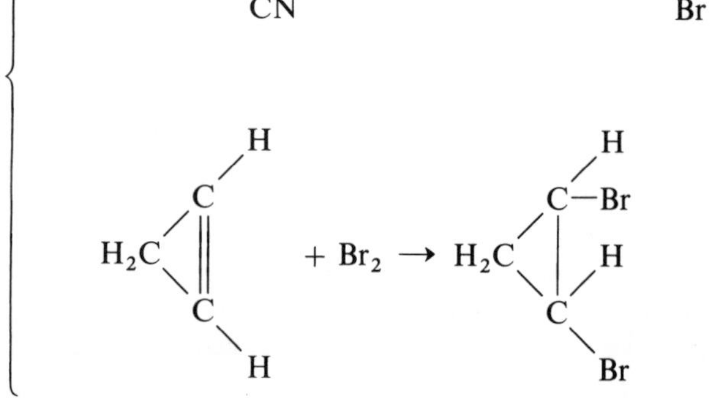

Hydrohalogenation

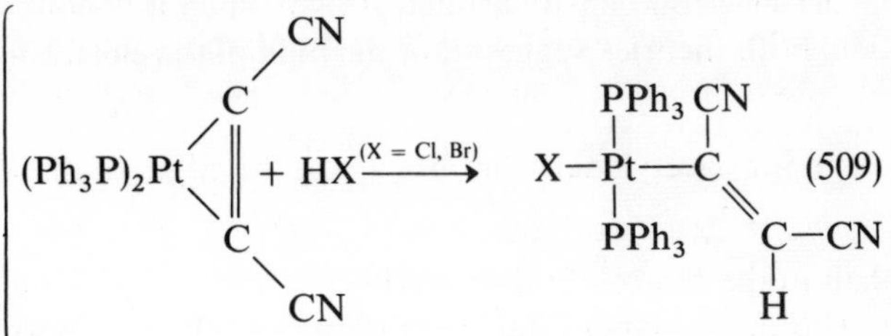

$$(Ph_3P)_2Pt \begin{array}{c} CN \\ C \\ \| \\ C \\ CN \end{array} + HX \xrightarrow{(X = Cl, Br)} X-Pt-C \begin{array}{c} PPh_3 \ CN \\ \| \\ C-CN \\ H \end{array} \quad (509)$$

$$(CH_3)_2C \begin{array}{c} CH_3 \\ C \\ \| \\ C \\ H \end{array} + HCl \rightarrow Cl-C-C \begin{array}{c} CH_3 \ CH_3 \\ \| \\ CH_3 \quad C-H \\ H \end{array} \quad (510)$$

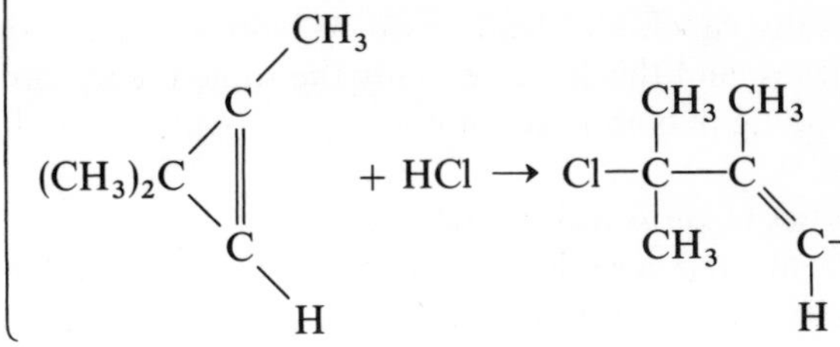

bromination and hydrohalogenation reactions are not sufficiently well understood to know if the structural and mechanistic implications of the valence bond formulation are valid.

However, the valence bond description is hard pressed to account for the following experimental observations:

(a) The angle α in Fig. 41 is between 8° and 14°. To account for this it is necessary to assume that the eclipsed position of the substituents on the multiple bond is sufficiently unstable to cause the sp^3 hybrid orbitals to rotate towards the staggered position.[236] However, these substituents are sufficiently remote from each other for this to be rather unlikely.

(b) The small values observed for β (Fig. 41 and Table 113) suggest that

the three-membered $Pt\begin{array}{c} C \\ | \\ C \end{array}$ ring would be very strained. However, since the

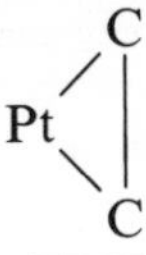

melting-points of the $[(PPh_3)_2Pt(olefin)]$ complexes vary from 122–5°C for the ethylene complex to 268–70°C for the TCNE complex[237] there would not appear to be as much strain present as is suggested by β values of around 40°.

(c) Although the C–C–Pt angles are about 70° whereas the optimum angle for the formation of strong σ-bonds is 109° there does not appear to be undue

strain in the $Pt\begin{array}{c} C \\ | \\ C \end{array}$ ring.

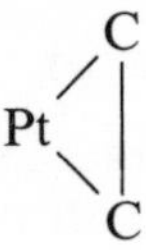

(iii) Equivalence of the molecular-orbital and valence-bond descriptions
The molecular-orbital and valence-bond descriptions are equivalent in that both predict that the metal atom has less than 10 d-electrons (i.e. is in a positive oxidation state) and that the co-ordinated unsaturated ligand has a

higher electron density around it than when it is not co-ordinated. In addition, both theories suggest that the bulk of the electron density in the metal–olefin (or –acetylene) bond lies along the edges of the Pt $\diagdown\!\!\!\!\!\underset{\displaystyle C}{\overset{\displaystyle C}{\big|}}$ triangle rather than in the centre of this triangle.

The molecular-orbital description has three important advantages over the valence-bond description: firstly it can readily explain all the experimental observations reported to date for these complexes, secondly, it is potentially capable of being used to interpret the electronic spectra of the complexes, and thirdly, essentially the same theory can be used to describe the bonding in complexes of both the zerovalent and divalent metals.

Properties of zerovalent complexes

The olefin and acetylene complexes of the zerovalent metals are stable, generally white, crystalline materials with fairly high decomposition points (Table 114). The importance of electron-withdrawing substituents attached

TABLE 114
Decomposition points of complexes of the zerovalent metals

	Decomposition points (°C)			
Complex	*Pt*	*Reference*	*Pd*	*Reference*
$[(PPh_3)_2M(TCNE)]$	268–270	204	256–257	209
$[(PPh_3)_2M(C_2F_4)]$	218–220	234		
$[(PPh_3)_2M(CF_3C\equiv CCF_3)]$	215–216	238	194–195	238
$[(PPh_3)_2M(CCOOCH_3)_2)]$	188–190	239	195–197	239
$[(PPh_3)_2M(C_2H_4)]$	122–125	212		

to the multiple bond is indicated by the decreasing thermal stability of $[(PPh_3)_2Pt(R_2C=CR_2)]$ in the order $R = CN > F > H$.

Both kinetic and equilibrium measurements have been made for the replacement of one acetylene (ac) co-ordinated to platinum(0) by another (ac') (equilibrium 511).[240,241] The rate is proportional to the concentration of the platinum(0)–acetylene complex but independent of the concentration of the reacting acetylene (ac'), suggesting that the rate-determining step

$$[(PPh_3)_2Pt(ac)] + ac' \rightleftharpoons [(PPh_3)_2Pt(ac')] + ac \qquad (511)$$

involves loss of the co-ordinated acetylene (ac). Thus the rate coefficient is a measure of the platinum(0)–acetylene bond strength. Both the rate and equilibrium constants for equilibrium 511, where $ac = PhC\equiv CH$ and $ac' =$ substituted-$PhC\equiv CH$, coupled with the chemical shifts of the acetylenic protons in the n.m.r. spectra of the free acetylenes[242] indicate that the π-acceptor capacity of the acetylene is more important than its σ-donor capacity for the formation of a stable platinum(0)–acetylene bond.

This is consistent with the bonding scheme discussed above. From equilibrium studies of the displacement of one olefin or acetylene by another according to equilibrium 511 a stability order of $PhC{\equiv}CPh > PhC{\equiv}CH >$ alkyl$C{\equiv}C$alkyl $> HC{\equiv}CH >$ olefins was established,[243] and a stability order of $TCNE > NCC{\equiv}CCN > PhC{\equiv}CH$ has been established from rate studies of the displacement of one ligand by another.[204,205] The weakness of the olefin complexes, except those of tetracyanoethylene, relative to the acetylene complexes of platinum(0) is particularly noteworthy as the reverse is true for the platinum(II) complexes.

Spectroscopic properties of zerovalent complexes

(i) Infra-red spectra

Complexes of the zerovalent metals with olefins exhibit no infra-red bands assignable to the olefinic double-bond. Acetylenes, however, exhibit a lowering of the $C{\equiv}C$ stretching frequency of about $450\ cm^{-1}$ on co-ordination to the zerovalent metals (Table 115). The resulting stretching frequency is

TABLE 115
$C{\equiv}C$ stretching frequencies in complexes with zerovalent platinum and palladium

	$v_{C{\equiv}C}\ (cm^{-1})$					
Complex	Free acetylene	Reference	Pt	Reference	Pd	Reference
$[(PPh_3)_2M(CF_3C{\equiv}CCF_3)]$	2300	244	1775	238	1811, 1838	239
$[(PPh_3)_2M(CCOOCH_3)_2]$	2150	245	1782, 1765^a	239	1845, 1830^a	239

^{a} Shoulder.

only slightly higher than the typical value for an unco-ordinated double-bond, which is consistent with the mode of binding discussed above for these complexes. The slightly greater lowering of the $C{\equiv}C$ stretching frequency on co-ordination to platinum(0) than palladium(0) (Table 115) is consistent with the greater ability of platinum than palladium to donate electrons to the acetylene π^*-(antibonding)-orbital.

(ii) Nuclear magnetic resonance spectra

As with the divalent complexes n.m.r. has given valuable information on the internal rotations of zerovalent olefin and acetylene complexes as well as on the nature of the bonding.

(a) *Internal rotations*. The n.m.r. spectra of $[(PPh_3)_2Pt(ac)]$, where ac = $PhC{\equiv}CCH_3$ or a hydroxy–acetylene, show coupling between the methyl protons and both *cis*- and *trans*-phosphorus atoms, implying that the rate of rotation of the acetylene about the platinum(0)–acetylene bond is slow relative to the n.m.r. time scale (i.e. less than $1.2\ sec^{-1}$), indicating a fairly high activation energy for this rotation.[246,247] Similarly, the phosphorus atoms in $[(PPh_3)_2Pt(CF_2{=}CFX)]$, where X = Cl or CF_3, were found to be non-equivalent, indicating the absence of free rotation about the platinum(0)–olefin bond.[248]

(b) *Coupling constants.* The geminal fluorine–fluorine coupling constant in the n.m.r. spectrum of $[(PPh_3)_2Pt(CF_3CF=CF_2)]$ is of the same order as found in the saturated fluorocarbon.[234] This is consistent with the molecular-orbital bonding scheme, given above, which suggests that the very strong π-donation of charge from platinum to the π^*-(antibonding)-orbitals of the olefin and the relatively weak σ-donation of charge from the π-orbitals of the olefin to platinum reduces the olefin bond approximately to a single bond.

The ^{13}C–^1H coupling constants in organic molecules have been shown to vary linearly with the fractional s character of the ^{13}C–H bond.[249,250] On co-ordination to platinum(0) there is a decrease in the ^{13}C–^1H coupling constant of olefins and acetylenes (Table 116) that is consistent with the

TABLE 116
^{13}C–^1H coupling constants for free and co-ordinated ligands (data from ref. 244)

Complex	$J_{^{13}C-^1H}$ (Hz)	Free ligand	$J_{^{13}C-^1H}$ (Hz)
trans-$[(PPh_3)_2PtCH_3X]$	131 ± 2^a	'CH$_3$'	125^b
$[(PPh_3')_2Pt(C_2H_4)]^c$	146·5	C$_2$H$_4$	156
$[(PPh_3')_2Pt(C_2H_2)]^c$	210	C$_2$H$_2$	250

^{a} Variation of ± 2 due to change of X.
^{b} Value for a typical methyl group in an organic compound.
^{c} PPh$_3'$ = fully deuterated PPh$_3$.

idea of a reduction in the carbon–carbon bond order on co-ordination.[251] The decrease in the ^{13}C–^1H coupling constant for acetylene is greater than that for ethylene on co-ordination, which is consistent with a greater reduction in the triple-bond order than the double-bond order. This is not un-expected since acetylene forms a stronger complex with platinum(0) than ethylene (see above p. 407).

Chemical reactions of zerovalent complexes
(i) Exchange reactions
The exchange of one unsaturated ligand for another has been studied extensively and has been shown to occur by initial formation of $[Pt(PPh_3)_2]$[213,241,252] (reactions 512 and 513). The rate of dissociation of an acetylene (i.e. k_{-B}) is much slower than the rate of dissociation of an

$$[(PPh_3)_2Pt(C_2H_4)] \underset{}{\overset{k_A}{\rightleftharpoons}} [Pt(PPh_3)_2] + C_2H_4 \qquad (512)$$

$$[Pt(PPh_3)_2] + PhC\equiv CH \underset{k_{-B}}{\overset{}{\rightleftharpoons}} [(PPh_3)_2Pt(PhC\equiv CH)] \qquad (513)$$

olefin (i.e. k_A). Thus if an acetylene is bubbled into a benzene solution of $[Pt(PPh_3)_3]$ the n.m.r. spectrum shows two well-separated signals; one at higher field due to the free acetylene and the other at lower field due to the co-ordinated acetylene. By contrast when ethylene is bubbled into a similar solution the n.m.r. spectrum at room temperature shows a very broad signal due to the ethylenic protons, with a chemical shift that is dependent on the olefin concentration and the temperature.[253] This behaviour is typical of a very fast exchange of the free and co-ordinated ethylene.

(ii) Displacement of olefins by tertiary phosphines

Both olefins and acetylenes are displaced from $[(PPh_3)_2PtL]$, where L = olefin or acetylene, by excess triphenylphosphine. In the case of olefins this reaction occurs spontaneously but in the case of acetylenes it is generally necessary to remove the free acetylene, for example by blowing nitrogen through the solution if the acetylene concerned is gaseous.

(iii) Reaction of acids with acetylene complexes

Acids such as HCl, HBr, picric, thioacetic and trifluoroacetic acid react with $[(PPh_3)_2Pt(acetylene)]$ to give complexes such as cis-$[(PPh_3)_2PtCl_2]$ and the corresponding olefin.[254,255] The reaction is thought to proceed via an addition-elimination mechanism (eqn. 514). Although in this example,

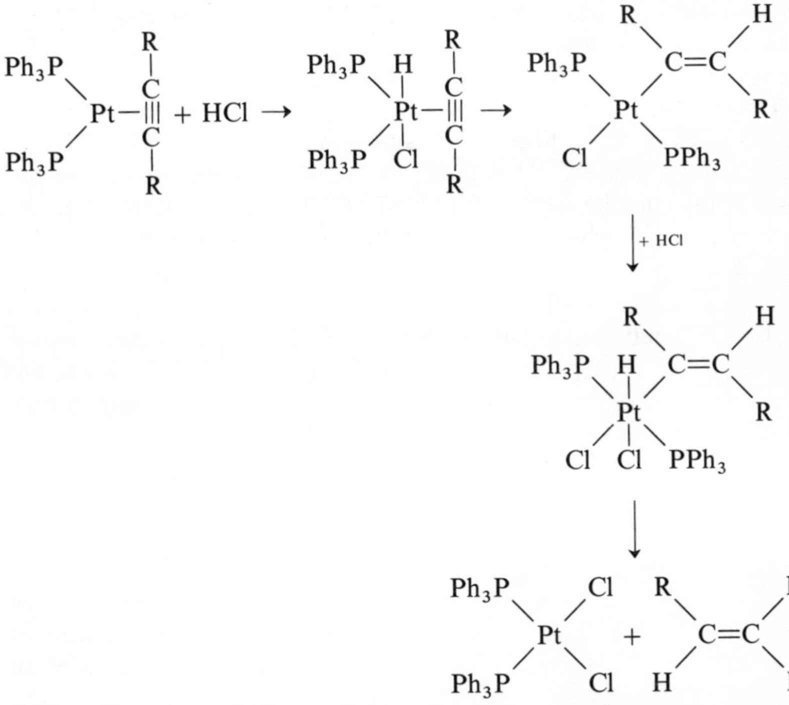

where the acetylene was a pure hydrocarbon, it was not possible to isolate the intermediate five-co-ordinate complexes, an analogous complex of tetracyanoethylene, $[(PEt_3)_2PtH(CN)(TCNE)]$ has been isolated,[216] and platinum(0)–acetylene complexes have been found to undergo oxidative addition reactions with tetrachloro-*o*-benzoquinone to give platinum(II) acetylene complexes[256] (reaction 515). Further support for the mechanism shown in eqn. 514 is obtained from the reaction of perfluoroacetylene

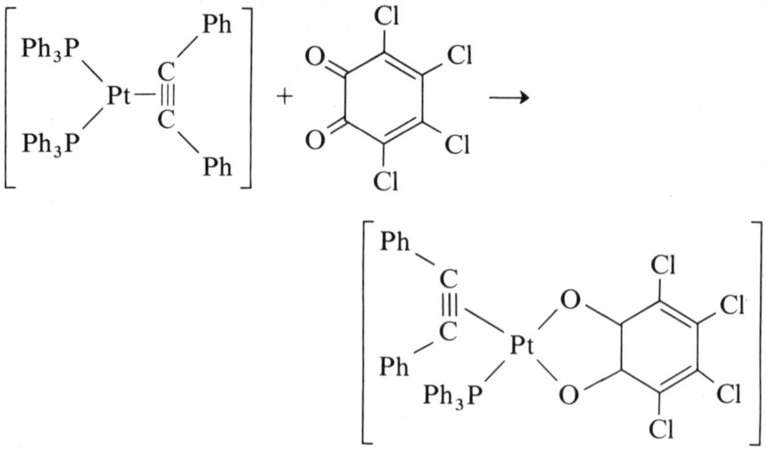

(515)

complexes such as $[(PPh_3)_2Pt(CF_3C{\equiv}CCF_3)]$ with acids such as $HCl^{[255]}$ or $CF_3COOH^{[254]}$ which yield the vinyl complexes $[(PPh_3)_2PtX\{C(CF_3){=}CHCF_3\}]$, where $X = Cl$ or CF_3COO, which are too stable to be decomposed to the free olefin.

The reactions just mentioned are almost the exact reverse of those found with TCNE. Thus TCNE reacts with $[(PR_3)_2PtHX]$, where $R = Et$ or Ph and $X = Cl$ or Br, to liberate HX and give $[(PR_3)_2Pt(TCNE)]$. Kinetic and mechanistic studies,[216] together with the isolation of the five-co-ordinate cyano-complex $[(PEt_3)_2PtH(CN)(TCNE)]$, have indicated a mechanism (reaction 516) which is closely similar to the reverse of reaction 514.

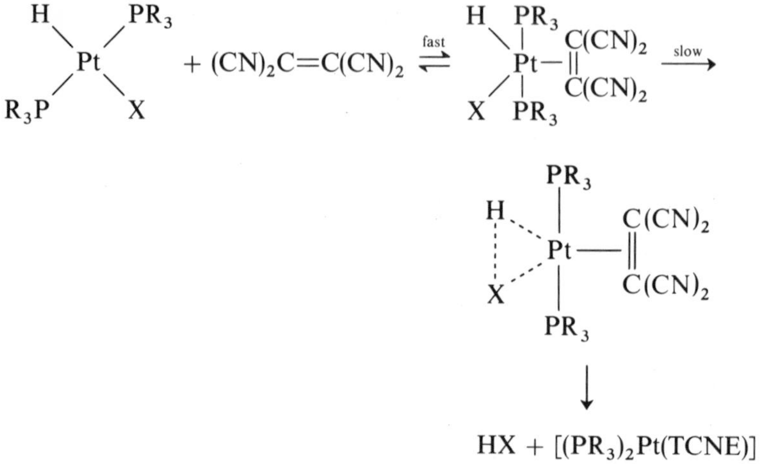

$$HX + [(PR_3)_2Pt(TCNE)] \quad (516)$$

(iv) Rearrangement of metal(0)–olefin complexes to metal(II)–vinyl complexes
As mentioned above (p. 398) perchloro– and perfluoro–olefin complexes of platinum(0) isomerise to vinyl complexes of platinum(II) on heating in a suitable polar solvent such as an alcohol.[206,248,257–9] The palladium(0)

complexes of these olefins isomerise to palladium(II)–vinyl complexes so readily that it is not possible to isolate the palladium(0)–olefin complex.[207–9] Kinetic and mechanistic studies of these isomerisations have indicated that in the case of tetrachloroethylene the isomerisation occurs by rate-determining S_N1 loss of chloride from the olefin (reaction 517), whereas for olefins such as $CHCl=CCl_2$ this mechanism competes with a simple intramolecular process[190] (reaction 518).

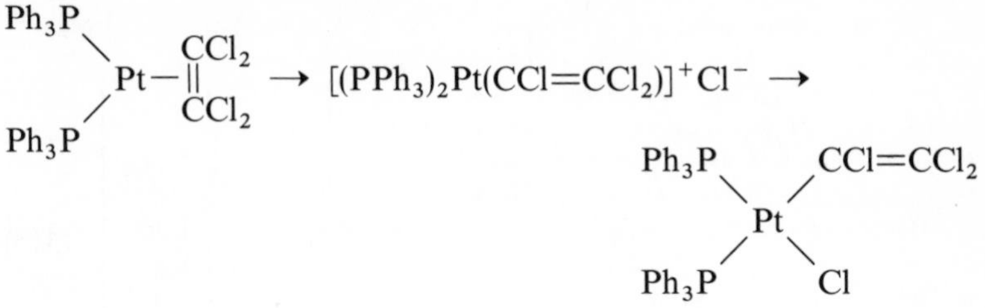

$$
\begin{array}{c}
Ph_3P \\
\diagdown \\
Pt-\underset{CCl_2}{\overset{CCl_2}{||}} \\
\diagup \\
Ph_3P
\end{array}
\longrightarrow [(PPh_3)_2Pt(CCl=CCl_2)]^+Cl^- \longrightarrow
$$

$$
\begin{array}{ccc}
Ph_3P & & CCl=CCl_2 \\
\diagdown & & \diagup \\
 & Pt & \\
\diagup & & \diagdown \\
Ph_3P & & Cl
\end{array}
\qquad (517)
$$

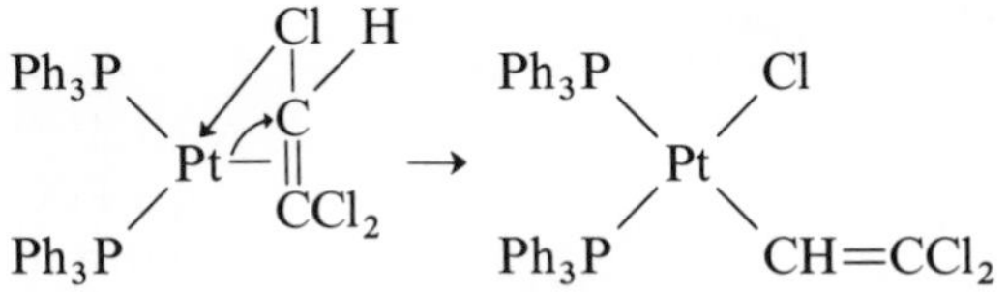

$$
\begin{array}{c}
\overset{Cl}{}\overset{H}{} \\
Ph_3P \\
\diagdown C \\
Pt-\underset{CCl_2}{||} \\
\diagup \\
Ph_3P
\end{array}
\longrightarrow
\begin{array}{ccc}
Ph_3P & & Cl \\
\diagdown & & \diagup \\
 & Pt & \\
\diagup & & \diagdown \\
Ph_3P & & CH=CCl_2
\end{array}
\qquad (518)
$$

(v) cis-trans-*isomerisation of olefins*

When $[(PPh_3)_2Pt(PhCH=CHPh)]$ is treated with a 1:1 mixture of *cis* and *trans* $CFCl=CFCl$ a mixture of the two isomers of $[(PPh_3)_2Pt(CClF=CClF)]$ is obtained.[213a] However, instead of obtaining a 1:1 mixture of the two isomers, a 23:1 mixture in favour of the *trans*-isomer is obtained although an examination of the unreacted olefin shows it to be still a 1:1 mixture, indicating that the *cis* to *trans* isomerisation of the olefin occurs on co-ordination to platinum(0). It has been suggested that this isomerisation involves an intermediate such as 172 and this is supported by the observation that the $CCl_2=C(CN)_2$ ligand binds to platinum(0) in an asymmetric manner with a much shorter $Pt-C(Cl_2)$ bond than $Pt-C(CN)_2$ bond[222]

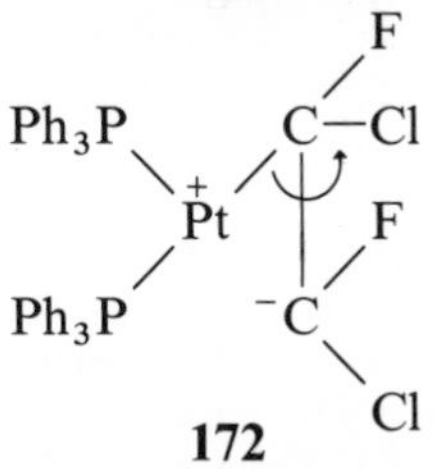

$$
\begin{array}{ccc}
 & & F \\
 & & \diagup \\
Ph_3P & & C-Cl \\
\diagdown & \diagup & \\
 & \overset{+}{Pt} & F \\
\diagup & & \diagdown\diagup \\
Ph_3P & & C \\
 & & \diagdown \\
 & 172 & Cl
\end{array}
$$

implying a contribution from the canonical structure 173.

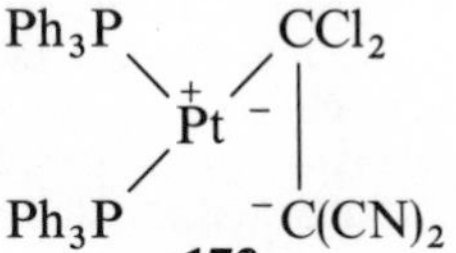

$$
\begin{array}{ccc}
Ph_3P & & CCl_2 \\
\diagdown & \diagup & | \\
 & \overset{+}{Pt} & \\
\diagup & & | \\
Ph_3P & & {}^-C(CN)_2 \\
 & 173 &
\end{array}
$$

REFERENCES

1 W. C. Zeise, *Mag. Pharm.*, **35** (1830) 105.
2 J. Smidt, *Chem. Ind. (London)*, (1962) 54.
3 J. C. Bailar, H. Itatani, M. J. Crespi and J. Geldard, *Adv. Chem. Ser.*, **62** (1967) 103.
4 F. R. Hartley, *Chem. Rev.*, **69** (1969) 799.
5 U. Belluco, B. Crociani, R. Pietropaolo and P. Uguagliati, *Inorg. Chim. Acta Rev.*, **3** (1969) 19.
6 R. Jones, *Chem. Rev.*, **68** (1968) 785.
7 M. L. H. Green, '*Organometallic Compounds*', by G. E. Coates, M. L. H. Green and K. Wade, Methuen, London, 3rd edn., 1968, vol. 2, chapter 1.
8 R. G. Guy and B. L. Shaw, *Adv. Inorg. and Radiochem.*, **4** (1962) 77.
9 J. H. Nelson and H. B. Jonassen, *Coord. Chem. Rev.*, **6** (1971) 27.
10 F. R. Hartley, *Angew. Chem., Int. Ed.*, **11** (1972) 596.
10a H. W. Quinn and J. H. Tsai, *Adv. Inorg. Radiochem.*, **12** (1969) 217.
11 F. R. Hartley, *Proceedings Fifth Int. Conf. on Organometallic Chemistry*, Moscow, August, 1971, abstract 331, vol. 2, p. 279.
12 R. G. Denning and L. M. Venanzi, *J. Chem. Soc.*, (1963) 3241.
13 K. A. Jensen, *Acta Chem. Scand.*, **7** (1953) 866.
14 R. D. Cramer, E. L. Jenner, R. V. Lindsey and U. G. Stolberg, *J. Amer. Chem. Soc.*, **85** (1963) 1691.
15 J. Chatt and M. L. Searle, *Inorg. Synth.*, **5** (1957) 210.
15a F. R. Hartley, *Organometal. Chem. Rev.*, **A6** (1970) 119.
16 W. M. MacNevin and S. A. Giddings, *Chem. Ind. (London)*, (1960) 1191.
17 R. A. Alexander, N. C. Baenziger, C. Carpenter and J. R. Doyle, *J. Amer. Chem. Soc.*, **82** (1960) 535.
18 F. R. Hartley, *Proceedings Third Inorg. Chim. Acta Symp.*, Venice, 1970, B3.
19 F. R. Hartley, *Inorg. Chim. Acta*, **5** (1971) 197.
20 J. R. Doyle, P. E. Slade and H. B. Jonassen, *Inorg. Synth.*, **6** (1960) 216.
21 J. S. Anderson, *J. Chem. Soc.*, (1936) 1042.
22 F. Basolo and R. G. Pearson, '*Mechanisms of Inorganic Reactions*', John Wiley, New York, 2nd edn., 1967, p. 377.
23 D. Banerjea, F. Basolo and R. G. Pearson, *J. Amer. Chem. Soc.*, **79** (1957) 4055.
24 R. M. Milburn and L. M. Venanzi, *Inorg. Chim. Acta*, **2** (1968) 97.
25 M. Green and C. J. Wilson, *Disc. Faraday Soc.*, **47** (1969) 110.
26 M. Green and C. J. Wilson, *Proceedings Third Inorg. Chim. Acta Symp.*, Venice, 1970, B2.
27 R. Pietropaolo, G. Dolcetti, M. Giustiniani and U. Belluco, *Inorg. Chem.*, **9** (1970) 549.
28 G. Dolcetti, R. Pietropaolo and U. Belluco, *Inorg. Chem.*, **9** (1970) 553.
29 W. Ostwald, *Phys. Z.*, **3** (1902) 313.
30 J. A. Wunderlich and D. P. Mellor, *Acta Cryst.*, **7** (1954) 130; **8** (1955) 57.
31 G. B. Bokii and G. A. Kukina, *Zh. Strukt. Khim.*, **5** (1965) 706.
32 M. Black, R. H. B. Mais and P. G. Owston, *Acta Cryst.*, **B25** (1969) 1753.
33 J. A. J. Jarvis, B. T. Kilbourn and P. G. Owston, *Acta Cryst.*, **B26** (1970) 876.
34 J. A. J. Jarvis, B. T. Kilbourn and P. G. Owston, *Acta Cryst.*, **B27** (1971) 366.
35 R. Spratley, K. Klanderman and W. C. Hamilton, quoted by L. Manojlovic-Muir, K. W. Muir and J. A. Ibers, *Disc. Faraday Soc.*, **47** (1969) 84.
36 N. C. Baenziger, J. R. Doyle and C. L. Carpenter, *Acta Cryst.*, **14** (1961) 303.
37 N. C. Baenziger, J. R. Doyle, G. F. Richards and C. L. Carpenter, in '*Advances in the Chemistry of the Coordination Compounds*', Ed. by S. Kirschner, Macmillan, New York, 1961, p. 131.
38 N. C. Baenziger, G. F. Richards and J. R. Doyle, *Acta Cryst.*, **18** (1965) 924.
39 C. E. Holloway, G. Hulley, B. F. G. Johnson and J. Lewis, *J. Chem. Soc. (A)*, (1970) 1653.
40 E. Benedetti, P. Corradini and C. Pedone, *J. Organometal. Chem.*, **18** (1969) 203.
41 C. Pedone and E. Benedetti, *J. Organometal. Chem.*, **29** (1971) 443.
42 G. R. Davies, W. Hewertson, R. H. B. Mais, P. G. Owston and C. G. Patel, *J. Chem. Soc. (A)*, (1970) 1873.

43 I. A. Zakharova, G. A. Kukina, T. S. Kuli-Zade, I. I. Moiseev, G. Yu. Pek and M. A. Porai-Koshits, *Russ. J. Inorg. Chem.*, **11** (1966) 1364.

44 C. V. Goebel, *Diss. Abs.*, **B28** (1967) 625.

45 N. C. Baenziger, R. C. Medrud and J. R. Doyle, *Acta Cryst.*, **18** (1965) 237.

46 J. R. Holden and N. C. Baenziger, *J. Amer. Chem. Soc.*, **77** (1955) 4987.

47 G. H. W. Milburn and M. R. Truter, *J. Chem. Soc. (A)*, (1966) 1609.

48 R. Mason and G. B. Robertson, *J. Chem. Soc. (A)*, (1969) 492.

49 V. C. Adam, J. A. J. Jarvis, B. T. Kilbourn and P. G. Owston, *Chem. Comm.*, (1971) 467.

50 *Tables of Interatomic Distances and Configuration in Molecules and Ions*, Ed. by L. E. Sutton, Chemical Society Special Publication nos. 11 and 18 (1958 and 1965).

51 P. R. H. Alderman, P. G. Owston and J. M. Rowe, *Acta Cryst.*, **13** (1960) 149.

52 J. Chatt and L. A. Duncanson, *J. Chem. Soc.*, (1953) 2939.

53 D. B. Powell and N. Sheppard, *Spectrochim. Acta*, **13** (1958) 69.

54 A. A. Babushkin, L. A. Gribov and A. D. Gel'man, *Dokl. Akad. Nauk. SSSR*, **123** (1958) 461; *Chem. Abs.*, **53** (1959) 3880c.

55 C. Smart, Ph.D. Thesis, London, 1962; quoted by D. M. Adams in *Metal-Ligand and Related Vibrations*, Arnold, London, (1967) p. 44.

56 H. P. Fritz and C. G. Kreiter, *Chem. Ber.*, **96** (1963) 1672.

57 A. D. Allen and T. Theophanides, *Can. J. Chem.*, **42** (1964) 1551.

58 H. P. Fritz and D. Sellmann, *J. Organomet. Chem.*, **6** (1966) 558.

59 M. J. Grogan and K. Nakamoto, *J. Amer. Chem. Soc.*, **88** (1966) 5454.

60 J. Pradilla-Sorzano and J. P. Fackler, *J. Mol. Spectry.*, **22** (1967) 80.

61 M. J. Grogan and K. Nakamoto, *J. Amer. Chem. Soc.*, **90** (1968) 918.

62 D. M. Adams and P. J. Chandler, *J. Chem. Soc. (A)*, (1969) 588.

63 J. Hiraishi, *Spectrochim. Acta*, **A25** (1969) 749.

64 J. Hiraishi, D. Finseth and F. A. Miller, *Spectrochim. Acta*, **A25** (1969) 1657.

65 F. R. Hartley, *D. Phil. Thesis*, Oxford, 1966.

66 F. R. Hartley, *Proceedings Third Inorg. Chim. Acta Symp.*, Venice, (1970) B3.

67 D. B. Powell and N. Sheppard, *J. Chem. Soc.*, (1960) 2519.

68 P. J. Hendra and D. B. Powell, *Spectrochim. Acta*, **17** (1961) 909.

69 E. O. Greaves and P. M. Maitlis, *J. Organometal. Chem.*, **6** (1966) 104.

70 L. W. Reeves, *Canad. J. Chem.*, **38** (1960) 736.

71 S. Maricic, C. R. Redpath and J. A. S. Smith, *J. Chem. Soc.*, (1963) 4905.

72 P. D. Kaplan, P. Schmidt and M. Orchin, *J. Amer. Chem. Soc.*, **89** (1967) 4537.

73 H. P. Fritz, K. E. Schwarzhans and D. Sellmann, *J. Organometal. Chem.*, **6** (1966) 551.

74 P. D. Kaplan and M. Orchin, *Inorg. Chem.*, **6** (1967) 1096.

75 B. F. G. Johnson, C. E. Holloway, G. Hulley and J. Lewis, *Chem. Comm.*, (1967) 1143.

76 C. E. Holloway, G. Hulley, B. F. G. Johnson and J. Lewis, *J. Chem. Soc. (A)*, (1969) 53.

77 F. R. Hartley, unpublished results.

78 A. R. Brause, F. Kaplan and M. Orchin, *J. Amer. Chem. Soc.*, **89** (1967) 2661.

79 P. D. Kaplan and M. Orchin, *Inorg. Chem.*, **4** (1965) 1393.

80 W. Partenheimer, *Diss. Abs.*, **29B** (1968) 524.

81 P. S. Braterman, *Inorg. Chem.*, **5** (1966) 1085.

82 T. Kinugasa, N. Nakamura, H. Yamada and A. Saika, *Inorg. Chem.*, **7** (1968) 2649.

83 R. Cramer, *Inorg. Chem.*, **4** (1965) 445.

84 G. Paiaro and A. Panunzi, *Ric. Sci., Rend. Sez. A*, **4** (1964) 601; *Chem. Abs.*, **62** (1965) 6127c.

85 C. E. Holloway and J. Fogelman, *Canad. J. Chem.*, **48** (1970) 3802.

86 P. D. Kaplan, P. Schmidt and M. Orchin, *J. Amer. Chem. Soc.*, **90** (1968) 4175.

87 P. D. Kaplan, P. Schmidt, A. Brause and M. Orchin, *J. Amer. Chem. Soc.*, **91** (1969) 85.

88 J. P. Yesinowski and T. L. Brown, *Inorg. Chem.*, **10** (1971) 1097.

89 Y. K. Syrkin, *Bull. Acad. Sci. USSR, Classe Sci. Chim.*, (1948) 69; *Chem. Abs.*, **42** (1948) 5368.

90 R. McWeeny, R. Mason and A. D. C. Towl, *Disc. Faraday Soc.*, **47** (1969) 20.

91 C. W. Fryer and J. A. S. Smith, *J. Chem. Soc. (A)*, (1970) 1029.

92 S. J. Lokken and D. S. Martin, *Inorg. Chem.*, **2** (1963) 562.

93   S. I. Shupack and M. Orchin, *J. Amer. Chem. Soc.*, **86** (1964) 586.

94   J. W. Moore, *Acta Chem. Scand.*, **20** (1966) 1154.

95   R. G. Denning, F. R. Hartley and L. M. Venanzi, *J. Chem. Soc. (A)*, (1967) 1322.

96   E. Premuzic and A. I. Scott, *Chem. Comm.*, (1967) 1078.

97   A. I. Scott and A. D. Wrixon, *Chem. Comm.*, (1969) 1184.

98   A. D. Wrixon, E. Premuzic and A. I. Scott, *Chem. Comm.*, (1968) 639.

99   F. Basolo and R. G. Pearson, *Mechanism of Inorganic Reactions*, John Wiley, New York, 2nd edn., 1967, p. 531.

100   R. G. Denning, F. R. Hartley and L. M. Venanzi, *J. Chem. Soc. (A)*, (1967) 324.

101   R. G. Denning, F. R. Hartley and L. M. Venanzi, *J. Chem. Soc. (A)*, (1967) 328.

102   F. R. Hartley and L. M. Venanzi, *J. Chem. Soc. (A)*, (1967) 330.

103   R. G. Denning and L. M. Venanzi, *J. Chem. Soc. (A)*, (1967) 336.

104   P. M. Henry, *J. Amer. Chem. Soc.*, **86** (1964) 3246.

105   I. I. Moiseev, M. N. Vargaftik and Ya. K. Syrkin, *Dokl. Akad. Nauk. SSSR*, **152** (1963) 147; *Chem. Abs.*, **60** (1964) 184d.

106   S. V. Pestrikov and I. I. Moiseev, *Izv. Akad. Nauk. SSSR, Ser. Khim.*, (1965) 349; *Chem. Abs.*, **62** (1965) 16018d.

107   S. V. Pestrikov, I. I. Moiseev and T. N. Romanova, *Zh. Neorgan. Khim.*, **10** (1965) 2203; *Chem. Abs.*, **63** (1965) 15836b.

108   S. V. Pestrikov, I. I. Moiseev and B. A. Tsivilikhoviskaya, *Zh. Neorgan. Khim.*, **11** (1966) 1742; *Chem. Abs.*, **65** (1966) 13757c.

109   S. V. Pestrikov, I. I. Moiseev and L. M. Sverzh, *Zh. Neorgan. Khim.*, **11** (1966) 2081; *Chem. Abs.*, **66** (1967) 45991.

110   J. R. Joy and M. Orchin, *J. Amer. Chem. Soc.*, **81** (1959) 310.

111   R. B. Turner, D. E. Nettleton and M. Perelman, *J. Amer. Chem. Soc.*, **80** (1958) 1430.

112   A. C. Cope, C. R. Ganellin, H. W. Johnson, T. V. Van Auken and H. J. S. Winkler, *J. Amer. Chem. Soc.*, **85** (1963) 3276.

113   G. Paiaro, P. Corradini, R. Palumbo and A. Panunzi, *Makromol. Chem.*, **71** (1964) 184.

114   G. Paiaro and A. Panunzi, *J. Amer. Chem. Soc.*, **86** (1964) 5148.

115   G. Paiaro and A. Panunzi, *Tetra. Letters*, (1965) 441.

116   P. Corradini, S. F. Mason, G. Paiaro, A. Panunzi and G. H. Searle, *J. Amer. Chem. Soc.*, **88** (1966) 2863.

117   A. Panunzi and G. Paiaro, *J. Amer. Chem. Soc.*, **88** (1966) 4843.

118   P. Ganis and C. Pedone, *Ric. Sci., Rend. Sez. A.*, **8** (1965) 1462; *Chem. Abs.*, **65** (1966) 1515h.

119   P. Corradini, P. Ganis and C. Pedone, *Acta Cryst.*, **20** (1966) 458.

120   R. Lazzaroni, P. Salvadori and P. Pino, *Tetra. Letters*, (1968) 2507.

121   A. C. Cope, C. R. Ganellin and H. W. Johnson, *J. Amer. Chem. Soc.*, **84** (1962) 3191.

122   A. C. Cope, J. K. Hecht, H. W. Johnson, H. Keller and H. J. S. Winkler, *J. Amer. Chem. Soc.*, **88** (1966) 761.

123   A. C. Cope, K. Banholzer, H. Keller, B. A. Pawson, J. J. Whang and H. J. S. Winkler, *J. Amer. Chem. Soc.*, **87** (1965) 3644.

124   A. C. Cope and M. W. Fordice, *J. Amer. Chem. Soc.*, **89** (1967) 6187.

125   A. C. Cope and B. A. Pawson, *J. Amer. Chem. Soc.*, **90** (1968) 636.

126   A. Panunzi, A. De Renzi and G. Paiáro, *J. Amer. Chem. Soc.*, **92** (1970) 3488.

127   G. Paiaro, A. Panunzi and A. De Renzi, *Tetra. Letters*, (1966) 3905.

128   A. Panunzi, A. De Renzi and G. Paiaro, *Inorg. Chim. Acta*, **1** (1967) 475.

129   R. Lazzaroni, P. Salvadori and P. Pino, *Chem. Comm.*, (1970) 1164.

130   Reference 99, p. 355.

131   M. J. S. Dewar, *Bull. Soc. Chim. France*, **18** (1951) C79.

132   D. P. Craig, A. Maccoll, R. S. Nyholm, L. E. Orgel and L. E. Sutton, *J. Chem. Soc.*, (1954) 332.

133   J. H. Nelson, K. S. Wheelock, L. C. Cusachs and H. B. Jonassen, *Chem. Comm.*, (1969) 1019.

134   J. H. Nelson, K. S. Wheelock, L. C. Cusachs and H. B. Jonassen, *J. Amer. Chem. Soc.*, **91** (1969) 7005.

135 D. M. Adams, J. Chatt, J. Gerratt and A. D. Westland, *J. Chem. Soc.*, (1964) 734.
136 R. J. Gillespie and R. S. Nyholm, *Quart. Rev.*, **11** (1957) 339.
137 A. C. Blizzard and D. P. Santry, *J. Amer. Chem. Soc.*, **90** (1968) 5749.
138 R. Mason, *Nature*, **217** (1968) 543.
139 J. N. Murrell, *Disc. Faraday Soc.*, **47** (1969) 59.
140 H. B. Gray and C. J. Ballhausen, *J. Amer. Chem. Soc.*, **85** (1963) 260.
141 J. K. Stille, R. A. Morgan, D. D. Whitehurst and J. R. Doyle, *J. Amer. Chem. Soc.*, **87** (1965) 3282.
142 J. K. Stille and R. A. Morgan, *J. Amer. Chem. Soc.*, **88** (1966) 5135.
143 M. Green and R. I. Hancock, *J. Chem. Soc. (A)*, (1967) 2054.
144 C. B. Anderson and B. J. Burreson, *J. Organometal. Chem.*, **7** (1967) 181.
145 C. B. Anderson and S. Winstein, *J. Org. Chem.*, **28** (1963) 605.
146 F. R. Hartley, *Nature*, **223** (1969) 615.
147 D. A. White, *J. Chem. Soc. (A)*, (1971) 145.
148 G. Carturan, L. Busetto, A. Palazzi and U. Belluco, *J. Chem. Soc. (A)*, (1971) 219.
149 J. Smidt, W. Hafner, R. Jira, R. Sieber, J. Sedlmeier and A. Sabel, *Angew. Chem. Int. Ed.*, **1** (1962) 80.
150 C. E. Moore, *Atomic Energy Levels*, vol. III, National Bureau of Standards, Circular 467, U.S. Government Printing Office, Washington, D.C., 1958.
151 F. Basolo, H. B. Gray and R. G. Pearson, *J. Amer. Chem. Soc.*, **82** (1960) 4200.
152 G. Booth and J. Chatt, *Proc. Chem. Soc.*, (1961) 67.
153 R. S. Nyholm and M. L. Tobe, *Experientia Suppl.*, **9** (1964) 112; *Chem. Abs.*, **61** (1964) 6612d.
154 P. M. Henry, *J. Amer. Chem. Soc.*, **88** (1966) 1595.
155 F. R. Hartley and M. Green, unpublished results.
156 B. L. Shaw, *Chem. Comm.*, (1968) 464.
157 J. Tsuji, *Adv. Org. Chem., Methods and Results*, **6** (1969) 109.
158 E. W. Stern and M. L. Spector, *Proc. Chem. Soc.*, (1961) 370.
159 H. Hirai, H. Sawai and S. Makishima, *Bull. Chem. Soc. Japan*, **43** (1970) 1148.
160 G. Paiaro, A. De Renzi and R. Palumbo, *Chem. Comm.*, (1967) 1150.
161 R. Palumbo, A. De Renzi, A. Panunzi and G. Paiaro, *J. Amer. Chem. Soc.*, **91** (1969) 3874.
162 R. N. Haszeldine, R. V. Parish and D. W. Robbins, *J. Organometal. Chem.*, **23** (1970) C33.
163 A. De Renzi, R. Palumbo and G. Paiaro, *J. Amer. Chem. Soc.*, **93** (1971) 880.
164 A. Panunzi, A. De Renzi, R. Palumbo and G. Paiaro, *J. Amer. Chem. Soc.*, **91** (1969) 3879.
165 Y. Fujiwara, I. Moritani, S. Danno, R. Asano and S. Teranishi, *J. Amer. Chem. Soc.*, **91** (1969) 7166 and references therein.
166 R. F. Heck, *J. Amer. Chem. Soc.*, **91** (1969) 6707.
167 I. Moritani, S. Danno, Y. Fujiwara and S. Teranishi, *Bull. Chem. Soc. Japan*, **44** (1971) 578 and references therein.
168 S. Danno, I. Moritani and Y. Fujiwara, *Tetrahedron*, **25** (1969) 4819.
169 S. Danno, I. Moritani, Y. Fujiwara and S. Teranishi, *Bull. Chem. Soc. Japan*, **43** (1970) 3966.
170 C. K. Ingold, *Structure and Mechanism in Organic Chemistry*, Cornell University Press, 2nd edn., 1969, p. 248.
171 M. A. Bennett, G. J. Erskine and R. S. Nyholm, *J. Chem. Soc. (A)*, 1967, 1260.
172 M. A. Bennett, *Symposium on Metal-Olefin and π-Allyl Complexes*, Sheffield, April, 1967.
173 D. I. Hall and J. D. Ling, personal communication.
174 J. C. Bailar, *Platinum Metals Review*, **15** (1971) 2.
175 A. J. Deeming, B. F. G. Johnson and J. Lewis, *Chem. Comm.*, (1970) 598.
176 A. J. Chalk and J. F. Harrod, *J. Amer. Chem. Soc.*, 1965, **87**, 16.
177 J. L. Speier, J. A. Webster and G. H. Barnes, *J. Amer. Chem. Soc.*, 1957, **79**, 974.
178 J. C. Saam and J. L. Speier, *J. Amer. Chem. Soc.*, **80** (1958) 4104.
179 J. C. Saam and J. L. Speier, *J. Amer. Chem. Soc.*, **83** (1961) 1351.
180 J. W. Ryan and J. L. Speier, *J. Amer. Chem. Soc.*, **86** (1964) 895.

181   R. H. Fish and H. G. Kuivila, *J. Org. Chem.*, **31** (1966) 2445.
182   L. Malatesta, G. Santarella, L. M. Vallarino and F. Zingales, *Angew. Chem.*, **72** (1960) 34.
183   P. M. Maitlis and M. L. Games, *Canad. J. Chem.*, **42** (1964) 183.
184   A. T. Blomquist and P. M. Maitlis, *J. Amer. Chem. Soc.*, **84** (1962) 2329.
185   R. Huettel and H. J. Neugebauer, *Tetra. Letters*, (1964) 3541.
186   P. M. Maitlis, D. Pollock, M. L. Games and W. J. Pryde, *Canad. J. Chem.*, **43** (1965) 470.
187   D. F. Pollock and P. M. Maitlis, *J. Organometal. Chem.*, **26** (1971) 407.
188   L. F. Dahl and W. E. Oberhansli, *Inorg. Chem.*, **4** (1965) 629.
189   F. Canziani, P. Chini, A. Quarta and A. Di Martino, *J. Organometal. Chem.*, **26** (1971) 285.
190   H. C. Longuet-Higgins and L. E. Orgel, *J. Chem. Soc.*, (1956) 1969.
191   R. C. Cookson and D. W. Jones, *J. Chem. Soc.*, (1965) 1881.
192   E. J. Wharton, *Inorg. Nucl. Chem. Lett.*, **7** (1971) 307.
193   K. A. Molodova, *Uch. Zap. Leningrad. Gos. Ped., Inst. im A. I. Gertsena*, **160** (1959) 151; *Chem. Abs.*, **55** (1961) 6236g.
194   J. Chatt, R. G. Guy, L. A. Duncanson and D. T. Thompson, *J. Chem. Soc.*, (1963) 5170.
195   S. V. Bukhovets and A. O. Sheveleva, *Zh. Neorg. Khim.*, **9** (1964) 471; *Chem. Abs.*, **60** (1964) 10182b.
196   A. O. Sheveleva and S. V. Bukhovets, *Zh. Neorg. Khim.*, **12** (1967) 965; *Chem. Abs.*, **67** (1967) 63902.
197   A. D. Allen and T. Theophanides, *Canad. J. Chem.*, **44** (1966) 2703.
198   A. D. Allen and T. Theophanides, *Canad. J. Chem.*, **43** (1965) 290.
199   T. Theophanides, personal communication.
200   J. Chatt, B. L. Shaw and A. A. Williams, *J. Chem. Soc.*, (1962) 3269.
201   S. Otsuka, A. Nakamura and K. Tani, *J. Organometal. Chem.*, **14** (1968) P30.
202   W. J. Bland and R. D. W. Kemmitt, *J. Chem. Soc. (A)*, (1968) 1278.
203   D. M. Roundhill and G. Wilkinson, *J. Chem. Soc. (A)*, (1968) 506.
204   W. H. Baddley and L. M. Venanzi, *Inorg. Chem.*, **5** (1966) 33.
205   G. L. McClure and W. H. Baddley, *J. Organometal. Chem.*, **27** (1971) 155.
206   J. A. Osborn, *Chem. Comm.*, (1968) 1231.
206a W. J. Bland, J. Burgess and R. D. W. Kemmitt, *J. Organometal. Chem.*, **14** (1968) 201.
207   C. D. Cook and G. S. Jauhal, *Canad. J. Chem.*, **45** (1967) 301.
208   S. Takahashi and N. Hagihara, *Nippon Kagaku Zasshi*, **88** (1967) 1306; *Chem. Abs.*, **69** (1968) 27514.
209   P. Fitton and J. E. McKeon, *Chem. Comm.*, (1968) 4.
210   P. Fitton, M. P. Johnson and J. E. McKeon, *Chem. Comm.*, (1968) 6.
211   S. Cenini, R. Ugo and G. La Monica, *J. Chem. Soc. (A)*, (1971) 409.
212   C. D. Cook and G. S. Jauhal, *Inorg. Nucl. Chem. Lett.*, **3** (1967) 31.
213   J. P. Birk, J. Halpern and A. L. Pickard, *Inorg. Chem.*, **7** (1968) 2672.
213a J. Ashley-Smith, M. Green and D. C. Wood, *J. Chem. Soc. (A)*, (1970) 1847.
214   D. M. Blake and C. J. Nyman, *Chem. Comm.*, (1969) 483.
215   D. M. Blake and C. J. Nyman, *J. Amer. Chem. Soc.*, **92** (1970) 5359.
216   P. Uguagliati and W. H. Baddley, *J. Amer. Chem. Soc.*, **90** (1968) 5446.
217   R. Van der Linde and R. O. de Jongh, *Chem. Comm.*, (1971) 563.
218   C. Panattoni, G. Bombieri, U. Belluco and W. H. Baddley, *J. Amer. Chem. Soc.*, **90** (1968) 798.
219   G. Bombieri, E. Forsellini, C. Panattoni, R. Graziani and G. Bandoli, *J. Chem. Soc.*, (1970) 1313.
220   C. D. Cook, C. H. Koo, S. C. Nyburg and M. T. Shiomi, *Chem. Comm.*, (1967) 426.
221   J. N. Francis, A. McAdam and J. A. Ibers, *J. Organometal. Chem.*, **29** (1971) 131.
222   A. McAdam, J. N. Francis and J. A. Ibers, *J. Organometal. Chem.*, **29** (1971) 149.
223   C. Panattoni and R. Graziani, *Progress in Coordination Chemistry*, ed. by M. Cais, Elsevier, Amsterdam, 1968, p. 310.
224   J. O. Glanville, J. M. Stewart and S. O. Grim, *J. Organometal. Chem.*, **7** (1967) P9.
225   C. D. Cook and G. S. Jauhal, *J. Amer. Chem. Soc.*, **90** (1968) 1464.
226   B. R. Penfold and W. N. Lipscomb, *Acta Cryst.*, **14** (1961) 589.

227 V. G. Albano, G. N. Basso Ricci and P. L. Bellon, *Inorg. Chem.*, **8** (1968) 2109.

228 V. G. Albano, P. L. Bellon and M. Sansoni, *Chem. Comm.*, (1969) 899.

229 H. G. M. Edwards, *D. Phil. Thesis, Oxford*, 1968.

230 J. C. Marriott, J. A. Salthouse, M. J. Ware and J. M. Freeman, *Chem. Comm.*, (1970) 595.

231 L. E. Orgel, *An Introduction to Transition Metal Chemistry*, Methuen, London, 1960, p. 65.

232 K. S. Wheelock, J. H. Nelson, L. C. Cusachs and H. B. Jonassen, *J. Amer. Chem. Soc.*, **92** (1970) 5110.

233 J. H. Nelson and H. B. Jonassen, *Coord. Chem. Rev.*, **6** (1971) 67.

234 M. Green, R. B. L. Osborn, A. J. Rest and F. G. A. Stone, *Chem. Comm.*, (1966) 502.

235 G. L. McClure and W. H. Baddley, *J. Organometal. Chem.*, **25** (1970) 261.

236 E. L. Eliel, *Stereochemistry of Carbon Compounds*, McGraw-Hill, New York, 1962, chapter 6.

237 W. H. Baddley, *Inorg. Chim. Acta Rev.*, **2** (1968) 7.

238 J. L. Boston, S. O. Grim and G. Wilkinson, *J. Chem. Soc.*, (1963) 3468.

239 E. O. Greaves and P. M. Maitlis, *J. Organometal. Chem.*, **6** (1966) 104.

240 A. D. Allen and C. D. Cook, *Canad. J. Chem.*, **41** (1963) 1235.

241 A. D. Allen and C. D. Cook, *Canad. J. Chem.*, **42** (1964) 1063.

242 C. D. Cook and S. S. Danyluk, *Tetrahedron*, **19** (1963) 177.

243 J. Chatt, G. A. Rowe and A. A. Williams, *Proc. Chem. Soc.*, (1957) 208.

244 F. A. Miller and R. P. Bauman, *J. Chem. Phys.*, **22** (1954) 1544.

245 E. Dallwigk, H. Paillard and E. Briner, *Helv. Chim. Acta*, **36** (1953) 1757.

246 E. O. Greaves, R. Bruce and P. M. Maitlis, *Chem. Comm.*, (1967) 860.

247 J. H. Nelson, J. J. R. Reed and H. B. Jonassen, *J. Organometal. Chem.*, **29** (1971) 163.

248 M. Green, R. B. L. Osborn, A. J. Rest and F. G. A. Stone, *J. Chem. Soc. (A)*, (1968) 2525.

249 N. Muller and D. E. Pritchard, *J. Chem. Phys.*, **31** (1959) 768.

250 N. Muller and D. E. Pritchard, *J. Chem. Phys.*, **31** (1959) 1471.

251 C. D. Cook and K. Y. Wan, *J. Amer. Chem. Soc.*, **92** (1970) 2595.

252 J. P. Birk, J. Halpern and A. L. Pickard, *J. Amer. Chem. Soc.*, **90** (1968) 4491.

253 R. Ugo, *Coord. Chem. Rev.*, **3** (1968) 319.

254 D. M. Barlex, R. D. W. Kemmitt and G. W. Littlecott, *Chem. Comm.*, (1969) 613.

255 P. B. Tripathy and D. M. Roundhill, *J. Amer. Chem. Soc.*, **92** (1970) 3825.

256 D. M. Barlex, R. D. W. Kemmitt and G. W. Littlecott, *Chem. Comm.*, (1971) 199.

257 W. J. Bland, J. Burgess and R. D. W. Kemmitt, *J. Organometal. Chem.*, **15** (1968) 217.

258 W. J. Bland, J. Burgess and R. D. W. Kemmitt, *J. Organometal. Chem.*, **18** (1969) 199.

259 A. J. Mukhedkar, M. Green and F. G. A. Stone, *J. Chem. Soc. (A)*, (1970) 947.

CHAPTER 14

π-Allyl Complexes

π-Allyl complexes, also known as π-enyl complexes, are formed when a three-carbon-atom system donates three electrons to a metal. The simplest is the π-allyl ligand, π-C_3H_5, and more complex π-allyl ligands have one or more of the carbon atoms substituted by groups other than hydrogen, or the terminal carbon atoms joined by a carbon chain, Fig. 45. Because the nature of the bonding in these complexes has been questioned (see pp. 429–433) the π-allyl to metal bond will be represented by a system of dashes (Fig. 45) which avoids precise structural commitment.

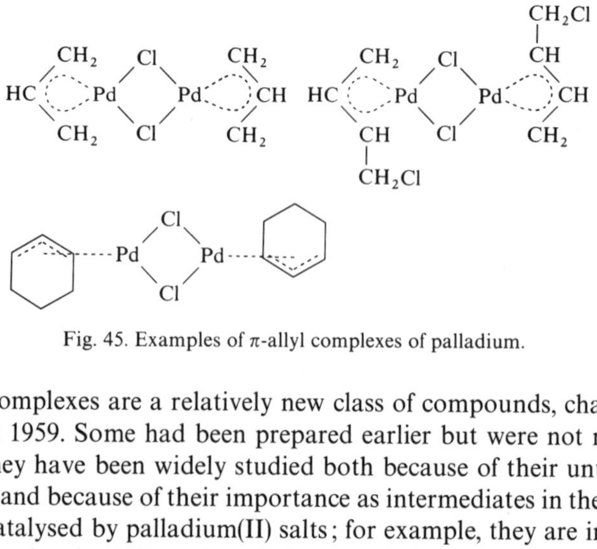

Fig. 45. Examples of π-allyl complexes of palladium.

π-Allyl complexes are a relatively new class of compounds, characterised since about 1959. Some had been prepared earlier but were not recognised as such. They have been widely studied both because of their unusual type of bonding and because of their importance as intermediates in the reactions of olefins catalysed by palladium(II) salts; for example, they are involved in the isomerisation (p. 448), carbonylation (pp. 445–447) and oligomerisation (pp. 443–444) of olefins. The number of reviews in this field (Table 117) is hardly commensurate with the volume of published material.

Palladium forms a wide range of π-allyl complexes whilst platinum forms many fewer, so that in considering the reactions of these two metals with olefins it is important to remember that whilst platinum will probably form an olefin complex of the type discussed in Chapter 13, palladium may well form a π-allyl complex. This is well illustrated by the reaction of allyl alcohol with the two metals (reactions 519 and 520) and is a point we shall return

TABLE 117
Reviews of π-allyl complexes

Title	Reference
Three-electron ligands	1
Allyl metal complexes	2
Olefin, acetylene and π-allylic complexes of transition metals	3

to (p. 441). Although most other ligands bond to more than one oxidation state of platinum and palladium, the π-allyl complexes so far prepared all involve the metals in the $+2$ oxidation state.

$$CH_2=CH-CH_2OH + Na_2PtCl_4 \rightarrow$$

$$Na^+[Pt(CH_2=CH-CH_2OH)Cl_3]^- + NaCl \quad (519)$$

$$2CH_2=CH-CH_2OH + 2Na_2PdCl_4 \rightarrow$$

$$[(\pi\text{-}C_3H_5)PdCl]_2 + 4NaCl \quad (520)$$

After considering the preparative routes to π-allyl complexes much of the emphasis in this chapter will be placed on a discussion of the bonding between the metal and the π-allylic ligand as deduced from the structural studies that have been reported. Particular attention will be devoted to n.m.r. because this has been the most important physical technique both for characterising π-allylic complexes and for studying their reactions in solution. The chapter concludes with a section on the chemical reactions of π-allyl complexes with the emphasis on the insertion of groups such as carbon monoxide and olefins into the π-allyl–metal bond since these are the reactions that are most likely to be of importance in the exploitation of π-allyl chemistry both for laboratory and industrial organic syntheses.

PREPARATION OF π-ALLYL COMPLEXES

The established preparative routes to π-allyl complexes are described below. There are many other reactions, particularly with palladium complexes, which have unexpectedly led to formation of a π-allyl complex, but these preparative routes are not described here.

(i) From allyl halides and allyl alcohol

Allyl halides, allyl alcohol and allyl silanes are frequently used to prepare π-allyl complexes (eqns. 521–4). The reaction between allyl chloride and

$$2CH_2=CH-CH_2-X + 2Na_2PdCl_4 \text{ (or PdCl}_2) \xrightarrow[X = Cl^{(4)},\, OH^{(5)},\, SiMe_3^{(6)}]{50\% \text{ aq. } CH_3COOH}$$

$$[(\pi\text{-}C_3H_5)PdCl]_2 \quad (521)$$

$$2C_3H_5Br + 2Pd \text{ sponge} \xrightarrow[\text{ref. 7}]{\text{reflux}} [(\pi\text{-}C_3H_5)PdBr]_2 \quad (522)$$

$$C_3H_5OH + PdBr_2 \xrightarrow[\text{ref. 8}]{\text{dil. HBr}} [(\pi\text{-}C_3H_5)PdBr_2PdBr_2Pd(\pi\text{-}C_3H_5)] \quad (523)$$

$$[Pt(PPh_3)_4] + C_3H_5X \xrightarrow[\text{ref. 9 and 10}]{X = Cl,\, Br} [(PPh_3)_2Pt(\pi\text{-}C_3H_5)]^+X^- + 2PPh_3 \quad (524)$$

palladium(II) chloride in 50% aqueous acetic acid (eqn. 521) is a particularly convenient way to prepare the parent π-allyl palladium complex $[(\pi\text{-}C_3H_5)PdCl]_2$, although a recent report[11] suggests that a solution of Na_2PdCl_4 and stannous chloride (1 equiv.) in methanol treated with allyl or substituted allyl halides almost quantitatively yields that complex. The reaction of allyl alcohol with platinum(II) and palladium(II) salts demonstrates an important difference between the two metals; platinum(II) salts give an olefin complex[11a] whereas palladium(II) salts give a π-allyl complex[5] (reaction 523). However allyl chloride reacts with platinum(II) dichloride in the presence of stannous chloride in an organic solvent such as acetone or tetrahydrofuran to give $[(\pi\text{-allyl})PtCl]_n$.[12]

(ii) From olefin complexes

π-Allylpalladium(II) complexes have been prepared by a number of routes that involve preliminary formation of an olefin complex. Thus olefins such as *iso*-butene react at room temperature with palladium(II) chloride in 50% aqueous acetic acid to give olefin complexes which on heating lose hydrogen chloride to give the more stable π-allylic complex[4] (reaction 525). The larger the alkyl groups of the olefins and the higher the degree of branching, the

$$2PdCl_2 + (CH_3)_2C{=}CH_2 \xrightarrow[20°C]{50\% \ CH_3COOH} [Pd\{(CH_3)_2C{=}CH_2\}Cl_2]_2 \xrightarrow[-HCl]{50-100°C}$$

$$[(\pi\text{-}CH_2{-}C(CH_3){-}CH_2)PdCl]_2 \quad (525)$$

greater is the proportion of π-allyl complex formed. These are of course just the requirements for the formation of an olefin complex of low stability (p. 376). Alternatively a solution of sodium acetate in acetic acid can be used as the solvent.[13] In the absence of sodium acetate the olefin complex is in equilibrium with a small amount of π-allyl complex (eqn. 526) but sodium

$$2(CH_3)_2C{=}CH_2 + 2PdCl_4^{2-} \rightleftharpoons$$

$$2Cl^- + 2[\{(CH_3)_2C{=}CH_2\}PdCl_3]^- \overset{NaOAc}{\rightleftharpoons}$$

$$\left[\begin{array}{c} H_3C{-}C \overset{CH_2}{\underset{CH_2}{\diagup\diagdown}} Pd \overset{Cl}{\diagdown} \end{array} \right]_2 + 2H^+ + 4Cl^- \quad (526)$$

acetate shifts this equilibrium completely to the right. Here the role of sodium acetate as a base is similar to that of dimethylformamide in the reaction of $CH_3(CH_2)_nCH{=}CH_2$ (where $n = 1, 2, 3$ or 5) with $PdCl_2$ to give π-allyl complexes, since the complex $[(\text{dimethylformamide})_2H]_2[Pd_2Cl_6]$ can be isolated from such mixtures.[14] Similarly, sodium carbonate in chloroform will convert the olefin–palladium(II) complexes of propene and butene into π-allyl complexes.[15] No reactions have been reported in which olefin–platinum(II) complexes are converted to π-allyl–platinum(II) complexes. This is consistent with the tendency of platinum(II) to form olefin complexes in contrast to that of palladium(II) to form π-allyl complexes.

Olefin complexes have been used in reactions in which the original olefin is displaced by a group such as vinylcyclopropane, to form π-allyl complexes[16] (reaction 527). Here the olefin complex merely acts as a palladium

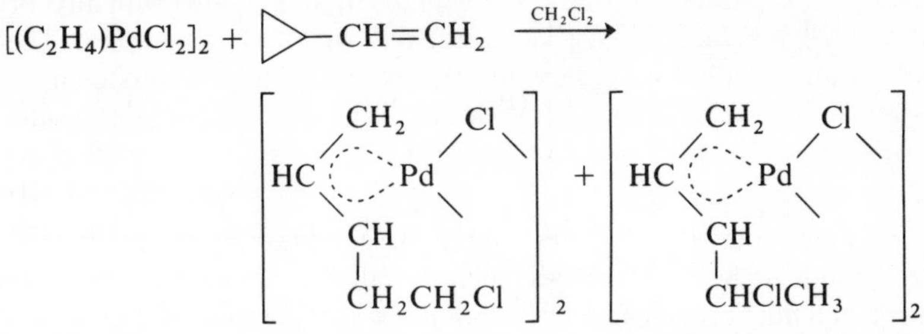

$$[(C_2H_4)PdCl_2]_2 + \triangleright\!\!-CH\!=\!CH_2 \xrightarrow{CH_2Cl_2}$$

$$\left[\begin{array}{c} CH_2 \quad Cl \\ HC \stackrel{\cdots}{\underset{\cdots}{\bigcirc}} Pd \\ CH \\ | \\ CH_2CH_2Cl \end{array}\right]_2 + \left[\begin{array}{c} CH_2 \quad Cl \\ HC \stackrel{\cdots}{\underset{\cdots}{\bigcirc}} Pd \\ CH \\ | \\ CHClCH_3 \end{array}\right]_2 \quad (527)$$

complex with a good leaving group (ethylene, a gas). Bis(benzonitrile)-dichloropalladium(II) has been used equally successfully in the preparation of π-allylic complexes from substituted vinylcyclopropanes[17] and methylenecyclopropanes.[18]

(iii) From conjugated dienes
When conjugated dienes, as opposed to simple olefins, are treated with palladium(II) chloride an extra group must be added to the diene if a π-allyl complex is to be formed. Thus the product of the reaction of bis(benzonitrile)-dichloropalladium(II) with butadiene depends on the solvent. In benzene, chlorine is incorporated into the π-allyl ligand[19] (reaction 528) whereas in methanol the methoxy derivative is formed[19] (reaction 529) and in glacial

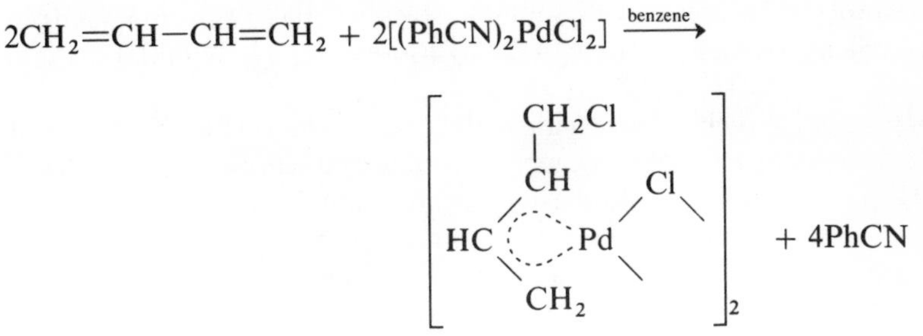

$$2CH_2\!=\!CH\!-\!CH\!=\!CH_2 + 2[(PhCN)_2PdCl_2] \xrightarrow{benzene}$$

$$\left[\begin{array}{c} CH_2Cl \\ | \\ CH \quad Cl \\ HC \stackrel{\cdots}{\underset{\cdots}{\bigcirc}} Pd \\ CH_2 \end{array}\right]_2 + 4PhCN \quad (528)$$

$$2CH_2\!=\!CH\!-\!CH\!=\!CH_2 + 2[(PhCN)_2PdCl_2] \xrightarrow{CH_3OH}$$

$$\left[\begin{array}{c} CH_2OCH_3 \\ | \\ CH \quad Cl \\ HC \stackrel{\cdots}{\underset{\cdots}{\bigcirc}} Pd \\ CH_2 \end{array}\right]_2 + 2HCl + 4NaCl \quad (529)$$

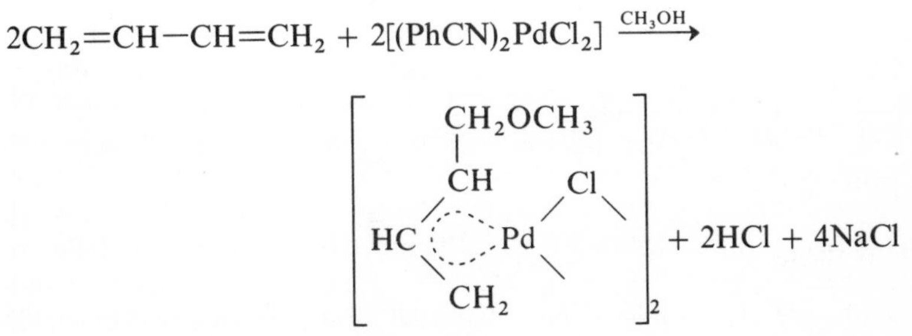

acetic acid containing sodium acetate, the acetate group is incorporated into the π-allyl complex[20] (reaction 530).

$$2CH_3COOH + 2CH_2{=}CH{-}CH{=}CH_2 + PdCl_2 \xrightarrow{CH_3COONa}$$

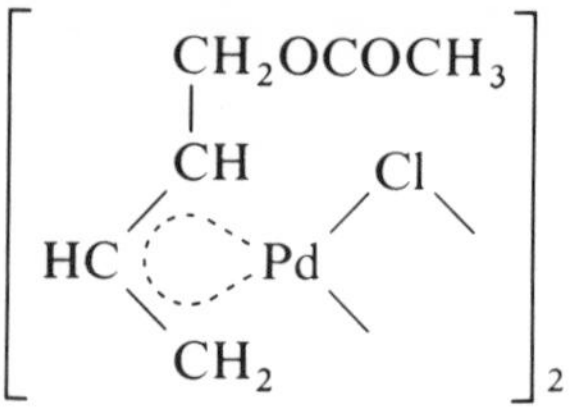

$$\left[\begin{array}{c} CH_2OCOCH_3 \\ | \\ CH \quad\quad Cl \\ HCPd \\ CH_2 \end{array} \right]_2 + 2HCl \quad (530)$$

Hydrido-platinum(II) complexes containing good leaving groups such as nitrate ligands react with dienes such as allene, butadiene, 1,5-cyclo-octadiene and norbornadiene to give cationic π-allyl complexes[21] (reaction 531), probably by initial substitution of the nitrate ligand by the diene

$$trans\text{-}[(PEt_3)_2PtH(NO_3)] + CH_2{=}CH{-}CH{=}CH_2 \longrightarrow$$

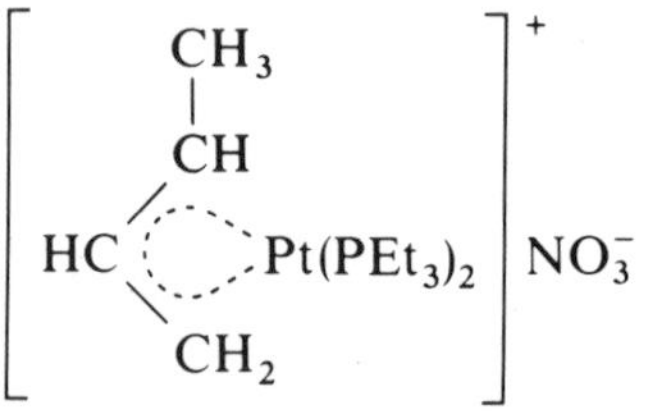

$$\left[\begin{array}{c} CH_3 \\ | \\ CH \\ HCPt(PEt_3)_2 \\ CH_2 \end{array} \right]^{+} NO_3^{-} \quad (531)$$

followed by intramolecular nucleophilic attack of hydride on the co-ordinated olefin. In agreement with this, treatment of *trans*-$[(PEt_3)_2PtH(NO_3)]$ in methanol in the presence of sodium tetraphenylborate with the mono-olefin ethylene gave a precipitate of the hydrido-olefin–platinum(II) complex *trans*-$[(PEt_3)_2Pt(C_2H_4)H]^{+}BPh_4^{-}$.

When hexa-2,4-diene reacts with palladium(II) chloride at $-40°C$ in an aprotic solvent two diastereoisomers are obtained since *cis-cis* and *trans-trans*-hexa-2,4-diene give one diastereoisomer and *cis-trans*-hexa-2,4-diene gives the other.[22] On warming to room temperature the two diastereo-isomers equilibrate. The *cis-cis* and the *trans-trans* isomers both give the same diastereoisomeric complex because *syn*-substituted π-allyl complexes are thermodynamically more stable than *anti*-complexes so that the *cis-cis* isomer is isomerised on co-ordination to give the *trans-trans*-isomer. Thus when the original hexa-2,4-diene is regenerated from the π-allylic complex by adding a co-ordinating ligand such as triphenyl-phosphine or dimethyl-sulphoxide all the *cis-cis*-hexa-2,4-diene is found to have been converted to the *trans-trans*-isomer.

(iv) From allenes

Like conjugated dienes, allenes react with palladium(II) salts to give π-allyl complexes that necessarily contain an additional substituent incorporated into the π-allylic group during the reaction; the substituent depends on the polarity of the solvent (Scheme 6). The addition of allene to palladium(II) chloride can be reversed by treating the resultant π-allyl complexes with a

SCHEME 6

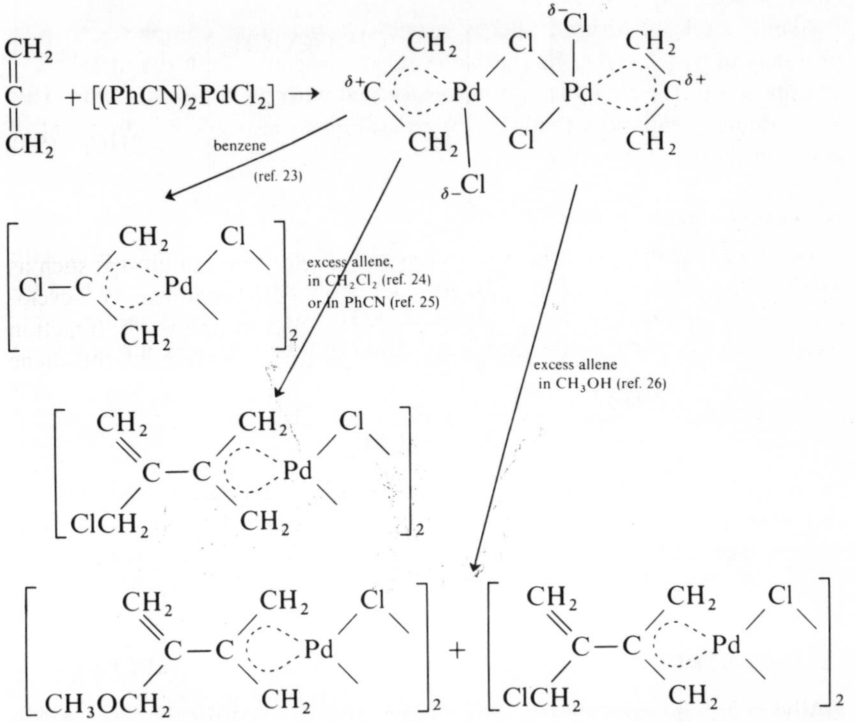

$$CH_2=\!\!=\!\!C=\!\!=CH_2 + [(PhCN)_2PdCl_2] \xrightarrow[\text{(ref. 23)}]{\text{benzene}}$$

tertiary phosphine[24] (reaction 532). When palladium(II) acetate is treated

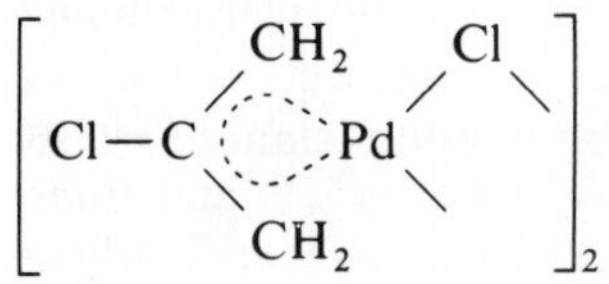

$$+ 4PPh_3 \rightarrow 2[(PPh_3)_2PdCl_2] + 2CH_2=\!\!=\!\!C=\!\!=CH_2 \quad (532)$$

with allene, three molecules of allene are condensed together to give a dimeric product in which two π-allyl–palladium bonds are formed by the one organic ligand (reaction 533). This bidentate ligand is cleaved to give a

$$2Pd(OAc)_2 + 3CH_2=\!\!=\!CH=\!\!=CH_2 \rightarrow$$

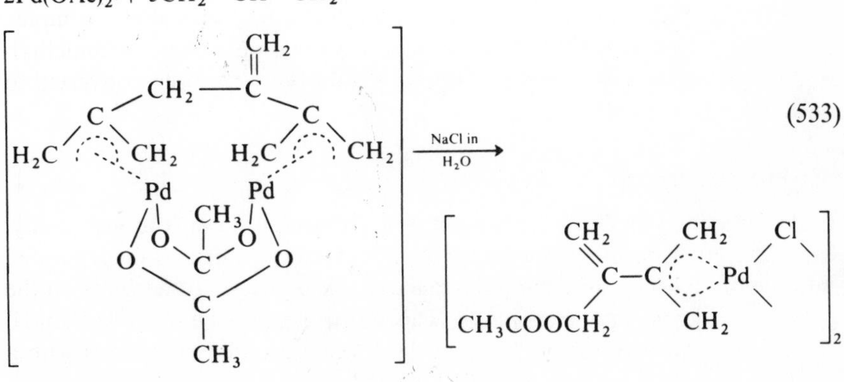

(533)

unidentate π-allylic ligand when the product is treated with aqueous sodium chloride solution[27] (reaction 533).

Allenes and substituted allenes react with platinum complexes to give products in which only one double-bond is associated with the metal,[28–31] as opposed to the π-allylic complexes formed with palladium(II) salts. This again demonstrates the tendency of palladium to π-allylic bonding and of platinum to olefin bonding.

(v) From Grignard reagents

Although Grignard reagents such as allyl magnesium bromide react with halide complexes of platinum(II) and palladium(II) to give π-allyl complexes[32] (reaction 534), they have rarely been used. There are two possible reasons for this: first π-allyl complexes can often be prepared directly from

$$2C_3H_5MgX + MX_2 \rightarrow [M(\pi\text{-}C_3H_5)_2] + 2MgX_2 \qquad (534)$$

$$(M = Pd, Pt)$$

more readily available compounds such as allyl chloride or allyl alcohol, and secondly allylmagnesium halide solutions are difficult to prepare because without suitable precautions the partly formed Grignard reacts with unreacted allyl halide to give biallyl.

THE STRUCTURE OF π-ALLYLPALLADIUM COMPLEXES

π-Allylpalladium complexes have been widely investigated by x-ray diffraction (Tables 118, 119 and 120). Four independent studies of $[(\pi\text{-}C_3H_5)PdCl]_2$, ranging from preliminary results to a detailed three-dimensional analysis at $-140°C$,[33–6] indicate a planar $(PdCl)_2$ bridge and planar C_3H_5 groups, with the C_3H_5 and $(PdCl)_2$ places intersecting at $111.5° \pm 0.9°$ (Fig. 46). At $-140°C$ the palladium–carbon distances are all

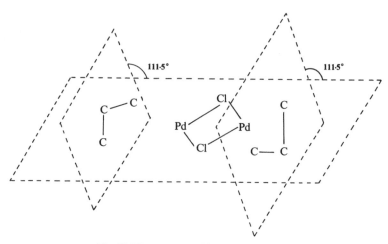

Fig. 46. The structure of $[(\pi\text{--}C_3H_5)PdCl)]_2$.

TABLE 118

Structural details for dimeric π-allyl–palladium complexes

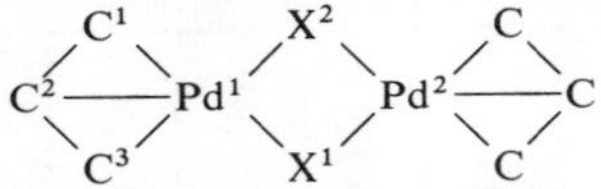

$$C^1 \diagdown \underset{C^2}{} Pd^1 \underset{C^3}{\overset{X^2}{\diamond}} Pd^2 \diagup \overset{C}{\underset{C}{}}$$

π-Allyl ligand	X	Planarity of PdX₂Pd	Conformation of π-allyl ligands	C¹–C²	C²–C³	Pd–C¹	Pd–C²	Pd–C³	Pd–X	No. of structure in Appendix II	Reference
							Bond-lengths (Å)				
CH₂–CH–CH₂	Cl	Planar	Anti	1·38	1·38	2·12	2·11	2·12	2·41	36	36
CH₃ $\vert$ CH₂–C–CH₂	Cl	Planar	Anti	1·36	1·36	2·08	2·12	2·06	2·40	40	37, 41
(CH₃)₂C–CH–C(CH₃)₂	Cl	Planar	Anti	1·41	1·41	2·12	2·14	2·14	2·42	60	38, 41
OH O $\vert$ $\Vert$ CH₂–C–CH–C–OC₂H₅	Cl	Planar	Anti	1·42	1·34	2·08	2·16	2·11	2·40 (Pd–Cl¹) 2·41 (Pd–Cl²)	51	39, 40
t-C₄H₉ $\vert$ CH₂–C–CH₂	Cl	Planar	Anti	1·39	1·39	2·14	2·15	2·14	—	—	42
Endo- and exo- 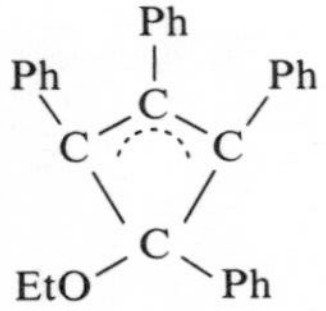 (Ph–C–C–C–Ph structure, Ph and EtO substituents)	Cl	Planar	Anti	1·47 (av.)	1·47 (av.)	2·17 (av.)	2·12 (av.)	2·14 (av.)	2·42 (av.)	90	43
CH₃–CH–CH–CH–CH₃	Cl	Non-planar (Pd¹Cl¹Cl²/Pd²Cl¹Cl² angle = 150°)	Syn	—	—	2·14	2·07	2·04	2·40	48	44
CH₂(CH₃)(CH₂)₂CH₂	Br	Non-planar (Pd¹Br¹Br²/Pd²Br¹Br² angle = 139°)	Syn	1·40	1·38	2·15	2·10	2·14	2·53	59	45

equal (mean 2·12 Å), whereas at room temperature the palladium–central carbon distance may be marginally longer (*ca.* 0·1 Å) than the other two palladium–carbon distances, although this may be apparent rather than real due to the uncertainties introduced by thermal motion. The carbon–carbon bond lengths are all equal (mean 1·38 Å) as are the palladium–chlorine distances (mean 2·41 Å). Although within the limits of accuracy of the measurement all the hydrogen atoms appear to be coplanar with the plane of the π-allyl ligand, x-ray diffraction of π-2-methylallylpalladium chloride indicates that the methyl group is bent out of the allyl plane towards the palladium atom.[37] This distortion together with a reduction in the C–C–C bond angle from 119·8° ± 0·9° in the π-allyl complex[36] to 112·4° ± 1·6° in the π-2-methylallyl complex indicates that the palladium atom enables the σ and π electrons in the allyl ligand to mix. This σ/π-electron mixing is confirmed by an x-ray diffraction study of dimeric π-1,1,3,3-tetramethylallyl-palladium chloride[38] in which the stereochemistry of the terminal carbon atoms is neither trigonal nor tetrahedral.

The dimeric complex π-1-ethoxycarbonyl-2-hydroxylallylpalladium chloride, although retaining a planar $PdCl_2Pd$ bridge, has a very slightly asymmetric structure[39,40] (see Table 118) probably due to the inherent asymmetry of the electron density in the orbitals of this substituted allyl ligand (see discussion on bonding below, pp. 432–433). The asymmetry in monomeric π-allylcomplexes which have two different ligands bound to palladium *trans* to the π-allyl ligand is different. This is exemplified by the two π-2-methylallylcomplexes[46,47] in Table 119 where the palladium–carbon bond lengths *trans* to the ligand of higher *trans*-effect (PPh_3 and S respectively) are longer than those *trans* to the ligand of lower *trans*-effect (Cl and O respectively). In the complex $[(\pi-C_3H_5)Pd(PPh_3)(SnCl_3)]$, where the two ligands triphenylphosphine and trichlorotin have comparable *trans*-effects, the two palladium–terminal carbon bond lengths are effectively equal.[48] The unequal palladium–carbon bond lengths in $[(\pi-C_4H_7)PdCl(PPh_3)]$ have been used to argue that in this complex the π-allyl group is bound to palladium by both a palladium–carbon σ-bond and a palladium–olefin bond[46] (see Fig. 54, p. 433), although this is now thought to be unlikely (see pp. 432–433). In this connection it is noteworthy that in the monothiobenzoylmethanato-π-2-methylallylpalladium(II) complex, where the palladium–terminal carbon bond lengths *trans* to oxygen and sulphur are very different, the carbon–carbon bond lengths in the π-allyl ligand are effectively equal,[47] strongly supporting the idea of a delocalised bond system in the π-allyl ligand.

All the π-allyl complexes described so far have either been monomeric and square-planar or dimeric with planar $PdCl_2$ bridges with the two π-allyl groups in an 'anti' configuration (Fig. 47). However, two structures of dimeric complexes have been reported in which the π-allyl groups are 'syn' (Fig. 47) and the palladium–halogen bridge is not planar. In $[(\pi-CH_3(CH)_3CH_3)PdCl]_2$ the π-1,3-dimethylallyl groups are 'syn' and the angle between the two $PdCl_2$ planes is 150°,[44] a result in agreement with the predictions from the dipole moment of this complex. The dimeric

TABLE 119

Structural details for monomeric π-allyl–palladium complexes

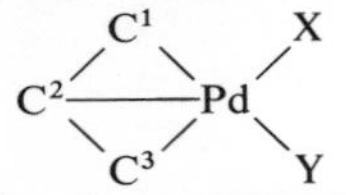

$$
\begin{array}{c}
\text{C}^1 \quad \text{X} \\
\diagdown \quad \diagup \\
\text{Pd} \\
\diagup \quad \diagdown \\
\text{C}^2 \quad \text{Y} \\
\diagdown \\
\text{C}^3
\end{array}
$$

π-Allyl ligand	X	Y	Bond lengths (Å)					No. of structure in Appendix II	Reference
			Pd–C^1	Pd–C^2	Pd–C^3	C^1–C^2	C^2–C^3		
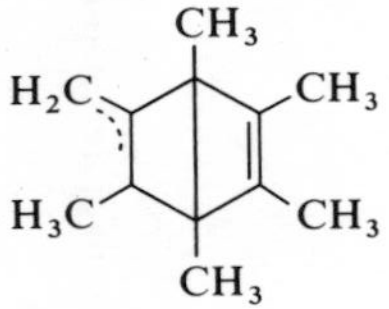 (cyclic ligand with CH$_3$, CH$_3$, CH$_3$, H$_2$C, H$_3$C, CH$_3$)	Acetylacetonato		2·12	2·12	2·12	1·43	1·43	64	49
CH$_2$–C–CH$_2$ with CH$_3$	Cl	PPh$_3$	2·28	2·22	2·14	1·40	1·47	75	46
CH$_2$–C–CH$_2$ with CH$_3$	O^a	S^a	2·21	2·15	2·08	1·40	1·42	69	47
CH$_2$–CH–CH$_2$	SnCl$_3$	PPh$_3$	2·19	2·12	2·20	1·35	1·38	74	48

^{a} O and S are oxygen and sulphur in the chelating ligand $HS—C(C_6H_5)=CH—C—C_6H_5$.
$$\underset{\parallel}{\quad\quad\quad\quad\quad\quad\quad\quad} O$$

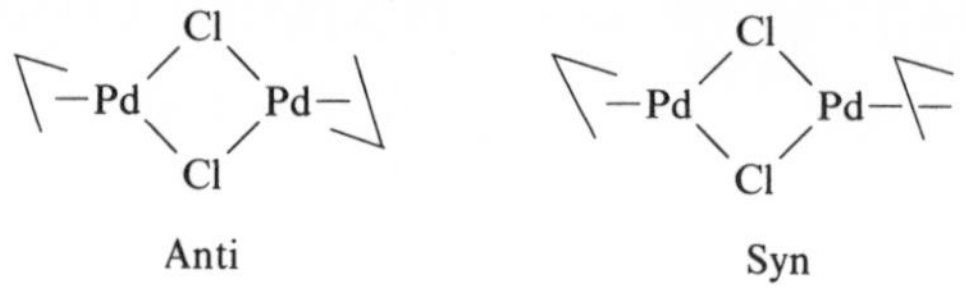

Fig. 47. 'Syn' and 'anti' configurations in dimeric π-allyl complexes.

complex π-cycloheptenylpalladium bromide[45] also has 'syn' π-cycloheptenyl groups and a non-planar bridge with an angle of 139° between
the two $PdBr_2$ planes. The unusual structure of this latter complex compared
with the symmetrical planar bridged complexes discussed earlier is reflected
in the n.m.r. spectrum where the central proton of the π-allylic group appears
at a higher $τ$ value (5·2 p.p.m.) than the other allylic protons[38] ($τ = 4·96$),
which is exactly the reverse of the order normally found (see pp. 434–435).
It has been suggested that the lack of symmetry in the binding of the π-
allylic groups in the non-planar complexes implies that all π-allylic complexes
are inherently non-rigid, a suggestion partly supported by the n.m.r. data, (pp.
435–436), and that the conformation in the solid state is largely determined by
the packing in the crystals.[44,45] Other palladium π-allyl structures that
have been determined include the dimeric $[(π-C_3H_5)Pd(CH_3COO)]_2$ in
which the two π-allyl groups are relatively close together[50] (Fig. 48) and

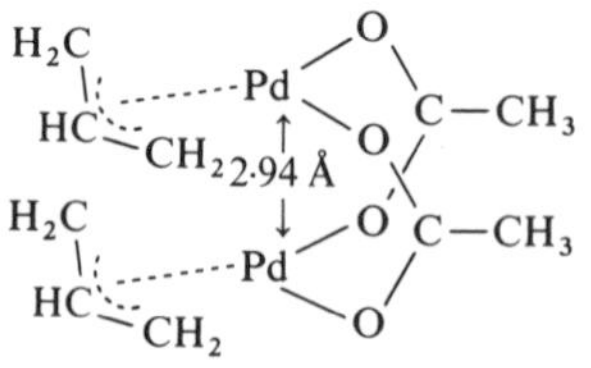

Fig. 48. The structure of $[(π-C_3H_5)Pd(CH_3COO)]_2$.

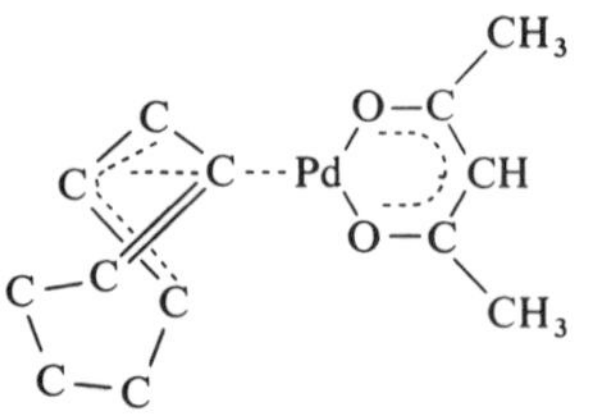

Fig. 49. The structure of $[(π-cyclo-octa-2,4-dienyl)Pd(acac)]$.

acetylacetonato(cyclo-octa-2,4-dienyl)palladium(II) (Fig. 49) which is of
interest because although the cyclo-octa-2,4-diene ligand has five adjacent
essentially sp^2 hybridised carbon atoms only three of these take part in
metal–ligand bonding, the other two forming a localised double-bond.[51,52]

THE STRUCTURE OF ALLYLPLATINUM COMPLEXES

Only two allylplatinum complexes have been studied by x-ray diffraction. Both have allyl groups bridging between two platinum atoms rather than the π-allylic structure found with palladium. Treatment of $[Pt(C_3H_5)_2]$ with dry hydrogen chloride gives $[Pt(C_3H_5)Cl]_4$, which has the structure[53] shown in Fig. 50, and on treatment with thallium(I) acetylacetonate this

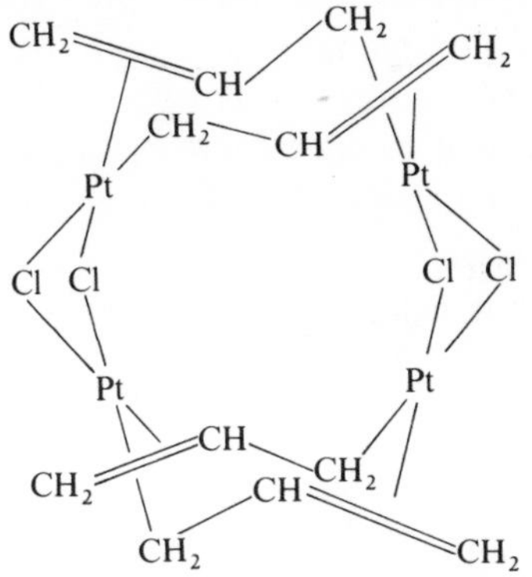

Fig. 50. The structure of $[Pt(C_3H_5)Cl]_4$.

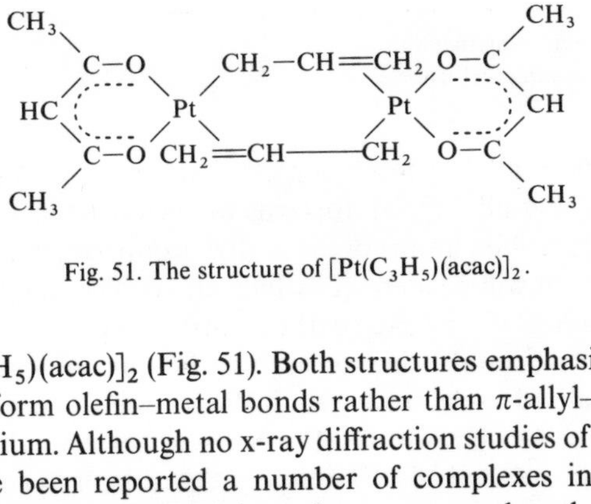

Fig. 51. The structure of $[Pt(C_3H_5)(acac)]_2$.

gives[54] $[Pt(C_3H_5)(acac)]_2$ (Fig. 51). Both structures emphasise the tendency of platinum to form olefin–metal bonds rather than π-allyl–metal bonds as found for palladium. Although no x-ray diffraction studies of π-allylplatinum complexes have been reported a number of complexes in which a platinum(II)–π-allyl bond is formed have been prepared and their structures established by comparison of their infra-red and n.m.r. spectra with those of the corresponding palladium(II) complexes.[9,10,55,56]

BONDING IN π-ALLYL COMPLEXES

Any bonding scheme that is put forward for π-allyl complexes must take account of the following experimental facts:

(i) In complexes containing a symmetrical π-allyl ligand bound to a palladium atom which has two further identical ligands co-ordinated to it, the three carbon atoms of the π-allyl group are essentially equidistant from the metal. Further the carbon–carbon bond lengths in such a π-allyl ligand are equal.

(ii) The dihedral angle between the plane of the π-allyl group and the square-plane containing palladium is generally about $116 \pm 10°$ (Table 120).

TABLE 120
Dihedral angle between π-allylic group and square-plane surrounding palladium

Complex	Dihedral angle	Reference
$[(\pi\text{-}C_3H_5)PdCl]_2$	111·5°	36
$[(\pi\text{-}C_3H_5)Pd(CH_3COO)]_2$	117° (average)	50
$[(\pi\text{-}C_8H_{11})Pd(acac)]$	121°	51, 52
$[(\pi\text{-}C_4H_7)PdCl]_2$	111·6°	37
$[\{\pi\text{-}(CH_3)_2CCHC(CH_3)_2\}PdCl]_2$	121·5°	38
$[\{\pi\text{-}CH_2C(OH)CHCOOC_2H_5\}PdCl]_2$	108°	39
$[\{\pi\text{-}CH_3(CH)_3CH_3\}PdCl]_2$	125° (average)	44
$[(\pi\text{-cycloheptenyl})PdBr]_2$	121° (average)	45
$[(\pi\text{-}C_3H_5)Pd(SnCl_3)(PPh_3)]$	114·5°	48
$[(\pi\text{-}C_4H_7)PdCl(PPh_3)]$	116°	46
$[(\pi\text{-}C_4H_7)Pd(S\text{-}DBM)]^a$	115·8°	47
$[\{\pi\text{-}CH_2C(t\text{-}C_4H_9)CH_2\}PdCl]_2$	119·5°	42
$[(\pi\text{-}DHMB)Pd(acac)]^b$	121°	49, 57
endo- and *exo-*$[\{\pi\text{-}C(Ph)C(Ph)C(Ph)C(Ph)(OEt)\}PdCl]_2$	95°	43

Average value 116°

^{a} S-DBM = thiodibenzoylmethanato.
^{b} DHMB = Dewarhexamethylbenzene.

(iii) Where the π-allyl group has methyl substituents, these are bent out of the plane of the π-allyl group towards the metal atom.

Since the three carbon atoms of the π-allyl group form an isosceles triangle it is likely that each will be identically hybridised. This can be accounted for, as in the allyl radical,[58] by each carbon atom being sp^2 hybridised with the remaining three p-orbitals (one on each carbon atom) interacting together to give a delocalised π-electron system. When the three p-orbitals interact they form three molecular orbitals (Fig. 52), which are bonding, non-bonding and antibonding respectively. If initially the dihedral angle between the π-allyl and $PdCl_2$ planes is considered to be 90°, then using the axes of Fig. 53, the ψ_1 and ψ_3 orbitals have the correct symmetry to overlap with the s, p_z, p_y, d_{yz} and d_{z^2} orbitals, and the ψ_2 orbital to overlap with the palladium p_x and d_{xz} orbitals. The palladium d_{xy} and $d_{x^2-y^2}$ orbitals can to a good approximation be considered as non-bonding orbitals, since first they lie in a plane parallel to but displaced horizontally from the π-allyl ligands and secondly, being d-orbitals, they are inherently diffuse. The palladium orbitals likely to overlap with ψ_1 can be separated from those likely to overlap with ψ_3 by pairing off those with coincident nodes. In this way the pairing scheme for the formation of π-allyl–palladium molecular orbitals becomes that shown in Table 121. A study of the overlap integrals of each of these possible pairs

Bonding

$$\psi_1 = \tfrac{1}{2}[\varphi_1 + \sqrt{2}\varphi_2 + \varphi_3]$$

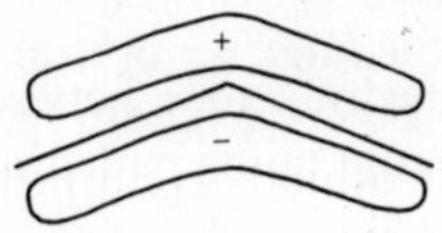

Non-Bonding

$$\psi_2 = \frac{1}{\sqrt{2}}(\varphi_1 - \varphi_3)$$

Antibonding

$$\psi_3 = \tfrac{1}{2}[\varphi_1 - \sqrt{2}\varphi_2 + \varphi_3]$$

Fig. 52. $p\pi$-Molecular orbitals of the π-allyl group. (The three carbon atoms lie in a plane (the xy plane of Fig. 53) which is perpendicular to the plane of the page (the xz plane of Fig. 53) with the terminal carbon atoms above and the central carbon atom below the plane of the page.)

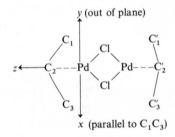

Fig. 53. The x, y and z axes referred to in discussing π-allyl complexes.

TABLE 121

Pairing scheme for the formation of
π-allyl palladium molecular orbitals

π-Allyl orbital	Palladium orbital
ψ_1	$5s, 5p_z, 4d_{z^2}$
ψ_2	$5p_x, 4d_{xz}$
ψ_3	$5p_y, 5d_{yz}$

suggests that only the overlap integrals for ψ_1 with d_{z^2} and ψ_2 with d_{xz} are appreciable and that all the other overlap integrals are very small.[59] Thus the principal cause of bonding in π-allyl–metal complexes is donation of electron density from the filled bonding π-orbital (ψ_1) on the allyl group to the empty metal d_{z^2} orbital complemented by back-donation of electron density from the filled d_{xz} orbital on the metal to the half-filled non-bonding π-orbital (ψ_2) on the allyl group. It should be emphasised that whilst these two molecular orbitals are principally responsible for bonding between the metal and π-allyl group a complete description of the bonding must involve all the orbitals listed in Table 121. The bonding/back bonding scheme suggested here is, of course, analogous to that found in complexes of olefins and carbon monoxide.

So far, only the first of the experimental observations mentioned at the beginning of this section has been used. To explain the dihedral angle between the π-allyl and $PdCl_2$ planes of about 116°, it is necessary to look closely at the criteria for maximum overlap of ψ_1 with d_{z^2} and ψ_2 with d_{xz}. Qualitatively it is apparent by considering the location of each of these orbitals in space that overlap of both the d_{z^2} with ψ_1 and d_{xz} with ψ_2 orbitals will be increased by tipping the plane of the π-allyl group in such a way that the central carbon atom is moved away from the metal, but that if the plane is tipped too far the overlaps of these orbitals will decrease again. When this qualitative argument is analysed by an approximately quantitative approach it is found that the d_{z^2}/ψ_1 and d_{xz}/ψ_2 overlaps are optimised when the dihedral angles are 112° and 102° respectively,[59] which in view of the approximations made during the calculation is remarkably close to the average value observed (116°—Table 120).

The displacement of methyl substituents on the allyl group out of the plane of this group towards the metal atom implies that the metal atom facilitates mixing of the σ- and π-electrons of the π-allyl ligand.[37] This indicates the crudeness of the above bonding scheme, since in this scheme the σ- and π-bonding orbitals of the allyl ligand are rigidly separated.

Before concluding this discussion of the bonding in π-allyl complexes it is important to consider the distortion that arises in the $(\pi\text{-}C_3H_5)Pd$ geometry on going from the symmetrical complex $[(\pi\text{-}C_3H_5)PdCl]_2$ where all the palladium–carbon bond lengths are essentially equal (Table 118) to $[(\pi\text{-}C_3H_5)PdCl(PPh_3)]$ where the palladium–carbon bond *trans* to chlorine is 2·14 Å and that *trans* to triphenylphosphine is 2·28 Å (Table 119). This asymmetry has led to the suggestion that the bonding in this latter complex is best described as a $\sigma + \pi$ bond (drawn as in Fig. 54) rather than a π-allyl bond.[46] This suggestion, although it has received some support[60] is extremely unlikely for two reasons:

(i) If the palladium atom lies in a suitable position to form a strong π-olefin bond it is far away from the optimum location to form a σ-bond to an sp^3 hybrid orbital on the other carbon atom and vice versa.

(ii) The $\sigma + \pi$-bonding scheme (Fig. 54) is quantum mechanically unattractive because if all three carbon atoms of the π-allyl group are bound to the metal, the π-allyl–metal molecular orbitals will all interact together rather than forming two separate groups.

A far more reasonable explanation for the definite, but relatively small disymmetry which results from the presence of two different ligands X and Y

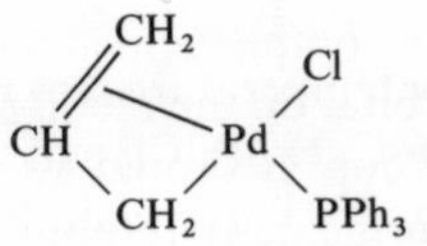

Fig. 54. Schematic representation for $\sigma + \pi$-bonding suggested for $[(\pi\text{-}C_3H_5)PdCl(PPh_3)]$.

in $[(\pi\text{-allyl})PdXY]$ complexes is that the different σ-donor abilities of X and Y perturb the molecular orbital bonding scheme discussed above for symmetrical π-allyl complexes. Recent molecular orbital calculations[61] on the *trans*-influence of L in *trans*-$[PtCl_2L(NH_3)]$ have shown that as the σ-donor power of L increases the covalent character of the *trans* platinum–nitrogen bond decreases. Further, this decrease arises from a decrease in the platinum $5d$ and $6s$ contributions whilst that from the platinum $6p$ actually increases. These calculations have received experimental support.[62,63] When this is applied to $[(\pi\text{-allyl})PdXY]$ it is not surprising to find that the palladium–carbon bond *trans* to the powerful σ-donor triphenylphosphine is lengthened, since the length of this bond is critically dependent on the overlap of the allyl ψ_2 and palladium d_{xz} orbitals. In addition, two ligands with very different σ-donor abilities will alter the coefficients of the p-orbitals for the different carbon atoms of the allyl ligand in the allyl molecular orbitals. Depending on the relative σ-donor abilities of X and Y in $[(\pi\text{-allyl})PdXY]$ three situations may arise (see Table 119). (i) Both the Pd–C_1 and Pd–C_3 as well as C_1–C_2 and C_2–C_3 bond lengths may be different (e.g. X = Cl, Y = PPh_3[46]); (ii) the Pd–C_1 and Pd–C_3 bond lengths can be different whilst the C_1–C_2 and C_2–C_3 bond lengths remain effectively equal (e.g. when X and Y are the oxygen and sulphur atoms of the bidentate ligand

$$HS\text{—}C(C_6H_5)\text{=}CH\text{—}\overset{\overset{\displaystyle O}{\|}}{C}\text{—}C_6H_5{}^{(47)});$$ (iii) the Pd–C_1 and Pd–C_3 as well as C_1–C_2 and C_2–C_3 bond lengths respectively can be effectively equal (e.g. X = $SnCl_3$, Y = PPh_3[48]).

NUCLEAR MAGNETIC RESONANCE STUDIES

The n.m.r. spectra of π-allyl complexes have been very extensively studied as they are potentially capable of yielding a great deal of information on the nature of the allyl group. The spectra fall into five main classes, of which the fifth (bis π-allyl complexes) is a special sub-group of the second (AA′BB′X π-allyl spectra).

(i) σ-Allyl spectra
From their formulae $(M\text{—}CH_2\text{—}CH\text{=}CH_2)$ σ-allyl complexes would be expected to exhibit three bands—two due to the terminal protons and one from the central proton. However, in $[PdCl(DMSO)_2(CH_2\text{—}CCl\text{=}CH_2)]^{(64)}$ and $[PdCl(PPh_3)_2(CH_2\text{—}C(CH_3)\text{=}CH_2)]^{(60)}$ the terminal protons give rise to only one band due to a rapid equilibrium (eqn. 534) being set up.

$$CH_2\text{=}CH\text{—}\overset{*}{C}H_2\text{—}Pd \rightleftharpoons \overset{*}{C}H_2\text{=}CH\text{—}CH_2\text{—}Pd \qquad (534)$$

(ii) AA'BB'X π-allyl spectra

A typical low resolution low temperature n.m.r. spectrum for a symmetrical π-allyl complex such as $[(\pi\text{-}C_3H_5)PdCl]_2$ in a non-co-ordinating solvent (e.g. $CDCl_3$) is shown in Fig. 55. Three main bands are observed due to the

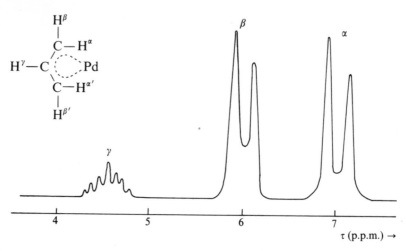

Fig. 55. AA'BB'X π-allyl–palladium spectrum. ($[(\pi\text{-}C_3H_5)PdCl]_2$ in $CDCl_3$ at room temperature).

proton bound to the central carbon atom (H_γ), the terminal *anti* protons (H_β) and the terminal *syn* protons (H_α). Typical τ values are H_γ 4·55, H_β 5·98 and H_α 7·05, which are observed for $[(\pi\text{-}C_3H_5)PdCl]_2$ in $CDCl_3$ at ambient temperature.[65] The phrase 'low temperature' in connection with these spectra varies considerably, decreasing to well below room temperature in

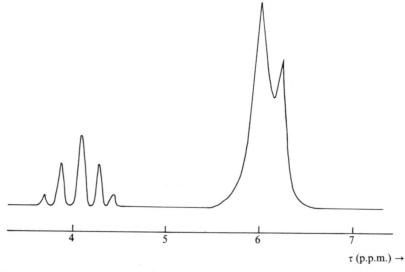

Fig. 56. A_4X π-allyl–palladium NMR spectrum ($[(\pi\text{-}C_3H_5)PdCl]_2$ + PPh_3 in $CDCl_3$ at room temperature).

the presence of a co-ordinating solvent or added ligand. In an inert solvent such as $CHCl_2CHCl_2$ the spectrum shown in Fig. 55 persists up to 150°C, when decomposition of the complex sets in.

(iii) A_4X π-allyl spectra

At high temperature (which, as with low temperature, is a relative term and varies considerably from one system to the next) in the presence of a co-ordinating solvent or added ligand the AA'BB'X spectrum of Fig. 55 collapses to an A_4X spectrum by coalescence of the H_α and H_β peaks into a single peak (see Fig. 56), indicating that all the terminal protons (i.e. H_α and H_β) are equivalent.

(iv) ABCDX π-allyl spectra

At very low temperatures asymmetric π-allyl complexes such as $[(\pi\text{-}C_3H_5)PdCl(PPh_3)]^{[60]}$ and $[(\pi\text{-}C_3H_5)Pd(PPh_3)(SnCl_3)]^{[67]}$ exhibit ABCDX n.m.r. spectra due to the non-equivalence of H_α and $H_{\alpha'}$ as well as H_β and $H_{\beta'}$ protons (see Fig. 55) which results from their being *trans* to two different ligands.

(v) Bis π-allyl complexes

The proton n.m.r. spectrum of $[Pd(\pi\text{-}C_3H_5)_2]$ at low temperature consists of two AB_2C_2 spectra present in unequal amounts. This has been interpreted in terms of an equilibrium mixture of *syn*- and *anti*-isomers[68] (Fig. 57). The

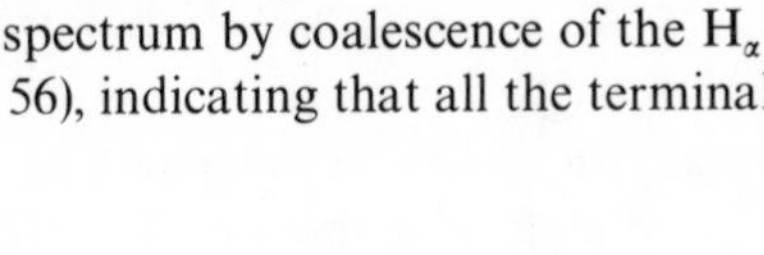

Fig. 57. Syn- and anti-isomers of $[M(\pi\text{-}C_3H_5)_2]$.

100 MHz spectrum of the platinum analogue $[Pt(\pi\text{-}C_3H_5)_2]$ appeared to indicate a different system,[68] but this well illustrates the dangers inherent in the interpretation of any n.m.r. spectrum since this apparent difference between the platinum and palladium complexes is simply the result of accidental coincidences, which are removed at 220 MHz.[69]

Temperature dependence of the nuclear magnetic resonance spectra

The temperature dependence of the n.m.r. spectra of π-allyl complexes both in the presence and absence of co-ordinating ligands has been studied extensively. Four main types of motion have been put forward to account for the various n.m.r. spectra that have been observed (in accordance with a recently introduced phrase, π-allyl complexes may be described as 'stereo-chemically non-rigid').

   (i) A momentary conversion of the π-allyl complex to a σ-allyl complex can enable free rotation to occur about the palladium–carbon or adjacent carbon–carbon bond followed by reformation of the π-allyl complex[64,70–5] (eqn. 535). This would enable H_β and H_α to exchange, hence giving an A_4B spectrum.

$$
\begin{array}{c}
H^\beta \\
| \\
C-H^\alpha \\
\diagup \quad \vdots \\
H-C \quad \quad Pd \\
\diagdown \quad \vdots \\
C-H^{\alpha'} \\
| \\
H^{\beta'}
\end{array}
\rightleftharpoons
\begin{array}{c}
H \quad \quad H \\
\diagdown \quad \diagup \\
H_2C \quad \quad C \\
\diagdown \quad \quad \diagup \\
C \\
| \\
H \quad Pd
\end{array}
\tag{535}
$$

(ii) Rotation of the allyl group in the plane of its equilibrium position relative to the metal (i.e. in a plane at about 116° to the metal–π-allyl bond) would enable H_α and $H_{\alpha'}$ to become equivalent (they are not equivalent in unsymmetrical complexes such as $[(\pi\text{-}C_3H_5)PdCl(PPh_3)]$ unless some such rotation occurs) but would still keep H_α and H_β non-equivalent.[71,76,77] This would convert an ABCDE spectrum at low temperature into an A_2B_2C spectrum at high temperature.

(iii) Rotation of the terminal CH_2 groups about the carbon–carbon bonds to the central carbon atom whilst the allyl group as a whole retains the π-configuration relative to the metal would enable H_α and H_β as well as $H_{\alpha'}$ and $H_{\beta'}$ to become equivalent whilst still keeping H_α and $H_{\alpha'}$ non-equivalent.[78] This rotation would also convert an ABCDE spectrum at low temperature into an A_2B_2C spectrum at a higher temperature. Although this rotation appears to be present in $[Zr(\pi\text{-}C_3H_5)_4]$, it now appears unlikely that it occurs in palladium–π-allyl complexes.[74]

(iv) A flip motion in which the allyl group rotates backwards (i.e. eqn. 536) would also enable H_α and H_β as well as $H_{\alpha'}$ and $H_{\beta'}$ to become equivalent

$$
\begin{array}{ccc}
\diagdown\!\!\!\!-M & \rightleftharpoons & \diagdown\!\!\!\!-M & \rightleftharpoons & \diagup\!\!\!\!-M
\end{array}
\tag{536}
$$

whilst still keeping H_α and $H_{\alpha'}$ non-equivalent,[66] again converting an ABCDE spectrum at low temperature into an A_2B_2C spectrum at a higher temperature.

It is extremely difficult with the evidence currently available to distinguish between those suggestions which provide satisfactory interpretations of the observed phenomena and those which are actually correct. It would appear that two main lines of corroborative work are needed, firstly independent non-n.m.r. support for the postulated motion and secondly a repetition of the vital experiments on a 220 MHz n.m.r. spectrometer to minimise the likelihood that any accidental coincidences which distort the observed n.m.r. spectra are present.

There is some independent support for rotation (i) (equilibrium 535), since above $-20°C$ the $(-)$-diastereoisomer of chloro(3-acetyl-2-methyl-allyl)-(S-α-phenethylamine)palladium(II) epimerises rapidly to a 1:1 mixture of $(+)$- and $(-)$-diastereoisomers.[79–81] This can be understood in terms of eqn. 537. Further, the isomerisation of (π-crotyl)(2-picoline)palladium(II)

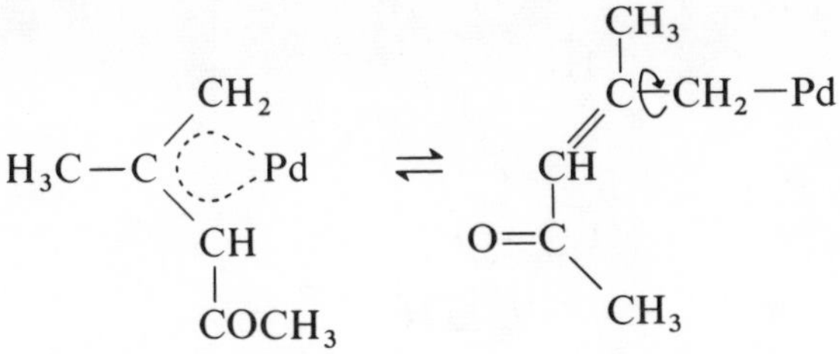

$$
\begin{array}{ccc}
& CH_2 & \\
H_3C-C{\Big\langle}\underset{\cdots}{::}{\Big\rangle}Pd & \rightleftharpoons & \\
& CH & \\
& | & \\
& COCH_3 &
\end{array}
\qquad
\begin{array}{c}
CH_3 \\
| \\
C{\overset{+}{\rightleftharpoons}}CH_2-Pd \\
\| \\
CH \\
| \\
O=C \\
\backslash \\
CH_3
\end{array}
\qquad (537)
$$

chloride[82] (eqn. 538) and isomerisation and epimerisation of (1-acetyl-2-methylallyl)(amine)palladium(II) chloride are now thought to occur via a σ-allyl intermediate with rotation about the palladium–carbon σ-bond,[81,83] although reaction 538 was originally thought to occur via the dimer

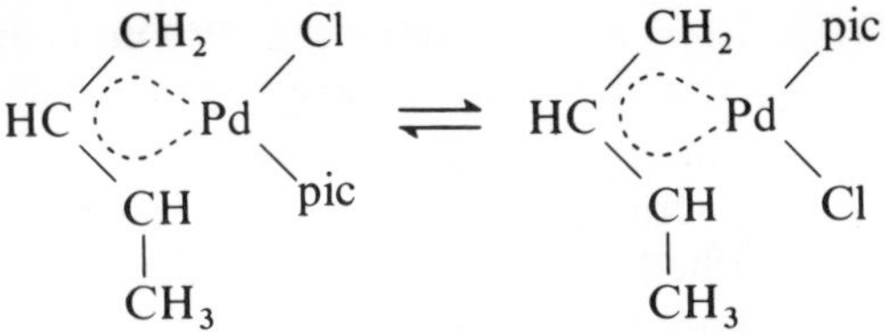

$$
\begin{array}{cc}
CH_2 & Cl \\
HC{\Big\langle}\underset{\cdots}{::}{\Big\rangle}Pd & \\
CH & pic \\
| & \\
CH_3 &
\end{array}
\rightleftharpoons
\begin{array}{cc}
CH_2 & pic \\
HC{\Big\langle}\underset{\cdots}{::}{\Big\rangle}Pd & \\
CH & Cl \\
| & \\
CH_3 &
\end{array}
\qquad (538)
$$

[(π-C$_4$H$_7$)PdCl]$_2$ formed by loss of 2-picoline.[82] The *anti*-π-allylic complex is formed by treating 2,4,4-trimethylpent-2-ene with palladium chloride in acetic acid containing sodium acetate (eqn. 539), because of the conforma-

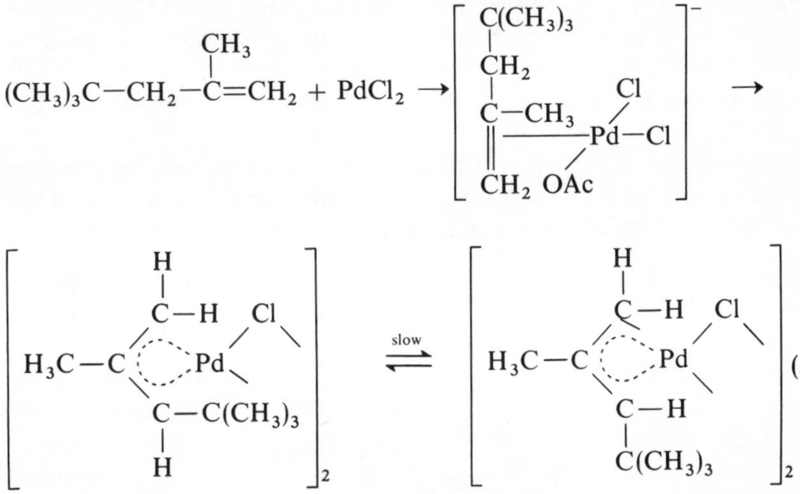

$$
(CH_3)_3C-CH_2-\underset{\underset{CH_3}{|}}{C}=CH_2 + PdCl_2 \rightarrow
\left[
\begin{array}{c}
C(CH_3)_3 \\
| \\
CH_2 \\
| \quad\quad Cl \\
C-CH_3 \quad / \\
\| \quad\quad Pd-Cl \\
CH_2 \quad OAc
\end{array}
\right]^-
\rightarrow
$$

$$
\left[
\begin{array}{cc}
H & \\
| & \\
C-H & Cl \\
H_3C-C{\Big\langle}\underset{\cdots}{::}{\Big\rangle}Pd & \\
C-C(CH_3)_3 & \\
| & \\
H &
\end{array}
\right]_2
\overset{slow}{\rightleftharpoons}
\left[
\begin{array}{cc}
H & \\
| & \\
C-H & Cl \\
H_3C-C{\Big\langle}\underset{\cdots}{::}{\Big\rangle}Pd & \\
C-H & \\
| & \\
C(CH_3)_3 &
\end{array}
\right]_2
\qquad (539)
$$

tion of the intermediate π-olefin complex, which is dictated by steric hindrance between the chloride ligands and the *t*-butyl groups.[84] Prolonged heating or addition of a co-ordinating ligand gives isomerisation of the *anti*-to the *syn*-isomer, which strongly suggests the formation of a labile σ-allyl intermediate with free rotation about the palladium–carbon σ-bond. The observation that on heating *syn*-1-isopropylallylpalladium(II) chloride (**174**) in *o*-dichlorobenzene both the isopropyl and the H_{3a} and H_{3s} doublets

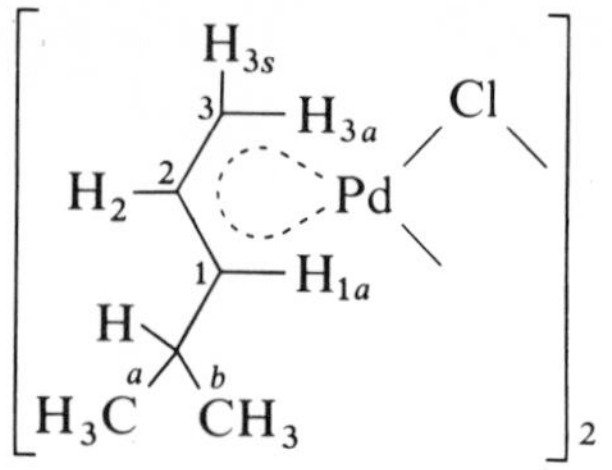

$$\left[\begin{array}{c} H_{3s} \\ \underset{3}{\overset{}{}} H_{3a} \quad Cl \\ H_2 \underset{2}{\overset{}{}} \quad Pd \\ \underset{1}{\overset{}{}} H_{1a} \\ H \\ \overset{a}{H_3C} \overset{b}{CH_3} \end{array} \right]_2$$

174

collapse with the same free energy of activation (*ca.* 19·5 kcal/mol) is consistent with rotation (i) occurring.[85]

Rotation (i) is also supported by studies of the rearrangement of π-allyl–palladium complexes which contain two chemically identical but stereochemically non-equivalent groups. Two types of complex have been used. The first, $[(\pi\text{-}C_3H_5)Pd(PMe_2Ph)Cl]$, contains the two non-equivalent methyl groups in the tertiary phosphine ligand[74] and the second (**175**) contains the

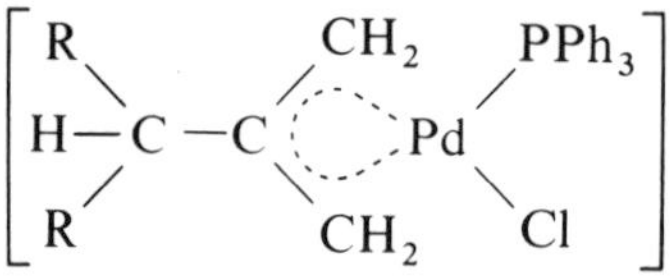

$$\left[\begin{array}{c} R \qquad\qquad CH_2 \quad PPh_3 \\ H-C-C \quad Pd \\ R \qquad\qquad CH_2 \quad Cl \end{array} \right]$$

175

two non-equivalent groups in the π-allylic ligand itself.[86] When the n.m.r. spectra of both types of complex were recorded it was found that the non-equivalent nuclei became equivalent as the temperature was raised, indicating that the complexes had attained a plane of symmetry not initially present.[87] This is consistent with the intermediate formation of a σ-allyl complex (i.e. rotation *i*).

The complexity of the n.m.r. spectra of π-allyl complexes in the presence of added ligands is hardly surprising since, in addition to the motions of the allyl group outlined above, five equilibria (eqns. 540–4) are also set up.[60]

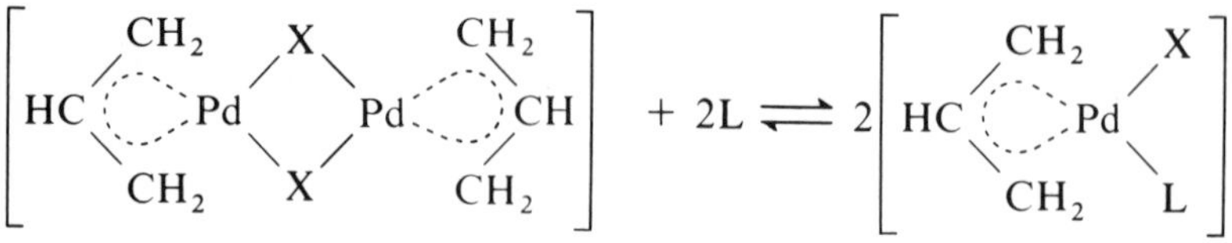

$$\left[\begin{array}{c} CH_2 \quad X \quad CH_2 \\ HC \quad Pd \quad Pd \quad CH \\ CH_2 \quad X \quad CH_2 \end{array} \right] + 2L \rightleftharpoons 2 \left[\begin{array}{c} CH_2 \quad X \\ HC \quad Pd \\ CH_2 \quad L \end{array} \right] \quad (540)$$

$$\left[\begin{array}{c} CH_2 \quad X \\ HC \quad Pd \\ CH_2 \quad L \end{array} \right] + \tfrac{1}{2} \left[\begin{array}{c} CH_2 \quad X \\ HC \quad Pd \\ CH_2 \end{array} \right]_2 \rightleftharpoons$$

$$\left[\begin{array}{c} CH_2{=}CH-H_2C \qquad X \qquad CH_2 \\ Pd \quad Pd \quad CH \\ L \qquad X \qquad CH_2 \end{array} \right] \quad (541)$$

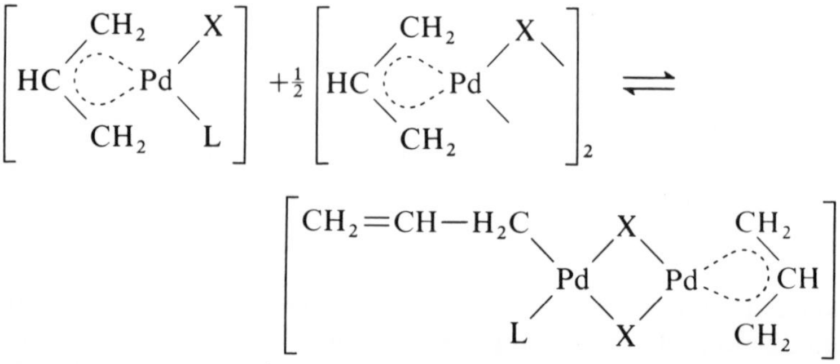

$$\begin{bmatrix} \begin{array}{c} CH_2 \\ HC \end{array} \begin{array}{c} X \\ Pd \\ CH_2 \end{array} \begin{array}{c} \\ \\ L \end{array} \end{bmatrix} + L \; \rightleftharpoons \; \begin{bmatrix} \begin{array}{c} CH_2 \\ HC \end{array} \begin{array}{c} L \\ Pd-X \\ CH_2 \end{array} \begin{array}{c} \\ \\ L \end{array} \end{bmatrix} \qquad (542)$$

$$\begin{bmatrix} \begin{array}{c} CH_2 \\ HC \end{array} \begin{array}{c} L \\ Pd-X \\ CH_2 \end{array} \begin{array}{c} \\ \\ L \end{array} \end{bmatrix} \; \rightleftharpoons \; \begin{bmatrix} CH_2{=}CH{-}CH_2 \qquad X \\ Pd \\ L \qquad\qquad L \end{bmatrix} \qquad (543)$$

$$\begin{bmatrix} CH_2{=}CH{-}CH_2 \qquad X \\ Pd \\ L \qquad\qquad L \end{bmatrix} \; \rightleftharpoons \; \begin{bmatrix} \begin{array}{c} CH_2 \\ HC \end{array} \begin{array}{c} L \\ Pd \\ CH_2 \end{array} \begin{array}{c} \\ \\ X \end{array} \end{bmatrix} + L \qquad (544)$$

The concentration of each of the species shown in these equilibria depends on the temperature, the nature of the π-allylic group present, the halide ligand (X) and the added ligand (L) which can be PPh_3, $AsPh_3$, $SbPh_3$, CO or DMSO.

VIBRATIONAL SPECTROSCOPY

The principal use of the infra-red and Raman spectra of π-allyl complexes is to demonstrate the absence of a band due to the double-bond present in the free allyl ligand. Thus neither $[(\pi\text{-}C_3H_5)PdCl]_2$ nor its 1:1 adducts with triphenylphosphine or dimethylsulphoxide show bands in their laser Raman spectra due to a free double bond,[88] which is consistent with the n.m.r.[72,76] and electronic[89] spectra and suggests that dimethylsulphoxide cleaves the halogen-bridge to give a monomeric π-allyl system (reaction 545). Complete

$$[(\pi\text{-}C_3H_5)PdCl]_2 + 2DMSO \rightarrow [(\pi\text{-}C_3H_5)PdCl(DMSO)] \qquad (545)$$

infra-red and Raman studies of π-allyl complexes have been reported,[90–4] although there is very little agreement about their assignment. Indeed there is not even agreement about the number of bands expected to arise from the π-allyl–palladium stretching mode. One recent paper predicts three such vibrations, one symmetric and two asymmetric,[94] whereas another does not predict any particular number, but finds two, one of which is split in the solid by crystal field effects.[93] It is clear that a great deal more work in this field is required.

ELECTRONIC SPECTROSCOPY

A recent investigation by the present author[89] of the electronic spectra of $[(\pi\text{-}C_3H_5)PdCl]_2$ in eight different solvents indicates that there are five bands in the region 20,000–50,000 cm^{-1} (Fig. 58) rather than three as suggested by earlier workers.[95,96] The considerable solvent dependence of these spectra is generally consistent with the positions of the solvents in the spectrochemical series except where a specific reaction of the solvent is

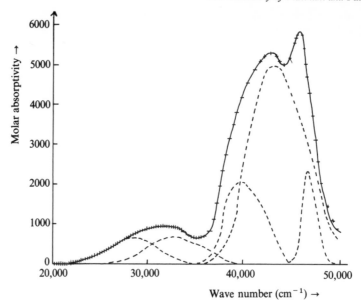

Fig. 58. The ultra-violet and visible spectrum of π-allylpalladium chloride in ethanol (reproduced with permission from ref. 89).

known to occur[89] (e.g. DMSO). The spectrum of $[(\pi\text{-}C_3H_5)PdCl]_2$ in aqueous 4 M potassium chloride suggests that in solution equilibrium 546 lies well to the left although addition of a suitable large cation leads to the precipitation of the anion $[(\pi\text{-}C_3H_5)PdCl_2]^-$.[97] The assignment of the bands

$$[(\pi\text{-}C_3H_5)PdCl]_2 + 2K^+Cl^- \overset{H_2O}{\rightleftharpoons} 2K^+[(\pi\text{-}C_3H_5)PdCl_2]^- \qquad (546)$$

in the electronic spectra of π-allyl–palladium complexes to particular electronic transitions is extremely complex and little progress has been made to date.

THERMODYNAMIC STUDIES

The quantitative decomposition of π-allyl palladium chloride to palladium metal and allyl chloride (reaction 547) which occurs at 180°C has been studied by differential scanning calorimetry giving values for the enthalpies

$$[(\pi\text{-}C_3H_5)PdCl]_2 \longrightarrow 2Pd + 2C_3H_5Cl \qquad (547)$$

of formation of crystalline and gaseous π-allylpalladium chloride of -25.7 and $+4.3$ kcal/mol respectively.[98] This gives a minimum value of about 55.7 kcal/mol for the palladium(II)–π-allyl bond dissociation energy.

PROPERTIES OF π-ALLYL COMPLEXES

The platinum– and palladium–π-allyl complexes are yellow, or occasionally red (only in the case of palladium) crystalline complexes the majority of which are easily handled in air at room temperature. Although the stability of the platinum complexes to water has not been reported, the palladium complexes are hydrolysed readily at room temperature[4,99] (eqn. 556, p. 442).

One of the most interesting and as yet virtually unanswered problems is: 'Why does palladium show such a great preference for forming π-allyl complexes whereas platinum prefers to form olefin complexes?' It is still too early to give a definite answer to this question, as the few relevant comments that can be made indicate:

(i) Palladium–olefin complexes are far less stable than platinum–olefin complexes both to displacement of the olefin by a halide ion (i.e. equilibrium 458, p. 375) and to reaction of the complex with a variety of reagents including water and most nucleophiles.

(ii) It has been stated that for the bis-π-allyl complexes $[M(\pi\text{-}C_3H_5)_2]$ the platinum complex is more stable than the palladium complex, although the word 'stable' was not defined.[32]

(iii) Several workers have suggested that the dearth of π-allylplatinum complexes may be the result of insufficient experimental effort in this field coupled with the greater stability of the olefin complexes of platinum than palladium.

CHEMICAL PROPERTIES OF π-ALLYL COMPLEXES

In the following account the chemical reactions of π-allyl complexes have been divided into four groups.

(i) Bridge cleavage reactions

A number of reagents will cleave the bridge of dimeric π-allylpalladium complexes whilst leaving the palladium–π-allyl bond intact although perhaps slightly modified (see pp. 432–433). A number of these reactions, which need no further comment, are shown in eqns. 548–53. The reaction of triphenylphosphine is a little more complex. With an equimolar concentration of

$$[(\pi\text{-allyl})Pd \underset{Cl}{\overset{Cl}{\diagdown\diagup}} Pd(\pi\text{-allyl})]$$

$$\xrightarrow[\text{HCl}]{\text{Na}^+\text{C}_5\text{H}_5^-} \quad [(\pi\text{-allyl})Pd(\pi\text{-C}_5H_5)] \qquad (548)$$
(ref. 100)

$$\xrightarrow[\text{acetone}]{\text{LiBr in}} [(\pi\text{-allyl})PdBr]_2 \qquad (549)$$
(ref. 100)

$$\xrightarrow[\text{(ref. 100)}]{\text{Tl}^+ \text{acac}^-} [(\pi\text{-allyl})Pd(\text{acac})] \qquad (550)$$

$$\xrightarrow[\text{(ref. 100)}]{\text{pyr.}} [(\pi\text{-allyl})PdCl(\text{pyr})] \qquad (551)$$

$$\xrightarrow[\text{(ref. 101)}]{\text{Fe}_2(\text{CO})_9} [(\pi\text{-allyl})PdCl\{Fe(CO)_4\}] \qquad (552)$$

$$\xrightarrow[\substack{+\text{Ph}_4\text{P}^+\text{Cl}^- \\ \text{(ref. 97)}}]{\text{Aq. KCl}} \text{Ph}_4\text{P}^+[(\pi\text{-allyl})PdCl_2]^- \qquad (553)$$

triphenylphosphine the dimeric complex $[(\pi\text{-allyl})PdCl]_2$ is split giving $[(\pi\text{-allyl})PdCl(PPh_3)]^{(70)}$ (the bonding in this complex is discussed on pp. 432–433). The product of the action of a further mole of triphenyl-phosphine depends on the polarity of the solvent. In a non-polar solvent such as chloroform the σ-allyl complex $[(PPh_3)_2PdCl(\sigma\text{-allyl})]$ is formed,[60] whereas in a more polar solvent, such as aqueous acetone, the cationic π-allyl complex $[(PPh_3)_2Pd(\pi\text{-allyl})]^+Cl^-$ is formed.[102] The tendency of ligands to form such cationic species decreases in the order $PMe_2Ph \sim PEt_2Ph \sim PEt_3 > PPh_3 > AsPh_3 > SbPh_3 >$ pyridine. Treatment of $[(\pi\text{-allyl})PdCl]_2$ in ethanol with 5–6 mol of triphenylphosphine not only

$$[(\pi\text{-allyl})PdCl]_2 + 6PPh_3 \rightarrow 2[Pd(PPh_3)_4] + 2[allylPPh_3]^+Cl^- \quad (554)$$

cleaves the halogen bridge, but also cleaves the palladium–π-allyl bond[102] (reaction 554). A similar reaction occurs for π-allyl–platinum complexes (reaction 555), although in this case the reaction is reversible.[9,10]

$$[(\pi\text{-allyl})Pt(PPh_3)_2]^+Cl^- + 2PPh_3 \rightleftharpoons [Pt(PPh_3)_4] + allylCl \quad (555)$$

(ii) π-Allyl–metal bond cleavage
(a) Hydrolysis
π-Allylpalladium chloride dimers are readily hydrolysed by dilute alkali at room temperature[4,99] (reaction 556). This reaction does not occur by a simple hydrolysis of the π-allyl–palladium bond, but rather occurs by initial formation of an olefin from one π-allyl group through abstraction of a hydrogen atom from the other π-allyl group, followed by addition of water

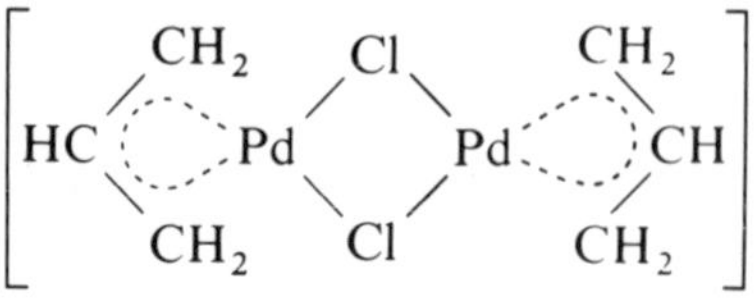

$$CH_2=CH-CHO + CH_2=CH-CH_3 + 2Pd + 2HCl \quad (556)$$

to this second π-allyl group.[103] This complex mechanism is based on fact that in $NaOD/D_2O$ the olefin formed is largely (83 %) undeuterated, whereas direct hydrolytic cleavage of the π-allyl–palladium bond should yield mainly monodeuterated olefin.

(b) Other reagents
No systematic study of reagents that cleave π-allyl–metal bonds has been made. Lead tetra-acetate cleaves π-allyl–palladium bonds to give allyl-acetate and palladium(II) acetate.[104] Cationic π-allyl–platinum complexes are cleaved by hydrogen to give alkanes (reaction 557) and by hydrogen chloride to give olefins[105] (reaction 558).

$$[(PPh_3)_2Pt(\pi\text{-}C_3H_5)]^+Cl^- \xrightarrow[CHCl_3]{H_2 \text{ in}}$$

$$CH_3CH_2CH_3 + cis\text{-}[(PPh_3)_2PtCl_2] \quad (557)$$

$$[(PPh_3)_2Pt(\pi\text{-}C_3H_5)]^+Cl^- + HCl \xrightarrow{CHCl_3}$$

$$CH_3CH{=}CH_2 + cis\text{-}[(PPh_3)_2PtCl_2] \quad (558)$$

(iii) Insertion into the π-allyl–metal bond

The insertions of both dienes and carbon monoxide into the π-allyl–palladium bond have been investigated in some detail because the reactions are of potential commercial importance.

(a) Dienes

Dienes react with π-allyl–palladium complexes to give the diene insertion product[105a—109] (reactions 559 and 560). The rate of diene insertion depends

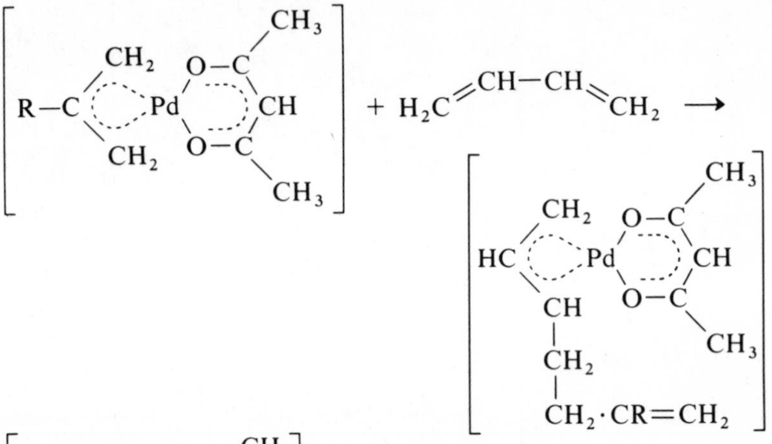

$$(559)$$

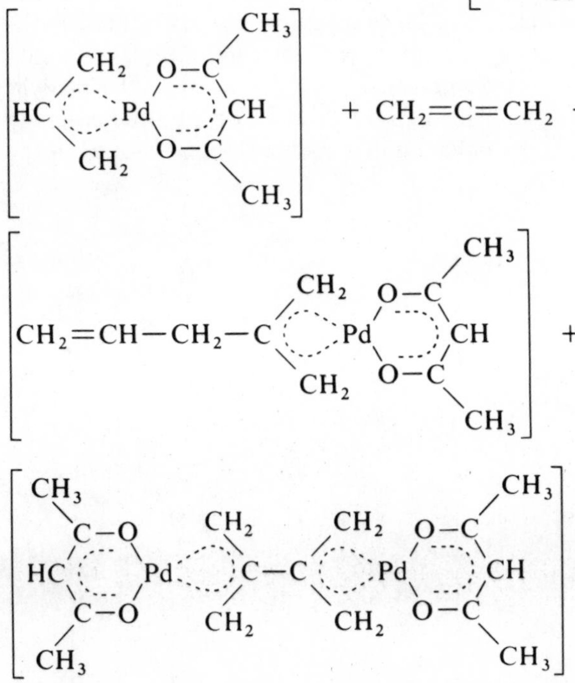

$$+ \text{ 1,5-hexadiene} \quad (560)$$

on the π-allyl group, the diene, the other ligands bound to palladium, the nature of the solvent and any extra ligand that may be added. Thus treatment of π-allylpalladium acetate with butadiene in benzene gives a linear trimer of butadiene, whereas in the presence of added triphenylphosphine a dimer of butadiene is the main product. When alcohols (or carboxylic acids) are used as solvents, unsaturated ethers (or esters) of the type $ROCH_2$-$CH=CHCH_2(CH_2-CH=CH-CH_2)_nCH_2CH_2CH=CH_2$ ($n = 0$, 1 or 2) are formed.[110] Deuteration studies on the insertion of allene into π-allyl-palladium acetate showed that the reaction (reaction 561) involves insertion

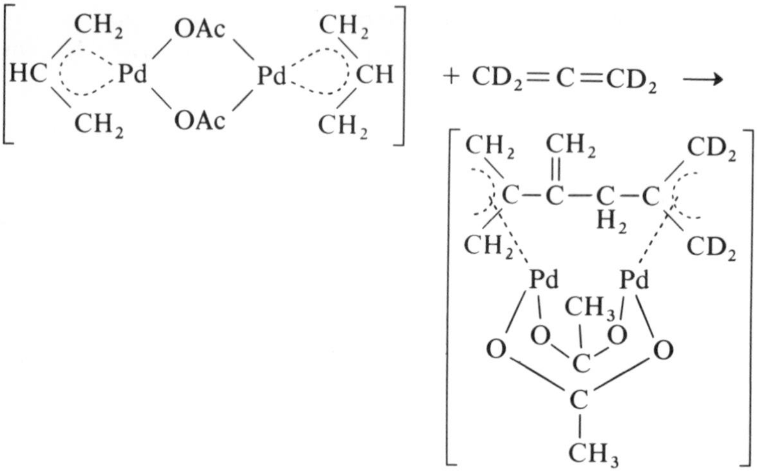

$$(561)$$

of allene into the π-allyl–palladium bond[111] and not trimerisation of allene.

An n.m.r. study of the kinetics of diene insertion into palladium–π-allyl bonds has suggested that the reaction involves an initial rapid pre-equilib-rium in which the diene co-ordinates to the metal converting the π-allylic complex to a σ-allylic complex. This step is followed by the rate determining insertion of the diene into the palladium(II)–carbon σ-bond formed in the first stage[108] (reaction 562).

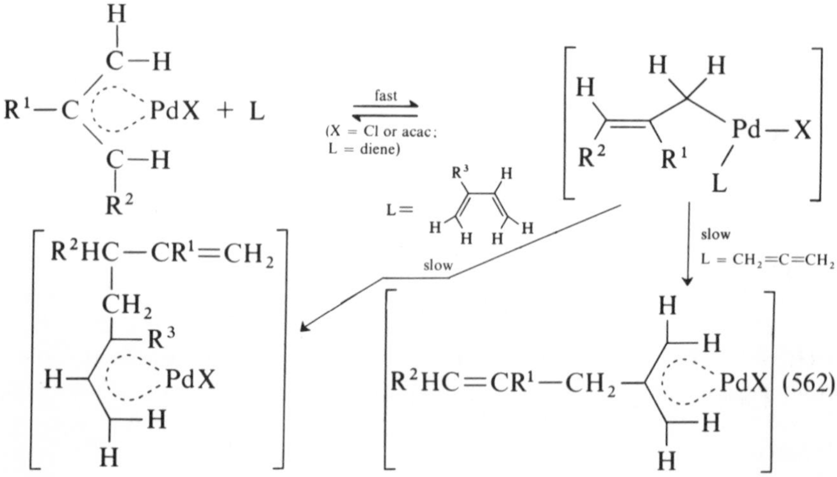

$$(562)$$

(b) Carbon Monoxide

The overall product obtained from insertion of carbon monoxide into the π-allyl–palladium bond depends on the solvent. In an inert solvent the product is an acid chloride, whereas in a solvent such as ethanol an ester is formed[112,113] (reaction 563). As it is not necessary to prepare the π-allyl

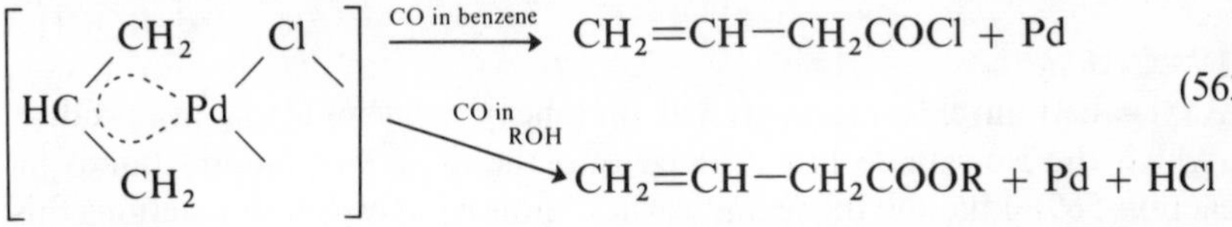

$$
\begin{bmatrix} CH_2 & Cl \\ HC \quad Pd & \\ CH_2 & \end{bmatrix}
\xrightarrow{\text{CO in benzene}} CH_2{=}CH{-}CH_2COCl + Pd
$$

$$\xrightarrow[\text{ROH}]{\text{CO in}} CH_2{=}CH{-}CH_2COOR + Pd + HCl \tag{563}$$

complex first, it is possible to carbonylate allylic compounds in the presence of catalytic amounts of palladium(II) chloride.[114] In this carbonylation of allylic compounds such as allyl halides, alcohols, ethers and esters the carbon monoxide is always found in the allylic position[114] (reactions 564–7). The results of the carbonylation of a range of allylic compounds have been summarised in a recent review.[115]

$$CH_2{=}CH{-}CH_2Cl + CO \xrightarrow[\text{cat.}]{PdCl_2} CH_2{=}CH{-}CH_2COCl \tag{564}$$

$$CH_2{=}CH{-}CH_2OR + CO \xrightarrow[\text{cat.}]{PdCl_2} CH_2{=}CH{-}CH_2COOR \tag{565}$$

$$CH_2{=}CH{-}CH_2OCOCH_3 + CO \xrightarrow[\text{cat.}]{PdCl_2}$$

$$CH_2{=}CH{-}CH_2{-}\overset{\overset{O}{\|}}{C}{-}O{-}\overset{\overset{O}{\|}}{C}{-}CH_3 \tag{566}$$

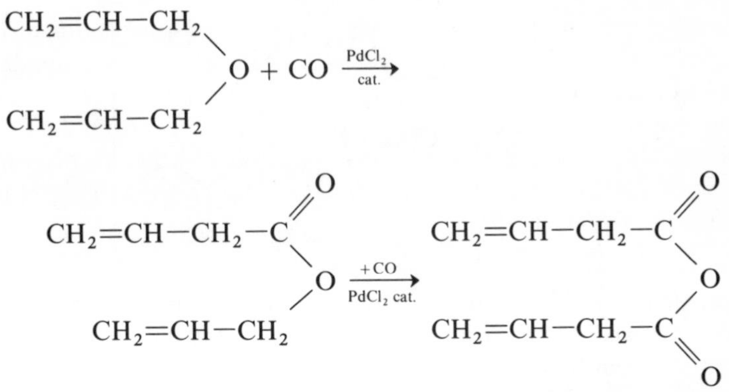

$$
\begin{array}{c} CH_2{=}CH{-}CH_2 \\ \qquad\qquad\qquad O + CO \xrightarrow[\text{cat.}]{PdCl_2} \\ CH_2{=}CH{-}CH_2 \end{array}
$$

(567)

$$
\begin{array}{c} CH_2{=}CH{-}CH_2{-}C \overset{O}{\diagdown}_O \\ \qquad\qquad\qquad CH_2{=}CH{-}CH_2 \end{array}
\xrightarrow[\text{PdCl}_2\text{ cat.}]{+CO}
\begin{array}{c} CH_2{=}CH{-}CH_2{-}C \overset{O}{\diagdown}_O \\ CH_2{=}CH{-}CH_2{-}C \overset{}{\diagup}_O \end{array}
$$

A mechanism analogous to the diene insertion mechanism has recently been suggested for the carbonylation of π-allylic complexes[116] (Fig. 59). The first stage involves cleavage of the chloride bridge of the π-allyl complex to give a monomeric carbonyl complex, which then picks up a molecule of the allyl halide to give a five-co-ordinate complex in the rate-determining step. The five-co-ordinate complex then rearranges by insertion of carbon monoxide into the π-allyl–palladium bond to give a four-co-ordinate acyl complex, which rapidly accepts a halogen atom from the co-ordinated olefin before the vinylacetyl halide group is displaced by further carbon monoxide. When π-allyl–platinum complexes are carbonylated the acyl complexes formed are more stable than their palladium analogues and can be iso-lated[105] (reaction 568).

$$[(PPh_3)_2Pt(\pi\text{-}C_3H_5)]^+Cl^- + CO \rightarrow$$

$$1:1 \text{ mixture of } cis\text{- and } trans\text{-}[(PPh_3)_2PtCl(COCH=CHCH_3)] \quad (568)$$

(c) Isocyanide

π-Allyl palladium chloride reacts with cyclohexyl isocyanide to give a product in which the isocyanide has inserted into the π-allyl–palladium bond[117] (reaction 569). Like the diene and carbon monoxide insertion reactions this

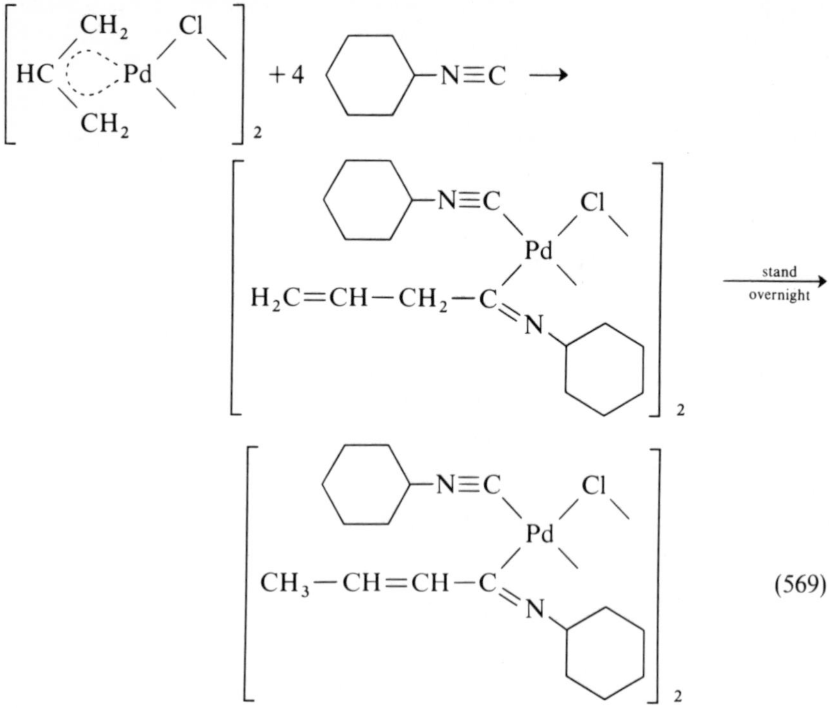

reaction involves chloride-bridge cleavage with co-ordination of the iso-cyanide followed by insertion of the co-ordinated isocyanide into the π-allyl–palladium bond.

(d) Sulphur dioxide

Sulphur dioxide reacts in a similar manner to carbon monoxide with π-allyl–platinum complexes[105] (reaction 570).

$$[(PPh_3)_2Pt(\pi\text{-}C_3H_5)]^+ + SO_2 \rightarrow$$

$$cis\text{- and } trans\text{-}[(PPh_3)_2PtCl(SO_2CH=CH-CH_3)] \quad (570)$$

(iv) Substitution reactions on co-ordinated π-allyl ligands

π-Allyl–palladium complexes with electron-withdrawing substituents (alkoxy or halogeno groups) in the side chain are readily attacked by alcohols in the

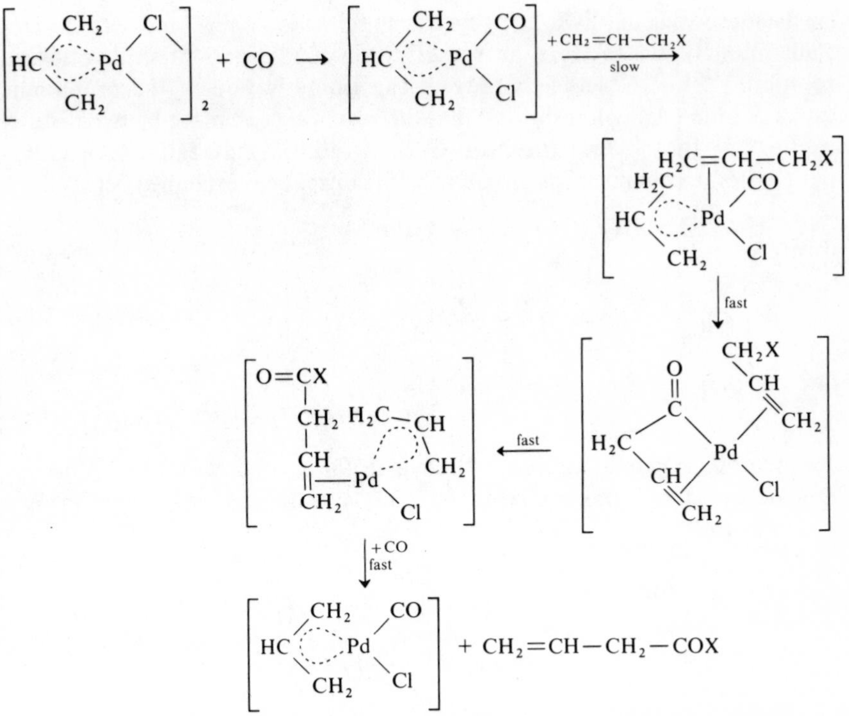

Fig. 59. Mechanism for the carbonylation of π-allylic complexes.

presence of hydrogen chloride[95,100] (reaction 571). These extremely easy alkoxy-exchanges are much faster when R (in eqn. 571) is CH_3 than H, suggesting the intermediate formation of an allyl carbonium ion, possibly stabilised by interaction with palladium.

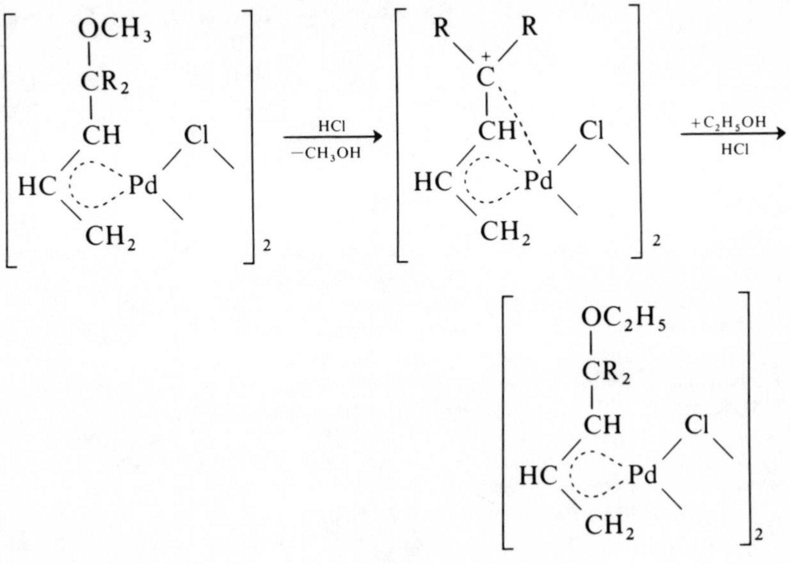

(571)

(v) Isomerisation of allylic compounds

Palladium(II) salts catalyse the isomerisation of allylic acetates in acetic acid solution.[118,119] A detailed study of the isomerisation of the propionate esters of butenyl alcohol showed that the reaction took place by two distinct routes.[120] In the first (reaction 572) an acetoxypalladation–deacetoxypalladation mechanism was involved with consequent exchange between the

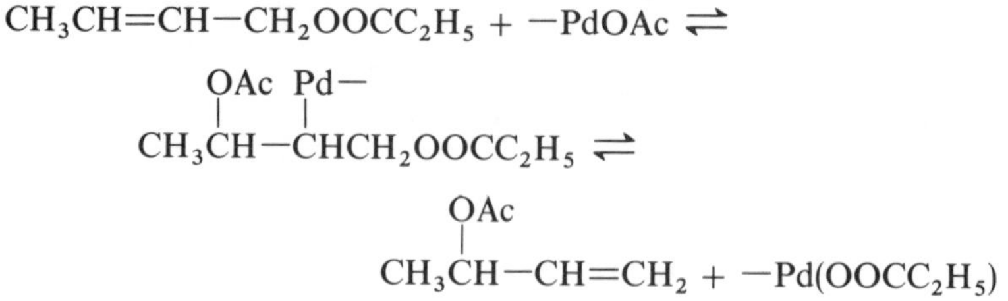

$$CH_3CH{=}CH{-}CH_2OOCC_2H_5 + {-}PdOAc \rightleftharpoons$$

$$\overset{\displaystyle OAc}{\underset{|}{CH_3CH}}{-}\overset{\displaystyle Pd-}{\underset{|}{CH}}CH_2OOCC_2H_5 \rightleftharpoons$$

$$\overset{\displaystyle OAc}{\underset{|}{CH_3CH}}{-}CH{=}CH_2 + {-}Pd(OOCC_2H_5) \quad (572)$$

acetate and propionate groups, whereas in the second route the isomerisation occurred with no exchange between acetate and propionate groups (reaction 573).

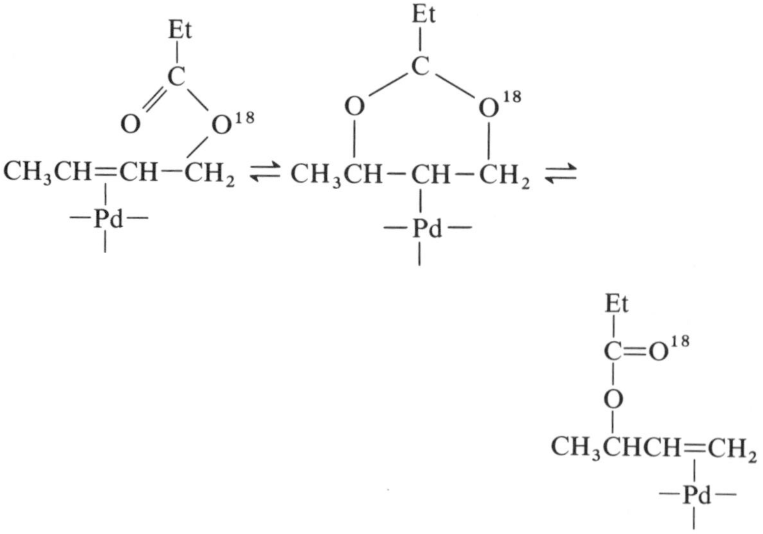

$$CH_3CH{=}CH{-}CH_2 \rightleftharpoons CH_3CH{-}CH{-}CH_2 \rightleftharpoons$$

$$\overset{\displaystyle Et}{\underset{|}{C}}{=}O^{18}$$
$$CH_3CHCH{=}CH_2 \quad (573)$$

REFERENCES

1 M. L. H. Green, *Organometallic Compounds*, by G. E. Coates, M. L. H. Green and K. Wade, Methuen, London, 3rd edn., 1968, vol. 2, chapter 2.
2 M. L. H. Green and P. L. I. Nagy, *Adv. Organometal. Chem.*, **2** (1964) 325.
3 R. G. Guy and B. L. Shaw, *Adv. Inorg. Radiochem.*, **4** (1962) 77.
4 R. Hüttel, J. Kratzer and M. Bechter, *Chem. Ber.*, **94** (1961) 766.
5 J. Smidt and W. Hafner, *Angew. Chem.*, **71** (1959) 284.
6 J. M. Kliegman, *J. Organometal. Chem.*, **29** (1971) 73.
7 E. O. Fisher and G. Bürger, *Z. Naturforsch.*, **16B** (1961) 702; *Chem. Abs.*, **57** (1962) 378b.
8 I. I. Moiseev, M. N. Vargaftik and Ya. K. Syrkin, *Izvest. Akad. Nauk. SSSR., Ser. Khim.*, (1964) 775; *Chem. Abs.*, **61** (1964) 3146f.
9 H. C. Volger and K. Vrieze, *J. Organometal. Chem.*, **6** (1966) 297.
10 H. C. Volger and K. Vrieze, *J. Organometal. Chem.*, **9** (1967) 527.

11   M. Sakakibara, Y. Takahashi, S. Sakai and Y. Ishii, *Chem. Comm.*, (1969) 396.

11a   J. Chatt, R. G. Guy, L. A. Duncanson and D. T. Thompson, *J. Chem. Soc.*, (1963) 5170.

12   J. H. Lukas and J. E. Blom, *J. Organometal Chem.*, **26** (1971) C25.

13   H. C. Volger, *Rec. Trav. Chim. Pays-Bas*, **88** (1969) 225.

14   D. Morelli, R. Ugo, F. Conti and M. Donati, *Chem. Comm.*, (1967) 801

15   A. D. Ketley and J. A. Braatz, *Chem. Comm.*, (1968) 169.

16   A. D. Ketley and J. A. Braatz, *J. Organometal. Chem.*, **9** (1967) P5.

17   T. Shono, T. Yoshimura, Y. Matsumura and R. Oda, *J. Org. Chem.*, **33** (1968) 876.

18   R. Noyori and H. Takaya, *Chem. Comm.*, (1969) 525.

19   B. L. Shaw, *Chem. Ind. (London)*, (1962) 1190.

20   J. M. Rowe and D. A. White, *J. Chem. Soc. (A)*, (1967) 1451.

21   A. J. Deeming, B. F. G. Johnson and J. Lewis, *Chem. Comm.*, (1970) 598.

22   J. Lukas, P. W. N. M. van Leeuwen, H. C. Volger and P. Kramer, *Chem. Comm.*, (1970) 799.

23   M. S. Lupin and B. L. Shaw, *Tetra. Letters*, (1964) 883.

24   M. S. Lupin, J. Powell and B. L. Shaw, *J. Chem. Soc. (A)*, (1966) 1687.

25   R. G. Schultz, *Tetra. Letters*, (1964) 301.

26   R. G. Schultz, *Tetrahedron*, **20** (1964) 2809.

27   T. Okamoto, Y. Sakakibara and S. Kunichika, *Bull. Chem. Soc. Japan*, **43** (1970) 2658.

28   S. Otsuka, A. Nakamura and K. Tani, *J. Organometal. Chem.*, **14** (1968) P30.

29   J. A. Osborn, *Chem. Comm.*, (1968) 1231.

30   T. G. Hewitt, K. Anzenhofer and J. J. De Boer, *J. Organometal. Chem.*, **18** (1968) P19.

31   K. Vrieze, H. C. Volger and A. P. Praat, *J. Organometal. Chem.*, **21** (1970) 467.

32   G. Wilke, B. Bogdanovic, P. Hardt, P. Heimbach, W. Kreim, M. Kröner, W. Oberkirch, K. Tanaka, E. Steinrücke, D. Walter and H. Zimmermann, *Angew. Chem. Int. Ed.*, **5** (1966) 151.

33   J. M. Rowe, *Proc. Chem. Soc.*, (1962) 66.

34   V. F. Levdik and M. A. Porai-Koshits, *Zh. Strukt. Khim.*, **3** (1962) 472; *Chem. Abs.*, **58** (1963) 7452e.

35   W. E. Oberhansli and L. F. Dahl, *J. Organometal. Chem.*, **3** (1965) 43.

36   A. E. Smith, *Acta. Cryst.*, **18** (1965) 331.

37   R. Mason and A. G. Wheeler, *J. Chem. Soc. (A)*, (1968) 2549.

38   R. Mason and A. G. Wheeler, *J. Chem. Soc. (A)*, (1968) 2543.

39   K. Oda, N. Yasuoka, T. Ueki, N. Kasai, M. Kakudo, Y. Tezuka, T. Ogura and S. Kawaguchi, *Chem. Comm.*, (1968) 989.

40   K. Oda, N. Yasuoka, T. Ueki, N. Kasai and M. Kakudo, *Bull. Chem. Soc. Japan.*, **43** (1970) 362.

41   R. Mason and A. G. Wheeler, *Nature*, **217** (1968) 1253.

42   Result quoted in reference 47.

43   L. F. Dahl and W. E. Oberhansli, *Inorg. Chem.*, **4** (1965) 629.

44   G. R. Davies, R. H. B. Mais, S. O'Brien and P. G. Owston, *Chem. Comm.*, (1967) 1151.

45   B. T. Kilbourn, R. H. B. Mais and P. G. Owston, *Chem. Comm.*, (1968) 1438.

46   R. Mason and D. R. Russell, *Chem. Comm.*, (1966) 26.

47   S. J. Lippard and S. M. Morehouse, *J. Amer. Chem. Soc.*, **91** (1969) 2504.

48   R. Mason and P. O. Whimp, *J. Chem. Soc. (A)*, (1969) 2709.

49   J. F. Malone and W. S. McDonald, *J. Chem. Soc. (A)*, (1970) 3124.

50   M. R. Churchill and R. Mason, *Nature*, **204** (1964) 777.

51   M. R. Churchill, *Chem. Comm.*, (1965) 625.

52   M. R. Churchill, *Inorg. Chem.*, **5** (1966) 1608.

53   G. Raper and W. S. McDonald, *Chem. Comm.*, (1970) 655.

54   W. S. McDonald, B. E. Mann, G. Raper, B. L. Shaw and G. Shaw, *Chem. Comm.*, (1969) 1254.

55   K. Vrieze and H. C. Volger, *J. Organometal. Chem.*, **9** (1967) 537.

56   J. N. Crosby and R. D. W. Kemmitt, *J. Organometal. Chem.*, **26** (1971) 277.

57   J. F. Malone, W. S. McDonald, B. L. Shaw and G. Shaw, *Chem. Comm.*, (1968) 869.

58   F. A. Cotton, *Chemical Applications of Group Theory*, John Wiley, New York, 1963, p. 156.

59   S. F. A. Kettle and R. Mason, *J. Organometal. Chem.*, **5** (1966) 573.
60   J. Powell and B. L. Shaw, *J. Chem. Soc. (A)*, (1967) 1839.
61· S. S. Zumdahl and R. S. Drago, *J. Amer. Chem. Soc.*, **90** (1968) 6669.
62   A. Pidcock, R. E. Richards and L. M. Venanzi, *J. Chem. Soc. (A)*, (1966) 1707.
63   M. J. Church and M. J. Mays, *J. Chem. Soc. (A)*, (1970) 1938.
64   J. C. W. Chien and H. C. Dehm, *Chem. Ind. (London)*, (1961) 745.
65   B. L. Shaw and N. Sheppard, *Chem. Ind. (London)*, (1961) 517.
66   F. A. Cotton, J. W. Faller and A. Musco, *Inorg. Chem.*, **6** (1967) 179.
67   M. Sakakibara, Y. Takahashi, S. Sakai and Y. Ishii, *J. Organometal. Chem.*, **27** (1967) 139.
68   J. K. Becconsall, B. E. Job and S. O'Brien, *J. Chem. Soc. (A)*, (1967) 423.
69   S. O'Brien, *J. Chem. Soc. (A)*, (1970) 9.
70   J. Powell, S. D. Robinson and B. L. Shaw, *Chem. Comm.*, (1965) 78.
71   K. Vrieze, C. Maclean, P. Cossee and C. W. Hilbers, *Rec. Trav. Chim. Pays-Bas*, **85** (1966) 1077.
72   G. L. Statton and K. C. Ramey, *J. Amer. Chem. Soc.*, **88** (1966) 1327.
73   K. Vrieze, A. P. Praat and P. Cossee, *J. Organometal. Chem.*, **12** (1968) 533.
74   P. W. N. M. van Leeuwen and A. P. Praat, *Chem. Comm.*, (1970) 365.
75   D. L. Tibbetts and T. L. Brown, *J. Amer. Chem. Soc.*, **92** (1970) 3031.
76   K. C. Ramey and G. L. Statton, *J. Amer. Chem. Soc.*, **88** (1966) 4387.
77   K. Vrieze, P. Cossee, A. P. Praat and C. W. Hilbers, *J. Organometal. Chem.*, **11** (1968) 353.
78   J. K. Becconsall and S. O'Brien, *Chem. Comm.*, (1966) 302.
79   P. Corradini, G. Maglio, A. Musco and G. Paiaro, *Chem. Comm.*, (1966) 618.
80   F. De Candia, G. Maglio, A. Musco and G. Paiaro, *Inorg. Chim. Acta*, **3** (1968) 233.
81   J. W. Faller and M. E. Thomsen, *J. Amer. Chem. Soc.*, **91** (1969) 6871.
82   J. W. Faller and M. J. Incorvia, *J. Organometal. Chem.*, **19** (1969) P13.
83   J. W. Faller, M. E. Thomsen and M. J. Mattina, *J. Amer. Chem. Soc.*, **93** (1971) 2642.
84   J. Lukas, S. Coren and J. E. Blom, *Chem. Comm.*, (1969) 1303.
85   C. W. Alexander, W. R. Jackson and R. Spratt, *J. Amer. Chem. Soc.*, **92** (1970) 4990.
86   P. W. N. M. van Leeuwen, A. P. Praat and M. van Diepen, *J. Organometal. Chem.*, **24** (1970) C31.
87   P. W. N. M. Leeuwen, A. P. Praat and M. van Diepen, *J. Organometal. Chem.*, **29** (1971) 433.
88   L. A. Leites, V. T. Aleksanyan, S. S. Bukalov and A. Z. Rubezhov, *Chem. Comm.*, (1971) 265.
89   F. R. Hartley, *J. Organometal. Chem.*, **21** (1970) 227.
90   H. P. Fritz, *Chem. Ber.*, **94** (1961) 1217.
91   W. T. Dent, R. Long and A. J. Wilkinson, *J. Chem. Soc.*, (1964) 1585.
92   M. S. Lupin, J. Powell and B. L. Shaw, *J. Chem. Soc. (A)*, (1966) 1410.
93   K. Shobatake and K. Nakamoto, *J. Amer. Chem. Soc.*, **92** (1970) 3339.
94   D. M. Adams and A. Squire, *J. Chem. Soc. (A)*, (1970) 1808.
95   S. D. Robinson and B. L. Shaw, *J. Chem. Soc.*, (1963) 4806.
96   B. F. Hegarty and W. Kitching, *J. Organometal. Chem.*, **6** (1966) 578.
97   R. J. Goodfellow and L. M. Venanzi, *J. Chem. Soc. (A)*, (1966) 784.
98   S. J. Ashcroft and C. T. Mortimer, *J. Chem. Soc. (A)*, (1971) 781.
99   R. Hüttel, J. Kratzer and M. Bechter, *Angew. Chem.*, **71** (1959) 456.
100 S. D. Robinson and B. L. Shaw, *J. Chem. Soc.*, (1964) 5002.
101 A. N. Nesmeyanov, S. P. Gubin and A. Z. Rubezhov, *J. Organometal. Chem.*, **16** (1969) 163.
102 J. Powell and B. L. Shaw, *J. Chem. Soc. (A)*, (1968) 774.
103 T. A. Schenach and F. F. Caserio, *J. Organometal. Chem.*, **18** (1969) P17.
104 S. Winstein, personal communication to R. F. Heck, *J. Amer. Chem. Soc.*, **90** (1968) 5542.
105 H. C. Volger and K. Vrieze, *J. Organometal Chem.*, **13** (1968) 495.
105a Y. Takahashi, S. Sakai and Y. Ishii, *Chem. Comm.*, (1967) 1092.
106 Y. Takahashi, S. Sakai and Y. Ishii, *J. Organometal. Chem.*, **16** (1969) 177.
107 R. P. Hughes and J. Powell, *J. Organometal. Chem.*, **20** (1969) P17.
108 R. P. Hughes and J. Powell, *Chem. Comm.*, (1971) 275.

109   V. N. Sokolov, G. M. Khvostic, I. Ya. Poddubnyi and G. P. Kondratenkov, *J. Organo-metal. Chem.*, **29** (1971) 313.

110   D. Medema and R. van Helden, *Abstr. 157th Amer. Chem. Soc. Meeting*, April 1969, Petr. 36.

111   T. Okamoto, *Chem. Comm.*, (1970) 1126.

112   J. Tsuji, J. Kiji and M. Morikawa, *Tetra. Letters*, (1963) 1811.

113   W. T. Dent, R. Long and G. H. Whitfield, *J. Chem. Soc.*, (1964) 1588.

114   J. Tsuji, J. Kiji, S. Imamura and M. Morikawa, *J. Amer. Chem. Soc.*, **86** (1964) 4350.

115   J. Tsuji, *Adv. Org. Chem.*, **6** (1969) 109 (especially pp. 166–71).

116   D. Medema, R. van Helden and C. F. Kohll, *Inorg. Chim. Acta.*, **3** (1969) 255.

117   T. Kajimoto, H. Takahashi and J. Tsuji, *J. Organometal. Chem.*, **23** (1970) 275.

118   W. Kitching, Z. Rappoport, S. Winstein and W. G. Young, *J. Amer. Chem. Soc.*, **88** (1966) 1054.

119   T. Matsuda, T. Mitsuyasu and Y. Nakamura, *Kogyo Kagaku Zasshi*, **72** (1969) 1751; *Chem. Abs.*, **72** (1970) 42693.

120   P. M. Henry, *Chem. Comm.*, (1971) 328.

Starting Materials for the Preparation of Complexes of Platinum and Palladium

This appendix describes the preparation and properties of those compounds of platinum and palladium that are most commonly used as starting materials for the preparation of complexes of these two elements. The compounds described were selected by studying the prominent journals in this field from 1960 to 1969 and analysing the starting materials used. Properties such as the solubilities in solvents are given for all the compounds described, whereas the preparations of only those compounds not commercially available are given in detail. However, because of delivery delays and other difficulties, literature references have been cited for the commercially available compounds to enable them to be prepared if they are urgently required.

A. COMPLEXES OF THE ZEROVALENT METALS

The general properties of the zerovalent tertiary phosphine complexes, which are the only zerovalent complexes widely used as starting materials, are given in Table 122. More detailed chemical properties are given in Chapter 3. The detailed procedures for the preparation of individual commonly used zerovalent starting materials are as follows:

(i) $[Pt(PPh_3)_4]$ by hydrazine hydrate reduction of cis-$[(PPh_3)_2PtCl_2]$.[1] A suspension of cis-$[(PPh_3)_2PtCl_2]$ (1 equiv.) and triphenylphosphine (3 equiv.) in ethanol is warmed to 70–80°C and then treated slowly with a solution of 10% aqueous ethanolic hydrazine. The suspended platinum(II) complex dissolves with evolution of nitrogen and after a few minutes $[Pt(PPh_3)_4]$ begins to settle out as a yellow crystalline precipitate which is filtered off and can be recrystallised from benzene.

(ii) $[Pt(PPh_3)_4]$ by borohydride reduction of K_2PtCl_4.[2] When an aqueous alcoholic solution of excess triphenylphosphine and K_2PtCl_4 (1 equiv.) is treated with an aqueous solution of sodium borohydride (3 equiv.), $[Pt(PPh_3)_4]$ is precipitated. The precipitate is washed first with warm ethanol, then cold water and finally ethanol.

(iii) $[Pt(PPh_3)_4]$ from K_2PtCl_4 using potassium hydroxide.[1,3] When a saturated aqueous solution of K_2PtCl_4 (1 equiv.) is slowly added with

TABLE 122

Properties of zerovalent platinum and palladium complexes

Complex	Colour	M.p. (°C)	Soluble in	Insoluble in	Remarks
[Pt(PPh₃)₄]	Ivory-yellow	118 (decomp.)	CHCl₃; C₆H₆ with extensive dissociation (see p. 31)	C₂H₅OH; hexane	Stable in air for some hours and completely stable in nitrogen. Reacts with CCl₄ to give cis-[Pt(PPh₃)₂Cl₂]
[Pt(PPh₃)₃]	Yellow	125–130 (decomp.)	(CH₃)₂CO; CHCl₃; C₆H₆ with some dissociation (see p. 31)	C₂H₅OH; hexane	As for [Pt(PPh₃)₄]
[(PPh₃)₂PtO₂]	Yellow	130–132	C₂H₅OH; CHCl₃; toluene (only slightly)	C₆H₆; hexane	Stable in air.
[(PPh₃)₂Pt(C₂H₄)]	Off-white	122–125	CHCl₃; C₆H₆	H₂O; C₂H₅OH; hexane	Stable in air.
[Pd(PPh₃)₄]	Yellow	100–105 (decomp.)	C₂H₅OH; C₆H₆ with extensive dissociation (see p. 31)		Stable in air for only a very short time, but stable in nitrogen.

vigorous stirring to a warm (50–60°C) saturated ethanolic solution of triphenylphosphine (5 equiv.) containing potassium hydroxide (2 equiv.), yellow crystals of [Pt(PPh₃)₄] separate almost immediately. After keeping the solution on a water-bath for a few minutes, the precipitate is filtered off and washed as in (ii) above.

(iv) *[Pt(PPh₃)₃] via hydrazine reduction of* cis-[(PPh₃)₂PtCl₂]. When the reaction described in (i) above is carried out in the absence of triphenylphosphine,[1] or for better yields in the presence of only one equivalent of triphenylphosphine,[4] [Pt(PPh₃)₃] is formed. It can be recrystallised from acetone.

(v) *[Pt(PPh₃)₃] from [Pt(PPh₃)₄].*[3] [Pt(PPh₃)₄] is suspended in absolute ethanol under a nitrogen atmosphere and boiled with stirring for 2 hours. After filtering the hot suspension, the precipitate of [Pt(PPh₃)₃] is washed with cold ethanol and dried in vacuo. The yield is 66%.

(vi) *[(PPh₃)₂PtO₂].*[5,6] When oxygen is passed through a saturated solution of [Pt(PPh₃)₄] in benzene the oxygen complex [(PPh₃)₂PtO₂] is rapidly precipitated. The excess triphenylphosphine is oxidised during the course of the reaction to triphenylphosphine oxide. After passing oxygen for 20 minutes the precipitated product, which contains one molecule of benzene of crystallisation, is filtered off and washed with benzene and hexane. The yield is 48%.

(vii) *[(PPh₃)₂Pt(C₂H₄)].*[6] [(PPh₃)₂PtO₂] (12 g) is dissolved in ethanol (200 ml) and the solution saturated with ethylene. Dropwise addition of a 0·1 M solution of sodium borohydride in ethanol (80 ml) over a period of 20 minutes accompanied by further passage of ethylene leads to the precipitation of the off-white crystalline compound [(PPh₃)₂Pt(C₂H₄)] which is filtered off, washed with water, alcohol and hexane and dried in air. The yield is 10·4 g.

(viii) *[Pd(PPh₃)₄] from palladium nitrate.*[7] Treat powdered palladium nitrate (1 equiv.) with an excess of triphenylphosphine (10 equiv.) in hot benzene. After a vigorous reaction in which a considerable amount of nitrous oxide is evolved a solid mass is obtained on cooling. The crude product is recrystallised from ethanol leaving triphenylphosphine oxide in the mother liquid.

(ix) *[Pd(PPh₃)₄] by borohydride reduction of K₂PdCl₄.*[2] The experimental details are identical to those given for [Pt(PPh₃)₄] in paragraph (ii) above.

B. COMPLEXES OF THE DIVALENT METALS CONTAINING INORGANIC LIGANDS

The solubility properties of the salts and complexes of the divalent metals with inorganic ligands are given in Table 123. Since all these compounds are

TABLE 123

Solubilities of salts and complexes of the divalent metals with inorganic ligands

Compound	Water	Ethanol	Acetone	Diethyl ether	Other
			Solvent		
$PtCl_2$	v.sl.s.	v.sl.s.	v.sl.s.	i	s. HCl (to give H_2PtCl_4) s. aq. NH_4OH (to give $[Pt(NH_3)_4]Cl_2$)
$PtBr_2$	i	i	i	i	s. HBr (to give H_2PtBr_4) s. MBr (to give M_2PtBr_4, M = Na, K, etc.)
H_2PtCl_4	s				
Li_2PtCl_4	s			i	s. Dimethylformamide
Na_2PtCl_4	s	sl.s.	v.sl.s.	i	sl. s. CH_3COOH v. sl. s. $CH_3COOC_2H_5$ sl. s. THF
K_2PtCl_4	s	i	i	i	
$(NH_4)_2PtCl_4$	s	i	i	i	
$[Pt(NH_3)_4]Cl_2$	s	v.sl.s.	i	i	
$K_2Pt(CN)_4$	sl.s. (cold) s. (hot)	sl.s.	sl.s.	sl.s.	
$PdCl_2$	sl.s	v.sl.s.	v.sl.s.	i	s. HCl (to give H_2PdCl_4) v.sl.s. 50% CH_3COOH v.sl.s. $CH_3COOC_2H_5$
$PdBr_2$	i	s		i	s. HBr (to give H_2PdBr_4)
PdI_2	i	i	i	i	s. KI (to give K_2PdI_4) i. dil. HCl
$PdSO_4$	v.s. (decomp. in hot water)				
$Pd(NO_3)_2$	s. (with decomp.)				s. HNO_3

$Pd(CH_3COO)_2$	i	i (cold; de-
		  comp. on
		  warming)
			s	s	s. HCl (with decomp.)
					s. $CHCl_3$; s. CH_2Cl_2;
					s. CH_3CN

H_2PdCl_4	s				
Na_2PdCl_4	v.s.	s	sl.s.	i	sl.s. CH_3COOH sl.s. $CH_3COOC_2H_5$
K_2PdCl_4	sl.s. (cold) s. (hot)	i	i	i	
$(NH_4)_2PdCl_4$	s	v.sl.s.		i	
K_2PdBr_4	s		i	i	
$K_2Pd(CN)_4$	s	s			precipitates $Pd(CN)_2$ in dilute acids

s = soluble; sl.s. = slightly soluble; v.sl.s. = very slightly soluble; i = insoluble.

either commercially available or very simply prepared from commercially available materials no preparative details are given here, although references for the preparations of all of them are given in Table 124.

A new reagent, $(Ph_2I^+)_2[PtCl_4]^{2-}$ has been advocated for the preparation of platinum(II) complexes.[31] On heating this reagent in co-ordinating solvents such as dimethylsulphoxide or pyridine, complexes of the type

TABLE 124
Preparation of salts and complexes of the divalent metals with inorganic ligands

Compound	Starting material	Reference
PtCl$_2$	H$_2$PtCl$_6$	8, 9
PtBr$_2$	H$_2$PtBr$_4$ or H$_2$PtBr$_6$	10
H$_2$PtCl$_4$	PtCl$_2$	9
	H$_2$PtCl$_6$	8
Li$_2$PtCl$_4$	PtCl$_2$	11
M$_2$PtCl$_4$ (where M = Na or K)	H$_2$PtCl$_6$ or M$_2$PtCl$_6$	12–15
(NH$_4$)$_2$PtCl$_4$	PtCl$_2$	16
[Pt(NH$_3$)$_4$]Cl$_2$	H$_2$PtCl$_6$ or H$_2$PtCl$_4$	17
	PtCl$_2$	18
K$_2$Pt(CN)$_4$	PtCl$_2$ or K$_2$PtCl$_4$	19
PdCl$_2$	Pd sponge	20
PdBr$_2$	Pd sponge	21
PdI$_2$	Pd(CH$_3$COO)$_2$	22
	PdCl$_2$	23
PdSO$_4$	Pd(CH$_3$COO)$_2$	24
	PdO	25
Pd(NO$_3$)$_2$	PdO	25
Pd(CH$_3$COO)$_2$	Pd sponge or Pd(NO$_3$)$_2$	22
H$_2$PdCl$_4$	Pd sponge	26
M$_2$PdCl$_4$ (where M = Na or K)	Pd sponge or H$_2$PdCl$_4$	26
	PdCl$_2$	27
(NH$_4$)$_2$PdCl$_4$	PdCl$_2$	27
K$_2$PdBr$_4$	K$_2$PdCl$_4$ or K$_2$PdBr$_6$	28
	Pd sponge	15, 29
K$_2$Pd(CN)$_4$	PdCl$_2$	30

[Pt(solvent)$_2$Cl$_2$] are formed. The main advantage claimed for the new reagent is that purification of the complexes is simple because the by-products of these reactions are not salts, as would be obtained from K$_2$PtCl$_4$, but chloro- and iodo-benzene.

C. COMPLEXES OF THE DIVALENT METALS CONTAINING ORGANIC LIGANDS

There are four main groups of complexes containing organic ligands bound to the divalent metals that are widely used as starting materials for the preparation of platinum and palladium complexes. These contain respectively tertiary phosphines, benzonitrile, ethylene and amines as the organic ligands. None of these complexes are available commercially in spite of their widespread use.

1. Tertiary phosphine complexes
(a) *Preparation of mononuclear tertiary phosphine complexes—[(PR$_3$)$_2$MX$_2$]*
Both *cis-* and *trans-*isomers of the bis(tertiary phosphine)platinum(II) complexes [(PR$_3$)$_2$PtX$_2$], where X is a halogen, are known, whereas the corresponding palladium(II) complexes are always *trans*,[32] reflecting the

much greater rate of *cis-trans* isomerisation found for palladium(II) complexes in general. Treatment of potassium tetrachloroplatinate(II) with a trialkylphosphine gives a precipitate of $[Pt(PR_3)_4]^{2+}[PtCl_4]^{2-}$ which is transformed into a mixture of *cis*- and *trans*-$[(PR_3)_2PtCl_2]$ on standing at room temperature for several weeks, or more rapidly on heating.[33] The stability of the intermediate ionic compounds decreases with increasing size of the alkyl group and when the alkyl group is *n*-butyl or larger the ionic compound is not formed at all. Although the tertiary phosphine complexes themselves are stable in air they must be prepared in the absence of oxygen as tertiary alkylphosphines are readily oxidised to tertiary alkylphosphine oxides. The detailed procedures for the preparation of commonly used tertiary phosphine complexes are as follows:

(i) Cis-$[(PR_3)_2PtCl_2]$, where R = alkyl but not methyl.[34] Shake vigorously K_2PtCl_4 (1 equiv.) in water with the tertiary phosphine (2 equiv.) for about three hours until no further lightening of the colour of the supernatant liquid occurs. In the case of the lower alkyl groups warm or allow to stand at room temperature for several weeks to allow the initially formed $[Pt(PR_3)_4]^{2+}[PtCl_4]^{2-}$ to be converted to a mixture of *cis*- and *trans*-$[(PR_3)_2PtCl_2]$. In the case of *n*-butyl and higher alkyl groups proceed immediately to scrape up the waxy salmon coloured material, wash it with water and dry it in vacuo at 30°C for about one hour. Extract this dried material with ice-cold light petroleum (b.p. 30–60°C) to remove the bulk of the *trans*-isomer, then heat the insoluble material in boiling light petroleum (b.p. 30–60°C) and add 95% ethanol dropwise with stirring until the solid just dissolves. Cool in an ice-salt bath to deposit white crystals. Repeat the recrystallisation until snow-white crystals of the pure product are obtained.

(ii) Trans-$[(PR_3)_2PtCl_2]$ where R = alkyl but not methyl.[34] Prepare the mixture of *cis*- and *trans*-$[(PR_3)_2PtCl_2]$ as described above in section (i) and heat this mixture in an evaporating basin on an oil-bath just above its melting point for one hour. Dissolve the residue in the minimum volume of warm 95% ethanol and cool for several minutes in an ice-salt bath. Filter off the yellow crystals and recrystallise from warm ethanol.

(iii) Cis-$[(PMe_3)_2PtCl_2]$.[35] Trimethylphosphine is very difficult to isolate and store as a pure compound owing to its very high volatility (b.p. 37·8°C),[36] its very rapid oxidation to trimethylphosphine oxide and the fact that it is spontaneously inflammable in air. However, it can readily be handled by preparing the 1:1 adduct with silver nitrate, which is a white crystalline complex that is indefinitely stable away from bright light.[35]

Stir together a solution of $AgNO_3.PMe_3$ (2 equiv.) and Na_2PtCl_4 (1 equiv.) in water for three hours. Filter off the solid and extract with dichloromethane to give a crude product which is recrystallised from methanol to give pure *cis*-$[(PMe_3)_2PtCl_2]$ in 75% yield.

(iv) Cis-$[(PPh_3)_2PtCl_2]$.[1] A saturated solution of triphenylphosphine (3 equiv.) in xylene is refluxed with K_2PtCl_4 (1 equiv.) to give white crystals of *cis*-$[(PPh_3)_2PtCl_2]$. The product can be recrystallised from benzene.

(v) Trans-$[(PPh_3)_2PtCl_2]$.[37] Irradiate cis-$[(PPh_3)_2PtCl_2]$ at 366 mμ for 4 hours (a mercury emission lamp is ideal as the presence of wavelengths higher than 366 mμ should be avoided, since light of such wavelengths decomposes the cis-complex). The resulting bright yellow solution contains a mixture of approximately 60% trans- and 40% cis-$[(PPh_3)_2PtCl_2]$. After evaporating the residue to dryness in vacuo, it is extracted with benzene to give a yellow solution and a white solid. On evaporating the yellow solution and adding methanol a yellow precipitate of trans-$[(PPh_3)_2PtCl_2]$ is formed which can be recrystallised from benzene. The yield is 42%.

(vi) $[(PR_3)_2PdCl_2]$ from $(NH_4)_2PdCl_4$ where R = alkyl or aryl but not methyl.[38] A solution of $(NH_4)_2PdCl_4$ in water is shaken vigorously with the tertiary phosphine to give yellow crystals of $[(PR_3)_2PdCl_2]$ and a brown scum which is separated off. The yellow crystals are then filtered off and recrystallised from ethanol.

(vii) $[(PR_3)_2PdCl_2]$ from Na_2PdCl_4 where R = alkyl or aryl but not methyl.[38] On shaking an ethanolic solution of the tertiary phosphine with Na_2PdCl_4 the initial red colour lightens to yellow and after filtering off the precipitated sodium chloride, evaporation of the solution yields $[(PR_3)_2PdCl_2]$. As an alternative hydrated palladium(II) chloride can be used instead of Na_2PdCl_4.

(viii) $[(PMe_3)_2PdCl_2]$.[35] This is prepared from $AhNO_3.PMe_3$ (2 equiv.) and $Na_2PdCl_4.3H_2O$ (1 equiv.) using identical conditions to those described in section (iii) above for the preparation of cis-$[(PMe_3)_2PtCl_2]$.

(ix) Complexes with tertiary phosphines and halogens other than chlorine. These complexes are prepared from starting materials containing the desired halogen. These starting materials are generally prepared in situ by treating the tetrachloroplatinate(II) or tetrachloropalladate(II) complex with potassium bromide or potassium iodide.

(b) Properties of mononuclear tertiary phosphine complexes
Both the cis- and trans-isomers of tertiary phosphine complexes such as $[(PBu_3^n)_2PtCl_2]$ are soluble in water containing excess tertiary phosphine (where $[Pt(PR_3)_4]^{2+}(Cl^-)_2$ is formed), benzene, carbon disulphide, ethanol and dimethylformamide.[34] However, the cis-isomer is considerably less soluble in all these solvents than the trans-isomer. In particular the cis-isomer is almost insoluble in diethyl ether and low boiling light petroleum, both of which dissolve the trans-isomer. Further, whereas the cis-isomer is slightly soluble in pure water, the trans-isomer is insoluble in this medium. The colours and melting-points of the commonest chloro-complexes are given in Table 125 from which it can be seen that the cis-dichloro complexes are all white and the trans-dichloro complexes, which of course include all the palladium complexes, are yellow. The corresponding bromo-complexes of platinum are similar except for the triphenylphosphine derivative where the cis-complex is orange. The iodo-complexes of platinum are pale yellow

TABLE 125
Colours and melting-points of the tertiary phosphine complexes
[(PR$_3$)$_2$MCl$_2$]

Complex	Melting point (°C).	Colour
cis-[(PMe$_3$)$_2$PtCl$_2$]	324–326 (with decomp.)	White
cis-[(PEt$_3$)$_2$PtCl$_2$]	191–192	White
trans-[(PEt$_3$)$_2$PtCl$_2$]	142–143	Yellow
cis-[(PPr$_3^n$)$_2$PtCl$_2$]	149–150	White
trans-[(PPr$_3^n$)$_2$PtCl$_2$]	85–86	Yellow
cis-[(PBu$_3^n$)$_2$PtCl$_2$]	144	White
trans-[(PBu$_3^n$)$_2$PtCl$_2$]	65–66	Yellow
cis-[(PPh$_3$)$_2$PtCl$_2$]	310 (with decomp.)	White
trans-[(PPh$_3$)$_2$PtCl$_2$]	306–309	Yellow
[(PMe$_3$)$_2$PdCl$_2$]	282 (with decomp.)	Yellow
[(PEt$_3$)$_2$PdCl$_2$]	139	Yellow
[(PPr$_3^n$)$_2$PdCl$_2$]	96	Yellow
[(PBu$_3^n$)$_2$PdCl$_2$]	66	Yellow
[(PPh$_3$)$_2$PdCl$_2$]	270 (with decomp.)	Yellow

(*cis*) and dark yellow (*trans*). In the case of the palladium complexes the bromides and iodides are generally deep yellow and orange respectively. All the tertiary phosphine complexes are stable in air.

(c) *Preparation of halogen bridged tertiary phosphine complexes* [(PR$_3$)$_2$M$_2$X$_4$]
The halogen bridged tertiary phosphine complexes [(PR$_3$)$_2$M$_2$X$_4$], where M = Pt or Pd, PR$_3$ = tertiary phosphine and X = halogen, all have[39] the *trans*-halogen bridged structure **176**. For a number of years the most satisfactory method for preparing the platinum(II) complexes consisted of fusing together the mononuclear complexes [(PR$_3$)$_2$PtX$_2$] and the platinum(II)

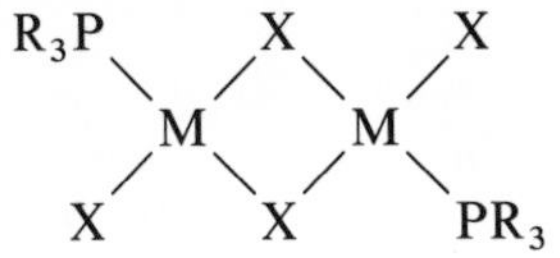

176

halide PtX$_2$.[40,41] However, the success of this method depends on maintaining the mixture in a molten state, which with high melting compounds demands a high degree of thermal stability from both reactants and products. Accordingly, the method cannot be used where the products decompose on melting or with [(PPh$_3$)$_2$PtCl$_2$] or [(PMe$_3$)$_2$PtCl$_2$] both of which decompose before melting. However, this difficulty can be overcome by heating a finely powdered slurry of the reactants either in a high-boiling hydrocarbon solvent[42] or in tetrachloroethane.[43] The solvent assists the reaction in three ways, first it keeps the reactants mobile, secondly it lowers their melting-points and thirdly it may dissolve them slightly. The halogen bridged tertiary phosphine palladium(II) complexes are prepared rather more easily by boiling a mixture of the mononuclear bis(tertiary phosphine) complex $[(PR_3)_2PdCl_2]$ in a suitable solvent with sodium tetrachloropalladate(II) in ethanol.[44] The detailed procedures for the preparation of the commonly used halogen bridged tertiary phosphine complexes are as follows:

(i) $[(PR_3)_2Pt_2Cl_4]$.[42] Finely grind a mixture of $[(PR_3)_2PtCl_2]$ (1 equiv.) and $PtCl_2$ (1·05 equiv.) and transfer it to a boiling tube. Add the hydrocarbon solvent to make a light slurry (about 10 ml of hydrocarbon is required for every 2–3 g of reactants). Heat the mixture in an oil-bath at the required temperature with occasional stirring. After cooling, the mixture is placed in a soxhlet thimble and extracted continuously with light petroleum (b.p. 40–60°C). The residue is then extracted with dichloromethane and recrystallised from either dichloromethane or chloroform. The hydrocarbon solvents, reaction temperatures and approximate yields for the different tertiary phosphines are as follows: PMe_3, xylene plus sufficient naphthalene to give a reflux temperature of 160°C, 63% yield. PEt_3, xylene, 140°C, 82% yield. PPh_3, naphthalene, 160°C, 70% yield.

(ii) $[(PR_3)_2Pd_2Cl_4]$.[44] A solution of sodium tetrachloropalladate(II) (1·1 equiv.) in ethanol and $[(PR_3)_2PdCl_2]$ (1 equiv.) in either ethanol (for $R = C_2H_5$) or chloroform (for $R = C_6H_5$) is boiled under reflux for an hour. In the case of the triphenylphosphine complex, $[(PPh_3)_2Pd_2Cl_4]$ is obtained as a precipitate, whereas the triethylphosphine complex is obtained by by evaporating the solution to dryness in vacuo. The triethylphosphine complex is washed with water, dried and recrystallised from ethanol to give $[(PEt_3)_2Pd_2Cl_4]$ in 81% yield. The triphenylphosphine complex is washed with water, ethanol and chloroform and dried before being recrystallised by dissolving it in a large volume of hot chloroform and precipitating it with light petroleum (b.p. 40–60°C) to give $[(PPh_3)_2Pd_2Cl_4]$ in 98% yield.

(iii) $[(PR_3)_2Pt_2X_4]$ where $X = Br$ or I.[45] These complexes are prepared analogously to the chloro-complexes described in section (i) above using $[(PR_3)_2PtX_2]$ and PtX_2 as the starting materials. They can be recrystallised from dichloromethane.

(iv) $[(PR_3)_2Pd_2X_4]$ where $X = Br$ or I.[44] The preparations of these complexes are analogous to those described above for the chloro-complex (section ii) except that a mixture of either sodium bromide or lithium iodide and sodium tetrachloropalladate(II) is used instead of sodium tetrachloropalladate(II) alone.

(d) *Properties of the halogen bridged complexes*
The complexes $[(PR_3)_2M_2X_4]$ are stable in air, soluble in organic solvents such as ethanol, chloroform and dichloromethane, but insoluble in water and light petroleum (b.p. 40–60°). Their colours and melting-points are given in Table 126.

TABLE 126

Colours, melting-points and solubilities of the chloro-bridged tertiary phosphine complexes [(PR₃)₂M₂Cl₄]

Complex	Colour	M.p. (°C)	Solubility[a]						
			H_2O	EtOH	$CHCl_3$	Pet. ether	CH_2Cl_2	Benzene	Acetone
$[(PMe_3)_2Pt_2Cl_4]$	Orange	200–210 (with decomp.)	i	s	s	i	s	s	s
$[(PEt_3)_2Pt_2Cl_4]$	Orange	224–225	i	s	s	i	s	s	s
$[(PPh_3)_2Pt_2Cl_4]$	Deep-orange	270–280	i	s	s	i	s	sl.s.	
$[(PEt_3)_2Pd_2Cl_4]$	Orange	230	i	s	s	i	s	s	s
$[(PPr^n_3)_2Pd_2Cl_4]$	Orange	189–191	i	s	s	i	s	s	s
$[(PPh_3)_2Pd_2Cl_4]$	Brownish-orange	250–270 (with decomp.)	i	i	sl.s. (hot)	i	sl.s.		

[a] s = soluble; sl.s. = slightly soluble; i = insoluble.

2. Bis(benzonitrile) complexes

The bis(benzonitrile) complexes $[M(PhCN)_2X_2]$, where X is halogen, have been widely used as starting materials for the preparation of organometallic complexes of platinum and palladium because of their ready solubility in organic solvents and their ease of preparation. The palladium complexes have the *trans*-configuration whereas the platinum complexes are *cis*.[46]

(a) Preparation

(i) $[Pt(PhCN)_2X_2]$ *where* $X = Cl$ *or* Br.[47] The appropriate platinum(II) halide (i.e. $PtCl_2$ or $PtBr_2$) is dissolved in the minimum volume of hot (100°C) benzonitrile. After filtering, the solution is cooled to yield a heavy precipitate of $[Pt(PhCN)_2X_2]$ which is collected by filtration. A further crop of crystals can be obtained by diluting the filtrate with light petroleum (b.p. 40–60°C). The yield is virtually quantitative. The product can be recrystallised from benzene.

Two other preparations of $[Pt(PhCN)_2X_2]$, both starting from potassium tetrachloroplatinate(II) have been described,[49,50] but neither gives such a good yield as the preparation described above.

(ii) $[Pd(PhCN)_2X_2]$ *where* $X = Cl$ *or* Br.[48,51] Palladium(II) halide (2 g of either $PdCl_2$ or $PdBr_2$) is suspended in benzonitrile (50 ml) and the mixture heated to 100°C for 20 minutes, by which time the majority of the palladium(II) halide has dissolved to give a red solution. The solution is filtered while still warm and the filtrate poured into low boiling light petroleum (300 ml). The light yellow precipitate is filtered off, washed with light petroleum and dried to give about 93 % yield of $[Pd(PhCN)_2X_2]$. The product can be recrystallised from benzene.[52]

(b) Properties

The bis(benzonitrile) complexes are all stable in air and soluble in non-polar organic solvents such as chloroform and benzene. Details of their colours, melting-points (where published) and solubilities are given in Table 127.

3. Ethylene complexes

Both the monomeric ethylene complex $K[Pt(C_2H_4)Cl_3].H_2O$ and its dimer $[Pt_2(C_2H_4)_2Cl_4]$ have been widely used as starting materials for the preparation of other olefin complexes ever since Anderson[53] showed that a less volatile olefin can displace a more volatile olefin from co-ordination to platinum(II). Ethylene complexes of palladium(II) are too unstable to be of use as starting materials for the preparation of other complexes.[54]

(a) Preparation

(i) $K[Pt(C_2H_4)Cl_3].H_2O$.[55] The method described here is a modification by the present author of the stannous chloride catalysed addition of ethylene to potassium tetrachloroplatinate(II) suggested by Cramer.[56] A mixture of K_2PtCl_4 (1 equiv.) and stannous chloride (0·05 equiv.) is ground up finely and dissolved in 1 N hydrochloric acid (lower concentrations of hydrochloric

TABLE 127
Colours, melting-points and solubilities of benzonitrile complexes

Complex	Colour	M.p. (°C)	Solubility[a]										
			C_6H_6	$CHCl_3$	$CHBr_3$	$(CH_3)_2CO$	$PhCN$	H_2O	$EtOH$	$MeOH$	Et_2O	P^b	C^c
[Pt(PhCN)$_2$Cl$_2$]	Pale-yellow	220 (with decomp.)	s	s	s	sl.s.	sl.s. (hot)	i	i	i	i	i	i
[Pt(PhCN)$_2$Br$_2$]	Yellow	223 (with decomp.)	s	s	s	sl.s.	sl.s. (hot)	i	i	i	i	i	i
[Pd(PhCN)$_2$Cl$_2$]	Pale-yellow		s	s	s	sl.s.	sl.s.	i	i	i	i	i	i
[Pd(PhCN)$_2$Br$_2$]	Yellow		s	s	s	sl.s.	sl.s.	i	i	i	i	i	i

[a] s = soluble; sl.s. = slightly soluble; i = insoluble.
[b] P = low boiling petroleum ether (b.p. 40–60°C).
[c] C = cyclohexane.

acid result in decomposition of the ethylene complex and deposition of platinum metal, higher concentrations lead to oxidation of K_2PtCl_4 to K_2PtCl_6). The solution is transferred to a two-necked flask provided with a PTFE covered bar magnet. The flask is fitted with a narrow gas inlet tube that conveys the incoming gas below the liquid level at the side of the flask remote from the region of the rotating bar magnet. The solution is vigorously stirred on a magnetic stirrer whilst a brisk stream of 'oxygen-free' nitrogen is bubbled through to ensure complete deoxygenation of the solution. After half-an-hour, whilst stirring and maintaining a steady flow of nitrogen, a brisk stream of ethylene is introduced into the flask for a few minutes and this is then reduced to a slow stream that is passed for 5 hours whilst the solution is maintained at room temperature. The initially dark-red solution gradually becomes golden-yellow. After 5 hours the ethylene and nitrogen streams are stopped and the solution evaporated *in vacuo* in a desiccator containing 18 M sulphuric acid and potassium hydroxide pellets. After being ground up the residue is extracted with absolute alcohol and the suspension filtered to give a residue of potassium chloride and unreacted potassium tetrachloroplatinate(II) and a filtrate containing $K[Pt(C_2H_4)Cl_3]$. After evaporating the ethanolic solution to dryness *in vacuo* the resulting yellow crystalline material is dissolved in 1 N hydrochloric acid, treated with animal charcoal to remove any traces of platinum metal, filtered and then evaporated slowly *in vacuo* in a desiccator containing 18 M sulphuric acid and potassium hydroxide pellets to give long yellow needles of $K[Pt(C_2H_4)Cl_3].H_2O$ in 84% yield.

(ii) $[Pt_2(C_2H_4)_2Cl_4]$.[57] $K[Pt(C_2H_4)Cl_3].H_2O$ (prepared as described above) is dissolved in ethanol. Concentrated hydrochloric acid (twice the amount needed to effect the reaction $K[Pt(C_2H_4)Cl_3] + HCl \rightarrow H[Pt(C_2H_4)Cl_3] + KCl$) is added, and after shaking for a few minutes the precipitated potassium chloride is filtered off and the yellow filtrate evaporated to dryness *in vacuo* keeping the temperature to 20°C until the colour deepens to yellow-orange and then raising it slowly to 60°C. On shaking, the viscous orange liquid suddenly crystallises to give $[Pt_2(C_2H_4)_2Cl_4]$ in 90–95% yield. The product can be recrystallised from boiling toluene, but should only be boiled for about two minutes to avoid excessive decomposition. The addition of animal charcoal during recrystallisation, followed by filtration, ensures complete removal of the colloidal platinum formed during the decomposition.

(b) *Properties*

(i) $K[Pt(C_2H_4)Cl_3].H_2O$. Potassium trichloro(ethylene)platinum(II) is stable in air, soluble in water (with slow decomposition to platinum metal and acetaldehyde), soluble in ethanol (with slow decomposition to platinum metal) and soluble in acetone (without decomposition). It is insoluble in non-polar solvents such as hexane, light petroleum and diethyl ether.

(ii) $[Pt_2(C_2H_4)_2Cl_4]$. $[Pt_2(C_2H_4)_2Cl_4]$ is sensitive to moisture and darkens after exposure to air for a few hours and must therefore be stored in an atmosphere of dry nitrogen. For preference, it is prepared as required from $K[Pt(C_2H_4)Cl_3]$ and not isolated from solution but used *in situ*. When isolated it exists either as long pale-orange needles, or larger, darker orange granular crystals. The granular form is slightly more stable. $[Pt_2(C_2H_4)_2Cl_4]$ is only slightly soluble in most organic solvents, although it dissolves readily in acetone or ethanol. The ethanol solution, however, decomposes slowly at room temperature and more rapidly on warming to deposit platinum metal. $[Pt_2(C_2H_4)_2Cl_4]$ decomposes in boiling water with the formation of acetaldehyde.[58]

4. Amine complexes

The amine complexes most widely used as starting materials for the preparation of platinum and palladium complexes have the general formula $[M(am)_2Cl_2]$, where am represents any neutral primary, secondary or tertiary amine. The platinum(II) complexes exist as *cis*- and *trans*-isomers, whereas the palladium(II) complexes are always *trans*. The situation is analogous to that for the tertiary phosphine complexes of these two metals. Typical preparations for both platinum(II) and palladium(II) complexes are illustrated below.

(a) Preparation
(i) Cis-$[Pt(C_5H_5N)_2Cl_2]$.[59] A solution of pyridine (2 equiv.) in water is added slowly, with vigorous stirring, to a solution of potassium tetrachloroplatinate(II) (1 equiv.) in water. After a few minutes a yellow precipitate begins to form. Stirring is continued for several hours before placing the mixture in a refrigerator for 24 hours to allow the precipitation to complete. The yellow precipitate is then filtered off, washed with ice-cold water to remove potassium chloride and dried at 100°C for 1 hour to give cis-$[Pt(C_5H_5N)_2Cl_2]$ in 89% yield.

(ii) Trans-$[Pt(C_5H_5N)_2Cl_2]$.[59] Cis-$[Pt(C_5H_5N)_2Cl_2]$ (5 g) is dissolved in excess 40% aqueous pyridine (containing 20 ml pure pyridine) by heating and stirring on a water bath to yield a colourless solution of $[Pt(C_5H_5N)_4]Cl_2$. After filtering, the solution is evaporated to dryness on a steam-bath, treated with 50 ml of concentrated hydrochloric acid and then evaporated to dryness again on a steam-bath to give a yellow residue. If any unchanged white $[Pt(C_5H_5N)_4]Cl_2$ remains, evaporation to dryness with concentrated hydrochloric acid is repeated. The pale-yellow crystals are then put into a filter funnel, washed with ice-cold water to remove hydrochloric acid, and then washed with ethanol and diethyl ether. The yield of pale-yellow crystalline trans-$[Pt(C_5H_5N)_2Cl_2]$ is virtually quantitative. It can be recrystallised from boiling chloroform, treating the mother liquor from which the bulk of the product has been obtained with diethyl ether to recover the last traces.

(iii) Trans-$[Pd(PhEtNH)_2Cl_2]$.[60] A suspension of palladium(II) chloride (1 g) and phenylethylamine (2 ml) in chloroform (50 ml) is heated for 2 hours to give a deep-yellow solution. The solution is evaporated to dryness to give

a cream-coloured solid which is recrystallised from a mixture of carbon tetrachloride and light petroleum (b.p. 40–60°C).

(b) Properties

The complexes [M(am)$_2$Cl$_2$] are all pale-yellow crystalline solids that are stable in air. They are virtually insoluble in water although *cis*-platinum(II) complexes do have a slight solubility. By contrast the *cis*-platinum(II) complexes are less soluble in organic solvents than their *trans*-analogues, as found with the bis(tertiary phosphine) complexes. Thus the *trans*-complexes, which include both platinum(II) and palladium(II) complexes, are soluble in chloroform, carbon tetrachloride, acetone, dimethylformamide and benzene, whereas the *cis*-complexes, whilst soluble in the first four solvents, are insoluble in benzene. Both the *cis*- and the *trans*-isomers of [M(am)$_2$Cl$_2$] are insoluble in light petroleum (b.p. 40–60°C).

D. COMPLEXES OF THE TETRAVALENT METALS

The compounds and complexes of the tetravalent metals that are commonly used as starting materials for the preparation of further complexes of platinum and palladium are all available commercially. Accordingly as for the corresponding divalent materials (section B) only references are given for the preparation of these compounds (see Table 128). The solubility properties of these materials are given in Table 129.

REFERENCES

1   L. Malatesta and C. Cariello, *J. Chem. Soc.*, (1958) 2323.
2   D. T. Rosevear and F. G. A. Stone, *J. Chem. Soc. (A)*, (1968) 164.
3   R. Ugo, F. Cariati and G. La Monica, *Inorg. Synth.*, **11** (1968) 105.
4   C. D. Cook and G. S. Jauhal, *Canad. J. Chem.*, **45** (1967) 301.
5   C. D. Cook and G. S. Jauhal, *Inorg. Nucl. Chem. Letters*, **3** (1967) 31.
6   C. D. Cook and G. S. Jauhal, *J. Amer. Chem. Soc.*, **90** (1968) 1464.
7   L. Malatesta and M. Angoletta, *J. Chem. Soc.*, (1957) 1186.
8   W. E. Cooley and D. H. Busch, *Inorg. Synth.*, **5** (1957) 208.
9   A. J. Cohen, *Inorg. Synth.*, **6** (1960) 209.
10 N. V. Sidgwick, *The Chemical Elements and their Compounds*, Oxford University Press, 1950, vol. 2, p. 1582.
11 E. Lodewijk and D. Wright, *J. Chem. Soc. (A)*, (1968) 119.
12 R. N. Keller, *Inorg. Synth.*, **2** (1946) 247.
13 G. B. Kauffman and D. O. Cowan, *Inorg. Synth.*, **7** (1963) 239.
14 G. L. Johnson, *Inorg. Synth.*, **8** (1966) 242.
15 S. E. Livingstone, *Synth. in Inorg. and Metal-Organic Chem.*, **1** (1971) 1.
16 Reference 10, p. 1605.
17 R. N. Keller, *Inorg. Synth.*, **2** (1946) 250.
18 W. E. Cooley and D. H. Busch, *Inorg. Synth.*, **5** (1957) 209.
19 H. L. Grube in *Handbook of Preparative Inorganic Chemistry*, Ed. by G. Brauer, Academic Press, London, 2nd edn., 1965, vol. 2, p. 1576.
20 F. Puche, *Ann. Chim. (Paris)*, **9** (1938) 233; *Chem. Abs.*, **32** (1938) 5322.
21 J. H. Canterford and R. Colton, *Halides of the Second and Third Row Transition Metals*, John Wiley, New York, 1968, p. 368.
22 T. A. Stephenson, S. M. Morehouse, A. R. Powell, J. P. Heffer and G. Wilkinson, *J. Chem. Soc.*, (1965) 3632.

TABLE 128

Preparation of salts and complexes of the tetravalent metals

Compound	Starting material	References
$PtCl_4$	H_2PtCl_6	61
$PtBr_4$	Pt metal or H_2PtBr_6	62
	Pt metal	63
PtI_4	Pt metal	64, 65
H_2PtCl_6	Pt metal	66, 67
M_2PtCl_6 (where M = Li, Na or K)	Pt metal or H_2PtCl_6	68
$(NH_4)_2PtCl_6$	Pt metal or H_2PtCl_6	67, 68
H_2PtBr_6	Pt metal	69
	$PtCl_4$	70
M_2PtBr_6 (where M = Na, K or NH_4)	H_2PtBr_6	69, 71
M_2PtI_6 (where M = Na or K)	H_2PtCl_6	72
	PtI_4	73
	H_2PtCl_4	74
M_2PdCl_6 (where M = Na, K or NH_4)	Pd sponge	20

TABLE 129

Solubilities of salts and complexes of the tetravalent metals

Compound	Solubility[a]				
	Water	Ethanol	Acetone	Diethyl ether	Other
$PtCl_4$	s	sl.s.	s	i	s HCl (to give H_2PtCl_6)
$PtBr_4$	sl.s.	s	s	i	s HBr (to give H_2PtBr_6)
PtI_4	s (with decomp.)	s	s	i	s HI (to give H_2PtI_6)
H_2PtCl_6	v.s.	v.s.	v.s.	v.s.	v.s. CH_3COOH; v.s. $CH_3COOC_2H_5$
Li_2PtCl_6	v.s.	v.s.		i	
Na_2PtCl_6	v.s.	s	s	i	s MeOH; sl.s. CH_3COOH; v.sl.s. $CH_3COOC_2H_5$
K_2PtCl_6	s	i		i	
$(NH_4)_2PtCl_6$	v.sl.s.	v.sl.s.		i	
H_2PtBr_6	s	s		s	
Na_2PtBr_6	v.s.	v.s.		i	
K_2PtBr_6	s	i		i	
$(NH_4)_2PtBr_6$	sl.s.			i	
Na_2PtI_6	s	s		i	
K_2PtI_6	v.s. (decomp. on heating)	sl.s.		i	
Na_2PdCl_6	v.s.	s	s	v.sl.s.	sl.s. CH_3COOH; v.sl.s. $CH_3COOC_2H_5$
K_2PdCl_6	sl.s. (decomp.)	i		i	sl.s. HCl
$(NH_4)_2PdCl_6$	v.sl.s.	v.sl.s.		i	

[a] v.s. = very soluble; s = soluble; sl.s. = slightly soluble; v.sl.s. = very slightly soluble; i = insoluble.

23 S. A. Shchukarev, T. A. Tolmacheva and Yu. L. Pazukhina, *Russ. J. Inorg. Chem.*, **9** (1964) 1354.

24 J. M. Davidson and C. Triggs, *J. Chem. Soc. (A)*, (1968) 1324.

25 F. A. Cotton and G. Wilkinson, *Advanced Inorganic Chemistry*, Interscience, New York, 2nd edn., 1966, p. 1029.

26 G. B. Kauffman and J. H. Tsai, *Inorg. Synth.*, **8** (1966) 234.

27 Reference 19, p. 1584.

28 E. Biilmann and A. C. Andersen, *Chem. Ber.*, **36** (1903) 1565.

29 C. M. Harris, S. E. Livingstone and I. H. Reece, *J. Chem. Soc.*, (1959) 1505.

30 J. H. Brigelow, *Inorg. Synth.*, **2** (1946) 245.

31 R. A. Krause, *Inorg. Chem.*, **4** (1965) 1227.

32 G. Booth, *Adv. Inorg. Radiochem.*, **6** (1964) 1.

33 K. A. Jensen, *Z. anorg. allgem. Chem.*, **229** (1936) 237.

34 G. B. Kauffman and L. A. Teter, *Inorg. Synth.*, **7** (1963) 245.

35 J. G. Evans, P. L. Goggin, R. J. Goodfellow and J. G. Smith, *J. Chem. Soc. (A)*, (1968) 464.

36 E. J. Rosenbaum and G. R. Sandberg, *J. Amer. Chem. Soc.*, **62** (1940) 1622.

37 S. H. Mastin and P. Haake, *Chem. Comm.*, (1970) 202.

38 F. G. Mann and D. Purdie, *J. Chem. Soc.*, (1935) 1549.

39 R. J. Goodfellow, P. L. Goggin and L. M. Venanzi, *J. Chem. Soc. (A)*, (1967) 1897 and references therein.

40 J. Chatt, *J. Chem. Soc.*, (1951) 652.

41 J. Chatt and L. M. Venanzi, *J. Chem. Soc.*, (1955) 2787.

42 R. J. Goodfellow and L. M. Venanzi, *J. Chem. Soc.*, (1965) 7533.

43 A. C. Smithies, M. Rycheck and M. Orchin, *J. Organometal. Chem.*, **12** (1968) 199.

44 J. Chatt and L. M. Venanzi, *J. Chem. Soc.*, (1957) 2351.

45 R. J. Goodfellow, J. G. Evans, P. L. Goggin and D. A. Duddell, *J. Chem. Soc. (A)*, (1968) 1604.

46 J. M. Jenkins and J. G. Verkade, *Inorg. Synth.*, **11** (1968) 108.

47 M. S. Kharasch and T. A. Ashford, unpublished results quoted in reference 48.

48 M. S. Kharasch, R. C. Seyler and F. R. Mayo, *J. Amer. Chem. Soc.*, **60** (1938) 882.

49 K. A. Hofmann and G. Bugge, *Chem. Ber.*, **40** (1907) 1772.

50 L. Ramberg, *Chem. Ber.*, **40** (1907) 2578.

51 J. R. Doyle, P. E. Slade and H. B. Jonassen, *Inorg. Synth.*, **6** (1960) 218.

52 E. Kuljian and H. Frye, *Z. Naturforsch.*, **B19** (1964) 651.

53 J. S. Anderson, *J. Chem. Soc.*, (1936) 1042.

54 F. R. Hartley, *Chem. Rev.*, **69** (1969) 799.

55 F. R. Hartley, *Organometal. Chem. Rev.*, **A6** (1970) 119.

56 R. D. Cramer, E. L. Jenner, R. V. Lindsey and U. G. Stolberg, *J. Amer. Chem. Soc.*, **85** (1963) 1691.

57 J. Chatt and M. L. Searle, *Inorg. Synth.*, **5** (1957) 210.

58 J. S. Anderson, *J. Chem. Soc.*, (1934) 971.

59 G. B. Kauffman, *Inorg. Synth.*, **7** (1963) 249.

60 B. Bosnich, *J. Chem. Soc. (A)*, (1966) 1394.

61 R. N. Keller, *Inorg. Synth.*, **2** (1946) 253.

62 L. Wöhler and F. Müller, *Z. anorg. allgem. Chem.*, **149** (1925) 377; *Chem. Abs.*, **20** (1926) 718.

63 S. A. Shchukarev, T. A. Tolmacheva, M. A. Oranskaya and L. V. Komandrovskaya, *Zhur. Neorg. Khim.*, **1** (1956) 8; *Chem. Abs.*, **50** (1956) 9833g.

64 G. R. Argue and J. J. Banewicz, *J. Inorg. Nucl. Chem.*, **25** (1963) 923.

65 S. A. Shchukarev, T. A. Tolmacheva and G. M. Slavutskaya, *Russ. J. Inorg. Chem.*, **9** (1964) 1351.

66 D. C. Giedt and C. J. Nyman, *Inorg. Synth.*, **8** (1966) 239.

67 G. B. Kauffman, *Inorg. Synth.*, **9** (1967) 182.

68 G. B. Kauffman and L. A. Teter, *Inorg. Synth.*, **7** (1963) 232.

69 A. Gutbier and F. Bauriedel, *Chem. Ber.*, **42** (1909) 4243.

70 C. Duval in *Nouveau Traité de Chimie Minerale*, Ed. by P. Pascal, Masson, Paris, 1958, vol. 19, p. 753.
71 A. Gutbier and A. Krell, *Chem. Ber.*, **38** (1905) 2385.
72 R. L. Datta, *J. Amer. Chem. Soc.*, **35** (1913) 1185.
73 Reference 70, p. 755.
74 G. A. Shagisultanova, *Russ. J. Inorg. Chem.*, **6** (1961) 904.

Structural Data on Palladium and Platinum Compounds

In this appendix an attempt has been made to summarise the structural data available for palladium and platinum complexes. No attempt has been made to give the full crystal data, but the more important molecular parameters are recorded and in all cases a reference to the original literature is given. The compounds are arranged in alphabetical order under the headings, palladium compounds that do not contain carbon, palladium compounds that contain carbon, platinum compounds that do not contain carbon, platinum compounds that contain carbon.

Unless it is specifically mentioned to the contrary all metal(II) complexes are square-planar and metal(IV) complexes are octahedral. Where authors have given estimates of standard deviations these are given in brackets as follows:

$$2\cdot044(3)\ \text{Å} \quad \text{means } 2\cdot044 \pm 0\cdot003\ \text{Å}$$
$$2\cdot044(30)\ \text{Å} \quad \text{means } 2\cdot044 \pm 0\cdot03\ \text{Å}$$
$$96\cdot5(3)° \quad \text{means } 96\cdot5 \pm 0\cdot3°$$
$$96\cdot5(14)° \quad \text{means } 96\cdot5 \pm 1\cdot4°$$

A. PALLADIUM COMPOUNDS THAT DO NOT CONTAIN CARBON

1. *Pd*   *Palladium Metal*
   (W. B. Pearson, *Lattice Spacings and Structures of Metals and Alloys*, Pergamon Press, London, 1957).
   Co-ordination number 12 (face centre cubic). Pd–Pd = 2·7511 Å at 25°C.

2. *PdCl$_2$*   *Palladium Dichloride*
   (A. F. Wells, *Z. Krist.*, **100** (1939) 189).

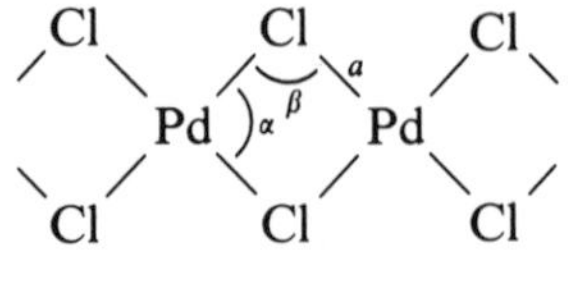

$$a = 2\cdot31\ \text{Å}$$
$$\alpha = 87°$$
$$\beta = 93°$$

3. *PdCl$_4^{2-}$*   in *K$_2$[PdCl$_4$]*
   (W. Theilacker, *Z. anorg. allgem. Chem.*, **234** (1937) 161).
   Pd square-planar. Pd–Cl = 2·30 Å.

4. $PdCl_4^{2-}$ in $(NH_4)_2[PdCl_4]$
   (J. D. Bell, D. Hall and T. N. Waters, *Acta Cryst.*, **21** (1966) 440).
   Pd square-planar. Pd–Cl = 2·299(4) Å.
5. $PdCl_4^{2-}$ in $[Pd(NH_3)_4][PdCl_4]$
   (J. R. Miller, *Proc. Chem. Soc.*, (1960) 318).
   Pd–Pd = 3·25 Å.
6. $PdCl_6^{2-}$ in $(NH_4)_2[PdCl_6]$
   (J. D. Bell, D. Hall and T. N. Waters, *Acta Cryst.*, **21** (1966) 440).
   Pd octahedral. Pd–Cl = 2·300(7) Å.
7. PdF_2 *Palladium Difluoride*
   (N. Bartlett and R. Maitland, *Acta Cryst.*, **11** (1958) 747).
   Rutile type ionic lattice. Four Pd–F = 2·155 Å. Two Pd–F = 2·171 Å.
8. PdF_3 *Palladium Trifluoride*
   (M. A. Hepworth, K. H. Jack, R. D. Peacock and G. J. Westland, *Acta Cryst.*, **10** (1957) 63; N. Bartlett and P. R. Rao, *Proc. Chem. Soc.*, (1964) 393; H. Henkel and R. Hoppe, *Z. anorg. allgem. Chem.*, **359** (1968) 160).
   Hepworth *et al.* reported Pd octahedral with Pd–F = 2·04 Å. Bartlett *et al.* later reported it to be $Pd^{2+}[PdF_6]^{2-}$. Pd–F bond length in $[PdF_6]^{2-}$ = 2·04 Å (Henkel *et al.*).
9. PdF_6^{2-} in $M_2[PdF_6]$
   (R. Hoppe and W. Klem, *Z. anorg. allgem. Chem.*, **268** (1952) 364; H. Henkel and R. Hoppe, *Z. anorg. allgem. Chem.*, **359** (1968) 160).
   Pd octahedral. Hoppe *et al.*: Pd–F = 1·89 Å (Cs); Pd–F = 1·80 Å (Rb); Henkel *et al.*: Pd–F = 1·98 Å (Mg); 2·22 Å (Ca); 2·08 Å (Cd); 2·07 Å (Zn).
10. $PdH_6N_2O_6S_2^{2-}$ in trans-$[Pd(SO_3)_2(NH_3)_2]^{2-}$, $2Na^+.6H_2O$
    (M. V. Capparelli and L. N. Becka, *J. Chem. Soc. (A)*, (1969) 260).

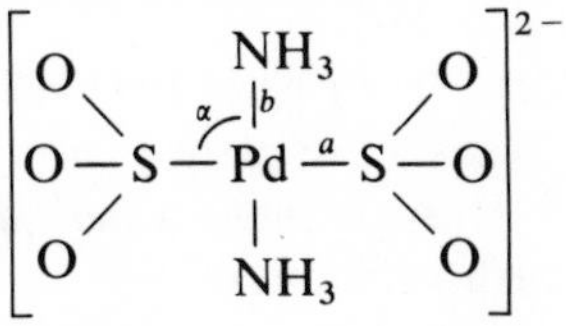

$a = 2·294(6)$ Å
$b = 2·060(9)$ Å
$\alpha = 90·9(3)°$

11. $PdH_9N_3O_3S$ $[Pd(SO_3)(NH_3)_3]$
    (M. A. Spinnler and L. N. Becka, *J. Chem. Soc. (A)*, (1967) 1194).

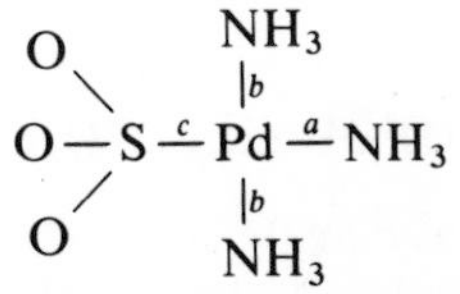

$a = 2·103(20)$ Å
$b = 2·107(20)$ Å (mean)
$c = 2·254(6)$ Å

12. $PdH_9N_4O_2$ in $[Pd(NO_2)(NH_3)_3]_2[Pd(NH_3)_4](NO_3)_4$
    (F. P. Boer, V. B. Carter and J. W. Turley, *Inorg. Chem.*, **10** (1971) 651).

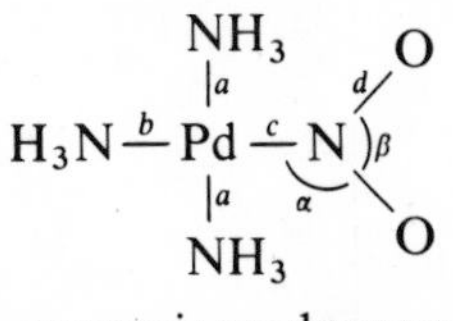

$a = 2·034(4)$ Å $d = 1·104(7)$ Å
$b = 2·053(8)$ Å $\alpha = 122·4(5)°$
$c = 1·984(8)$ Å $\beta = 155·2(9)°$

Plane of NO_2 group is coplanar with square-planar about Pd.

13. $PdH_{12}N_4$ in $[Pd(NH_3)_4][PtCl_4]$
(J. R. Miller, *Proc. Chem. Soc.*, (1960) 318).
Pd–Pt = 3·25 Å.
14. $PdH_{12}N_4$ in $[Pd(NH_3)_4][Pd(NO_2)(NH_3)_3]_2(NO_3)_4$
(F. P. Boer, V. B. Carter and J. W. Turley, *Inorg. Chem.*, **10** (1971) 651).
Pd–N = 2·044(3) Å. Co-ordination about Pd is square-planar.
15. $PdN_4O_8^{2-}$ in $K_2[Pd(NO_2)_4]$
(M. A. Porai-Koshits, *Dokl. Akad. Nauk. SSSR*, **58** (1947) 603).
Pd square-planar. Pd–N = 2·10 Å.
16. *PdO*   PdO
(W. J. Moore and L. Pauling, *J. Amer. Chem. Soc.*, **63** (1941) 1392).

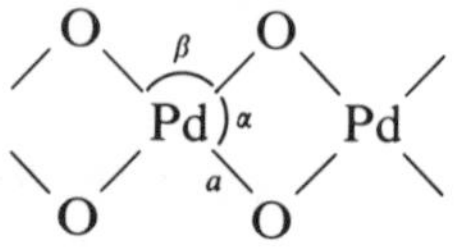

$a = 2\cdot01(1)$ Å
$\alpha = 82°$
$\beta = 98°$

Co-ordination about Pd is approximately planar. Co-ordination about O is approximately tetrahedral.
17. PdP_2 *Palladium diphosphide*
(W. H. Zachariasen, *Acta Cryst.*, **16** (1963) 1253).

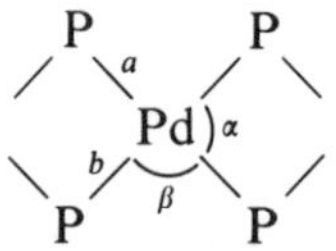

$a = 2\cdot341(10)$ Å $\alpha = 91\cdot1°$
$b = 2\cdot335(10)$ Å $\beta = 88\cdot9°$

Pd surrounded by four P in approximately square-planar arrangement.
18. PdS_2 *Palladium disulphide*
(F. Grønwold and E. Røst, *Acta Cryst.*, **10** (1957) 329).
Pd surrounded by four S in square-planar arrangement. Pd–S = 2·30 Å.
(R. A. Munson and J. S. Kasper, *Inorg. Chem.*, **8** (1969) 1198 report that at high pressures PdS_2 is orthorhombic).
19. $PdSe_2$ *Palladium diselenide*
(F. Grønwold and E. Røst, *Acta Cryst.*, **10** (1957) 329).
Pd surrounded by four Se in square-planar arrangement. Pd–Se = 2·44 Å.

B. PALLADIUM COMPOUNDS THAT CONTAIN CARBON

20. $C_2H_8N_2O_6PdS_4^{2-}$ in $[Pd(en)_2][Pd(S_2O_3)_2en]$
(S. Baggio, L. M. Amzel and L. N. Becka, *Acta Cryst.*, **B26** (1970) 1698).

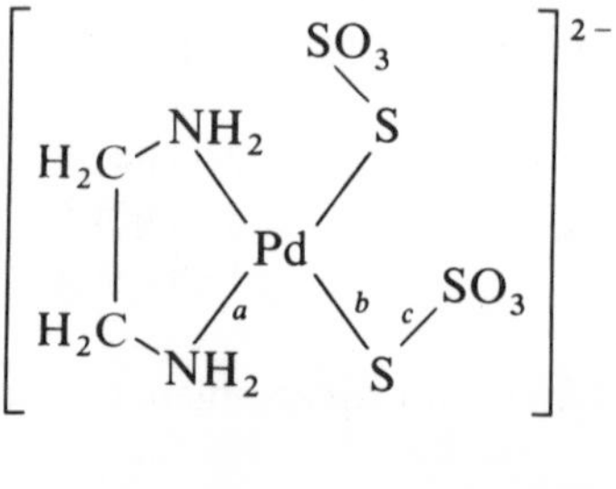

$a = 2\cdot071(7)$ Å (mean)
$b = 2\cdot297(4)$ Å (mean)
$c = 2\cdot066(6)$ Å (mean)

21. $C_4H_6O_4Pd$ *Palladium(II) acetate*
(A. C. Skapski and M. L. Smart, *Chem. Comm.*, (1970) 658).

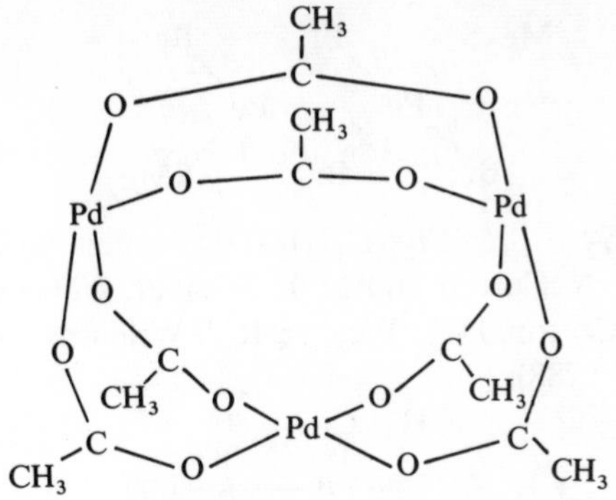

$$Pd-O = 1.973-2.014 \pm 0.009 \text{ Å (mean } 1.992 \text{ Å)}$$
$$Pd-Pd = 3.105-3.203 \pm 0.001 \text{ Å (mean } 3.109 \text{ Å)}$$

22. $C_4H_6N_2O_2Pd$ *Bis(glyoximato)palladium(II)*
(M. Calleri, G. Ferraris and D. Viterbo, *Inorg. Chim. Acta*, **1** (1967) 297).
Unit cell contains two crystallographically non-equivalent molecules.

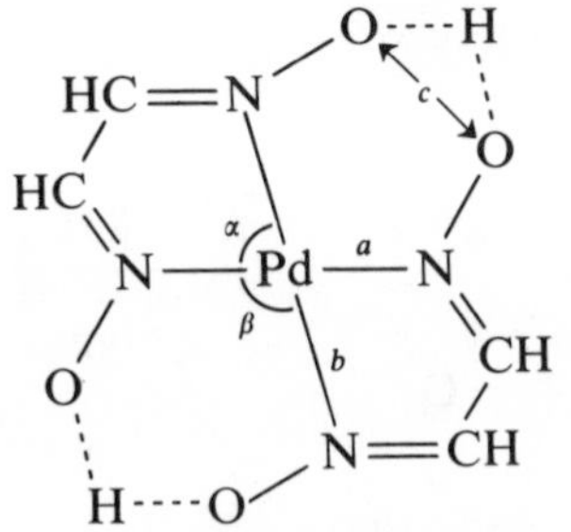

	Molecule 1	Molecule 2
a	1·958(7) Å	1·953(9) Å
b	1·991(8) Å	1·957(7) Å
c	2·659(12) Å	2·599(12) Å
α	80·3(4)°	80·3(4)°
β	99·7(4)°	99·7(4)°

23. $C_4H_8Cl_4Pd_2$ $[PdCl_2(C_2H_4)]_2$
(J. N. Dempsey and N. C. Baenziger, *J. Amer. Chem. Soc.*, **77** (1955) 4984).
Crystal contains two crystallographically non-equivalent molecules. C_2H_4 is perpendicular to $PdCl_3$ plane.

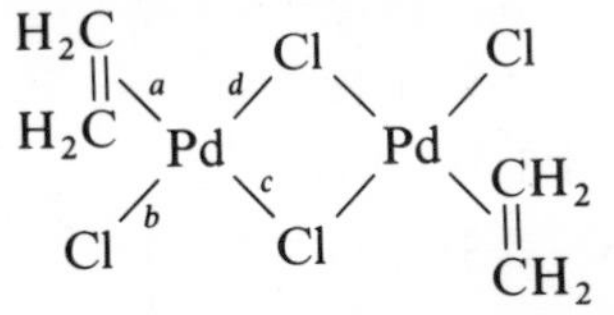

	Molecule 1	Molecule 2
a	2·28 Å	2·34 Å
b	2·31 Å	2·18 Å
c	2·43 Å	2·32 Å
d	2·33 Å	2·41 Å

24. $C_4H_{10}Cl_2N_4Pd$ $[PdCl_2\{(CH_3)_2C_2N_4H_4\}]$
(A. Burke, A. L. Balch and J. H. Enemark, *J. Amer. Chem. Soc.*, **92** (1970) 2555).

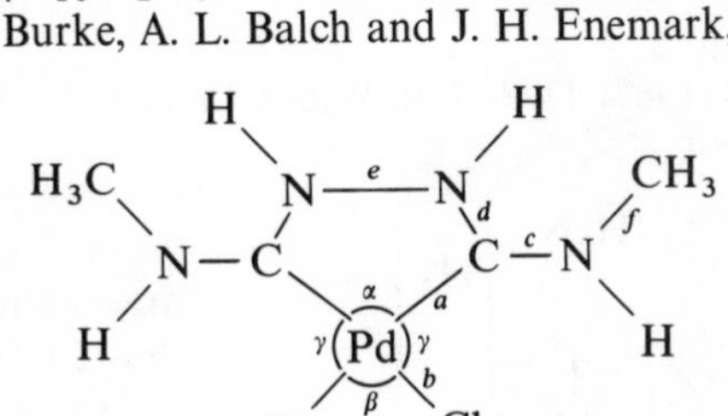

$a = 1.86(3)$ Å	$\alpha = 78°$
$b = 2.38(1)$ Å	$\beta = 94°$
$c = 1.45(4)$ Å	$\gamma = 94°$
$d = 1.38(4)$ Å	
$e = 1.45(5)$ Å	
$f = 1.44(4)$ Å	

25. $C_4H_{12}Br_4Pd_2S_2$ $[PdBr_2\{(CH_3)_2S\}]_2$
(D. L. Sales, J. Stokes and P. Woodward, *J. Chem. Soc. (A)*, (1968) 1852).

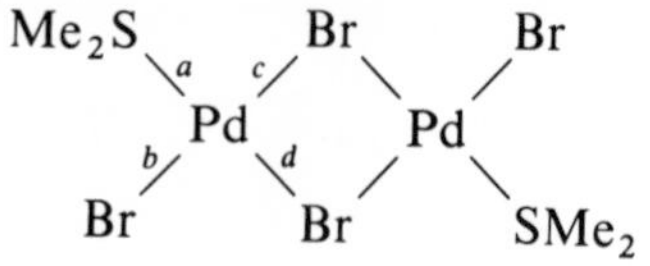

$a = 2\cdot30(2)$ Å
$b = 2\cdot404(4)$ Å
$c = 2\cdot429(4)$ Å
$d = 2\cdot447(11)$ Å

26. $C_4H_{12}Cl_2O_2PdS_2$ $[PdCl_2\{(CH_3)_2SO\}_2]$
(M. J. Bennett, F. A. Cotton and D. L. Weaver, *Nature*, **212** (1966) 286;
M. J. Bennett, F. A. Cotton, D. L. Weaver, R. J. Williams and W. H. Watson,
Acta Cryst., **23** (1967) 788).

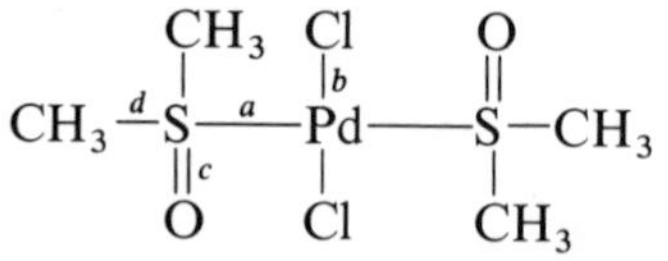

$a = 2\cdot298$ Å
$b = 2\cdot287$ Å
$c = 1\cdot475$ Å
$d = 1\cdot778$ Å

27. $C_4H_{12}N_2O_8PdS_2$ $[Pd(NO_3)_2\{(CH_3)_2SO\}_2]$
(D. A. Langs, C. R. Hare and R. G. Little, *Chem. Comm.*, (1967) 1080).

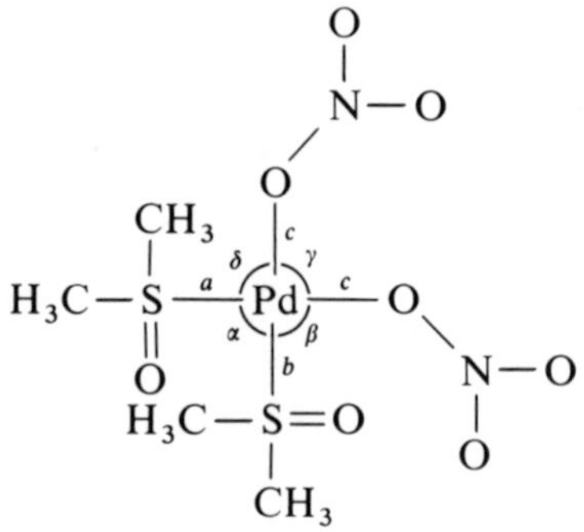

$a = 2\cdot231$ Å $\alpha = 90\cdot5°$
$b = 2\cdot253$ Å $\beta = 95\cdot4°$
$c = 2\cdot066$ Å $\gamma = 83\cdot0°$
                       $\delta = 91\cdot6°$

28. $C_4H_{12}PdS_4$ Bis(dithiobiureto)palladium(II)
(R. L. Girling and E. L. Amma, *Chem. Comm.*, (1968) 1487).

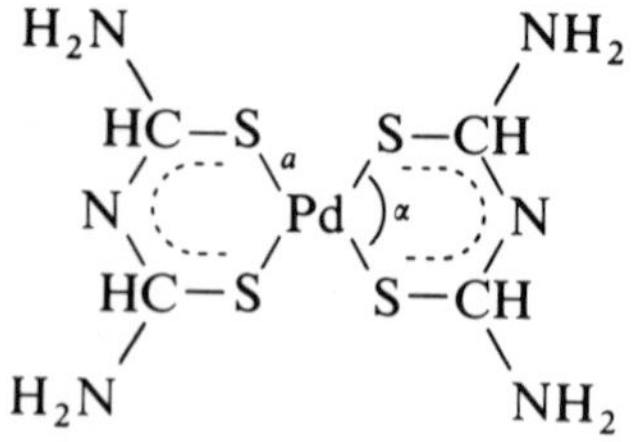

$a = 2\cdot295(4)$ Å (mean)
$\alpha = 92\cdot8°$

29. $C_4H_{16}N_4Pd^{2+}$ $[Pd(en)_2]Cl_2$
(J. R. Wiesner, *Diss. Abs.*, **27B** (1967) 4298; J. R. Wiesner and E. C. Lingafelter, *Inorg. Chem.*, **5** (1966) 1770).

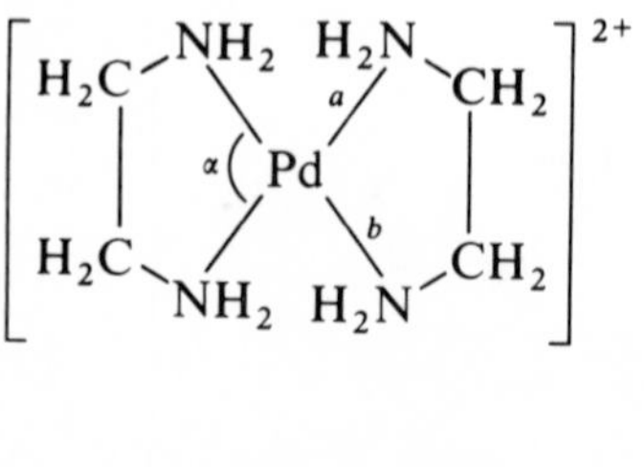

$a = 2\cdot030(6)$ Å
$b = 2\cdot043(7)$ Å
$\alpha = 83\cdot6°$

30. $C_4H_{16}N_4Pd^{2+}$ $[Pd(en)_2][Pd(en)(S_2O_3)_2]$
   (S. Baggio, L. M. Amzel and L. N. Becka, *Acta Cryst.*, **B26** (1970) 1698).
   For figure see structure 29.

$$a = 2\cdot074(13)\,\text{Å}$$
$$b = 2\cdot085(15)\,\text{Å}$$
$$\alpha = 94\cdot7°$$

31. $C_4H_{16}N_8PdS_4^{2+}$ $[Pd\{SC(NH_2)_2\}_4]Cl_2$
   (S. Oi, T. Kawase, K. Nakatsu and H. Kuroya, *Bull. Chem. Soc. Japan*,
   33 (1960) 861; D. A. Berta, W. A. Spofford, P. Boldrini and E. L. Amma,
   Inorg. Chem., **9** (1970) 136).

   PdS_4 is approximately square-planar but slightly distorted towards
   tetrahedral. Mean Pd–S = 2·33(1) Å. Two Cl⁻ essentially axial at 3·594(3) Å
   and 3·791(3) Å

32. $C_4N_4PdS_4$ $K_2[Pd(SCN)_4]$
   (A. Mawby and G. E. Pringle, *Chem. Comm.*, (1970) 385).

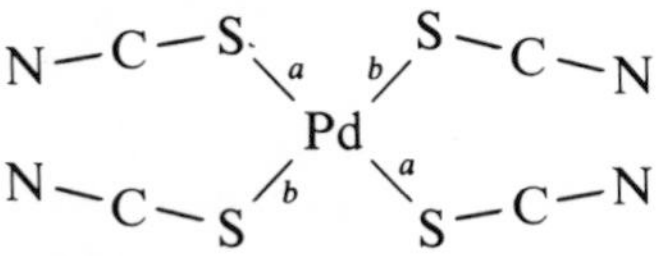

PdS_4 exactly planar
$a = 2\cdot312(96)$ Å
$b = 2\cdot392(90)$ Å—this S is associated with
a second Pd at a distance
of 3·66 Å.

33. $C_5H_9Cl_2NOPdS$ $[Pd(C_3H_5NH.C(OMe)S)Cl_2]$
   (P. Porta, *J. Chem. Soc. (A)*, (1971) 1217).

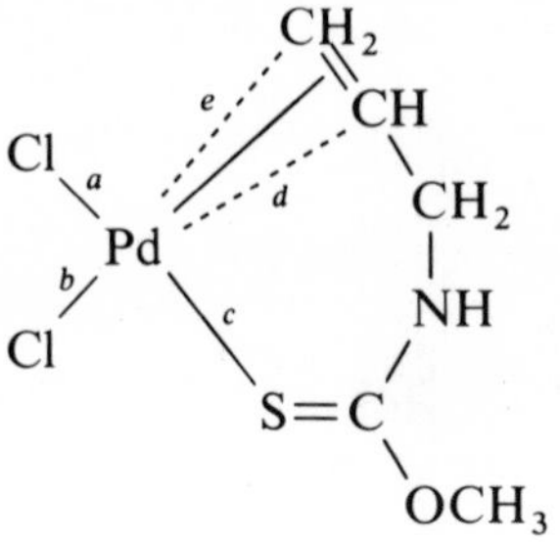

$a = 2\cdot33$ Å $d = 2\cdot15$ Å
$b = 2\cdot34$ Å $e = 2\cdot14$ Å
$c = 2\cdot30$ Å $f = 1\cdot304$ Å

$PdCl_2S$ are coplanar and double bond is perpendicular to that plane and
situated symmetrically about it.

34. $C_5H_{11}Cl_2NO_2PdS$ $[Pd(methionine-H)Cl_2]$
   (R. C. Warren, J. F. McConnell and N. C. Stephenson, *Acta Cryst.*, **B26**
   (1970) 1402).

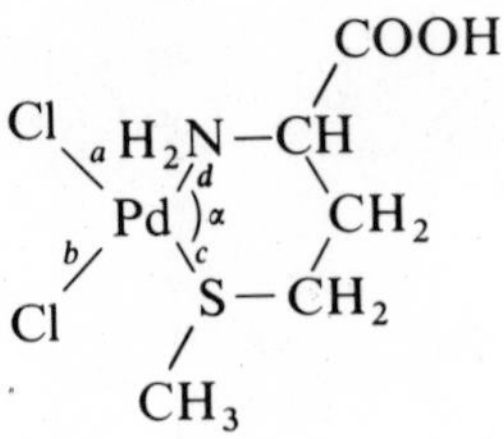

$a = 2\cdot332(4)$ Å
$b = 2\cdot308(4)$ Å
$c = 2\cdot265(4)$ Å
$d = 2\cdot061(14)$ Å
$\alpha = 97°$

35. $C_6H_{10}Cl_2Pd$ $[Pd(CH_2=CH-(CH_2)_2-CH=CH_2)Cl_2]$

(I. A. Zakharova, G. A. Kukina, T. S. Kuli-Zade, I. I. Moiseev, G. Yu. Pek and M. A. Porai-Koshits, *Russ. J. Inorg. Chem.*, **11** (1966) 1364).

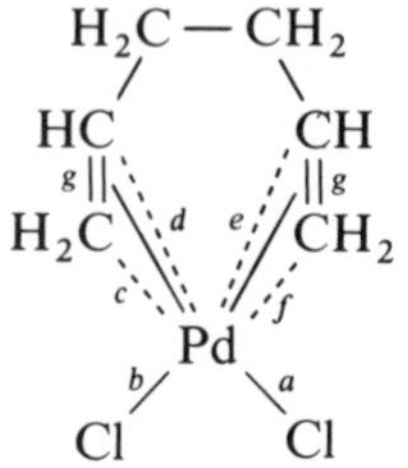

$$
\begin{array}{l}
a = 2\cdot30(1)\,\text{Å} \quad e = 2\cdot24(3)\,\text{Å} \\
b = 2\cdot31(1)\,\text{Å} \quad f = 2\cdot10(3)\,\text{Å} \\
c = 2\cdot20(3)\,\text{Å} \quad g = 1\cdot31(5)\,\text{Å} \\
d = 2\cdot24(3)\,\text{Å}
\end{array}
$$

Double-bonds are perpendicular to and symmetrical about the square-plane containing Pd.

36. $C_6H_{10}Cl_2Pd_2$ $[(\pi\text{-}C_3H_5)PdCl]_2$

(J. M. Rowe, *Proc. Chem. Soc.*, (1962) 66; A. E. Smith, *Acta Cryst.*, **18** (1965) 331; W. E. Oberhansli and L. F. Dahl, *J. Organometal Chem.*, **3** (1965) 43).

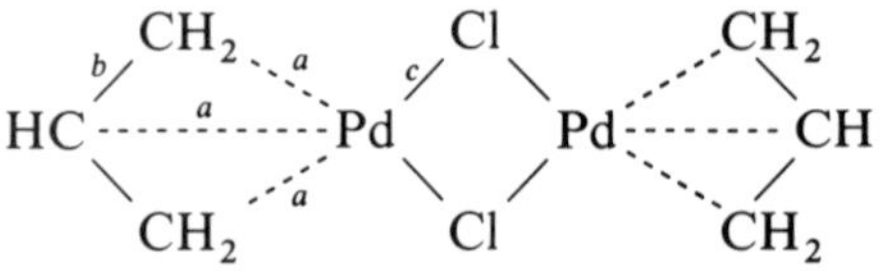

$$
\begin{array}{l}
a = 2\cdot12\,\text{Å} \\
b = 1\cdot38\,\text{Å} \\
c = 2\cdot41\,\text{Å}
\end{array}
$$

Angle between C_3 plane and $PdCl_2$ plane $= 111\cdot5 \pm 0\cdot90°$.

37. $C_6H_{14}Cl_2N_2Pd$ $[Pd(dimethylpiperazine)Cl_2]$

(O. Hassel and B. F. Pedersen, *Proc. Chem. Soc.*, (1959) 394).

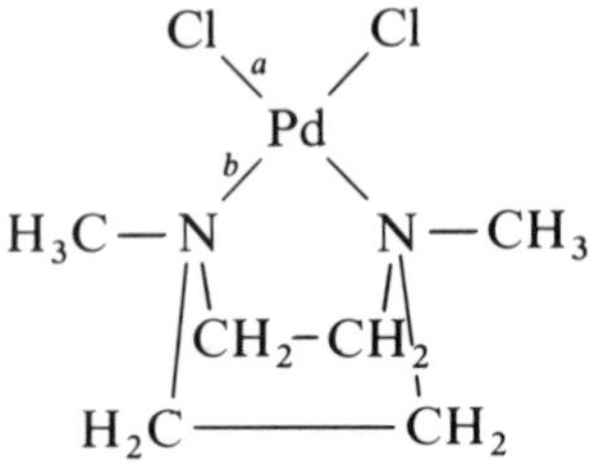

$$
\begin{array}{l}
a = 2\cdot30\,\text{Å} \\
b = 2\cdot00\,\text{Å}
\end{array}
$$

38. $C_6H_{14}PdS_2$ $[Pd(SC_3H_7)_2]$

(N. R. Kunchur, *Acta Cryst.*, **B24** (1968) 1623).

Molecule is hexameric with a puckered ring of six Pd atoms. Each Pd is approximately square-planar.

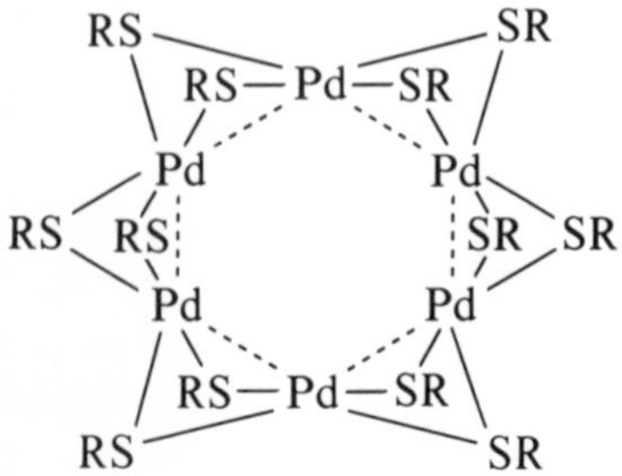

$R = n\text{-}C_3H_7$
Pd–S between $2\cdot282\,\text{Å}$ and $2\cdot454\,\text{Å}$

39. $C_6H_{18}As_2Br_4Pd_2$ $[Pd\{(CH_3)_3As\}Br_2]_2$
(A. F. Wells, *Proc. Roy. Soc.*, **A167** (1938) 169).

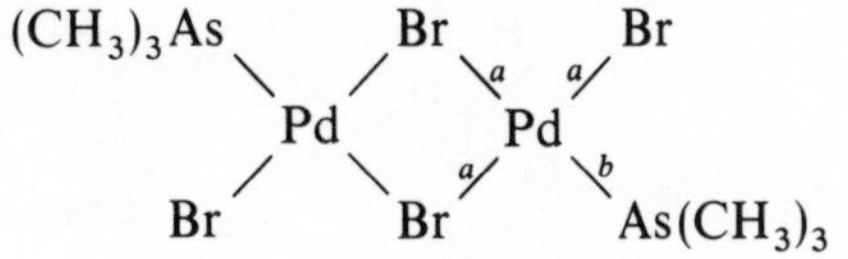

$a = 2.45\,\text{Å}$
$b = 2.50\,\text{Å}$

40. $C_7H_8Cl_2Pd$ $[Pd(norbornadiene)Cl_2]$
(N. C. Baenziger, J. R. Doyle and C. L. Carpenter, *Acta Cryst.*, **14** (1961) 303; N. C. Baenziger, J. R. Doyle, G. F. Richards and C. L. Carpenter in *Advances in the Chemistry of the Coordination Compounds*, Macmillan, New York, 1961, p. 131; N. C. Baenziger, G. F. Richards and J. R. Doyle, *Acta Cryst.*, **18** (1965) 924).

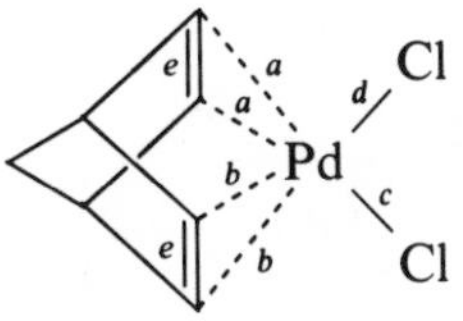

$a = 2.159(4)\,\text{Å}$
$b = 2.166(4)\,\text{Å}$
$c = 2.323(1)\,\text{Å}$
$d = 2.310(1)\,\text{Å}$
$e = 1.366(10)\,\text{Å}$

41. $C_8H_8Cl_2Pd$ $[(Cyclo\text{-}octatetraene)PdCl_2]$
(C. V. Goebel, *Diss. Abs.*, **B28** (1967) 625).

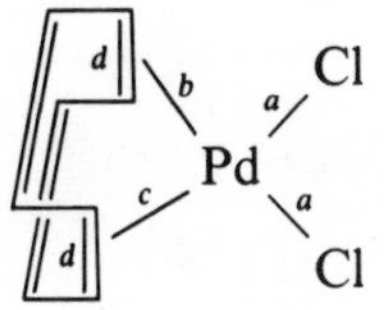

$a = 2.29\,\text{Å}$ (average)
$b = 2.12\,\text{Å}$
$c = 2.04\,\text{Å}$
$d = 1.42\,\text{Å}$ (average)

42. $C_8H_{14}Cl_2Pd$ $[(\pi\text{-}C_4H_7)PdCl]_2$
(R. Mason and A. G. Wheeler, *J. Chem. Soc. (A)*, (1968) 2549).

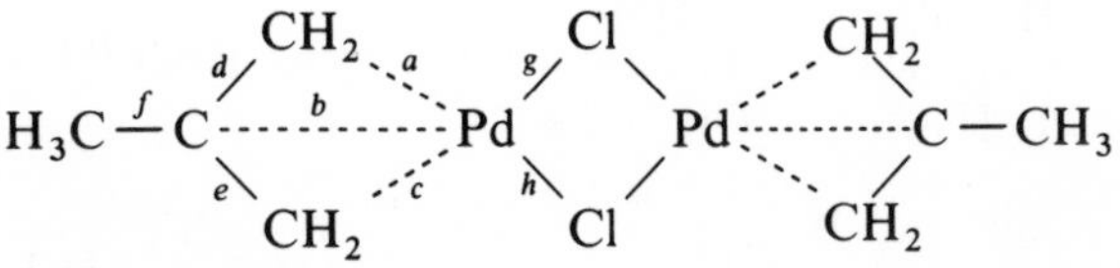

$a = 2.082(19)\,\text{Å}$ $e = 1.346(28)\,\text{Å}$
$b = 2.102(16)\,\text{Å}$ $f = 1.542(26)\,\text{Å}$
$c = 2.062(18)\,\text{Å}$ $g = 2.413(4)\,\text{Å}$
$d = 1.369(26)\,\text{Å}$ $h = 2.395(5)\,\text{Å}$

CH_3 group is bent out of π-allyl plane towards Pd by $0.29\,\text{Å}$. $PdCl_2Pd$ unit is planar.

43. $C_8H_{14}N_4O_4Pd$ Bis(dimethylglyoxime)palladium(II)

(C. Panattoni, E. Frasson and R. Zanetti, *Gazz. Chim. Ital.*, **89** (1959) 2132; D. E. Williams, G. Wohlauer and R. E. Rundle, *J. Amer. Chem. Soc.*, **81** (1959) 755).

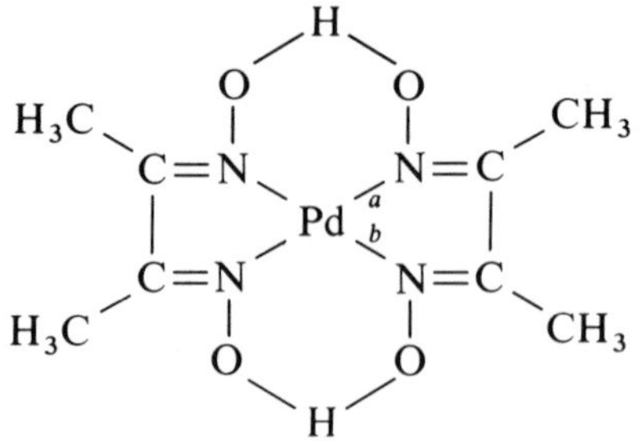

$a = 1{\cdot}93$ Å
$b = 1{\cdot}99$ Å

44. $C_8H_{18}Cl_2PdSe_2$ $[Pd(^iC_3H_7SeCH_2CH_2Se^iC_3H_7)Cl_2]$

(H. J. Whitfield, *J. Chem. Soc.* (*A*), (1970) 113).

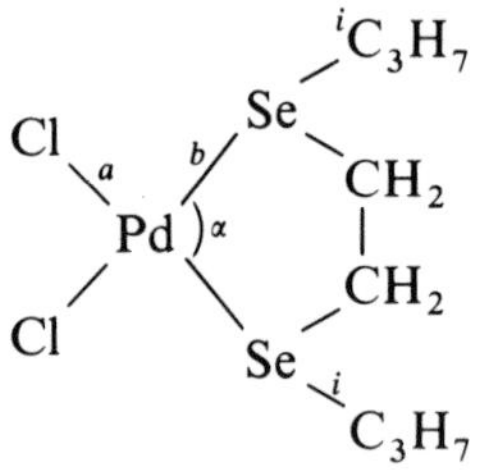

$a = 2{\cdot}31$ Å
$b = 2{\cdot}38$ Å
$\alpha = 89{\cdot}5°$

45. $C_8H_{20}Cl_2PdSe_2$ Trans-$[Pd\{(C_2H_5)_2Se\}_2Cl_2]$

(P. E. Skakke and S. E. Rasmussen, *Acta Chem. Scand.*, **24** (1970) 2634).

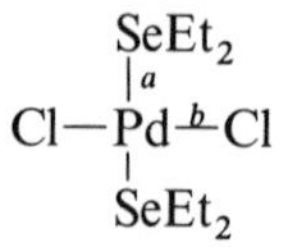

$a = 2{\cdot}424(7)$ Å
$b = 2{\cdot}266(9)$ Å

46. $C_{10}H_{16}Cl_2N_2O_8Pd$ $[Pd(H_4EDTA)Cl_2]$

(D. J. Robinson and C. H. L. Kennard, *Chem. Comm.*, (1967) 1236; D. J. Robinson and C. H. L. Kennard, *J. Chem. Soc.* (*A*), (1970) 1008).

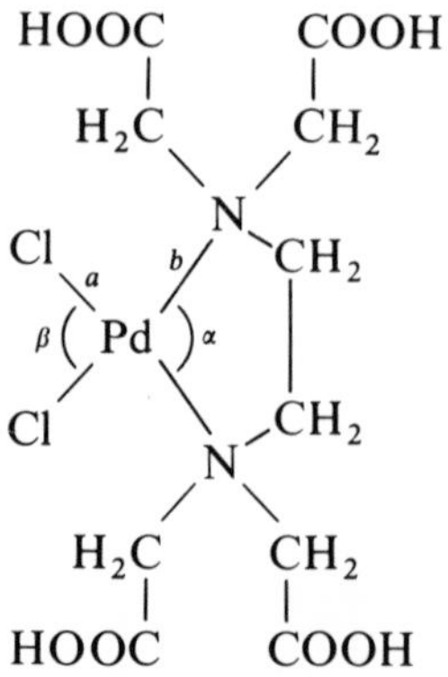

$a = 2{\cdot}31(1)$ Å
$b = 2{\cdot}09(2)$ Å
$\alpha = 84{\cdot}5°$
$\beta = 90{\cdot}6°$

47. $C_{10}H_{16}O_4Pd_2$ $[(\pi\text{-}C_3H_5)Pd(CH_3COO)]_2$
(M. R. Churchill and R. Mason, *Nature*, **204** (1964) 777).

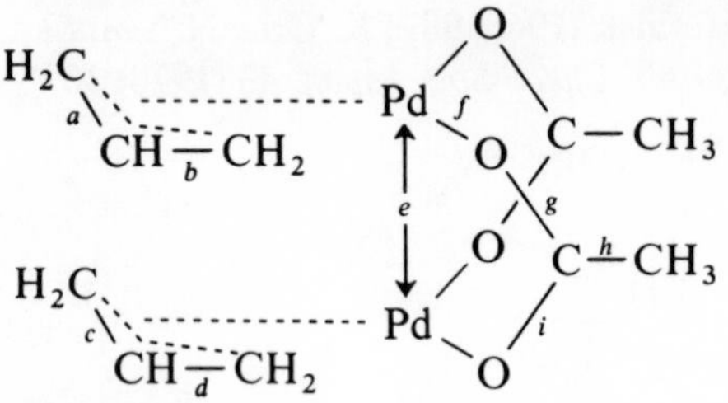

$a = 1.34$ Å	$f = 2.11$ Å (mean)
$b = 1.37$ Å	$g = 1.20$ Å (mean)
$c = 1.42$ Å	$h = 1.57$ Å (mean)
$d = 1.33$ Å	$i = 1.27$ Å (mean)
$e = 2.94$ Å	

48. $C_{10}H_{18}Cl_2Pd$ $[Pd(\pi\text{-}1,3\text{-}dimethylallyl)Cl]_2$
(G. R. Davies, R. H. B. Mais, S. O'Brien and P. G. Owston, *Chem. Comm.*, (1967) 1151).

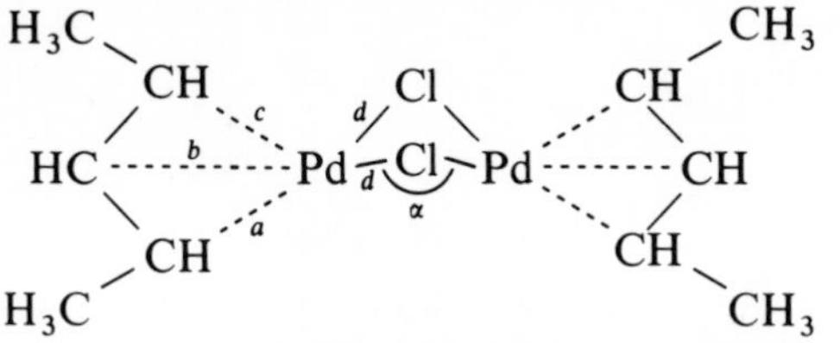

$a = 2.14(4)$ Å
$b = 2.07(4)$ Å
$c = 2.04(3)$ Å
$d = 2.40$ Å
$\alpha = 150°$

Angle between π-allyl plane and $PdCl_2$ plane $= 125°$ (mean).

49. $C_{11}H_{27}As_3BrPd^+$ $[Pd(triars)Br]^+ Br^-$
(G. A. Mair, H. M. Powell and D. E. Henn, *Proc. Chem. Soc.*, (1960) 415).

triars = $(CH_3)_2As(CH_2)_3As(CH_2)_3As(CH_3)_2$ with a CH_3 on the central As

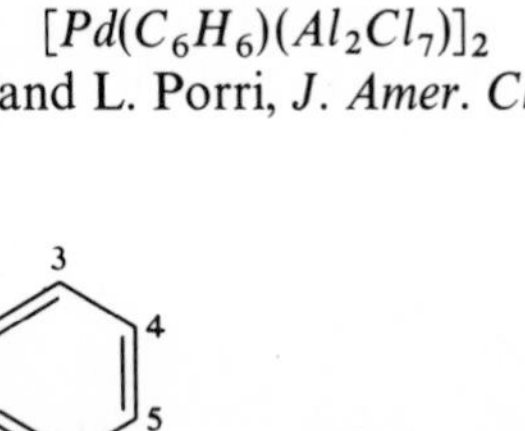

$\alpha = 10°$

50. $C_{12}H_{12}Al_4Cl_{14}Pd_2$ $[Pd(C_6H_6)(Al_2Cl_7)]_2$
(G. Allegra, A. Immirzi and L. Porri, *J. Amer. Chem. Soc.*, **87** (1965) 1394).

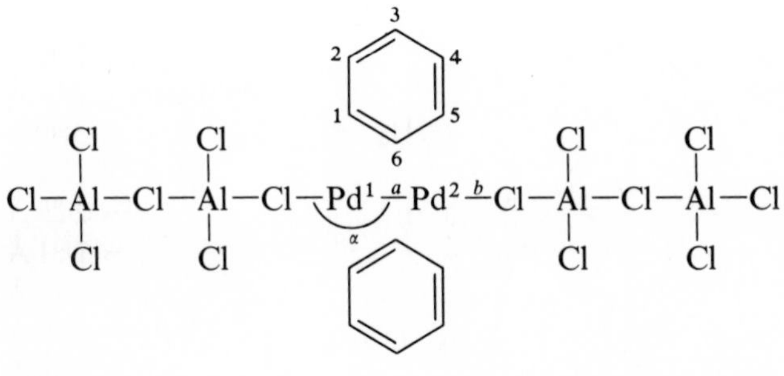

$a = 2.58(1)$ Å
$b = 2.46(2)$ Å

$\left.\begin{array}{l} Pd^1\text{-}C^1 \\ Pd^1\text{-}C^2 \\ Pd^2\text{-}C^4 \\ Pd^2\text{-}C^5 \end{array}\right\}$ 2.34 Å (mean)

$\left.\begin{array}{l} Pd^1\text{-}C^3 \\ Pd^1\text{-}C^6 \\ Pd^2\text{-}C^3 \\ Pd^2\text{-}C^6 \end{array}\right\}$ 2.94 Å (mean)

$\alpha = 176(1)°$

51. $C_{12}H_{18}Cl_2O_6Pd$ [$Pd(\pi$-1-*ethoxycarbonyl-2-hydroxyallyl)Cl*]$_2$
(K. Oda, N. Yasuoka, T. Ueki, N. Kasai, M. Kakudo, Y. Tezuka, T. Ogura and S. Kawaguchi, *Chem. Comm.*, (1968) 989; K. Oda, N. Yasuoka, T. Ueki, N. Kasai and M. Kakudo, *Bull. Chem. Soc. Japan*, **43** (1970) 362).

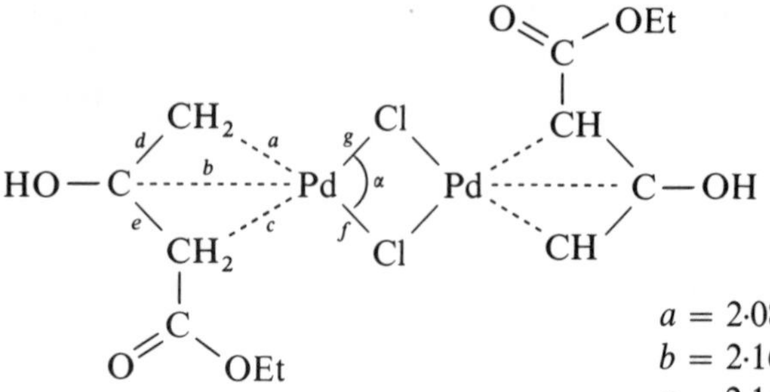

$a = 2.08(5)$ Å
$b = 2.16(5)$ Å } mean $2.12(5)$ Å
$c = 2.11(5)$ Å
$d = 1.42(7)$ Å
$e = 1.34(7)$ Å $g = 2.41(1)$ Å
$f = 2.40(1)$ Å $\alpha = 89.4(4)°$

Dihedral angle between π-allyl plane and PdCl$_2$Pd plane $= 108°$. –COOEt bent out of plane of π-allyl group towards Pd.

52. $C_{12}H_{22}Cl_2N_2O_2Pd$ [$Pd(C_6H_{10}NOH)_2Cl_2$]
(M. Tanimura, T. Mizushima and Y. Kinoshita, *Bull. Chem. Soc. Japan*, **40** (1967) 2777).

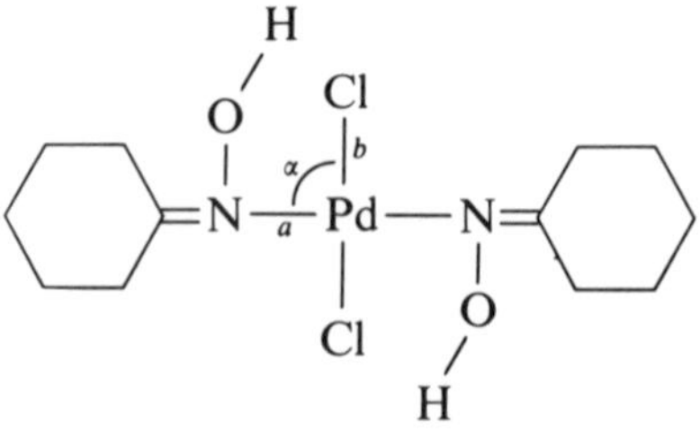

$a = 2.08$ Å
$b = 2.24$ Å
$\alpha = 87°$

The oxime hydrogen of one complex is hydrogen-bonded to the chloride ligand of the complex stacked vertically above it.

53. $C_{12}H_{24}Pd_3S_9$ [$Pd(2,2'$-*dimercaptodiethylsulphide*)]$_3$
(E. M. McPartlin and N. C. Stephenson, *Acta Cryst.*, **B25** (1969) 1659).

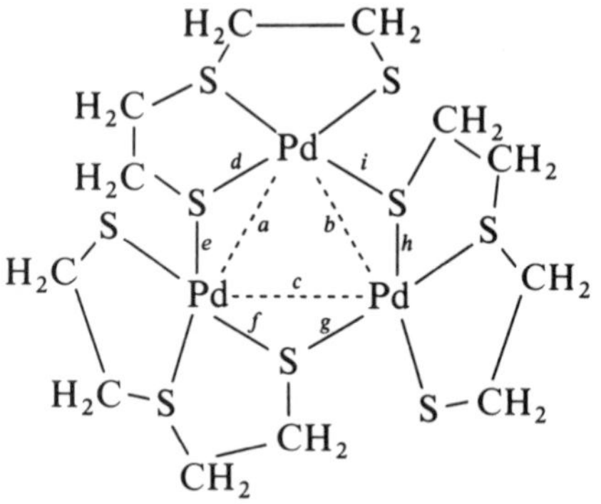

Pd–Pd { $a = 3.41$ Å
$b = 3.66$ Å
$c = 3.49$ Å

Pd–S { $d = 2.29$ Å
$e = 2.31$ Å
$f = 2.41$ Å
$g = 2.30$ Å
$h = 2.31$ Å
$i = 2.36$ Å

Alternative view emphasising
the approximately square-planar
co-ordination about each Pd.

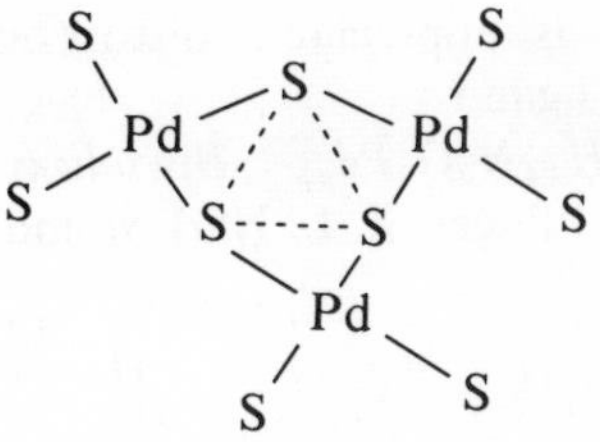

54. $C_{13}H_{18}O_2Pd$ (π-*cyclo-octadienyl*)*palladium acetylacetonate*
   (M. R. Churchill, *Chem. Comm.*, (1965) 625; M. R. Churchill, *Inorg. Chem.*,
   5 (1966) 1608).

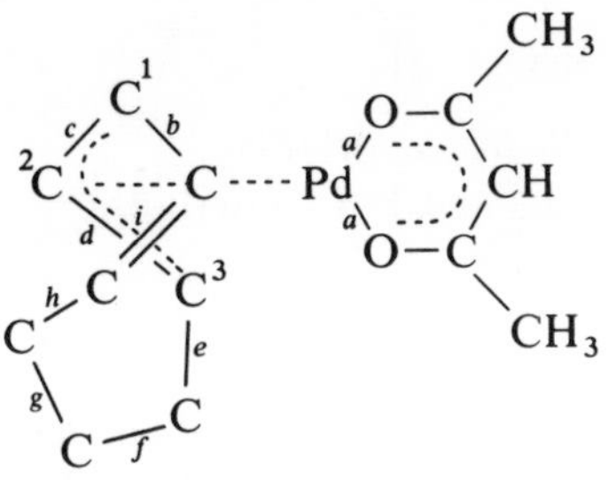

$$\left.\begin{array}{l} Pd-C^1 \\ Pd-C^2 \\ Pd-C^3 \end{array}\right\} 2{\cdot}11 \text{ Å (mean)}$$

$a = 2{\cdot}08$ Å	$d = 1{\cdot}42$ Å	$g = 1{\cdot}51$ Å
$b = 1{\cdot}47$ Å	$e = 1{\cdot}50$ Å	$h = 1{\cdot}46$ Å
$c = 1{\cdot}46$ Å	$f = 1{\cdot}44$ Å	$i = 1{\cdot}38$ Å

55. $C_{14}H_{10}Cl_2N_2Pd$ $[Pd(C_6H_5CN)_2Cl_2]$
   (J. R. Holden and N. C. Baenziger, *Acta Cryst.*, **9** (1956) 194).
   Pd–Cl (approx.) = 2·35 Å.
56. $C_{14}H_{10}PdS_4$ *Bis(dithiobenzoato)palladium*
   (M. Bonamico and G. Dessy, *Chem. Comm.*, (1968) 483).

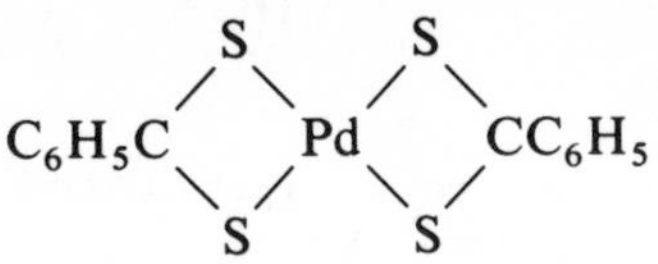

Stereochemistry around Pd is distorted tetragonal. Pd–S = 2·32–2·34 Å
(there are two types of molecule in the unit cell with slightly different bond
lengths).
57. $C_{14}H_{12}N_4O_4Pd$ $[Pd(2,9\text{-}dimethyl\text{-}1,10\text{-}phenanthroline)(NO_2)_2]$
   (L. F. Power, *Inorg. Nucl. Chem. Lett.*, **6** (1970) 791).

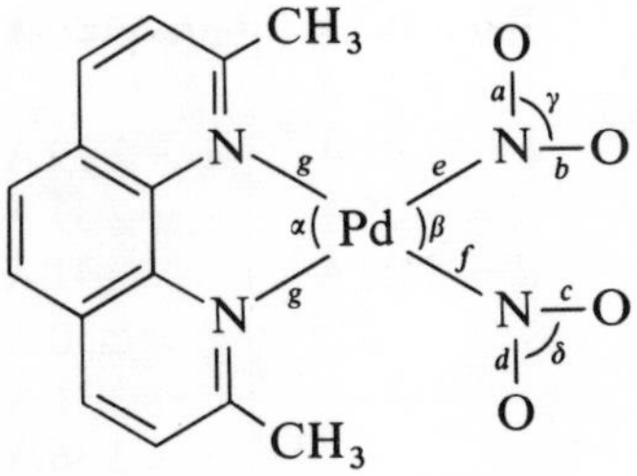

$a = 1{\cdot}24(2)$ Å	$e = 1{\cdot}98$ Å	$\alpha = 83°$
$b = 1{\cdot}17(2)$ Å	$f = 2{\cdot}02$ Å	$\beta = 76°$
$c = 1{\cdot}30$ Å	$g = 2{\cdot}09$ Å	$\gamma = 120°$
$d = 1{\cdot}16$ Å		$\delta = 133°$

There is appreciable distortion within the 2,9-dimethyl-1,10-phenanthroline ligand.

58. $C_{14}H_{14}N_2O_2Pd$ *Bis(salicylaldoximato)palladium*
(C. E. Pfluger, R. L. Harlow and S. H. Simonsen, *Acta Cryst.*, **B26** (1970) 1631).

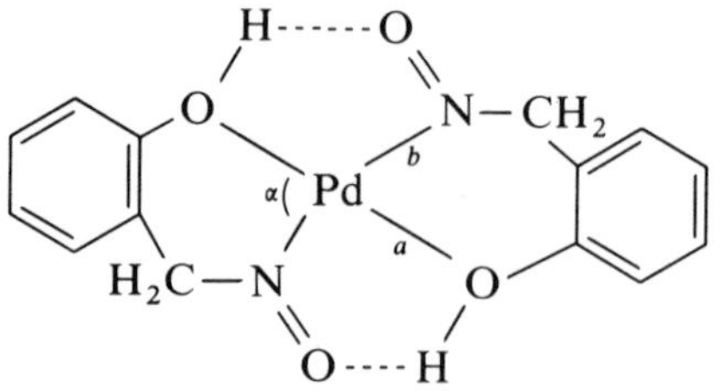

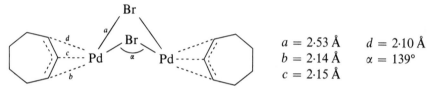

$a = 1.982(10)$ Å
$b = 1.961(11)$ Å
$\alpha = 92.5(7)$ Å

59. $C_{14}H_{22}Br_2Pd_2$ *[(π-Cycloheptenyl)PdBr]₂*
(B. T. Kilbourn, R. H. B. Mais and P. G. Owston, *Chem. Comm.*, (1968) 1438).

$a = 2.53$ Å $d = 2.10$ Å
$b = 2.14$ Å $\alpha = 139°$
$c = 2.15$ Å

Angle between π-C_3 plane and $PdBr_2$ plane = 118°.

60. $C_{14}H_{26}Cl_2Pd_2$ *[(π-1,1,3,3-tetramethylallyl)PdCl]₂*
(R. Mason and A. G. Wheeler, *J. Chem. Soc. (A)*, (1968) 2543).

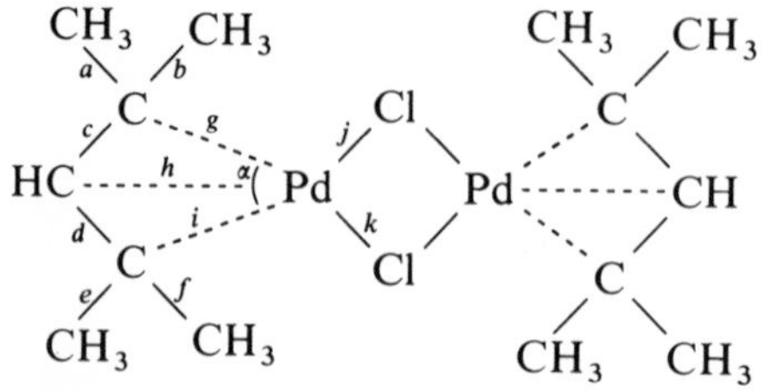

$a = 1.61$ Å $g = 2.12$ Å
$b = 1.54$ Å $h = 2.14$ Å
$c = 1.41$ Å $i = 2.14$ Å
$d = 1.42$ Å $j = 2.43$ Å
$e = 1.60$ Å $k = 2.42$ Å
$f = 1.56$ Å $\alpha = 69.5°$

61. $C_{15}H_{27}N_3O_9Pd_3$ *[Pd(CH₃COO)({CH₃}₂CNO)]₃*
(A. Mawby and G. E. Pringle, *Chem. Comm.*, (1970) 560).

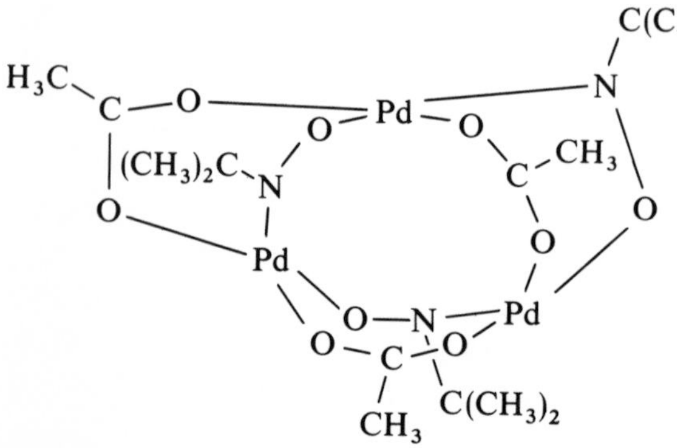

Each Pd is square-planar
$a = 2.04$ Å
$b = 1.97$ Å
$c = 2.05$ Å
$d = 2.00$ Å

62. $C_{16}H_{16}Cl_4Pd_2$ $[Pd(C_6H_5CH{=}CH_2)Cl_2]$
(J. R. Holden and N. C. Baenziger, *J. Amer. Chem. Soc.*, **77** (1955) 4987).

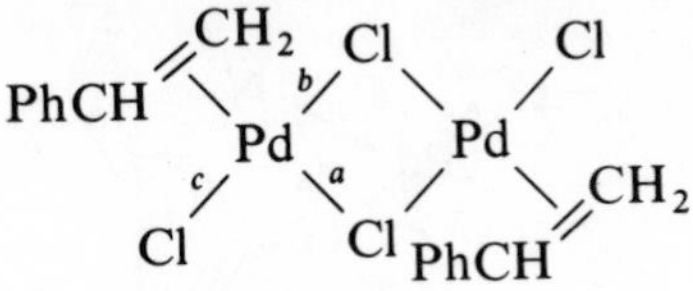

$a = 2{\cdot}41$ Å
$b = 2{\cdot}32$ Å
$c = 2{\cdot}27$ Å

Plane of olefinic double-bond is perpendicular to $PdCl_3$ plane.

63. $C_{16}H_{22}I_2Pd$ Trans-$[(PMe_2Ph)_2PdI_2]$
(N. A. Bailey, J. M. Jenkins, R. Mason and B. L. Shaw, *Chem. Comm.*, (1965) 237 and 296; N. A. Bailey and R. Mason, *J. Chem. Soc. (A)*, (1968) 2594).

(a) *Red Isomer*

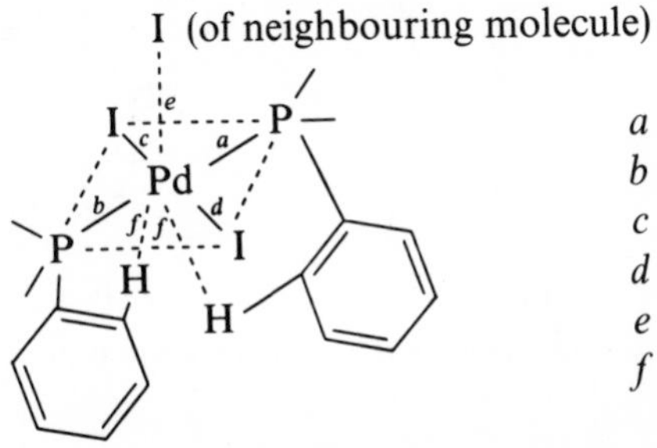

I (of neighbouring molecule)

$a = 2{\cdot}638(3)$ Å
$b = 2{\cdot}619(3)$ Å
$c = 2{\cdot}327(7)$ Å
$d = 2{\cdot}333(8)$ Å
$e = 3{\cdot}290(3)$ Å
$f = 2{\cdot}84$ Å

(b) *Yellow Isomer*

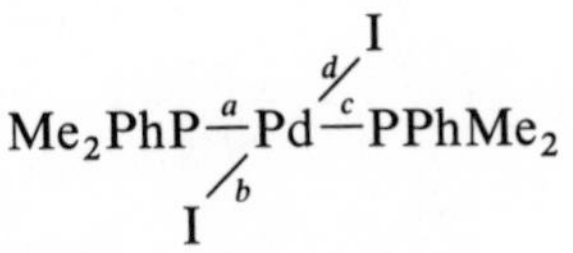

$a = c = 2{\cdot}592(3)$ Å
$b = d = 2{\cdot}333(7)$ Å

Pd–H of a phenyl bound to a neighbouring Pd $= 3{\cdot}28$ Å.

64. $C_{17}H_{26}O_2Pd$ (π-*Pentamethyl(bicyclo(2,2,0)hexa-2,5-dienylmethyl) palladium acetylacetonate*
(J. F. Malone, W. S. McDonald, B. L. Shaw and G. Shaw, *Chem. Comm.*, (1968) 869; J. F. Malone and W. S. McDonald, *J. Chem. Soc. (A)*, (1970) 3124).

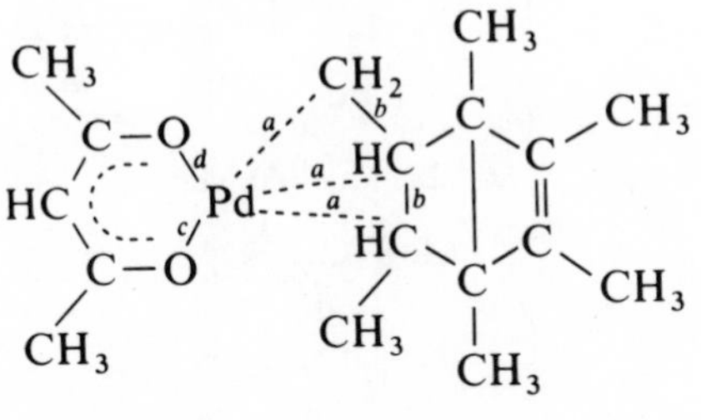

$a = 2{\cdot}12(2)$ Å
$b = 1{\cdot}433(20)$ Å
$c = 2{\cdot}050(10)$ Å
$d = 2{\cdot}077(9)$ Å

65. $C_{18}H_{12}N_2O_2Pd$ Bis(8-hydroxyquinolato)palladium(II)
   (C. K. Prout and A. G. Wheeler, *J. Chem. Soc.* (*A*), (1966) 1286).

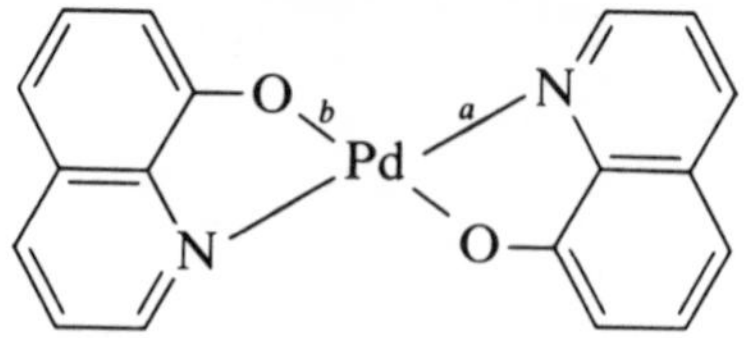

$a = 2.02(2)$ Å
$b = 2.02(2)$ Å
Pd is square-planar

66. $C_{18}H_{20}N_2O_2Pd$ Bis(N-ethylsalicylaldiminato)palladium(II)
   (E. Frasson, C. Panattoni and L. Sacconi, *Acta Cryst.*, **17** (1964) 85).

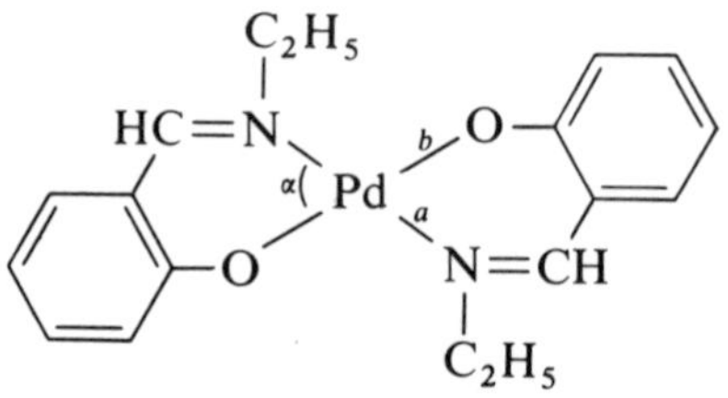

$a = 1.86$ Å
$b = 1.94$ Å
$\alpha = 91°$

67. $C_{18}H_{21}BrN_2OPd$ Trans-[PdBr(pyr)$_2$(5-methoxynortricycl-3-enyl)]
   (E. Forsellini, G. Bombieri, B. Crociani and T. Boschi, *Chem. Comm.*,
   (1970) 1203).

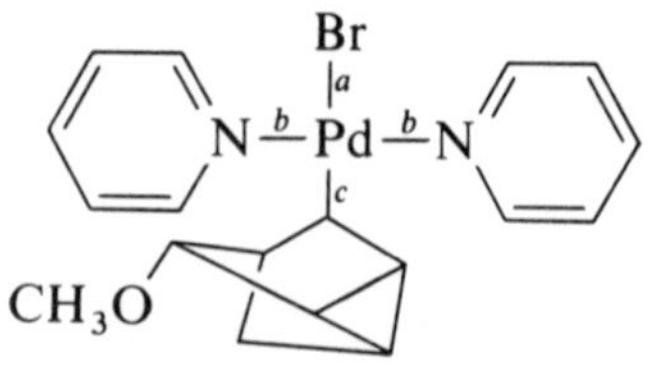

$a = 2.58$ Å
$b = 2.085$ Å (mean)
$c = 2.16$ Å

68. $C_{18}H_{26}As_2I_2PdS_2$ Trans-[Pd(dimethyl-o-methylthiophenylarsine)$_2$I$_2$]
   (J. P. Beale and N. C. Stephenson, *Acta Cryst.*, **B26** (1970) 1655).

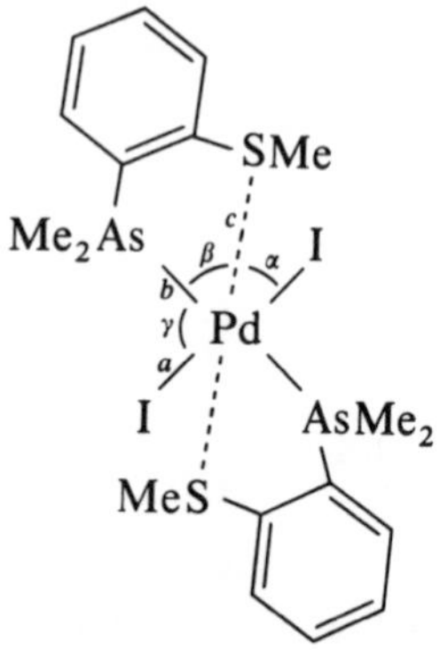

$a = 2.58$ Å $\alpha = 95°$
$b = 2.39$ Å $\beta = 56°$
$c = 3.84$ Å $\gamma = 92°$

69. $C_{19}H_{18}OPdS$ $[(\pi\text{-}C_4H_7)Pd(C_6H_5\text{-}\overset{\overset{O}{\|}}{C}\text{-}CH\text{-}\overset{\overset{S}{\|}}{C}\text{-}C_6H_5)]$
(S. J. Lippard and S. M. Morehouse, *J. Amer. Chem. Soc.*, **91** (1969) 2504).

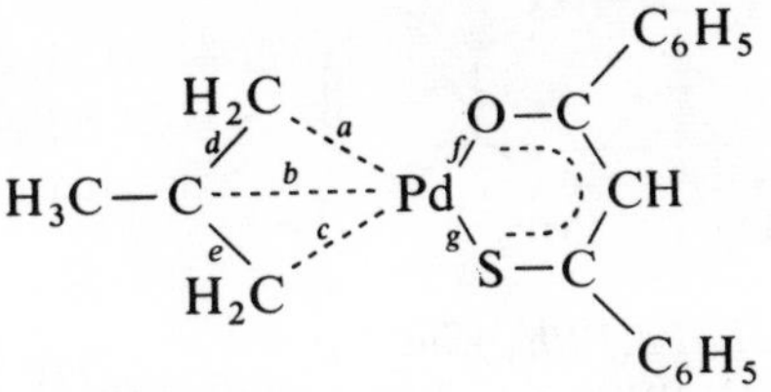

$a = 2\cdot21$ Å	$e = 1\cdot42$ Å
$b = 2\cdot15$ Å	$f = 2\cdot07$ Å
$c = 2\cdot08$ Å	$g = 2\cdot29$ Å
$d = 1\cdot40$ Å	

70. $C_{19}H_{22}N_3PPdS_2$ $[\{Ph_2PCH_2CH_2CH_2N(CH_3)_2\}Pd(NCS)(SCN)]$
(G. R. Clark, G. J. Palenik and D. W. Meek, *J. Amer. Chem. Soc.*, **92** (1970) 1077).

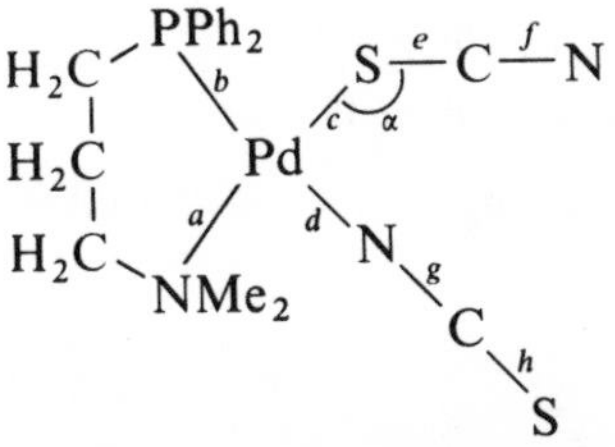

$a = 2\cdot148$ Å	$e = 1\cdot658$ Å
$b = 2\cdot243$ Å	$f = 1\cdot146$ Å
$c = 2\cdot295$ Å	$g = 1\cdot136$ Å
$d = 2\cdot063$ Å	$h = 1\cdot611$ Å
	$\alpha = 107\cdot3°$

71. $C_{20}H_{16}N_6Pd$ *Bis(2,2'-dipyridyliminato)palladium(II)*
(H. C. Freeman, J. F. Geldard, F. Lions and M. R. Snow, *Proc. Chem. Soc.*, (1964) 258; H. C. Freeman and M. R. Snow, *Acta Cryst.*, **18** (1965) 843).

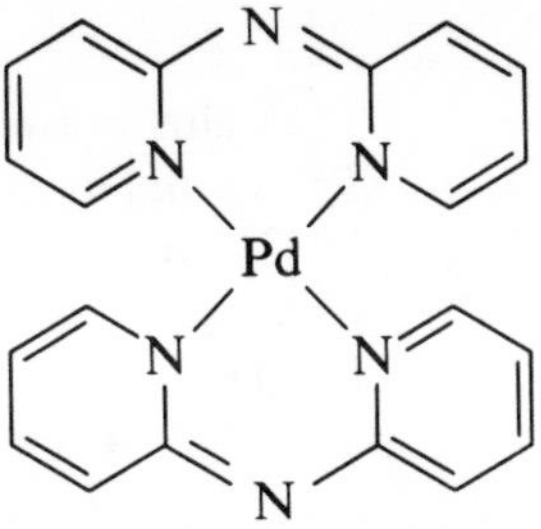

Pd–N = $2\cdot02(1)$ Å (mean)

72. $C_{20}H_{18}O_4Pd$ *(1-Phenyl-1,3-butanedionato)palladium(II)*
(P.-K. Hon, C. E. Pfluger and R. L. Belford, *Inorg. Chem.*, **6** (1967) 730).

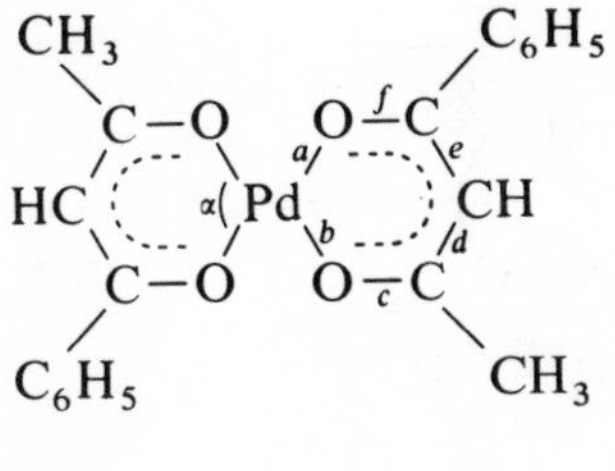

$a = 1\cdot976(14)$ Å	$d = 1\cdot470(30)$ Å
$b = 1\cdot965(15)$ Å	$e = 1\cdot375(28)$ Å
$c = 1\cdot233(28)$ Å	$f = 1\cdot307(23)$ Å
	$\alpha = 93\cdot3°$

73. $C_{20}H_{32}As_4I_2Pd$ $[Pd\{o\text{-}C_6H_4(AsMe_2)_2\}_2]I_2$
(C. M. Harris, R. S. Nyholm and N. C. Stephenson, *Nature*, **177** (1956) 1127).

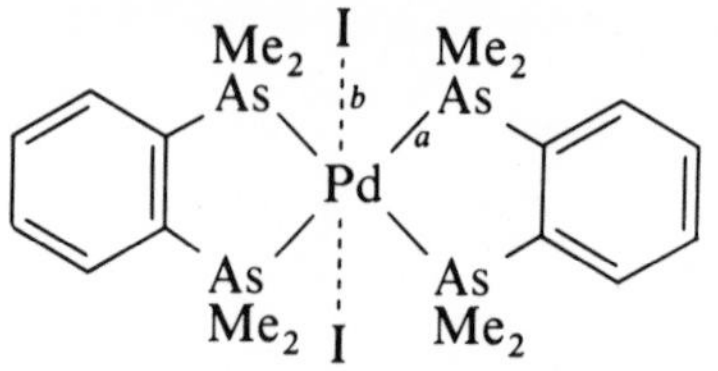

$$a = 2\cdot38 \text{ Å}$$
$$b = 3\cdot52 \text{ Å}$$

74. $C_{21}H_{20}Cl_3PPdSn$ $[(\pi\text{-}C_3H_5)Pd(PPh_3)(SnCl_3)]$
(R. Mason, G. B. Robertson, P. O. Whimp and D. A. White, *Chem. Comm.*, (1968) 1655; R. Mason and P. O. Whimp, *J. Chem. Soc.* (*A*), (1969) 2709).

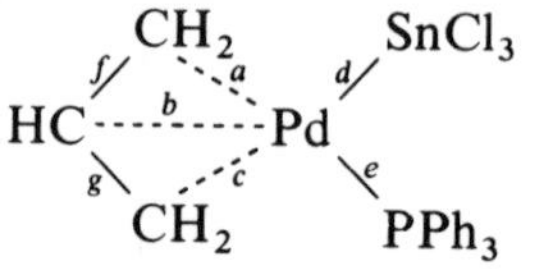

$a = 2\cdot193(12)$ Å	$e = 2\cdot317(3)$ Å
$b = 2\cdot116(21)$ Å	$f = 1\cdot347(24)$ Å
$c = 2\cdot201(16)$ Å	$g = 1\cdot379(26)$ Å
$d = 2\cdot563(1)$ Å	

75. $C_{22}H_{22}ClPPd$ $[(\pi\text{-}C_4H_7)PdCl(PPh_3)]$
(R. Mason and D. R. Russell, *Chem. Comm.*, (1966) 26).

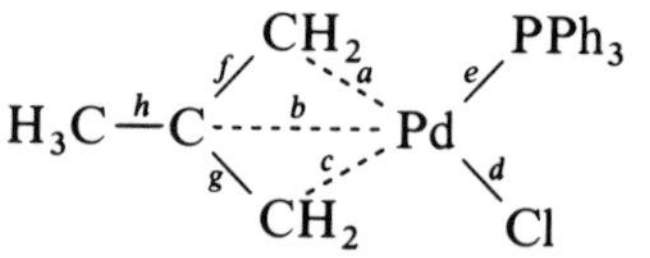

$a = 2\cdot14$ Å	$e = 2\cdot31$ Å
$b = 2\cdot22$ Å	$f = 1\cdot47$ Å
$c = 2\cdot28$ Å	$g = 1\cdot40$ Å
$d = 2\cdot38$ Å	$h = 1\cdot56$ Å

76. $C_{22}H_{28}N_2O_2Pd$ Bis-(N-*isopropyl*-3-*methylsalicylaldiminato*)-*palladium(II)*
(P. C. Jain and E. C. Lingafelter, *Acta Cryst.*, **23** (1967) 127).

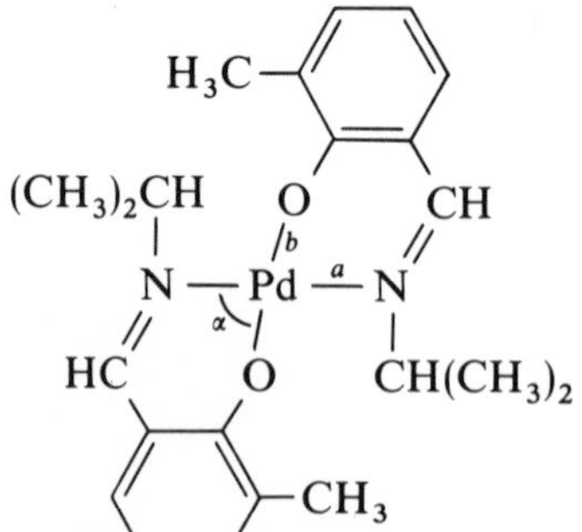

	Occurs in two crystalline forms	
	Tetragonal form	Monoclinic form
a	$2\cdot032(3)$ Å	$2\cdot019(8)$ Å
b	$1\cdot987(3)$ Å	$1\cdot988(4)$ Å
α	$91\cdot1(1)°$	$91\cdot3(3)°$

77. $C_{22}H_{28}N_2O_2Pd$ Bis(N-*h-butylsalicylaldiminato*)*palladium(II)*
(E. Frasson, C. Panattoni and L. Sacconi, *Acta Cryst.*, **17** (1964) 477).

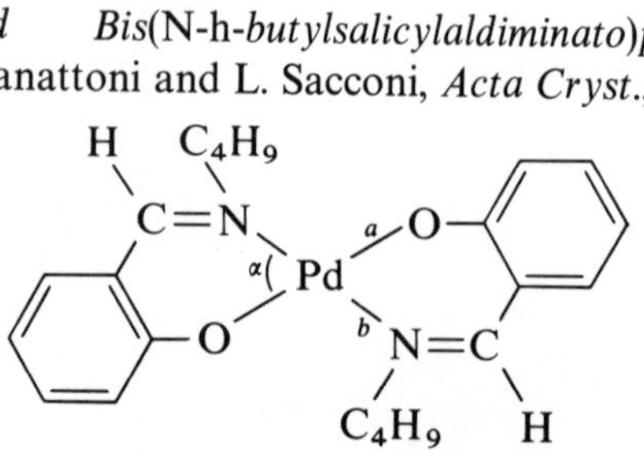

$$a = 2\cdot00 \text{ Å}$$
$$b = 2\cdot01 \text{ Å}$$
$$\alpha = 104°$$

78. $C_{22}H_{28}N_2O_2Pd$ *Bis(N-t-butylsalicylaldiminato)palladium(II)*

  (V. W. Day, M. D. Glick and J. L. Hoard, *J. Amer. Chem. Soc.*, **90** (1968) 4803).

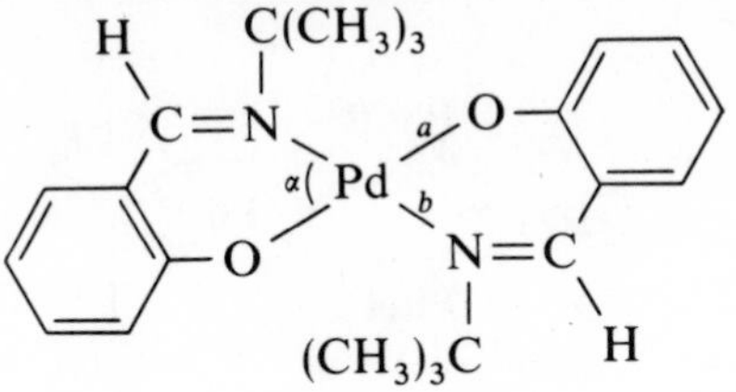

$a = 1.982(7)$ Å
$b = 2.059(9)$ Å
$\alpha = 88.3°$

Molecule is stepped

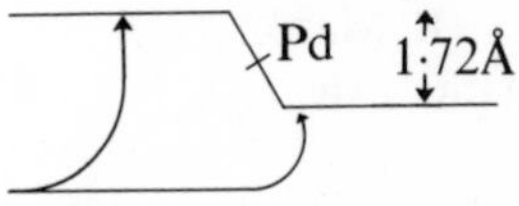

Salicylaldimine planes

Pd 1.72Å

79. $C_{24}H_{30}As_3ClPd^+$ $[Pd(TPAs)Cl]^+ClO_4^-$

  (T. L. Blundell and H. M. Powell, *J. Chem. Soc. (A)*, (1967) 1650).

TPAs =

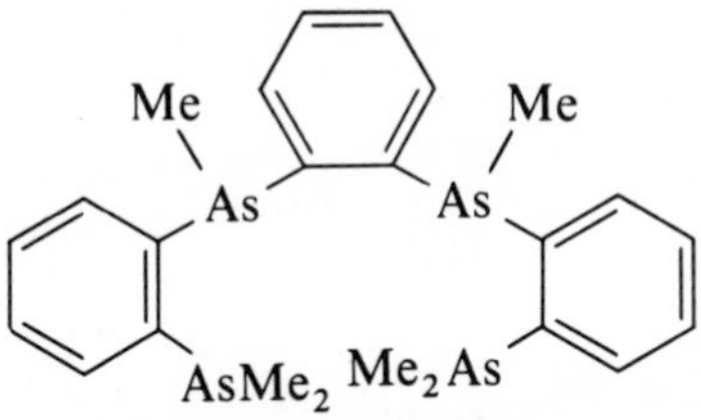

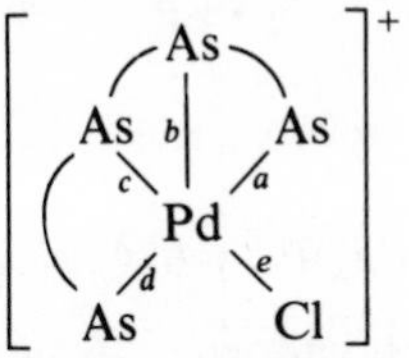

$a = 2.408(7)$ Å
$b = 2.860(7)$ Å
$c = 2.331(7)$ Å
$d = 2.375(7)$ Å
$e = 2.33(4)$ Å

Stereochemistry around Pd is square pyramidal.

80. $C_{24}H_{32}N_2O_2Pd$ *Bis-(N-isopropyl-3-ethylsalicylaldiminato)-*
                              palladium(II)

  (R. L. Braun and E. C. Lingafelter, *Acta Cryst.*, **22** (1967) 787).

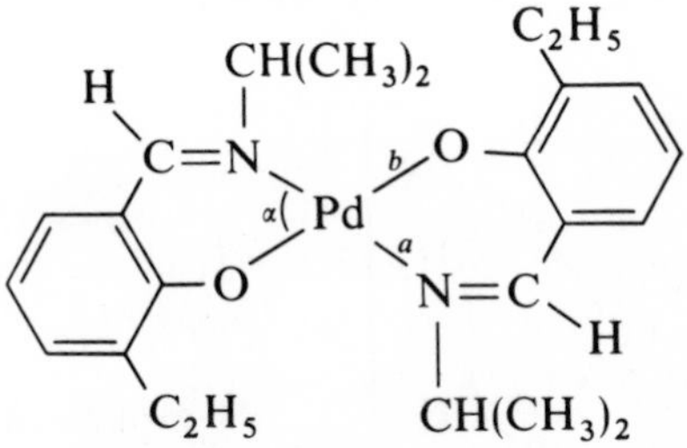

$a = 2.031$ Å
$b = 1.991$ Å
$\alpha = 91.3°$

Molecule is stepped

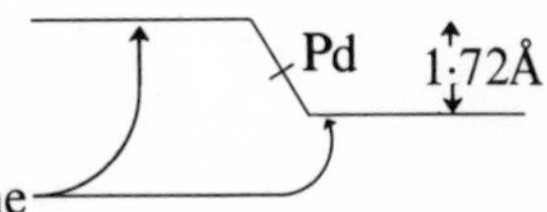

Salicylaldimine plane

Pd 1.72Å

81. $C_{26}H_{25}Cl_2NP_2Pd$ *Dichloro-bis(diphenylphosphino)ethylamine-palladium(II)*

(D. S. Payne, J. A. A. Mokuolo and J. C. Speakman, *Chem. Comm.*, (1965) 599).

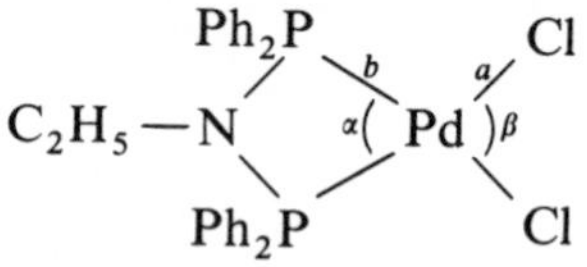

$$a = 2\cdot367(3) \text{ Å}$$
$$b = 2\cdot224(3) \text{ Å}$$
$$\alpha = 71\cdot43(15)°$$
$$\beta = 94\cdot77(17)°$$

82. $C_{28}H_{24}N_2P_2PdS_2$ $[(Ph_2PCH_2CH_2PPh_2)Pd(NCS)(SCN)]$

(G. Beran and G. J. Palenik, *Chem. Comm.*, (1970) 1354).

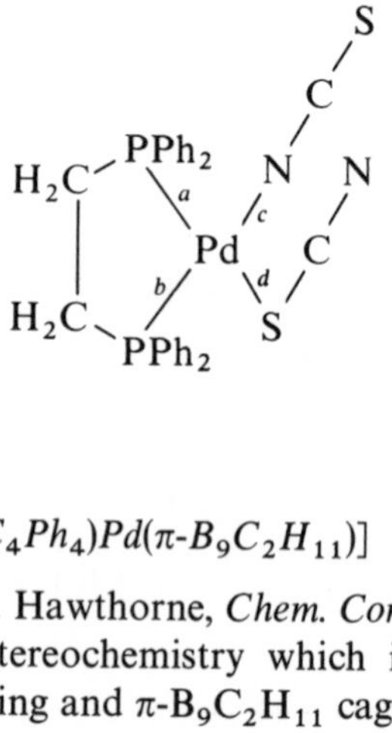

$$a = 2\cdot260 \text{ Å}$$
$$b = 2\cdot245 \text{ Å}$$
$$c = 2\cdot000 \text{ Å}$$
$$d = 2\cdot364 \text{ Å}$$

83. $C_{30}H_{31}B_9Pd$ $[(\pi\text{-}C_4Ph_4)Pd(\pi\text{-}B_9C_2H_{11})]$

(P. A. Wegner and M. F. Hawthorne, *Chem. Comm.*, (1966) 861).
Only reports general stereochemistry which involves Pd atom sandwiched between $\pi\text{-}C_4Ph_4$ ring and $\pi\text{-}B_9C_2H_{11}$ cage (see Fig. 14, p. 68).

84. $C_{32}H_{40}P_4Pd^{2+}$ $[Pd(C_6H_5CH_3PCH_2CH_2PCH_3C_6H_5)_2]Cl_2$

(P. Groth, *Acta Chem. Scand.*, **24** (1970) 2785).

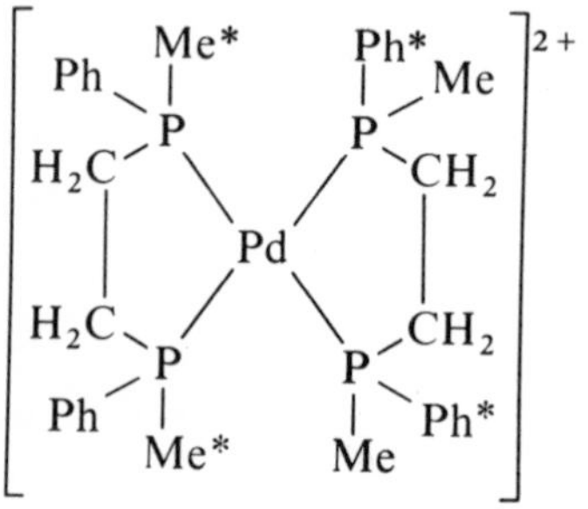

* groups above square-plane, i.e. molecule has a centre of symmetry.
$$Pd–P = 2\cdot333 \text{ Å}$$
$$Pd–Cl = 4\cdot090 \text{ Å}$$

85. $C_{38}H_{46}N_4O_8Pd$ *Bis(4,4'-dicarboxyethyl-3,3',5,5'-tetramethyl-dipyrromethene)palladium(II)*

(F. C. March, D. A. Couch, K. Emerson, J. E. Fergusson and W. T. Robinson, *J. Chem. Soc.* (*A*), (1971) 440).

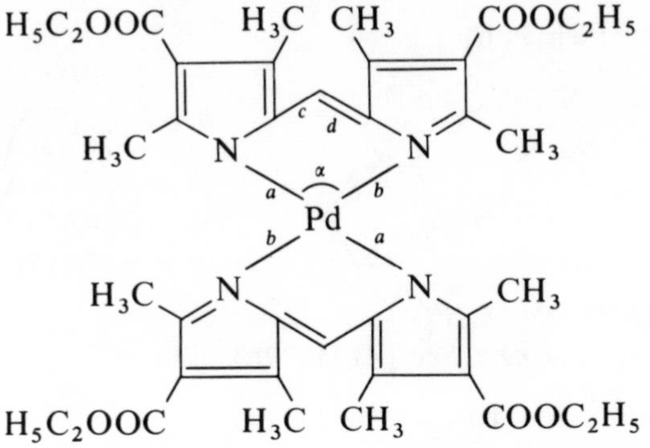

$a = 2.030(10)$ Å
$b = 2.026(09)$ Å
$c = 1.38(2)$ Å
$d = 1.40(2)$ Å
$\alpha = 84.2(5)°$

PdN$_4$ is planar but rest of ligand is bent 44° out of this plane

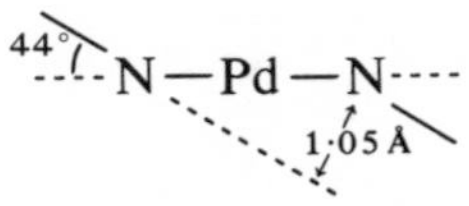

86. $C_{39}H_{28}Cl_2Fe_2O_8P_2Pd_2$ *Di-μ-chloro-bis(μ-diphenylphosphidotetra-carbonylironpalladium)*

(B. T. Kilbourn and R. H. B. Mais, *Chem. Comm.*, (1968) 1507).

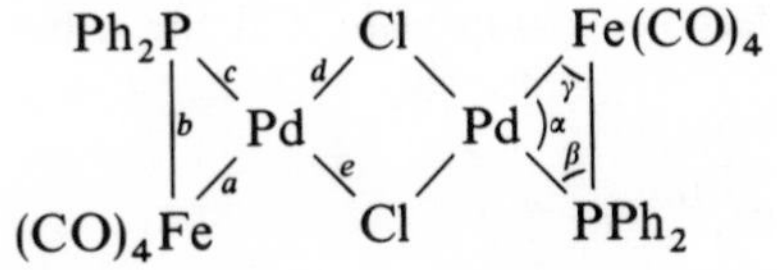

$a = 2.59(1)$ Å
$b = 2.24(2)$ Å $\alpha = 55.5(4)°$
$c = 2.15(1)$ Å $\beta = 52.2(4)°$
$d = 2.42(1)$ Å $\gamma = 72.3(4)°$
$e = 2.45(1)$ Å

87. $C_{42}H_{39}Br_2P_3Pd$

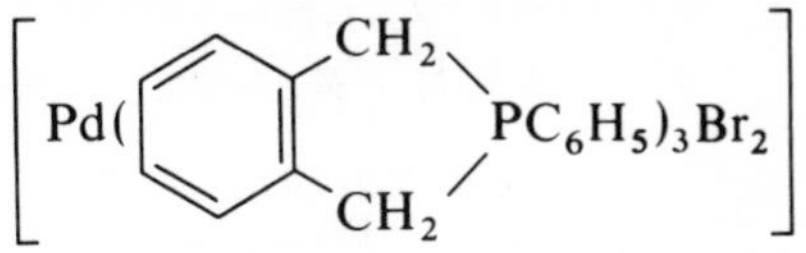

(J. W. Collier, F. G. Mann, D. G. Watson and H. R. Watson, *J. Chem. Soc.*, (1964) 1803).

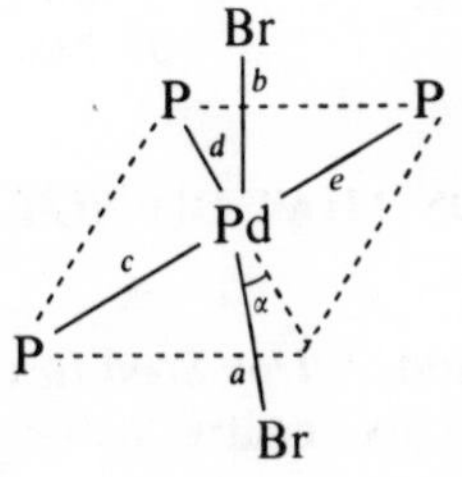

$a = 2.52$ Å $d = 2.19$ Å
$b = 2.93$ Å $e = 2.28$ Å
$c = 2.30$ Å $\alpha = 15°$

88. $C_{48}H_{39}ClN_2P_2Pd$ Trans-$[(PPh_3)_2PdCl(-C_6H_4N{=}NC_6H_5)]$
   (R. W. Siekman and D. L. Weaver, *Chem. Comm.*, (1968) 1021; D. L. Weaver, *Inorg. Chem.*, **9** (1970) 2250).

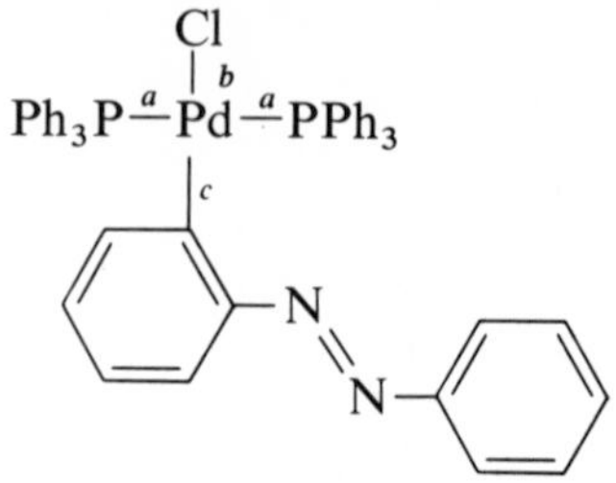

$a = 2{\cdot}306(5)$ Å
$b = 2{\cdot}382(5)$ Å
$c = 1{\cdot}994(15)$ Å

89. $C_{60}H_{30}F_{20}P_2Pd_2S_4$ $[(PPh_3)Pd(SC_6F_5)_2]_2$
   (R. H. Fenn and G. R. Segrott, *J. Chem. Soc. (A)*, (1970) 3197).

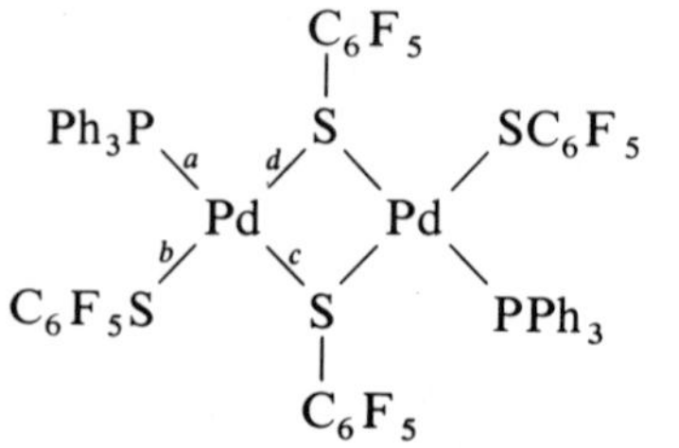

$a = 2{\cdot}294(11)$ Å
$b = 2{\cdot}311(11)$ Å
$c = 2{\cdot}378(11)$ Å
$d = 2{\cdot}363(11)$ Å

Stereochemistry about Pd is square-planar, but the two $PdPS_3$ planes are perpendicular.

90. $C_{60}H_{50}Cl_2O_2Pd_2$ $[(\pi\text{-}C_4Ph_4OEt)PdCl]_2$
   (L. F. Dahl and W. E. Oberhansli, *Inorg. Chem.*, **4** (1965) 629).
   Reports two structures, one with ethoxy group *endo* and the other *exo* with respect to palladium.

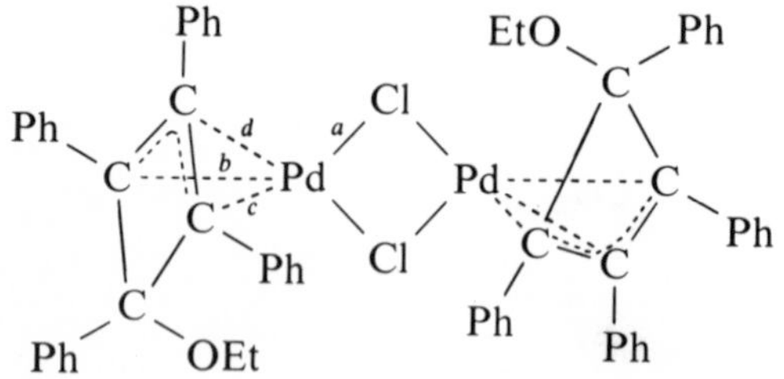

	Endo-isomer	Exo-isomer
a	2·43 Å	2·42 Å
b	2·16 Å	2·18 Å
c	2·15 Å	2·13 Å
d	2·09 Å	2·14 Å

C. PLATINUM COMPOUNDS THAT DO NOT CONTAIN CARBON

91. *Pt*   *Platinum Metal*
   (H. J. Goldschmidt and T. Land, *J. Iron Steel Inst.*, **155** (1947) 221).
   Co-ordination number 12 (face centre cubic). Pt–Pt = 2·7746 Å at 20°C.

92. $PtClH_{15}N_5^{3+}$ $[Pt(NH_3)_5Cl]Cl_3.H_2O$
(G. B. Bokii and L. A. Popova, *Dokl. Akad. Nauk. SSSR*, **64** (1949) 73;
Chem. Abs., **43** (1949) 7287b).

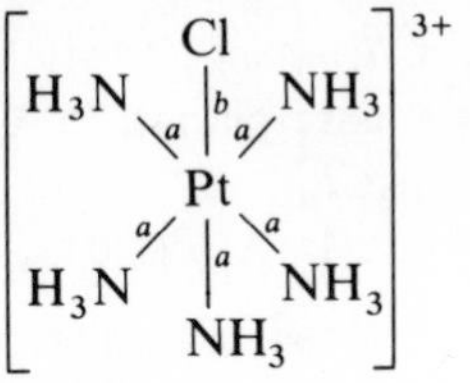

$$
\left[
\begin{array}{c}
\mathrm{Cl} \\
\mathrm{H_3N} \underset{a}{\diagdown} \overset{\mid}{\underset{}{b}} \underset{a}{\diagup} \mathrm{NH_3} \\
\mathrm{Pt} \\
\underset{a}{\diagup} \quad \underset{a}{\diagdown} \\
\mathrm{H_3N} \quad \overset{\mid}{\underset{}{a}} \quad \mathrm{NH_3} \\
\mathrm{NH_3}
\end{array}
\right]^{3+}
$$

$a = 2\cdot00\,\text{Å}$
$b = 2\cdot30\,\text{Å}$

93. $PtCl_2H_6N_2$ Cis-$[Pt(NH_3)_2Cl_2]$
(G. H. W. Milburn and M. R. Truter, *J. Chem. Soc. (A)*, (1966) 1609).

$$
\begin{array}{ccc}
\mathrm{H_3N} & & \mathrm{Cl} \\
\diagdown_{b} & \diagup_{a} & \\
& \mathrm{Pt} & \\
\diagup & \diagdown & \\
\mathrm{H_3N} & & \mathrm{Cl}
\end{array}
$$

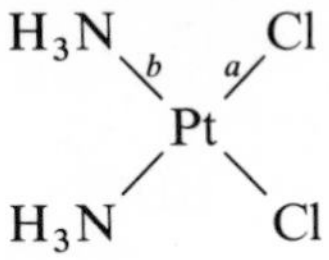

$a = 2\cdot33(1)\,\text{Å}$
$b = 2\cdot01(4)\,\text{Å}$

94. $PtCl_2H_6N_2$ Trans-$[Pt(NH_3)_2Cl_2]$
(M. A. Porai-Koshits, *Trudy Inst. Krist. Acad. Nauk. SSSR*, **9** (1954) 229;
G. H. W. Milburn and M. R. Truter, *J. Chem. Soc. (A)*, (1966) 1609).

$$
\begin{array}{ccc}
\mathrm{H_3N} & & \mathrm{Cl} \\
\diagdown & \diagup_{a} & \\
& \mathrm{Pt} & \\
\diagup & \diagdown_{b} & \\
\mathrm{Cl} & & \mathrm{NH_3}
\end{array}
$$

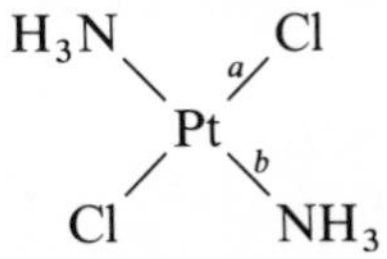

$a = 2\cdot32(1)\,\text{Å}$
$b = 2\cdot05(4)\,\text{Å}$

95. $PtCl_2H_{12}N_4^{2+}$ $[Pt(NH_3)_4Cl_2]Cl_2$
(G. B. Bokii and M. A. Porai-Koshits, *Dokl. Akad. Nauk. SSSR*, **64** (1949)
337; *Chem. Abs.*, **43** (1949) 4535i).

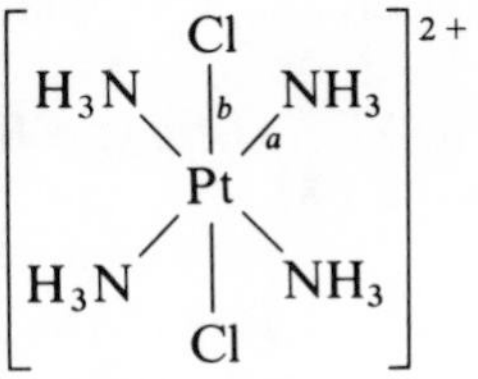

$$
\left[
\begin{array}{c}
\mathrm{Cl} \\
\mathrm{H_3N} \underset{}{\diagdown} \overset{\mid}{\underset{b}{}} \underset{a}{\diagup} \mathrm{NH_3} \\
\mathrm{Pt} \\
\diagup \quad \mid \quad \diagdown \\
\mathrm{H_3N} \quad \mid \quad \mathrm{NH_3} \\
\mathrm{Cl}
\end{array}
\right]^{2+}
$$

$a = 2\cdot03\,\text{Å}$
$b = 2\cdot25\,\text{Å}$

96. $PtCl_3H_3N^-$ $K[Pt(NH_3)Cl_3].H_2O$
(Y. P. Jeannin and D. R. Russell, *Inorg. Chem.*, **9** (1970) 778).

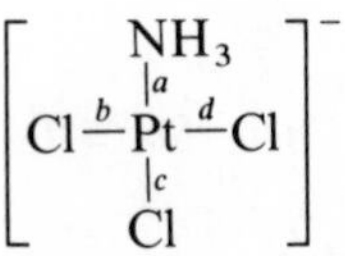

$$
\left[
\begin{array}{c}
\mathrm{NH_3} \\
\mid a \\
\mathrm{Cl} \overset{b}{-} \mathrm{Pt} \overset{d}{-} \mathrm{Cl} \\
\mid c \\
\mathrm{Cl}
\end{array}
\right]^-
$$

$a = 2\cdot06(2)\,\text{Å}$
$b = 2\cdot31(1)\,\text{Å}$
$c = 2\cdot321(7)\,\text{Å}$ } Equal within experimental
$d = 2\cdot32(3)\,\text{Å}$ } error

97. $PtCl_4$ $PtCl_4$
(M. T. Falqui, *Ann. Chim. (Rome)*, **48** (1958) 1160).
Assumed tetrahedral, Pt–Cl = 2·26 Å. This is most unlikely to be correct.

98. $PtCl_4^{2-}$ $[Pd(NH_3)_4][PtCl_4]$
(J. R. Miller, *Proc. Chem. Soc.*, (1960) 318).
Pd–Pt = 3·25 Å.

99. $PtCl_4^{2-}$ $[Cu(NH_3)_4][PtCl_4]$

(M. Bukovska and M. A. Porai–Koshits, *Kristallografia*, **5** (1960) 137).
Cu–Pt = 3·23 and 3·22 Å.

100. $PtCl_4^{2-}$ K_2PtCl_4

(R. G. Dickinson, *J. Amer. Chem. Soc.*, **44** (1922) 2404).
Co-ordination about Pt is square-planar. Pt–Cl = 2·33 Å.

101. $PtCl_4H_6N_2$ $[Pt(NH_3)_2Cl_4]$

(N. V. Belov, G. B. Bokii and L. A. Popova, *Izvest. Akad. Nauk. SSSR*, **8** (1947) 249).

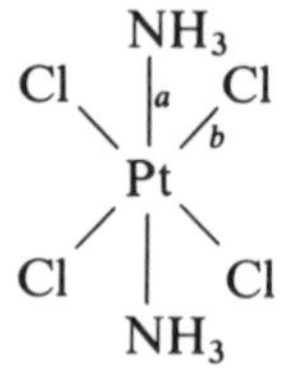

$a = 2·35$ Å
$b = 2·31$ Å

102. $PtCl_6^{2-}$ M_2PtCl_6 ($M = Cs, K, NH_4, Rb, Tl$)

(G. Engel, *Z. Krist.*, **90** (1935) 341).

Pt is octahedrally surrounded by six Cl. Pt–Cl = 2·29 Å (Tl), 2·32 Å (Rb), 2·33 Å (K), 2·34 Å (Cs), 2·36 Å (NH₄).

103. $PtCl_{15}Sn_5^{3-}$ $(PMe_3Ph^+)_3[Pt(SnCl_3)_5]^{3-}$

(R. D. Cramer, R. V. Lindsey, C. T. Prewitt and U. G. Stolberg, *J. Amer. Chem. Soc.*, **87** (1965) 658).

Pt is surrounded by 5 $SnCl_3^-$ ligands in a trigonal bipyramidal arrangement. Mean Pt–Sn = 2·54 Å.

104. PtF_6 PtF_6

(B. Weinstock, H. H. Claasen and J. G. Malm, *J. Amer. Chem. Soc.*, **79** (1957) 5832).

PtF_6 is isostructural with WF_6, OsF_6 and IrF_6. By extrapolation from this series Pt–F estimated to be 1·83 Å.

105. PtF_6^- $O_2^+[PtF_6]^-$

(N. Bartlett and D. H. Lohmann, *Proc. Chem. Soc.*, (1962) 115; N. Bartlett and D. H. Lohmann, *J. Chem. Soc.*, (1962) 5253; J. A. Ibers and W. C. Hamilton, *J. Chem. Phys.*, **44** (1965) 1748 (neutron diffraction)).
Pt–F = 1·74 Å
O–O = 1·13 Å

106. PtF_6^- $(XeF_5)^+[PtF_6]^-$

(N. Bartlett, F. Einstein, D. F. Stewart and J. Trotter, *J. Chem. Soc. (A)*, (1967) 1190).

PtF_6^- is octahedral. Two fluorides bridge between Pt and Xe and have Pt–F distance of 1·94 Å (mean), the remainder have Pt–F distance of 1·85 Å. (See Fig. 4, p. 23).

107. PtF_6^{2-} K_2PtF_6

(D. P. Mellor and N. C. Stephenson, *Austral. J. Sci. Res.*, **A4** (1951) 406).
Pt is octahedral. Pt–F = 1·91 Å.

108. $PtH_2N_4S_4$ *Bis-(tetrathionitrosyl)platinum*
   (I. Lindqvist and J. Weiss, *J. Inorg. Nucl. Chem.*, **6** (1958) 184; I. Lindqvist and R. Rosenstein, *J. Inorg. Nucl. Chem.*, **7** (1958) 421).

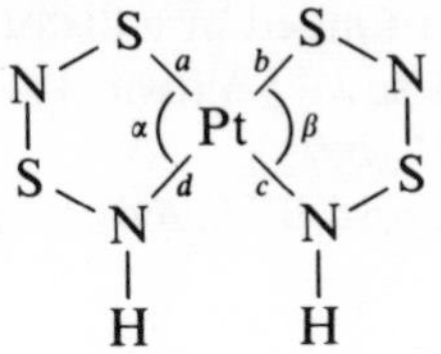

$a = 2\cdot23$ Å
$b = 2\cdot25$ Å $\alpha = 89\cdot9°$
$c = 2\cdot15$ Å $\beta = 86\cdot9°$
$d = 2\cdot03$ Å

109. $PtH_6O_6^{2-}$ $K_2Pt(OH)_6$
   (C. O. Björling, *Arkiv. Kemi Min. Geol.*, **15** (1941) no. 2).
   Pt is octahedral. Pt–O = 1·96 Å.
110. $PtN_4O_8^{2-}$ $K_2Pt(NO_2)_4$
   (H. Lambot, *Bull. Soc. Roy. Sci. Liège*, **12** (1943) 463, 541).
   Pt is square-planar. Pt–N = 2·02 Å.
111. *PtO*   *PtO*
   (W. J. Moore and L. Pauling, *J. Amer. Chem. Soc.*, **63** (1941) 1392).

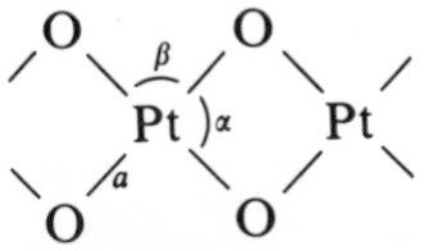

$a = 2\cdot02(2)$ Å
$\alpha = 82°$
$\beta = 98°$

Co-ordination about Pt is approximately planar. Co-ordination about O is approximately tetrahedral.
112. PtO_2 PtO_2
   (N. A. Shishakov, *Kristallografiya*, **2** (1957) 689; *Chem. Abs.*, **53** (1959) 7712).
   Pt–O = 1·9 Å.
113. PtP_2 PtP_2
   (E. Dahl, *Acta Chem. Scand.*, **23** (1969) 2677).
   Pt is surrounded by 6P at corners of a slightly distorted octahedron. Pt–P = 2·391 Å (mean).
114. PtS_{15}^{2-} $(NH_4)_2PtS_{15}.2H_2O$
   (P. E. Jones and L. Katz, *Chem. Comm.*, (1967) 842; P. E. Jones and L. Katz, *Acta Cryst.*, **B25** (1969) 745).

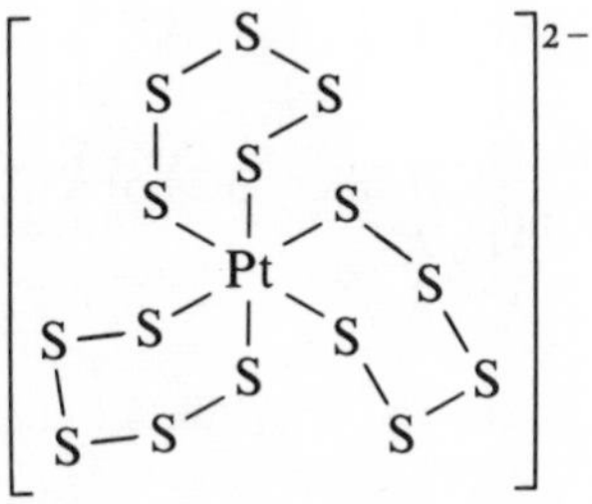

Co-ordination around Pt is only approximately octahedral. Pt–S in the range 2·365(7)–2·428(8) Å. Mean Pt–S = 2·39 Å.

115. Pt_2Br_6 $[PtBr_4][PtBr_2]$

(G. Thiele and P. Woditsch, *Angew. Chem. Int. Ed.*, **8** (1969) 672).

$PtBr_2$ units are hexameric, i.e. Pt_6Br_{12} with the 6Pt atoms forming a distorted octahedron. Pt–Br = 2·447(18) Å.

$PtBr_4$ units involve Pt surrounded by a distorted octahedron of Br atoms. Pt–Br = 2·57 Å, 2·52 Å, 2·48 Å, 2·47 Å (twice) and 2·44 Å.

116. $Pt_2Br_6^{2-}$ $(NEt_4^+)_2[Pt_2Br_6^{2-}]$

(N. C. Stephenson, *Acta Cryst.*, **17** (1964) 587).

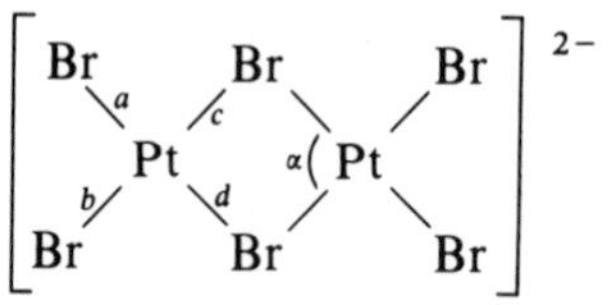

$a = 2·446(6)$ Å
$b = 2·421(5)$ Å
$c = 2·413(7)$ Å
$d = 2·451(7)$ Å
$\alpha = 86·19°$

117. $Pt_2Br_6H_{12}N_4$ $[Pt^{II}(NH_3)_2Br_2][Pt^{IV}(NH_3)_2Br_4]$

(C. Brosset, *Arkiv. Kemi. Min. Geol.*, **A25** (1948) no. 19; D. Hall and P. P. Williams, *Acta Cryst.*, **11** (1958) 624; J. Wallen, C. Brosset and N.-G. Vannerberg, *Arkiv. Kemi.*, **18** (1962) 541).

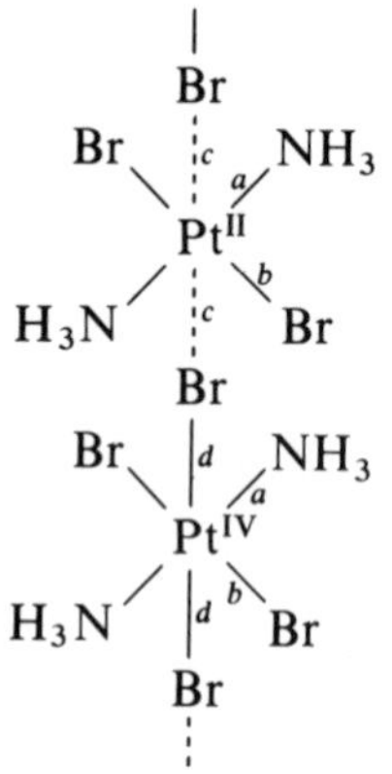

$a = 2·16$ Å
$b = 2·63$ Å
$c = 3·02$ Å
$d = 2·44$ Å

118. $Pt_2Cl_4H_{12}N_4$ $[Pt(NH_3)_4^{2+}][PtCl_4^{2-}]$

(S. Yamada, *J. Amer. Chem. Soc.*, **73** (1951) 1579; M. Atoji, J. W. Richardson and R. E. Rundle, *J. Amer. Chem. Soc.*, **79** (1957) 3017; J. R. Miller, *Proc. Chem. Soc.*, (1960) 318).

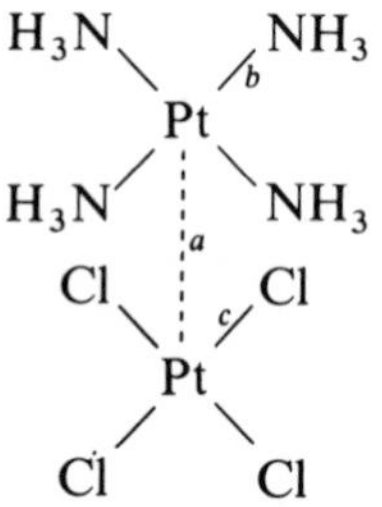

$a = 3·23$ (Miller), 3·24(5) Å (Atoji *et al.*)
$b = 2·06$ Å
$c = 2·34$ Å

119. Pt_2Cl_6 $[PtCl_4][PtCl_2]$
(U. Wiese, H. Schäfer, H. G. v. Schnering, C. Brendel and K. Rinke, *Angew. Chem. Int. Ed.*, **9** (1970) 158).
Same structure as Pt_2Br_6 (no. 115 above).
Pt^{II}–Cl = 2·32 Å.
Pt^{IV}–Cl = 2·37 Å (twice) and 2·29 Å (four times).

D. PLATINUM COMPLEXES THAT CONTAIN CARBON

120. $C_2H_4Br_3Pt^-$ $K^+[Pt(C_2H_4)Br_3]^-$
(G. B. Bokii and G. A. Kukina, *Kristallografiya*, **2** (1957) 400; *Chem. Abs.*, **52** (1958) 3455h).

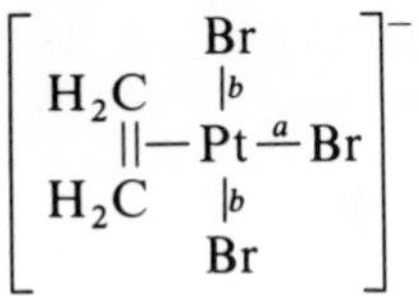

$$\begin{bmatrix} & Br \\ H_2C & |b \\ ||-Pt\,{}^a\,Br \\ H_2C & |b \\ & Br \end{bmatrix}^-$$

a = 2·42 Å
b = 2·50 Å
    Plane of C=C is perpendicular to PtBr$_3$ plane.

121. $C_2H_4Cl_3Pt^-$ $K^+[Pt(C_2H_4)Cl_3]^-.H_2O$
(J. A. Wunderlich and D. P. Mellor, *Acta Cryst.*, **7** (1954) 130; **8** (1955) 57; G. B. Bokii and G. A. Kukina, *Krystallografiya*, **2** (1957) 400; M. Black, R. H. B. Mais and P. G. Owston, *Acta Cryst.*, **B25** (1969) 1753; J. A. J. Jarvis, B. T. Kilbourn and P. G. Owston, *Acta Cryst.*, **B26** (1970) 876; J. A. J. Jarvis, B. T. Kilbourn and P. G. Owston, *Acta Cryst.*, **B27** (1971) 366).

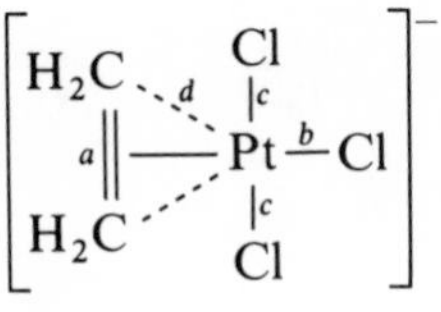

$$\begin{bmatrix} H_2C & Cl \\ & {}_d & |c \\ {}^a|| & Pt\,{}^b\,Cl \\ H_2C & |c \\ & Cl \end{bmatrix}^-$$

a = 1·37(3) Å c = 2·305(7) Å
                                  (mean)
b = 2·327(5) Å d = 2·127(19) Å
                                  (mean)
    Plane of C=C makes an angle of 84° with PtCl$_3$ plane.

122. $C_2H_6N_4PtS_2$ Cis-$[Pt(NH_3)_2(SCN)_2]$
(Ia. Ia. Bleidelis, *Kristallografiya*, **2** (1957) 278; *Chem. Abs.*, **51** (1957) 14364b).

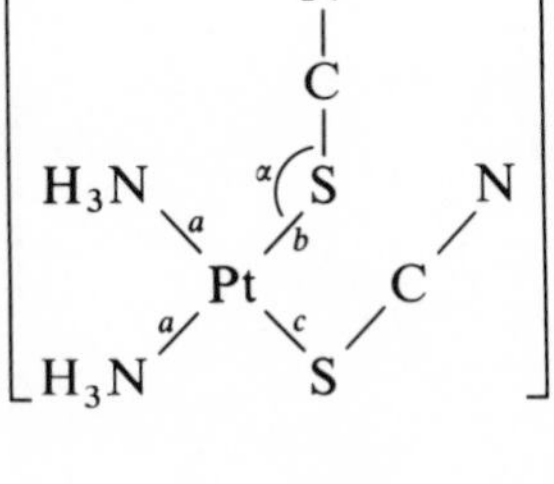

$$\begin{bmatrix} & & N \\ & & | \\ & & C \\ & & | \\ H_3N & {}^a\,S \\ & {}_a & {}_b & N \\ & Pt & C \\ H_3N & {}^c & S \end{bmatrix}$$

a = 2·0 Å
b = 2·29 Å
c = 2·31 Å
$\alpha \sim 110°$

123. $C_2H_6N_4PtS_2$ Trans-$[Pt(NH_3)_2(SCN)_2]$

(Ia. Ia. Bleidelis and G. B. Bokii, *Kristallografiya*, **2** (1957) 281; *Chem. Abs.*, **51** (1957) 14363i).

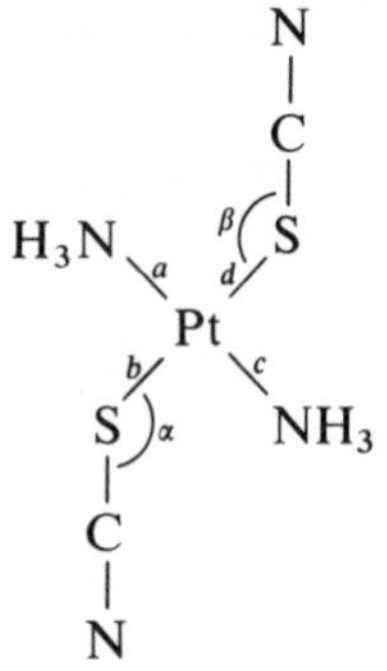

$$N$$
$$|$$
$$C$$
$$|$$

H$_3$N, S

Pt

S, NH$_3$

C

N

$a = 2 \cdot 13$ Å
$b = 2 \cdot 26$ Å $\alpha \sim 78°$
$c = 2 \cdot 10$ Å $\beta \sim 102°$
$d = 2 \cdot 27$ Å

124. $C_2H_8Br_3N_2Pt$ $[Pt(en)Br_3]$

(T. D. Ryan and R. E. Rundle, *J. Amer. Chem. Soc.*, **83** (1961) 2814).

Contains $[Pt(en)Br_2]$ and $[Pt(en)Br_4]$ units.

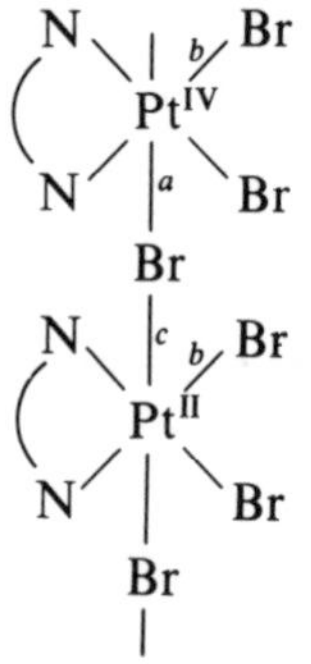

N, Br

Pt^{IV}

N, Br

Br

N, Br

Pt^{II}

N, Br

Br

$a = 2 \cdot 48$ Å
$b = 2 \cdot 51$ Å
$c = 3 \cdot 125$ Å

125. $C_2H_8Cl_2N_2Pt$ $[Pt(en)Cl_2]$

(D. S. Martin, R. A. Jacobson, L. D. Hunter and J. E. Benson, *Inorg. Chem.*, **9** (1970) 1276).

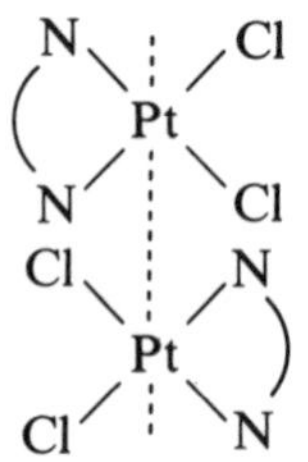

N, Cl

Pt

N, Cl

Cl, N

Pt

Cl, N

Pt–Pt $= 3 \cdot 39$ Å

126. $C_4O_8Pt^{2-}$ $K_2[Pt(oxalate)_2].2H_2O$

(R. Mattes and K. Krogmann, *Z. anorg. allgem. Chem.*, **332** (1964) 247).

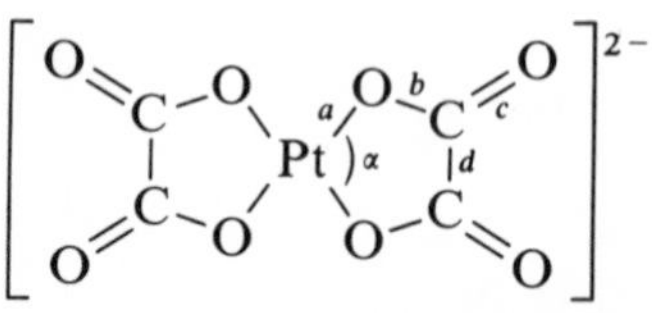

$$\left[\begin{array}{c} O=C-O \quad O-C=O \\ | \quad Pt \quad | \\ C-O \quad O-C \\ O \qquad\qquad O \end{array} \right]^{2-}$$

$a = 2 \cdot 00$ Å
$b = 1 \cdot 29$ Å
$c = 1 \cdot 22$ Å
$d = 1 \cdot 54$ Å
$\alpha = 83°$

127. $C_4H_6Cl_6Pt_2^{2-}$ $(NMe_3Et^+)_2[Cl_3PtCH_2=CH-CH=CH_2PtCl_3]^{2-}$
(V. C. Adam. J. A. J. Jarvis, B. T. Kilbourn and P. G. Owston, *Chem. Comm.*, (1971) 467).

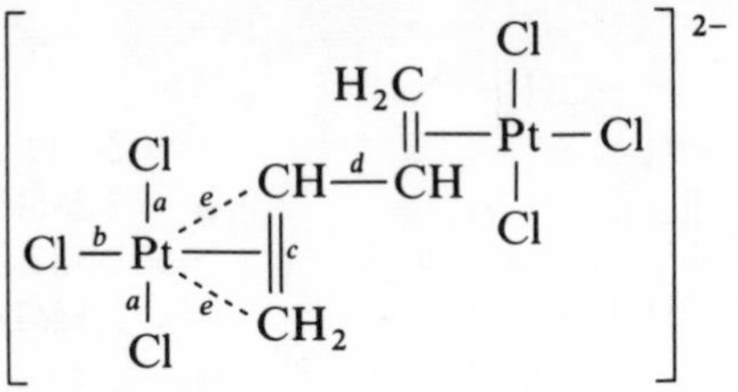

$a = 2\cdot290(7)$ Å
$b = 2\cdot320(7)$ Å
$c = 1\cdot514(28)$ Å
$d = 1\cdot360(28)$ Å
$e = 2\cdot188(25)$ Å

Plane of $-CH=CH_2$ bond is perpendicular to $PtCl_3$ plane.

128. $C_4H_6N_4O_4Pt$ *Bis(glyoximato)platinum(II)*
(G. Ferraris and D. Viterbo, *Acta Cryst.*, **B25** (1969) 2066).

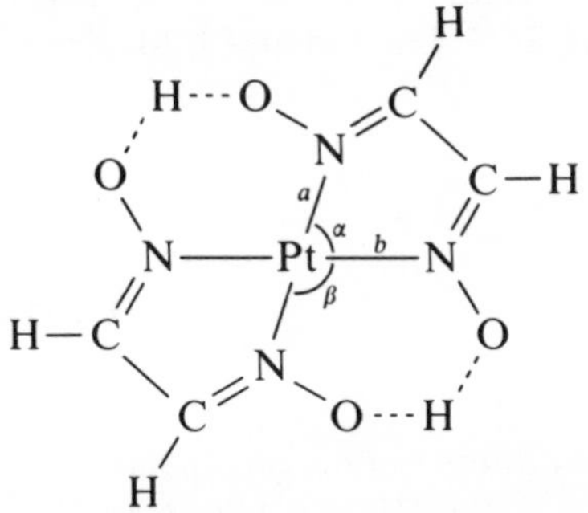

$a = 1\cdot968$ Å
$b = 2\cdot013$ Å
$\alpha = 78\cdot0°$
$\beta = 102\cdot0°$

129. $C_4H_8N_2O_4Pt$ *Trans-bis(glycinato)platinum(II)*
(H. C. Freeman and M. L. Golomb, *Acta Cryst.*, **B25** (1969) 1203).

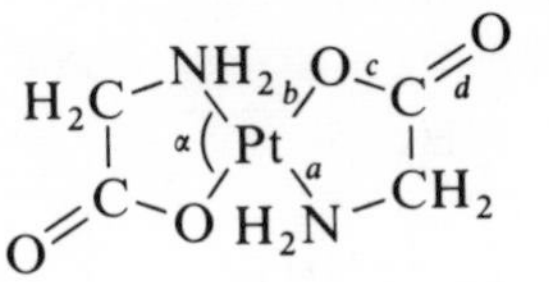

$a = 2\cdot037(4)$ Å
$b = 2\cdot002(4)$ Å
$c = 1\cdot290(6)$ Å
$d = 1\cdot231(7)$ Å
$\alpha = 82\cdot5°$

130. $C_4H_{10}Cl_3NPt$ $[Pt(cis-but-2-enylammonium)Cl_3]$
(R. Spagna, L. M. Venanzi and L. Zambonelli, *Inorg. Chim. Acta*, **4** (1970) 475).

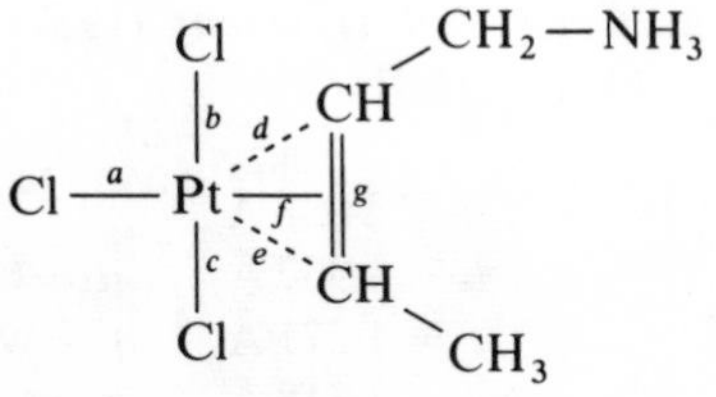

$a = 2\cdot320(6)$ Å $e = 2\cdot18(2)$ Å
$b = 2\cdot301(6)$ Å $f = 2\cdot05(2)$ Å
$c = 2\cdot293(6)$ Å $g = 1\cdot40(4)$ Å
$d = 2\cdot15(2)$ Å

Double bond lies at 85·5° to $PtCl_3$ plane. The four carbon atoms of the olefin are almost coplanar.

131. $C_4H_{10}Cl_3NPt$ [Pt(trans-but-2-enylammonium)Cl_3]
(R. Spagna, L. M. Venanzi and L. Zambonelli, *Inorg. Chim. Acta*, **4** (1970) 283).

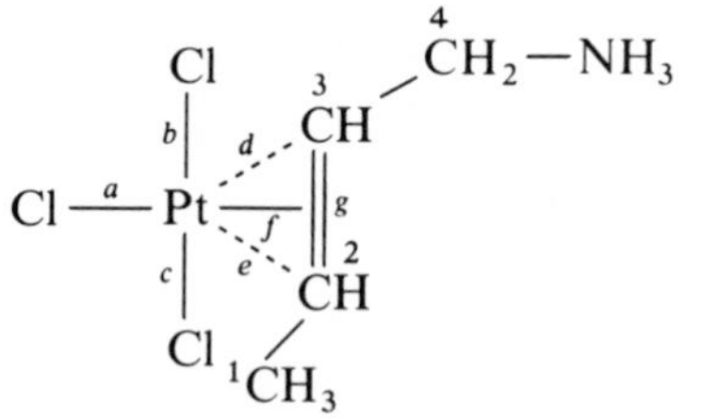

$a = 2\cdot339(7)$ Å	$e = 2\cdot16(3)$ Å
$b = 2\cdot288(7)$ Å	$f = 2\cdot02(3)$ Å
$c = 2\cdot307(6)$ Å	$g = 1\cdot40(4)$ Å
$d = 2\cdot12(3)$ Å	

Double-bond lies at 81·5° to $PtCl_3$ plane. C^1, C^2, C^3 and C^4 are not coplanar (cf. *cis*-complex). C^1 and C^4 are bent out of plane of double-bond away from Pt. The internal rotation angle around $-C\!=\!C-$ bond = 143.2°.

132. $C_4H_{11}Cl_2NPt$ Trans-[$Pt(C_2H_4)(NHMe_2)Cl_2$]
(P. R. H. Alderman, P. G. Owston and J. M. Rowe, *Acta Cryst.*, **13** (1960) 149).

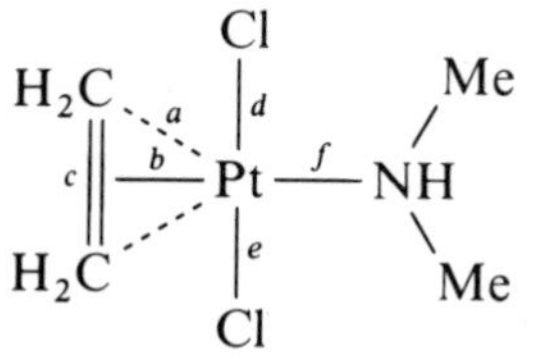

$a = 2\cdot21$ Å	$d = 2\cdot30(4)$ Å
$b = 2\cdot09$ Å	$e = 2\cdot33(4)$ Å
$c = 1\cdot47(18)$ Å	$f = 2\cdot02(19)$ Å

Plane of $CH_2\!=\!CH_2$ double-bond is perpendicular to $PtCl_2N$ plane.

133. $C_4H_{12}N_8Pt_2S_4$ [$Pt(NH_3)_4$]$^{2+}$[$Pt(SCN)_4$]$^{2-}$
(J. R. Miller, *J. Chem. Soc.*, (1961) 4452).
Pt–Pt = 3·35 Å.

134. $C_4H_{16}Cl_2N_4Pt^{2+}$ Cis-[$Pt(en)_2Cl_2$]Cl_2
(C. F. Liu and J. A. Ibers, *Inorg. Chem.*, **9** (1970) 773).

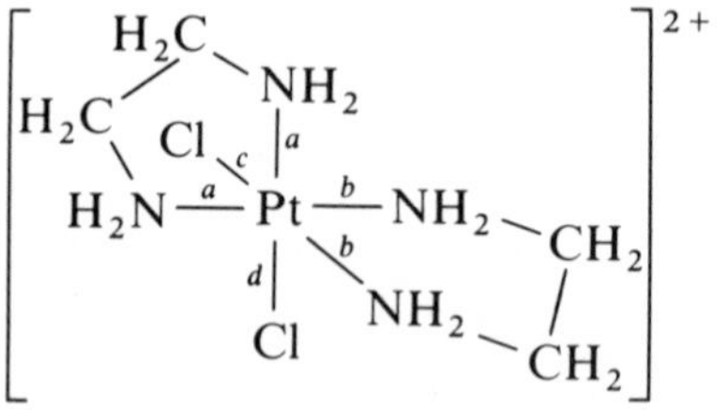

$a = 2\cdot07$ Å	
$b = 2\cdot05$ Å	
$c = 2\cdot30$ Å	
$d = 2\cdot31$ Å	

Paper also reports the absolute configuration

135. $C_4H_{18}N_6Pt^{2+}$ [$Pt(NH_3)_4(CH_3CN)_2$]$Cl_2.H_2O$
(C. M. Harris and N. C. Stephenson, *Chem. Ind.* (London), (1957) 426; N. C. Stephenson, *J. Inorg. Nucl. Chem.*, **24** (1962) 801).

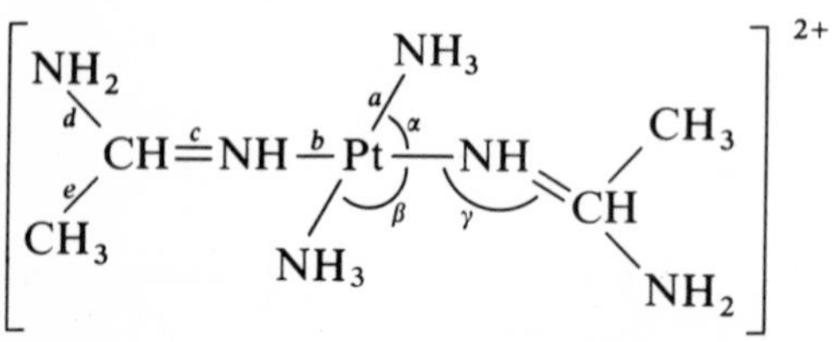

$a = 2\cdot038$ Å	
$b = 1\cdot957$ Å	$\alpha = 83\cdot8°$
$c = 1\cdot271$ Å	$\beta = 96\cdot2°$
$d = 1\cdot318$ Å	$\gamma = 130\cdot2°$
$e = 1\cdot533$ Å	

136. $C_4H_{20}Cl_4N_4Pt_2$ $[Pt(CH_3NH_2)_4]^{2+}[PtCl_4]^{2-}$
   (S. Yamada and R. Tsuchida, *Bull. Chem. Soc. Japan*, **31** (1958) 813; J. R. Miller, *Proc. Chem. Soc.*, (1960) 318).
   Pt–Pt = 3·25 Å.

137. $C_4H_{20}N_4Pt^{2+}$ Trans-$[Pt(Me_2NH)_2(NH_3)_2]Cl_2$
   (J. S. Anderson, J. W. Carmichael and A. W. Cordes, *Inorg. Chem.*, **9** (1970) 143).

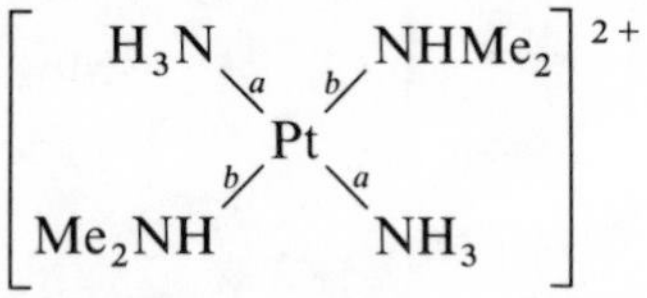

$$\left[\begin{array}{c} H_3N \quad NHMe_2 \\ Pt \\ Me_2NH \quad NH_3 \end{array} \right]^{2+}$$

   $a = 2·04$ Å
   $b = 2·04$ Å

138. $C_4N_4Pt^{2-}$ $[Pt(CN)_4]^{2-}$
   (W. Krogmann, *Angew. Chem. Int. Ed.*, **8** (1969) 35).
   Reports Pt–Pt distance for $[Pt(CN)_4]^{2-}$ with a range of cations.

139. $C_5N_5Pt^{2-}$ $K_2[Pt(CN)_5].3H_2O$
   (A. Piccinin and J. Toussaint, *Bull. Soc. Roy. Sci. Liège*, (1967) 122; *Chem. Abs.*, **68** (1968) 73213).

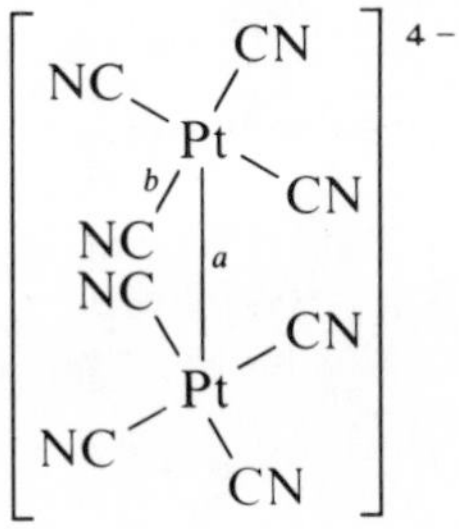

$$\left[\begin{array}{c} NC \quad CN \\ Pt \\ CN \\ NC \\ NC \quad CN \\ Pt \\ NC \quad CN \end{array} \right]^{4-}$$

   $a = 2·92$ Å
   $b = 2·02$ Å
   $C \equiv N = 1·18$ Å
   The fifth cyanide ion is bound as an anion within the crystal lattice.

140. $C_5H_{11}Cl_2NO_2PtS$ $[Pt(L\text{-}methionineH)Cl_2]$
   (H. C. Freeman and M. L. Golomb, *Chem. Comm.*, (1970) 1523).

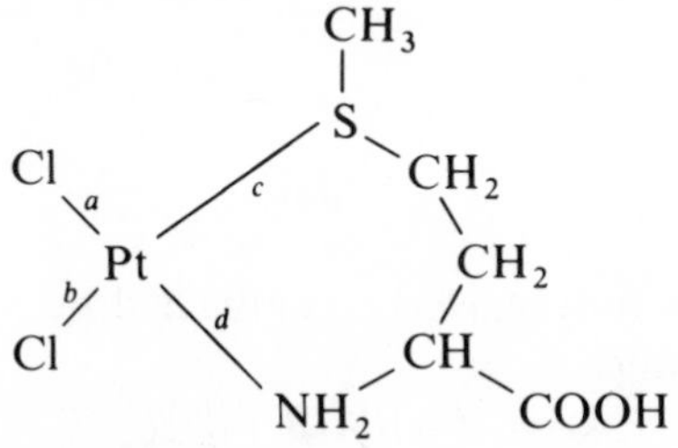

$$\begin{array}{c} CH_3 \\ | \\ S \\ Cl \quad CH_2 \\ Pt \quad CH_2 \\ Cl \quad CH \\ NH_2 \quad COOH \end{array}$$

   $a = 2·31$ Å
   $b = 2·32$ Å
   $c = 2·26$ Å
   $d = 2·03$ Å

141. $C_6N_6PtSe_6^{2-}$ $K_2[Pt(SeCN)_6]$
   (G. V. Tsintsadze, V. V. Skopenko and A. E. Shvelashvili, *Tr. Gruz. Politekh. Inst.*, (1968) 19; *Chem. Abs.*, **70** (1969) 100619).
   $K_2[Pt(SeCN)_6]$ is isostructural with $K_2[Pt(SCN)_6]$. Pt octahedrally surrounded by six Se. Pt–Se = 2·8 Å; C–Se = 1·9 Å; C–N = 1·1 Å.

142. $C_6H_{18}As_2Cl_4Pt_2$ $[(Me_3As)PtCl_2]_2$
(S. F. Watkins, *Chem. Comm.*, (1968) 504; S. F. Watkins, *J. Chem. Soc. (A)*, (1970) 168).

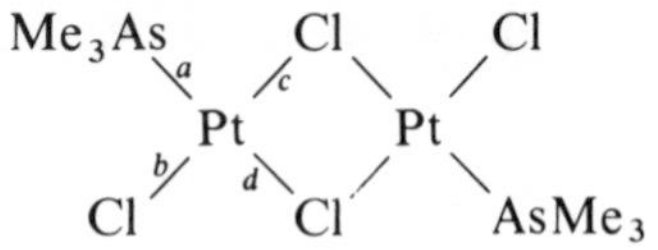

$a = 2\cdot308$ Å
$b = 2\cdot268(6)$ Å
$c = 2\cdot394(6)$ Å
$d = 2\cdot312(5)$ Å

143. $C_6H_{18}Cl_2P_2Pt$ Cis-$[(PMe_3)_2PtCl_2]$
(G. G. Messmer, E. L. Amma and J. A. Ibers, *Inorg. Chem.*, **6** (1967) 725).

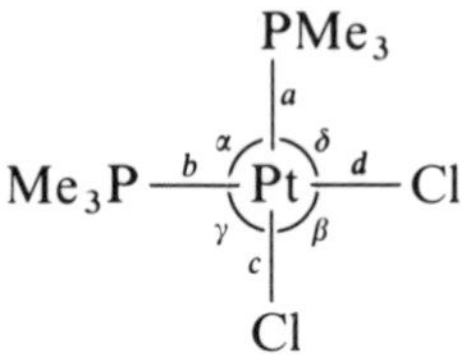

$a = 2\cdot239(6)$ Å $\alpha = 96\cdot3(3)°$
$b = 2\cdot256(8)$ Å $\beta = 87\cdot7(3)°$
$c = 2\cdot364(8)$ Å $\gamma = 85\cdot1(4)°$
$d = 2\cdot388(9)$ Å $\delta = 91\cdot3(3)°$

Configuration about Pt is slightly distorted from square-planar towards tetrahedral.

144. $C_7H_{13}ClN_2O_3PtS$ $[Pt(glycyl$-L$-methionineH)Cl]$
(H. C. Freeman and M. L. Golomb, *Chem. Comm.*, (1970) 1523).

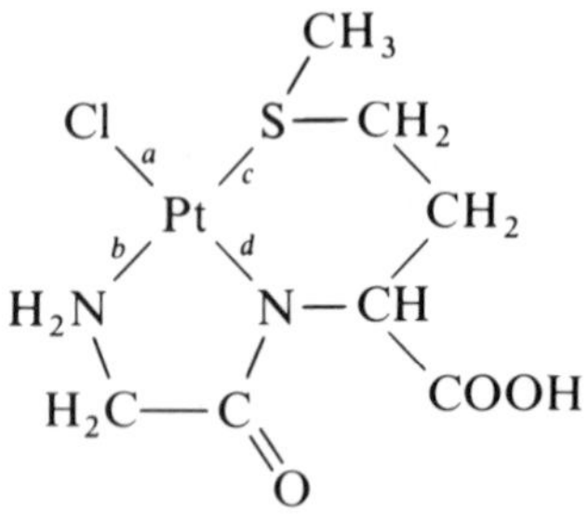

$a = 2\cdot30$ Å
$b = 2\cdot07$ Å
$c = 1\cdot98$ Å
$d = 2\cdot25$ Å

Neutron diffraction study shows loss of the hydrogen originally bound to the peptide nitrogen atom.

145. $C_7H_{15}Cl_2OPPt$ cis-$[(PEt_3)Pt(CO)Cl_2]$
(E. M. Badley and G. A. Sim, to be published).

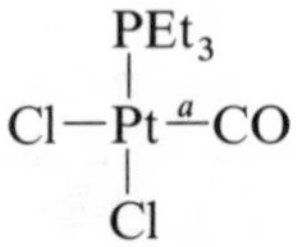

$a = 1\cdot75$ Å

146. $C_8H_{14}N_4O_4Pt$ *Bis-(dimethylglyoximato)platinum(II)*
(C. Panattoni, E. Frasson and R. Zannetti, *Acta Cryst.*, **12** (1959) 1027).

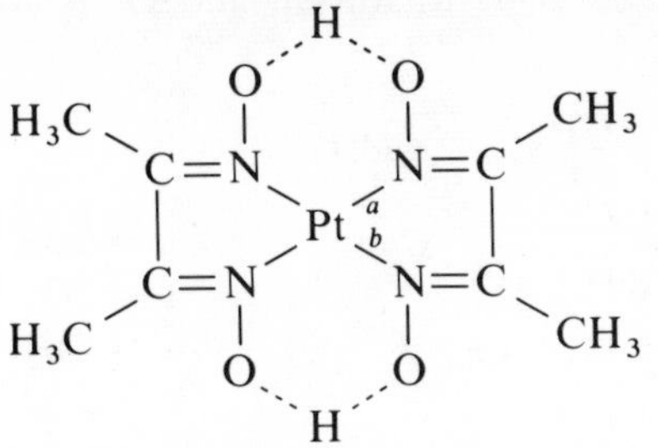

$a = 1.93$ Å
$b = 1.95$ Å

147. $C_8H_{14}Pt$ $[(\pi\text{-}C_5H_5)Pt(CH_3)_3]$
(V. A. Semion, A. Z. Rubeznov, Yu. T. Struckov and S. P. Gubin, *Russ. J. Struct. Chem.*, **10** (1969) 144; G. W. Adamson and J. C. J. Bart, reported by K. W. Egger, *J. Organometal. Chem.*, **24** (1970) 501).

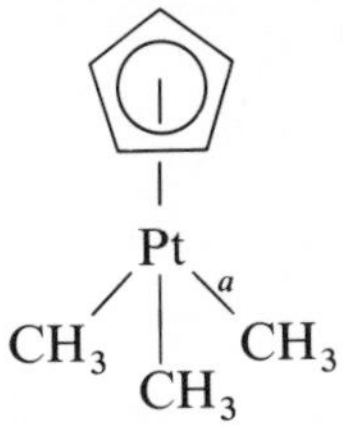

$a = 2.11$ Å
Pt–C of $C_5H_5 = 2.32$ Å
C–C in $C_5H_5 = 1.43$ Å
Pt lies below centre of C_5H_5 ring.

148. $C_8H_{20}Br_4Pt_2S_2$ $[(Et_2S)PtBr_2]_2$
(D. L. Sales, J. Stokes and P. Woodward, *J. Chem. Soc. (A)*, (1968) 1852).

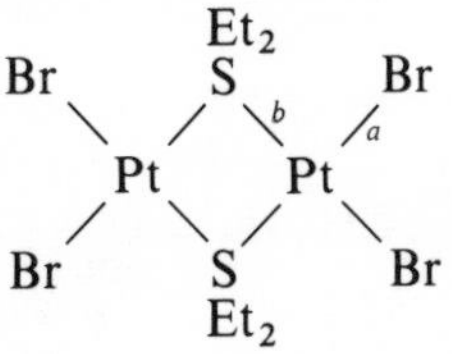

$a = 2.388(5)$ Å (mean)
$b = 2.218(10)$ Å
Co-ordination around Pt is square-planar.

149. $C_{10}H_{14}ClO_4Pt^-$ $K[PtCl(acac)_2]$
(B. N. Figgis, J. Lewis, R. F. Long, R. Mason, R. S. Nyholm, P. J. Pauling and G. B. Robertson, *Nature*, **195** (1962) 1278; R. Mason, G. B. Robertson and P. J. Pauling, *J. Chem. Soc. (A)*, (1969) 485).

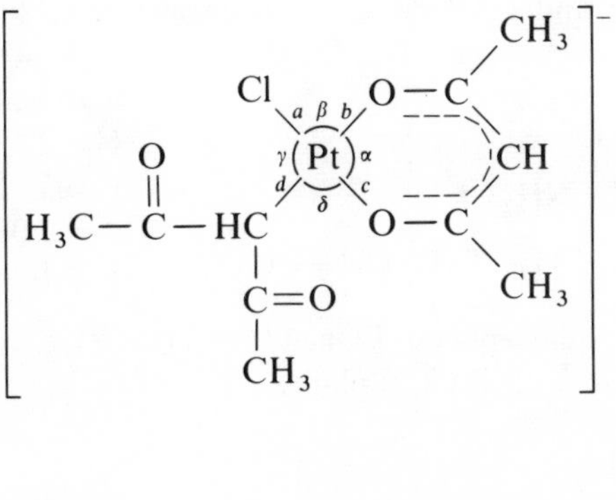

$a = 2.76(5)$ Å	$\alpha = 95°$
$b = 2.07$ Å	$\beta = 85°$
$c = 1.97$ Å	$\gamma = 96°$
$d = 2.11$ Å	$\delta = 84°$

150. $C_{10}H_{14}Cl_2O_3Pt$ $[Pt(C_{10}H_{14}O_3)Cl_2]$

(D. Gibson, C. Oldham, J. Lewis, D. Lawton, R. Mason and G. B. Robertson, *Nature*, **208** (1965) 580; R. Mason and G. B. Robertson, *J. Chem. Soc. (A)*, (1969) 492).

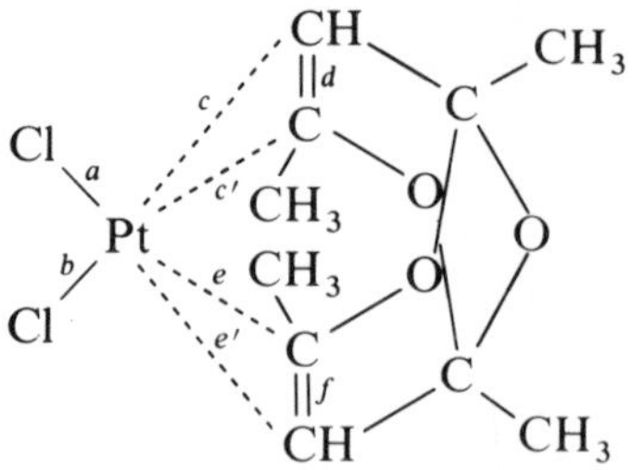

$a = 2\cdot30$ Å	$d = 1\cdot39(3)$ Å
$b = 2\cdot31$ Å	$e = 2\cdot33$ Å
$c = 2\cdot04$ Å	$e' = 2\cdot13$ Å
$c' = 2\cdot32$ Å	$f = 1\cdot44(3)$ Å

151. $C_{10}H_{16}Cl_2Pt$ $[Pt(dipentene)Cl_2]$

(N. C. Baenziger, R. C. Medrud and J. R. Doyle, *Acta Cryst.*, **18** (1965) 237).

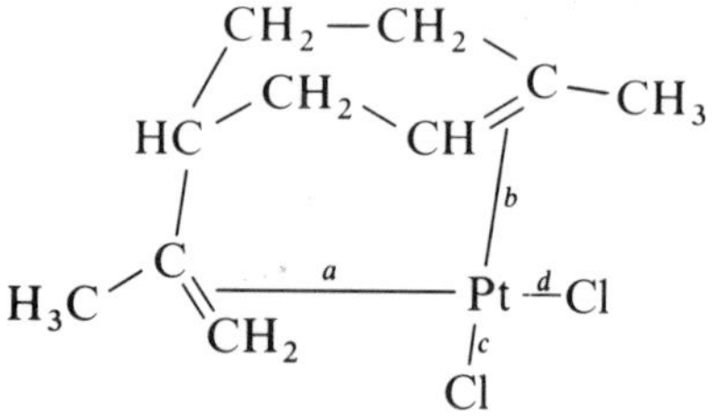

$a = 2\cdot14(3)$ Å
$b = 2\cdot13(3)$ Å
$c = 2\cdot32(1)$ Å
$d = 2\cdot34(1)$ Å

152. $C_{10}H_{23}N_4O_2Pt^+$ *Bis-(2-amino-2-methyl-3-butanoneoximato)-platinum(II) chloride*

(E. O. Schlemper, *Inorg. Chem.*, **8** (1969) 2740).

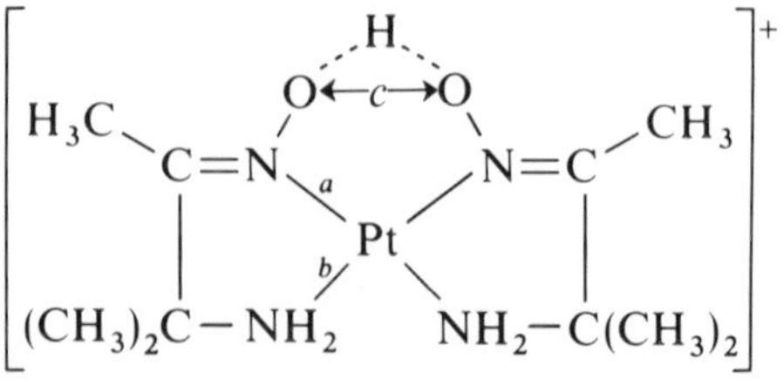

$a = 1\cdot98$ Å
$b = 2\cdot03$ Å
$c = 2\cdot48$ Å

Co-ordination around Pt is square-planar.

153. $C_{12}H_{19}Cl_2NPt$ Cis-$[Pt(trans\text{-}CH_3CH\!=\!CHCH_3)\{(S)\text{-}\alpha\text{-}phen\text{-}ethylamine\}Cl_2]$

(E. Benedetti, P. Corradini and C. Pedone, *J. Organometal. Chem.*, **18** (1969) 203).

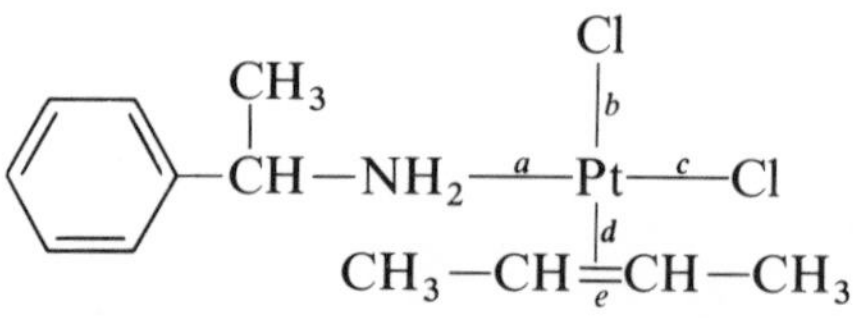

$a = 2\cdot14(3)$ Å
$b = 2\cdot35(2)$ Å
$c = 2\cdot31(2)$ Å
$d = 2\cdot05(2)$ Å
$e = 1\cdot36(6)$ Å

Co-ordination around Pt is square-planar. Plane of $-CH\!=\!CH-$ double bond makes an angle of $77 \pm 4°$ with $PtCl_2N$ plane.

154. $C_{12}H_{19}Cl_2NPt$ Cis[Pt{(S)-1-CH_3CH_2CH=CH_2}{(S)-α-phen-
ethylamine}Cl_2]

(C. Pedone and E. Benedetti, *J. Organometal. Chem.*, **29** (1971) 443).

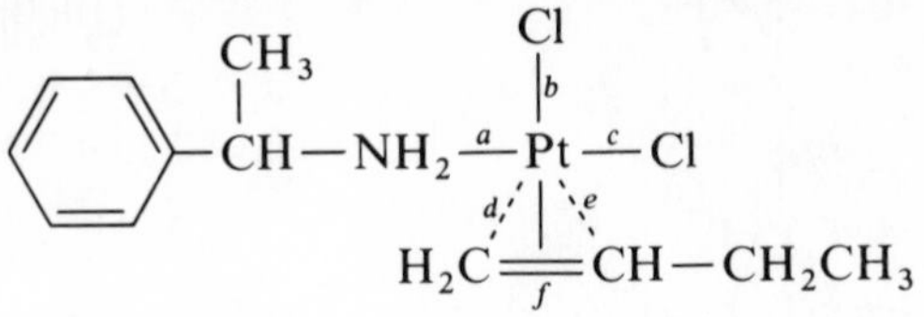

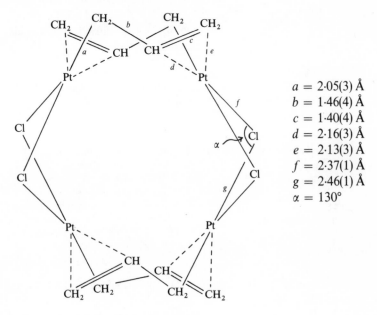

$a = 2\cdot076(15)$ Å
$b = 2\cdot361(5)$ Å
$c = 2\cdot289(7)$ Å
$d = 2\cdot163(25)$ Å
$e = 2\cdot173(23)$ Å
$f = 1\cdot350(3)$ Å

Plane of double bond makes angle of 79° with square-plane about Pt.

155. $C_{12}H_{20}Cl_4Pt_4$ [$Pt_4(C_3H_5)_4Cl_4$]

(G. Raper and W. S. MacDonald, *Chem. Comm.*, (1970) 655).

$a = 2\cdot05(3)$ Å
$b = 1\cdot46(4)$ Å
$c = 1\cdot40(4)$ Å
$d = 2\cdot16(3)$ Å
$e = 2\cdot13(3)$ Å
$f = 2\cdot37(1)$ Å
$g = 2\cdot46(1)$ Å
$\alpha = 130°$

156. $C_{12}H_{30}Br_2P_2Pt$ Trans-[$(PEt_3)_2PtBr_2$]

(G. G. Messmer and E. L. Amma, *Inorg. Chem.*, **5** (1966) 1775).

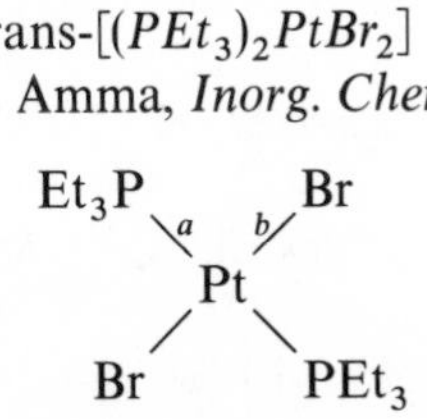

$a = 2\cdot315(4)$ Å
$b = 2\cdot428(2)$ Å

157. $C_{12}H_{30}Cl_2P_2Pt$ Trans-[$(PEt_3)_2PtCl_2$]

(G. G. Messmer and E. L. Amma, *Inorg. Chem.*, **5** (1966) 1775).

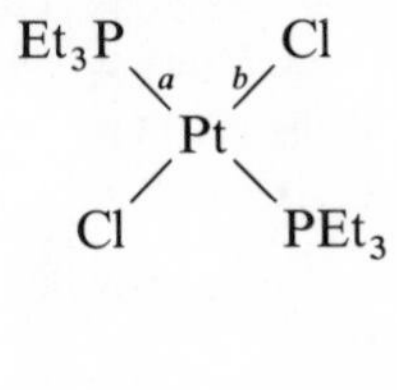

$a = 2\cdot300(19)$ Å
$b = 2\cdot294(90)$ Å

158. $C_{12}H_{30}Cl_4PPt$ Trans-$[(PEt_3)_2PtCl_4]$

(L. Aslanov, R. Mason, A. G. Wheeler and P. O. Whimp, *Chem. Comm.*, (1970) 30).

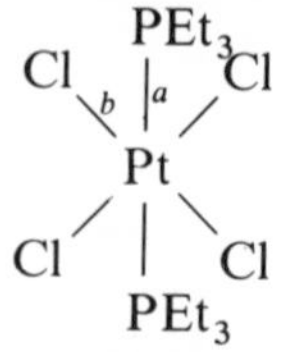

$$
\begin{array}{c}
PEt_3 \\
Cl \underset{b}{\diagdown}\ |a\ \diagup Cl \\
Pt \\
Cl\diagup\ |\ \diagdown Cl \\
PEt_3
\end{array}
$$

$a = 2\cdot393(5)$ Å
$b = 2\cdot332(5)$ Å

159. $C_{12}H_{31}BrP_2Pt$ Trans-$[(PEt_3)_2PtHBr]$

(P. G. Owston, J. M. Partridge and J. M. Rowe, *Acta Cryst.*, **13** (1960) 246).

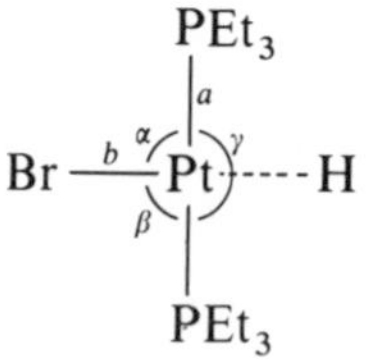

$$
\begin{array}{c}
PEt_3 \\
|a \\
Br \overset{\alpha}{\underset{\beta}{\diagdown}} Pt \overset{\gamma}{\diagup}\text{----}H \\
| \\
PEt_3
\end{array}
$$

$a = 2\cdot26$ Å $\alpha = 93\cdot7°$
$b = 2\cdot56$ Å $\beta = 94\cdot1°$
$\gamma = 172\cdot2°$

Note: x-ray diffraction study did not locate hydridic proton.

160. $C_{12}H_{31}ClP_2Pt$ Trans-$[(PEt_3)_2PtHCl]$

(I. R. Beattie and K. M. S. Livingston, *J. Chem. Soc. (A)*, (1969) 2201).

Single crystal Raman study demonstrated that hydridic proton was in the PtP_2Cl plane.

161. $C_{12}H_{36}Cl_4Pt_4$ $[(CH_3)_3PtCl]_4$

(R. E. Rundle and J. H. Sturdivant, *J. Amer. Chem. Soc.*, **69** (1947) 1561).

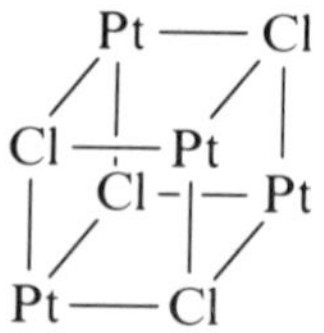

$$
\begin{array}{c}
Pt \text{----} Cl \\
\diagup |\ \ \diagup | \\
Cl \text{----} Pt\ | \\
|\ \ Cl \text{-}|\text{---} Pt \\
|\diagup\ \ |\diagup \\
Pt \text{----} Cl
\end{array}
$$

Pt–Cl = $2\cdot48$ Å (mean)
Pt–Pt = $3\cdot73$ Å (mean)
∠ Pt–Cl–Pt = 99°
∠ Cl–Pt–Cl = 81°

162. $C_{12}H_{36}I_4Pt$ $[(CH_3)_3PtI]_4$

(G. Donnay, L. B. Coleman, N. G. Krieghoff and D. O. Cowan, *Acta Cryst.*, **B24** (1968) 157).

Same basic structure as $[(CH_3)_3PtCl]_4$ (no. 161). Pt–I = $2\cdot83$ Å (mean).

163. $C_{12}H_{40}O_4Pt_4$ $[(CH_3)_3Pt(OH)]_4$

(X-ray diffraction: T. G. Spiro, D. H. Templeton and A. Zalkin, *Inorg. Chem.*, **7** (1968) 2165;

Neutron diffraction: H. S. Preston, J. C. Mills and C. H. L. Kennard, *J. Organometal. Chem.*, **14** (1968) 447).

Same basic structure as $[(CH_3)_3PtCl]_4$ (no. 161).

	X-ray	Neutron	
Pt–C	2·04 Å	2·02 Å	
Pt–O	2·22 Å	2·20 Å	
Pt–Pt	3·43 Å	3·41 Å	
∠ O–Pt–O	77·6°	O–H 0·93 Å	
∠ C–Pt–C	87°	C–H 1·12 Å	} two different C–H bond lengths
∠ Pt–O–Pt	101·2°	1·06 Å	

164. $C_{12}H_{41}B_9P_2PtS$ $[(PEt_3)_2Pt(H)(B_9H_{10}S)]$
(A. R. Kane, L. J. Guggenberger and E. L. Muetterties, *J. Amer. Chem. Soc.*, **92** (1970) 2571).

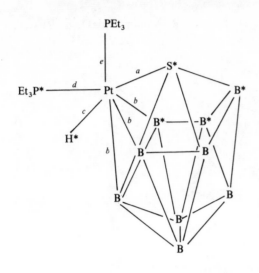

$a = 2.43$ Å
$b = 2.20$–2.25 Å
$c = 1.66$ Å
$d = 2.36$ Å
$e = 2.39$ Å
* atoms are approx. in one plane.

165. $C_{12}H_{42}N_6Pt_2$ $[(en)Pt(CH_3)_3\text{-}en\text{-}Pt(CH_3)_3(en)]$
(M. R. Truter and E. G. Cox, *J. Chem. Soc.*, (1956) 948).

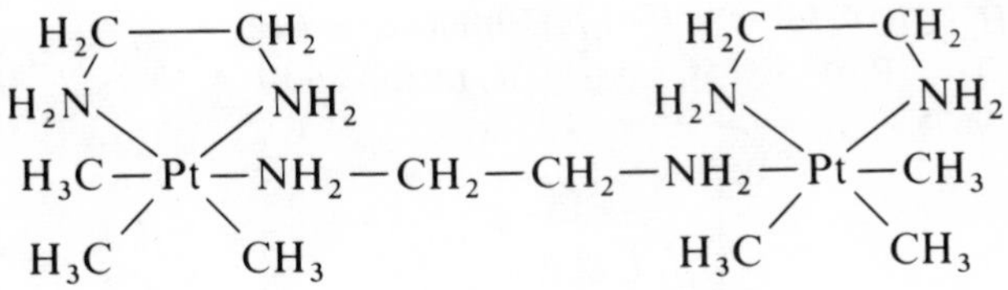

$$H_2C \text{———} CH_2$$
$$H_2N \diagup \qquad \diagdown NH_2$$
$$H_3C \text{—} Pt \text{—} NH_2 \text{—} CH_2 \text{—} CH_2 \text{—} NH_2 \text{—} Pt \text{—} CH_3$$
$$H_3C \diagup \qquad \diagdown CH_3$$

No accurate interatomic distances quoted, but crystallographic observations establish the diagrammatic formula shown.

166. $C_{13}H_{16}Cl_2N_2Pt$ $[(C_3H_6)Pt(pyr)_2Cl_2]$
(N. A. Bailey, R. D. Gillard, M. Keeton, R. Mason and D. R. Russell, *Chem. Comm.*, (1966) 396).

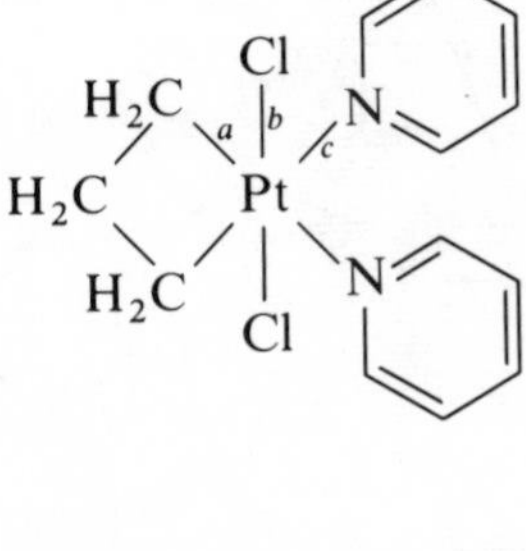

$a = 2.15$ Å (mean)
$b = 2.225$ Å (mean)
$c = 2.20$ Å (mean)

167. $C_{13}H_{16}Cl_4N_2Pt$ $[Pt(pyr)\left(CH\begin{smallmatrix}C_2H_5\\ \\pyr\end{smallmatrix}\right)Cl_4]$

(N. A. Bailey, R. D. Gillard, M. Keeton, R. Mason and D. R. Russell, *Chem. Comm.*, (1966) 396).

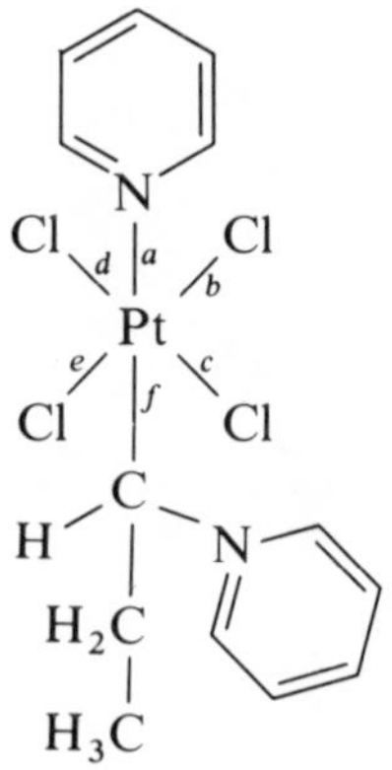

$$
a = 2{\cdot}17\,\text{Å}
$$
$$
b = 2{\cdot}31\,\text{Å}
$$
$$
c = 2{\cdot}28\,\text{Å}
$$
$$
d = 2{\cdot}34\,\text{Å}
$$
$$
e = 2{\cdot}33\,\text{Å}
$$
$$
f = 2{\cdot}05\,\text{Å}
$$

168. $C_{13}H_{20}Cl_2NPPt$ Cis-$[(PEt_3)Pt(CNPh)Cl_2]$

(E. M. Badley and G. A. Sim, to be published; quoted by E. M. Badley, J. Chatt and R. L. Richards, *J. Chem. Soc.* (*A*), (1971) 21).

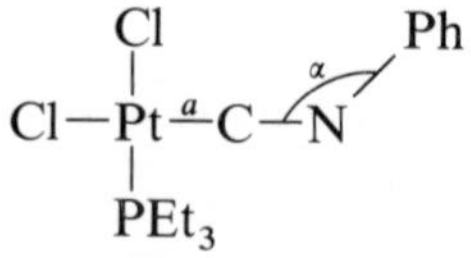

$$
a = 1{\cdot}87\,\text{Å}
$$
$$
\alpha = 165{\cdot}5°
$$

169. $C_{13}H_{30}ClOP_2Pt^+$ $[(PEt_3)_2Pt(CO)Cl]^+ BF_4^-$

(H. C. Clark, P. W. R. Corfield, K. R. Dixon and J. A. Ibers, *J. Amer. Chem. Soc.*, **89** (1967) 3360).

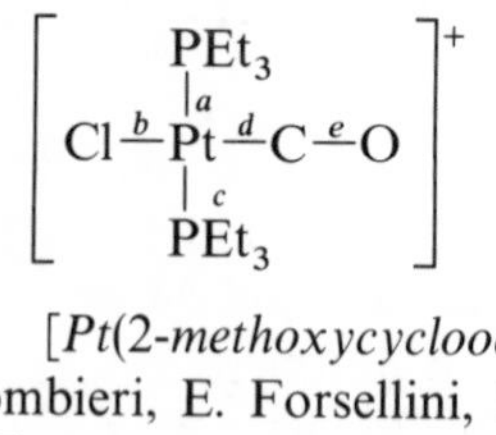

$a = 2{\cdot}35\,\text{Å}$	$d = 1{\cdot}78\,\text{Å}$
$b = 2{\cdot}30\,\text{Å}$	$e = 1{\cdot}14\,\text{Å}$
$c = 2{\cdot}34\,\text{Å}$	

170. $C_{14}H_{20}ClNOPt$ $[Pt(2\text{-}methoxycycloocta\text{-}1,5\text{-}dienyl)(pyr)Cl]$

(C. Panattoni, G. Bombieri, E. Forsellini, B. Crociani and U. Belluco, *Chem. Comm.*, (1969) 187).

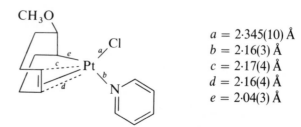

$$
a = 2{\cdot}345(10)\,\text{Å}
$$
$$
b = 2{\cdot}16(3)\,\text{Å}
$$
$$
c = 2{\cdot}17(4)\,\text{Å}
$$
$$
d = 2{\cdot}16(4)\,\text{Å}
$$
$$
e = 2{\cdot}04(3)\,\text{Å}
$$

171. $C_{14}H_{24}Cl_4Pt_2$ [(1,1,3,3-*tetramethylallene*)$PtCl_2$]$_2$

(T. G. Hewitt, K. Anzenhofer and J. J. De Boer, *J. Organometal. Chem.*, **18** (1969) P19; T. G. Hewitt and J. J. De Boer, *J. Chem. Soc. (A)*, (1971) 817).

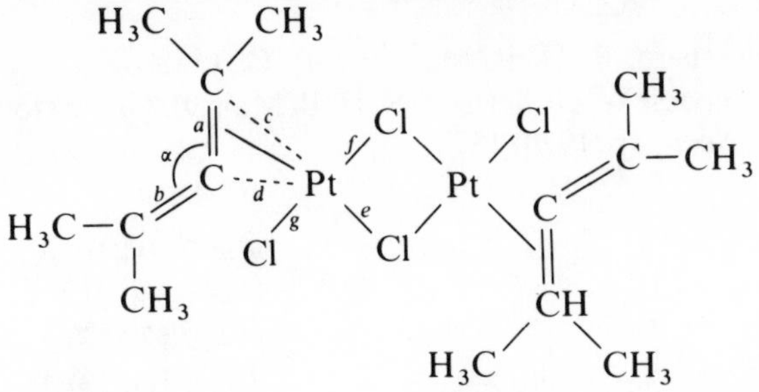

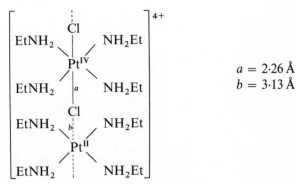

$a = 1·37$ Å $e = 2·38$ Å
$b = 1·36$ Å $f = 2·34$ Å
$c = 2·25$ Å $g = 2·27$ Å
$d = 2·07$ Å $\alpha = 151°$

CH_3–C–CH_3 planes at opposite ends of allene are perpendicular to each other.

172. $C_{15}H_{26}Cl_2NOPPt$ Cis-[(PEt_3)Pt$\left(-C\begin{smallmatrix}\diagup OC_2H_5\\\diagdown NHPh\end{smallmatrix}\right)Cl_2$]

(E. M. Badley, J. Chatt, R. L. Richards and G. A. Sim, *Chem. Comm.*, (1969) 1322).

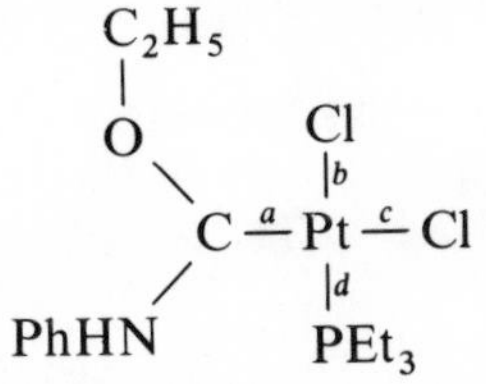

$a = 1·98(2)$ Å
$b = 2·368(7)$ Å
$c = 2·355(5)$ Å
$d = 2·240(8)$ Å

173. $C_{16}H_{56}Cl_6N_8Pt_2$ [$Pt^{IV}(C_2H_5NH_2)_4Cl_2$]$^{2+}$[$Pt^{II}(C_2H_5NH_2)_4$]$^{2+}$-
(Cl^-)$_4$

(B. M. Craven and D. Hall, *Acta Cryst.*, **14** (1961) 475).

$a = 2·26$ Å
$b = 3·13$ Å

174. $C_{17}H_{27}Cl_2NPt$ Trans-$[Pt(Me_3C-C\equiv C-CMe_3)-$

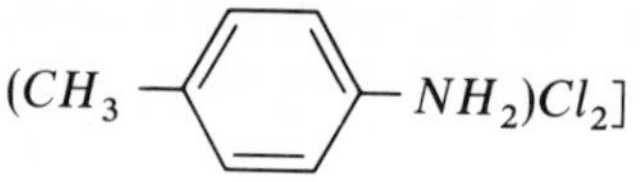

$(CH_3$ ⬡ $-NH_2)Cl_2]$

(G. R. Davies, W. Hewertson, R. H. B. Mais and P. G. Owston, *Chem. Comm.*, (1967) 423; G. R. Davies, W. Hewertson, R. H. B. Mais, P. G. Owston and C. G. Patel, *J. Chem. Soc. (A)*, (1970) 1873).

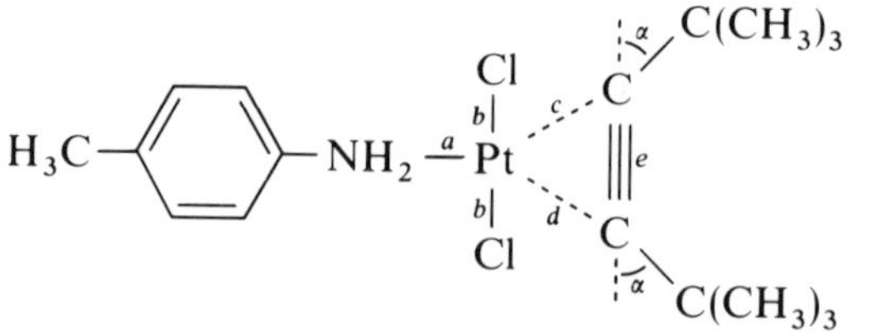

$a = 2\cdot104(10)$ Å
$b = 2\cdot300(2)$ Å
$c = 2\cdot138(13)$ Å
$d = 2\cdot180(13)$ Å
$e = 1\cdot24$ Å
$\alpha = 163\cdot5°$ (mean)

Plane of C≡C is at an angle of 85° to $PtCl_2N$ plane.

175. $C_{18}H_{24}N_2O_2Pt$ $[(CH_3)_3Pt(acac)(bipyr)]$
(A. G. Swallow and M. R. Truter, *Proc. Chem. Soc.*, (1961) 166; A. G. Swallow and M. R. Truter, *Proc. Roy. Soc.*, **A266** (1962) 527).

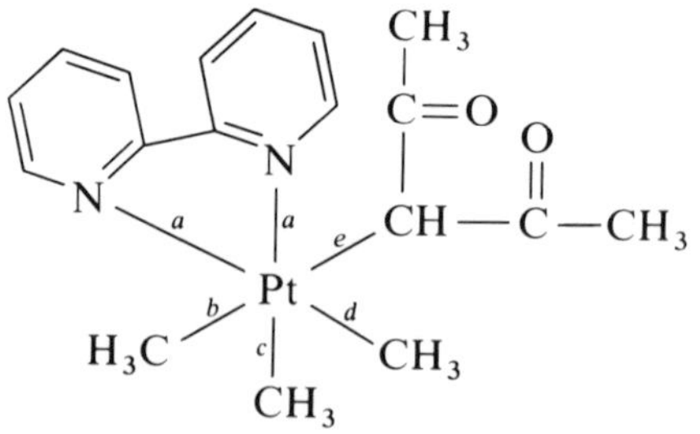

$a = 2\cdot15(1)$ Å
$b = 2\cdot05(2)$ Å
$c = 2\cdot07(2)$ Å
$d = 2\cdot03(2)$ Å
$e = 2\cdot15(3)$ Å

176. $C_{18}H_{26}O_6Pt_2$ $[(CH_3)_3Pt(CH_3COCHCOOEt)]_2$
(A. C. Hazell and M. R. Truter, *Proc. Roy. Soc.*, **A254** (1960) 218).

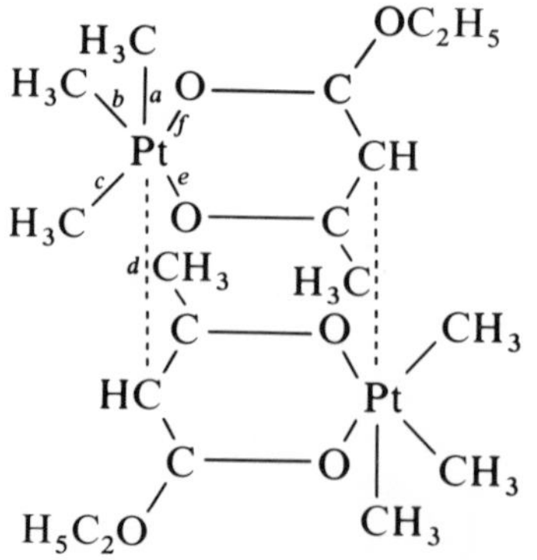

$a = 2\cdot17(10)$ Å
$b = 1\cdot96(7)$ Å
$c = 2\cdot14(9)$ Å
$d = 2\cdot56(9)$ Å
$e = 2\cdot19(5)$ Å
$f = 2\cdot18(5)$ Å

177. $C_{18}H_{40}N_2O_4Pt_2$ $[(acac)(CH_3)_3Pt\text{-}en\text{-}Pt(CH_3)_3(acac)]$
(A. Robson and M. R. Truter, *J. Chem. Soc.*, (1965) 630).

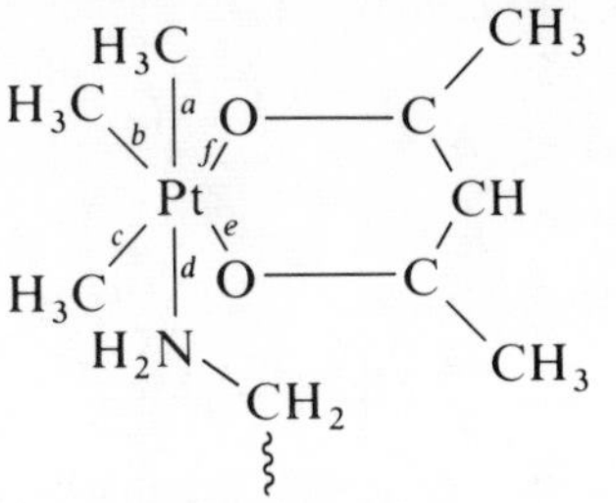

$a = 2.03(7)$ Å
$b = 2.17(8)$ Å
$c = 2.12(8)$ Å
$d = 2.31(6)$ Å
$e = 2.18(4)$ Å
$f = 2.16(4)$ Å

178. $C_{18}H_{42}Cl_4P_2Pt_2$ $[(P^nPr_3)PtCl_2]_2$
(M. Black, R. H. B. Mais and P. G. Owston, *Acta Cryst.*, **B25** (1969) 1760).

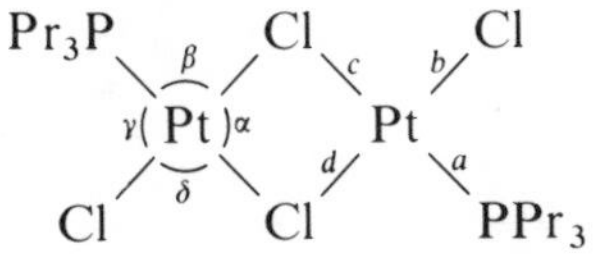

$a = 2.230(9)$ Å $\alpha = 83.6°$
$b = 2.279(9)$ Å $\beta = 96.5°$
$c = 2.425(8)$ Å $\gamma = 88.4°$
$d = 2.315(8)$ Å $\delta = 91.5°$

179. $C_{20}H_{28}O_4Pt_2$ $[(CH_3)_3Pt(salicylaldehyde)]_2$
(J. E. Lydon, M. R. Truter and R. C. Watling, *Proc. Chem. Soc.*, (1964) 193;
M. R. Truter and R. C. Watling, *J. Chem. Soc. (A)*, (1967) 1955).

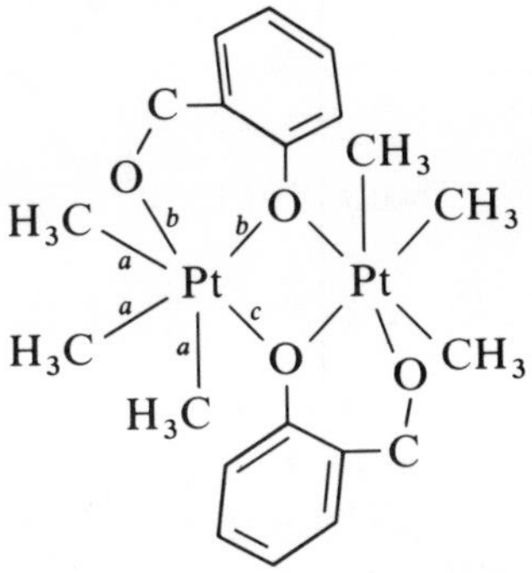

$a = 2.03(3)$ Å (mean)
$b = 2.24(3)$ Å (mean)
$c = 2.20(3)$ Å (mean)

180. $C_{20}H_{33}ClP_2PtSi$ Trans-$[(PPhMe_2)_2Pt(CH_2SiMe_3)Cl]$
(B. Jovanovic, Lj. Manojlovic-Muir and K. W. Muir, unpublished results).

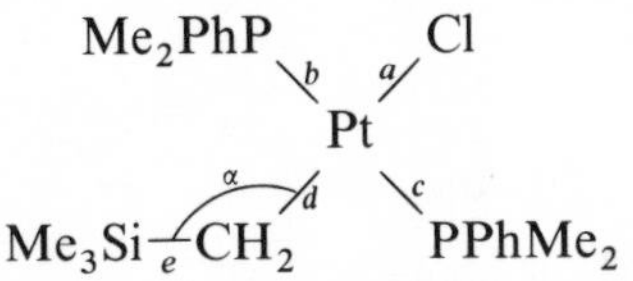

$a = 2.415(5)$ Å $\alpha = 114(2)°$
$b = 2.294(5)$ Å
$c = 2.292(5)$ Å
$d = 2.079(14)$ Å
$e = 1.844(19)$ Å

181. $C_{20}H_{36}As_4Pt^{2+}$ *Bis-(o-phenylenebisdimethylarsine)-*
platinum(II)dichloride
(N. C. Stephenson and G. A. Jeffrey, *Proc. Chem. Soc.*, (1963) 173; N. C.
Stephenson, *Acta Cryst.*, **17** (1964) 1517).

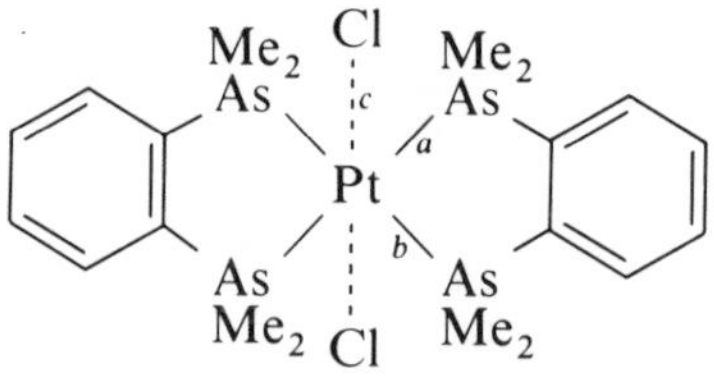

$a = 2\cdot37$ Å
$b = 2\cdot38$ Å
$c = 4\cdot16$ Å

182. $C_{20}H_{42}Cl_2N_2P_2Pt_2S_2$ $[(P^nPr_3)ClPt(SCN)_2PtCl(P^nPr_3)]$
(P. G. Owston and J. M. Rowe, *Acta Cryst.*, **13** (1960) 253; U. A. Gregory,
J. A. J. Jarvis, B. T. Kilbourn and P. G. Owston, *J. Chem. Soc. (A)*, (1970)
2770).

α-isomer   β-isomer

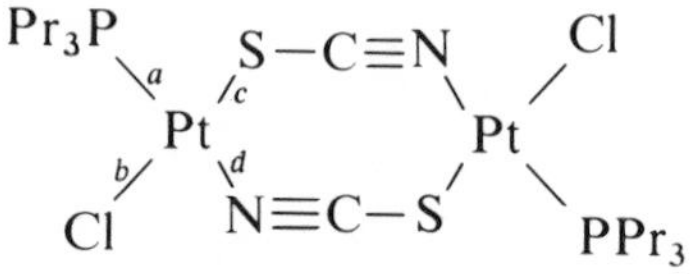

$a = 2\cdot244(4)$ Å
$b = 2\cdot304(4)$ Å
$c = 2\cdot327(5)$ Å
$d = 2\cdot08(1)$ Å

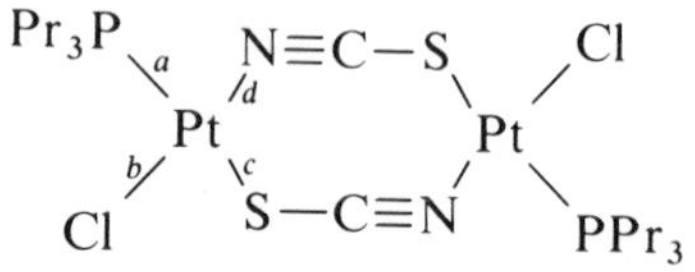

$a = 2\cdot262(4)$ Å
$b = 2\cdot277(4)$ Å
$c = 2\cdot408(4)$ Å
$d = 1\cdot97(1)$ Å

183. $C_{21}H_{29}Cl_2N_2PPt$ Trans-

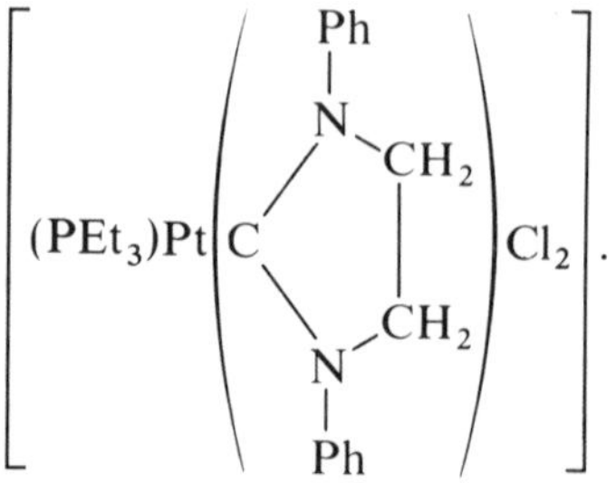

(D. J. Cardin, B. Cetinkaya, M. F. Lappert, L. J. Manojlovic-Muir and
K. W. Muir, *Chem. Comm.*, (1971) 400; L. J. Manojlovic-Muir and K. W.
Muir, personal communication).

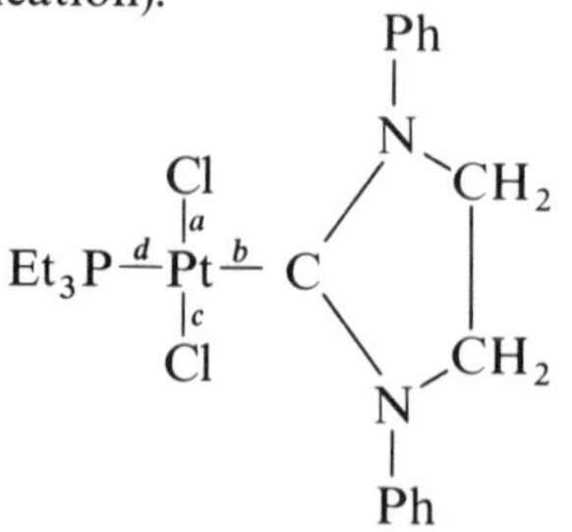

$a = 2\cdot311(6)$ Å
$b = 2\cdot02(16)$ Å
$c = 2\cdot292(6)$ Å
$d = 2\cdot291(4)$ Å

N–C(Pt)–N plane is at 70° to PtCCl$_2$P plane.

184. $C_{22}H_{30}Cl_2O_2Pt_2$ $[Pt(dicyclopentadieneOMe)Cl]_2$
(W. A. Whitla, H. M. Powell and L. M. Venanzi, *Chem. Comm.*, (1966) 310).

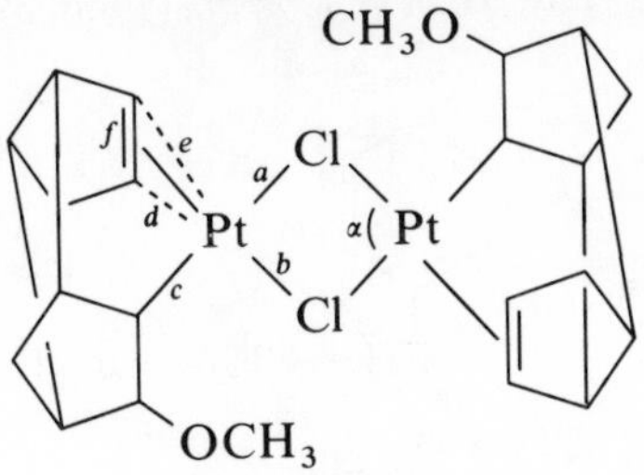

$$a = 2\cdot51\ \text{Å}$$
$$b = 2\cdot34\ \text{Å}$$
$$c = 2\cdot07\ \text{Å}$$
$$d = 2\cdot11\ \text{Å}$$
$$e = 2\cdot23\ \text{Å}$$
$$f = 1\cdot37\ \text{Å}$$
$$\alpha = 85°$$

185. $C_{24}H_{30}N_2O_2Pt_2$ $[(CH_3)_3Pt(8\text{-}quinolinol)]_2$
(J. E. Lydon, M. R. Truter and R. C. Watling, *Proc. Chem. Soc.*, (1964) 193;
J. E. Lydon and M. R. Truter, *J. Chem. Soc.*, (1965) 6899).

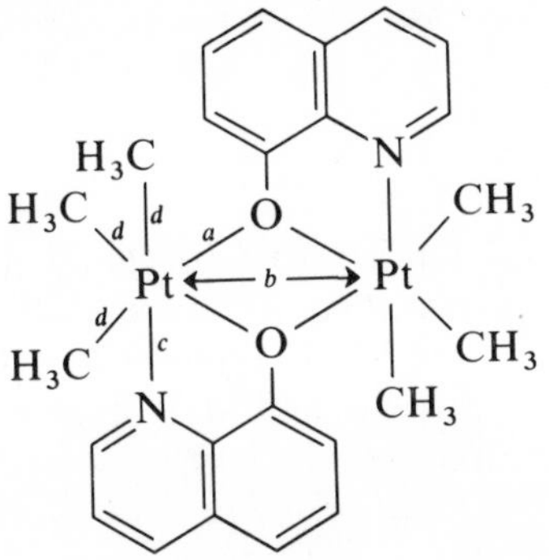

$$a = 2\cdot24(3)\ \text{Å}$$
$$b = 3\cdot384(3)\ \text{Å}$$
$$c = 2\cdot13(5)\ \text{Å}$$
$$d = 2\cdot06(5)\ \text{Å (mean)}$$

186. $C_{24}H_{34}Cl_2Pt_2$

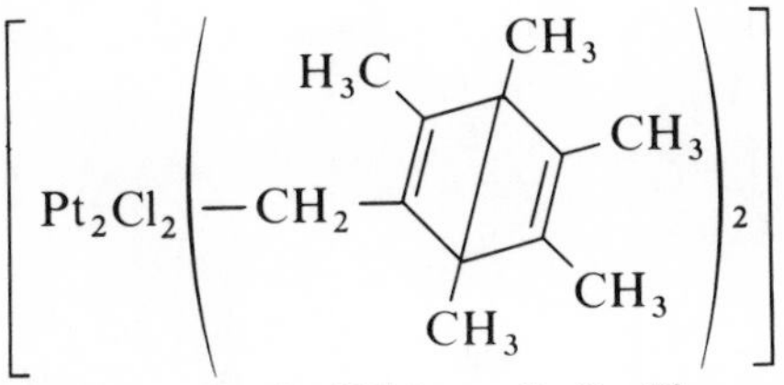

(R. Mason, G. B. Robertson, P. O. Whimp, B. L. Shaw and G. Shaw,
Chem. Comm., (1968) 868; R. Mason, G. B. Robertson and P. O. Whimp,
J. Chem. Soc. (A), (1970) 535).

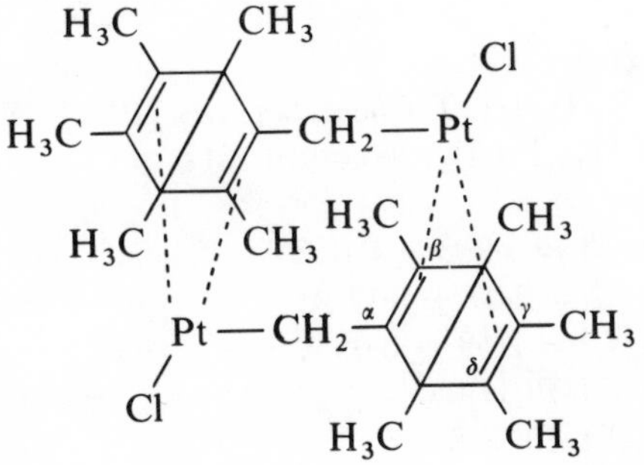

$$\text{Pt–Cl} = 2\cdot308(7)\ \text{Å}$$
$$\text{Pt–C}\alpha = 2\cdot18\ \text{Å}$$
$$\text{Pt–C}\beta = 2\cdot17\ \text{Å}$$
$$\text{Pt–C}\gamma = 2\cdot31\ \text{Å}$$
$$\text{Pt–C}\delta = 2\cdot36\ \text{Å}$$
$$\text{Pt–CH}_2 = 2\cdot07\ \text{Å}$$

187. $C_{24}H_{35}As_2Br_3OPt$ [Pt(o-allylphenyldimethylarsineOEt)Br₃]
(M. A. Bennett, G. J. Erskine, J. Lewis, R. Mason, R. S. Nyholm, G. B. Robertson and A. D. C. Towl, *Chem. Comm.*, (1966) 395).

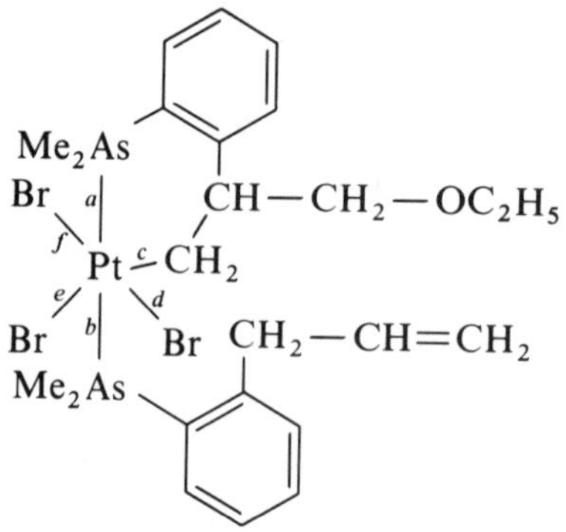

$a = 2.43$ Å
$b = 2.46$ Å
$c = 1.95$ Å
$d = 2.46$ Å
$e = 2.55$ Å
$f = 2.48$ Å

188. $C_{24}H_{48}O_4Pt_2$ [(CH₃)₃Pt(C₃H₇COCHCOC₃H₇)]₂
(A. G. Swallow and M. R. Truter, *Proc. Roy. Soc.*, **A254** (1960) 205, (x-ray diffraction); R. N. Hargreaves and M. R. Truter, *Chem. Comm.*, (1968) 473 (neutron diffraction); R. N. Hargreaves and M. R. Truter, *J. Chem. Soc. (A)*, (1969) 2282 (neutron diffraction).)

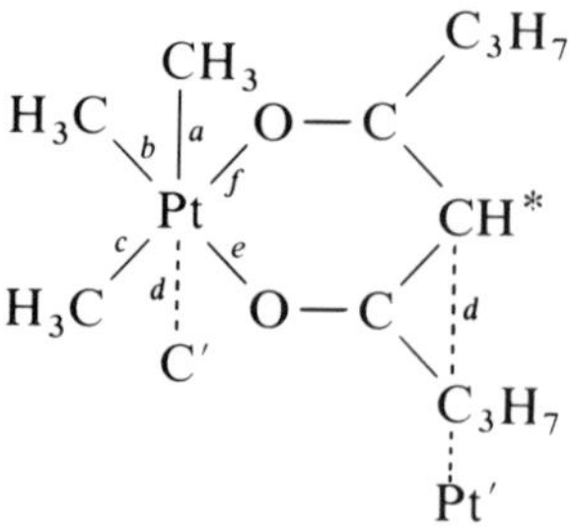

$a = 2.00(3)$ Å $d = 2.39(3)$ Å
$b = 2.04(2)$ Å $e = 2.14(2)$ Å
$c = 2.02(4)$ Å $f = 2.16(2)$ Å
H* occupies a site sterically suitable for much larger atoms.

189. $C_{24}H_{60}Cl_4Pt_4$ [(C₂H₅)₃PtCl]₄
(R. N. Hargreaves and M. R. Truter, *J. Chem. Soc. (A)*, (1971) 90).
Same basic structure as [(CH₃)₃PtCl]₄ (structure 161).

Pt–Pt $= 3.903 \pm 0.015$ Å
Pt–C $= 2.07 \pm 0.11$ Å
Pt–Cl $= 2.58 \pm 0.05$ Å
∠ Pt–Cl–Pt $= 98°$
∠ Cl–Pt–Cl $= 81°$

190. $C_{29}H_{47}NO_3P_2Pt$ $[\{P(o\text{-}CH_3C_6H_4)(^tBu)_2\}Pt\{P(C_6H_4)(^tBu)_2\}\text{-}(NO_3)]$

(R. Countryman and W. S. McDonald, personal communication).

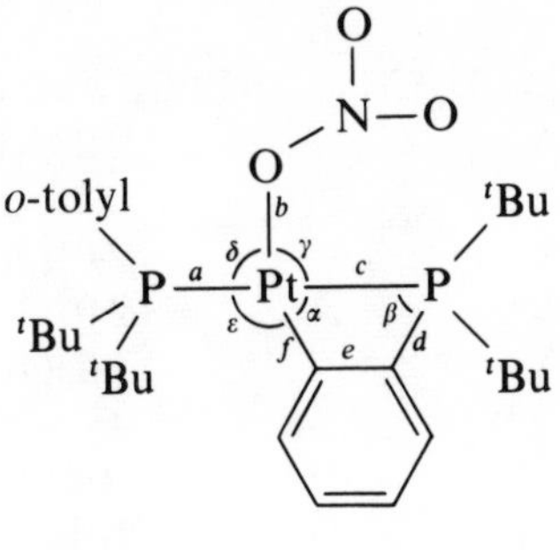

$a = 2\cdot32(1)\,\text{Å}$ $a = 68(1)°$
$b = 2\cdot14(3)\,\text{Å}$ $\beta = 81(1)°$
$c = 2\cdot30(1)\,\text{Å}$ $\gamma = 96(1)°$
$d = 1\cdot90(5)\,\text{Å}$ $\delta = 95(1)°$
$e = 1\cdot26(10)\,\text{Å}$ $\varepsilon = 100(1)°$
$f = 2\cdot03(4)\,\text{Å}$

191. $C_{28}H_{31}ClP_2Pt$ Trans-$[(PPh_2Et)_2PtHCl]$
(R. Eisenberg and J. A. Ibers, *Inorg. Chem.*, **4** (1965) 773).

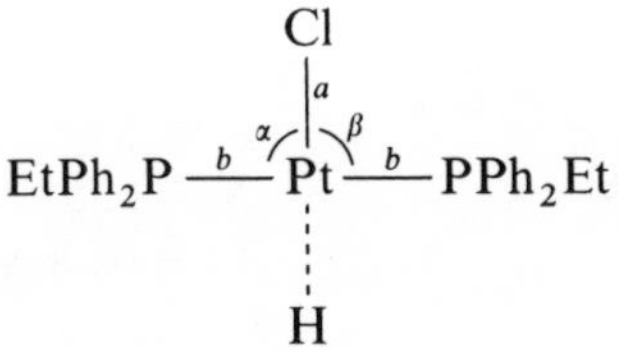

$a = 2\cdot422\,\text{Å}$ $\alpha = 92\cdot6°$
$b = 2\cdot268\,\text{Å}$ $\beta = 94\cdot5°$

192. $C_{29}H_{35}ClP_2PtSi$ Trans-$[(PPhMe_2)Pt(SiPh_2Me)Cl]$
(R. McWeeney, R. Mason and A. D. C. Towl, *Disc. Faraday Soc.*, **47** (1969) 20; R. Mason, personal communication).
Pt–Si = 2.29 Å   Pt–Cl (*trans* to Si) = 2.45(1) Å.

193. $C_{32}H_{16}N_8Pt$ $[Pt(phthalocyanin)]$
(J. M. Robertson and I. Woodward, *J. Chem. Soc.*, (1940) 36; C. J. Brown, *J. Chem. Soc. (A)*, (1968) 2494).

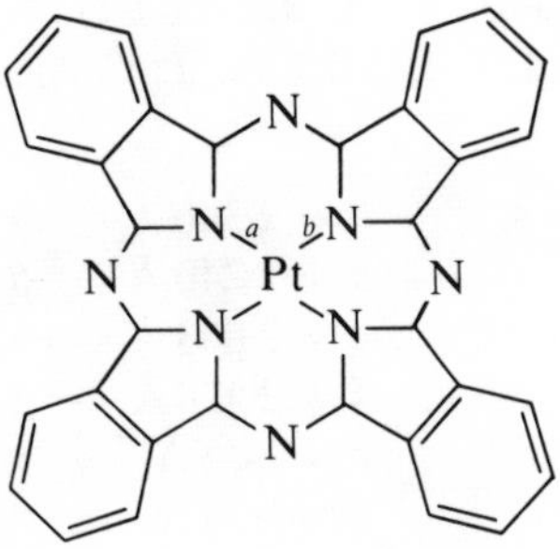

	α-polymorph	γ-polymorph
a	1·97 Å	2·02 Å
b	1·99 Å	1·95 Å

194. $C_{36}H_{30}O_2P_2Pt$ $[(PPh_3)_2PtO_2]$

(C. D. Cook, P.-T. Cheng and S. C. Nyburg, *J. Amer. Chem. Soc.*, **91** (1969) 2123 ($[(PPh_3)_2PtO_2].C_6H_5CH_3$); T. Kashiwagi, N. Yasuoka, N. Kasai, M. Kakudo, S. Takahashi and N. Hagihara, *Chem. Comm.*, (1969) 743 ($[(PPh_3)_2PtO_2].1.5\,C_6H_6$)).

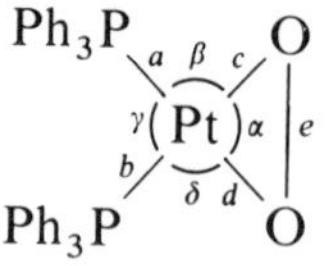

	$[(PPh_3)_2PtO_2].C_6H_5CH_3$	$[(PPh_3)_2PtO_2].1.5\,C_6H_6$
a	2·28 Å	2·282(11) Å
b	2·22 Å	2·253(12) Å
c	1·90 Å	2·01(3) Å
d	1·99 Å	2·01(2) Å
e	1·26 Å	1·45(4) Å
α	38°	43°
β	109°	108·5°
γ	100°	101·2°
δ	113°	107·8°
P_2PtO_2	Virtually coplanar	Virtually coplanar

195. $C_{36}H_{54}F_{10}P_2PtS_2$ Trans-$[(PBu_3)_2Pt(SC_6F_5)_2]$

(R. H. Fenn and G. R. Segrott, *J. Chem. Soc. (A)*, (1970) 2781).

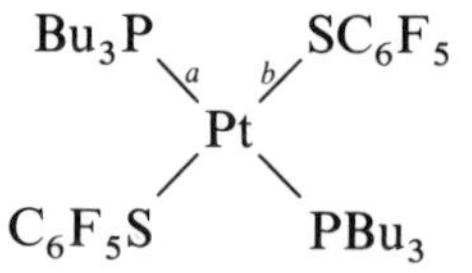

$a = 2.329$ Å
$b = 2.355$ Å
Pt is 0·28 Å out of S_2P_2 plane.

196. $C_{37}H_{30}O_3P_2Pt$ $[(PPh_3)_2Pt(CO_3)]$

(F. Cariati, R. Mason, G. B. Robertson and R. Ugo, *Chem. Comm.*, (1967) 408).

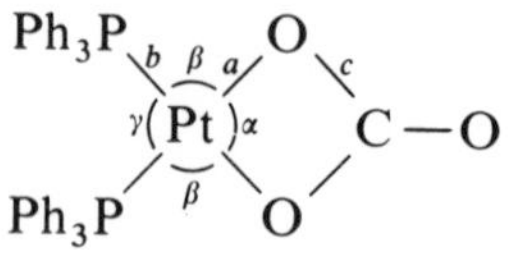

$a = 2.07$ Å $\alpha = 64°$
$b = 2.24$ Å $\beta = 98.8°$
$c = 1.28$ Å $\gamma = 98.2°$

197. $C_{37}H_{30}P_2PtS_2$ $[(PPh_3)_2Pt(CS_2)]$

(M. C. Baird, G. Hartwell, R. Mason, A. I. M. Rae and G. Wilkinson, *Chem. Comm.*, (1967) 92; R. Mason and A. I. M. Rae, *J. Chem. Soc. (A)*, (1970) 1767).

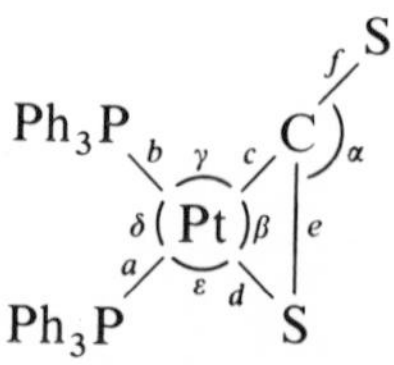

$a = 2.346$ Å $\alpha = 136.2°$
$b = 2.240$ Å $\beta = 45.5°$
$c = 2.063$ Å $\gamma = 99.7°$
$d = 2.328$ Å $\delta = 107.1°$
$e = 1.72$ Å $\varepsilon = 107.9°$
$f = 1.54$ Å

198. $C_{37}H_{44}O_5P_4Pt_4$ $[Pt_4(PPhMe_2)_4(CO)_5]$
(R. G. Vranka, L. F. Dahl, P. Chini and J. Chatt, *J. Amer. Chem. Soc.*, **91** (1969) 1574).

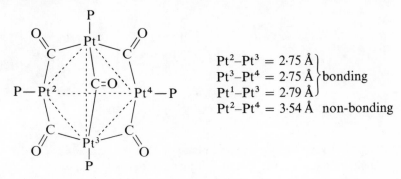

$Pt^2–Pt^3 = 2\cdot75$ Å
$Pt^3–Pt^4 = 2\cdot75$ Å bonding
$Pt^1–Pt^3 = 2\cdot79$ Å
$Pt^2–Pt^4 = 3\cdot54$ Å non-bonding

199. $C_{38}H_{30}Cl_4P_2Pt$ $[(PPh_3)_2Pt(CCl_2{=}CCl_2)]$
(J. N. Francis, A. McAdam and J. A. Ibers, *J. Organometal. Chem.*, **29** (1971) 131).

$a = 2\cdot278(8)$ Å	$f = 1\cdot77(3)$ Å
$b = 2\cdot292(7)$ Å	$g = 1\cdot69(3)$ Å
$c = 2\cdot02(3)$ Å	$h = 1\cdot81(3)$ Å
$d = 2\cdot05(3)$ Å	$i = 1\cdot73(3)$ Å
$e = 1\cdot62(3)$ Å	$\alpha = 47\cdot1(1\cdot0)°$

Angle between CCl_2 plane and plane of double bond $= 40\cdot6(3\cdot0)°$.
Angle between PtP_2 plane and PtC_2 plane $= 12\cdot3(1\cdot5)°$.

200. $C_{38}H_{30}Cl_2F_2P_2Pt$ $[(PPh_3)_2Pt(CCl_2{=}CF_2)]$
(J. N. Francis, A. McAdam and J. A. Ibers, *J. Organometal. Chem.*, **29** (1971) 131).

$a = 2\cdot303(6)$ Å	$d = {\sim}2\cdot00$ Å
$b = 2\cdot314(5)$ Å	$e = 1\cdot68(3)$ Å
$c = 2\cdot05(3)$ Å	$f = 1\cdot77(3)$ Å

201. $C_{39}H_{36}O_3P_2Pt$ $[(PPh_3)_2Pt(O_3CMe_2)]$
(R. Ugo, F. Conti, S. Cenini, R. Mason and G. B. Robertson, *Chem. Comm.*, (1968) 1498).

$a = 2\cdot27$ Å	
$b = 2\cdot26$ Å	$\alpha = 83\cdot2°$
$c = 2\cdot00$ Å	$\beta = 87\cdot0°$
$d = 2\cdot01$ Å	$\gamma = 98\cdot5°$
$e = 1\cdot32$ Å	$\delta = 91\cdot4°$
$f = 1\cdot40$ Å	
$g = 1\cdot46$ Å	

202. $C_{40}H_{30}Cl_2N_2P_2Pt$ $[(PPh_3)_2Pt(CCl_2{=}C(CN)_2)]$
(A. McAdam, J. N. Francis and J. A. Ibers, *J. Organometal. Chem.*, **29** (1971) 149).

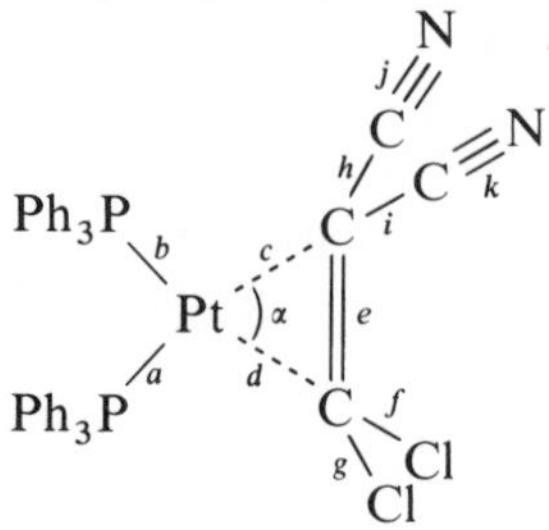

$$a = 2\cdot260(6)\text{ Å}$$
$$b = 2\cdot339(6)\text{ Å}$$
$$c = 2\cdot10(2)\text{ Å}$$
$$d = 2\cdot00(2)\text{ Å}$$
$$e = 1\cdot42(3)\text{ Å}$$
$$f = 1\cdot75(2)\text{ Å}\}\,1\cdot80(2)\text{ Å}$$
$$g = 1\cdot85(2)\text{ Å}$$

$$h = 1\cdot48(3)\text{ Å}\}\,1\cdot44(3)\text{ Å}$$
$$i = 1\cdot40(2)\text{ Å}$$
$$j = 1\cdot08(3)\text{ Å}\}\,1\cdot13(3)\text{ Å}$$
$$k = 1\cdot19(3)\text{ Å}$$
$$\alpha = 40\cdot6(9)°$$

Angle between CCl_2 plane and plane of double-bond = 48·4(2·3)°.
Angle between $C(CN)_2$ plane and plane of double-bond = 69·9(3·3)°.
Angle between PtP_2 plane and PtC_2 plane = 1·9(1·3)°.

203. $C_{40}H_{30}N_2P_2Pt$ $[(PPh_3)_2Pt(NCC{\equiv}CCN)]$
(C. Panattoni and R. Graziani in *Progress in Coordination Chemistry*, Ed. by M. Cais, Elsevier, 1968, p. 310).

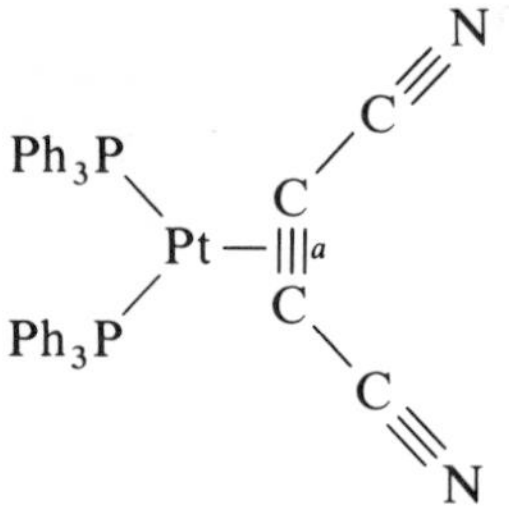

$$a = 1\cdot40\text{ Å}$$

Angle between PtP_2 plane and PtC_2 plane = 8°.

204. $C_{40}H_{31}F_6N_3P_2Pt$

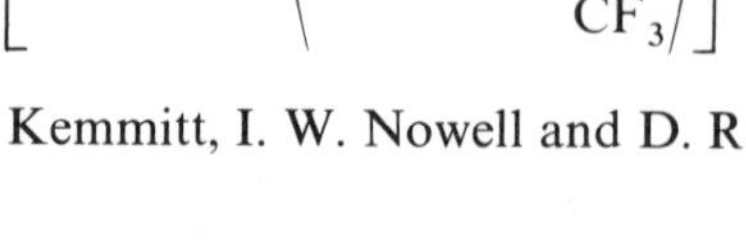

(W. D. Bland, R. D. W. Kemmitt, I. W. Nowell and D. R. Russell, *Chem. Comm.*, (1968) 1065).

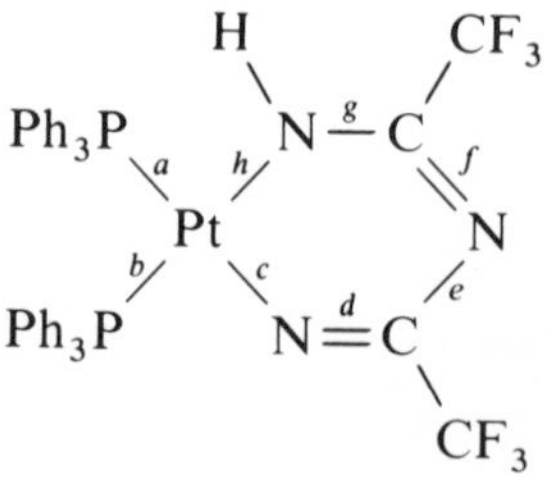

$$a = 2\cdot31\text{ Å}\qquad e = 1\cdot32\text{ Å}$$
$$b = 2\cdot28\text{ Å}\qquad f = 1\cdot32\text{ Å}$$
$$c = 2\cdot01\text{ Å}\qquad g = 1\cdot36\text{ Å}$$
$$d = 1\cdot24\text{ Å}\qquad h = 2\cdot02\text{ Å}$$

205. $C_{40}H_{32}N_2P_2Pt$ $[(PPh_3)_2Pt(NCCH{=}CHCN)]$

(C. Panattoni, R. Graziani, U. Belluco and W. H. Baddley, private communication to L. Manojlovic-Muir, K. W. Muir and J. A. Ibers, *Disc. Faraday Soc.*, **47** (1969) 84).

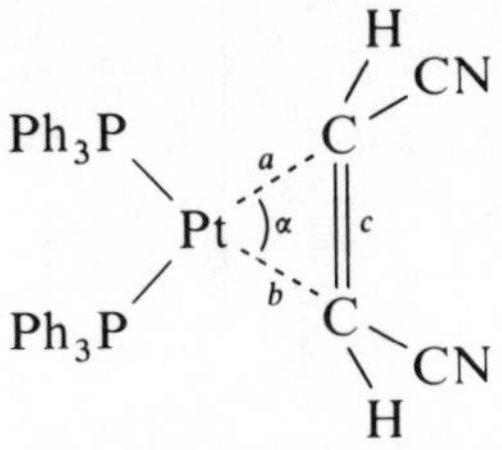

$$
\begin{array}{ll}
Ph_3P & \\
& \diagdown \\
& Pt \\
& \diagup \\
Ph_3P &
\end{array}
\qquad
\begin{array}{l}
H \\
\diagup \quad CN \\
C \\
\| \\
C \\
\diagdown \quad CN \\
H
\end{array}
$$

$a = 2{\cdot}025(6)$ Å
$b = 2{\cdot}162(6)$ Å
$c = 1{\cdot}525(8)$ Å
$\alpha = 43 \pm 2°$

Angle between PtP_2 and PtC_2 planes $= 6°$.

206. $C_{42}H_{30}N_4P_2Pt$ $[(PPh_3)_2Pt\{(NC)_2C{=}C(CN)_2\}]$

(C. Panattoni, G. Bombieri, U. Belluco and W. H. Baddley, *J. Amer. Chem. Soc.*, **90** (1968) 798; G. Bombieri, E. Forsellini, C. Panattoni, R. Graziani and G. Bandoli, *J. Chem. Soc. (A)*, (1970) 1313).

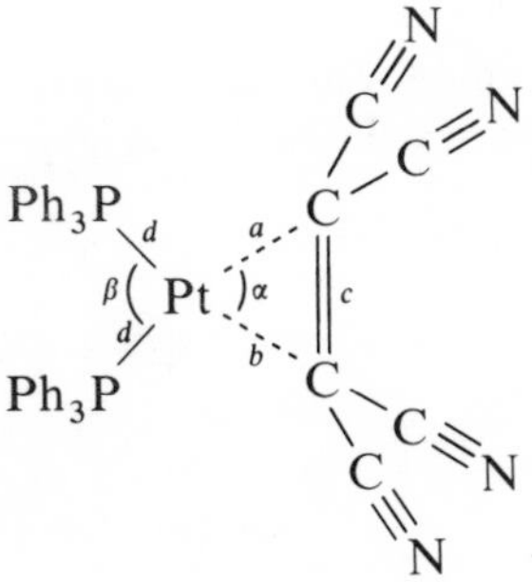

$a = 2{\cdot}10$ Å $\alpha = 41{\cdot}5°$
$b = 2{\cdot}12$ Å $\beta = 101{\cdot}4°$
$c = 1{\cdot}49$ Å
$d = 2{\cdot}29$ Å

Angle between PtP_2 and PtC_2 planes $= 8{\cdot}3°$.

207. $C_{50}H_{40}P_2Pt$ $[(PPh_3)_2Pt(C_6H_5C{\equiv}CC_6H_5)]$

(J. O. Glanville, J. M. Stewart and S. O. Grim, *J. Organometal. Chem.*, **7** (1967) P9).

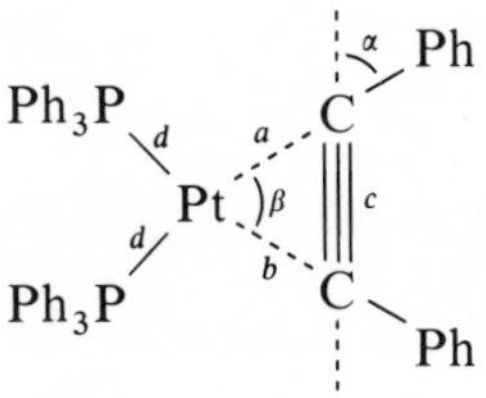

$a = 2{\cdot}06$ Å $\alpha = 40°$
$b = 2{\cdot}01$ Å $\beta = 39°$
$c = 1{\cdot}32(9)$ Å
$d = 2{\cdot}275(8)$ Å (mean)

Angle between PtP_2 and PtC_2 planes $= 14°$.

208. $C_{54}H_{42}As_4IPt^+$ $[Pt(QAs)I]^+(BPh_4^-)$

(G. A. Mair, H. M. Powell and L. M. Venanzi, *Proc. Chem. Soc.*, (1961) 170; G. A. Mair, *D. Phil. Thesis*, Oxford, 1961).

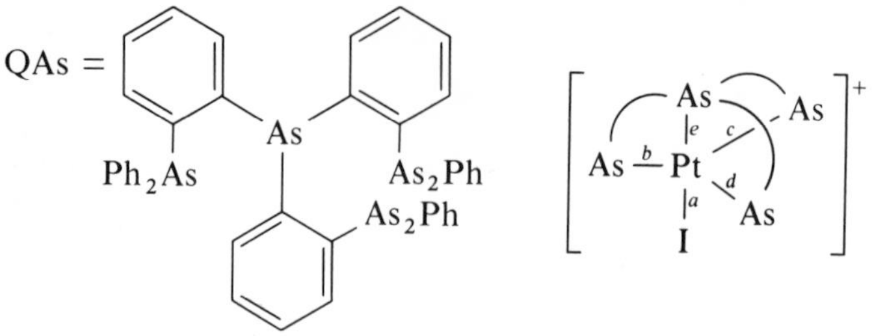

QAs =

$a = 2.65$ Å
$b = 2.49$ Å
$c = 2.45$ Å
$d = 2.43$ Å
$e = 2.30$ Å

Co-ordination about Pt is trigonal bipyramidal with Pt slightly displaced out of trigonal plane towards iodine.

209. $C_{54}H_{45}P_3Pt$ $[Pt(PPh_3)_3]$

(V. G. Albano, P. L. Bellon and V. Scatturin, *Chem. Comm.*, (1966) 507).

Molecule is approximately planar with Pt distorted 0·1 Å out of P_3 plane due to packing of phenyl rings. Pt–P = 2·25(1)–2·28(1) Å.

210. $C_{55}H_{45}OP_3Pt$ $[(PPh_3)_3Pt(CO)]$

(V. G. Albano, P. L. Bellon and M. Sansoni, *Chem. Comm.*, (1969) 899; V. G. Albano, G. M. Basso Ricci and P. L. Bellon, *Inorg. Chem.*, **8** (1969) 2109).

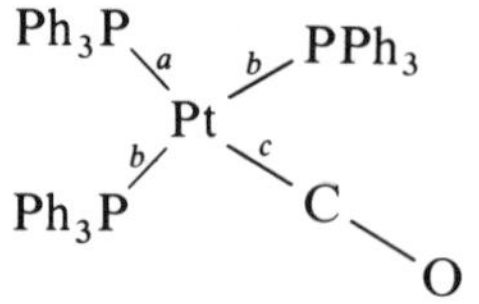

$a = 2.352(8)$ Å
$b = 2.334(7)$ Å
$c = 1.86(3)$ Å

Co-ordination about Pt is tetrahedral.

211. $C_{55}H_{45}OP_3Pt_2S$ $[(PPh_3)_2PtSPt(CO)(PPh_3)]$

(A. C. Skapski and P. G. H. Troughton, *Chem. Comm.*, (1969) 170; A. C. Skapski and P. G. H. Troughton, *J. Chem. Soc. (A)*, (1969) 2772).

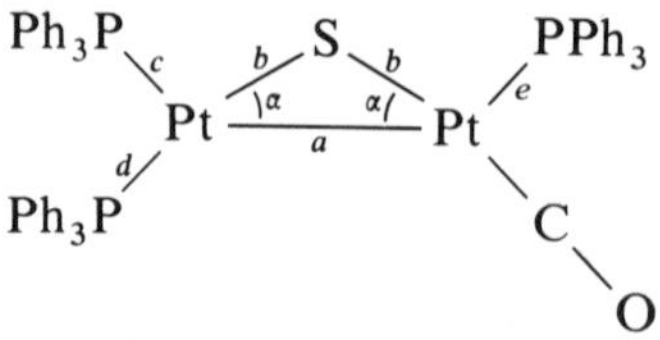

$a = 2.647$ Å
$b = 2.22$ Å
$c = 2.226$ Å
$d = 2.267$ Å
$e = 2.318$ Å
$\alpha = 53.5°$

212. $C_{72}H_{63}N_3P_4Pt_2$ $[(PPh_3)_2Pt(N_2H)(NH_2)Pt(PPh_3)_2]$

(G. C. Dobinson, R. Mason, G. B. Robertson, R. Ugo, F. Conti, D. Morelli, S. Cenini and F. Bonati, *Chem. Comm.*, (1967) 739).

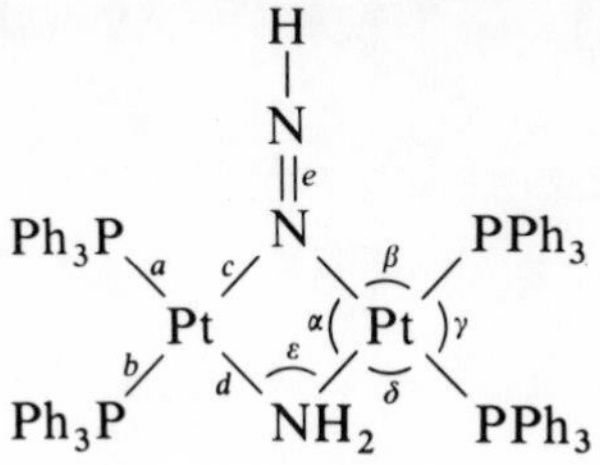

$$
\begin{array}{c}
H \\
| \\
N \\
\parallel e \\
\end{array}
$$

Ph$_3$P, a c N β PPh$_3$

Pt α (Pt)γ

b d ε δ

Ph$_3$P   NH$_2$ PPh$_3$

$a = 2{\cdot}27(2)$ Å	$\alpha = 70°$
$b = 2{\cdot}29(2)$ Å	$\beta = 98°$
$c = 1{\cdot}97(5)$ Å	$\gamma = 96°$
$d = 2{\cdot}09(5)$ Å	$\delta = 96°$
$e = 1{\cdot}18(9)$ Å	$\varepsilon = 110°$

Index

Page numbers followed by an S in brackets refer to crystal structures in Appendix II.

'S' olefin complex, 361–2
Salicylaldehyde complexes, 509(S)
Salicylaldimine complexes, 119–20, 484(S), 486–7(S)
Salicylaldoxime complexes, 119–20, 482(S)
Schiff's base, 209, 391
    complexes, 151, 209
Selenides
    divalent metals, 166–8
    tetravalent metals, 270–2, 472(S)
Selenocyanate complexes
    divalent metals, 220, 297
    tetravalent metals, 275, 499(S)
Selenoether complexes
    divalent metals, 175–9, 478(S)
    tetravalent metals, 270–2
Selenoethers in organometallics, 325
Semiconductivity, 72
Semiconductor, 72, 167–8, 268–9
Silanes, addition to olefins, 95, 394
Silicon (*see under* 'Group IVB elements')
Silicon complexes
    divalent metals, 75–99, 513(S)
    tetravalent metals, 252, 257–8, 394
Single crystal n.m.r. of olefin complexes, 369–70
Six co-ordinate complexes of divalent metals, 18–9, 127–8, 148, 236
Soft acid character, 13–5, 143, 218–9, 297
    palladium(0), 15
    palladium(II), 13–5, 143, 217–9, 230, 238–9
    platinum(0), 15
    platinum(II), 13–5, 143, 217–9, 230, 238–9
    platinum(IV), 15, 278
Solubilities of starting materials, 452–67
Solvation of square-planar complexes, 308
Solvent in substitution reactions, 307–8
Soybean oil, 393
Square-planar complexes
    isomerisation, 175, 208, 313–5
    mechanism of substitution reactions, 292–312
Square-planar geometry and crystal field splitting, 19–20, 402
Square-pyramidal complexes, 20, 147–8, 479(S), 487(S)
Square-pyramidal intermediate in substitution reactions, 310–1
Stability constants of
    acetylene complexes, 406–7
    amine complexes, 115–6
    olefin complexes, 374–8

Stability of
    metal–carbon σ-bonds, 331–4
    olefin complexes, 364–5
Stannous chloride, 261
    catalyst for formation of
        π-allyl complexes, 420
        olefin complexes, 362–3
Starting materials for further preparations, 452–67
Stereochemical course of olefin reactions, 380–1, 386–9
Stereochemical non-rigidity in π-allyl complexes, 434–9
Stereochemistry
    divalent complexes, 17–9
    tetravalent complexes, 21
    zerovalent complexes, 16–7
Stereospecific reactions, 264, 380–1
Steric effect in substitution reactions, 294, 305–6
Steric hindrance in substitution reactions, 294, 305–6
Stibine complexes of divalent metals
    bonding, 143–6
    dipole moments, 144
    five co-ordinate complexes, 148–9
    preparation, 129–30
    properties, 132–4
Stibines in organometallics, 325
Structural data, 470–519(S)
Substitution reactions
    divalent complexes, 292–313
        acid catalysis, 311
        anomalous ligand exchange reactions, 311–2
        associative mechanism, 293–6, 308–11
        cis-effect, 303–5
        electrophilic catalysis, 311
        entering ligand, 117, 294–8
        general mechanism, 293–6
        intimate mechanism, 308–11
        isotope exchange reactions, 311–2
        leaving group, 306–7
        non-labile ligand, 299–306
        rate law, 293–4
        solvent, 294, 307–8
        steric effect, 305–6
        steric hindrance, 294, 306
        substrate, 298–9
        trans-effect, 299–301
        trans-influence, 301–3
        trigonal bipyramidal complexes, 312–3
    tetravalent metals, 315–9
        bridging ligands, 316–7
        catalysis, 315–6